Optics in Magnetic Multilayers and Nanostructures

OPTICAL SCIENCE AND ENGINEERING

1. Electron and Ion Microscopy and Microanalysis: Principles and Applications, *Lawrence E. Murr*
2. Acousto-Optic Signal Processing: Theory and Implementation, *edited by Norman J. Berg and John N. Lee*
3. Electro-Optic and Acousto-Optic Scanning and Deflection, *Milton Gottlieb, Clive L. M. Ireland, and John Martin Ley*
4. Single-Mode Fiber Optics: Principles and Applications, *Luc B. Jeunhomme*
5. Pulse Code Formats for Fiber Optical Data Communication: Basic Principles and Applications, *David J. Morris*
6. Optical Materials: An Introduction to Selection and Application, *Solomon Musikant*
7. Infrared Methods for Gaseous Measurements: Theory and Practice, *edited by Joda Wormhoudt*
8. Laser Beam Scanning: Opto-Mechanical Devices, Systems, and Data Storage Optics, *edited by Gerald F. Marshall*
9. Opto-Mechanical Systems Design, *Paul R. Yoder, Jr.*
10. Optical Fiber Splices and Connectors: Theory and Methods, *Calvin M. Miller with Stephen C. Mettler and Ian A. White*
11. Laser Spectroscopy and Its Applications, *edited by Leon J. Radziemski, Richard W. Solarz, and Jeffrey A. Paisner*
12. Infrared Optoelectronics: Devices and Applications, *William Nunley and J. Scott Bechtel*
13. Integrated Optical Circuits and Components: Design and Applications, *edited by Lynn D. Hutcheson*
14. Handbook of Molecular Lasers, *edited by Peter K. Cheo*
15. Handbook of Optical Fibers and Cables, *Hiroshi Murata*
16. Acousto-Optics, *Adrian Korpel*
17. Procedures in Applied Optics, *John Strong*
18. Handbook of Solid-State Lasers, *edited by Peter K. Cheo*
19. Optical Computing: Digital and Symbolic, *edited by Raymond Arrathoon*
20. Laser Applications in Physical Chemistry, *edited by D. K. Evans*
21. Laser-Induced Plasmas and Applications, *edited by Leon J. Radziemski and David A. Cremers*

22. Infrared Technology Fundamentals, *Irving J. Spiro and Monroe Schlessinger*
23. Single-Mode Fiber Optics: Principles and Applications, Second Edition, Revised and Expanded, *Luc B. Jeunhomme*
24. Image Analysis Applications, *edited by Rangachar Kasturi and Mohan M. Trivedi*
25. Photoconductivity: Art, Science, and Technology, *N. V. Joshi*
26. Principles of Optical Circuit Engineering, *Mark A. Mentzer*
27. Lens Design, *Milton Laikin*
28. Optical Components, Systems, and Measurement Techniques, *Rajpal S. Sirohi and M. P. Kothiyal*
29. Electron and Ion Microscopy and Microanalysis: Principles and Applications, Second Edition, Revised and Expanded, *Lawrence E. Murr*
30. Handbook of Infrared Optical Materials, *edited by Paul Klocek*
31. Optical Scanning, *edited by Gerald F. Marshall*
32. Polymers for Lightwave and Integrated Optics: Technology and Applications, *edited by Lawrence A. Hornak*
33. Electro-Optical Displays, *edited by Mohammad A. Karim*
34. Mathematical Morphology in Image Processing, *edited by Edward R. Dougherty*
35. Opto-Mechanical Systems Design: Second Edition, Revised and Expanded, *Paul R. Yoder, Jr.*
36. Polarized Light: Fundamentals and Applications, *Edward Collett*
37. Rare Earth Doped Fiber Lasers and Amplifiers, *edited by Michel J. F. Digonnet*
38. Speckle Metrology, *edited by Rajpal S. Sirohi*
39. Organic Photoreceptors for Imaging Systems, *Paul M. Borsenberger and David S. Weiss*
40. Photonic Switching and Interconnects, *edited by Abdellatif Marrakchi*
41. Design and Fabrication of Acousto-Optic Devices, *edited by Akis P. Goutzoulis and Dennis R. Pape*
42. Digital Image Processing Methods, *edited by Edward R. Dougherty*
43. Visual Science and Engineering: Models and Applications, *edited by D. H. Kelly*
44. Handbook of Lens Design, *Daniel Malacara and Zacarias Malacara*
45. Photonic Devices and Systems, *edited by Robert G. Hunsberger*
46. Infrared Technology Fundamentals: Second Edition, Revised and Expanded, *edited by Monroe Schlessinger*
47. Spatial Light Modulator Technology: Materials, Devices, and Applications, *edited by Uzi Efron*
48. Lens Design: Second Edition, Revised and Expanded, *Milton Laikin*
49. Thin Films for Optical Systems, *edited by Francoise R. Flory*
50. Tunable Laser Applications, *edited by F. J. Duarte*

51. Acousto-Optic Signal Processing: Theory and Implementation, Second Edition, *edited by Norman J. Berg and John M. Pellegrino*
52. Handbook of Nonlinear Optics, *Richard L. Sutherland*
53. Handbook of Optical Fibers and Cables: Second Edition, *Hiroshi Murata*
54. Optical Storage and Retrieval: Memory, Neural Networks, and Fractals, *edited by Francis T. S. Yu and Suganda Jutamulia*
55. Devices for Optoelectronics, *Wallace B. Leigh*
56. Practical Design and Production of Optical Thin Films, *Ronald R. Willey*
57. Acousto-Optics: Second Edition, *Adrian Korpel*
58. Diffraction Gratings and Applications, *Erwin G. Loewen and Evgeny Popov*
59. Organic Photoreceptors for Xerography, *Paul M. Borsenberger and David S. Weiss*
60. Characterization Techniques and Tabulations for Organic Nonlinear Optical Materials, *edited by Mark G. Kuzyk and Carl W. Dirk*
61. Interferogram Analysis for Optical Testing, *Daniel Malacara, Manuel Servin, and Zacarias Malacara*
62. Computational Modeling of Vision: The Role of Combination, *William R. Uttal, Ramakrishna Kakarala, Spiram Dayanand, Thomas Shepherd, Jagadeesh Kalki, Charles F. Lunskis, Jr., and Ning Liu*
63. Microoptics Technology: Fabrication and Applications of Lens Arrays and Devices, *Nicholas Borrelli*
64. Visual Information Representation, Communication, and Image Processing, *edited by Chang Wen Chen and Ya-Qin Zhang*
65. Optical Methods of Measurement, *Rajpal S. Sirohi and F. S. Chau*
66. Integrated Optical Circuits and Components: Design and Applications, *edited by Edmond J. Murphy*
67. Adaptive Optics Engineering Handbook, *edited by Robert K. Tyson*
68. Entropy and Information Optics, *Francis T. S. Yu*
69. Computational Methods for Electromagnetic and Optical Systems, *John M. Jarem and Partha P. Banerjee*
70. Laser Beam Shaping, *Fred M. Dickey and Scott C. Holswade*
71. Rare-Earth-Doped Fiber Lasers and Amplifiers: Second Edition, Revised and Expanded, *edited by Michel J. F. Digonnet*
72. Lens Design: Third Edition, Revised and Expanded, *Milton Laikin*
73. Handbook of Optical Engineering, *edited by Daniel Malacara and Brian J. Thompson*

74. Handbook of Imaging Materials: Second Edition, Revised and Expanded, *edited by Arthur S. Diamond and David S. Weiss*
75. Handbook of Image Quality: Characterization and Prediction, *Brian W. Keelan*
76. Fiber Optic Sensors, *edited by Francis T. S. Yu and Shizhuo Yin*
77. Optical Switching/Networking and Computing for Multimedia Systems, *edited by Mohsen Guizani and Abdella Battou*
78. Image Recognition and Classification: Algorithms, Systems, and Applications, *edited by Bahram Javidi*
79. Practical Design and Production of Optical Thin Films: Second Edition, Revised and Expanded, *Ronald R. Willey*
80. Ultrafast Lasers: Technology and Applications, *edited by Martin E. Fermann, Almantas Galvanauskas, and Gregg Sucha*
81. Light Propagation in Periodic Media: Differential Theory and Design, *Michel Nevière and Evgeny Popov*
82. Handbook of Nonlinear Optics, Second Edition, Revised and Expanded, *Richard L. Sutherland*
83. Polarized Light: Second Edition, Revised and Expanded, *Dennis Goldstein*
84. Optical Remote Sensing: Science and Technology, *Walter Egan*
85. Handbook of Optical Design: Second Edition, *Daniel Malacara and Zacarias Malacara*
86. Nonlinear Optics: Theory, Numerical Modeling, and Applications, *Partha P. Banerjee*
87. Semiconductor and Metal Nanocrystals: Synthesis and Electronic and Optical Properties, edited by *Victor I. Klimov*
88. High-Performance Backbone Network Technology, *edited by Naoaki Yamanaka*
89. Semiconductor Laser Fundamentals, *Toshiaki Suhara*
90. Handbook of Optical and Laser Scanning, *edited by Gerald F. Marshall*
91. Organic Light-Emitting Diodes: Principles, Characteristics, and Processes, *Jan Kalinowski*
92. Micro-Optomechatronics, *Hiroshi Hosaka, Yoshitada Katagiri, Terunao Hirota, and Kiyoshi Itao*
93. Microoptics Technology: Second Edition, *Nicholas F. Borrelli*
94. Organic Electroluminescence, *edited by Zakya Kafafi*
95. Engineering Thin Films and Nanostructures with Ion Beams, *Emile Knystautas*
96. Interferogram Analysis for Optical Testing, Second Edition, *Daniel Malacara, Manuel Sercin, and Zacarias Malacara*
97. Laser Remote Sensing, *edited by Takashi Fujii and Tetsuo Fukuchi*
98. Passive Micro-Optical Alignment Methods, *edited by Robert A. Boudreau and Sharon M. Boudreau*

99. Organic Photovoltaics: Mechanism, Materials, and Devices, *edited by Sam-Shajing Sun and Niyazi Serdar Saracftci*
100. Handbook of Optical Interconnects, *edited by Shigeru Kawai*
101. GMPLS Technologies: Broadband Backbone Networks and Systems, *Naoaki Yamanaka, Kohei Shiomoto, and Eiji Oki*
102. Laser Beam Shaping Applications, *edited by Fred M. Dickey, Scott C. Holswade and David L. Shealy*
103. Electromagnetic Theory and Applications for Photonic Crystals, *Kiyotoshi Yasumoto*
104. Physics of Optoelectronics, *Michael A. Parker*
105. Opto-Mechanical Systems Design: Third Edition, *Paul R. Yoder, Jr.*
106. Color Desktop Printer Technology, *edited by Mitchell Rosen and Noboru Ohta*
107. Laser Safety Management, *Ken Barat*
108. Optics in Magnetic Multilayers and Nanostructures, *Štefan Višňovský*

Optics in Magnetic Multilayers and Nanostructures

Štefan Višňovský

A CRC title, part of the Taylor & Francis imprint, a member of the Taylor & Francis Group, the academic division of T&F Informa plc.

Published in 2006 by
CRC Press
Taylor & Francis Group
6000 Broken Sound Parkway NW, Suite 300
Boca Raton, FL 33487-2742

Printed in the United States of America on acid-free paper
10 9 8 7 6 5 4 3 2 1

International Standard Book Number-10: 0-8493-3686-4 (Hardcover)
International Standard Book Number-13: 978-0-8493-3686-7 (Hardcover)

Library of Congress Cataloging-in-Publication Data

Catalog record is available from the Library of Congress

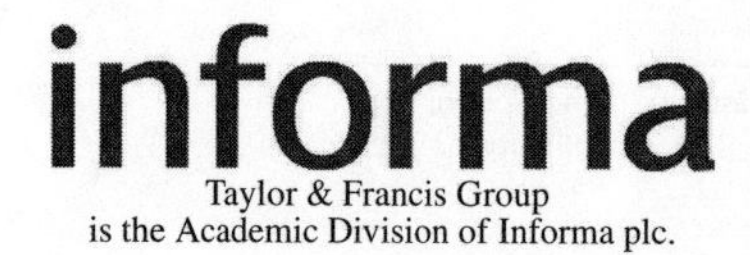

Taylor & Francis Group
is the Academic Division of Informa plc.

**Visit the Taylor & Francis Web site at
http://www.taylorandfrancis.com**

**and the CRC Press Web site at
http://www.crcpress.com**

To Milada and Daniel

Preface

Multilayers with individual layer thickness in the nanometer range, and nanostructures in general, represent a new type of material with properties controlled by interfaces and patterning. They find an increased number of applications in information technology, engineering, chemistry, biology, and medicine. In their characterization, the techniques based on the multilayer or nanostructure response to incident electromagnetic or material waves play an important role. A special class is formed by nanostructures with parameters periodic in space, typical examples being optical devices, semiconductor memories, magnetic, and magnetooptic (MO) recording media. In microwave and optical systems, the periodic nanostructures can perform frequency selective functions. The most recent activities in this area are related to so-called *photonic band* (PBG) materials or *photonic crystals*.

The expanding research field attracts specialists from various branches of science and technology as well as students. The motivation of this book arises from the need for an accessible and timesaving introduction to the subject of optics in anisotropic magnetic multilayers and nanostructures. The electromagnetic response of nanostructures containing magnetic layers or elements depends on their magnetic state, which can be controlled by external magnetic fields. This feature enables optical *sensing* of the magnetic state with a diffraction limited lateral resolution. It may also be used to *control* the light. Practical applications include magnetic domain observations using magnetooptic microscopy, reading the information recorded in magnetooptic disks, and magnetooptic light modulation. The wave propagation can be made *unidirectional, i.e., nonreciprocal*, and exploited in devices like isolators or circulators.

The simultaneous presence of natural crystallographic and induced anisotropy including magnetic ordering requires each individual layer to be characterized by its own permittivity tensor, which may be of the most general form. This is necessary in order to treat the crystallographic (or preference) axes and the direction of magnetization as mutually independent. The response analysis can be performed using the 4×4 matrix Yeh's formalism [1] originally developed for the birefringent layered media characterized by real symmetric permittivity tensors with orthogonal linear proper polarizations. In absorbing magnetic crystals, the permittivity tensor is no more symmetric and consists of Hermitian and anti-Hermitian parts. In multilayers, the tensor is assumed homogeneous inside each layer. Then the response of a medium with continuous change of the permittivity tensor parallel to an axis normal to the interfaces can be approximated by splitting the medium to an appropriate number of homogeneous layers. This is so-called *multilayer* or *staircase* approximation.

The formalism can be extended to anisotropic multilayered nanostructures either with one-dimensional (1D) periodicity of the tensor parallel to an arbitrary axis in a layer or with a two-dimensional (2D) tensor periodicity in the interface plane [2–5]. These situations occur in photonic crystals. The latter may contain magnetic elements [6] and perform the nonreciprocal functions in *magnetophotonics*. The incorporation of three-dimensional (3D) periodicity into the permittivity tensors for multilayer systems makes the problem rather general. In addition to the photonic crystals, this opens ways to fundamental problems on spatial dispersion and crystal physics.

The generalized formalism includes many topics of the electromagnetic optics as special cases. For example, propagation in homogeneous and nonhomogeneous[1] isotropic media, reflection and transmission at interfaces, thin film and multilayer optics, dielectric optical waveguides, metal optics, crystal optics, anisotropic multilayers, interference and Fraunhofer diffraction, *etc*., can be explained from a unified viewpoint.

The understanding of magnetic multilayer optics, which employs the notions of both *polar* (electric field, electric dipole moment) and *axial* (magnetization, magnetic flux density, *etc*.) vectors, is helpful in the development of a physical insight in both optics and magnetism. The formal approaches employed in the multilayer and nanostructure optics display much analogy with quantum theory of (periodic) *quantum well problems*. Thanks to this circumstance, quantum solutions can often be illustrated on formally similar but less abstract situations in optics.

In recent years, several books and reviews dealing with magnetooptics appeared [7–11]. Here, we focus on those aspects that received less attention in those volumes but which have been found useful in the research of magnetic multilayers and nanostructures. The book assumes familiarity with the ellipsometry and the representation of the polarized light using the Jones formalism as presented, *e.g*., in the book by Azzam and Bashara [12]. An independent part on microscopic semiclassical model, not necessary for the understanding of the rest of the book, assumes an acquaintance with ligand fields [13,14]. The treatment is restricted to a *linear* response to plane electromagnetic waves (Fraunhofer approximation). It may be used as an introductory reading before the study of nonlinear magnetooptics [15] and near-field magnetooptics [16,17].

---

[1] The nonhomogeneity is understood in a restricted sense; the permittivity changes, in an arbitrary way, on an axis normal to the coplanar medium boundaries.

The emphasis is on those aspects that can be characterized analytically. It should, however, be emphasized that even in the simplest situations, the evaluation of the optical response in anisotropic absorbing multilayers and nanostructures requires a direct use of numerical methods because of the complex number nature of the parameters involved. Nevertheless, the analytical formulae are indispensable in tracing and testing the complicated phase relationship and the employed sign conventions. In addition, the analytical representations provide a physical insight useful in new problems. The formulae of the optics in anisotropic media are always complicated, and their derivation requires effort and time. Because of want of space, only final results are presented in the literature. This may make the reading less comfortable and more time consuming to the community of nonspecialists in the field. To address this community, we prefer to provide detailed explanations. The possibility to follow derivations step by step helps us to understand the procedure and to develop approximations. The detailed treatment also helps to avoid the confusion related to sign conventions and those arising from misprints and errors.

The first chapter briefly recaps the history and uses of magnetooptics in magnetic media. It introduces basic concepts of magnetooptics in media that would be isotropic in the absence of magnetization. At nonzero magnetization, an originally scalar permittivity of the medium transforms to a tensor of specific symmetry. By substitution of the tensor into Maxwell equations, the basic formulae for Faraday and Voigt effects and magnetooptic Kerr effects follow. The changes in an originally isotropic medium induced by the axial vector of magnetization enable a unique and practically important function, a nonreciprocal (or unidirectional) wave propagation. The final part of the chapter is devoted to the first encounter with a multilayer problem illustrated on the multilayer response at polar magnetization to a normally incident polarized wave. This treatment, based on the simplest form of the permittivity tensor in magnetic medium, is already sufficient in many practical situations.

A more involved approach should include the effect of crystalline symmetry. In the second chapter we thus analyze the simultaneous effect of crystalline symmetry and arbitrarily oriented magnetization on the form of permittivity tensors. We start with the development of the permittivity into power series of magnetization. We consider terms of the zero, first, and second degrees and define corresponding magnetization independent tensors, *i.e.*, a second rank permittivity tensor, a third rank linear magnetooptic tensor, and a fourth rank quadratic magnetooptic tensor. For understanding the magnetooptics, the first two terms in the development are indispensable. However, these two terms cannot distinguish between amorphous media and those with cubic symmetry. The distinction only becomes possible starting from the terms quadratic magnetization.

In order to gain some idea of the spectral dependence of the permittivity, the discussion of two simple models follows. The simplest microscopic interpretation of the optical and magnetooptic properties of magnetically ordered materials employs the classical Lorentz and Drude theories. More sophisticated semiclassical theories consider the linear response of a magnetic quantum system, *i.e.*, magnetically ordered crystal, to classical electromagnetic monochromatic waves. In media with spontaneous magnetization, the semiclassical model should incorporate spin-orbit coupling and exchange interactions responsible for magnetic ordering. However, a rigorous account for the exchange interactions from first principles would be too difficult, and we replace their effect by an effective exchange field. This step simplifies the problem to that of the *anomalous Zeeman effect*, with the action of the effective field restricted to spin magnetic moments. The model provides a microscopic basis for the optical response studies in magnetic media. This aspect is not limited to fundamental academic research but presents a practical importance. For example, a plausible spectroscopic response in magnetic multilayers and nanostructures should allow a decomposition into a sum of elementary Lorentz or Voigt lines [18]. These lines originate from materials forming the structure, and their weights in the sum contain valuable information.

General treatment of the electromagnetic response in planar layered absorbing anisotropic media displaying magnetic order is given in the third chapter. Optical interactions in a multilayer system are described with the Yeh's 4×4 matrix formalism extended to absorbing anisotropic magnetic media. This approach is universal yet simple and based on classical optical terms. In particular, crystal optics, electrooptics, elastooptics, magnetooptics, optics of thin films, and waveguiding in planar isotropic or anisotropic multilayer structures are included as special cases. The subsequent three chapters discuss the multilayer response for three special orientations of the magnetization vector with respect to the plane of incidence and the interface plane, *i.e.*, polar, longitudinal, and transverse ones. The analysis starts with a permittivity tensor suitable for the treatment of the media of orthorhombic symmetry and magnetized parallel to one of the orthorhombic axes. No restrictions are therefore imposed on the three diagonal and two off-diagonal elements (symmetrically positioned with respect to the tensor diagonal) for the three orientations of the magnetization

$$\begin{pmatrix} \varepsilon_{xx}^{(n)} & \varepsilon_{xy}^{(n)} & 0 \\ \varepsilon_{yx}^{(n)} & \varepsilon_{yy}^{(n)} & 0 \\ 0 & 0 & \varepsilon_{zz}^{(n)} \end{pmatrix}, \quad \begin{pmatrix} \varepsilon_{xx}^{(n)} & 0 & \varepsilon_{xz}^{(n)} \\ 0 & \varepsilon_{yy}^{(n)} & 0 \\ \varepsilon_{zx}^{(n)} & 0 & \varepsilon_{zz}^{(n)} \end{pmatrix}, \quad \begin{pmatrix} \varepsilon_{xx}^{(n)} & 0 & 0 \\ 0 & \varepsilon_{yy}^{(n)} & \varepsilon_{yz}^{(n)} \\ 0 & \varepsilon_{zy}^{(n)} & \varepsilon_{zz}^{(n)} \end{pmatrix}. \quad (1)$$

In this way, the analysis can treat the multilayer response in a broad class of low symmetry media. The 4×4 matrix formalism is then applied to the reflection on an anisotropic film sandwiched between isotropic ambient and the anisotropic substrate.[2]

The fourth chapter deals extensively with the multilayers at polar magnetization using the general approach developed in the third chapter. The analytical representations greatly simplify at normal incidence. Thanks to this, analytical formulae can be written down for systems of several layers and for periodic multilayers. This simple configuration is at the same time the most important one as it is the most widely employed in practice. In the fifth and sixth chapters, we apply the general approach of the third chapter to both exact and approximate treatments of multilayers at longitudinal and transverse orientations of the magnetization vector. The case of normal or nearly normal incidence on multilayers with arbitrarily oriented magnetizations and general anisotropies in individual layers forms the subject of the seventh chapter. The restriction to the magnetizations either normal or parallel to interfaces allows significant simplifications. The final part of the chapter is devoted to the analysis of the practical configuration for the studies of quadratic MO effects in reflection.

The case of an arbitrary angle of incidence and an arbitrary oriented magnetization in multilayers consisting of layers with the thicknesses in nanometer range can be reasonably presented analytically. With a small loss in accuracy, the permittivity tensor is chosen in the form that accounts for the response linear in magnetization only. Consequently, in the response characteristics, the difference in the diagonal tensor elements is ignored and the terms containing the off-diagonal tensor elements in the second and higher power are neglected. This approach is justified in most practical situations dealing with absorbing thin films. The magnetizations in individual layers are assumed uniform within each layer but need not be necessarily of the same orientation. The resulting formulae provide a basis for *magnetooptic vector magnetometry*. In particular, the case of chiral magnetization profiles can be treated in this way by splitting the medium into a stack of layers (multilayer or staircase approximation). These topics form the subject of the eighth chapter. The Fraunhofer or far field diffraction in laterally periodic multilayer anisotropic structures, treated in the ninth chapter, represents a generalization of the 4×4 matrix formalism. The chapter can be employed as an introduction to magnetophotonic crystals.

This work is based on the experience acquired in the experimental research of magnetic metallic and oxide thin film and multilayers performed since 1973 in cooperation with Ramanathan Krishnan of Laboratoire

[2] The film and the substrate are characterized by the permittivity tensor of the same orthorhombic symmetry with one orthorhombic axis perpendicular to the interfaces. The remaining two axes are perpendicular and parallel with respect to the plane of incidence. The film and substrate magnetizations are parallel to each other.

de magnétisme, Centre national de la recherche scientifique in Meudon-Bellevue. The advice from Katsu Rokushima, professor emeritus at the University of Osaka Prefecture, Václav Prosser of Charles University, Tomuo Yamaguchi of Shizuoka University, Kiyotoshi Yasumoto of Kyushu University, Pierre Beauvillain of Paris XI University, and Jaromír Pištora of Ostrava Technical University is acknowledged. Kamil Postava of Ostrava Technical University performed the simulations of the optical response in magnetic multilayers employed in this book.

The main part of this subject was developed in frame of Czechoslovak State Project of Fundamental Research supported by Czechoslovak Federal Ministry for the Development in Technology and Investment and Czech Ministry for Education. The cooperation with CNRS Bellevue Laboratoire de Magnetisme, Université Paris-Sud (Jacques Ferré, Danielle Renard), Université de Versailles (Marcel Guyot, Niels Keller), Shizuoka University, Kaiserslautern University (Burkard Hillebrands), Cambridge University (J. Antony C. Bland), and Queen's University of Belfast (Ronald Atkinson) is also acknowledged. The numerical aspects of the matrix approach discussed in the book have been developed and applied to original experimental data by Miroslav Nývlt, Radek Lopušnik, Radovan Urban, Jan Mistrík, Jaroslav Hamrle, and Roman Antoš in their Ph.D. theses. The theoretical and experimental aspects of the optics in magnetic multilayers and nanostructures were studied in frame of several research projects, in particular: PECO (Magnetic properties of novel magnetic films), NATO (International Scientific Program: Priority Area on High Technology), Grant Agency of Czech Republic #202/06/0531, Grant Agency of Charles University #314/2004/B – FYZ/MFF as well as the projects supported by Czech Ministry of Education Youth and Sports #MSM 0021 620834. Thanks go to Radovan Urban of California Institute of Technology for discussion and useful suggestions.

Kiyotoshi Yasumoto provided support and encouragement essential for the accomplishment of this work. Finally, I would like to thank Taisuke Soda and his coworkers Theresa Del Forn, Takisha Jackson, Jacqueline Callahan, and James Miller of Taylor & Francis for their help.

References

1. P. Yeh, "Optics of anisotropic layered media: a new 4 × 4 matrix algebra," Surf. Sci. **96**, 41–53, 1980; P. Yeh, *Optical Waves in Layered Media* (John Wiley & Sons, New York, 1988), Chapter 9.
2. K. Rokushima and J. Yamakita, "Analysis of anisotropic dielectric gratings," J. Opt. Soc. Am. **73**, 901–908 (1983).

3. E. N. Glytsis and T. K. Gaylord, "3-dimensional (vector) rigorous coupled-wave analysis of anisotropic grating diffraction," J. Opt. Soc. Am. A **7**, 1399–1420 (1990).
4. L. Li, "Fourier modal method for crossed anisotropic gratings with arbitrary permittivity and permeability tensors," J. Optics A: Pure Appl. Opt. **5**, 345–355 (2003).
5. M. Nevière and E. Popov, *Light Propagation in Periodic Media: Differential Theory and Design* (Marcel Dekker, Inc., New York, 2003).
6. A. Figotin and I. Vitebsky, "Nonreciprocal magnetic photonic crystals," Phys. Rev. E **63**, 066609(17) (2001).
7. J. Schoenes, in *Materials Science and Technology, Vol. 3A: Electronic and Magnetic Properties of Metals and Ceramics—Part I (Volume Editor K. H. J. Buschow)* R. W. Cahn, P. Haasen, and E. J. Kramer, eds. (Verlag Chemie, Weinheim, 1992), p. 147.
8. A. K. Zvezdin and V. A. Kotov, *Modern Magnetooptics and Magnetooptical Materials* (Institute of Physics Publishing, Bristol and Philadelphia, 1997).
9. P. M. Oppeneer, Magnetooptical Kerr Spectra, in *Handbook of Magnetic Materials*, K. H. J. Buschow, ed. (Elsevier, Amsterdam, 2001), Vol. 13, pp. 229–422.
10. B. Heinrich and J. A. C. Bland, eds., *Ultrathin Magnetic Structures* (Springer Verlag, Berlin, 1994).
11. M. Mansuripur, *The Physical Principles of Magnetooptical Recording* (Cambridge University Press, London, 1996).
12. R. M. A. Azzam and N. M. Bashara, *Ellipsometry and Polarized Light* (Elsevier, Amsterdam, 1987).
13. C. J. Ballhausen, *Introduction to Ligand Field Theory* (McGraw-Hill Book Company, New York, 1962).
14. S. Sugano, Y. Tanabe, and H. Kamimura, *Multiplets of Transition-Metal Ions in Crystals* (Academic Press, New York and London, 1970).
15. K. H. Bennemann, *Nonlinear Optics in Metals* (Oxford University Press, Oxford, 1999).
16. C. Hermann, V. A. Kosobukin, G. Lampel, J. Peretti, V. I. Safarov, and P. Bertrand, "Surface-enhanced magnetooptics in metallic multilayer films," Phys. Rev. B **64**, 235422 (2001).
17. P. Fumagalli, A. Rosenberger, G. Eggers *et al.*, "Scanning near-field magnetooptic microscopy using illuminated fiber tips," Ultramicroscopy **71**, 249–256 (1998).
18. B. Di Bartolo, *Optical Interactions in Solids* (J. Wiley, New York, 1968).

The Author

Štefan Višňovský received his master's degree (Ing.) in applied physics in 1965 from Slovak Technical University in Bratislava, Czechoslovakia. In 1966, he joined the Research Institute for Telecommunications in Prague, where he developed tunable microwave filters using single-crystals of yttrium iron garnets. In 1969, he joined the Institute of Physics at Charles University in Prague to study optical properties of magnetic garnets. He received his Ph.D. (CSc.) and D.Sc. (DrSc.) degrees in 1976 and 1991, respectively, from Charles University. In 1997, he received a full professor position at Charles University.

Višňovský's research focuses on optics of magnetic and anisotropic media, multilayer optics, and physics of magnetic oxides and ultrathin metallic films. He has published over 100 articles in reviewed international journals and 50 contributions for international conferences, including 9 invited talks. He has been an invited lecturer at several universities in Japan, France, Germany, the United Kingdom, and the United States, including Université Paris XI, Kyoto University, Cambridge University, and California Institute of Technology.

He is a member of the International Advisory Committees of Magneto-Optical Recording International Symposium (since 1996) and the International Colloquium on Magnetic Films and Surfaces (since 1997). At Charles University, Višňovský taught courses on electromagnetism, optics, atomic, statistics and quantum physics, thin film optics, optics of anisotropic multilayered structures, integrated optics, and fiber optics. He was a partner in the NATO International Scientific Program (priority area on high technology), a partner in the Program of Commission of European Communities (magnetic properties of novel ultrathin films), and a partner in the European Science Foundation Network (nanomagnetism and growth on vicinal surfaces). He has been responsible for projects financed by the Scientific Foundation of Czech Republic, Czech Ministry of Education, Youth and Sports, and Grant Agency of Charles University.

Contents

List of Figures

List of Tables

1

Introduction

Optical response in magnetic media depends on their magnetic state, and this dependence manifests itself in magnetooptic (MO) effects [1]. Under normal circumstances, the action of magnetization even in media with nearly perfect magnetic ordering is weak, and special measures are required to make them observable. Usually a full advantage of the vectorial nature of the electromagnetic plane wave is taken by using the probing radiation with a defined (*e.g.*, linear or circular) polarization state. In this introductory chapter, we give a brief overview of magnetooptics in magnetic materials. The main part that follows deals with the bases of magnetooptics from the phenomenological viewpoint of classical wave optics. The final part is devoted to the MO effects in multilayers. The information provided in this chapter allows the reader to immediately approach a broad class of MO problems encountered in practice.

1.1 History

1.1.1 Early Studies

The magnetooptic effects played an important role in the early history of electromagnetism [2]. They provided an experimental support to the electromagnetic theory of light (Faraday effect), as well as to both classical (*e.g.*, Zeeman effect) and quantum theory of matter including the notions of electron spin and spin-orbit coupling (*e.g.*, anomalous Zeeman effect).

The first magnetooptic effect was observed in 1845 by Michael Faraday. Faraday found that the azimuth of a linearly polarized light is rotated after travelling parallel to a magnetic field and to the axis of a flint glass rod with polished faces. The azimuth rotation is proportional to the applied magnetic field and to the length of the glass rod. The phenomenon is known as *Faraday rotation*. In 1876, Reverend John Kerr announced azimuth rotations, however much weaker, upon the reflection of a linearly polarized light on a perpendicularly (*i.e.*, in the polar direction) magnetized air-iron interface. The effect is labeled *MO polar*

Kerr rotation.[1] Later, it was found that the azimuth transformation of the incident wave in the Faraday or MO polar Kerr effects is more involved, often being accompanied by the transformation of the wave ellipticity. This is a manifestation of circular birefringence and dichroism induced by magnetization parallel to the propagation vector and a consequence of the dispersion relations.

In the magnetooptic effects observed in reflection, three special (or pure) configurations of magnetization vector with respect to the interface plane and the incident wave are distinguished by symmetry. In the polar geometry mentioned above, the magnetization is oriented normal to the interface. The MO longitudinal Kerr effect (discovered in 1878) takes place when the magnetization is parallel both to the interface and to the plane of incidence. Again, upon reflection, a linearly polarized incident wave is transformed to an elliptically polarized one, with the major axis of the polarization ellipse rotated with respect to the incident azimuth.

The MO transverse Kerr effect (the magnetization lies in the plane of interface and normal to the plane of incidence) was predicted by Cornelis Harm Wind in 1897 and observed by Peter Zeeman in 1898. Here, only the wave with its electric field vector polarized in the plane of incidence (called p wave or TM wave) displays changes.[2] These manifest themselves by transformations of both the amplitude and phase of the TM reflected wave linear (or more generally, odd) in magnetization.

The three Kerr effects belong to the category of magnetooptic effects that take place upon the reflection at surfaces and interfaces. As a rule, they are much weaker than the effects that arise from the propagation in bulk media, *e.g.*, Faraday effect. In fact, the interface MO effects, which take place in both reflection and transmission, are much easier to detect in media with spontaneous magnetization established below their Curie or Néel temperatures like ferromagnets, ferrimagnets, weak ferromagnets, *etc*. Moreover, even in materials with spontaneous magnetization, the interface effects are easy to detect in the spectral regions with a strong MO activity, only, *i.e.*, close to resonances of the electron transitions where the material absorbs strongly.

The Faraday effect or magnetic circular birefringence and dichroism (MCB and MCD) in transmission and magnetooptic Kerr effect (MCB and

---

[1] This should not be confused with another discovery of this scientist: the electrooptical Kerr effect, which occurs when the polarized light travels across a medium perpendicularly to an applied electric field vector; this is a transmission effect even (quadratic) in the applied electric field.

[2] Note that this behavior is due to the dominant contributions in the optical region from the electric dipole electron transition. The transverse effect would be observed on s polarized or TE waves in the spectral (*e.g.*, microwave) regions where the magnetic dipole electron transitions dominate. At a nonzero magnetization, the latter are characterized by a magnetic permeability tensor and a scalar permittivity.

MCD in reflection) are classified as *linear magnetooptic effects*, *i.e.*, the effects linear (or more precisely, odd) in magnetization and consequently changing sign upon the reversal of magnetization. This property is essential for the applications in *magnetooptic recording*, where the two polarities of magnetization represent logic zero and one.

In a magnetized but otherwise optically isotropic medium, the circularly polarized (CP) waves travel parallel to the magnetization vector, $\boldsymbol{M}$, with different velocities assigned to them in a fixed point of space according to the sense of wave field vector rotation as the time elapses with respect to the orientation of $\boldsymbol{M}$. Note that the definition common in the ellipsometry classifies the CP polarization according to the sense of rotation of the wave field vectors with respect to the sense of wave propagation: In a fixed point of space, an observer looking against the coming wave perceives the right and left circularly polarized (RCP and LCP) waves with their field vectors rotating clockwise and counterclockwise, respectively [3]. This magnetization sensitive propagation of CP waves explains unidirectional behavior of the Faraday rotation. From the practical point of view, the phenomenon of the unidirectional, *i.e.*, nonreciprocal, wave propagation represents an important unique property, not achievable by other means.

It was Fresnel who made the first important step to understanding the principle of the unidirectional propagation. In his natural optical activity studies, he realized that a linearly polarized (LP) wave can be decomposed into two oppositely circularly polarized waves of equal amplitudes. The azimuth of the LP wave is governed by the difference in the initial phases of the CP waves. The idea was later found useful for the explanation of the propagation of the linearly polarized wave parallel to the magnetization manifested by the Faraday rotation. The unidirectional nature of the linearly polarized waves propagating parallel to the magnetization was discovered in 1885 by Lord Rayleigh, John William Strutt. He wrote, "*... I drew attention to a peculiarity of magnetic rotation of the plane of polarisation arising from the circumstance that the rotation is in the same absolute direction whichever way the light may be travelling*" [4]. The phenomenon of unidirectional propagation allows the construction of isolators and circulators, devices widely used at microwave frequencies. These devices are currently developed for optical communication frequencies in integrated optoelectronics versions to perform the same functions, in particular, the protection of sources (semiconductor lasers) against back reflections occurring in transmission lines (optical fibers).

In 1897, W. Voigt observed linear birefringence induced by the magnetic field oriented normal to the propagation vector of a linearly polarized beam. The same scientist wrote the first book on electrooptics and magnetooptics, which appeared in 1908 [5]. The manifestations of the Voigt effect, magnetic linear birefringence and accompanying magnetic linear dichroism, abbreviated as MLB and MLD, respectively, are insensitive to the magnetic field polarity. These transmission effects are, therefore, classified as *quadratic MO effects, i.e.*, the MO effects quadratic (or more precisely, even) in magnetization. There also exist MLB and MLD in reflection labelled simply even (or quadratic) MO effects in reflection.

The long-sought effect of magnetic field on light emission was finally discovered by Peter Zeeman in 1897. The microscopic model was almost immediately provided by Lorentz with his electron theory. The difficulties in the explanation with the classical Lorentz electron theory of the later discovered *anomalous Zeeman effect* stimulated the progress in the development of quantum theory. The notions like electron spin angular momentum or spin-orbit coupling had to be introduced in order to interpret the effect.

1.1.2 Twentieth Century

1.1.2.1 Metals

Ferromagnetic metals have been the subject of the extensive optical and magnetooptic studies since the beginning of the 20th century [2,6]. Most of the earlier results, however, are of little value for the present multilayer studies because of contamination of their highly reactive surfaces and excessive levels of impurities in the samples. The optical response of magnetic thin metallic films covered by dielectric thin film structures was widely studied since the early 1960s, with the aim to enhance the MO response in the multilayer system.

Already at that time, the interest was motivated by the potential exploitation of ferromagnetic metallic thin films in an optically accessed high-density (nonvolatile rewritable) information storage. Their high-saturation magnetization stabilized the in-plane orientation of the magnetization vector. For optical reading of the recorded binary information, represented by the magnetization vector $\pm M$, in the plane of incidence, a rather weak MO longitudinal Kerr effect was considered. An improvement was expected from multiple reflections in dielectric layers on the magnetic films. It was soon realized that MO azimuth or ellipticity enhancement with overlayers, which was not difficult to achieve, did not solve the problem. As a rule, the MO enhancement was usually accompanied with reduced irradiance of the output beam. In order to achieve the minimum bit size and consequently the maximum recording density, it is not sufficient to maximize the difference in the response azimuths of the reflected beams corresponding to the state one and zero of the information bit. We have to maximize the difference in energy delivered to the detector from the accessed information bit in the state zero and one. An appropriate design of the overlayer structure, a kind of a wave impedance matching network, optimizes the MO response ultimately determined by the intrinsic magnetic material parameters. To improve the MO response, we have, therefore, to look for new materials with strong MO active electron transitions in the spectral region of interest. Remarkable examples of the high MO efficiency are bismuth or rare-earth doped iron garnet films.

In the 1980s, the research in MO recording media was extended to magnetic multilayers, planar structures composed of alternating ultrathin[3] magnetic and nonmagnetic films. The properties of these media are controlled by interfaces. The highest perfection was achieved in sandwiches with ferromagnetic elemental metallic films epitaxially deposited on single-crystal substrates. Using extremely flat substrate surfaces, the film thickness could be adjusted with the precision down to a single atomic layer. Layer by layer growth with sharp interfaces on areas with linear dimensions as large as hundreds of atomic spacing was achieved. The reduced dimensionality, specific local coordination imposed by the substrate, and electron coupling to the substrate or to adjacent layers resulted in ultrathin film properties often remarkably different from the corresponding bulk phases [7]. Note that the progress was made possible by previous achievements in the ultrahigh vacuum technology developed primarily for the semiconductor industry.

The availability of these new materials artificially prepared on an atomic scale offered the opportunity for academic research in theoretical and experimental magnetism. A new and deeper insight into the nature of magnetic ordering from the fundamental point of view resulted from the confrontation between experiment and models. Good examples are range of exchange interactions, origin of interface induced and magnetocrystalline anisotropies, magnetization switching, exchange coupling between adjacent layers, *etc*.

Magnetooptic techniques developed into sensitive tools in diagnostics of structures containing magnetic ultrathin (*i.e.*, only a few atomic monolayers thick) layers [7]. Most extensively employed are the MO effects linear in magnetization; *i.e.*, the Faraday effect in light transmission and the MO Kerr effects in reflection. As mentioned above, thanks to advanced film engineering, magnetic multilayers can display a precisely defined composition profile. For example, they can be produced with an interface roughness of the order of 10^{-3} of the visible radiation wavelength. This makes them remarkable objects for the confrontation between the theory of electromagnetic plane waves in magnetic multilayers and MO experiments.

Compared to bulk materials, ultrathin magnetic structures possess specific optical properties for at least two reasons. First, the thicknesses are comparable to or smaller than the penetration depth (20–40 nm in metals), and both propagation and interface contributions to the magnetization dependent optical response must be accounted for. Moreover, because of the reduced magnetic thickness and the presence of adjacent media, the MO properties of the ultrathin magnetic layer more or less depart from those of

[3] The thickness of ultrathin films is in the nanometer range.

corresponding bulk (or thin film) medium and may become nonuniform across its thickness. We note that a clear manifestation of reduced dimension can only be achieved in ultrathin film magnetic sandwiches with extremely flat interfaces. All these effects should be accounted for in the MO response simulations, which usually start from the optical and MO data for bulk metals.[4]

In magnetic multilayers, the control of the profile is restricted to one dimension, *i.e.*, to the control of the ultrathin film thickness. As a natural next step, the research focused on the lateral patterning and sensing in nanometer scale. The main motivation came from the magnetic recording industry. Concurrent large-scale activities take place in the semiconductor industry and also in the research field of nanophotonics. The latter is focused on frequency selective photonic devices. The incorporation of magnetic elements into these nanostructures offers the feasibility of nonreciprocal propagation or magnetically controlled functions. This forms the subject of the new field of *magnetophotonics*.

1.1.2.2 Nonmetals

Among nonmetals, most attention was focused on ferrimagnetic oxides with garnet structure, the optical transparency of which was discovered in 1959 [8,10,11]. In the 1970s and 1980s, thin films of magnetic garnets prepared by liquid phase epitaxy (LPE) on single crystal gadolinium gallium garnet (GGG) substrates were the subject of extensive research in leading world laboratories. The research was motivated by magnetic bubble domain memory, a concept using nonmechanical access to the stored information. These films are optically transparent for visible light and exhibit large Faraday rotation. Thanks to this unique property, theoretically predicted domain structures could be verified using MO microscopy. This circumstance enabled a progress in fundamental understanding of magnetic domain static and dynamic properties. Since then, many magnetic oxide films with garnet, spinel, and perovskite structures were prepared by sputtering or pulsed laser deposition. The latter technique was found useful for the preparation of high-quality films of manganite perovskites with the thickness ranging between 20 to 40 nm. The material displays the phenomenon of colossal magnetoresistance. An interesting correlation between electron transport properties and MO spectra in these films was observed.

MO techniques were also applied to many other nonmetallic materials, often of lower than cubic crystalline symmetry [12]. We only mention in

---

[4] The correlation in the optical data cannot be straightforward, as the differences between the ultrathin film and bulk data do not originate exclusively from the interface effects in ultrathin films. Due to the small volume growth of ultrathin films in ultrahigh vacuum conditions, a high level of perfection and purity is much easier to achieve than in their bulk or thick film counterparts.

passing an extensive research in semiconductor magnetooptics where the main activity was focused on the effects observed in transmission [13,14]. Note that most semiconductors lack spontaneous magnetization. Here, the magnetooptic effects are in general weaker and linearly dependent on the magnitude of the applied magnetic field. Recently, much attention was paid to by Mn doped gallium arsenide, a magnetic semiconductor with the Curie temperature in the range of 100 Kelvin investigated for spin electronics applications.

1.1.2.3 Magnetooptic Vector Magnetometry

Magnetooptic techniques with highly coherent monochromatic (laser) radiation provide a sensitive, nondestructive, noninvasive, and noncontact tool in the metrology of magnetic multilayers and nanostructures. For example, magnetic characteristics in ferromagnetic ultrathin films with thicknesses approaching a single atomic layer with a laser beam diameter close to the diffraction limit can be sensed magnetooptically. Even deeply buried magnetic layers can be detected with magnetooptics. Remember that the thickness of ultrathin magnetic films falls in the region where the magnetooptic effects originating at interfaces are comparable to those originating from the wave propagation. From the response to an incident wave of defined polarization obtained experimentally and compared with the model, valuable information on the multilayer parameters can be deduced, *e.g.*, magnetization vector in-depth profile, magnetic anisotropy, magnetic hysteresis, individual layer thickness, detection of alloying or mixing at interfaces, *etc*. In this procedure, the nanostructure parameters are deduced from the solution of the inverse problem. The magnetooptic vector magnetometry, both *ex situ* and *in situ* in a wide temperature range, was successfully exploited in the development of magnetic recording media, magnetic random access memory structures, in the studies of giant or colossal magnetoresistance, spin electronics, *etc*. [24,25].

In the single wavelength MO (vector) magnetometry, the magnetic information on the magnetic state in a specimen is deduced from electron transitions betweens the ground and excited states. In optical regions, energy differences between ground and excited states exceed the energy required for magnetic ordering by several orders of magnitude. The MO response depends on the nature of these electron transitions, whereas the magnetic characteristic are determined by the electron ground state and by the electron states separated from the ground state by energies much smaller than those involved in the optical electron transitions. Strictly speaking, the MO magnetometry conveys only indirect information on the magnetic properties. In most cases, this does not present difficulties. But there are situations, *e.g.*, in temperature dependence studies, where the relations between ground and excited states play some role. These effects can be accounted for with the help of MO spectroscopy.

1.1.2.4 Magnetooptic Spectroscopic Ellipsometry

In the MO spectroscopic ellipsometry, the reduced irradiance from available quasimonochromatic light sources (*e.g.*, high-pressure Xe arc lamps followed by prism or grating monochromators) with respect to laser sources makes the signal-to-noise ratio somewhat inferior compared to that achieved with polarized laser radiation. Consequently, the minimum area to which the quasimonochromatic radiation can be focused on sample surfaces is larger and the lateral resolution lower. In addition, the *in situ* and temperature dependence measurements are more difficult to realize.

The MO spectroscopic ellipsometry, in a wide range of photon energies from the near infrared to near ultraviolet part of the spectrum, preferably combined with the classical spectroscopic ellipsometry, provides fundamental information on the energy levels of magnetic electrons. The inverse problem solutions developed in the classical spectroscopic ellipsometry for isotropic surfaces can be generalized for use in the MO spectroscopic ellipsometry. One of the most important achievements of the classical spectroscopic ellipsometry is the capability to detect surface layers of submonolayer thicknesses. The experimentally obtained information is practically never sufficient to describe the studied surface in a unique way. A usually applied procedure consists of confrontation of experimental results with judiciously chosen models. Through mathematical evaluation, the models are compared with the experimentally observed ellipsometric parameters, and the most probable picture of the surface is proposed. Complementary information acquired by other methods of analysis, like normal incidence reflectance spectroscopy, atomic force microscopy, x-ray, electron and neutron diffraction, *etc.*, proved to be particularly helpful in this procedure.

Magnetooptic spectroscopic ellipsometry represents a special case of the spectroscopic ellipsometry of anisotropic surfaces and follows similar lines. As already mentioned, magnetic ultrathin films and multilayers now available with practically ideally sharp planar interfaces represent unique objects for this kind of computer modelling and analysis. The optical anisotropy introduced by atomic order (two-dimensional crystal structure) and magnetization (giving rise to the magnetooptic effects) increases the number of experimentally observable effects. This extends the volume of optical data available from the system response. Correspondingly, the number of the system characteristics to be determined is also increased. In addition, as already mentioned, at thicknesses approaching one or a few atomic layers, the material parameters more or less change with respect to those of bulk phase. Then, in multilayer with a high level of perfection, quantum well effects become observable. Here also, in view of the complex nature of magnetic ultrathin sandwiches and multilayer systems and lack of sufficiently precise knowledge of optical parameters, the interpretation is hardly possible without the recourse to modelling. An appropriate choice of adequate models requires a good deal of insight and experience with similar magnetic multilayer systems. At present, efficient procedures have been developed in the optical metrology of magnetic nanostructures, and significant progress, motivated by practical applications, can be expected.

Compared to the classical ellipsometry, the MO techniques are easier to implement. First, the experiments can often be performed close to the normal light incidence. Second, the experimental data are obtained as a difference in photoelectronic signals at two opposite applied magnetic field polarities without the necessity of mechanical movement. Although the sensitivity of the classical ellipsometry is restricted to the atomic planes close to the sample surface, the MO ellipsometry can add information even on deeply buried magnetic layers. The input spectroscopic MO data for the inverse problem solution in a given magnetic nanostructure are, in principle, richer than those obtained with the fixed wavelength MO magnetometry. This allows the magnetic nanostructure parameters to be extracted with improved reliability.

1.1.2.5 Practical Applications

As already mentioned, the importance of the optical effects in magnetic multilayers goes beyond the material research field. The magnetization-dependent optical response provides the basis for construction of several kinds of MO devices, *e.g.*, MO memories, MO magnetic field sensors, MO modulators, and integrated optoelectronic (or magnetophotonic) devices for fiber communication wavelengths, like circulators and isolators. Among practical applications, the most important position from the market viewpoint belongs to thermally assisted MO recording.

The concept of a memory with thermomagnetic recording and MO sensing of recorded information using nonmechanical access was proposed in 1965 [15]. The writing threshold was achieved using an anomalous temperature dependence of the coercive force in the region of either the Curie temperature or the ferrimagnetic compensation temperature. The writing uses a localized heating with a laser beam of a selected information unit (bit) and a coincident magnetic field pulse (leaving the data written on the adjacent bits unchanged thanks to a high coercive force in the unheated areas held at the ambient temperature). The magnetic easy axis is chosen normal to the film plane, the configuration favorable for high recording densities (perpendicular recording) as well as for easy magnetooptic sensing at the normal incidence of a focused beam.

Several kinds of materials were considered as magnetooptic recording films before the magnetooptic memories (with thermomagnetic recording) become commercially available. Ferrimagnetic iron garnets and amorphous 3d-transition metal-4f-rare-earth metal alloys were the most investigated candidates for more than 30 years; the latter were finally developed to the level of practical applications. Present effort remains focused on the increase of recording densities, *e.g.*, using multilayer design. The recording density after several decades of systematic improvement approaches the physical limits of current materials. New approaches based on deeper physical insight are needed. A remarkable chemical stability was achieved. Magnetooptic storage data stability is guaranteed for 50 years, the expected one reaches even 100 years. Note that the access remains all the time mechanical, using a rotating disk. The random access MO memory with non-mechanical beam addressing remains a challenge for the future.

1.2 Magnetized Medium

To understand the origin of magnetooptic effects on a phenomenological basis, it is not necessary to go into details of microscopic models describing the response of magnetic media to electromagnetic plane waves. We start from Maxwell equations assuming magnetization-dependent material tensors. Their form follows from symmetry arguments. In the simplest case of an originally isotropic medium subjected to an applied uniform time-independent magnetic flux density field, $\boldsymbol{B}_{\mathrm{appl}}$, the symmetry of the permittivity tensor is determined by the symmetry group of the axial vector $\boldsymbol{B}_{\mathrm{appl}}$ alone. A uniform magnetization, $\boldsymbol{M}$, induced by $\boldsymbol{B}_{\mathrm{appl}}$ is parallel to $\boldsymbol{B}_{\mathrm{appl}}$ and displays the symmetry of $\boldsymbol{B}_{\mathrm{appl}}$. The magnetization, $\boldsymbol{M}$, *i.e.*, the magnetic dipole moment density, can be characterized by an equivalent current density. For the present purpose it is thus sufficient to consider the symmetry of a current (*i.e.*, with a specified preferred sense) circular loop. The loop at its center generates magnetic dipole moment normal to the loop plane. The sense of the dipole moment is related to the sense of the current in the loop by the right-hand rule.

The choice of the material tensors depends on the spectral range of concern. In the optical region of electromagnetic spectrum, we can describe the medium either by complex conductivity or by complex electric susceptibility tensors. We choose the electric susceptibility tensor that is simply related to the relative permittivity tensor in Maxwell equations. In the optic spectral region, magnetic susceptibility assumes very small values monotonously dependent on the radiation frequency. It is, therefore, reasonable to set its value to a real scalar constant equal to a very small number or zero. In practice, the optical characteristics of the magnetic material are assumed to be completely contained in the electric susceptibility tensor. This is so-called *electric dipole approximation*. In most situations, we can also ignore the effect of crystalline structure in a medium. This leads to the simplest form of the material tensor in a magnetic medium. The substitution of the tensor into Maxwell equations allows us to demonstrate all relevant magnetooptic effects (either linear or quadratic in magnetization), *i.e.*, the

Faraday and Voigt effects observed in transmission and the magnetooptic polar and longitudinal Kerr and transverse effects observed in reflection.

We first derive the form of the electric susceptibility tensor $\boldsymbol{\chi}$ in a linear[5] homogeneous medium. Note that tensor $\boldsymbol{\chi}$ characterizes the medium local response to a monochromatic electromagnetic plane wave and relates the electric field vector, $\boldsymbol{E}(\boldsymbol{r})$, in the medium to its induced electric dipole moment density, $\boldsymbol{P}(\boldsymbol{r})$, at the same position specified by the position vector $\boldsymbol{r}$. In Cartesian coordinates, using SI units this relation assumes the form ($\varepsilon_{\mathrm{vac}}$ is the vacuum permittivity)

$$\begin{pmatrix} P_x \\ P_y \\ P_z \end{pmatrix} = \varepsilon_{\mathrm{vac}} \begin{pmatrix} \chi_{xx} & \chi_{xy} & \chi_{xz} \\ \chi_{yx} & \chi_{yy} & \chi_{yz} \\ \chi_{zx} & \chi_{zy} & \chi_{zz} \end{pmatrix} \begin{pmatrix} E_x \\ E_y \\ E_z \end{pmatrix}. \tag{1.1}$$

The symmetry point group of an originally isotropic medium subjected to a uniform magnetic flux density field contains all proper rotations about the principal axis and the rotations times reflection in a plane normal to this axis. In the Schönflies notation, it is denoted as $C_{\infty h}$ [16,17]. Its international symbol is $\frac{\infty}{m}$. The symmetry group operations leave current carrying circular loops fixed in planes normal to the principal axis unchanged.

Let us choose the principal axis parallel to the z axis of our Cartesian system. The symmetry requires that the electric susceptibility tensor is invariant under the operation of the group represented by matrix transformations

$$\begin{pmatrix} \cos\varphi & \sin\varphi & 0 \\ -\sin\varphi & \cos\varphi & 0 \\ 0 & 0 & \pm 1 \end{pmatrix} \begin{pmatrix} \chi_{xx} & \chi_{xy} & \chi_{xz} \\ \chi_{yx} & \chi_{yy} & \chi_{yz} \\ \chi_{zx} & \chi_{zy} & \chi_{zz} \end{pmatrix} \begin{pmatrix} \cos\varphi & -\sin\varphi & 0 \\ \sin\varphi & \cos\varphi & 0 \\ 0 & 0 & \pm 1 \end{pmatrix}. \tag{1.2}$$

Here, φ is an arbitrary angle of rotation about the magnetization z axis and the $\pm$ signs indicate the proper (+) and improper (−) rotations, respectively. The resulting complex electric susceptibility tensor takes the form

$$\begin{pmatrix} \chi_{xx} & \chi_{xy} & 0 \\ -\chi_{xy} & \chi_{xx} & 0 \\ 0 & 0 & \chi_{zz} \end{pmatrix}. \tag{1.3}$$

We have focused on the complex electric susceptibility tensor. The same arguments are valid for the complex permittivity tensor and for the complex conductivity tensor with the same results. In a nonabsorbing medium, the electric susceptibility tensor should be Hermitian ($\chi_{ij} = \chi_{ji}^*$, $i, j = x, y,$ and z). The asterisk superscript (*) indicates the complex conjugate. In this case, the diagonal tensor elements should be real and the off-diagonal

[5] The linearity means that the tensor is independent of the wave electric field, $\boldsymbol{E}$, and the wave magnetic flux density, $\boldsymbol{B}$.

ones imaginary pure [18]. In a more general case of dispersive and absorbing magnetic media, the complex electric susceptibility tensor consists of frequency-dependent Hermitian and anti-Hermitian parts.

The symmetry operations of reflection in the planes containing the principal axis do not belong to the symmetry group of uniform magnetization. In fact, they produce the magnetization reversal

$$\begin{pmatrix} -\cos 2\beta & \sin 2\beta & 0 \\ \sin 2\beta & \cos 2\beta & 0 \\ 0 & 0 & 1 \end{pmatrix} \begin{pmatrix} \chi_{xx} & \chi_{xy} & 0 \\ -\chi_{xy} & \chi_{xx} & 0 \\ 0 & 0 & \chi_{zz} \end{pmatrix} \begin{pmatrix} -\cos 2\beta & \sin 2\beta & 0 \\ \sin 2\beta & \cos 2\beta & 0 \\ 0 & 0 & 1 \end{pmatrix}. \quad (1.4)$$

The result of the transformation does not depend on the angle β, which specifies the orientation of the reflection planes. We obtain the following rules for the effect of magnetization reversal on the tensor elements

$$\chi_{xx}(M_z) = \chi_{xx}(-M_z) \quad (1.5a)$$

$$\chi_{zz}(M_z) = \chi_{zz}(-M_z) \quad (1.5b)$$

$$\chi_{xy}(M_z) = -\chi_{xy}(-M_z) = -\chi_{yx}(M_z). \quad (1.5c)$$

We arrive at the conclusion that in a magnetized medium, which was isotropic before switching the magnetization on, the diagonal and off-diagonal elements of the complex electric susceptibility tensor are even and odd functions of the magnetization, respectively. This is consistent with a more general, so-called, Onsager relation [19,20], which for an arbitrary orientation of the magnetization vector, $\boldsymbol{M}$, requires

$$\chi_{ij}(\boldsymbol{M}) = \chi_{ji}(-\boldsymbol{M}). \quad (1.6)$$

In this form, the Onsager relation is restricted to electric dipole effects. This is sufficient, as the medium response is assumed local and completely characterized by an electric susceptibility tensor.

The Cartesian electric susceptibility tensor (1.3) can be transformed into an electric susceptibility tensor in circular representation using a unitary matrix and its inverse as follows

$$\frac{1}{\sqrt{2}} \begin{pmatrix} 1 & -\mathrm{j} & 0 \\ 1 & \mathrm{j} & 0 \\ 0 & 0 & \sqrt{2} \end{pmatrix} \begin{pmatrix} \chi_{xx} & \chi_{xy} & 0 \\ -\chi_{xy} & \chi_{xx} & 0 \\ 0 & 0 & \chi_{zz} \end{pmatrix} \frac{1}{\sqrt{2}} \begin{pmatrix} 1 & 1 & 0 \\ \mathrm{j} & -\mathrm{j} & 0 \\ 0 & 0 & \sqrt{2} \end{pmatrix}, \quad (1.7)$$

where j represents the imaginary unit. In the circular representation, the material relation (1.3) becomes

$$\begin{pmatrix} P_x - \mathrm{j}P_y \\ P_x + \mathrm{j}P_y \\ P_z \end{pmatrix} = \frac{1}{2} \begin{pmatrix} \chi_{xx} + \mathrm{j}\chi_{xy} & 0 & 0 \\ 0 & \chi_{xx} - \mathrm{j}\chi_{xy} & 0 \\ 0 & 0 & 2\chi_{zz} \end{pmatrix} \begin{pmatrix} E_x - \mathrm{j}E_y \\ E_x + \mathrm{j}E_y \\ E_z \end{pmatrix}. \quad (1.8)$$

1.3 Propagation Parallel to Magnetization

The information just obtained on the electric susceptibility tensor is sufficient for a demonstration of the magnetooptic effects. We take the Maxwell curl equations in the form

$$\nabla \times \boldsymbol{B} = c^{-2} \varepsilon \frac{\partial \boldsymbol{E}}{\partial t} \tag{1.9a}$$

$$\nabla \times \boldsymbol{E} = -\frac{\partial \boldsymbol{B}}{\partial t}, \tag{1.9b}$$

where $c = (\varepsilon_{\text{vac}} \mu_{\text{vac}})^{-1/2}$ denotes the electromagnetic wave phase velocity in a vacuum and ε is the relative permittivity tensor defined as a sum of the unit diadic $\mathbf{1}$, the identity tensor, and the electric susceptibility tensor

$$\varepsilon = \mathbf{1} + \chi. \tag{1.10}$$

In a Cartesian system, the differential operator ∇ takes the form

$$\nabla = \hat{x}\frac{\partial}{\partial x} + \hat{y}\frac{\partial}{\partial y} + \hat{z}\frac{\partial}{\partial z}, \tag{1.11}$$

$\hat{x}$, $\hat{y}$, and $\hat{z}$ being the Cartesian unit vectors. The time derivatives are denoted with $\frac{\partial}{\partial t}$. Eliminating $\boldsymbol{B}$, the magnetic flux density vector, we arrive at the wave equation for the electric field vector

$$\nabla \times (\nabla \times \boldsymbol{E}) = -c^{-2} \varepsilon \frac{\partial^2 \boldsymbol{E}}{\partial t^2}. \tag{1.12}$$

We have assumed that the medium is infinite, homogeneous, time-invariant, and linear, *i.e.*, characterized by ε independent of $\boldsymbol{E}$. We now apply the relation

$$\nabla \times (\nabla \times \boldsymbol{A}) = \nabla (\nabla \cdot \boldsymbol{A}) - \nabla^2 \boldsymbol{A} \tag{1.13}$$

valid for a vector $\boldsymbol{A}$. This allows us to write the wave equation in a more convenient form

$$\nabla^2 \boldsymbol{E} - \nabla (\nabla \cdot \boldsymbol{E}) = c^{-2} \varepsilon \frac{\partial^2 \boldsymbol{E}}{\partial t^2}. \tag{1.14}$$

The solution can be found for a monochromatic plane wave with $\boldsymbol{E}$, harmonic both in time, t, and position, $\boldsymbol{r}$. Expressed in the complex (or phasor) notation, $\boldsymbol{E}$ becomes

$$\boldsymbol{E} = \boldsymbol{E}_0 \exp\left[\mathrm{j}\,(\omega t - \gamma \cdot \boldsymbol{r})\right], \tag{1.15}$$

where the quantities conjugated with t and $\boldsymbol{r}$ are the angular frequency, ω and the propagation vector γ, respectively. The constant complex amplitude of the electric field vector, $\boldsymbol{E}_0$, specifies the wave polarization.

The simplest and the most important situation takes place when the monochromatic plane wave propagates parallel to $\boldsymbol{M}$. For $\boldsymbol{M} = \hat{\mathbf{z}}M$ this corresponds to the propagation vector parallel to the z axis

$$\gamma = \gamma\hat{\mathbf{z}} = \frac{\omega}{c}N_z\hat{\mathbf{z}}. \tag{1.16}$$

The solution is independent of the coordinates x and y and the corresponding derivatives in Eq. (1.14) are zero

$$\begin{pmatrix} \frac{\partial^2}{\partial z^2} & 0 & 0 \\ 0 & \frac{\partial^2}{\partial z^2} & 0 \\ 0 & 0 & 0 \end{pmatrix} \begin{pmatrix} E_x \\ E_y \\ E_z \end{pmatrix} = c^{-2} \begin{pmatrix} \varepsilon_{xx} & \varepsilon_{xy} & \varepsilon_{xz} \\ \varepsilon_{yx} & \varepsilon_{yy} & \varepsilon_{yz} \\ \varepsilon_{zx} & \varepsilon_{zy} & \varepsilon_{zz} \end{pmatrix} \frac{\partial^2}{\partial t^2} \begin{pmatrix} E_x \\ E_y \\ E_z \end{pmatrix}.$$

The space and time derivatives may be replaced by the factors according to $\frac{\partial^2}{\partial z^2} \rightarrow -\gamma^2$ and $\frac{\partial^2}{\partial z^2} \rightarrow -\omega^2$. Then, the partial differential wave equation transforms to an algebraic one

$$\begin{pmatrix} \gamma^2 E_x \\ \gamma^2 E_y \\ 0 \end{pmatrix} = \frac{\omega^2}{c^2} \begin{pmatrix} \varepsilon_{xx} & \varepsilon_{xy} & \varepsilon_{xz} \\ \varepsilon_{yx} & \varepsilon_{yy} & \varepsilon_{yz} \\ \varepsilon_{zx} & \varepsilon_{zy} & \varepsilon_{zz} \end{pmatrix} \begin{pmatrix} E_x \\ E_y \\ E_z \end{pmatrix}. \tag{1.17}$$

Here, we have assumed that Eq. (1.1) relates the Fourier components of the electric dipole moment density and electric field of the wave as

$$\boldsymbol{P}(\omega) = \varepsilon_{\text{vac}}\boldsymbol{\chi}(\omega, \boldsymbol{M})\boldsymbol{E}(\omega). \tag{1.18}$$

The response of a magnetic medium to a monochromatic plane wave of Eq. (1.15) may be expressed with Eq. (1.10)

$$\boldsymbol{\varepsilon}(\omega, \boldsymbol{M}) = \mathbf{1} + \boldsymbol{\chi}(\omega, \boldsymbol{M}), \tag{1.19}$$

where we have emphasized the dependence on ω and $\boldsymbol{M}$. As the only nonzero component of the propagation vector is parallel to the z axis, we may write

$$\frac{\omega}{c}N_z = \pm\frac{\omega}{c}N, \tag{1.20}$$

where N denotes the complex index of refraction. The positive and negative signs account for the two senses of propagation. For the space and time dependence chosen in Eq. (1.15), the complex index of refraction takes the form

$$N = n - \mathrm{j}k, \tag{1.21}$$

where n is the real index of refraction and k is the extinction coefficient. By convention, $n > 0$ and $k > 0$.

We next look for the propagation vectors and polarizations consistent with the wave equation and expressed in terms of the medium parameters, *i.e.*, in terms of the components of the permittivity tensor. Substituting for the permittivity tensor in a medium magnetized parallel to the z axis according to Eq. (1.3) and Eq. (1.10) into Eq. (1.17),

$$\begin{pmatrix} \gamma^2 E_x \\ \gamma^2 E_y \\ 0 \end{pmatrix} = \frac{\omega^2}{c^2} \begin{pmatrix} \varepsilon_{xx} & \varepsilon_{xy} & 0 \\ -\varepsilon_{xy} & \varepsilon_{yy} & 0 \\ 0 & 0 & \varepsilon_{zz} \end{pmatrix} \begin{pmatrix} E_x \\ E_y \\ E_z \end{pmatrix}, \tag{1.22}$$

and making use of Eq. (1.16), we get the system of three equations for the Cartesian components of the vector field amplitude

$$\left(N_z^2 - \varepsilon_{xx}\right) E_{0x} - \varepsilon_{xy} E_{0y} = 0, \tag{1.23a}$$

$$\varepsilon_{xy} E_{0x} + \left(N_z^2 - \varepsilon_{xx}\right) E_{0y} = 0, \tag{1.23b}$$

$$\varepsilon_{zz} E_{0z} = 0. \tag{1.23c}$$

To get nontrivial solutions for the amplitude components E_{0x}, E_{0y}, and E_{0z}, the determinant of the equation system (1.23) should be zero, *i.e.*,

$$\varepsilon_{zz}\left[N_z^4 - 2\varepsilon_{xx} N_z^2 + \left(\varepsilon_{xx}^2 + \varepsilon_{xy}^2\right)\right] = 0. \tag{1.24}$$

We obtain a complex quadratic equation in N_z^2. This is a proper value equation with the solutions conventionally distinguished by $\pm$ signs as

$$(N_{z\pm})^2 = \varepsilon_{xx} \pm \mathrm{j}\varepsilon_{xy}. \tag{1.25}$$

The proper values of the complex index of refraction are independent of the sense of propagation. From Eq. (1.5c), it follows $\varepsilon_{xy}(M_z) = -\varepsilon_{xy}(-M_z)$. Consequently, upon the magnetization reversal, the two proper values are exchanged. This behavior is typical for the MO effect linear (or more precisely odd) in magnetization.

Substituting for $N_z^{\pm}$, the proper values, either into Eq. (1.23a) or Eq. (1.23b), we can specify the proper polarizations in terms of the relations between the amplitudes E_{0x} and E_{0y}

$$E_{0y} = \pm\mathrm{j}E_{0x} = \pm|E_{0x}|\exp\left[\mathrm{j}\left(\phi_{0x} \pm \frac{\pi}{2}\right)\right], \tag{1.26}$$

where ϕ_{0x} and $|E_{0x}|$ are the initial phase and the absolute value of the complex amplitude E_{0x}. The upper and lower signs correspond to the waves that propagate with $(\omega/c)\, N_+$ and $(\omega/c)\, N_-$, respectively.

Let us first consider the upper (+) sign, N_+, in Eq. (1.25). After the substitution of Eq. (1.26) with the upper sign (+) into Eq. (1.15) with Eq. (1.16), we can determine the real parts of the complex field components

$$\Re\{E_x\} = |E_{0x}| \cos\left[\omega t \mp N_+ \left(\frac{\omega}{c}\right) z + \phi_{0x}\right], \tag{1.27a}$$

$$\Re\{E_y\} = -|E_{0x}| \sin\left[\omega t \mp N_+ \left(\frac{\omega}{c}\right) z + \phi_{0x}\right], \tag{1.27b}$$

where the $\mp$ signs distinguish the sense of propagation, the forward one for the upper (−) sign and the backward (retrograde) one for the lower (+) sign. The symbol $\Re$ denotes the real part of the complex numbers representing the electric field of the waves. In a fixed plane normal to the z axis of our coordinate system fixed to the medium and the same for the forward and backward going waves, the endpoint of the real field vector traces, as time, t, elapses, a circle in the sense from the positive y axis to the positive x axis irrespective of the sense of propagation. At a fixed t, the endpoints of the vectors representing the fields travelling with $(\omega/c)\, N_+$ in different points on the z axis form the helices that differ in their chiralities: a right-handed helix for the forward-going wave and a left-handed helix for the backward-going wave.

For the lower (−) sign, N_-, in Eq. (1.25) the real parts of the complex field components become

$$\Re\{E_x\} = |E_{0x}| \cos\left[\omega t \mp N_- \left(\frac{\omega}{c}\right) z + \phi_{0x}\right], \tag{1.28a}$$

$$\Re\{E_y\} = |E_{0x}| \sin\left[\omega t \mp N_- \left(\frac{\omega}{c}\right) z + \phi_{0x}\right]. \tag{1.28b}$$

Now, in a fixed plane normal to the z axis the endpoint of the real field vector traces, as time elapses, a circle in the sense from the positive x axis to the positive y axis irrespective of the sense of propagation. At a fixed t, the endpoints of the vectors representing the fields travelling with $(\omega/c)\, N_-$ in different points on the z axis form the helices that differ in their chiralities: a left-handed helix for the forward-going wave and a right-handed helix for the backward-going wave.

According to their proper polarizations, the propagation vectors can be divided into two groups,

$$\frac{E_{0y}}{E_{0x}} = \mathrm{j}, \tag{1.29a}$$

$$\frac{\omega}{c} N_+ = \frac{\omega}{c} (\varepsilon_{xx} + \mathrm{j}\varepsilon_{xy})^{1/2}, \tag{1.29b}$$

$$-\frac{\omega}{c} N_+ = -\frac{\omega}{c} (\varepsilon_{xx} + \mathrm{j}\varepsilon_{xy})^{1/2}, \tag{1.29c}$$

and

$$\frac{E_{0y}}{E_{0x}} = -\mathrm{j}, \tag{1.30a}$$

$$\frac{\omega}{c} N_{-} = \frac{\omega}{c} (\varepsilon_{xx} - \mathrm{j}\varepsilon_{xy})^{1/2}, \tag{1.30b}$$

$$-\frac{\omega}{c} N_{-} = -\frac{\omega}{c} (\varepsilon_{xx} - \mathrm{j}\varepsilon_{xy})^{1/2}. \tag{1.30c}$$

We conclude that in a medium magnetized parallel to the wave propagation direction, the wave equation admits four circularly polarized proper polarizations (or proper polarization modes). Any solution to the wave equation can then be expressed in terms of four complex scalar amplitudes, which we shall denote as $\mathrm{E}_0^{(j)}$ with $j = 1, \ldots, 4$.

In the complex (*i.e.*, phasor) representation, the forward and retrograde CP waves propagating with the propagation vectors $\pm(\omega/c)N_{+}$ may be expressed as

$$\boldsymbol{E}_1 = 2^{-1/2} \left(\hat{\boldsymbol{x}} + \mathrm{j}\hat{\boldsymbol{y}}\right) \mathrm{E}_0^{(1)} \exp\left[\mathrm{j}\left(\omega t - N_{+}\frac{\omega}{c} z\right)\right], \tag{1.31a}$$

$$\boldsymbol{E}_2 = 2^{-1/2} \left(\hat{\boldsymbol{x}} + \mathrm{j}\hat{\boldsymbol{y}}\right) \mathrm{E}_0^{(2)} \exp\left[\mathrm{j}\left(\omega t + N_{+}\frac{\omega}{c} z\right)\right]. \tag{1.31b}$$

The wave $\boldsymbol{E}_1$ ($\boldsymbol{E}_2$) propagates parallel to the z axis in the positive (negative) sense. The sense of rotation of the field vectors $\boldsymbol{E}$ and $\boldsymbol{B}$ of a circular polarization in a plane normal to the propagation vector remains undetermined until we define the sense of the normal to the plane. In the classical ellipsometry, the rotation sense is defined with respect to the sense of wave propagation [3]. Then, the two solutions 1 and 2 pertain to opposite circular polarizations.[6] According to the classical ellipsometry, we would have to classify $\boldsymbol{E}_1$ (the forward wave) as right circularly polarized (RCP) and $\boldsymbol{E}_2$ (the backward wave) as left circularly polarized (LCP). For the forward and retrograde CP waves propagating with the propagation vectors $\pm(\omega/c)N_{-}$, we have

$$\boldsymbol{E}_3 = 2^{-1/2} \left(\hat{\boldsymbol{x}} - \mathrm{j}\hat{\boldsymbol{y}}\right) \mathrm{E}_0^{(3)} \exp\left[\mathrm{j}\left(\omega t - N_{-}\frac{\omega}{c} z\right)\right], \tag{1.32a}$$

$$\boldsymbol{E}_4 = 2^{-1/2} \left(\hat{\boldsymbol{x}} - \mathrm{j}\hat{\boldsymbol{y}}\right) \mathrm{E}_0^{(4)} \exp\left[\mathrm{j}\left(\omega t + N_{-}\frac{\omega}{c} z\right)\right], \tag{1.32b}$$

where the solution $\boldsymbol{E}_3$ ($\boldsymbol{E}_4$) corresponds to an LCP forward and RCP retrograde (backward) wave, respectively. The propagation of the CP waves in an isotropic medium could be described in the same way with $N_{+} = N_{-}$, which, in our choice of the coordinate system, would correspond to $\varepsilon_{xy} = 0$.

---

[6] The convention on CP polarizations is illustrated in Appendix A.

This behavior is basically different from that observed in media displaying natural optical activity. The natural optical activity manifested as circular birefringence (CB) and accompanying circular dichroism (CD) also originates from different phase velocities and different attenuation of the two CP waves, respectively. Upon the reflection the wave of a definite circular polarization (RCP or LCP) returns with an opposite circular polarization (LCP or RCP) but with a different phase velocity and a different attenuation.

The origin of the phenomenon consists of the chiral nature of the optically active medium. We note that an object (crystal, molecule) is said to be chiral if it cannot be brought into congruence with its mirror image by translation and rotation. A simple example that belongs to this category is a helix. In particular, the symmetry operations represented by the lower sign in Eq. (1.2) (leaving the oriented planar loop invariant) change the chirality of the helix to an opposite one and therefore do not belong to the helix symmetry group. A CP wave of a defined handedness (RCP or LCP) travelling along the helix in one direction changes its handedness upon the reflection at the boundary and propagates in the opposite direction at different conditions.

As we have just shown, this is not the case for a CP wave travelling parallel to the magnetization in a medium characterized by the electric susceptibility tensor of Eq. (1.3). Here, both the phase velocities and absorption coefficients do not differ for the forward and backward propagation. The propagation vectors are split into two groups according to their handedness (*i.e.*, the sense of tracing the circle) with respect to the magnetization vector instead of the wave propagation direction. This magnetization-related handedness is of much theoretical and practical importance.

To demonstrate it, let us consider the propagation of CP waves parallel to $\boldsymbol{M}$ in a slab with plane parallel ideally reflecting walls. Both the propagation vectors and the magnetization in the slab are assumed perpendicular to the planar boundaries. The phase difference between the CP wave that travels with the propagation vector $(\omega/c)\, N_+$ and that which travels in the same sense with the propagation vector $(\omega/c)\, N_-$ equals $(N_+ - N_-)\,(\omega d/c)$, where d is the thickness of the slab. It cannot depend on the propagation sense of the two CP waves. As a result, the pair of waves, after multiple reflections at the boundaries, acquires the phase difference given by the product $p\,(N_+ - N_-)\,(\omega d/c)$. Here, p denotes the number of times the waves travelled the distance d across the slab. The phase accumulation after multiple passes is a unique feature of the wave propagation in magnetic media. It provides the physical basis for unidirectional (or nonreciprocal) microwave and optical devices like isolators, circulators, *etc.*

Alternatively, the distinction between MO rotation and natural optical activity can be approached using simple analogies. We have considered symmetry of an originally isotropic medium with the magnetization induced by a circular current loop. We usually generate the magnetic field by a coil that may have a right-handed or a left-handed winding. For the sense of the induced magnetization, the handedness of the winding does not matter. What really matters is the sense of the current in the circle created by the projection of the coil on a plane perpendicular to its axis and to the magnetization vector.

Similarly, the sense of rotation of the field vectors $\boldsymbol{E}$ and $\boldsymbol{B}$ of a CP wave, conventionally specified with respect to the sense of the wave propagation, does not matter in the determination of its complex refractive index (or its phase velocity and attenuation). What really matters here is the sense of rotation of the field vectors $\boldsymbol{E}$ and $\boldsymbol{B}$ (with respect to the sense of the generating current or the equivalent magnetization current in the loop) in a circle created by the projection of the endpoints of the field vectors to a plane normal to the magnetization and the propagation vector. The sense of rotation of $\boldsymbol{E}$ (and $\boldsymbol{B}$) may be either the same (congruent) or opposite (incongruent) with respect to the sense of the equivalent magnetization current. We can then assign one value of the index of refraction to the CP wave with $\boldsymbol{E}$ rotating in the same sense as the generating current and another value of the index of refraction to the CP wave with $\boldsymbol{E}$ rotating in the opposite sense, irrespective of the sense of the wave propagation. The reversal of the magnetization corresponds to the current reversal (equivalent to the reversal of t) and exchanges the roles of the CP polarizations.

The optically active medium does not possess the symmetry of the closed circular loop discussed above. Its symmetry corresponds to the symmetry of a helix. The CP wave of a given handedness, defined with respect to the sense of propagation, travels with the same value of the complex index of refraction across an optically active medium irrespective of the sense of propagation. Let us now consider the propagation of a CP wave in a slab with plane parallel ideally reflecting walls displaying natural optical activity. Upon reflection at the walls, the CP wave changes handedness and travels back with a different value of the refractive index. Therefore, no multiple pass phase accumulation is possible in an optically active medium. The snapshot of the endpoints of $\boldsymbol{E}$ ($\boldsymbol{B}$) forms a helix in the space. Then, the refractive index determining the propagation conditions of a CP wave depend on the chirality of the CP wave with respect to the chirality of the helix representing the symmetry of the optically active medium. There is no time reversal effect on the refractive index here. If we rotate by 180 deg the plate about an axis parallel to the plate interfaces (and exchange the entrance and exit faces of the plate in this way), the symmetry of the helix remains unchanged, and the propagation of the CP waves is not affected by this rotation.

In summary, the propagation of the CP waves parallel to vect M is classified according to the *temporal* dependence of $\boldsymbol{E}$ in a fixed plane normal to $\boldsymbol{M}$ (clockwise or counterclockwise rotation of $\boldsymbol{E}$ with respect to $\boldsymbol{M}$). The propagation of CP waves in a naturally optically active chiral *medium* is classified according to the *spatial* dependence of **E**. The medium displaying natural optical activity can be dextrorotary or levorotary, and the snapshot of a CP wave may have a chirality congruent or incongruent with that of the medium.

1.3.1 Faraday Effect

We now describe the propagation of polarized waves in a magnetized medium neglecting the effect of interfaces. The magnetization and the propagation vectors remain parallel to each other and to the z axis. Let us consider a monochromatic plane wave propagating in the direction of the positive z axis linearly polarized at $z = 0$, its electric field vector being parallel to the x axis, $\mathbf{E}(0) = \hat{x}E_0$. In terms of the azimuth θ and the ellipticity ϵ, the wave polarization at $z = 0$ may be characterized by $\theta = 0$ (for the azimuth origin in the x axis) and $\epsilon = 0$. As first realized by Fresnel, a linearly polarized wave can be understood as a sum of RCP and LCP waves of equal amplitudes. Its azimuth is determined by the difference in the initial phases of the CP waves.

After travelling the distance d parallel to the magnetization, the electric field of the output wave can be written as a sum of the forward going proper polarization modes $\mathbf{E}_1$ and $\mathbf{E}_3$ given by Eq. (1.31a) and Eq. (1.32a)

$$\mathbf{E}(d) = \frac{1}{2}E_0 e^{\mathrm{j}\omega t}\left\{\begin{pmatrix}1\\ \mathrm{j}\end{pmatrix}\exp(-\mathrm{j}N_+\frac{\omega}{c}d) + \begin{pmatrix}1\\ -\mathrm{j}\end{pmatrix}\exp\left(-\mathrm{j}N_-\frac{\omega}{c}d\right)\right\}. \quad (1.33)$$

Here, the sum of two circular polarizations was expressed with help of the normalized CP Jones 2×1 column vectors in Cartesian representation

$$2^{-1/2}\left(\hat{x} \pm \mathrm{j}\hat{y}\right) = 2^{-1/2}\begin{pmatrix}1\\ \pm\mathrm{j}\end{pmatrix}. \quad (1.34)$$

For the waves propagating along the positive z axis, the upper and lower signs correspond to the RCP and LCP waves [3]. We denote $\beta_\pm = N_\pm \omega d/c$ and obtain

$$\mathbf{E}(d) = \begin{pmatrix}E_{0x}(d)\\ E_{0y}(d)\end{pmatrix} = E_0 e^{\mathrm{j}\omega t} e^{-\mathrm{j}(\beta_+ + \beta_-)/2}\begin{pmatrix}\cos[(\beta_+ - \beta_-)/2]\\ \sin[(\beta_+ - \beta_-)/2]\end{pmatrix}. \quad (1.35)$$

Thus, the polarization state of the input wave ($\theta = \epsilon = 0$) corresponds to the linear polarization parallel to the x axis. After travelling the distance d, the wave transforms in such a way that its azimuth is rotated by a complex angle

$$\chi_F = (\beta_+ - \beta_-)/2 = \frac{\omega}{2c}(N_+ - N_-)d = \frac{\omega}{2c}\left[(n_+ - n_-) - \mathrm{j}(k_+ - k_-)\right]d. \quad (1.36)$$

We have used the definition for the complex indices of refraction $N_+ = n_+ - jk_+$ and $N_- = n_- - jk_-$, where n_+ and n_- are the real indices of refraction and k_+ (k_-) are the extinction coefficients for the RCP and LCP waves, respectively.[7] The complex angle χ_F can be approximately expressed in terms of the permittivity tensor components as

$$\chi_F = \mathrm{j}\frac{\omega}{2c}\frac{\varepsilon_{xy}}{\sqrt{\varepsilon_{xx}}}d. \tag{1.37}$$

We employed Eq. (1.29b) and Eq. (1.30b) with the assumption $\varepsilon_{xy} \ll \varepsilon_{xx}$.

To express the polarization state of the output wave $\boldsymbol{E}(d)$ in terms of θ, the azimuth and ϵ, the ellipticity, we employ the Cartesian Jones vector representation [3]

$$\begin{aligned} \begin{pmatrix} E_{0x} \\ E_{0y} \end{pmatrix} &= E_0 \exp(\mathrm{j}\phi_0) \left\{ \begin{pmatrix} \cos\theta\cos\epsilon - \mathrm{j}\sin\theta\sin\epsilon \\ \sin\theta\cos\epsilon + \mathrm{j}\cos\theta\sin\epsilon \end{pmatrix} \right\} \\ &= 2^{-1/2} E_0 \exp(\mathrm{j}\phi_0) \begin{pmatrix} 1 & 1 \\ \mathrm{j} & -\mathrm{j} \end{pmatrix} \left\{ 2^{-1/2} \begin{pmatrix} (\cos\epsilon + \sin\epsilon)\exp(-\mathrm{j}\theta) \\ (\cos\epsilon - \sin\epsilon)\exp(\mathrm{j}\theta) \end{pmatrix} \right\}. \end{aligned} \tag{1.38}$$

E_0 is the real amplitude and ϕ_0 the initial phase. In the second row of Eq. (1.38), the Cartesian Jones vector is expressed as a product of the transformation matrix and the corresponding Jones vector in circular (CP) representation. The curly brackets indicate *normalized* Cartesian and circular (CP) Jones vectors, respectively. The RCP and LCP polarizations are obtained for $\theta = 0$ and $\epsilon = \pm\pi/4$. The linear polarizations parallel to $\hat{x}$ and $\hat{y}$ are obtained for $\theta = 0$ and $\theta = \pi/2$ with $\epsilon = 0$, respectively.

We define the azimuth of the output wave as the Faraday rotation (or magnetic circular birefringence) [8]

$$\theta_F = \frac{\omega}{2c}(n_+ - n_-)d. \tag{1.39}$$

The corresponding Faraday ellipticity is defined as

$$\tan\epsilon_F = -\tanh\left[\frac{\omega d}{2c}(k_+ - k_-)\right] = \tanh\left[\frac{d}{4}(\alpha_- - \alpha_+)\right]. \tag{1.40}$$

It is simply related to magnetic circular dichroism, defined as a difference $(\alpha_+ - \alpha_-)$ of the absorption coefficients

$$\alpha_\pm = \frac{2\omega}{c}k_\pm \tag{1.41}$$

for the RCP and LCP waves.

---

[7] For $(n_+ - n_-) > 0$ the azimuth of the wave travelling parallel to the z axis is rotated from the positive x axis at $z = 0$ towards the positive y axis at $z > 0$, according to Eq. (1.35).

For an incident wave of arbitrary polarization state, the output wave follows from

$$\begin{pmatrix} E_{0x}(d) \\ E_{0y}(d) \end{pmatrix} = e^{-j(\beta_+ + \beta_-)/2} \begin{pmatrix} \cos[(\beta_+ - \beta_-)/2] & -\sin[(\beta_+ - \beta_-)/2] \\ \sin[(\beta_+ - \beta_-)/2] & \cos[(\beta_+ - \beta_-)/2] \end{pmatrix} \begin{pmatrix} E_{0x}(0) \\ E_{0y}(0) \end{pmatrix}, \tag{1.42}$$

where the 2×2 Cartesian Jones matrix for the rotation by a complex angle $(\beta_+ - \beta_-)/2$ represents the Faraday effect.[8]

The picture of LCP and RCP proper polarization waves travelling with different velocities and absorption coefficients does not work in the media with reduced symmetry, as realized by Tabor and Chen [21]. They studied the propagation in magnetic crystals parallel to $\boldsymbol{M}$ fixed to one of the orthorhombic axes. There, the azimuth oscillates with the distance travelled by the wave as opposed to the situation in isotropic media with induced magnetization, where it linearly increases. This is a consequence of the fact that the proper polarization modes in lower symmetry media are no more circularly polarized than the elliptically polarized ones. A similar transition to the elliptical proper polarization modes takes place when the radiation modes are to be replaced by the guided ones in planar structures with the magnetization in the plane of the zigzag travelling wave. Here, the anisotropy originates from the oblique incidence at interfaces resulting in the LP proper modes classified as TE and TM modes at $\boldsymbol{M} = 0$.

1.3.2 Normal Incidence Polar Kerr Effect

Among the magnetooptic effects that take place when the linearly polarized wave is reflected on the plane interface between an ambient and magnetic medium, the most important is the polar Kerr effect at the normal incidence. The propagation vector and magnetization are parallel as in the case of the Faraday effect. In the Cartesian representation, the incident and reflected fields are related by the Cartesian Jones reflection matrix, which must be invariant with respect to the principal axis normal to the interface and parallel to the magnetization

$$\begin{pmatrix} E_x^{(r)} \\ E_y^{(r)} \end{pmatrix} = \begin{pmatrix} r_{xx} & r_{xy} \\ -r_{xy} & r_{xx} \end{pmatrix} \begin{pmatrix} E_x^{(i)} \\ E_x^{(i)} \end{pmatrix}. \tag{1.43}$$

The circularly polarized RCP and LCP plane waves of equal amplitudes, which together form a linearly polarized wave, have different amplitude

[8] The common phase $(\beta_+ + \beta_-)$ is difficult to determine experimentally. In the Jones formalism, we therefore usually set the factor $e^{-j(\beta_+ + \beta_-)/2} = 1$.

reflection coefficients[9] at the interface with vacuum [8]

$$r_\pm = \frac{1 - N_\pm}{1 + N_\pm}. \tag{1.44}$$

We therefore transform the Cartesian Jones reflection matrix into the CP representation

$$2^{-1/2}\begin{pmatrix}1 & -\mathrm{j}\\ 1 & \mathrm{j}\end{pmatrix}\begin{pmatrix}r_{xx} & r_{xy}\\ -r_{xy} & r_{xx}\end{pmatrix}2^{-1/2}\begin{pmatrix}1 & 1\\ \mathrm{j} & -\mathrm{j}\end{pmatrix}$$
$$= \begin{pmatrix}r_{xx} + \mathrm{j}r_{xy} & 0\\ 0 & r_{xx} - \mathrm{j}r_{xy}\end{pmatrix} \equiv \begin{pmatrix}r_+ & 0\\ 0 & r_-\end{pmatrix}. \tag{1.45}$$

We also transform to the CP representation the Jones vectors of Eq. (1.43) and obtain

$$\begin{pmatrix}E_+^{(r)}\\ E_-^{(r)}\end{pmatrix} = 2^{-1/2}\begin{pmatrix}E_x^{(r)} - \mathrm{j}E_y^{(r)}\\ E_x^{(r)} + \mathrm{j}E_y^{(r)}\end{pmatrix} = \begin{pmatrix}r_+ & 0\\ 0 & r_-\end{pmatrix}2^{-1/2}\begin{pmatrix}E_x^{(i)} - \mathrm{j}E_y^{(i)}\\ E_x^{(i)} + \mathrm{j}E_y^{(i)}\end{pmatrix}. \tag{1.46}$$

With the Jones formalism in circular (CP) representation, the wave reflected at the interface is described by

$$\begin{pmatrix}E_+^{(r)}\\ E_-^{(r)}\end{pmatrix} = 2^{-1/2}\begin{pmatrix}r_+ & 0\\ 0 & r_-\end{pmatrix}\begin{pmatrix}1\\ 1\end{pmatrix}E_x^{(i)}. \tag{1.47}$$

Here, the incident wave of the amplitude $E_x^{(i)}$ is assumed polarized along the x axis (*i.e.*, with the zero azimuth, $\theta^{(\mathrm{pol})} = 0$, and zero ellipticity, $\tan\epsilon^{(\mathrm{pol})} = 0$), $E_\pm^{(r)}$ are the amplitudes of the reflected RCP and LCP waves.

The Jones vector of the reflected wave transformed back into the Cartesian representation becomes

$$\begin{pmatrix}E_x^{(r)}\\ E_y^{(r)}\end{pmatrix} = 2^{-1/2}\begin{pmatrix}1 & 1\\ \mathrm{j} & -\mathrm{j}\end{pmatrix}\begin{pmatrix}E_+^{(r)}\\ E_-^{(r)}\end{pmatrix} = \frac{1}{2}\begin{pmatrix}r_+ + r_-\\ \mathrm{j}(r_+ - r_-)\end{pmatrix}E^{(i)}. \tag{1.48}$$

The Cartesian complex number representation for the polarization state of the reflected wave, $\chi_P^{(xy)}$ in MO polar Kerr effect, follows from the ratio of the complex components of the Cartesian Jones vector

$$\chi_r^{(\mathrm{pol})} = \frac{E_y^{(r)}}{E_x^{(r)}} = \frac{\mathrm{j}(r_+ - r_-)}{r_+ + r_-} = \mathrm{j}\frac{N_+ - N_-}{N_+N_- - 1}. \tag{1.49}$$

[9] The expressions for $r_\pm$ are analogous to that of the Fresnel reflection coefficient at the normal light incidence and follow from a simple consideration of the field continuity at the interface. More details will be given in Chapter 3 and Chapter 4.

The incident wave with the zero azimuth and the zero ellipticity, corresponding to the origin of the Cartesian complex plane representation of polarization states, is transformed to a wave with the polarization state characterized by a point $\chi^{(xy)}$ in the plane. The transformation described by $\chi_r^{(\mathrm{pol})}$ thus defines the normal incidence MO complex polar Kerr effect. The complex number representation of the polarization state, $\chi^{(xy)}$, contains the information on the associated azimuth rotation and ellipticity given by the general expression, which follows from Eq. (1.38) for the Cartesian representation [3]

$$\chi^{(xy)} = \frac{E_{0y}}{E_{0x}} = \frac{\tan\theta + \mathrm{j}\tan\epsilon}{1 - \mathrm{j}\tan\theta\tan\epsilon}. \tag{1.50}$$

According to this result, the MO polar Kerr rotation follows from the equation

$$\tan 2\theta^{(\mathrm{pol})} = \frac{2\Re\left(\chi_r^{(\mathrm{pol})}\right)}{1 - |\chi_r^{(\mathrm{pol})}|^2}, \tag{1.51}$$

and the corresponding MO polar Kerr ellipticity $\tan\epsilon^{(\mathrm{pol})}$ from the equation

$$\sin 2\epsilon^{(\mathrm{pol})} = \frac{2\Im\left(\chi_r^{(\mathrm{pol})}\right)}{1 + |\chi_r^{(\mathrm{pol})}|^2}. \tag{1.52}$$

Here, $\Re(\chi_r^{(\mathrm{pol})})$ and $\Im(\chi_r^{(\mathrm{pol})})$ denote the real and imaginary part of the complex number $\chi_r^{(\mathrm{pol})}$, respectively. In many practical situations, normal incidence MO polar Kerr azimuth rotations and ellipticities are small, and it is justified to take a small ellipsometric angle (small azimuth and ellipticity) approximation

$$\chi_r^{(\mathrm{pol})} \approx \theta^{(\mathrm{pol})} + \mathrm{j}\epsilon^{(\mathrm{pol})}. \tag{1.53}$$

To express $\chi_r^{(\mathrm{pol})}$ in terms of the relative permittivity tensor components, we make use of Eq. (1.29b) and Eq. (1.30b) with the assumption $\varepsilon_{xy} \ll \varepsilon_{xx}$ and find out $N_+ - N_- \approx \mathrm{j}\varepsilon_{xy}\varepsilon_{xx}^{-1/2}$ and $N_+N_- \approx \varepsilon_{xx} = N^2 = (n - \mathrm{j}k)^2$. Then, the substitution into Eq. (1.49), always for the interface with vacuum, gives

$$\chi_r^{(\mathrm{pol})} \approx \theta^{(\mathrm{pol})} + \mathrm{j}\epsilon^{(\mathrm{pol})} \approx \frac{\varepsilon_{xy}^{(1)}}{\varepsilon_{xx}^{(1)1/2}(1 - \varepsilon_{xx}^{(1)})}. \tag{1.54}$$

In more detail, omitting the magnetic medium index (1)

$$\theta^{(\mathrm{pol})} \approx \frac{\Re\left(\varepsilon_{xy}\right)\left(3nk^2 - n^3 + n\right) + \Im(\varepsilon_{xy})\left(3n^2k - k^3 - k\right)}{\left(n^2 + k^2\right)\left[\left(n^2 + k^2\right) - 2n^2 + 2k^2 + 1\right]}, \tag{1.55a}$$

$$\epsilon^{(\mathrm{pol})} \approx \frac{\Im(\varepsilon_{xy})\left(3nk^2 - n^3 + n\right) - \Re(\varepsilon_{xy})\left(3n^2k - k^3 - k\right)}{\left(n^2 + k^2\right)\left[\left(n^2 + k^2\right) - 2n^2 + 2k^2 + 1\right]}. \tag{1.55b}$$

Here, $\Re\left(\varepsilon_{xy}\right)$ and $\Im\left(\varepsilon_{xy}\right)$ stand for the real and imaginary part of the off-diagonal permittivity tensor element ε_{xy}, respectively.

It is instructive to express the MO Kerr azimuth rotation and ellipticity using the amplitudes and phase angles of the CP reflection coefficients

$$r_{\pm} = |r_{\pm}| \mathrm{e}^{\mathrm{j}\varrho_{\pm}}. \tag{1.56}$$

We transform the Cartesian representation of the polarization state in Eq. (1.38) to the circular one and compare the resulting Jones vector with that representing the polarization state of the wave reflected on a surface with polar magnetization as given in Eq. (1.47)

$$\begin{aligned}
\begin{pmatrix} E_{+}^{(r)} \\ E_{-}^{(r)} \end{pmatrix} &= 2^{-1/2} \begin{pmatrix} 1 & -\mathrm{j} \\ 1 & \mathrm{j} \end{pmatrix} \begin{pmatrix} \cos\theta^{(\mathrm{pol})} \cos\epsilon^{(\mathrm{pol})} - \mathrm{j} \sin\theta^{(\mathrm{pol})} \sin\epsilon^{(\mathrm{pol})} \\ \sin\theta^{(\mathrm{pol})} \cos\epsilon^{(\mathrm{pol})} + \mathrm{j} \cos\theta^{(\mathrm{pol})} \sin\epsilon^{(\mathrm{pol})} \end{pmatrix} E^{(r)} \\
&= 2^{-1/2} \begin{pmatrix} \mathrm{e}^{-\mathrm{j}\theta^{(\mathrm{pol})}} \left(\cos\epsilon^{(\mathrm{pol})} + \sin\epsilon^{(\mathrm{pol})}\right) \\ \mathrm{e}^{\mathrm{j}\theta^{(\mathrm{pol})}} \left(\cos\epsilon^{(\mathrm{pol})} - \sin\epsilon^{(\mathrm{pol})}\right) \end{pmatrix} E^{(r)} \\
&= \begin{pmatrix} r_{+} \\ r_{-} \end{pmatrix} E_{x}^{(i)},
\end{aligned} \tag{1.57}$$

always for the incident wave of the zero azimuth and ellipticity. The complex plane circular representation of polarization states is obtained from the ratio of LCP to RCP fields [3]

$$\chi^{(\mathrm{CP})} = \frac{E_{-}}{E_{+}} = \mathrm{e}^{2\mathrm{j}\theta} \tan\left(\frac{\pi}{4} - \epsilon\right). \tag{1.58}$$

We can express the polarization state as a ratio of the circular (CP) Jones vector components

$$\chi_{(\mathrm{CP})}^{(\mathrm{pol})} = \frac{r_{-}}{r_{+}}. \tag{1.59}$$

According to Eq. (1.56), the MO Kerr azimuth rotation is given by the phase difference

$$\theta^{(\mathrm{pol})} = \frac{1}{2}(\varrho_{-} - \varrho_{+}), \tag{1.60}$$

whereas the MO Kerr ellipticity depends on the relative difference of the amplitude reflection coefficients for RCP and LCP waves

$$\tan\epsilon^{(\mathrm{pol})} = \frac{|r_{+}| - |r_{-}|}{|r_{+}| + |r_{-}|}. \tag{1.61}$$

1.4 Voigt Effect

The Voigt effect is observed when a polarized wave travels across the medium in the direction normal to the magnetization. This is a magnetooptic effect quadratic in magnetization. Putting the magnetization parallel to the x axis and allowing the wave to propagate in the z direction, we have

$$\begin{pmatrix} \gamma^2 E_x \\ \gamma^2 E_y \\ 0 \end{pmatrix} = \frac{\omega^2}{c^2} \begin{pmatrix} \varepsilon_{xx} & 0 & 0 \\ 0 & \varepsilon_{yy} & \varepsilon_{yz} \\ 0 & -\varepsilon_{yz} & \varepsilon_{yy} \end{pmatrix} \begin{pmatrix} E_x \\ E_y \\ E_z \end{pmatrix}. \tag{1.62}$$

After dividing by the vacuum propagation constant (ω/c), we obtain the equation system

$$\left(N^2 - \varepsilon_{xx}\right) E_{0x} = 0, \tag{1.63a}$$

$$\left(N^2 - \varepsilon_{yy}\right) E_{0y} - \varepsilon_{yz} E_{0z} = 0, \tag{1.63b}$$

$$\varepsilon_{yz} E_{0y} - \varepsilon_{yy} E_{0z} = 0. \tag{1.63c}$$

The proper value equation for the complex index of refraction, related to the proper value of the propagation vector, follows from the zero determinant condition

$$\left(N^2 - \varepsilon_{xx}\right) \left(\varepsilon_{yy} N^2 - \varepsilon_{yy}^2 - \varepsilon_{yz}^2\right) = 0. \tag{1.64}$$

In a simplified picture, the wave with the electric field polarized parallel to the magnetization propagates as if there were no magnetization in the medium.[10] This can be deduced from a model of a free charged particle subjected to electric and **B**, magnetic flux density, fields parallel to each other. According to the Lorentz force law, the **B** field does not affect particles moving parallel to **B**. The complex index of refraction of the wave polarized with the electric field parallel to **B** assumes the value

$$(N_{\parallel})^2 = \epsilon_{xx}. \tag{1.65}$$

On the other hand, the wave polarized normal to the magnetization is affected by the magnetization, as expected from the Lorentz force law

$$(N_{\perp})^2 = \epsilon_{yy} + \frac{\epsilon_{yz}^2}{\epsilon_{yy}}. \tag{1.66}$$

The proper values of the propagation vector do not depend on the sense of propagation. They also show no dependence on the sense of magnetization; *i.e.*, the magnetization reversal does not produce any effect on the wave propagation. This behavior is typical for the MO effects quadratic (or more precisely even) in the magnetization. The proper modes polarized parallel to the magnetization are linearly polarized. The proper modes polarized

[10] The simplification neglects the electron redistribution induced by magnetization, which affects the dipole moment both in the direction of **M** and in the plane normal to **M**. More details on the permittivity tensor symmetry at $\boldsymbol{M} \neq 0$ can be found in Chapter 2.

perpendicularly to the magnetization have a small longitudinal field component independent of the sense of propagation

$$E_{1,2\parallel} = \hat{x}\, E_{0\parallel} \exp\left[\mathrm{j}(\omega t \mp \frac{\omega}{c} N_{\parallel} z)\right], \tag{1.67a}$$

$$E_{3,4\perp} = \left(\hat{y} + \hat{z}\,\frac{\varepsilon_{yz}}{\varepsilon_{yy}}\right) E_{0\perp} \exp\left[\mathrm{j}(\omega t \mp \frac{\omega}{c} N_{\perp} z)\right]. \tag{1.67b}$$

The longitudinal component of $E_{3,4\perp}$ is proportional to the off-diagonal element of the permittivity tensor. This changes sign upon the reversal of the magnetization from $\hat{x}M$ to $-\hat{x}M$, *i.e.*, $\varepsilon_{yz}(M_x) = -\varepsilon_{yz}(-M_x)$. Magnetic linear birefringence (MLB) is defined in terms of the proper values of real indices of refraction [8]

$$\Re(N_{\perp} - N_{\parallel}) = n_{\perp} - n_{\parallel}. \tag{1.68}$$

Magnetic linear dichroism (MLD) depends on the imaginary part of the magnetic complex birefringence $(N_{\perp} - N_{\parallel})$, *i.e.*,

$$\Im(N_{\perp} - N_{\parallel}) = k_{\parallel} - k_{\perp}. \tag{1.69}$$

MLD is defined as a difference of the absorption coefficients for the two proper polarization modes

$$(\alpha_{\parallel} - \alpha_{\perp}) = \frac{2\omega}{c}(k_{\parallel} - k_{\perp}). \tag{1.70}$$

An effect analogous to the Voigt effect in transmission can be observed in reflection. The effect is due to the difference in the normal reflection coefficients at an interface between a vacuum and the magnetic medium magnetized in the plane of the interface. For the wave linearly polarized normal to the magnetization, we have

$$r_{\perp} = \frac{1 - N_{\perp}}{1 + N_{\perp}}. \tag{1.71}$$

When the wave is linearly polarized parallel to the magnetization, the result is

$$r_{\parallel} = \frac{1 - N_{\parallel}}{1 + N_{\parallel}}. \tag{1.72}$$

1.5 Propagation in Anisotropic Media

In the reflection cases treated so far, the propagation vector of the incident wave was always normal to the reflecting interface. A more complete treatment of the MO effects in reflection requires the extension to arbitrary angles of incidence. We first consider the proper values of the propagation vector and the proper electric field polarizations of monochromatic plane waves propagating in an anisotropic medium characterized by a relative permittivity tensor ε of the most general form. The wave equation (1.14) written in components becomes

$$\begin{pmatrix} \left(\frac{\partial^2}{\partial y^2}+\frac{\partial^2}{\partial z^2}\right) & -\frac{\partial^2}{\partial x \partial y} & -\frac{\partial^2}{\partial x \partial z} \\ -\frac{\partial^2}{\partial y \partial x} & \left(\frac{\partial^2}{\partial z^2}+\frac{\partial^2}{\partial x^2}\right) & -\frac{\partial^2}{\partial y \partial z} \\ -\frac{\partial^2}{\partial z \partial x} & -\frac{\partial^2}{\partial z \partial y} & \left(\frac{\partial^2}{\partial x^2}+\frac{\partial^2}{\partial y^2}\right) \end{pmatrix} \begin{pmatrix} E_x \\ E_y \\ E_z \end{pmatrix}$$
$$= c^{-2} \begin{pmatrix} \varepsilon_{xx} & \varepsilon_{xy} & \varepsilon_{xz} \\ \varepsilon_{yx} & \varepsilon_{yy} & \varepsilon_{yz} \\ \varepsilon_{zx} & \varepsilon_{zy} & \varepsilon_{zz} \end{pmatrix} \frac{\partial^2}{\partial t^2} \begin{pmatrix} E_x \\ E_y \\ E_z \end{pmatrix}. \tag{1.73}$$

We look for monochromatic plane wave solutions represented in Eq. (1.15), where $\gamma = \hat{x}\gamma_x + \hat{y}\gamma_y + \hat{z}\gamma_z$

$$\begin{pmatrix} (\gamma_y^2+\gamma_z^2) & -\gamma_x\gamma_y & -\gamma_x\gamma_z \\ -\gamma_y\gamma_x & (\gamma_z^2+\gamma_x^2) & -\gamma_y\gamma_z \\ -\gamma_z\gamma_x & -\gamma_z\gamma_y & (\gamma_x^2+\gamma_y^2) \end{pmatrix} \begin{pmatrix} E_x \\ E_y \\ E_z \end{pmatrix} = \frac{\omega^2}{c^2} \begin{pmatrix} \varepsilon_{xx} & \varepsilon_{xy} & \varepsilon_{xz} \\ \varepsilon_{yx} & \varepsilon_{yy} & \varepsilon_{yz} \\ \varepsilon_{zx} & \varepsilon_{zy} & \varepsilon_{zz} \end{pmatrix} \begin{pmatrix} E_x \\ E_y \\ E_z \end{pmatrix}. \tag{1.74}$$

We express the propagation vector as $\gamma = \frac{\omega}{c}\left(\hat{x}N_x + \hat{y}N_y + \hat{z}N_z\right)$ and write the wave equation for the monochromatic plane wave solutions as

$$\begin{pmatrix} (N_y^2+N_z^2)-\varepsilon_{xx} & -N_xN_y-\varepsilon_{xy} & -N_xN_z-\varepsilon_{xz} \\ -N_yN_x-\varepsilon_{yx} & (N_z^2+N_x^2)-\varepsilon_{yy} & -N_yN_z-\varepsilon_{yz} \\ -N_zN_x-\varepsilon_{zx} & -N_zN_y-\varepsilon_{zy} & (N_x^2+N_y^2)-\varepsilon_{zz} \end{pmatrix} \begin{pmatrix} E_{0x} \\ E_{0y} \\ E_{0z} \end{pmatrix} = 0. \tag{1.75}$$

We wish to find propagation vectors and corresponding proper polarization modes. The propagation vectors follow from the condition for zero determinant of the wave equation (1.75) expressed as

$$\begin{aligned} &\left(\varepsilon_{xx}N_x^2+\varepsilon_{yy}N_y^2+\varepsilon_{zz}N_z^2\right)\left(N_x^2+N_y^2+N_z^2\right) \\ &\quad + \left[\left(\varepsilon_{yz}+\varepsilon_{zy}\right)N_yN_z + \left(\varepsilon_{xy}+\varepsilon_{yx}\right)N_xN_y + \left(\varepsilon_{zx}+\varepsilon_{xz}\right)N_zN_x\right] \\ &\quad \times \left(N_x^2+N_y^2+N_z^2\right) - \left(\varepsilon_{yy}\varepsilon_{zz}-\varepsilon_{yz}\varepsilon_{zy}\right)\left(N_y^2+N_z^2\right) \end{aligned}$$

$$
\begin{aligned}
&-\left(\varepsilon_{zz}\varepsilon_{xx}-\varepsilon_{zx}\varepsilon_{xz}\right)\left(N_z^2+N_x^2\right)-\left(\varepsilon_{xx}\varepsilon_{yy}-\varepsilon_{xy}\varepsilon_{yx}\right)\left(N_x^2+N_y^2\right)\\
&-\left[\varepsilon_{xx}\left(\varepsilon_{yz}+\varepsilon_{zy}\right)-\left(\varepsilon_{xy}\varepsilon_{zx}+\varepsilon_{yx}\varepsilon_{xz}\right)\right]N_yN_z\\
&-\left[\varepsilon_{yy}\left(\varepsilon_{zx}+\varepsilon_{xz}\right)-\left(\varepsilon_{yz}\varepsilon_{xy}+\varepsilon_{zy}\varepsilon_{yx}\right)\right]N_zN_x\\
&-\left[\varepsilon_{zz}\left(\varepsilon_{xy}+\varepsilon_{yx}\right)-\left(\varepsilon_{zx}\varepsilon_{yz}+\varepsilon_{xz}\varepsilon_{zy}\right)\right]N_xN_y\\
&+\varepsilon_{xx}\varepsilon_{yy}\varepsilon_{zz}+\varepsilon_{xy}\varepsilon_{yz}\varepsilon_{zx}+\varepsilon_{yx}\varepsilon_{zy}\varepsilon_{xz}\\
&-\varepsilon_{xx}\varepsilon_{yz}\varepsilon_{zy}-\varepsilon_{yy}\varepsilon_{zx}\varepsilon_{xz}-\varepsilon_{zz}\varepsilon_{xy}\varepsilon_{yx}=0.
\end{aligned}
\tag{1.76}
$$

The monochromatic plane wave solutions are strictly valid only in infinite homogeneous media. However, they may serve as approximate solutions in bounded regions. We consider a homogeneous anisotropic medium characterized by a relative permittivity tensor ε occupying the half space $z > z^{(0)}$. The half space $z < z^{(0)}$ is occupied by a homogeneous isotropic nonmagnetic nonabsorbing medium[11] characterized by a real scalar permittivity $\varepsilon^{(0)}$. We choose the plane of wave incidence normal to the x axis. There is no restriction to the problem generality, as we can always transform our tensor ε of the general form to new axes. The transformation will, of course, change the meaning of the tensor components but does not reduce their number, leaving the tensor completely general. We assume that a monochromatic plane wave with the propagation vector

$$
\gamma^{(0)}=\frac{\omega}{c}\left(\hat{\mathbf{y}}\sqrt{\varepsilon^{(0)}}\sin\varphi^{(0)}+\hat{\mathbf{z}}\sqrt{\varepsilon^{(0)}}\cos\varphi^{(0)}\right)
\tag{1.77}
$$

impinges at the interface $z = z_0$ at an arbitrary angle of incidence, $\varphi^{(0)}$, with electric field component proportional to $\exp[\mathrm{j}(\omega t - \sqrt{\varepsilon^{(0)}}\frac{\omega}{c}z\cos\varphi^{(0)}$ $-\sqrt{\varepsilon^{(0)}}\frac{\omega}{c}y\sin\varphi^{(0)})]$ from the region $z < z^{(0)}$ of the isotropic half space. Snell law requires the continuity of the propagation vector components parallel to the interfaces. In our choice, $N_x^{(0)} = 0$ and $N_y^{(0)} = \sqrt{\varepsilon^{(0)}}\frac{\omega}{c}y\sin\varphi^{(0)}$. Then, the propagation vector components $\frac{\omega}{c}N_x = 0$ and $\frac{\omega}{c}N_y = \sqrt{\varepsilon^{(0)}}$ $\frac{\omega}{c}\sin\varphi^{(0)}$ in the anisotropic medium with $\varepsilon^{(1)}$ are already fixed, and it remains to determine the values of the z component, $\frac{\omega}{c}N_z^{(1)}$, of the propagation vector in the anisotropic medium

$$
\gamma=\frac{\omega}{c}(N_y\hat{\mathbf{y}}+N_z^{(1)}\hat{\mathbf{z}}).
\tag{1.78}
$$

[11] A generalization to an absorbing isotropic medium characterized by scalar complex permittivity and magnetic permeability presents no difficulty.

To this purpose, we transform the wave equation system (1.75) to a simpler form

$$\begin{pmatrix} (N_y^2 + N_z^{(1)2}) - \varepsilon_{xx}^{(1)} & -\varepsilon_{xy}^{(1)} & -\varepsilon_{xz}^{(1)} \\ -\varepsilon_{yx}^{(1)} & N_z^{(1)2} - \varepsilon_{yy}^{(1)} & -N_y N_z^{(1)} - \varepsilon_{yz}^{(1)} \\ -\varepsilon_{zx}^{(1)} & -N_z^{(1)} N_y - \varepsilon_{zy}^{(1)} & N_y^2 - \varepsilon_{zz}^{(1)} \end{pmatrix} \begin{pmatrix} E_{0x}^{(1)} \\ E_{0y}^{(1)} \\ E_{0z}^{(1)} \end{pmatrix} = 0. \tag{1.79}$$

Then, Eq. (1.76) becomes a complex fourth degree algebraic equation in the reduced normal component of the propagation vector, N_z,

$$\begin{aligned} &\varepsilon_{zz}^{(1)} N_z^{(1)4} + \left(\varepsilon_{yz}^{(1)} + \varepsilon_{zy}^{(1)}\right) N_y N_z^{(3)3} \\ &\quad - \left[\left(\varepsilon_{yy}^{(1)}\varepsilon_{zz}^{(1)} - \varepsilon_{yz}^{(1)}\varepsilon_{zy}^{(1)}\right) + \left(\varepsilon_{zz}^{(1)}\varepsilon_{xx}^{(1)} - \varepsilon_{zx}^{(1)}\varepsilon_{xz}^{(1)}\right) - \left(\varepsilon_{yy}^{(1)} + \varepsilon_{zz}^{(1)}\right) N_y^2\right] N_z^{(1)2} \\ &\quad - \left[\left(\varepsilon_{yz}^{(1)} + \varepsilon_{zy}^{(1)}\right)\left(\varepsilon_{xx}^{(1)} - N_y^2\right) - \left(\varepsilon_{xy}^{(1)}\varepsilon_{zx}^{(1)} + \varepsilon_{yx}^{(1)}\varepsilon_{xz}^{(1)}\right)\right] N_y N_z^{(1)} \\ &\quad + \varepsilon_{yy}^{(1)} N_y^4 - \left(\varepsilon_{yy}^{(1)}\varepsilon_{zz}^{(1)} - \varepsilon_{yz}^{(1)}\varepsilon_{zy}^{(1)}\right) N_y^2 - \left(\varepsilon_{xx}^{(1)}\varepsilon_{yy}^{(1)} - \varepsilon_{xy}^{(1)}\varepsilon_{yx}^{(1)}\right) N_y^2 \\ &\quad + \varepsilon_{xx}^{(1)}\varepsilon_{yy}^{(1)}\varepsilon_{zz}^{(1)} + \varepsilon_{xy}^{(1)}\varepsilon_{yz}^{(1)}\varepsilon_{zx}^{(1)} + \varepsilon_{yx}^{(1)}\varepsilon_{zy}^{(1)}\varepsilon_{xz}^{(1)} \\ &\quad - \varepsilon_{xx}^{(1)}\varepsilon_{yz}^{(1)}\varepsilon_{zy}^{(1)} - \varepsilon_{yy}^{(1)}\varepsilon_{zx}^{(1)}\varepsilon_{xz}^{(1)} - \varepsilon_{zz}^{(1)}\varepsilon_{xy}^{(1)}\varepsilon_{yx}^{(1)} = 0. \end{aligned} \tag{1.80}$$

In general, there are four different solutions, the proper values $N_{zj}^{(1)}$, $j = 1, 2, 3$, and 4. The substitution of $N_{zj}^{(1)}$ into the wave equation system (1.75) gives the corresponding, in general different, proper polarizations $\boldsymbol{e}_j^{(1)}$.

1.6 Reflection at an Arbitrary Angle of Incidence

We consider a homogeneous plane wave in an isotropic ambient propagating with an index of refraction $N^{(0)}$ towards the interface $z = z_0$. We specify the orientation of its propagation vector by direction cosines $\alpha_y^{(0)} = \sin\varphi^{(0)}$ and $\alpha_z^{(0)} = \cos\varphi^{(0)}$ with $\alpha_x^{(0)} = 0$. This choice sets the plane of incidence normal to the x axis. The four propagating plane wave solutions of Eq. (1.79) correspond to two waves propagating in the forward direction of the positive z axis (out of the interface $z = z_0$), indexed 1 and 3, and two waves propagating in the backward direction of the negative z axis (towards the interface $z = z_0$), indexed 2 and 4. The latter two solutions (2 and 4) need not be considered in the problem of a single interface $z = z_0$ between two semi-infinite media once we have accepted the solutions 1 and 3 as the forward propagating proper polarization modes.

The propagation vectors of the incident (i), reflected (r), and transmitted (t) waves can be written as

$$\gamma^{(i)} = N^{(0)}\frac{\omega}{c}\left(\alpha_y^{(0)}\hat{y} + \alpha_z^{(0)}\hat{z}\right), \tag{1.81a}$$

$$\gamma^{(r)} = N^{(0)}\frac{\omega}{c}\left(\alpha_y^{(0)}\hat{y} - \alpha_z^{(0)}\hat{z}\right), \tag{1.81b}$$

$$\gamma_{1,3}^{(t)} = \frac{\omega}{c}\left(N^{(0)}\alpha_y^{(0)}\hat{y} + N_{z1,3}^{(1)}\hat{z}\right). \tag{1.81c}$$

We then have for the electric fields of the waves

$$\boldsymbol{E}^{(i)} = \left[E_{0s}^{(i)}\hat{x} + E_{0p}^{(i)}(\alpha_z^{(0)}\hat{y} - \alpha_y^{(0)}\hat{z})\right]\exp\left[\mathrm{j}\omega t - \mathrm{j}\frac{\omega}{c}N^{(0)}\left(\alpha_y^{(0)}y + \alpha_z^{(0)}z\right)\right], \tag{1.82a}$$

$$\boldsymbol{E}^{(r)} = \left[E_{0s}^{(r)}\hat{x} + E_{0p}^{(r)}(\alpha_z^{(0)}\hat{y} + \alpha_y^{(0)}\hat{z})\right]\exp\left[\mathrm{j}\omega t - \mathrm{j}\frac{\omega}{c}N^{(0)}\left(\alpha_y^{(0)}y - \alpha_z^{(0)}z\right)\right], \tag{1.82b}$$

$$\boldsymbol{E}^{(t)} = \sum_{j=1,3}\left(E_{0xj}^{(1)}\hat{x} + E_{0yj}^{(1)}\hat{y} + E_{0zj}^{(1)}\hat{z}\right)\exp\left[\mathrm{j}\omega t - \mathrm{j}\frac{\omega}{c}\left(N^{(0)}\alpha_y^{(0)}y + N_{zj}^{(1)}z\right)\right]. \tag{1.82c}$$

Here, $E_{0s}^{(i)}$, $E_{0s}^{(r)}$, $E_{0p}^{(i)}$, and $E_{0p}^{(r)}$ are the amplitudes of the incident (i) and reflected (r) waves in the isotropic medium (0) polarized normal s (from German *senkrecht*, also called TE waves) and parallel p (also called TM waves) with respect to the plane of incidence (normal to $\hat{x}$). The last equation (1.82c) characterizes the transmitted wave induced in the medium (1) on the other side of the planar interface and characterized by the relative permittivity tensor ε. The corresponding magnetic fields of the waves follow from the Maxwell equations (1.9)

$$\boldsymbol{H}^{(i)} = N^{(0)}\left[-E_{0p}^{(i)}\hat{x} + E_{0s}^{(i)}\left(\alpha_z^{(0)}\hat{y} - \alpha_y^{(0)}\hat{z}\right)\right] \times\left(\frac{\varepsilon_{\mathrm{vac}}}{\mu_{\mathrm{vac}}}\right)^{1/2}\exp\left[\mathrm{j}\omega t - \mathrm{j}\frac{\omega}{c}N^{(0)}\left(\alpha_y^{(0)}y + \alpha_z^{(0)}z\right)\right] \tag{1.83a}$$

$$\boldsymbol{H}^{(r)} = N^{(0)}\left[E_{0p}^{(r)}\hat{x} + E_{0s}^{(r)}\left(-\alpha_z^{(0)}\hat{y} - \alpha_y^{(0)}\hat{z}\right)\right] \times\left(\frac{\varepsilon_{\mathrm{vac}}}{\mu_{\mathrm{vac}}}\right)^{1/2}\exp\left[\mathrm{j}\omega t - \mathrm{j}\frac{\omega}{c}N^{(0)}\left(\alpha_y^{(0)}y - \alpha_z^{(0)}z\right)\right] \tag{1.83b}$$

$$H^{(t)} = \sum_{j=1,3} \left[\left(N^{(0)} \alpha_y^{(0)} E_{0zj}^{(1)} - N_{zj}^{(1)} E_{0yj}^{(1)} \right) \hat{\boldsymbol{x}} + N_{zj}^{(1)} E_{0xj}^{(1)} \hat{\boldsymbol{y}} - N^{(0)} E_{0xj}^{(1)} \hat{\boldsymbol{z}} \right]$$
$$\times \left(\frac{\varepsilon_{\text{vac}}}{\mu_{\text{vac}}} \right)^{1/2} \exp\left[j\omega t - j\frac{\omega}{c} \left(N^{(0)} \alpha_y^{(0)} y + N_{zj}^{(1)} z \right) \right]. \tag{1.83c}$$

Here, $\sum_{j=1,3}$ denotes the sum over the fields of two refracted waves, and ε_{vac} and μ_{vac} are the permittivity and magnetic permeability in a vacuum, respectively.

The requirements of the continuity for the electric and magnetic field components parallel to the interface plane $z = 0$ provide four equations. We obtain from Eq. (1.82) and Eq. (1.83) [9]

E_x continuous:

$$-E_{0s}^{(r)} + E_{0x}^{(1)} + E_{0x}^{(3)} = E_{0s}^{(i)}. \tag{1.84a}$$

H_y continuous:

$$N^{(0)} \alpha_z^{(0)} E_{0s}^{(r)} + N_{z1} E_{0x}^{(1)} + N_{z3} E_{0x}^{(3)} = N^{(0)} \alpha_z^{(0)} E_{0s}^{(i)}. \tag{1.84b}$$

E_y continuous:

$$-\alpha_z^{(0)} E_{0p}^{(r)} + E_{0y}^{(1)} + E_{0y}^{(3)} = \alpha_z^{(0)} E_{0p}^{(i)}. \tag{1.84c}$$

H_x continuous:

$$N^{(0)} E_{0p}^{(r)} - N^{(0)} \alpha_y^{(0)} \left(E_{0z}^{(1)} + E_{0z}^{(3)} \right) + N_{z1} E_{0y}^{(1)} + N_{z3} E_{0y}^{(3)} = N^{(0)} E_{0p}^{(i)}. \tag{1.84d}$$

The wave equation (1.79) provides the relations between the field components. Then, the modal amplitudes $E_{0x}^{(1)}$, $E_{0y}^{(1)}$, and $E_{0z}^{(1)}$ can be expressed with help of one of them. Similarly, the amplitudes $E_{0x}^{(3)}$, $E_{0y}^{(3)}$, $E_{0z}^{(3)}$ can be expressed with help of one of them. For example, from the first and third equations of the system (1.79), we obtain for $j = 1, 3$

$$E_{0x}^{(j)} = p_j^{(1)} E_{0y}^{(j)} \tag{1.85a}$$

$$E_{0z}^{(j)} = q_j^{(1)} E_{0y}^{(j)}, \tag{1.85b}$$

where[12]

$$p_j^{(1)} = \frac{-\varepsilon_{xy}^{(1)}\left(\varepsilon_{zz}^{(1)} - N_y^2\right) + \varepsilon_{xz}^{(1)}\left(\varepsilon_{zy}^{(1)} + N_y N_{zj}^{(1)}\right)}{\left(\varepsilon_{xx}^{(1)} - N_y^2 - N_{zj}^{(1)2}\right)\left(\varepsilon_{zz}^{(1)} - N_y^2\right) - \varepsilon_{zx}^{(1)}\varepsilon_{xz}^{(1)}}, \tag{1.86a}$$

$$q_j^{(1)} = \frac{-\left(\varepsilon_{xx}^{(1)} - N_y^2 - N_{zj}^{(1)2}\right)\left(\varepsilon_{zy}^{(1)} + N_y N_{zj}^{(1)}\right) + \varepsilon_{zx}^{(1)}\varepsilon_{xy}^{(1)}}{\left(\varepsilon_{xx}^{(1)} - N_y^2 - N_{zj}^{(1)2}\right)\left(\varepsilon_{zz}^{(1)} - N_y^2\right) - \varepsilon_{zx}^{(1)}\varepsilon_{xz}^{(1)}}. \tag{1.86b}$$

The solutions for the reflection of the system (1.84) provide the amplitudes $E_{0s}^{(r)}$ and $E_{0p}^{(r)}$ in terms of the amplitudes of the incident waves $E_{0s}^{(i)}$ and $E_{0p}^{(i)}$

$$\begin{pmatrix} -1 & 0 & p_1^{(1)} & p_3^{(1)} \\ N^{(0)}\alpha_z^{(0)} & 0 & N_{z1}^{(1)} p_1^{(1)} & N_{z3}^{(1)} p_3^{(1)} \\ 0 & -\alpha_z^{(0)} & 1 & 1 \\ 0 & N^{(0)} & N_{z1}^{(1)} - q_1^{(1)} N^{(0)}\alpha_y^{(0)} & N_{z3}^{(1)} - q_3^{(1)} N^{(0)}\alpha_y^{(0)} \end{pmatrix} \begin{pmatrix} E_{0s}^{(r)} \\ E_{0p}^{(r)} \\ E_{0y1}^{(1)} \\ E_{0y3}^{(1)} \end{pmatrix}$$
$$= \begin{pmatrix} E_{0s}^{(i)} \\ N^{(0)}\alpha_z^{(0)} E_{0s}^{(i)} \\ \alpha_z^{(0)} E_{0p}^{(i)} \\ N^{(0)} E_{0p}^{(i)} \end{pmatrix}. \tag{1.87}$$

It is convenient to solve the system (1.84) or (1.87) for either $E_{0s}^{(i)} \neq 0$ (and $E_{0p}^{(i)} = 0$) or $E_{0p}^{(i)} \neq 0$ (and $E_{0s}^{(i)} = 0$), separately. The determinant of this system becomes

$$\begin{vmatrix} -1 & 0 & p_1^{(1)} & p_3^{(1)} \\ N^{(0)}\alpha_z^{(0)} & 0 & N_{z1}^{(1)} p_1^{(1)} & N_{z3}^{(1)} p_3^{(1)} \\ 0 & -\alpha_z^{(0)} & 1 & 1 \\ 0 & N^{(0)} & N_{z1}^{(1)} - q_1^{(1)} N^{(0)}\alpha_y^{(0)} & N_{z3}^{(1)} - q_3^{(1)} N^{(0)}\alpha_y^{(0)} \end{vmatrix}$$
$$= N^{(0)}\alpha_z^{(0)2}\mathcal{P}^{(1)} + N^{(0)2}\alpha_z^{(0)}\mathcal{Q}^{(1)} + \alpha_z^{(0)}\mathcal{S}^{(1)} + N^{(0)}\mathcal{T}^{(1)}, \tag{1.88}$$

---

[12] In simpler situations characterized by a permittivity tensor with $\varepsilon_{xy}^{(1)} = \varepsilon_{yx}^{(1)} = \varepsilon_{zx}^{(1)} = \varepsilon_{xz}^{(1)} = 0$, *i.e.*, in the absence of the coupling between the wave fields polarized perpendicular and parallel to the plane of incidence $x = 0$, this procedure may lead to indefinite expressions. It is then more convenient to return to the wave equation and solve it for this special form of the permittivity tensor.

where $N^{(0)}\alpha_y^{(0)} = N_y$ and

$$\mathcal{P}^{(1)} = \left[N_y\left(p_1^{(1)}q_3^{(1)} - p_3^{(1)}q_1^{(1)}\right) + \left(N_{z1}^{(1)}p_3^{(1)} - N_{z3}^{(1)}p_1^{(1)}\right)\right], \tag{1.89a}$$

$$\mathcal{Q}^{(1)} = \left(p_3^{(1)} - p_1^{(1)}\right), \tag{1.89b}$$

$$\mathcal{S}^{(1)} = \left[N_y\left(N_{z1}^{(1)}p_1^{(1)}q_3^{(1)} - N_{z3}^{(1)}p_3^{(1)}q_1^{(1)}\right) + N_{z1}^{(1)}N_{z3}^{(1)}\left(p_3^{(1)} - p_1^{(1)}\right)\right], \tag{1.89c}$$

$$\mathcal{T}^{(1)} = \left(N_{z3}^{(1)}p_3^{(1)} - N_{z1}^{(1)}p_1^{(1)}\right). \tag{1.89d}$$

We have for $E_{0p}^{(i)} = 0$

$$E_{0s}^{(r)} = \begin{vmatrix} 1 & 0 & p_1^{(1)} & p_3^{(1)} \\ N^{(0)}\alpha_z^{(0)} & 0 & N_{z1}^{(1)}p_1^{(1)} & N_{z3}^{(1)}p_3^{(1)} \\ 0 & -\alpha_z^{(0)} & 1 & 1 \\ 0 & N^{(0)} & N_{z1}^{(1)} - q_1^{(1)}N^{(0)}\alpha_y^{(0)} & N_{z3}^{(1)} - q_3^{(1)}N^{(0)}\alpha_y^{(0)} \end{vmatrix} \times \frac{E_{0s}^{(i)}}{N^{(0)}\alpha_z^{(0)2}\mathcal{P}^{(1)} + N^{(0)2}\alpha_z^{(0)}\mathcal{Q}^{(1)} + \alpha_z^{(0)}\mathcal{S}^{(1)} + N^{(0)}\mathcal{T}^{(1)}}, \tag{1.90}$$

$$E_{0p}^{(r)} = \begin{vmatrix} -1 & 1 & p_1^{(1)} & p_3^{(1)} \\ N^{(0)}\alpha_z^{(0)} & N^{(0)}\alpha_z^{(0)} & N_{z1}^{(1)}p_1^{(1)} & N_{z3}^{(1)}p_3^{(1)} \\ 0 & 0 & 1 & 1 \\ 0 & 0 & N_{z1}^{(1)} - q_1^{(1)}N^{(0)}\alpha_y^{(0)} & N_{z3}^{(1)} - q_3^{(1)}N^{(0)}\alpha_y^{(0)} \end{vmatrix} \times \frac{E_{0s}^{(i)}}{N^{(0)}\alpha_z^{(0)2}\mathcal{P}^{(1)} + N^{(0)2}\alpha_z^{(0)}\mathcal{Q}^{(1)} + \alpha_z^{(0)}\mathcal{S}^{(1)} + N^{(0)}\mathcal{T}^{(1)}}, \tag{1.91}$$

and for $E_{0p}^{(i)} = 0$

$$E_{0s}^{(r)} = \begin{vmatrix} 0 & 0 & p_1^{(1)} & p_3^{(1)} \\ 0 & 0 & N_{z1}^{(1)}p_1^{(1)} & N_{z3}^{(1)}p_3^{(1)} \\ \alpha_z^{(0)} & -\alpha_z^{(0)} & 1 & 1 \\ N^{(0)} & N^{(0)} & N_{z1}^{(1)} - q_1^{(1)}N^{(0)}\alpha_y^{(0)} & N_{z3}^{(1)} - q_3^{(1)}N^{(0)}\alpha_y^{(0)} \end{vmatrix} \times \frac{E_{0p}^{(i)}}{N^{(0)}\alpha_z^{(0)2}\mathcal{P}^{(1)} + N^{(0)2}\alpha_z^{(0)}\mathcal{Q}^{(1)} + \alpha_z^{(0)}\mathcal{S}^{(1)} + N^{(0)}\mathcal{T}^{(1)}}, \tag{1.92}$$

$$E_{0p}^{(r)} = \begin{vmatrix} -1 & 0 & p_1^{(1)} & p_3^{(1)} \\ N^{(0)}\alpha_z^{(0)} & 0 & N_{z1}^{(1)} p_1^{(1)} & N_{z3}^{(1)} p_3^{(1)} \\ 0 & \alpha_z^{(0)} & 1 & 1 \\ 0 & N^{(0)} & N_{z1}^{(1)} - q_1^{(1)} N^{(0)} \alpha_y^{(0)} & N_{z3}^{(1)} - q_3^{(1)} N^{(0)} \alpha_y^{(0)} \end{vmatrix}$$
$$\times \frac{E_{0p}^{(i)}}{N^{(0)}\alpha_z^{(0)2}\mathcal{P}^{(1)} + N^{(0)2}\alpha_z^{(0)}\mathcal{Q}^{(1)} + \alpha_z^{(0)}\mathcal{S}^{(1)} + N^{(0)}\mathcal{T}^{(1)}}. \tag{1.93}$$

We define the generalized Fresnel reflection coefficients for an interface between isotropic half space (0) and a half space (1) characterized by a general permittivity tensor

$$r_{ss}^{(01)} = \frac{E_{0s}^{(r)}}{E_{0s}^{(i)}} = \frac{N^{(0)}\alpha_z^{(0)2}\mathcal{P}^{(1)} + N^{(0)2}\alpha_z^{(0)}\mathcal{Q}^{(1)} - \alpha_z^{(0)}\mathcal{S}^{(1)} - N^{(0)}\mathcal{T}^{(1)}}{N^{(0)}\alpha_z^{(0)2}\mathcal{P}^{(1)} + N^{(0)2}\alpha_z^{(0)}\mathcal{Q}^{(1)} + \alpha_z^{(0)}\mathcal{S}^{(1)} + N^{(0)}\mathcal{T}^{(1)}}, \quad (E_{0p}^{(i)} = 0), \tag{1.94a}$$

$$r_{ps}^{(01)} = \frac{E_{0p}^{(r)}}{E_{0s}^{(i)}} = \frac{2N^{(0)}\alpha^{(0)}\left[\left(N_{z1}^{(1)} - N_{z3}^{(1)}\right) - N_y\left(q_1^{(1)} - q_3^{(1)}\right)\right]}{N^{(0)}\alpha_z^{(0)2}\mathcal{P}^{(1)} + N^{(0)2}\alpha_z^{(0)}\mathcal{Q}^{(1)} + \alpha_z^{(0)}\mathcal{S}^{(1)} + N^{(0)}\mathcal{T}^{(1)}}, \quad (E_{0p}^{(i)} = 0), \tag{1.94b}$$

$$r_{sp}^{(01)} = \frac{E_{0s}^{(r)}}{E_{0p}^{(i)}} = \frac{2N^{(0)}\alpha^{(0)} p_1^{(1)} p_3^{(1)}\left(N_{z3}^{(1)} - N_{z1}^{(1)}\right)}{N^{(0)}\alpha_z^{(0)2}\mathcal{P}^{(1)} + N^{(0)2}\alpha_z^{(0)}\mathcal{Q}^{(1)} + \alpha_z^{(0)}\mathcal{S}^{(1)} + N^{(0)}\mathcal{T}^{(1)}}, \quad (E_{0s}^{(i)} = 0), \tag{1.94c}$$

$$r_{pp}^{(01)} = \frac{E_{0p}^{(r)}}{E_{0p}^{(i)}} = \frac{-N^{(0)}\alpha_z^{(0)2}\mathcal{P}^{(1)} + N^{(0)2}\alpha_z^{(0)}\mathcal{Q}^{(1)} - \alpha_z^{(0)}\mathcal{S}^{(1)} + N^{(0)}\mathcal{T}^{(1)}}{N^{(0)}\alpha_z^{(0)2}\mathcal{P}^{(1)} + N^{(0)2}\alpha_z^{(0)}\mathcal{Q}^{(1)} + \alpha_z^{(0)}\mathcal{S}^{(1)} + N^{(0)}\mathcal{T}^{(1)}}, \quad (E_{0s}^{(i)} = 0). \tag{1.94d}$$

In order to write $\mathcal{P}^{(1)}$, $\mathcal{Q}^{(1)}$, $\mathcal{S}^{(1)}$, and $\mathcal{T}^{(1)}$ as well as $r_{ps}^{(01)}$ and $r_{sp}^{(01)}$ explicitly, we make use of the following expressions

$$\begin{aligned} p_1^{(1)} p_3^{(1)} &= \mathcal{G}^{-1}\left\{\left[-\varepsilon_{xy}^{(1)}\left(\varepsilon_{zz}^{(1)} - N_y^2\right) + \varepsilon_{xz}^{(1)}\varepsilon_{zy}^{(1)}\right]^2 \right. \\ &\left. + \varepsilon_{xz}^{(1)} N_y\left(N_{z1}^{(1)} + N_{z3}^{(1)}\right)\left[-\varepsilon_{xy}^{(1)}\left(\varepsilon_{zz}^{(1)} - N_y^2\right) + \varepsilon_{xz}^{(1)}\varepsilon_{zy}^{(1)}\right] + \varepsilon_{xz}^{(1)2} N_y^2\left(N_{z1}^{(1)} N_{z3}^{(1)}\right)\right\}, \end{aligned} \tag{1.95a}$$

$$
\begin{aligned}
& q_1^{(1)} - q_3^{(1)} \\
& = \left(N_{z1}^{(1)} - N_{z3}^{(1)}\right) \mathcal{G}^{-1} \\
& \times \left\{ N_y \left[\left(N_{z1}^{(1)2} + N_{z3}^{(1)2} \right) - \left(\varepsilon_{xx}^{(1)} - N_y^2 \right) \right] \left[\left(\varepsilon_{xx}^{(1)} - N_y^2 \right) \left(\varepsilon_{zz}^{(1)} - N_y^2 \right) - \varepsilon_{zx}^{(1)} \varepsilon_{xz}^{(1)} \right] \right. \\
& - \varepsilon_{zx}^{(1)} \varepsilon_{xz}^{(1)} \left[\varepsilon_{zy}^{(1)} \left(N_{z1}^{(1)} + N_{z3}^{(1)} \right) + N_y \left(N_{z1}^{(1)} N_{z3}^{(1)} \right) \right] - N_y \left(\varepsilon_{zz}^{(1)} - N_y^2 \right) \\
& \left. \times \left(N_{z1}^{(1)} N_{z3}^{(1)} \right)^2 \right\},
\end{aligned} \tag{1.95b}
$$

$$
\begin{aligned}
& \left(N_{z1}^{(1)} - N_{z3}^{(1)}\right) - N_y \left(q_1^{(1)} - q_3^{(1)}\right) \\
& \quad = \left(N_{z1}^{(1)} - N_{z1}^{(3)}\right) \mathcal{G}^{-1} \\
& \quad \times \left\{ \left[\left(\varepsilon_{xx}^{(1)} - N_y^2 \right) \left(\varepsilon_{zz}^{(1)} - N_y^2 \right) - \varepsilon_{zx}^{(1)} \varepsilon_{xz}^{(1)} \right] \left[\varepsilon_{zz}^{(1)} \left(\varepsilon_{xx}^{(1)} - N_y^2 \right) - \varepsilon_{zx}^{(1)} \varepsilon_{xz}^{(1)} - \varepsilon_{zz}^{(1)} \right. \right. \\
& \quad \times \left. \left(N_{z1}^{(1)2} + N_{z3}^{(1)2} \right) \right] + \varepsilon_{zz}^{(1)} \left(\varepsilon_{zz}^{(1)} - N_y^2 \right) \left(N_{z1}^{(1)} N_{z3}^{(1)} \right)^2 \\
& \quad \left. + \varepsilon_{zx}^{(1)} \varepsilon_{xz}^{(1)} N_y^2 \left(N_{z1}^{(1)} N_{z3}^{(1)} \right) + \varepsilon_{zx}^{(1)} \varepsilon_{xz}^{(1)} \varepsilon_{zy}^{(1)} N_y \left(N_{z1}^{(1)} + N_{z3}^{(1)} \right) \right\},
\end{aligned} \tag{1.95c}
$$

$$
\begin{aligned}
\mathcal{Q}^{(1)} & = p_3^{(1)} - p_1^{(1)} \\
& = \left(N_{z1}^{(1)} - N_{z3}^{(1)}\right) \mathcal{G}^{-1} \left\{ -\varepsilon_{xz}^{(1)} N_y \left(\varepsilon_{zz}^{(1)} - N_y^2 \right) \left(N_{z1}^{(1)} N_{z3}^{(1)} \right) \right. \\
& + \left(\varepsilon_{zz}^{(1)} - N_y^2 \right) \left[\varepsilon_{xy}^{(1)} \left(\varepsilon_{zz}^{(1)} - N_y^2 \right) - \varepsilon_{xz}^{(1)} \varepsilon_{zy}^{(1)} \right] \left(N_{z1}^{(1)} + N_{z3}^{(1)} \right) \\
& \left. - \ \varepsilon_{xz}^{(1)} N_y \left[\left(\varepsilon_{xx}^{(1)} - N_y^2 \right) \left(\varepsilon_{zz}^{(1)} - N_y^2 \right) - \varepsilon_{zx}^{(1)} \varepsilon_{xz}^{(1)} \right] \right\},
\end{aligned} \tag{1.95d}
$$

$$
\begin{aligned}
\mathcal{T}^{(1)} & = N_{z3}^{(1)} p_3^{(1)} - N_{z1}^{(1)} p_1^{(1)} \\
& = \left(N_{z1}^{(1)} - N_{z3}^{(1)}\right) \mathcal{G}^{-1} \left\{ -\left(\varepsilon_{zz}^{(1)} - N_y^2 \right) \left[-\varepsilon_{xy}^{(1)} \left(\varepsilon_{zz}^{(1)} - N_y^2 \right) + \varepsilon_{xz}^{(1)} \varepsilon_{zy}^{(1)} \right] \right. \\
& \times \left(N_{z1}^{(1)} N_{z3}^{(1)} \right) - \varepsilon_{xz}^{(1)} N_y \left[\left(\varepsilon_{xx}^{(1)} - N_y^2 \right) \left(\varepsilon_{zz}^{(1)} - N_y^2 \right) - \varepsilon_{zx}^{(1)} \varepsilon_{xz}^{(1)} \right] \left(N_{z1}^{(1)} + N_{z3}^{(1)} \right) \\
& \left. - \left[-\varepsilon_{xy}^{(1)} \left(\varepsilon_{zz}^{(1)} - N_y^2 \right) + \varepsilon_{xz}^{(1)} \varepsilon_{zy}^{(1)} \right] \left[\left(\varepsilon_{xx}^{(1)} - N_y^2 \right) \left(\varepsilon_{zz}^{(1)} - N_y^2 \right) - \varepsilon_{zx}^{(1)} \varepsilon_{xz}^{(1)} \right] \right\},
\end{aligned} \tag{1.95e}
$$

$$
\begin{aligned}
& N_{z1}^{(1)} p_3^{(1)} - N_{z3}^{(1)} p_1^{(1)} \\
& \quad = \left(N_{z1}^{(1)} - N_{z3}^{(1)}\right) \mathcal{G}^{-1} \left\{ \left[-\varepsilon_{xy}^{(1)} \left(\varepsilon_{zz}^{(1)} - N_y^2 \right) + \varepsilon_{xz}^{(1)} \varepsilon_{zy}^{(1)} \right] \right.
\end{aligned}
$$

$$\times \left[\left(\varepsilon_{xx}^{(1)} - N_y^2\right)\left(\varepsilon_{zz}^{(1)} - N_y^2\right) - \varepsilon_{zx}^{(1)}\varepsilon_{xz}^{(1)}\right]$$
$$- \left(N_{z1}^{(1)2} + N_{z1}^{(1)}N_{z3}^{(1)} + N_{z3}^{(1)2}\right)\left(\varepsilon_{zz}^{(1)} - N_y^2\right)\left[-\varepsilon_{xy}^{(1)}\left(\varepsilon_{zz}^{(1)} - N_y^2\right) + \varepsilon_{xz}^{(1)}\varepsilon_{zy}^{(1)}\right]$$
$$-\varepsilon_{xz}^{(1)}N_y\left(\varepsilon_{zz}^{(1)} - N_y^2\right)\left(N_{z1}^{(1)} + N_{z3}^{(1)}\right)\left(N_{z1}^{(1)}N_{z3}^{(1)}\right)\Big\}, \qquad (1.95f)$$

$$p_1^{(1)}q_3^{(1)} - p_3^{(1)}q_1^{(1)}$$
$$= \left(N_{z1}^{(1)} - N_{z3}^{(1)}\right)\mathcal{G}^{-1}\Big\{ - \varepsilon_{xy}^{(1)}N_y\left[\left(\varepsilon_{xx}^{(1)} - N_y^2\right)\left(\varepsilon_{zz}^{(1)} - N_y^2\right) - \varepsilon_{zx}^{(1)}\varepsilon_{xz}^{(1)}\right]$$
$$- \varepsilon_{zy}^{(1)}\left(N_{z1}^{(1)} + N_{z3}^{(1)}\right)\left[-\varepsilon_{xy}^{(1)}\left(\varepsilon_{zz}^{(1)} - N_y^2\right) + \varepsilon_{xz}^{(1)}\varepsilon_{zy}^{(1)}\right]$$
$$- N_y\left(N_{z1}^{(1)2} + N_{z3}^{(1)2}\right)\left[-\varepsilon_{xy}^{(1)}\left(\varepsilon_{zz}^{(1)} - N_y^2\right) + \varepsilon_{xz}^{(1)}\varepsilon_{zy}^{(1)}\right]$$
$$- N_y\left(N_{z1}^{(1)}N_{z3}^{(1)}\right)\left[-\varepsilon_{xy}^{(1)}\left(\varepsilon_{zz}^{(1)} - N_y^2\right) + 2\varepsilon_{xz}^{(1)}\varepsilon_{zy}^{(1)}\right]$$
$$- \varepsilon_{xz}^{(1)}N_y^2\left(N_{z1}^{(1)}N_{z3}^{(1)}\right)\left(N_{z1}^{(1)} + N_{z3}^{(1)}\right)\Big\}, \qquad (1.95g)$$

$$N_{z1}^{(1)}p_1^{(1)}q_3^{(1)} - N_{z3}^{(1)}p_3^{(1)}q_1^{(1)}$$
$$= \left(N_{z1}^{(1)} - N_{z3}^{(1)}\right)\mathcal{G}^{-1}\Big\{\left[-\varepsilon_{xy}^{(1)}\left(\varepsilon_{zz}^{(1)} - N_y^2\right) + \varepsilon_{zy}^{(1)}\varepsilon_{xz}^{(1)}\right]$$
$$\times\left[-\varepsilon_{zy}^{(1)}\left(\varepsilon_{xx}^{(1)} - N_y^2\right) + \varepsilon_{xy}^{(1)}\varepsilon_{zx}^{(1)}\right]$$
$$+\varepsilon_{xz}^{(1)}N_y\left[-\varepsilon_{zy}^{(1)}\left(\varepsilon_{xx}^{(1)} - N_y^2\right) + \varepsilon_{xy}^{(1)}\varepsilon_{zx}^{(1)}\right]\left(N_{z1}^{(1)} + N_{z3}^{(1)}\right)$$
$$- \varepsilon_{xz}^{(1)}N_y^2\left(\varepsilon_{xx}^{(1)} - N_y^2\right)\left(N_{z1}^{(1)}N_{z3}^{(1)}\right) - \varepsilon_{zy}^{(1)}\left[-\varepsilon_{xy}^{(1)}\left(\varepsilon_{zz}^{(1)} - N_y^2\right) + \varepsilon_{xz}^{(1)}\varepsilon_{zy}^{(1)}\right]$$
$$\times\left(N_{z1}^{(1)}N_{z3}^{(1)}\right) - N_y\left[-\varepsilon_{xy}^{(1)}\left(\varepsilon_{zz}^{(1)} - N_y^2\right) + \varepsilon_{zy}^{(1)}\varepsilon_{xz}^{(1)}\right]$$
$$\times\left(N_{z1}^{(1)}N_{z3}^{(1)}\right)\left(N_{z1}^{(1)} + N_{z3}^{(1)}\right) - \varepsilon_{xz}^{(1)}N_y^2\left(N_{z1}^{(1)}N_{z3}^{(1)}\right)^2\Big\}, \qquad (1.95h)$$

where the denominator $\mathcal{G}$ of the expressions in Eq. (1.95) can be found from Eq. (1.86). We have

$$\mathcal{G} = \left[\left(\varepsilon_{xx}^{(1)} - N_y^2\right)\left(\varepsilon_{zz}^{(1)} - N_y^2\right) - \varepsilon_{zx}^{(1)}\varepsilon_{xz}^{(1)}\right]^2$$
$$- \left(\varepsilon_{zz}^{(1)} - N_y^2\right)\left(N_{z1}^{(1)2} + N_{z3}^{(1)2}\right)\left[\left(\varepsilon_{xx}^{(1)} - N_y^2\right)\left(\varepsilon_{zz}^{(1)} - N_y^2\right) - \varepsilon_{zx}^{(1)}\varepsilon_{xz}^{(1)}\right]$$
$$+ \left(\varepsilon_{zz}^{(1)} - N_y^2\right)^2\left(N_{z1}^{(1)}N_{z3}^{(1)}\right)^2. \qquad (1.96)$$

From Eq. (1.95f) and Eq. (1.95g), we determine $\mathcal{P}^{(1)}$ using

$$\begin{aligned}
\mathcal{P}^{(1)} &= N_y\left(p_1^{(1)}q_3^{(1)} - p_3^{(1)}q_1^{(1)}\right) + \left(N_{z1}^{(1)}p_3^{(1)} - N_{z3}^{(1)}p_1^{(1)}\right) \\
&= \left(N_{z1}^{(1)} - N_{z3}^{(1)}\right)\mathcal{G}^{-1}\Big\{\left(-\varepsilon_{xy}^{(1)}\varepsilon_{zz}^{(1)} + \varepsilon_{xz}^{(1)}\varepsilon_{zy}^{(1)}\right) \\
&\times\left[\left(\varepsilon_{xx}^{(1)} - N_y^2\right)\left(\varepsilon_{zz}^{(1)} - N_y^2\right) - \varepsilon_{zx}^{(1)}\varepsilon_{xz}^{(1)}\right] \\
&- \varepsilon_{zy}^{(1)}N_y\left[-\varepsilon_{xy}^{(1)}\left(\varepsilon_{zz}^{(1)} - N_y^2\right) + \varepsilon_{xz}^{(1)}\varepsilon_{zy}^{(1)}\right]\left(N_{z1}^{(1)} + N_{z3}^{(1)}\right) \\
&+ \left[\varepsilon_{zz}^{(1)}\varepsilon_{xy}^{(1)}\left(\varepsilon_{zz}^{(1)} - N_y^2\right) - \varepsilon_{xz}^{(1)}\varepsilon_{zy}^{(1)}\left(\varepsilon_{zz}^{(1)} + N_y^2\right)\right]\left(N_{z1}^{(1)}N_{z3}^{(1)}\right) \\
&- \varepsilon_{zz}^{(1)}\left[-\varepsilon_{xy}^{(1)}\left(\varepsilon_{zz}^{(1)} - N_y^2\right) + \varepsilon_{xz}^{(1)}\varepsilon_{zy}^{(1)}\right]\left(N_{z1}^{(1)2} + N_{z3}^{(1)2}\right) \\
&- \varepsilon_{xz}^{(1)}\varepsilon_{zz}^{(1)}N_y\left(N_{z1}^{(1)}N_{z3}^{(1)}\right)\left(N_{z1}^{(1)} + N_{z3}^{(1)}\right)\Big\}.
\end{aligned} \tag{1.97}$$

From Eq. (1.95d) and Eq. (1.95h), we further establish the expression for $\mathcal{S}^{(1)}$

$$\begin{aligned}
\mathcal{S}^{(1)} &= N_{z1}^{(1)}N_{z3}^{(1)}\left(p_3^{(1)} - p_1^{(1)}\right) + N_y\left(N_{z1}^{(1)}p_1^{(1)}q_3^{(1)} - N_{z3}^{(1)}p_3^{(1)}q_1^{(1)}\right) \\
&= \left(N_{z1}^{(1)} - N_{z3}^{(1)}\right)\mathcal{G}^{-1}\Big\{-\varepsilon_{xz}^{(1)}\varepsilon_{zz}^{(1)}N_y\left(N_{z1}^{(1)}N_{z3}^{(1)}\right)^2 \\
&- \varepsilon_{zz}^{(1)}\left[-\varepsilon_{xy}^{(1)}\left(\varepsilon_{zz}^{(1)} - N_y^2\right) + \varepsilon_{xz}^{(1)}\varepsilon_{zy}^{(1)}\right]\left(N_{z1}^{(1)} + N_{z3}^{(1)}\right)\left(N_{z1}^{(1)}N_{z3}^{(1)}\right) \\
&+ \varepsilon_{xz}^{(1)}N_y^2\left[-\varepsilon_{zy}^{(1)}\left(\varepsilon_{xx}^{(1)} - N_y^2\right) + \varepsilon_{xy}^{(1)}\varepsilon_{zx}^{(1)}\right]\left(N_{z1}^{(1)} + N_{z3}^{(1)}\right) \\
&- \varepsilon_{zy}^{(1)}N_y\left[-\varepsilon_{xy}^{(1)}\left(\varepsilon_{zz}^{(1)} - N_y^2\right) + \varepsilon_{xz}^{(1)}\varepsilon_{zy}^{(1)}\right]\left(N_{z1}^{(1)}N_{z3}^{(1)}\right) \\
&- \varepsilon_{xz}^{(1)}N_y\left[\varepsilon_{zz}^{(1)}\left(\varepsilon_{xx}^{(1)} - N_y^2\right) - \varepsilon_{zx}^{(1)}\varepsilon_{xz}^{(1)}\right]\left(N_{z1}^{(1)}N_{z3}^{(1)}\right) \\
&+ N_y\left[-\varepsilon_{xy}^{(1)}\left(\varepsilon_{zz}^{(1)} - N_y^2\right) + \varepsilon_{zy}^{(1)}\varepsilon_{xz}^{(1)}\right]\left[-\varepsilon_{zy}^{(1)}\left(\varepsilon_{xx}^{(1)} - N_y^2\right) + \varepsilon_{xy}^{(1)}\varepsilon_{zx}^{(1)}\right]\Big\}.
\end{aligned} \tag{1.98}$$

The relations between the incident and reflected electric field amplitudes can be conveniently expressed with help of the 2 × 2 Cartesian Jones reflection matrix [9]

$$\begin{pmatrix} E_{0s}^{(r)} \\ E_{0p}^{(r)} \end{pmatrix} = \begin{pmatrix} r_{ss}^{(01)} & r_{sp}^{(01)} \\ r_{ps}^{(01)} & r_{pp}^{(01)} \end{pmatrix}\begin{pmatrix} E_{0s}^{(i)} \\ E_{0p}^{(i)} \end{pmatrix}, \tag{1.99}$$

which provides the response of an anisotropic interface to an incident wave with an arbitrary polarization. In this way, we have solved completely the problem of the reflection at an interface between a half space characterized by a scalar permittivity and a half space characterized by a general permittivity tensor.

We now focus on important limiting cases. Our purpose is to describe the reflection at an arbitrary angle of incidence of a polarized monochromatic plane wave travelling from vacuum toward the interface, assuming three special orientations of the magnetization with respect to the plane of incidence and to the plane of interface. These correspond to the geometries of the magnetooptic Kerr effects. We express results in terms of generalized Fresnel amplitude coefficients, which form the Cartesian Jones reflection matrix defined in Eq. (1.99). They follow from Eq. (1.94) for a particular choice of the permittivity tensor in Eq. (1.95) to Eq. (1.97).

In the cases where the magnetization is induced in an originally isotropic medium parallel to one of the Cartesian axes, the permittivity tensor (pertinent to the same coordinate axes) takes its simplest form, *e.g.*, that employed in the wave equation (1.22) or (1.62). Then, useful formulas can be derived in a transparent way under the following simplifying assumptions on the relative values of the ε components, *i.e.*,

$$\varepsilon_{ij} \ll \varepsilon_{jj}, \tag{1.100}$$

and

$$\varepsilon_{ii} \approx \varepsilon_{jj}, \tag{1.101}$$

for $i, j = x, y, z$ with $i \neq j$. With the simplifying assumptions of Eq. (1.100) and Eq. (1.101), justified in most situations, the diagonal elements of the Jones reflection matrix, defined in Eq. (1.94), do not differ from the classical Fresnel equations for an interface between two isotropic media (0) and (1)

$$r_{ss}^{(01)} = \frac{N^{(0)}\alpha_z^{(0)} - N_z^{(1)}}{N^{(0)}\alpha_z^{(0)} + N_z^{(1)}} = \frac{N^{(0)}\cos\varphi^{(0)} - N_z^{(1)}}{N^{(0)}\cos\varphi^{(0)} + N_z^{(1)}}, \tag{1.102a}$$

$$r_{pp}^{(01)} = \frac{N^{(0)}N_z^{(1)} - \varepsilon_0^{(1)}\alpha_z^{(0)}}{N^{(0)}N_z^{(1)} + \varepsilon_0^{(1)}\alpha_z^{(0)}} = \frac{N^{(0)}N_z^{(1)} - \varepsilon_0^{(1)}\cos\varphi^{(0)}}{N^{(0)}N_z^{(1)} + \varepsilon_0^{(1)}\cos\varphi^{(0)}}, \tag{1.102b}$$

where $(\omega N_z^{(1)}/c)$ is the component of the propagation vector of the transmitted wave normal to the interface and $\varepsilon_0^{(1)} \approx \varepsilon_{jj}^{(1)}$, where $j = x, y, z$ stands for any of the diagonal elements of the permittivity tensor[13] in the magnetic medium (1). The elements of the reflection matrices for the magnetooptic Kerr effects can be written rather concisely thanks to the experimental notion, expressed by Eq. (1.100) and Eq. (1.101), that the off-diagonal elements in the permittivity tensor of a magnetized medium are much smaller than the diagonal ones with the diagonal elements differing very little, by very

[13] In this section, we may skip the region superscript (1) in the magnetic medium for brevity.

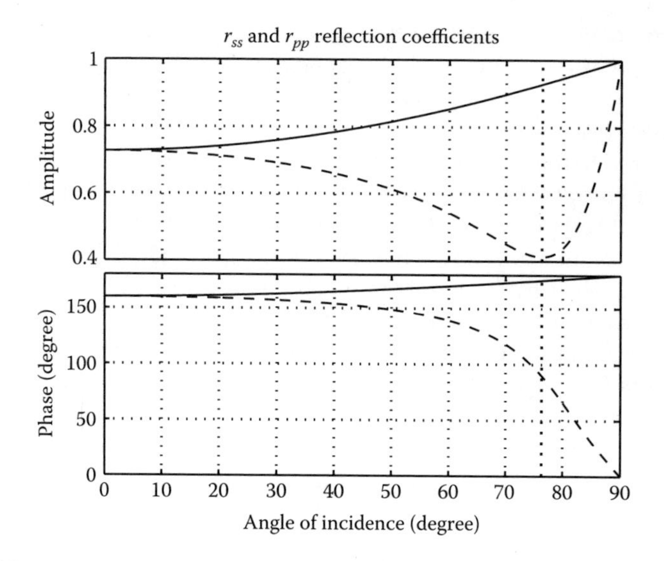

FIGURE 1.1
The amplitude and phase of the diagonal elements of the reflection matrix at an air-iron interface as a function of the angle of incidence computed for the complex index of refraction $N^{(Fe)} = 2.922 - j3.070$, corresponding to the scalar permittivity $\varepsilon_0^{(Fe)} = -0.8845 - j17.938$. The dotted vertical line indicates the principal angle of incidence, corresponding to the phase of $\pi/2$ of the reflection coefficient, r_{pp}.

small terms proportional to the even orders of $\boldsymbol{M}$, from each other. At the oblique incidence, we usually consider the off-diagonal elements up to the first order and neglect completely the differences among the diagonal elements. In this section, we limit ourselves, as before, to the magnetized media that had been isotropic before the magnetization was established. The dependence of the reflection matrix elements on the angle of incidence will be illustrated in the case of an air-iron interface for a chosen set of values of the diagonal and off-diagonal permittivity tensor elements pertinent to the wavelength of 632.8 nm [22]. Figure 1.1 displays this dependence for the diagonal elements.

1.6.1 MO Polar Kerr Effect

Magnetooptic polar Kerr effect takes place in the geometry where the magnetization in the magnetic medium (1) is oriented in the direction normal to the plane interface (*i.e.*, parallel to the z axis of our coordinate system) between the isotropic region (0) and region (1). The medium (1) is characterized by the relative permittivity tensor of the form, which follows from Eq. (1.3) and Eq. (1.10). It was already used in Eq. (1.22). We skip the medium index (1). The choice of the coordinate system for the polar magnetization is denoted by the subscript P, *i.e.*,

$$\varepsilon_P = \begin{pmatrix} \varepsilon_{xx} & \varepsilon_{xy} & 0 \\ -\varepsilon_{xy} & \varepsilon_{yy} & 0 \\ 0 & 0 & \varepsilon_{zz} \end{pmatrix} \approx \begin{pmatrix} \varepsilon_0 & \mathrm{j}\varepsilon_1 & 0 \\ -\mathrm{j}\varepsilon_1 & \varepsilon_0 & 0 \\ 0 & 0 & \varepsilon_0 \end{pmatrix}. \tag{1.103}$$

The substitution into the wave equation system (1.79) gives

$$\begin{pmatrix} (N_y^2 + N_z^2) - \varepsilon_0 & -\mathrm{j}\varepsilon_1 & 0 \\ \mathrm{j}\varepsilon_1 & N_z^2 - \varepsilon_0 & -N_y N_z \\ 0 & -N_z N_y & N_y^2 - \varepsilon_0 \end{pmatrix} \begin{pmatrix} E_{0x} \\ E_{0y} \\ E_{0z} \end{pmatrix} = 0, \tag{1.104}$$

and the proper value equation for the propagation vector at polar magnetization becomes a biquadratic one

$$\varepsilon_0 N_z^4 - 2\varepsilon_0 \left(\varepsilon_0 - N_y^2\right) N_z^2 + \left(\varepsilon_0 - N_y^2\right)\left(\varepsilon_0^2 - \varepsilon_0 N_y^2 - \varepsilon_1^2\right) = 0, \tag{1.105}$$

with the solutions

$$\left(N_z^2\right)^{\pm} = \left(\varepsilon_0 - N_y^2\right) \pm \left(\varepsilon_0 - N_y^2\right)^{1/2} \frac{\varepsilon_1}{\varepsilon_0^{1/2}}. \tag{1.106}$$

Applying a similar nomenclature as in the case of the Faraday effect in Section 1.3.1, we can classify the solutions, to first order in ε_1, as follows

$$N_{z1,3} = +\left(\varepsilon_0 - N_y^2\right)^{1/2} \pm \frac{\varepsilon_1}{2\varepsilon_0^{1/2}}, \qquad \text{forward modes,} \tag{1.107a}$$

$$N_{z2,4} = -\left(\varepsilon_0 - N_y^2\right)^{1/2} \mp \frac{\varepsilon_1}{2\varepsilon_0^{1/2}}, \qquad \text{backward modes.} \tag{1.107b}$$

We wish to find the proper polarization modes in the magnetized medium. The substitution of $N_{z1,3}$ from Eq. (1.107a) into the first and second equations of the system (1.104) provides

$$\pm \left(\varepsilon_0 - N_y^2\right)^{1/2} \frac{\varepsilon_1}{\varepsilon_0^{1/2}} E_{0x}^{(1,3)} - \mathrm{j}\varepsilon_1 E_{0y}^{(1,3)} = 0,$$

$$\mathrm{j}\varepsilon_1 E_{0x}^{(1,3)} \pm \left[\left(\varepsilon_0 - N_y^2\right)^{1/2} \frac{\varepsilon_1}{\varepsilon_0^{1/2}} \mp N_y^2\right] E_{0y}^{(1,3)} = N_y \left[\left(\varepsilon_0 - N_y^2\right)^{1/2} \pm \frac{\varepsilon_1}{2\varepsilon_0^{1/2}}\right] E_{0z}^{(1,3)}.$$

We get the relations between the field components for the forward-going polarization modes

$$E_{0y}^{(1,3)} = \mp\mathrm{j} \frac{\left(\varepsilon_0 - N_y^2\right)^{1/2}}{\varepsilon_0^{1/2}} E_{0x}^{(1,3)}, \tag{1.108a}$$

$$E_{0x}^{(1,3)} = \pm\mathrm{j} \frac{\varepsilon_0^{1/2}}{\left(\varepsilon_0 - N_y^2\right)^{1/2}} E_{0y}^{(1,3)}, \tag{1.108b}$$

$$E_{0z}^{(1,3)} = -\frac{N_y}{\left(\varepsilon_0 - N_y^2\right)^{1/2}} \frac{\left(\varepsilon_0 - N_y^2\right)^{1/2} \pm \dfrac{\varepsilon_1}{\varepsilon_0^{1/2}}}{\left(\varepsilon_0 - N_y^2\right)^{1/2} \pm \dfrac{\varepsilon_1}{2\varepsilon_0^{1/2}}} E_{0y}^{(1,3)}$$

$$\approx -\frac{N_y}{\left(\varepsilon_0 - N_y^2\right)} \left[\left(\varepsilon_0 - N_y^2\right)^{1/2} \pm \frac{\varepsilon_1}{2\varepsilon_0^{1/2}}\right] E_{0y}^{(1,3)}. \quad (1.108c)$$

From the expression for $E_{0z}^{(1,3)}$, the terms proportional to ε_1^2 were removed. At oblique angles of incidence $(N_y \neq 0)$, the proper polarization modes are elliptically polarized. As expected, at $N_y = 0$, they become circularly polarized waves. The proper polarization modes may be treated as approximately CP at small angles of incidence and at $\left(\varepsilon_0 - N_y^2\right)^{1/2} \gg \frac{\varepsilon_1}{\varepsilon_0^{1/2}}$. Then, $E_{0x}^{(1,3)2} \approx E_{0y}^{(1,3)2}$.

The fields of backward-going modes differ from the forward-going ones in sign of the E_{0z} components

$$E_{0y}^{(2,4)} = \mp j \frac{\left(\varepsilon_0 - N_y^2\right)^{1/2}}{\varepsilon_0^{1/2}} E_{0x}^{(2,4)}, \quad (1.109a)$$

$$E_{0x}^{(2,4)} = \pm j \frac{\varepsilon_0^{1/2}}{\left(\varepsilon_0 - N_y^2\right)^{1/2}} E_{0y}^{(2,4)}, \quad (1.109b)$$

$$E_{0z}^{(2,4)} \approx \frac{N_y}{\left(\varepsilon_0 - N_y^2\right)} \left[\left(\varepsilon_0 - N_y^2\right)^{1/2} \pm \frac{\varepsilon_1}{2\varepsilon_0^{1/2}}\right] E_{0y}^{(2,4)}. \quad (1.109c)$$

We now apply the conditions for the continuity of the field components parallel to the interface at the interface. The substitutions into Eq. (1.84) for the field amplitudes $E_{0x}^{(1,3)}$ and $E_{0z}^{(1,3)}$ according to Eq. (1.108b) and Eq. (1.108c) lead to an equation system

$$- E_{0s}^{(r)} + j\frac{\varepsilon_0^{1/2}}{N_{z0}} \left(E_{0y}^{(1)} - E_{0y}^{(3)}\right) = E_{0s}^{(i)}$$

$$N^{(0)}\alpha_z^{(0)} E_{0s}^{(r)} + j\frac{\varepsilon_1}{2N_{z0}} \left(E_{0y}^{(1)} + E_{0y}^{(3)}\right) + j\varepsilon_0^{1/2} \left(E_{0y}^{(1)} - E_{0y}^{(3)}\right) = N^{(0)}\alpha_z^{(0)} E_{0s}^{(i)}$$

$$-\alpha_z^{(0)} E_{0p}^{(r)} + \left(E_{0y}^{(1)} + E_{0y}^{(3)}\right) = \alpha_z^{(0)} E_{0p}^{(i)}$$

$$N^{(0)} E_{0p}^{(r)} + \frac{\varepsilon_0}{N_{z0}} \left(E_{0y}^{(1)} + E_{0y}^{(3)}\right) + \frac{\varepsilon_1 \varepsilon_0^{1/2}}{2N_{z0}^2} \left(E_{0y}^{(1)} - E_{0y}^{(3)}\right) = N^{(0)} E_{0p}^{(i)}$$

or in a matrix form

$$\begin{pmatrix} -1 & 0 & 0 & \mathrm{j}\dfrac{\varepsilon_0^{1/2}}{N_{z0}} \\ N^{(0)}\alpha_z^{(0)} & 0 & \mathrm{j}\dfrac{\varepsilon_1}{2N_{z0}} & \mathrm{j}\varepsilon_0^{1/2} \\ 0 & -\alpha_z^{(0)} & 1 & 0 \\ 0 & N^{(0)} & \dfrac{\varepsilon_0}{N_{z0}} & \dfrac{\varepsilon_1\varepsilon_0^{1/2}}{2N_{z0}^2} \end{pmatrix} \begin{pmatrix} E_{0s}^{(r)} \\ E_{0p}^{(r)} \\ E_{0y}^{(1)} + E_{0y}^{(3)} \\ E_{0y}^{(1)} - E_{0y}^{(3)} \end{pmatrix} = \begin{pmatrix} E_{0s}^{(i)} \\ N^{(0)}\alpha_z^{(0)} E_{0s}^{(i)} \\ \alpha_z^{(0)} E_{0p}^{(i)} \\ N^{(0)} E_{0p}^{(i)} \end{pmatrix}, \tag{1.110}$$

where $N_{z0} = \left(\varepsilon_0 - N_y^2\right)^{1/2}$. We compute the determinant of the 4×4 matrix

$$\begin{vmatrix} -1 & 0 & 0 & \mathrm{j}\dfrac{\varepsilon_0^{1/2}}{N_{z0}} \\ N^{(0)}\alpha_z^{(0)} & 0 & \mathrm{j}\dfrac{\varepsilon_1}{2N_{z0}} & \mathrm{j}\varepsilon_0^{1/2} \\ 0 & -\alpha_z^{(0)} & 1 & 0 \\ 0 & N^{(0)} & \dfrac{\varepsilon_0}{N_{z0}} & \dfrac{\varepsilon_1\varepsilon_0^{1/2}}{2N_{z0}^2} \end{vmatrix}$$

$$= \mathrm{j}\frac{\varepsilon_0^{1/2}}{N_{z0}^2}\left[\left(N^{(0)}\alpha_z^{(0)} + N_{z0}\right)\left(N^{(0)}N_{z0} + \varepsilon_0\alpha_z^{(0)}\right) - \frac{\varepsilon_1^2\alpha_z^{(0)}}{4N_{z0}}\right] \tag{1.111}$$

and recognize the product of denominators of the Fresnel equations (1.102). The $E_{0s}^{(r)}$ and $E_{0p}^{(r)}$ amplitudes are obtained with the following determinants

$$\begin{vmatrix} E_{0s}^{(i)} & 0 & 0 & \mathrm{j}\dfrac{\varepsilon_0^{1/2}}{N_{z0}} \\ N^{(0)}\alpha_z^{(0)} E_{0s}^{(i)} & 0 & \mathrm{j}\dfrac{\varepsilon_1}{2N_{z0}} & \mathrm{j}\varepsilon_0^{1/2} \\ \alpha_z^{(0)} E_{0p}^{(i)} & -\alpha_z^{(0)} & 1 & 0 \\ N^{(0)} E_{0p}^{(i)} & N^{(0)} & \dfrac{\varepsilon_0}{N_{z0}} & \dfrac{\varepsilon_1\varepsilon_0^{1/2}}{2N_{z0}^2} \end{vmatrix}$$

$$= \mathrm{j}\frac{\varepsilon_0^{1/2}}{N_{z0}^2}\left[\left(N^{(0)}\alpha_z^{(0)} - N_{z0}\right)\left(N^{(0)}N_{z0}^{(0)} + \varepsilon_0\alpha_z^{(0)}\right) E_{0s}^{(i)} - \mathrm{j}N^{(0)}\alpha_z^{(0)}\varepsilon_1 E_{0p}^{(i)}\right], \tag{1.112}$$

$$\begin{vmatrix} -1 & E_{0s}^{(i)} & 0 & \mathrm{j}\frac{\varepsilon_0^{1/2}}{N_{z0}} \\ N^{(0)}\alpha_z^{(0)} & N^{(0)}\alpha_z^{(0)}E_{0s}^{(i)} & \mathrm{j}\frac{\varepsilon_1}{2N_{z0}} & \mathrm{j}\varepsilon_0^{1/2} \\ 0 & \alpha_z^{(0)}E_{0p}^{(i)} & 1 & 0 \\ 0 & N^{(0)}E_{0p}^{(i)} & \frac{\varepsilon_0}{N_{z0}} & \frac{\varepsilon_1\varepsilon_0^{1/2}}{2N_{z0}^2} \end{vmatrix}$$

$$= \mathrm{j}\frac{\varepsilon_0^{1/2}}{N_{z0}^2}\left[\left(N^{(0)}\alpha_z^{(0)} + N_{z0}\right)\left(N^{(0)}N_{z0}^{(0)} - \varepsilon_0\alpha_z^{(0)}\right)E_{0s}^{(i)} + \mathrm{j}N^{(0)}\alpha_z^{(0)}\varepsilon_1 E_{0s}^{(i)}\right]. \tag{1.113}$$

According to Eqs. (1.111) to (1.113), there are terms proportional to ε_1^2 in the diagonal elements of the Cartesian Jones reflection matrix (r_{ss} and r_{pp}). In the subsequent step, these were removed in the spirit of our approximation. We observe that r_{ss} and r_{pp} are then approximately given by the Fresnel equations for an interface between isotropic media, Eq. (1.102), as expected. For the off-diagonal elements defined in (1.94), we obtain

$$r_{ps}^{(\mathrm{pol})} = -r_{sp}^{(\mathrm{pol})} = \frac{\mathrm{j}\varepsilon_1 N^{(0)}\alpha_z^{(0)}}{\left(N^{(0)}\alpha_z^{(0)} + N_{z0}\right)\left(N^{(0)}N_{z0} + \varepsilon_0\alpha_z^{(0)}\right)}$$
$$= \frac{\mathrm{j}\varepsilon_1 N^{(0)}\cos\varphi^{(0)}}{\left(N^{(0)}\cos\varphi^{(0)} + N_{z0}\right)\left(N^{(0)}N_{z0} + \varepsilon_0\cos\varphi^{(0)}\right)}. \tag{1.114}$$

We can define the MO polar Kerr effect for the *incident* linearly s and p polarized waves by the complex ratios

$$\chi_{rs}^{(\mathrm{pol})} = \frac{r_{ps}^{(\mathrm{pol})}}{r_{ss}} = \frac{\mathrm{j}\varepsilon_1 N^{(0)}\cos\varphi^{(0)}}{\left(N^{(0)}\cos\varphi^{(0)} - N_{z0}\right)\left(N^{(0)}N_{z0} + \varepsilon_0\cos\varphi^{(0)}\right)}, \tag{1.115a}$$

$$\chi_{rp}^{(\mathrm{pol})} = \frac{r_{sp}^{(\mathrm{pol})}}{r_{pp}} = \frac{-\mathrm{j}\varepsilon_1 N^{(0)}\cos\varphi^{(0)}}{\left(N^{(0)}\cos\varphi^{(0)} + N_{z0}\right)\left(N^{(0)}N_{z0} - \varepsilon_0\cos\varphi^{(0)}\right)}. \tag{1.115b}$$

It can be shown that the ratio of the off-diagonal to diagonal elements of the Cartesian Jones reflection matrix at the normal incidence ($\varphi^{(0)} = 0$) is consistent with $\chi_r^{(\mathrm{pol})}$ in Section 1.3.2. Figure 1.2 and Figure 1.3 display the dependence on the angle of incidence, $\varphi^{(0)}$, of Eq. (1.114) and Eq. (1.115). Note the plateau in $r_{ps}^{(\mathrm{pol})} = -r_{sp}^{(\mathrm{pol})}$ extended from zero over a wide $\varphi^{(0)}$ range. Here the $\varphi^{(0)}$ dependence of the MO polar Kerr rotation and ellipticity is predominantly due to the $\varphi^{(0)}$ dependence of r_{pp}, the diagonal reflection matrix element.

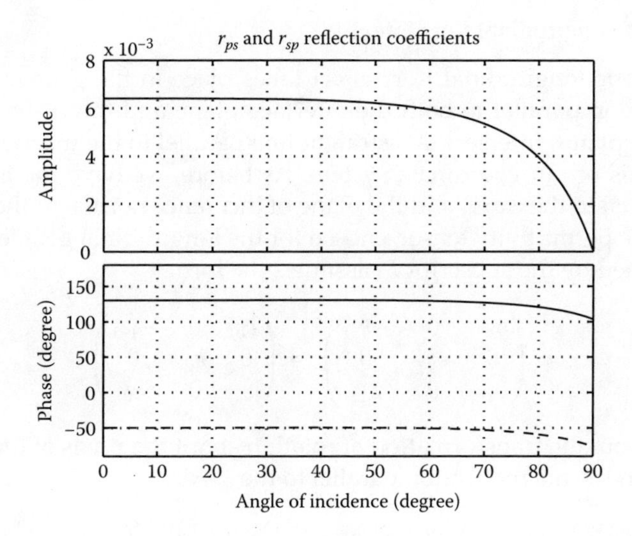

FIGURE 1.2
The amplitude and phase of the off-diagonal elements r_{ps} (solid lines) and r_{sp} (dashed line) of the Cartesian Jones reflection matrix, pertinent to the polar magnetization, at an air-iron interface as a function of the angle of incidence. The curves were computed using the diagonal and off-diagonal permittivity tensor elements given by $\varepsilon_{xx}^{(\mathrm{Fe})} = \varepsilon_0^{(\mathrm{Fe})} = -0.8845 - \mathrm{j}17.938$ and $\varepsilon_{xy}^{(\mathrm{Fe})} = \mathrm{j}\varepsilon_1^{(\mathrm{Fe})} = -0.6676 - \mathrm{j}0.08988$, respectively.

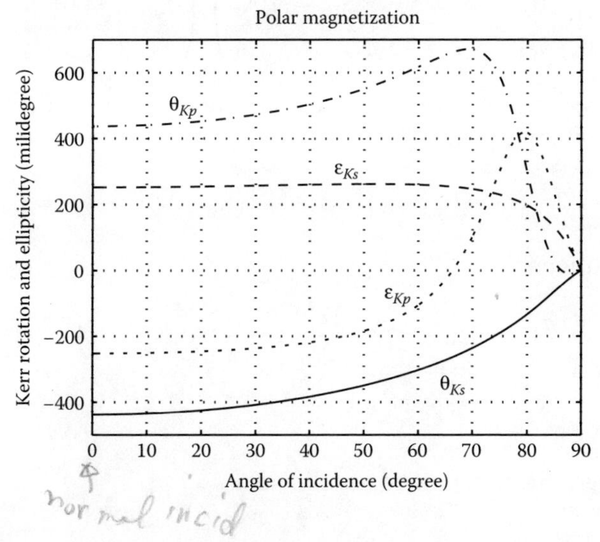

FIGURE 1.3
Complex magnetooptic polar Kerr rotation and ellipticity $\chi_{rs}^{(\mathrm{pol})} = \theta_{Ks} + \mathrm{j}\varepsilon_{Ks}$ and $\chi_{rp}^{(\mathrm{pol})} = \theta_{Kp} + \mathrm{j}\varepsilon_{Kp}$ as functions of the angle of incidence for the s and p polarized incident waves, respectively, at an air-iron interface. The curves of the magnetooptic azimuths θ_{Ks} and θ_{Kp} and the ellipticities ϵ_{Ks} and ϵ_{Kp} were computed using the data of Figure 1.1 and Figure 1.2.

1.6.2 MO Longitudinal Kerr Effect

Magnetooptic longitudinal Kerr effect takes place in the geometry with magnetization parallel to both the interface plane and the plane of incidence. The planes intersect in a straight line parallel to the magnetization in the y axis of our coordinate system. As before, we have the interface plane normal to the z axis and the plane of incidence normal to the x axis. The relative permittivity tensor, chosen for the longitudinal geometry and distinguished by the subscript L, assumes the form

$$\varepsilon_L = \begin{pmatrix} \varepsilon_{xx} & 0 & -\varepsilon_{zx} \\ 0 & \varepsilon_{yy} & 0 \\ \varepsilon_{zx} & 0 & \varepsilon_{zz} \end{pmatrix} \approx \begin{pmatrix} \varepsilon_0 & 0 & -\mathrm{j}\varepsilon_1 \\ 0 & \varepsilon_0 & 0 \\ \mathrm{j}\varepsilon_1 & 0 & \varepsilon_0 \end{pmatrix}, \tag{1.116}$$

resulting from the transformation of rotation about the x axis of the tensor (1.103) with the magnetization parallel to the z axis

$$\varepsilon_L = \begin{pmatrix} 1 & 0 & 0 \\ 0 & 0 & 1 \\ 0 & -1 & 0 \end{pmatrix} \begin{pmatrix} \varepsilon_0 & \mathrm{j}\varepsilon_1 & 0 \\ -\mathrm{j}\varepsilon_1 & \varepsilon_0 & 0 \\ 0 & 0 & \varepsilon_0 \end{pmatrix} \begin{pmatrix} 1 & 0 & 0 \\ 0 & 0 & -1 \\ 0 & 1 & 0 \end{pmatrix}. \tag{1.117}$$

The substitution into the wave equation system (1.79) gives

$$\begin{pmatrix} (N_y^2 + N_z^2) - \varepsilon_0 & 0 & \mathrm{j}\varepsilon_1 \\ 0 & N_z^2 - \varepsilon_0 & -N_y N_z \\ -\mathrm{j}\varepsilon_1 & -N_z N_y & N_y^2 - \varepsilon_0 \end{pmatrix} \begin{pmatrix} E_{0x} \\ E_{0y} \\ E_{0z} \end{pmatrix} = 0, \tag{1.118}$$

and the proper value equation for the propagation vector at the longitudinal magnetization becomes

$$\varepsilon_0 N_z^4 - \left[2\varepsilon_0 \left(\varepsilon_0 - N_y^2\right) - \varepsilon_1^2\right] N_z^2 + \varepsilon_0 \left[\left(\varepsilon_0 - N_y^2\right)^2 - \varepsilon_1^2\right] = 0, \tag{1.119}$$

with the solutions

$$\begin{aligned} \left(N_z^2\right)_\pm &= \left(\varepsilon_0 - N_y^2\right) - \frac{\varepsilon_1^2}{2\varepsilon_0} \pm \frac{\varepsilon_1}{2\varepsilon_0} \left(4\varepsilon_0 N_y^2 + \varepsilon_1^2\right)^{1/2} \\ &\approx \left(\varepsilon_0 - N_y^2\right) \pm \frac{\varepsilon_1 N_y}{\varepsilon_0^{1/2}}. \end{aligned} \tag{1.120}$$

The solutions may be classified as follows

$$N_{z1,3} = +\left(\varepsilon_0 - N_y^2\right)^{1/2} \pm \frac{\varepsilon_1 N_y}{2\varepsilon_0^{1/2} \left(\varepsilon_0 - N_y^2\right)^{1/2}}, \quad \text{forward modes,} \tag{1.121a}$$

$$N_{z2,4} = -\left(\varepsilon_0 - N_y^2\right)^{1/2} \mp \frac{\varepsilon_1 N_y}{2\varepsilon_0^{1/2} \left(\varepsilon_0 - N_y^2\right)^{1/2}}, \quad \text{backward modes.} \tag{1.121b}$$

We again express the electric fields of the forward-going modes in the magnetized medium in terms of one of them. The substitution of $N_{z1,3}$ from Eq. (1.121a) into the second and third equations of the system (1.118) provides

$$E_{0z}^{(1,3)} = \left[-\frac{N_y}{\left(\varepsilon_0 - N_y^2\right)^{1/2}} \pm \varepsilon_1 \frac{2\varepsilon_0 - N_y^2}{2\varepsilon_0^{1/2}\left(\varepsilon_0 - N_y^2\right)^{3/2}} \right] E_{0y}^{(1,3)}, \quad (1.122a)$$

$$E_{0x}^{(1,3)} = \pm \mathrm{j} \frac{\varepsilon_0^{1/2}}{\left(\varepsilon_0 - N_y^2\right)^{1/2}} E_{0y}^{(1,3)}. \quad (1.122b)$$

The continuity of the field components at the interface is expressed by the equation system

$$\begin{pmatrix} -1 & 0 & 0 & \mathrm{j}\dfrac{\varepsilon_0^{1/2}}{N_{z0}} \\ N^{(0)}\alpha_z^{(0)} & 0 & \mathrm{j}\dfrac{\varepsilon_1 N_y}{2N_{z0}^2} & \mathrm{j}\varepsilon_0^{1/2} \\ 0 & -\alpha_z^{(0)} & 1 & 0 \\ 0 & N^{(0)} & \dfrac{\varepsilon_0}{N_{z0}} & -\dfrac{\varepsilon_1 \varepsilon_0^{1/2} N_y}{2N_{z0}^3} \end{pmatrix} \begin{pmatrix} E_{0s}^{(r)} \\ E_{0p}^{(r)} \\ E_{0y}^{(1)} + E_{0y}^{(3)} \\ E_{0y}^{(1)} - E_{0y}^{(3)} \end{pmatrix} = \begin{pmatrix} E_{0s}^{(i)} \\ N^{(0)}\alpha_z^{(0)} E_{0s}^{(i)} \\ \alpha_z^{(0)} E_{0p}^{(i)} \\ N^{(0)} E_{0p}^{(i)} \end{pmatrix}, \quad (1.123)$$

where $N_{z0} = \left(\varepsilon_0 - N_y^2\right)^{1/2}$. We compute the determinant of the 4×4 matrix

$$\begin{vmatrix} -1 & 0 & 0 & \mathrm{j}\dfrac{\varepsilon_0^{1/2}}{N_{z0}} \\ N^{(0)}\alpha_z^{(0)} & 0 & \mathrm{j}\dfrac{\varepsilon_1 N_y}{2N_{z0}^2} & \mathrm{j}\varepsilon_0^{1/2} \\ 0 & -\alpha_z^{(0)} & 1 & 0 \\ 0 & N^{(0)} & \dfrac{\varepsilon_0}{N_{z0}} & -\dfrac{\varepsilon_1 N_y \varepsilon_0^{1/2}}{2N_{z0}^3} \end{vmatrix}$$

$$= \mathrm{j}\frac{\varepsilon_0^{1/2}}{N_{z0}^2} \left[\left(N^{(0)}\alpha_z^{(0)} + N_{z0}\right)\left(N^{(0)} N_{z0} + \varepsilon_0 \alpha_z^{(0)}\right) + \frac{\varepsilon_1^2 \alpha_z^{(0)} N_y^2}{4N_{z0}^3} \right]. \quad (1.124)$$

Using the procedure similar to that employed above for the polar magnetization, Eqs. (1.111) to (1.113), we obtain r_{ss} and r_{pp}, which again differ from those for the interface of isotropic media by terms proportional to ε_1^2, only. It remains to compute $r_{sp}^{(\mathrm{lon})}$ using the determinant

$$\begin{vmatrix} E_{0s}^{(i)} & 0 & 0 & \mathrm{j}\dfrac{\varepsilon_0^{1/2}}{N_{z0}} \\ N^{(0)}\alpha_z^{(0)} E_{0s}^{(i)} & 0 & \mathrm{j}\dfrac{\varepsilon_1 N_y}{2N_{z0}^2} & \mathrm{j}\varepsilon_0^{1/2} \\ \alpha_z^{(0)} E_{0p}^{(i)} & -\alpha_z^{(0)} & 1 & 0 \\ N^{(0)} E_{0p}^{(i)} & N^{(0)} & \dfrac{\varepsilon_0}{N_{z0}} & -\dfrac{\varepsilon_1 \varepsilon_0^{1/2} N_y}{2N_{z0}^3} \end{vmatrix}$$

$$= \mathrm{j}\frac{\varepsilon_0^{1/2}}{N_{z0}^2}\left[\left(N^{(0)}\alpha_z^{(0)} - N_{z0}\right)\left(N^{(0)}N_{z0}^{(0)} + \varepsilon_0\alpha_z^{(0)}\right) E_{0s}^{(i)} - \mathrm{j}N^{(0)}\alpha_z^{(0)}\frac{\varepsilon_1 N_y}{N_{z0}} E_{0p}^{(i)}\right], \tag{1.125}$$

with $E_{0s}^{(i)} = 0$, in agreement with the definition (1.94c), and $r_{ps}^{(\mathrm{lon})}$, using the determinant

$$\begin{vmatrix} -1 & E_{0s}^{(i)} & 0 & \mathrm{j}\dfrac{\varepsilon_0^{1/2}}{N_{z0}} \\ N^{(0)}\alpha_z^{(0)} & N^{(0)}\alpha_z^{(0)} E_{0s}^{(i)} & \mathrm{j}\dfrac{\varepsilon_1 N_y}{2N_{z0}^2} & \mathrm{j}\varepsilon_0^{1/2} \\ 0 & \alpha_z^{(0)} E_{0p}^{(i)} & 1 & 0 \\ 0 & N^{(0)} E_{0p}^{(i)} & \dfrac{\varepsilon_0}{N_{z0}} & -\dfrac{\varepsilon_1 \varepsilon_0^{1/2} N_y}{2N_{z0}^3} \end{vmatrix}$$

$$= \mathrm{j}\frac{\varepsilon_0^{1/2}}{N_{z0}^2}\left[\left(N^{(0)}\alpha_z^{(0)} + N_{z0}\right)\left(N^{(0)}N_{z0}^{(0)} - \varepsilon_0\alpha_z^{(0)}\right) E_{0p}^{(i)} - \mathrm{j}N^{(0)}\alpha_z^{(0)}\frac{\varepsilon_1 N_y}{N_{z0}} E_{0s}^{(i)}\right], \tag{1.126}$$

with $E_{0p}^{(i)} = 0$, in agreement with the definition (1.94b).

Again, the anisotropy of the medium induced by magnetization leads to nonzero off-diagonal reflection matrix elements

$$r_{ps}^{(\text{lon})} = r_{sp}^{(\text{lon})} = \frac{-j\varepsilon_1 N^{(0)}\alpha_z^{(0)} \dfrac{N_y}{N_{z0}}}{\left(N^{(0)}\alpha_z^{(0)} + N_{z0}\right)\left(N^{(0)}N_{z0} + \varepsilon_0\alpha_z^{(0)}\right)}$$

$$= \frac{-j\varepsilon_1 N^{(0)2}\cos\varphi^{(0)}\sin\varphi^{(0)}}{N_{z0}\left(N^{(0)}\cos\varphi^{(0)} + N_{z0}\right)\left(N^{(0)}N_{z0} + \varepsilon_0\cos\varphi^{(0)}\right)}. \quad (1.127)$$

We can define the MO longitudinal Kerr effect for linearly s and p polarized incident waves by ratios

$$\chi_{rs}^{(\text{lon})} = \frac{r_{ps}^{(\text{lon})}}{r_{ss}} = \frac{-j\varepsilon_1 N^{(0)2}\cos\varphi^{(0)}\sin\varphi^{(0)}}{N_{z0}\left(N^{(0)}\cos\varphi^{(0)} - N_{z0}\right)\left(N^{(0)}N_{z0} + \varepsilon_0\cos\varphi^{(0)}\right)}, \quad (1.128a)$$

$$\chi_{rp}^{(\text{lon})} = \frac{r_{sp}^{(\text{lon})}}{r_{pp}} = \frac{-j\varepsilon_1 N^{(0)2}\cos\varphi^{(0)}\sin\varphi^{(0)}}{N_{z0}\left(N^{(0)}\cos\varphi^{(0)} + N_{z0}\right)\left(N^{(0)}N_{z0} - \varepsilon_0\cos\varphi^{(0)}\right)}. \quad (1.128b)$$

In the present approximation, the diagonal elements r_{ss} and r_{pp} remain the same as for an interface of isotropic media, *i.e*, given by Eq. (1.102). Figure 1.4 and Figure 1.5 display the dependence on the angle of incidence of the off-diagonal reflection coefficients and MO longitudinal Kerr effect characterized by Eq. (1.127) and Eq. (1.128), respectively. Note that the computed $\chi_{rs}^{(\text{lon})}\left(\varphi^{(0)}\right)$ and $\chi_{rp}^{(\text{lon})}\left(\varphi^{(0)}\right)$ indicate the optimum $\varphi^{(0)}$ for the longitudinal Kerr rotation and ellipticity measurements.

1.6.3 MO Transverse Effect

In the last considered geometry, known as transverse, the magnetization is restricted to the planar interface being normal to the plane of incidence. In our choice of axes, it is parallel to the x axis. The permittivity tensor, chosen for the transverse geometry and distinguished by the subscript T, assumes the form

$$\varepsilon_T = \begin{pmatrix} \varepsilon_{xx} & 0 & 0 \\ 0 & \varepsilon_{yy} & \varepsilon_{yz} \\ 0 & -\varepsilon_{yz} & \varepsilon_{yy} \end{pmatrix} = \begin{pmatrix} \varepsilon_0 & 0 & 0 \\ 0 & \varepsilon_0 & j\varepsilon_1 \\ 0 & -j\varepsilon_1 & \varepsilon_0 \end{pmatrix}. \quad (1.129)$$

The substitution into the wave equation system (1.79) gives

$$\begin{pmatrix} \left(N_y^2 + N_z^2\right) - \varepsilon_3 & 0 & 0 \\ 0 & N_z^2 - \varepsilon_0 & -N_yN_z - j\varepsilon_1 \\ 0 & -N_zN_y + j\varepsilon_1 & N_y^2 - \varepsilon_0 \end{pmatrix} \begin{pmatrix} E_{0x} \\ E_{0y} \\ E_{0z} \end{pmatrix} = 0, \quad (1.130)$$

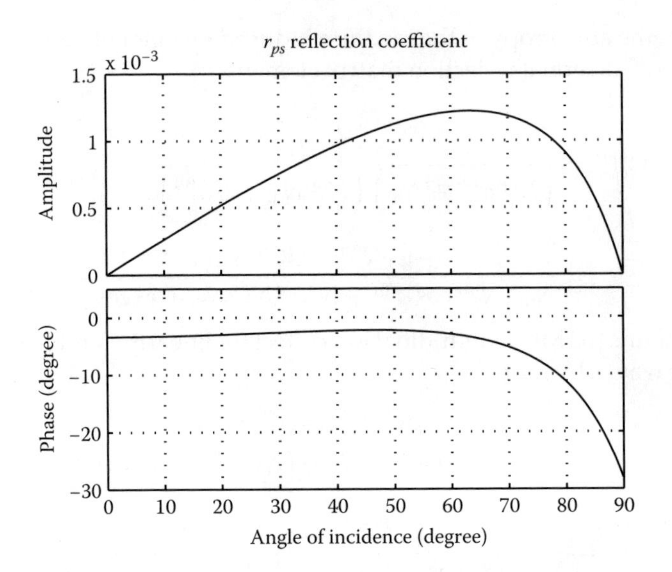

FIGURE 1.4
The amplitude and phase of the off-diagonal elements r_{ps} (solid lines) and r_{sp} (dashed line) of the Jones reflection matrix for the longitudinal magnetization at an air-iron interface as a function of the angle of incidence. The curves were computed using the values for the diagonal and off-diagonal permittivity tensor elements given by $\varepsilon_{xx}^{(\mathrm{Fe})} = \varepsilon_0^{(\mathrm{Fe})} = -0.8845 - \mathrm{j}17.938$ and $\varepsilon_{xy}^{(\mathrm{Fe})} = \mathrm{j}\varepsilon_1^{(\mathrm{Fe})} = -0.8845 - \mathrm{j}17.938$, respectively.

and the proper value equation for the propagation vector at the transverse magnetization becomes

$$\left[N_z^2 - \left(\varepsilon_3 - N_y^2\right)\right]\left[N_z^2 - \left(\varepsilon_0 - N_y^2\right) + \frac{\varepsilon_1^2}{\varepsilon_0}\right] = 0, \tag{1.131}$$

with the solutions for the waves polarized perpendicularly to the plane of incidence independent on the off-diagonal tensor elements ε_1 and the solutions for the waves polarized parallel to the plane of incidence perturbed by the term proportional to ε_1^2 of the second order in magnetization. The propagation vectors are always perpendicular to the magnetization, and we have

$$\gamma_{3,4} = \frac{\omega}{c}\left(N_y\hat{\mathbf{y}} + N_{z3,4}\hat{\mathbf{z}}\right) = \frac{\omega}{c}\left[N_y\hat{\mathbf{y}} \pm \left(\varepsilon_0 - N_y^2 - \frac{\varepsilon_1^2}{\varepsilon_0}\right)^{1/2}\hat{\mathbf{z}}\right]. \tag{1.132}$$

With the assumption

$$\frac{\varepsilon_1^2}{\varepsilon_0} \ll \left(\varepsilon_0 - N_y^2\right)^{1/2}, \tag{1.133}$$

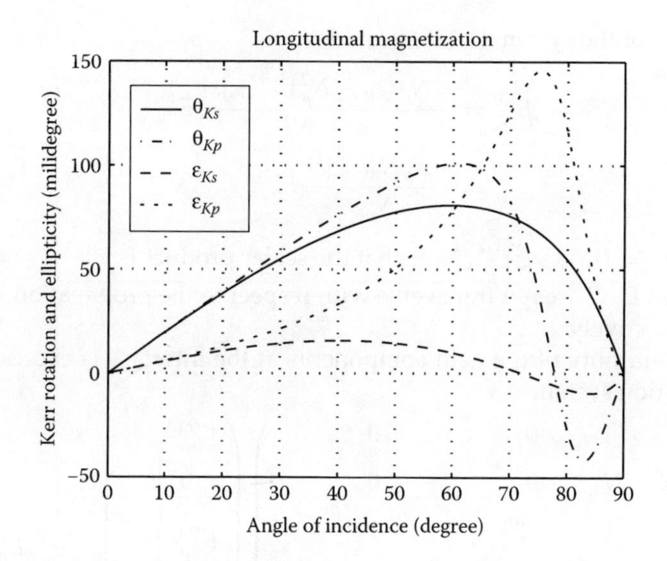

FIGURE 1.5
Complex longitudinal Kerr rotation and ellipticity $\chi_{rs}^{(\mathrm{lon})} = \theta_{Ks} + \mathrm{j}\epsilon_{Ks}$ and $\chi_{rp}^{(\mathrm{lon})} = \theta_{Kp} + \mathrm{j}\epsilon_{Kp}$ as functions of the angle of incidence for the s and p polarized incident waves, respectively, at an air-iron interface. The curves of the azimuths θ_{Ks} and θ_{Kp} and the ellipticities ϵ_{Ks} and ϵ_{Kp} were computed using the data of Figure 1.1 and Figure 1.4.

the proper modes may be classified as follows

$$N_{z1,2} = \pm\left(\varepsilon_3 - N_y^2\right)^{1/2}, \qquad s \text{ polarized,} \qquad (1.134a)$$

$$N_{z3,4} = \pm\left[\left(\varepsilon_0 - N_y^2\right)^{1/2} - \frac{\varepsilon_1^2}{2\varepsilon_0^{1/2}\left(\varepsilon_0 - N_y^2\right)^{1/2}}\right], \qquad p \text{ polarized.} \qquad (1.134b)$$

Strictly speaking, the second pair is only approximately p polarized as it contains a small field component parallel to the propagation vector and proportional to ε_1. For homogeneous plane waves propagating in originally isotropic media magnetized normal to the propagation vector, no MO effects linear in magnetization are possible. The linear magnetooptic effect at transverse magnetization can take place only at interfaces where a nonhomogeneous or leaky refracted wave is induced by a wave obliquely incident at the interface.

To first order in ε_1, the proper propagation vectors of the s and p polarized waves are parallel. We observe that the $E_{0x}^{(1)}$ component is not coupled to $E_{0y}^{(3)}$ and $E_{0z}^{(3)}$. The substitution for N_{z3} from Eq. (1.134b) into the third equations of the system (1.130) provides

$$E_{0z}^{(3)} = \frac{-N_y\left(\varepsilon_0 - N_y^2\right)^{1/2} + \mathrm{j}\varepsilon_1}{\varepsilon_0 - N_y^2} E_{0y}^{(3)} = \frac{-N_y N_{z0} + \mathrm{j}\varepsilon_1}{N_{z0}^2} E_{0y}^{(3)}, \tag{1.135}$$

where $N_{z0} = \left(\varepsilon_0 - N_y^2\right)^{1/2}$. Note that the scalar product $\boldsymbol{E}^{(3,4)} \cdot \gamma_{3,4} \neq 0$, *i.e.*, the waves $\boldsymbol{E}^{(3,4)}$ are not transverse with respect to the propagation vectors $\gamma_{3,4}$, respectively.

The continuity of the field components at the interface is expressed by the equation system

$$\begin{pmatrix} -1 & 1 & 0 & 0 \\ N^{(0)}\alpha_z^{(0)} & N_{z0} & 0 & 0 \\ 0 & 0 & -\alpha_z^{(0)} & 1 \\ 0 & 0 & N^{(0)} & \dfrac{1}{N_{z0}}\left(\varepsilon_0 - \dfrac{\mathrm{j}\varepsilon_1 N_y}{N_{z0}}\right) \end{pmatrix} \begin{pmatrix} E_{0s}^{(r)} \\ E_{0x}^{(1)} \\ E_{0p}^{(r)} \\ E_{0y}^{(3)} \end{pmatrix} = \begin{pmatrix} E_{0s}^{(i)} \\ N^{(0)}\alpha_z^{(0)} E_{0s}^{(i)} \\ \alpha_z^{(0)} E_{0p}^{(i)} \\ N^{(0)} E_{0p}^{(i)} \end{pmatrix}. \tag{1.136}$$

The element r_{ss} does not differ from that given by the Fresnel equation for an interface between isotropic media in Eq. (1.102a) apart from a small second order perturbation of the refractive index with respect to its value at zero magnetization. The off-diagonal elements are zero, $r_{sp} = r_{ps} = 0$. The effect of magnetization manifests itself in the reflected amplitude of the wave polarized parallel to the plane of incidence

$$\begin{aligned} E_{0p}^{(r)} &= \frac{N^{(0)} N_{z0} - \left(\varepsilon_0 - \dfrac{\mathrm{j}\varepsilon_1 N_y}{N_{z0}}\right)\alpha_z^{(0)}}{N^{(0)} N_{z0} + \left(\varepsilon_0 - \dfrac{\mathrm{j}\varepsilon_1 N_y}{N_{z0}}\right)\alpha_z^{(0)}} E_{0p}^{(i)} \\ &= \left[\frac{N^{(0)} N_{z0} - \varepsilon_0\alpha_z^{(0)}}{N^{(0)} N_{z0} + \varepsilon_0\alpha_z^{(0)}} + \frac{2\mathrm{j}\varepsilon_1 N_y N^{(0)}\alpha_z^{(0)}}{\left(N^{(0)} N_{z0} + \varepsilon_0\alpha_z^{(0)}\right)^2}\right] E_{0p}^{(i)} \\ &= \left[r_{pp} + \frac{\mathrm{j}\varepsilon_1 N_y}{2\varepsilon_0 N_{z0}}\left(1 - r_{pp}^2\right)\right] E_{0p}^{(i)}. \end{aligned} \tag{1.137}$$

We designate the amplitude reflection coefficient perturbed by the transverse magnetization as

$$\begin{aligned} r_{pp}^{(\mathrm{trans})} &= r_{pp} + \frac{\mathrm{j}\varepsilon_1 N_y}{2\varepsilon_0 N_{z0}}\left(1 - r_{pp}^2\right) \\ &= r_{pp} + \Delta r_{pp}^{(\mathrm{trans})}. \end{aligned} \tag{1.138}$$

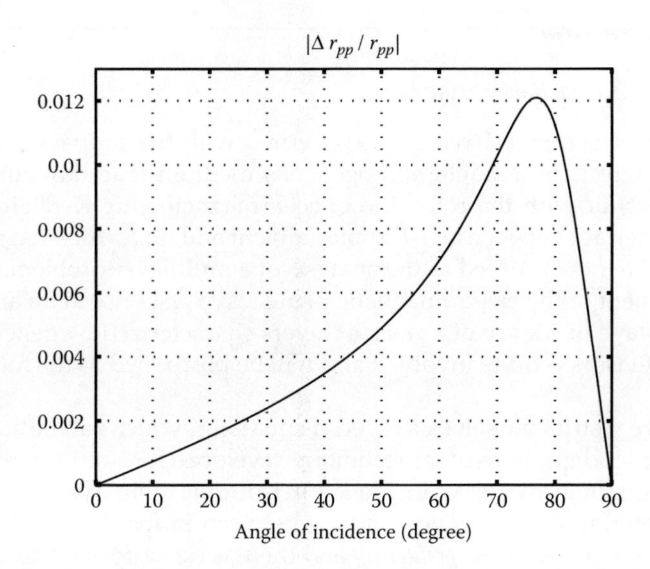

FIGURE 1.6
Perturbation $\Delta r_{pp}^{(\mathrm{trans})}$ to the diagonal element r_{pp}, induced by the transverse magnetization at an air-iron interface as a function of the angle of incidence. The curve $|\Delta r_{pp}/r_{pp}|$ was computed using the values for the diagonal and off-diagonal permittivity tensor elements given by $\varepsilon_{xx}^{(\mathrm{Fe})} = \varepsilon_0^{(\mathrm{Fe})} = -0.8845 - \mathrm{j}17.938$ and $\varepsilon_{xy}^{(\mathrm{Fe})} = \mathrm{j}\varepsilon_1^{(\mathrm{Fe})} = -0.8845 - \mathrm{j}17.938$, respectively.

Here, r_{pp} is given, to first order in $\boldsymbol{M}$, by Eq. (1.102b). Figure 1.6 displays the dependence of $|\Delta r_{pp}/r_{pp}|$ on the angle of incidence. In both the longitudinal and transverse geometry, the magnetization is parallel to the interface plane.

The MO longitudinal Kerr effect and transverse Kerr effect vanish at both the normal and grazing incidence. The magnetic medium is characterized by one of the tensors ε_P, ε_L, and ε_T given in Eq. (1.103), Eq. (1.116), and Eq. (1.129) (consistent with our choice of the interface plane normal to the z axis of the coordinate system). Upon the reflection at the interface between a vacuum and the magnetic medium, the azimuth and ellipticity of the incident wave are transformed. This transformation may be completely described by the corresponding Jones reflection matrices (independent of the choice of the coordinate system), the general form of which appears in Eq. (1.99). In the case of the polar and longitudinal effects, we usually express this transformation in terms of the MO polar and longitudinal Kerr rotation and ellipticity, with the reference to an incident wave linearly polarized perpendicular or parallel with respect to the plane of incidence.

1.7 Multilayer Response

Up to the present, we have been concerned with the propagation magnetooptic effects in a single homogeneous medium (Faraday effect and Voigt effect) or with the reflection effects (magnetooptic Kerr effects) at a single interface between an isotropic ambient and uniformly magnetized medium. We now proceed to the analysis of a multilayer problem. A general treatment of the electromagnetic plane wave response at an arbitrary angle of wave incidence of a stack of layers characterized by general permittivity tensors is quite involved and will be postponed to the following chapters.

Here, we restrict ourselves to special situations, which can be analyzed using simple adaptations of the technique developed for isotropic multilayers at oblique angles of wave incidence. In isotropic multilayers with plane parallel interfaces, there is no coupling between proper s and p modes. The modes propagate *independently* and there is no *mode coupling* at interfaces. Consequently, the multilayer response can be analyzed for each of the proper modes separately.

In magnetic multilayers, this feature, *i.e.*, the absence of mode coupling, takes place at normal incidence on a stack of layers with all magnetization vectors normal to the plane parallel interfaces. According to the wave equation (1.22), the proper polarization modes in each layer are circularly polarized waves. In isotropic media, which represent a special case here, the proper polarization modes can also be chosen as circularly polarized waves. The isotropic media form the half spaces sandwiching the multilayer and nonmagnetic layers in the magnetic multilayer stack. The global multilayer response will be odd in magnetization.

Another example of the independent propagation of the proper polarization modes without any mode coupling at interfaces represents the case of normal incidence at a multilayer with the magnetization vectors in the individual layers oriented all in the same direction parallel to the interface planes.[14] The proper modes follow from the wave equation (1.62). They are linearly polarized with the field vectors parallel to the interface plane, in the direction of the magnetization vector and perpendicular to it. The approach can be extended to the case of oblique incidence in multilayers with transverse magnetization. The response of proper polarization modes linearly polarized parallel to the plane of incidence and normal to the magnetization vector will be odd in magnetization. For illustration, we consider the most frequent case of multilayers with polar magnetization. In view of

[14] The susceptibility tensor in all layers displays the symmetry given in Eq. (2.28) with the magnetization component normal to interfaces set to zero. For the interface planes specified, *e.g.*, by $z^{(n)} = \text{const}$, this requires $M_z^{(n)} = 0$ in all layers.

the response linearity, the relations between the incident and transmitted or reflected waves can be represented by a linear matrix transformation.

We choose the z axis of the Cartesian system perpendicular to the multilayer interface planes situated at $z = z_n$, where $n = 0, 1..., \mathcal{N}; \mathcal{N}$ denotes the number of layers. The propagation vectors and magnetization (if nonzero) in all layers are parallel to the z axis. The media of all layers are characterized by the permittivity tensors of the same symmetry as in Eq. (1.22)

$$\varepsilon^{(n)} = \begin{pmatrix} \varepsilon_{xx}^{(n)} & \varepsilon_{xy}^{(n)} & 0 \\ -\varepsilon_{xy}^{(n)} & \varepsilon_{xx}^{(n)} & 0 \\ 0 & 0 & \varepsilon_{zz}^{(n)} \end{pmatrix}. \tag{1.139}$$

If a particular layer is isotropic, *i.e.*, it displays a zero magnetization, the off-diagonal tensor element is zero. We shall assume that in each of the layers we can take for the proper polarization modes those given by Eq. (1.31) and Eq. (1.32) obtained from the plane wave solutions of the wave equation (1.22) in corresponding unbound media. We therefore write the four circularly polarized proper modes as

$$\boldsymbol{E}_j^{(n)} = \boldsymbol{e}_j^{(n)} \mathrm{E}_{0j}^{(n)} \exp\left[\mathrm{j}\left(\omega t - \frac{\omega}{c} N_{zj}^{(n)} z\right)\right], \tag{1.140}$$

with $j = 1, \ldots, 4$ and $z^{(n-1)} \leq z \leq z^{(n)}$. Here, $\hat{\boldsymbol{e}}_{1,2}^{(n)} = 2^{-1/2}\left(\hat{x} + \mathrm{j}\hat{y}\right)$ and $\hat{\boldsymbol{e}}_{3,4}^{(n)} = 2^{-1/2}\left(\hat{x} - \mathrm{j}\hat{y}\right)$. We assign $N_{z1}^{(n)} = +N^{(n)+}$, $N_{z2}^{(n)} = -N^{(n)+}$, $N_{z3}^{(n)} = +N^{(n)-}$, and $N_{z4}^{(n)} = -N^{(n)-}$. We define a 4×1 column vector made of the complex amplitudes at a plane $z =$ const inside the n-th layer

$$\mathbf{E}_{0j}^{(n)}(z) = \mathrm{E}_{0j}^{(n)} \exp\left[\mathrm{j}\frac{\omega}{c} N_{zj}^{(n)} \left(z^{(n-1)} - z\right)\right] \tag{1.141}$$

as

$$\begin{pmatrix} \mathrm{E}_{01}^{(n)}(z) \\ \mathrm{E}_{02}^{(n)}(z) \\ \mathrm{E}_{03}^{(n)}(z) \\ \mathrm{E}_{04}^{(n)}(z) \end{pmatrix}. \tag{1.142}$$

The amplitudes in the isotropic half spaces $n = 0$ ($z < z^{(0)}$) and $\mathcal{N} + 1$ ($z > z^{(\mathcal{N})}$), which sandwich the multilayer, are related by the so-called *scattering matrix* $\mathbf{S}^{(\mathcal{N})}$ of the multilayer

$$\begin{pmatrix} \mathrm{E}_{01}^{(0)}(z^{(0)}) \\ \mathrm{E}_{02}^{(0)}(z^{(0)}) \\ \mathrm{E}_{03}^{(0)}(z^{(0)}) \\ \mathrm{E}_{04}^{(0)}(z^{(0)}) \end{pmatrix} = \begin{pmatrix} \mathcal{S}_{11}^{(\mathcal{N})} & \mathcal{S}_{12}^{(\mathcal{N})} & 0 & 0 \\ \mathcal{S}_{21}^{(\mathcal{N})} & \mathcal{S}_{22}^{(\mathcal{N})} & 0 & 0 \\ 0 & 0 & \mathcal{S}_{33}^{(\mathcal{N})} & \mathcal{S}_{34}^{(\mathcal{N})} \\ 0 & 0 & \mathcal{S}_{43}^{(\mathcal{N})} & \mathcal{S}_{44}^{(\mathcal{N})} \end{pmatrix} \begin{pmatrix} \mathrm{E}_{01}^{(\mathcal{N}+1)}(z^{(\mathcal{N})}) \\ \mathrm{E}_{02}^{(\mathcal{N}+1)}(z^{(\mathcal{N})}) \\ \mathrm{E}_{03}^{(\mathcal{N}+1)}(z^{(\mathcal{N})}) \\ \mathrm{E}_{04}^{(\mathcal{N}+1)}(z^{(\mathcal{N})}) \end{pmatrix}. \tag{1.143}$$

The amplitudes are pertinent to the planes $z = z^{(0)}$ and $z = z^{(\mathcal{N})}$ in sandwiching half spaces infinitessimally close to the multilayer boundaries. The scattering matrix is 2×2 block diagonal, reflecting the independence of the proper polarization modes specified by the polarizations defined by the unit vectors $\hat{e}_{1,2}^{(n)}$ with respect to those defined by $\hat{e}_{3,4}^{(n)}$.

For an interface between the $(n-1)$-th and n-th media, we have for one of the 2×2 blocks

$$\begin{pmatrix} E_{01}^{(n-1)}(z^{(n-1)}) \\ E_{02}^{(n-1)}(z^{(n-1)}) \end{pmatrix} = \begin{pmatrix} \mathcal{S}_{11} & \mathcal{S}_{12} \\ \mathcal{S}_{21} & \mathcal{S}_{22} \end{pmatrix} \begin{pmatrix} E_{01}^{(n)}(z^{(n-1)}) \\ E_{02}^{(n)}(z^{(n-1)}) \end{pmatrix}. \tag{1.144}$$

We now determine the elements of the scattering matrix. Let us first examine the situation when the only incident wave is in the medium $(n-1)$, *i.e.*, $E_{02}^{(n)}(z^{(n-1)}) = 0$. The expansion of the matrix equation (1.144) gives

$$E_{01}^{(n-1)}(z^{(n-1)}) = \mathcal{S}_{11} E_{01}^{(n)}(z^{(n-1)}) \tag{1.145a}$$

$$E_{02}^{(n-1)}(z^{(n-1)}) = \mathcal{S}_{21} E_{01}^{(n)}(z^{(n-1)}). \tag{1.145b}$$

We can thus define the single-interface transmission and reflection coefficients at the interface $z = z^{(n-1)}$ as

$$t_{+}^{(n-1,n)} = \frac{E_{01}^{(n)}(z^{(n-1)})}{E_{01}^{(n-1)}(z^{(n-1)})} = \mathcal{S}_{11}^{-1} \tag{1.146a}$$

$$r_{+}^{(n-1,n)} = \frac{E_{02}^{(n-1)}(z^{(n-1)})}{E_{01}^{(n-1)}(z^{(n-1)})} = \frac{\mathcal{S}_{21}}{\mathcal{S}_{11}}. \tag{1.146b}$$

For the wave incident from the medium n, now assuming $E_{01}^{(n-1)}(z^{(n-1)}) = 0$, we find

$$0 = \mathcal{S}_{11} E_{01}^{(n)}(z^{(n-1)}) + \mathcal{S}_{12} E_{02}^{(n)}(z^{(n-1)}) \tag{1.147a}$$

$$E_{02}^{(n-1)}(z^{(n-1)}) = \mathcal{S}_{21} E_{01}^{(n)}(z^{(n-1)}) + \mathcal{S}_{22} E_{02}^{(n)}(z^{(n-1)}). \tag{1.147b}$$

The substitution for $\mathcal{S}_{11}$ and $\mathcal{S}_{21}$, according to Eqs. (1.146), provides

$$\mathcal{S}_{11} = \left(t_{+}^{(n-1,n)}\right)^{-1} \tag{1.148a}$$

$$\mathcal{S}_{12} = -r_{+}^{(n,n-1)} \left(t_{+}^{(n-1,n)}\right)^{-1} \tag{1.148b}$$

$$\mathcal{S}_{21} = r_{+}^{(n-1,n)} \left(t_{+}^{(n-1,n)}\right)^{-1} \tag{1.148c}$$

$$\mathcal{S}_{22} = \left(t_{+}^{(n,n-1)} t_{+}^{(n-1,n)} - r_{+}^{(n-1,n)} r_{+}^{(n,n-1)}\right) \left(t_{+}^{(n-1,n)}\right)^{-1}. \tag{1.148d}$$

From the continuity of circularly polarized fields at the interface we find

$$r_{\pm}^{(n-1,n)} = \frac{N_{\pm}^{(n-1)} - N_{\pm}^{(n)}}{N_{\pm}^{(n-1)} + N_{\pm}^{(n)}} = -r_{\pm}^{(n,n-1)}, \tag{1.149a}$$

$$t_{\pm}^{(n-1,n)} = \frac{2N_{\pm}^{(n-1)}}{N_{\pm}^{(n-1)} + N_{\pm}^{(n)}} = 1 + r_{\pm}^{(n-1,n)}, \tag{1.149b}$$

and

$$t_{\pm}^{(n,n-1)} = \frac{2N_{\pm}^{(n)}}{N_{\pm}^{(n-1)} + N_{\pm}^{(n)}} = \frac{1 - \left(r_{\pm}^{(n,n-1)}\right)^2}{t_{\pm}^{(n-1,n)}}. \tag{1.150}$$

The scattering matrix of the interface consists of two 2×2 blocks, each pertinent to a specific circular polarization, *i.e.*,

$$\boldsymbol{I}^{(n-1,n)} = \begin{pmatrix} \left(t_{+}^{(n-1,n)}\right)^{-1} & r_{+}^{(n-1,n)}\left(t_{+}^{(n-1,n)}\right)^{-1} & 0 & 0 \\ r_{+}^{(n-1,n)}\left(t_{+}^{(n-1,n)}\right)^{-1} & \left(t_{+}^{(n-1,n)}\right)^{-1} & 0 & 0 \\ 0 & 0 & \left(t_{-}^{(n-1,n)}\right)^{-1} & r_{-}^{(n-1,n)}\left(t_{-}^{(n-1,n)}\right)^{-1} \\ 0 & 0 & r_{-}^{(n-1,n)}\left(t_{-}^{(n-1,n)}\right)^{-1} & \left(t_{-}^{(n-1,n)}\right)^{-1} \end{pmatrix}. \tag{1.151}$$

The propagation inside the n-th layer between the interfaces $z = z^{(n-1)}$ and $z = z^{(n)}$ is represented by the 4×4 matrix $\boldsymbol{L}_n$ defined in the following matrix relation

$$\begin{pmatrix} E_{01}^{(n)}(z^{(n-1)}) \\ E_{02}^{(n)}(z^{(n-1)}) \\ E_{03}^{(n)}(z^{(n-1)}) \\ E_{04}^{(n)}(z^{(n-1)}) \end{pmatrix} = \begin{pmatrix} e^{j\beta_{+}^{(n)}} & 0 & 0 & 0 \\ 0 & e^{-j\beta_{+}^{(n)}} & 0 & 0 \\ 0 & 0 & e^{j\beta_{-}^{(n)}} & 0 \\ 0 & 0 & 0 & e^{-j\beta_{-}^{(n)}} \end{pmatrix} \begin{pmatrix} E_{01}^{(n)}(z^{(n)}) \\ E_{02}^{(n)}(z^{(n)}) \\ E_{03}^{(n)}(z^{(n)}) \\ E_{04}^{(n)}(z^{(n)}) \end{pmatrix}, \tag{1.152}$$

where

$$\beta_{\pm}^{(n)} = \frac{\omega}{c} N_{z\pm}^{(n)} \left(z^{(n)} - z^{(n-1)}\right). \tag{1.153}$$

The multilayer scattering matrix is then represented by the matrix product

$$\mathbf{S}^{(\mathcal{N})} = \boldsymbol{I}^{(01)} \prod_{n=1}^{\mathcal{N}} \left(\boldsymbol{L}^{(n)} \boldsymbol{I}^{(n,n+1)}\right). \tag{1.154}$$

For example, the multilayer system consisting of alternating layers A and B on a substrate C

$$B(AB)^p C$$

will be represented by the matrix product

$$\mathbf{S}^{(\mathcal{N})} = \boldsymbol{I}^{(0B)}\boldsymbol{L}^{(B)}\boldsymbol{I}^{(BA)}\boldsymbol{L}^{(A)}\boldsymbol{I}^{(AB)}\boldsymbol{L}^{(B)}\ldots\boldsymbol{L}^{(B)}\boldsymbol{I}^{(BC)}\boldsymbol{L}^{(C)}\boldsymbol{I}^{(C,\mathcal{N}+1)}, \tag{1.155}$$

where $\mathcal{N} = 2(p+1)$ or

$$\mathbf{S}^{(\mathcal{N})} = \boldsymbol{I}^{(0B)}\left(\boldsymbol{L}^{(B)}\boldsymbol{I}^{(BA)}\boldsymbol{L}^{(A)}\boldsymbol{I}^{(AB)}\right)^p\boldsymbol{L}^{(B)}\boldsymbol{I}^{(BC)}\boldsymbol{L}^{(C)}\boldsymbol{I}^{(C,\mathcal{N}+1)}. \tag{1.156}$$

The global multilayer reflection and transmission coefficients for the waves incident from the medium $n = 0$ at the interface $z = z^{(0)}$ (*i.e.*, $\mathrm{E}_{02}^{(\mathcal{N})} = \mathrm{E}_{04}^{(\mathcal{N})} = 0$ at $z = z^{(\mathcal{N})}$) and exiting into the region $(\mathcal{N}+1)$ may be written as

$$r_+^{(0,\mathcal{N}+1)} = \frac{\mathrm{E}_{02}^{(0)}}{\mathrm{E}_{01}^{(0)}} = \frac{\mathcal{S}_{21}^{(\mathcal{N})}}{\mathcal{S}_{11}^{(\mathcal{N})}} \tag{1.157a}$$

$$r_-^{(0,\mathcal{N}+1)} = \frac{\mathrm{E}_{04}^{(0)}}{\mathrm{E}_{03}^{(0)}} = \frac{\mathcal{S}_{43}^{(\mathcal{N})}}{\mathcal{S}_{33}^{(\mathcal{N})}} \tag{1.157b}$$

$$t_+^{(0,\mathcal{N}+1)} = \frac{\mathrm{E}_{01}^{(\mathcal{N}+1)}}{\mathrm{E}_{01}^{(0)}} = \frac{1}{\mathcal{S}_{11}^{(\mathcal{N})}} \tag{1.157c}$$

$$t_-^{(0,\mathcal{N}+1)} = \frac{\mathrm{E}_{03}^{(\mathcal{N}+1)}}{\mathrm{E}_{03}^{(0)}} = \frac{1}{\mathcal{S}_{33}^{(\mathcal{N})}}. \tag{1.157d}$$

As an example, let us consider a system consisting of a magnetic film sandwiched between isotropic and magnetic half spaces represented by a matrix product

$$\mathbf{S} = \boldsymbol{I}^{(01)}\boldsymbol{L}^{(1)}\boldsymbol{I}^{(12)}. \tag{1.158}$$

It is sufficient to consider the product of 2×2 matrices

$$\begin{pmatrix} \mathcal{S}_{11\pm} & \mathcal{S}_{12\pm} \\ \mathcal{S}_{21\pm} & \mathcal{S}_{22\pm} \end{pmatrix} = \left(1 + r_\pm^{(01)}\right)^{-1}\left(1 + r_\pm^{(12)}\right)^{-1}\begin{pmatrix} 1 & r_\pm^{(01)} \\ r_\pm^{(01)} & 1 \end{pmatrix}\begin{pmatrix} \mathrm{e}^{\mathrm{j}\beta_\pm^{(1)}} & 0 \\ 0 & \mathrm{e}^{-\mathrm{j}\beta_\pm^{(1)}} \end{pmatrix}\begin{pmatrix} 1 & r_\pm^{(12)} \\ r_\pm^{(12)} & 1 \end{pmatrix}. \tag{1.159}$$

Using Eq. (1.157), we obtain for the reflected and transmitted amplitudes

$$\left(\mathrm{E}_{02}^{(0)}\right)_{\pm} = \frac{r_{\pm}^{(01)} + r_{\pm}^{(12)} \mathrm{e}^{2\mathrm{j}\beta_{\pm}^{(1)}}}{1 + r_{\pm}^{(01)} r_{\pm}^{(12)} \mathrm{e}^{2\mathrm{j}\beta_{\pm}^{(1)}}} \left(\mathrm{E}_{01}^{(0)}\right)_{\pm}, \tag{1.160a}$$

$$\left(\mathrm{E}_{01}^{(\mathcal{N}+1)}\right)_{\pm} = \frac{t_{01}^{\pm} t_{12}^{\pm} \mathrm{e}^{-\mathrm{j}\beta_{\pm}^{(1)}}}{1 + r_{\pm}^{(01)} r_{\pm}^{(12)} \mathrm{e}^{2\mathrm{j}\beta_{\pm}^{(1)}}} \left(\mathrm{E}_{01}^{(0)}\right)_{\pm}, \tag{1.160b}$$

with the help of Eq. (1.149b).

For a multilayer consisting of $\mathcal{N}$ layers, we have got the circular Jones (2×2) reflection and transmission matrices

$$\begin{pmatrix} r_{+}^{(0,\mathcal{N}+1)} & 0 \\ 0 & r_{-}^{(0,\mathcal{N}+1)} \end{pmatrix}, \quad \begin{pmatrix} t_{+}^{(0,\mathcal{N}+1)} & 0 \\ 0 & t_{-}^{(0,\mathcal{N}+1)} \end{pmatrix}. \tag{1.161}$$

The circular Jones matrices (1.161) can be transformed to the Cartesian ones with help of unitary matrices

$$\begin{aligned} &\begin{pmatrix} r_{xx}^{(0,\mathcal{N}+1)} & r_{xy}^{(0,\mathcal{N}+1)} \\ r_{yx}^{(0,\mathcal{N}+1)} & r_{xx}^{(0,\mathcal{N}+1)} \end{pmatrix} \\ &= 2^{-1/2} \begin{pmatrix} 1 & 1 \\ \mathrm{j} & -\mathrm{j} \end{pmatrix} \begin{pmatrix} r_{+}^{(0,\mathcal{N}+1)} & 0 \\ 0 & r_{-}^{(0,\mathcal{N}+1)} \end{pmatrix} 2^{-1/2} \begin{pmatrix} 1 & -\mathrm{j} \\ 1 & \mathrm{j} \end{pmatrix} \\ &= \frac{1}{2} \begin{pmatrix} r_{+}^{(0,\mathcal{N}+1)} + r_{-}^{(0,\mathcal{N}+1)} & -\mathrm{j}\left(r_{+}^{(0,\mathcal{N}+1)} - r_{-}^{(0,\mathcal{N}+1)}\right) \\ \mathrm{j}\left(r_{+}^{(0,\mathcal{N}+1)} - r_{-}^{(0,\mathcal{N}+1)}\right) & r_{+}^{(0,\mathcal{N}+1)} + r_{-}^{(0)} \end{pmatrix}, \end{aligned} \tag{1.162}$$

$$\begin{aligned} &\begin{pmatrix} t_{xx}^{(0,\mathcal{N}+1)} & t_{xy}^{(0,\mathcal{N}+1)} \\ t_{yx}^{(0,\mathcal{N}+1)} & t_{xx}^{(0,\mathcal{N}+1)} \end{pmatrix} \\ &= 2^{-1/2} \begin{pmatrix} 1 & 1 \\ \mathrm{j} & -\mathrm{j} \end{pmatrix} \begin{pmatrix} t_{+}^{(0,\mathcal{N}+1)} & 0 \\ 0 & t_{-}^{(0,\mathcal{N}+1)} \end{pmatrix} 2^{-1/2} \begin{pmatrix} 1 & -\mathrm{j} \\ 1 & \mathrm{j} \end{pmatrix} \\ &= \frac{1}{2} \begin{pmatrix} t_{+}^{(0,\mathcal{N}+1)} + t_{-}^{(0,\mathcal{N}+1)} & -\mathrm{j}\left(t_{+}^{(0,\mathcal{N}+1)} - t_{-}^{(0,\mathcal{N}+1)}\right) \\ \mathrm{j}\left(t_{+}^{(0,\mathcal{N}+1)} - t_{-}^{(0,\mathcal{N}+1)}\right) & t_{+}^{(0,\mathcal{N}+1)} + t_{-}^{(0,\mathcal{N}+1)} \end{pmatrix}. \end{aligned} \tag{1.163}$$

With these Cartesian Jones matrices, we can characterize the transformation of the polarization state of an incident wave upon transmission, *i.e.*, multilayer Faraday effect, and that upon reflection, *i.e.*, multilayer normal incidence "MO polar Kerr effect." In particular, if the incident wave is linearly polarized, we get, for example in the transmission case,

$$\begin{pmatrix} E_{0x}^{(\mathcal{N}+1)} \\ E_{0y}^{(\mathcal{N}+1)} \end{pmatrix} = \frac{1}{2} \begin{pmatrix} t_+^{(0,\mathcal{N}+1)} + t_-^{(0,\mathcal{N}+1)} & -\mathrm{j}\left(t_+^{(0,\mathcal{N}+1)} - t_-^{(0,\mathcal{N}+1)}\right) \\ \mathrm{j}\left(t_+^{(0,\mathcal{N}+1)} - t_-^{(0,\mathcal{N}+1)}\right) & t_+^{(0,\mathcal{N}+1)} + t_-^{(0,\mathcal{N}+1)} \end{pmatrix} \begin{pmatrix} E_{0x}^{(0)} \\ 0 \end{pmatrix}. \tag{1.164}$$

Then, the polarization state of the reflected and transmitted waves may be expressed in terms of ellipsometric complex number parameters in Cartesian (x, y) representation as

$$\chi_r^{(0,\mathcal{N}+1)} = \mathrm{j}\frac{r_+^{(0,\mathcal{N}+1)} - r_-^{(0,\mathcal{N}+1)}}{r_+^{(0,\mathcal{N}+1)} + r_-^{(0,\mathcal{N}+1)}}, \tag{1.165a}$$

$$\chi_t^{(0,\mathcal{N}+1)} = \mathrm{j}\frac{t_+^{(0,\mathcal{N}+1)} - t_-^{(0)}}{t_+^{(0,\mathcal{N}+1)} + t_-^{(0,\mathcal{N}+1)}}. \tag{1.165b}$$

Formally, these expressions correspond to the results developed earlier, *i.e.*, in Eq. (1.42) and Eq. (1.49). They, however, pertain to more general situations including the propagation effects, interface reflection and transmission effects, and the effect of multiple reflections. Useful approximations can be developed where the differences $t_+^{(0)} - t_-^{(0)}$ and $r_+^{(0)} - r_-^{(0)}$ can be considered small with respect to the corresponding sums $t_+^{(0)} + t_-^{(0)}$ and $r_+^{(0)} + r_-^{(0)}$. We return to these problems in Chapter 4 from a more general viewpoint, using the 4×4 Yeh matrix formalism capable of treating general anisotropic multilayers at arbitrary angles of wave incidence. The present treatment of the multilayer problem has made use of the formalism originally applied to linearly polarized modes in isotropic media at oblique angles of incidence [3].

References

1. M. J. Freiser, "A survey of magnetooptic effects," IEEE Trans. Magn. **4**, 152–61, 1968.
2. E. D. Palik and B. W. Henvis, "A bibliography of magnetooptics of solids," Appl. Opt. **6**, 603–630, 1967.
3. R. M. A. Azzam and N. M. Bashara, *Ellipsometry and Polarized Light* (Elsevier, Amsterdam, 1987).
4. Lord Rayleigh, "On the magnetic rotation of light and the second law of thermodynamics," Nature (London), 64, 577, Oct.10, 1901.

5. W. Voigt, *Magneto- und Elektro-Optik* (Teubner, Leipzig, 1908).
6. A. V. Sokolov, *Optical Properties of Metals* (Blackie, Glasgow and London, 1967).
7. B. Heinrich and J. A. C. Bland, eds., *Ultrathin Magnetic Structures* (Springer Verlag, Berlin, 1994).
8. J. F. Dillon, Jr., "Magnetooptics and its uses," J. Magn. Magn. Mater. **31-34**, Pt 1, 1–9, 1983.
9. R. P. Hunt, "Magneto-optic scattering from thin solid films," J. Appl. Phys. **37**, 1652–1671, 1967.
10. J. C. Suits, "Faraday and Kerr effect in magnetic compounds," IEEE Trans. Magn. **8**, 95, 1972.
11. W. Wettling, "Magneto-optics of ferrites," J. Magn. Magn. Mater. **3**, 147, 1976.
12. K. Sato, *Hikari to Jiki* (in Japanese, *Light and Magnetism—Introduction to Magnetooptics*, Asakura, Tokyo, 1989).
13. J. T. Devreese, ed., Proc. of Antwerp Advanced Study Institute on Theoretical Aspects and New Developments in Magneto-Optics, Universiteit Antwerpen (RUCA) Antwerp, July 16–July 28, 1979.
14. T. S. Moss, G. J. Burrell, and B. Ellis, *Semiconductor Optoelectronics* (Butterworths, London, 1973).
15. J. T. Chang, J. F. Dillon, and V. F. Gianola, "Magneto-optical variable memory based upon the properties of a trasparent ferrimagnetic garnet at its compensation temperature," J. Appl. Phys. **36**, 1110–1111, 1965.
16. B. Di Bartolo, *Optical Interactions in Solids* (John Wiley, New York, 1968).
17. R. R. Birss, *Symmetry and Magnetism* (North-Holland, Amsterdam, 1964).
18. L. D. Landau and E. M. Lifshitz, *Electrodynamics of Continuous Media* (Pergamon Press, 1960).
19. R. M. White, *Quantum Theory of Magnetism* (Springer, Berlin, 1983).
20. D. B. Melrose and R. C. McPhedran, *Electromagnetic Processes in Dispersive Media. A Treatment Based on Dielectric Tensor* (Cambridge University Press, Cambridge, 1991) pp. 83–86.
21. W. J. Tabor and F. S. Chen, "Electromagnetic propagation through materials possessing both Faraday rotation and birefringence: Experiments with ytterbium orthoferrite," J. Appl. Phys. **40**, 2760–2765, 1969.
22. Š. Višňovský, K. Postava, T. Yamaguchi, and R. Lopušník, "Magneto-optic ellipsometry in exchange-coupled films, Appl. Opt. **41**, 3950–3960, 2002.
23. P. Yeh, "Optics of anisotropic layered media: A new 4×4 matrix algebra," Surf. Sci. **96**, 41–53, 1980; P. Yeh, *Optical Waves in Layered Media* (John Wiley & Sons, New York, 1988) Chapter 9.
24. S. Maekawa and T. Shinjo, eds. *Spin Dependent Transport in Magnetic Nanostructures* (Taylor & Francis, London, 2002).
25. B. Heinrich, "Magnetic nanostructures. From physical principles to spintronics," Can. J. Phys. **78**, 161–199 (2000).

2

Material Tensors

2.1 Introduction

Using symmetry arguments, we have obtained the electric susceptibility or permittivity tensor for an originally isotropic medium with anisotropy induced by the magnetization. The form of the electric susceptibility tensor should be refined in media where the anisotropy induced by magnetization coexists with other kinds of anisotropy. In this chapter, the electric susceptibility tensors for crystals of all 32 crystal classes are derived for arbitrary orientations of the magnetization vector under the assumption that the effect of magnetic ordering on the tensor elements is small.[1] In most cases, it is sufficient to stay in the frame of electric dipole approximation and to characterize the magnetization effect on the medium optical response with the electric susceptibility tensor. The perturbation of the first and second order in magnetization to the electric susceptibility tensor at zero magnetization is expressed using *linear magnetooptic tensor* and *quadratic magnetooptic tensor*.

In Chapter 1, we were able to demonstrate the optical response in magnetic media including the magnetooptic effects using the electric susceptibility tensor deduced from symmetry arguments only, without employing any microscopic models. However, we have to resort to models to explain spectral dependence of the tensors and related optical and magnetooptic spectra. We consider two of them: the *classical Lorentz model* and the simple single electron *semiclassical* model.

The Lorentz model for an elastically coupled charged particle, a bound electron, provides the simplest idea of the origin of magnetooptic effects, *e.g.*, in magnetic dielectrics. It shows some analogy with the quantum picture of *interband electron transitions*. The Drude's version of the model pertains to the case of free electrons in an ideal metal; here, some analogy can be found with *intraband electron transitions*. The situation in magnetic

[1] From the microscopic point of view, the approximation is based on the assumption that the magnetization-dependent part of the electric dipole moment induced by the electric field of the electromagnetic wave is small with respect to the electric dipole moment induced by the wave field at zero magnetization.

metals includes both the Lorentz's and Drude's aspects. An important advantage of the Lorentz and Drude models consists of their capability to simply account for the simultaneous action on the electron of two forces of different symmetry: that originating from the dominating spherically symmetric central Coulomb potential and the Lorentz force produced by the magnetic field.

We start with the Hamiltonian of an electrically charged particle that is elastically coupled and subjected to a uniform time-independent magnetic field and the field of a monochromatic plane electromagnetic wave. From the solution of the Hamilton equations, we obtain the desired information on the frequency dependence of the electric susceptibility tensor. Two special cases are distinguished here by the choice of the natural frequency characterizing the strength of the coupling: the case of an elastically coupled charged particle (Lorentz model—nonzero resonant frequency) and the case of a free charged particle (Drude model—zero resonant frequency).

The semiclassical model, which follows, can provide a deeper insight. It treats the action of *classic* electromagnetic radiation on a *quantum* system. We confine ourselves to nonrelativistical treatment with *electron spin* and *spin orbit coupling*, two concepts essential for the explanation of electron transitions in magnetic crystals. We consider the quantum Hamiltonian, including the effect of dominating spherically symmetric central potential and crystal field potential. We have further to account for the effect of external magnetic flux density field, that of hypothetical exchange magnetic field, and the contribution of spin-orbit coupling. Our model should explain the behavior in the spectral regions of absorption in magnetic materials and must, therefore, include the relaxation effects. To this purpose, we employ the density matrix method.

We obtain the solution for an average current density vector induced by the field of the electromagnetic plane wave. This allows us to define a generalized susceptibility tensor dependent on both the radiation frequency and the propagation vector. As in the case of simpler classical models, we find the results to be consistent with the Onsager relation, the special case of which has been given in Eq. (1.6). Assuming the dimension of the quantum system much smaller than the radiation wavelength, which is often acceptable in the optical region of the spectrum, we can distinguish electric dipole, electric quadrupole, and magnetic dipole components of the susceptibility.

In the following, we limit ourselves to the electron transitions in the electric dipole approximation. We demonstrate the similarity of the results following from the Lorentz model with those of the semiclassical model limited to the s- to p-orbital electron state transitions with the effect of electron spin neglected. This simplest classical model could be extended to include the effect of reduced crystal symmetry. This refinement as well as the electron transitions involving orbital d- and f-states are usually discussed in frame of the semiclassical model. Here, we consider the effect of cubic and orthorhombic symmetry on the proper values of energy and corresponding proper states for s-, p-, and d-states and compute the susceptibility tensor spectra for simple situations.

We treat the tensors in the Cartesian representation. This enables us to analyze the response from a more general point of view, which includes the effect of magnetic ordering in media displaying lower symmetry. The latter effect is usually ignored in the treatment restricted to circularly polarized induced moments. The Cartesian tensor-based approach is also more fundamental. It not only includes the magnetooptic effects linear in magnetization, *e.g.*, Faraday rotation, but provides a broader view on the optical response in magnetic media. The focus is on the electric dipole approximation expressed in terms of the electric susceptibility tensor. Using the duality principle, the analysis can be modified to the case where the medium is characterized by a magnetic susceptibility tensor. Rarer cases where the medium should be characterized by both the electric and magnetic susceptibility tensors are more complicated. We note that the response can be handled analytically in special configurations with the propagation vector parallel to the magnetization or with a wave obliquely incident on a planar structure with transverse magnetization.

2.2 Tensors in Magnetic Crystals

2.2.1 Effect of Magnetization

So far, we have studied the effect of magnetization on the electric susceptibility or permittivity in originally isotropic media. The problem can be approached from a more general viewpoint still in the frame of the classical macroscopic electric dipole approximation. The volume density of electric dipole moments, $\boldsymbol{P}$, in a linear isotropic homogeneous medium induced by the electric field, $\boldsymbol{E}$, of the wave propagating in the medium, is given by the linear material relation

$$\boldsymbol{P} = \varepsilon_{\mathrm{vac}} \chi_e \boldsymbol{E}, \tag{2.1}$$

where $\varepsilon_{\mathrm{vac}}$ is the vacuum permittivity and χ_e denotes the scalar electric susceptibility. The $\boldsymbol{E}$ and $\boldsymbol{P}$ are classed as polar vectors [1], *i.e.*, they change their signs under the spatial inversion symmetry operation, which in Cartesian coordinates may be represented by a matrix

$$\bar{\mathbf{I}} = \begin{pmatrix} -1 & 0 & 0 \\ 0 & -1 & 0 \\ 0 & 0 & -1 \end{pmatrix}. \tag{2.2}$$

They are also said to be odd under the spatial inversion. At the same time, they are invariant with respect to the reversal of time, $t \rightarrow -t$. The two vectors are related by a polar (*i.e.*, invariant with respect to $\bar{\mathbf{I}}$) second rank tensor invariant with respect to the time reversal, which in a linear isotropic homogenous medium reduces to a true scalar in Eq. (2.1). This relation is modified when the medium becomes magnetized. Equation (2.1) takes a more general form

$$\boldsymbol{P}(\boldsymbol{M}) = \varepsilon_{\text{vac}} \boldsymbol{\chi}_e(\boldsymbol{M})\,\boldsymbol{E}. \tag{2.3}$$

We start from the electric susceptibility tensor

$$\boldsymbol{\chi}_e^{(0)} = \begin{pmatrix} \chi_{xx}^{(0)} & 0 & 0 \\ 0 & \chi_{xx}^{(0)} & 0 \\ 0 & 0 & \chi_{xx}^{(0)} \end{pmatrix}, \tag{2.4}$$

which in the present case is reduced to a scalar quantity independent of $\boldsymbol{M}$, and assume that the magnetization, induced in the medium, produces small changes of this electric susceptibility tensor. The magnetization $\boldsymbol{M}$ is an axial vector (*i.e.*, invariant with respect to $\bar{\mathbf{I}}$) changing the sign under the time reversal. In the account of magnetization effect, we consider the perturbations to the tensor in Eq. (2.4) as additional terms that are linear and quadratic in $\boldsymbol{M}$,

$$\boldsymbol{\chi}_e = \boldsymbol{\chi}_e^{(0)} + \boldsymbol{\chi}_e^{(1)} + \boldsymbol{\chi}_e^{(2)}, \tag{2.5}$$

neglecting the higher order terms. Because of the space and time reversal symmetry requirement, the terms $\boldsymbol{\chi}_e(\boldsymbol{M})$ and $\boldsymbol{\chi}_e(\boldsymbol{M}^2)$ should also be polar tensor invariant with respect to the time reversal. There is a single term linear in $\boldsymbol{M}$ that meets this requirement, *i.e*,

$$\alpha_{\text{tot}}^{(\text{ani})} \frac{\partial \boldsymbol{E}}{\partial t} \times \boldsymbol{M}, \tag{2.6}$$

where $\alpha_{\text{tot}}^{(\text{ani})}$ is a constant independent of $\boldsymbol{M}$. A complete list of the terms proportional to $\boldsymbol{M}^2$ and invariant under the spatial inversion and time reversal includes

$$\beta^{(\text{iso})} \boldsymbol{M}^2 \boldsymbol{E} + \gamma^{(\text{iso})} \boldsymbol{M}^2 \frac{\partial^2 \boldsymbol{E}}{\partial t^2} + \beta^{(\text{ani})} \boldsymbol{M} \times (\boldsymbol{E} \times \boldsymbol{M}) + \gamma^{(\text{ani})} \boldsymbol{M} \times \left(\frac{\partial^2 \boldsymbol{E}}{\partial t^2} \times \boldsymbol{M} \right), \tag{2.7}$$

where $\beta^{(\text{iso})}$ and $\gamma^{(\text{iso})}$ are the proportionality constants of the terms insensitive to the magnetization direction, and $\beta^{(\text{ani})}$ and $\gamma^{(\text{ani})}$ are those of the terms sensitive to the magnetization direction. We are interested in the response to monochromatic waves. We therefore write Eq. (2.3) as a relation between Fourier components at the angular frequency ω, *i.e.*,

$$\boldsymbol{P}(\omega, \boldsymbol{M}) = \varepsilon_{\text{vac}} \boldsymbol{\chi}_e(\omega, \boldsymbol{M})\,\boldsymbol{E}(\omega), \tag{2.8}$$

where we assume[2] that $\boldsymbol{E} = \boldsymbol{E}(\omega)\,\mathrm{e}^{-\mathrm{i}\omega t}$. Then,

$$\begin{aligned}\frac{1}{\varepsilon_{\text{vac}}}\boldsymbol{P} &= \chi_e^{(0)}\boldsymbol{E} - \mathrm{i}\omega\alpha_{\text{tot}}^{(\text{ani})}\left(\boldsymbol{E}\times\boldsymbol{M}\right) + \left[\left(\beta^{(\text{iso})} - \omega^2\gamma^{(\text{iso})}\right)\right]M^2\boldsymbol{E} \\ &\quad + \left[\left(\beta^{(\text{ani})} - \omega^2\gamma^{(\text{ani})}\right)\right]\left[M^2\boldsymbol{E} - \boldsymbol{M}\left(\boldsymbol{E}\cdot\boldsymbol{M}\right)\right], \end{aligned} \tag{2.9}$$

or in the Cartesian components

$$\begin{aligned}&\frac{1}{\varepsilon_{\text{vac}}}\left(P_x\hat{\boldsymbol{x}} + P_y\hat{\boldsymbol{y}} + P_z\hat{\boldsymbol{z}}\right) \\ &= \chi_e^{(0)}\left(E_x\hat{\boldsymbol{x}} + E_y\hat{\boldsymbol{y}} + E_z\hat{\boldsymbol{z}}\right) \\ &\quad -\mathrm{i}\omega\alpha_{\text{tot}}^{(\text{ani})}\left[\left(E_yM_z - E_zM_y\right)\hat{\boldsymbol{x}} + \left(E_zM_x - E_xM_z\right)\hat{\boldsymbol{y}} + \left(E_xM_y - E_yM_x\right)\hat{\boldsymbol{z}}\right] \\ &\quad +\beta_{\text{tot}}^{(\text{iso})}M^2\left(E_x\hat{\boldsymbol{x}} + E_y\hat{\boldsymbol{y}} + E_z\hat{\boldsymbol{z}}\right) + \beta_{\text{tot}}^{(\text{ani})}\left[M^2\left(E_x\hat{\boldsymbol{x}} + E_y\hat{\boldsymbol{y}} + E_z\hat{\boldsymbol{z}}\right)\right. \\ &\quad \left. - \left(M_x\hat{\boldsymbol{x}} + M_y\hat{\boldsymbol{y}} + M_z\hat{\boldsymbol{z}}\right)\left(E_xM_x + E_yM_y + E_zM_z\right)\right], \end{aligned} \tag{2.10}$$

where $\beta_{\text{tot}}^{(\text{iso})} = \beta^{(\text{iso})} - \omega^2\gamma^{(\text{iso})}$ and $\beta_{\text{tot}}^{(\text{ani})} = \beta^{(\text{ani})} - \omega^2\gamma^{(\text{ani})}$. For example, the terms of the x component of this equation can be grouped as follows

$$\begin{aligned}\frac{1}{\varepsilon_{\text{vac}}}P_x &= \left[\chi_e^{(0)} + \beta_{\text{tot}}^{(\text{iso})}\left(M_x^2 + M_y^2 + M_z^2\right) + \beta_{\text{tot}}^{(\text{ani})}\left(M_y^2 + M_z^2\right)\right]E_x \\ &\quad + \left(-\mathrm{i}\omega\alpha_{\text{tot}}^{(\text{ani})}M_z - \beta_{\text{tot}}^{(\text{ani})}M_xM_y\right)E_y + \left(\mathrm{i}\omega\alpha_{\text{tot}}^{(\text{ani})}M_z - \beta_{\text{tot}}^{(\text{ani})}M_xM_z\right)E_z \\ &= \chi_e^{(0)}E_x - \mathrm{i}\omega\alpha_{\text{tot}}^{(\text{ani})}\left(M_zE_y - M_zE_z\right) \\ &\quad + \left(\beta_{\text{tot}}^{(\text{iso})}M_x^2 + \left(\beta_{\text{tot}}^{(\text{iso})} + \beta_{\text{tot}}^{(\text{ani})}\right)\left(M_y^2 + M_z^2\right)\right]E_x \\ &\quad -\beta_{\text{tot}}^{(\text{ani})}\left(M_xM_yE_y + M_xM_zE_z\right). \end{aligned} \tag{2.11}$$

This suggests that Eq. (2.4) can be expressed in a matrix form as

$$\begin{aligned}\begin{pmatrix}\chi_{xx} & \chi_{xy} & \chi_{xz} \\ \chi_{yx} & \chi_{yy} & \chi_{yz} \\ \chi_{zx} & \chi_{zy} & \chi_{zz}\end{pmatrix} &= \begin{pmatrix}\chi_{xx}^{(0)} & 0 & 0 \\ 0 & \chi_{xx}^{(0)} & 0 \\ 0 & 0 & \chi_{xx}^{(0)}\end{pmatrix} - \mathrm{i}\omega\alpha_{\text{tot}}^{(\text{ani})}\begin{pmatrix}0 & M_z & -M_y \\ -M_z & 0 & M_x \\ M_y & -M_x & 0\end{pmatrix} + \beta_{\text{tot}}^{(\text{iso})}\begin{pmatrix}M_x^2 & 0 & 0 \\ 0 & M_y^2 & 0 \\ 0 & 0 & M_z^2\end{pmatrix} \\ &\quad + \left(\beta_{\text{tot}}^{(\text{iso})} + \beta_{\text{tot}}^{(\text{ani})}\right)\begin{pmatrix}M_y^2 + M_z^2 & 0 & 0 \\ 0 & M_z^2 + M_x^2 & 0 \\ 0 & 0 & M_x^2 + M_y^2\end{pmatrix} \\ &\quad -\beta_{\text{tot}}^{(\text{ani})}\begin{pmatrix}0 & M_xM_y & M_xM_z \\ M_yM_x & 0 & M_yM_z \\ M_zM_x & M_zM_y & 0\end{pmatrix}. \end{aligned} \tag{2.12}$$

---

[2] In this chapter we employ the convention with $\mathrm{e}^{-\mathrm{i}\omega t}$ with i representing the imaginary unit. The convention is established in most texts on microscopic models.

In the Cartesian tensor notation, this may be written as

$$\chi_{ij} = \chi_{ij}^{(0)}\delta_{ij} - \mathrm{i}\omega\alpha_{\mathrm{tot}}^{(\mathrm{ani})}\epsilon_{ijk}M_k + \left(\beta_{\mathrm{tot}}^{(\mathrm{iso})} + \beta_{\mathrm{tot}}^{(\mathrm{ani})}\right)\delta_{ij}\delta_{kl}M_kM_l - \beta_{\mathrm{tot}}^{(\mathrm{ani})}\delta_{ik}\delta_{jl}M_kM_l. \tag{2.13}$$

Here, δ_{ij} denotes the Kronecker delta, $\delta_{ij} = 1$ for $i = j$ and ϕ otherwise. The Levi-Civita permutation symbol ϵ_{ijk} changes its sign, when any two of its indices exchange their position, *i.e.*, $\epsilon_{jki} = \epsilon_{kij} = \epsilon_{ijk} = \epsilon_{kji} = \epsilon_{ikj} = \epsilon_{jik}$, the absolute value $|\epsilon_{ijk}| = 1$ when all three indices are different and zero otherwise, *e.g.*, $\epsilon_{jkk} = \epsilon_{kii} = \epsilon_{jjk} = \epsilon_{iii} = \epsilon_{jjj} = \epsilon_{kkk} = 0$. The shortened notation employed in Eq. (2.13) uses the so-called Einstein sum convention. This assumes the summation over pairs of repeated indices with summation signs skipped. Without the convention, the information contained in Eq. (2.13) would be expressed as

$$\begin{aligned}\chi_{ij}\left(M_x, M_y, M_z\right) = \chi_{ij}^{(0)}\delta_{ij} + \mathrm{i}\omega\alpha_{\mathrm{tot}}^{(\mathrm{ani})}\sum_k \epsilon_{ijk}M_k \\ + \left(\beta_{\mathrm{tot}}^{(\mathrm{iso})} + \beta_{\mathrm{tot}}^{(\mathrm{ani})}\right)\sum_{kl}\delta_{ij}\delta_{kl}M_kM_l \\ - \beta_{\mathrm{tot}}^{(\mathrm{ani})}\sum_{kl}\delta_{ik}\delta_{jl}M_kM_l.\end{aligned} \tag{2.14}$$

The tensor is invariant with respect to *simultaneous* reversal of time ($\mathrm{i} \to \mathrm{i}^* = -\mathrm{i}$, in the assumed temporal dependence $\mathrm{e}^{-\mathrm{i}\omega t}$) and space inversion of the magnetization vector, *i.e.*, magnetization reversal ($M_k \to -M_k$), *i.e.*,

$$\begin{aligned}\widetilde{\chi}_{ij}\left(-M_x, -M_y, -M_z\right) = \chi_{ij}^{(0)}\delta_{ij} + \mathrm{i}\omega\alpha_{\mathrm{tot}}^{(\mathrm{ani})}\epsilon_{ijk}\left(-M_k\right) \\ + \left(\beta_{\mathrm{tot}}^{(\mathrm{iso})} + \beta_{\mathrm{tot}}^{(\mathrm{ani})}\right)\delta_{ij}\delta_{kl}\left(-M_k\right)\left(-M_l\right) \\ - \beta_{\mathrm{tot}}^{(\mathrm{ani})}\delta_{ik}\delta_{jl}\left(-M_k\right)\left(-M_l\right),\end{aligned} \tag{2.15}$$

where the tilde in $\widetilde{\chi}_{ij}$ indicates the time reversal. This is consistent with the Onsager principle. The second rank tensor δ_{ij}, as well as the fourth rank tensors $\delta_{ij}\delta_{kl}$ and $\delta_{ik}\delta_{jl}$, are invariant under spatial inversion. The magnetization reversal operation alone, switching $\boldsymbol{M}$

$$\chi_{ij}(\boldsymbol{M}) = \chi_{ij}^{(0)}\delta_{ij} - \mathrm{i}\omega\alpha_{\mathrm{tot}}^{(\mathrm{ani})}\epsilon_{ijk}M_k + \left(\beta_{\mathrm{tot}}^{(\mathrm{iso})} + \beta_{\mathrm{tot}}^{(\mathrm{ani})}\right)\delta_{ij}\delta_{kl}M_kM_l - \beta_{\mathrm{tot}}^{(\mathrm{ani})}\delta_{ik}\delta_{jl}M_kM_l \tag{2.16}$$

to $-\boldsymbol{M}$, produces

$$\begin{aligned}\chi_{ij}(-\boldsymbol{M}) &= \chi_{ij}^{(0)}\delta_{ij} - \mathrm{i}\omega\alpha_{\mathrm{tot}}^{(\mathrm{ani})}\epsilon_{ijk}(-M_k) + \left(\beta_{\mathrm{tot}}^{(\mathrm{iso})} + \beta_{\mathrm{tot}}^{(\mathrm{ani})}\right)\delta_{ij}\delta_{kl}(-M_k)(-M_l) \\ &\quad -\beta_{\mathrm{tot}}^{(\mathrm{ani})}\delta_{ik}\delta_{jl}(-M_k)(-M_l) \\ &= \chi_{ij}^{(0)}\delta_{ij} - \mathrm{i}\omega\alpha_{\mathrm{tot}}^{(\mathrm{ani})}\epsilon_{jik}M_k + \left(\beta_{\mathrm{tot}}^{(\mathrm{iso})} + \beta_{\mathrm{tot}}^{(\mathrm{ani})}\right)\delta_{ij}\delta_{kl}M_kM_l \\ &\quad -\beta_{\mathrm{tot}}^{(\mathrm{ani})}\delta_{ik}\delta_{jl}M_kM_l \\ &= \chi_{ji}(\boldsymbol{M}),\end{aligned} \tag{2.17}$$

where the use has been made of the properties of Kronecker delta (*i.e.*, the symmetry $\delta_{ij} = \delta_{ji}$) and Levi-Civita permutation symbols (*i.e.*, the antisymmetry $\epsilon_{jik} = -\epsilon_{jik}$).

The axial second rank tensor $\epsilon_{ijk}M_k$ is invariant under spatial inversion. Indeed, the third rank tensor ϵ_{ijk} represents a pseudotensor that transforms according to

$$\epsilon'_{ijk} = \det[a_{ip}a_{jq}a_{kr}]a_{ip}a_{jq}a_{kr}\epsilon_{pqr} \tag{2.18}$$

and the magnetization as a pseudovector

$$M'_k = \det[a_{kr}]a_{kr}M_r, \tag{2.19}$$

where a_{ip}, a_{jq}, and a_{kr} are the spatial transformation matrices. For the spatial inversion $\bar{\mathbf{I}}$ of Eq. (2.2) $a_{ip} = -\delta_{ip}$, $a_{jq} = -\delta_{jq}$, $a_{kr} = -\delta_{kr}$, and $\det[a_{ip}a_{jq}a_{kr}] = -1$. As a result, the pseudotensor ϵ_{ijk} is invariant under all proper and improper rotations and is, therefore, classified as an *isotropic third rank tensor*. The pseudovector M_k is then invariant with respect to the spatial inversion, because $a_{kr} = -\delta_{kr}$ and $\det[a_{kr}] = -1$.

In addition to ϵ_{ijk}, the other tensors entering Eq. (2.13) are isotropic, as the only anisotropy comes from the magnetization vector. Here, δ_{ij} represents the isotropic second rank tensor (or a unit matrix). The general fourth rank isotropic tensor takes the form [2]

$$\begin{aligned}G_{ijkl} = \mathcal{A}\delta_{ij}\delta_{kl} + \mathcal{B}\left(\delta_{ik}\delta_{jl} + \delta_{jk}\delta_{il}\right) \\ +\mathcal{C}\left(\delta_{ik}\delta_{jl} - \delta_{jk}\delta_{il}\right).\end{aligned} \tag{2.20}$$

Eq. (2.12) and Eq. (2.13) indicate that our fourth rank tensor should be symmetric in both the first and second pair of indices, *i.e.*,

$$G_{ijkl} = G_{jikl} = G_{ijlk} = G_{jilk}. \tag{2.21}$$

Then, $\mathcal{C} = \pm\mathcal{B}$ and we choose $\mathcal{C} = \mathcal{B}$, without any loss in generality. From this, we can deduce

$$G_{iiii} = \mathcal{A} + 2\mathcal{B} = \beta_{\mathrm{tot}}^{(\mathrm{iso})}, \tag{2.22a}$$

$$G_{iijj} = \mathcal{A} = \beta_{\mathrm{tot}}^{(\mathrm{iso})} + \beta_{\mathrm{tot}}^{(\mathrm{ani})}, \tag{2.22b}$$

$$G_{ijij} = G_{iiii} - G_{iijj} = 2\mathcal{B} = -\beta_{\mathrm{tot}}^{(\mathrm{ani})}. \tag{2.22c}$$

Referring to Eq. (2.5), we observe that $\chi_e^{(0)}$ and $\chi_e^{(2)}$ are symmetric tensors while $\chi_e^{(1)}$ is an antisymmetric tensor. In a matrix form, the second rank tensors $\chi_e^{(1)}$ and $\chi_e^{(2)}$ can be expressed as matrix products with corresponding column vectors, of a *linear third rank magnetooptic tensor*

$$\begin{pmatrix} \chi_{yz}^{(1)} \\ \chi_{zx}^{(1)} \\ \chi_{xy}^{(1)} \\ \chi_{zy}^{(1)} \\ \chi_{xz}^{(1)} \\ \chi_{yx}^{(1)} \end{pmatrix} = \mathrm{i}\omega\alpha_{\text{tot}}^{(\text{ani})} \begin{pmatrix} 1 & 0 & 0 \\ 0 & 1 & 0 \\ 0 & 0 & 1 \\ -1 & 0 & 0 \\ 0 & -1 & 0 \\ 0 & 0 & -1 \end{pmatrix} \begin{pmatrix} M_x \\ M_y \\ M_z \end{pmatrix}, \tag{2.23}$$

and a *quadratic fourth rank magnetooptic tensor*

$$\begin{pmatrix} \chi_{xx}^{(2)} \\ \chi_{yy}^{(2)} \\ \chi_{zz}^{(2)} \\ \chi_{yz}^{(2)} \\ \chi_{zx}^{(2)} \\ \chi_{xy}^{(2)} \end{pmatrix} = \begin{pmatrix} (\mathcal{A}+2\mathcal{B}) & \mathcal{A} & \mathcal{A} & 0 & 0 & 0 \\ \mathcal{A} & (\mathcal{A}+2\mathcal{B}) & \mathcal{A} & 0 & 0 & 0 \\ \mathcal{A} & \mathcal{A} & (\mathcal{A}+2\mathcal{B}) & 0 & 0 & 0 \\ 0 & 0 & 0 & 2\mathcal{B} & 0 & 0 \\ 0 & 0 & 0 & 0 & 2\mathcal{B} & 0 \\ 0 & 0 & 0 & 0 & 0 & 2\mathcal{B} \end{pmatrix} \begin{pmatrix} M_x^2 \\ M_y^2 \\ M_z^2 \\ M_y M_z \\ M_z M_x \\ M_x M_y \end{pmatrix}. \tag{2.24}$$

After this introduction, we can approach the investigation of the magnetization effect in crystals. This approach still does not require any microscopic model. As will be seen later, the symmetry-based conclusions agree with those following from the microscopic models for the magnetooptic effects. In crystals, the tensors, in the expansions similar to Eq. (2.13), are no more isotropic, in general. The allowed proper and improper rotations are restricted by the point symmetry group of a crystal. In the sum of Eq. (2.5) generalized to crystals, the nonmagnetic electric susceptibility tensor, $\chi_e^{(0)}$, is assumed symmetric

$$\chi_e^{(0)} = \begin{pmatrix} \chi_{xx}^{(0)} & \chi_{xy}^{(0)} & \chi_{zx}^{(0)} \\ \chi_{xy}^{(0)} & \chi_{yy}^{(0)} & \chi_{yz}^{(0)} \\ \chi_{zx}^{(0)} & \chi_{yz}^{(0)} & \chi_{zz}^{(0)} \end{pmatrix}, \tag{2.25}$$

the linear magnetooptic tensor should remain antisymmetric

$$\chi_e^{(1)} = \begin{pmatrix} 0 & \chi_{xy}^{(1)} & -\chi_{zx}^{(1)} \\ -\chi_{xy}^{(1)} & 0 & \chi_{yz}^{(1)} \\ \chi_{zx}^{(1)} & -\chi_{yz}^{(1)} & 0 \end{pmatrix}, \tag{2.26}$$

and the quadratic MO tensor should remain symmetric

$$\chi_e^{(2)} = \begin{pmatrix} \chi_{xx}^{(2)} & \chi_{xy}^{(2)} & \chi_{zx}^{(2)} \\ \chi_{xy}^{(2)} & \chi_{yy}^{(2)} & \chi_{yz}^{(2)} \\ \chi_{zx}^{(2)} & \chi_{yz}^{(2)} & \chi_{zz}^{(2)} \end{pmatrix}. \tag{2.27}$$

We continue to use mere symmetry arguments and assume that the effect of magnetic ordering on the electric susceptibility tensor is small, as observed in most situations. We limit ourselves to the case of collinear magnetic order and expand the complex electric susceptibility tensor χ_{ij} into the McLaurin power series of the macroscopic magnetization. In the Cartesian representation, this provides [3–8]

$$\begin{aligned} \chi_{ij} &= \chi_{ij}^{(0)} + \left[\frac{\partial \chi_{ij}}{\partial M_k}\right]_{M=0} M_k + \frac{1}{2}\left[\frac{\partial^2 \chi_{ij}}{\partial M_k M_l}\right]_{M=0} M_k M_l + \cdots \\ &= \chi_{ij}^{(0)} + K_{ijk} M_k + \frac{1}{2} G_{ijkl} M_k M_l + \cdots \end{aligned} \tag{2.28}$$

Here, $\chi_{ij}^{(0)} = \chi_{ji}^{(0)}$ denotes the components of the susceptibility tensor when the magnetization $\boldsymbol{M} = 0$. The third rank axial tensor, with the components denoted as K_{ijk}, relates the optical susceptibility (a polar tensor invariant under spatial inversion and time reversal) to static magnetization (axial vector or pseudovector). Thus, these components K_{ijk} define the linear magnetooptic tensor. Consequently, the linear magnetooptic effects are allowed in the crystal point groups that include the spatial inversion operation. Note that this is not the case for the linear electrooptic effects, which are exclusively allowed in the point groups, which do not include the space inversion operation [9–10]

The components of the quadratic magnetooptic polar tensor are denoted by G_{ijkl}. The quadratic magnetooptic polar tensor [3] relates the optical susceptibility (a polar tensor) to the products $M_k M_l$ (which form another polar tensor). Its form does not differ from that which characterizes quadratic (or Kerr) electrooptic effect and photoelastic effect in crystals of corresponding symmetry. The indices i, j, k, and l denote the Cartesian x, y and z axes. We recall that the sum convention is understood. The Onsager principle requires [3], [11–14]

$$\chi_{ij}(\omega, \boldsymbol{k}, \boldsymbol{M}) = \chi_{ji}(\omega, -\boldsymbol{k}, -\boldsymbol{M}). \tag{2.29}$$

In the electric dipole approximation (corresponding to the zero propagation vector, $\boldsymbol{k} \approx 0$), this simplifies to

$$\chi_{ij}(\omega, \boldsymbol{M}) = \chi_{ji}(\omega, -\boldsymbol{M}). \tag{2.30}$$

For the terms of Eq. (2.28) linear in M_k, it follows

$$\chi_{ij}^{(1)} = K_{ijk}M_k = -K_{jik}M_k = -\chi_{ji}^{(1)} \tag{2.31}$$

or

$$K_{iik} = 0 \quad \text{and} \quad K_{ijk} = -K_{jik}. \tag{2.32}$$

This result has already been verified for the special case of media originally isotropic in the absence of magnetization. Assuming that the order of differentiation in the term proportional to products $M_k M_l$ of Eq. (2.28) can be exchanged, we have

$$\chi_{ij}^{(2)} = G_{ijkl}M_k M_l = G_{ijkl}(-M_k)(-M_l) = \chi_{ji}^{(2)}, \tag{2.33}$$

from which it follows

$$G_{ijkl} = G_{jikl} = G_{ijlk} = G_{jilk}. \tag{2.34}$$

In a crystal of arbitrary symmetry and crystallographic orientation, we can thus write the following general relations for the contributions to the electric susceptibility tensor linear and quadratic in magnetization

$$\begin{pmatrix} \chi_{yz}^{(1)} \\ \chi_{zx}^{(1)} \\ \chi_{xy}^{(1)} \\ \chi_{zy}^{(1)} \\ \chi_{xz}^{(1)} \\ \chi_{yx}^{(1)} \end{pmatrix} = \begin{pmatrix} K_{yzx} & K_{yzy} & K_{yzz} \\ K_{zxx} & K_{zxy} & K_{zxz} \\ K_{xyx} & K_{xyy} & K_{xyz} \\ -K_{yzx} & -K_{yzy} & -K_{yzz} \\ -K_{zxx} & -K_{zxy} & -K_{zxz} \\ -K_{xyx} & -K_{xyy} & -K_{xyz} \end{pmatrix} \begin{pmatrix} M_x \\ M_y \\ M_z \end{pmatrix} \tag{2.35}$$

or in a more concise form suggested by Eq. (2.31)

$$\begin{pmatrix} \chi_{yz}^{(1)} \\ \chi_{zx}^{(1)} \\ \chi_{xy}^{(1)} \end{pmatrix} = \begin{pmatrix} K_{yzx} & K_{yzy} & K_{yzz} \\ K_{zxx} & K_{zxy} & K_{zxz} \\ K_{xyx} & K_{xyy} & K_{xyz} \end{pmatrix} \begin{pmatrix} M_x \\ M_y \\ M_z \end{pmatrix} \tag{2.36}$$

and

$$\begin{pmatrix} \chi_{xx}^{(2)} \\ \chi_{yy}^{(2)} \\ \chi_{zz}^{(2)} \\ \chi_{yz}^{(2)} \\ \chi_{zx}^{(2)} \\ \chi_{xy}^{(2)} \end{pmatrix} = \begin{pmatrix} G_{xxxx} & G_{xxyy} & G_{xxzz} & G_{xxyz} & G_{xxzx} & G_{xxxy} \\ G_{yyxx} & G_{yyyy} & G_{yyzz} & G_{yyyz} & G_{yyzx} & G_{yyxy} \\ G_{zzxx} & G_{zzyy} & G_{zzzz} & G_{zzyz} & G_{zzzx} & G_{zzxy} \\ G_{yzxx} & G_{yzyy} & G_{yzzz} & G_{yzyz} & G_{yzzx} & G_{yzxy} \\ G_{zxxx} & G_{zxyy} & G_{zxzz} & G_{zxyz} & G_{zxzx} & G_{zxxy} \\ G_{xyxx} & G_{xyyy} & G_{xyzz} & G_{xyyz} & G_{xyzx} & G_{xyxy} \end{pmatrix} \begin{pmatrix} M_x^2 \\ M_y^2 \\ M_z^2 \\ M_y M_z \\ M_z M_x \\ M_x M_y \end{pmatrix}. \tag{2.37}$$

Under proper and improper rotations of the crystal point group, the tensor components transform according to [15]

$$\chi'_{ij} = a_{im}a_{jn}\chi_{mn} \tag{2.38a}$$

$$K'_{ijk} = a_{im}a_{jn}a_{ko}K_{mno} \tag{2.38b}$$

$$G'_{ijkl} = a_{im}a_{jn}a_{ko}a_{lp}G_{mnop}. \tag{2.38c}$$

Here, $a_{\mu\nu}$ $(\mu, \nu = x, y, z)$ are the elements of 3×3 matrices that represent the Cartesian vector transformations. They can be expressed as a product

$$\begin{pmatrix} \pm\cos\psi_c & \mp\sin\psi_c & 0 \\ \pm\sin\psi_c & \pm\cos\psi_c & 0 \\ 0 & 0 & \pm 1 \end{pmatrix} \begin{pmatrix} \cos\theta_c & 0 & \sin\theta_c \\ 0 & 1 & 0 \\ -\sin\theta_c & 0 & \cos\theta_c \end{pmatrix} \begin{pmatrix} \cos\varphi_c & -\sin\varphi_c & 0 \\ \sin\varphi_c & \cos\varphi_c & 0 \\ 0 & 0 & 1 \end{pmatrix}, \tag{2.39}$$

where φ_c, θ_c, and ψ_c are the Euler angles. The determinant of this matrix product is ± 1. The lower signs here, as in the matrix product (2.39), correspond to improper rotations (*e.g.*, to the spatial inversion or reflection in a plane times a proper rotation).

The symmetry operations of a particular crystal point group leave the material tensors (here χ_{ij}, $\chi_{ij}^{(0)}$, K_{ijk}, and G_{ijkl}) unchanged. In the following sections, we summarize the form of the tensors in the 32 crystal classes. The principal axes are set parallel to the Cartesian z axis, the twofold orthorhombic and tetragonal axes, and the three mutually perpendicular twofold (fourfold) axes of the cubic system are chosen parallel to the Cartesian axes.

In our demonstration of the linear and quadratic magnetooptic tensors, we use a suitably chosen set of point group symmetry operations. Note that this choice is not unique. It has been found that the form of the tensor assumes the same form for a set of point groups. We illustrate the derivation for one chosen point group of the set. Note that the matrix product representing rotations in Eq. (2.39) may also be employed to investigate the effect of crystallographic orientation in a particular crystal point group on the effects characterized by the tensor.

2.2.2 Susceptibility in Nonmagnetic Crystals

For the sake of completeness, we first list the electric susceptibility tensors $\chi_{ij}^{(0)}$ in seven crystallographic systems [15], which follows from Eq. (2.38a):
(a) Triclinic system

$$\begin{pmatrix} \chi_{xx}^{(0)} & \chi_{xy}^{(0)} & \chi_{zx}^{(0)} \\ \chi_{xy}^{(0)} & \chi_{yy}^{(0)} & \chi_{yz}^{(0)} \\ \chi_{zx}^{(0)} & \chi_{yz}^{(0)} & \chi_{zz}^{(0)} \end{pmatrix}. \tag{2.40a}$$

(b) Monoclinic system

$$\begin{pmatrix} \chi_{xx}^{(0)} & \chi_{xy}^{(0)} & 0 \\ \chi_{xy}^{(0)} & \chi_{yy}^{(0)} & 0 \\ 0 & 0 & \chi_{zz}^{(0)} \end{pmatrix}. \tag{2.40b}$$

(c) Orthorhombic system

$$\begin{pmatrix} \chi_{xx}^{(0)} & 0 & 0 \\ 0 & \chi_{yy}^{(0)} & 0 \\ 0 & 0 & \chi_{zz}^{(0)} \end{pmatrix}. \tag{2.40c}$$

(d) Tetragonal, trigonal, and hexagonal systems

$$\begin{pmatrix} \chi_{xx}^{(0)} & 0 & 0 \\ 0 & \chi_{xx}^{(0)} & 0 \\ 0 & 0 & \chi_{zz}^{(0)} \end{pmatrix}. \tag{2.40d}$$

(e) Cubic system

$$\begin{pmatrix} \chi_{xx}^{(0)} & 0 & 0 \\ 0 & \chi_{xx}^{(0)} & 0 \\ 0 & 0 & \chi_{xx}^{(0)} \end{pmatrix}. \tag{2.40e}$$

2.2.3 Linear Magnetooptic Tensor

The form of the linear magnetooptic tensor K_{ijk} is obtained from Eq. (2.38b) using a suitable set of symmetry operations of the crystal point group considered. Auld and Wilson [16] derived the tensor in cubic crystals showing that the linear magnetooptic effects (*e.g.*, Faraday or Kerr rotation) do not depend on the crystallographic orientation. The orthorhombic symmetry case was treated by Tabor and Chen [17] and by Kahn *et al.* [18].

Below we list the linear magnetooptic tensors in 32 crystallographic point symmetry groups (we employ both the international and Schönflies notation) [19].

(a) Classes $1 = C_1$ and $\overline{1} = C_i$ of the triclinic system. The symmetry operations of these groups set no limitations on the form of the tensor. They leave it in the most general form given by Eq. (2.35) or Eq. (2.36).

(b) Classes $2 = C_2$, $m = C_S$, C_{1h}, and $2/m = C_{2h}$ of the monoclinic system. The use will be made of the rotation by the angle π about the principal twofold axis represented by the matrix

$$C_{2z} = \begin{pmatrix} -1 & 0 & 0 \\ 0 & -1 & 0 \\ 0 & 0 & 1 \end{pmatrix}. \tag{2.41}$$

The substitution of the matrix elements for a_{im}, a_{jn}, and a_{ko} in Eq. (2.38b) gives

$$\begin{pmatrix} \chi_{yz}^{(1)} \\ \chi_{zx}^{(1)} \\ \chi_{xy}^{(1)} \end{pmatrix} = \begin{pmatrix} K_{yzx} & K_{yzy} & 0 \\ K_{zxx} & K_{zxy} & 0 \\ 0 & 0 & K_{xyz} \end{pmatrix} \begin{pmatrix} M_x \\ M_y \\ M_z \end{pmatrix}. \tag{2.42}$$

(c) Classes $222 = D_2$, $mm2 = C_{2v}$, and $mmm = D_{2h}$ of the orthorhombic system.
Adding to the above C_{2z} symmetry operation the rotations about twofold axes x and y,

$$C_{2x} = \begin{pmatrix} 1 & 0 & 0 \\ 0 & -1 & 0 \\ 0 & 0 & -1 \end{pmatrix} \tag{2.43}$$

and

$$C_{2y} = \begin{pmatrix} -1 & 0 & 0 \\ 0 & 1 & 0 \\ 0 & 0 & -1 \end{pmatrix}, \tag{2.44}$$

we obtain the linear magnetooptic tensor in the form

$$\begin{pmatrix} K_{yzx} & 0 & 0 \\ 0 & K_{zxy} & 0 \\ 0 & 0 & K_{xyz} \end{pmatrix}. \tag{2.45}$$

We observe that any rotation about the orthorhombic axes does not change the form of the linear magnetooptic tensor in orthorhombic crystals. This fact allows one to separate the crystallographic effect of orthorhombic symmetry (manifested in the rotations about the orthorhombic axes) from the linear MO effects.

(d) Classes of uniaxial point groups $4 = C_4$, $\bar{4} = S_4$ and $4/m = C_{4h}$ of the tetragonal system, classes $3 = C_3$ and $\bar{3} = C_{3i}$ of the trigonal system and classes $6 = C_6$, $\bar{6} = C_{3h}$, and $6/m = C_{6h}$ of the hexagonal system.
The rotation about the principal axis by the angle $\pm\varphi = \pi/2, 2\pi/3$, and $\pi/3$ represented by

$$\begin{pmatrix} \cos\varphi & \mp\sin\varphi & 0 \\ \pm\sin\varphi & \cos\varphi & 0 \\ 0 & 0 & 1 \end{pmatrix} \tag{2.46}$$

yields the same result for the above point uniaxial groups, *i.e.*,

$$\begin{pmatrix} K_{yzx} & K_{yzy} & 0 \\ -K_{yzy} & K_{yzx} & 0 \\ 0 & 0 & K_{xyz} \end{pmatrix}. \tag{2.47}$$

We observe that rotations about the principal axis do not affect the form of the linear magnetooptic MO tensor in the uniaxial point groups.

(e) Classes of uniaxial point groups $422 = D_4$, $4mm = C_{4v}$, $\bar{4}2m = D_{2d}$, and $4/mmm = D_{4h}$ of the tetragonal system; classes $32 = D_3$, $3m = C_{3v}$, and $\bar{3}m = D_{3d}$ of the trigonal system; and classes $622 = D_6$, $6mm = C_{6v}$, $\bar{6}2m = D_{3h}$, and $6/mmm = D_{6h}$ of the hexagonal system.
The inclusion of the rotation by the angle π about a secondary axis (a twofold axis normal to the principal axis) represented for the groups 422, 42m, *etc*., by C_{2x} in Eq. (2.43) simplifies the previous case to

$$\begin{pmatrix} K_{yzx} & 0 & 0 \\ 0 & K_{yzx} & 0 \\ 0 & 0 & K_{xyz} \end{pmatrix}. \tag{2.48}$$

The same result is obtained for the other groups of the tetragonal, trigonal, and hexagonal systems. To prove this, we would have to investigate the effect of the matrix

$$\begin{pmatrix} \pm\cos 2\varphi & \mp\sin 2\varphi & 0 \\ \mp\sin 2\varphi & \mp\cos 2\varphi & 0 \\ 0 & 0 & \mp 1 \end{pmatrix}. \tag{2.49}$$

It represents either the rotation about a twofold axis normal to the principal axis (upper signs) or the reflection in the plane containing the principal axis (lower signs).

(f) All crystal classes of the cubic system, *i.e.*, $23 = T$, $m3 = T_h$, $432 = O$, $\bar{4}3m = T_d$, and $m3m = O_h$.
To obtain the form of the linear magnetooptic tensor in this most common case, it is sufficient to extend the set of symmetry operations C_{2x}, C_{2y}, and C_{2z} employed in the case (c) of the orthorhombic system by the operation of the rotation about a threefold axis. Choosing, *e.g.*,

$$C_3 = \begin{pmatrix} 0 & 0 & 1 \\ 1 & 0 & 0 \\ 0 & 1 & 0 \end{pmatrix} \tag{2.50}$$

we obtain [3], [16]

$$\begin{pmatrix} K_{xyz} & 0 & 0 \\ 0 & K_{xyz} & 0 \\ 0 & 0 & K_{xyz} \end{pmatrix}. \tag{2.51}$$

Under the above assumptions expressed by the development in Eq. (2.28), the linear magnetooptic effects in cubic crystal do not depend on the crystallographic orientation. This coincides with the result for isotropic medium given in Eq. (2.23). We return to this conclusion later in our discussion of the semiclassical model of the susceptibility.

2.2.4 Quadratic Magnetooptic Tensor

The form of the quadratic magnetooptic tensor follows from Eq. (2.38c) using a suitably chosen set of symmetry operations. It is analogous to that of the quadratic electrooptic tensor and that of the photoelasticity tensor, which can be found in the books by Nye [15], Kaminoff [9], or Yariv and Yeh [20]. Because of the permutation symmetry in the two pairs of indices, expressed by Eq. (2.34), the contracted notation proposed by Voigt is often used in the literature [15]. The correspondence is the following: $1 = (11)$, $2 = (22)$, $3 = (33)$, $4 = (23) = (32)$, $5 = (13) = (31)$, and $6 = (12) = (21)$. The numbers in the parentheses index the Cartesian coordinates $x_1 = x, x_2 = y$, and $x_3 = z$.

(a) Triclinic system.
The form of the tensor is the most general and is given by Eq. (2.37).

(b) Monoclinic system.
The operation of the rotation about the twofold axis C_{2z} sets to zero all components containing an odd number of the indices z, *i.e.*,

$$\begin{pmatrix} G_{xxxx} & G_{xxyy} & G_{xxzz} & 0 & 0 & G_{xxxy} \\ G_{yyxx} & G_{yyyy} & G_{yyzz} & 0 & 0 & G_{yyxy} \\ G_{zzxx} & G_{zzyy} & G_{zzzz} & 0 & 0 & G_{zzxy} \\ 0 & 0 & 0 & G_{yzyz} & G_{yzzx} & 0 \\ 0 & 0 & 0 & G_{zxyz} & G_{zxzx} & 0 \\ G_{xyxx} & G_{xyyy} & G_{xyzz} & 0 & 0 & G_{xyxy} \end{pmatrix}. \tag{2.52}$$

(c) Orthorhombic system.
The operations of the rotation about the twofold orthorhombic axes C_{2x}, C_{2y}, and C_{2z} set to zero the components with an odd number of indices

x, y, or z, which gives

$$\begin{pmatrix} G_{xxxx} & G_{xxyy} & G_{xxzz} & 0 & 0 & 0 \\ G_{yyxx} & G_{yyyy} & G_{yyzz} & 0 & 0 & 0 \\ G_{zzxx} & G_{zzyy} & G_{zzzz} & 0 & 0 & 0 \\ 0 & 0 & 0 & G_{yzyz} & 0 & 0 \\ 0 & 0 & 0 & 0 & G_{zxzx} & 0 \\ 0 & 0 & 0 & 0 & 0 & G_{xyxy} \end{pmatrix}. \tag{2.53}$$

(d) Tetragonal system: crystal classes $4 = C_4$, $\bar{4} = S_4$, and $4/m = C_{4h}$.
We start with the result for the monoclinic system under (b) and make use of the rotation about the fourfold principal axis $C_{4z}^{\pm}$ represented by

$$C_{4z}^{\pm} = \begin{pmatrix} 0 & \mp 1 & 0 \\ \pm 1 & 0 & 0 \\ 0 & 0 & 1 \end{pmatrix}. \tag{2.54}$$

We then obtain

$$\begin{pmatrix} G_{xxxx} & G_{xxyy} & G_{xxzz} & 0 & 0 & G_{xxxy} \\ G_{xxyy} & G_{xxxx} & G_{xxzz} & 0 & 0 & -G_{xxxy} \\ G_{zzxx} & G_{zzxx} & G_{zzzz} & 0 & 0 & 0 \\ 0 & 0 & 0 & G_{yzyz} & G_{yzzx} & 0 \\ 0 & 0 & 0 & -G_{yzzx} & G_{yzyz} & 0 \\ G_{xyxx} & -G_{xyxx} & 0 & 0 & 0 & G_{xyxy} \end{pmatrix}. \tag{2.55}$$

(e) Tetragonal system: crystal classes $422 = D_4$, $4mm = C_{4v}$, $\bar{4}2m = D_{2d}$, and $4/mmm = D_{4h}$.
Adding to the previous case (d) the rotation about one of the twofold axes normal to the principal fourfold axis represented by Eq. (2.49)

$$C_{2a} = \begin{pmatrix} 0 & 1 & 0 \\ 1 & 0 & 0 \\ 0 & 0 & -1 \end{pmatrix} \tag{2.56}$$

gives

$$\begin{pmatrix} G_{xxxx} & G_{xxyy} & G_{xxzz} & 0 & 0 & 0 \\ G_{xxyy} & G_{xxxx} & G_{xxzz} & 0 & 0 & 0 \\ G_{zzxx} & G_{zzxx} & G_{zzzz} & 0 & 0 & 0 \\ 0 & 0 & 0 & G_{yzyz} & 0 & 0 \\ 0 & 0 & 0 & 0 & G_{yzyz} & 0 \\ 0 & 0 & 0 & 0 & 0 & G_{xyxy} \end{pmatrix}. \tag{2.57}$$

(f) Trigonal system: crystal classes $3 = C_3$ and $\overline{3} = C_{3i}$.
We apply the rotation by the angle $\pm 2\pi 3$ represented by

$$C_3^{\pm} = \begin{pmatrix} -1/2 & \mp\sqrt{3}/2 & 0 \\ \pm\sqrt{3}/2 & -1/2 & 0 \\ 0 & 0 & 1 \end{pmatrix} \tag{2.58}$$

to the general form given in Eq. (2.37), which results in the following tensor

$$\begin{pmatrix} G_{xxxx} & G_{xxyy} & G_{xxzz} & G_{xxyz} & G_{xxzx} & G_{xxxy} \\ G_{xxyy} & G_{xxxx} & G_{xxzz} & -G_{xxyz} & -G_{xxzx} & -G_{xxxy} \\ G_{zzxx} & G_{zzxx} & G_{zzzz} & 0 & 0 & 0 \\ G_{yzxx} & -G_{yzxx} & 0 & G_{yzyz} & G_{yzzx} & -G_{zxxx} \\ G_{zxxx} & -G_{zxxx} & 0 & -G_{yzzx} & G_{yzyz} & G_{yzxx} \\ -G_{xxxy} & G_{xxxy} & 0 & -G_{xxzx} & G_{xxyz} & \frac{1}{2}\left(G_{xxxx} - G_{xxyy}\right) \end{pmatrix}. \tag{2.59}$$

(g) Trigonal system: crystal classes $3m = C_{3v}$, $32 = D_3$, and $\overline{3}m = D_{3d}$.
We add the operation of the rotation about a twofold axis normal to the principal axis, represented, *e.g.*, by C_{2x} in Eq. (2.43), to the previous case and obtain [3]

$$\begin{pmatrix} G_{xxxx} & G_{xxyy} & G_{xxzz} & G_{xxyz} & 0 & 0 \\ G_{xxyy} & G_{xxxx} & G_{xxzz} & -G_{xxyz} & 0 & 0 \\ G_{zzxx} & G_{zzxx} & G_{zzzz} & 0 & 0 & 0 \\ G_{yzxx} & -G_{yzxx} & 0 & G_{yzyz} & 0 & 0 \\ 0 & 0 & 0 & 0 & G_{yzyz} & G_{yzxx} \\ 0 & 0 & 0 & 0 & G_{xxyz} & \frac{1}{2}\left(G_{xxxx} - G_{xxyy}\right) \end{pmatrix}. \tag{2.60}$$

(h) Hexagonal system: crystal classes $6 = C_6$, $\overline{6} = C_{3h}$, and $6/m = C_{6h}$.
We start with Eq. (2.59), the result for the trigonal system under (f), and add the symmetry operation of the rotation by an angle $\pm\pi$ about the principal axis, which provides

$$\begin{pmatrix} G_{xxxx} & G_{xxyy} & G_{xxzz} & 0 & 0 & G_{xxxy} \\ G_{xxyy} & G_{xxxx} & G_{xxzz} & 0 & 0 & -G_{xxxy} \\ G_{zzxx} & G_{zzxx} & G_{zzzz} & 0 & 0 & 0 \\ 0 & 0 & 0 & G_{yzyz} & G_{yzzx} & 0 \\ 0 & 0 & 0 & -G_{yzzx} & G_{yzyz} & 0 \\ -G_{xxxy} & G_{xxxy} & 0 & 0 & 0 & \frac{1}{2}\left(G_{xxxx} - G_{xxyy}\right) \end{pmatrix}. \tag{2.61}$$

(i) Hexagonal system: The crystal classes $622 = D_6$, $6mm = C_{6v}$, $\bar{6}2m = D_{3h}$, and $6/mmm = D_{6h}$.
The required result may be obtained by the application of the symmetry operation C_{2x} (the rotation about the twofold secondary axis normal to the principal axis), *i.e.*,

$$\begin{pmatrix} G_{xxxx} & G_{xxyy} & G_{xxzz} & 0 & 0 & 0 \\ G_{xxyy} & G_{xxxx} & G_{xxzz} & 0 & 0 & 0 \\ G_{zzxx} & G_{zzxx} & G_{zzzz} & 0 & 0 & 0 \\ 0 & 0 & 0 & G_{yzyz} & 0 & 0 \\ 0 & 0 & 0 & 0 & G_{yzyz} & 0 \\ 0 & 0 & 0 & 0 & 0 & \frac{1}{2}\left(G_{xxxx} - G_{xxyy}\right) \end{pmatrix}. \tag{2.62}$$

(j) Cubic system: crystal class $23 = T$ and $m3 = T_h$.
Starting with the orthorhombic symmetry case (c), we include the operation C_3 (the rotation about a threefold axis) and obtain

$$\begin{pmatrix} G_{xxxx} & G_{xxyy} & G_{yyxx} & 0 & 0 & 0 \\ G_{yyxx} & G_{xxxx} & G_{xxyy} & 0 & 0 & 0 \\ G_{xxyy} & G_{yyxx} & G_{xxxx} & 0 & 0 & 0 \\ 0 & 0 & 0 & G_{yzyz} & 0 & 0 \\ 0 & 0 & 0 & 0 & G_{yzyz} & 0 \\ 0 & 0 & 0 & 0 & 0 & G_{yzyz} \end{pmatrix}. \tag{2.63}$$

(k) Cubic system: crystal classes $432 = O$, $\bar{4}3m = T_d$, and $m3m = O_h$ [3–8,21]
Adding the operation of the rotation about a twofold axis, *e.g.*, normal to the z axis represented by

$$C_{2a} = \begin{pmatrix} 0 & 1 & 0 \\ 1 & 0 & 0 \\ 0 & 0 & -1 \end{pmatrix}, \tag{2.64}$$

makes the elements G_{xxyy} and G_{yyxx} of the previous case (j) equal to each other. We thus finally have for the case of the highest symmetry among crystalline media

$$\begin{pmatrix} G_{xxxx} & G_{xxyy} & G_{xxyy} & 0 & 0 & 0 \\ G_{xxyy} & G_{xxxx} & G_{xxyy} & 0 & 0 & 0 \\ G_{xxyy} & G_{xxyy} & G_{xxxx} & 0 & 0 & 0 \\ 0 & 0 & 0 & G_{yzyz} & 0 & 0 \\ 0 & 0 & 0 & 0 & G_{yzyz} & 0 \\ 0 & 0 & 0 & 0 & 0 & G_{yzyz} \end{pmatrix}. \tag{2.65}$$

Note that the tensor is characterized by three parameters, *i.e.*, G_{xxxx}, G_{xxyy}, and G_{xyxy}, whereas only two are needed in the originally isotropic medium, where $2G_{yzyz} = G_{xxxx} - G_{xxyy}$, as seen in Eq. (2.24). It is only in the second order in magnetization where the difference between an isotropic medium and crystal of cubic symmetry manifests itself. We recall that the tensors (2.63) and (2.65) pertain to the orientation of the three mutually perpendicular twofold (fourfold) axes parallel to the x, y, and z axes of the Cartesian system.

2.3 Rotation About an Axis

The crystal rotation about an axis in the space affects the form of the electric susceptibility tensor. We illustrate this effect on the cubic crystals belonging to the classes $432 = O$, $\bar{4}3m = T_d$, and $m3m = O_h$ [21]. We assume the electric susceptibility tensor characterized by the first three terms in the development in Eq. (2.28). The susceptibility of a nonmagnetic cubic crystal is a scalar invariant to any spatial rotation. Similarly, the linear magnetooptic tensor is invariant to any spatial rotation. The difference between an isotropic medium and a cubic crystal manifests itself starting from the quadratic magnetooptic tensor given in a Cartesian system with the axes parallel to the fourfold axes of the crystal in Eq. (2.65). Let us rotate the crystal in such a way that a threefold axis is parallel to the z axis. To do this, we first rotate the crystal about the fourfold axis parallel to the z axis by an angle of $\pi/4$, described by the transformation matrix

$$\frac{1}{2^{1/2}}\begin{pmatrix} 1 & 1 & 0 \\ -1 & 1 & 0 \\ 0 & 0 & 2^{1/2} \end{pmatrix}, \tag{2.66}$$

and then we rotate about the new y axis by an angle of $\arccos 3^{-1/2}$, described by

$$\frac{1}{3}\begin{pmatrix} 3^{1/2} & 0 & -6^{1/2} \\ 0 & 3 & 0 \\ 6^{1/2} & 0 & 6^{1/2} \end{pmatrix}. \tag{2.67}$$

The resulting transformation is characterized by

$$\frac{1}{18^{1/2}}\begin{pmatrix} 3^{1/2} & 3^{1/2} & -12^{1/2} \\ -3 & 3 & 0 \\ 6^{1/2} & 6^{1/2} & 6^{1/2} \end{pmatrix}. \tag{2.68}$$

Making use of the transformation rule given in Eq. (2.38c), we obtain for the quadratic MO tensor of the crystal with the threefold axis parallel to the z axis

$$\begin{pmatrix} G'_{xxxx} & G'_{xxyy} & G'_{xxzz} & 0 & -G'_{xyyz} & 0 \\ G'_{xxyy} & G'_{xxxx} & G'_{xxzz} & 0 & G'_{xyyz} & 0 \\ G'_{xxzz} & G'_{xxzz} & G'_{zzzz} & 0 & 0 & 0 \\ 0 & 0 & 0 & G'_{yzyz} & 0 & G'_{xyyz} \\ -G'_{xyyz} & G'_{xyyz} & 0 & 0 & G'_{yzyz} & 0 \\ 0 & 0 & 0 & G'_{xyyz} & 0 & G'_{xyxy} \end{pmatrix}, \tag{2.69}$$

where the nonzero tensor components have the following meaning

$$G'_{xxxx} = \frac{1}{2}\left(G_{xxxx} + G_{xxyy} + 2G_{yzyz}\right) \tag{2.70a}$$

$$G'_{xxyy} = \frac{1}{6}\left(G_{xxxx} + 5G_{xxyy} - 2G_{yzyz}\right) \tag{2.70b}$$

$$G'_{xxzz} = \frac{1}{3}\left(G_{xxxx} + 2G_{xxyy} - 2G_{yzyz}\right) \tag{2.70c}$$

$$G'_{zzzz} = \frac{1}{3}\left(G_{xxxx} + 2G_{xxyy} + 4G_{yzyz}\right) \tag{2.70d}$$

$$G'_{yzyz} = \frac{1}{3}\left(G_{xxxx} - G_{xxyy} + G_{yzyz}\right) \tag{2.70e}$$

$$G'_{xyxy} = \frac{1}{6}\left(G_{xxxx} - G_{xxyy} + 4G_{yzyz}\right) = \frac{1}{2}\left(G_{xxxx} - G_{xxyy}\right) \tag{2.70f}$$

$$G'_{xyyz} = \frac{1}{18^{1/2}}\left(G_{xxxx} - G_{xxyy} - 2G_{yzyz}\right). \tag{2.70g}$$

Under the rotation about the threefold axis described by

$$\begin{pmatrix} \cos\psi_c & \sin\psi_c & 0 \\ -\sin\psi_c & \cos\psi_c & 0 \\ 0 & 0 & 1 \end{pmatrix} \tag{2.71}$$

the tensor transforms as

$$\begin{pmatrix} G'_{xxxx} & G'_{xxyy} & G'_{xxzz} & -G''_{xyzx} & -G''_{xyyz} & 0 \\ G'_{xxyy} & G'_{xxxx} & G'_{xxzz} & G''_{xyzx} & G''_{xyyz} & 0 \\ G'_{xxzz} & G'_{xxzz} & G'_{zzzz} & 0 & 0 & 0 \\ -G''_{xyzx} & G''_{xyzx} & 0 & G'_{yzyz} & 0 & G''_{xyyz} \\ -G''_{xyyz} & G''_{xyyz} & 0 & 0 & G'_{yzyz} & G''_{xyzx} \\ 0 & 0 & 0 & G''_{xyyz} & G''_{xyzx} & G'_{xyxy} \end{pmatrix}, \tag{2.72}$$

where

$$G''_{xyyz} = G'_{xyyz}\cos 3\psi_c, \tag{2.73a}$$

$$G''_{xyzx} = G'_{xyzx}\sin 3\psi_c. \tag{2.73b}$$

Assuming the magnetization fixed parallel to the x axis, the electric susceptibility tensor becomes

$$\begin{pmatrix} \chi_{zz}^{(0)} + \chi_{xx}^{(2)} & 0 & \chi_{zx}^{(2)} \\ 0 & \chi_{zz}^{(0)} + \chi_{yy}^{(2)} & \chi_{(yz)}^{(1)} + \chi_{yz}^{(2)} \\ \chi_{zx}^{(2)} & -\chi_{yz}^{(1)} + \chi_{yz}^{(2)} & \chi_{zz}^{(0)} + \chi_{zz}^{(2)} \end{pmatrix}, \tag{2.74}$$

where

$$\chi_{yz}^{(1)} = -\chi_{zy}^{(1)} = K_{xyz} M_x \tag{2.75a}$$

$$\chi_{xx}^{(2)} = G'_{xxxx} M_x^2, \qquad \chi_{yy}^{(2)} = G'_{xxyy} M_x^2, \qquad \chi_{zz}^{(2)} G'_{xxzz} M_x^2, \tag{2.75b}$$

$$\chi_{zx}^{(2)} = -G'_{xyyz} M_x^2 \cos 3\psi_c, \qquad \chi_{yz}^{(2)} = -G'_{xyzx} M_x^2 \sin 3\psi_c. \tag{2.75c}$$

Next, we investigate the rotation of the crystal about a twofold axis with the magnetization fixed in a particular direction in a plane perpendicular to the twofold axis. We transform a twofold axis of the tensor in Eq. (2.65) to the direction of the z axis. The magnetization vector is confined to the plane normal to the z axis, and its orientation is determined by the spherical polar coordinate angles $\theta_u = \pi/2$ and φ_u. At $\varphi_u = 0$, the magnetization is parallel to the x axis. The contribution to the electric susceptibility tensor originating from the linear MO tensor of the third rank is given by the terms

$$\chi_{yz}^{(1)} = -\chi_{zy}^{(1)} = \chi^{(1)} \cos \varphi_u, \tag{2.76a}$$

$$\chi_{zx}^{(1)} = -\chi_{xz}^{(1)} = \chi^{(1)} \sin \varphi_u. \tag{2.76b}$$

The transformation of the tensor in Eq. (2.65) consists of a sequence of three rotations, the rotation about the z axis by an angle $\varphi_c = \pi/4$, represented by the matrix (2.66), the rotation by an angle of $\theta_c = \pi/2$ about the new y axis

$$\begin{pmatrix} 0 & 0 & 1 \\ 0 & 1 & 0 \\ -1 & 0 & 0 \end{pmatrix}, \tag{2.77}$$

and finally the rotation by an angle of ψ_c about the new z axis

$$\begin{pmatrix} \cos \psi_c & \sin \psi_c & 0 \\ -\sin \psi_c & \cos \psi_c & 0 \\ 0 & 0 & 1 \end{pmatrix}. \tag{2.78}$$

The sequence of the three rotations is represented by the matrix

$$\begin{pmatrix} -2^{-1/2} \sin \psi_c & 2^{-1/2} \sin \psi_c & \cos \psi_c \\ -2^{-1/2} \cos \psi_c & 2^{-1/2} \cos \psi_c & -\sin \psi_c \\ -2^{-1/2} & -2^{-1/2} & 0 \end{pmatrix}. \tag{2.79}$$

The transformed tensor becomes

$$\begin{pmatrix} G'_{xxxx} & G'_{xxyy} & G'_{xxzz} & 0 & 0 & G'_{xxxy} \\ G'_{xxyy} & G'_{yyyy} & G'_{yyzz} & 0 & 0 & G''_{yyxy} \\ G'_{xxzz} & G'_{yyzz} & G'_{zzzz} & 0 & 0 & G''_{yzzx} \\ 0 & 0 & 0 & G'_{yzyz} & G''_{yzzx} & 0 \\ 0 & 0 & 0 & G''_{yzzx} & G'_{zxzx} & 0 \\ G''_{xxxy} & G''_{yyxy} & G''_{yzzx} & 0 & 0 & G'_{xyxy} \end{pmatrix}, \tag{2.80}$$

where

$$G'_{xxxx} = G_{xxxx} - \frac{1}{2}\left(\sin^4\psi_c + 4\cos^2\psi_c\sin^2\psi_c\right)\left(G_{xxxx} - G_{xxyy} - 2G_{yzyz}\right), \tag{2.81a}$$

$$G'_{yyyy} = G_{xxxx} - \frac{1}{2}\left(\cos^4\psi_c + 4\cos^2\psi_c\sin^2\psi_c\right)\left(G_{xxxx} - G_{xxyy} - 2G_{yzyz}\right), \tag{2.81b}$$

$$G'_{zzzz} = \frac{1}{2}\left(G_{xxxx} - G_{xxyy} - 2G_{yzyz}\right), \tag{2.81c}$$

$$G'_{xxyy} = G_{xxyy} + \frac{3}{2}\cos^2\psi_c\sin^2\psi_c\left(G_{xxxx} - G_{xxyy} - 2G_{yzyz}\right), \tag{2.81d}$$

$$G'_{xxzz} = G_{xxyy} - \frac{1}{2}\sin^2\psi_c\left(G_{xxxx} - G_{xxyy} - 2G_{yzyz}\right), \tag{2.81e}$$

$$G'_{yyzz} = G_{xxyy} + \frac{1}{2}\cos^2\psi_c\left(G_{xxxx} - G_{xxyy} - 2G_{yzyz}\right), \tag{2.81f}$$

$$G'_{yzyz} = G_{yzyz} + \frac{1}{2}\cos^2\psi_c\left(G_{xxxx} - G_{xxyy} - 2G_{yzyz}\right), \tag{2.81g}$$

$$G'_{zxzx} = G_{yzyz} + \frac{1}{2}\sin^2\psi_c\left(G_{xxxx} - G_{xxyy} - 2G_{yzyz}\right), \tag{2.81h}$$

$$G'_{xyxy} = G_{yzyz} + \frac{3}{2}\cos^2\psi_c\sin^2\psi_c\left(G_{xxxx} - G_{xxyy} - 2G_{yzyz}\right), \tag{2.81i}$$

$$G'_{xxxy} = \frac{1}{2}\cos\psi_c\sin\psi_c\left(1 - 3\cos^2\psi_c\right)\left(G_{xxxx} - G_{xxyy} - 2G_{yzyz}\right), \tag{2.81j}$$

$$G'_{yyxy} = \frac{1}{2}\cos\psi_c\sin\psi_c\left(1 - 3\sin^2\psi_c\right)\left(G_{xxxx} - G_{xxyy} - 2G_{yzyz}\right), \tag{2.81k}$$

$$G'_{yzzx} = \frac{1}{2}\cos\psi_c\sin\psi_c\left(G_{xxxx} - G_{xxyy} - 2G_{yzyz}\right). \tag{2.81l}$$

For the magnetization oriented perpendicular to the twofold rotation axis and parallel to the x axis, corresponding to $\theta_u = \pi/2$ and $\varphi_u = 0$, the components of the electric susceptibility tensor are given by

$$\chi_{xx}^{(2)} = \left[G_{xxxx} - \frac{1}{2} \left(\sin^4 \psi_c + 4\cos^2 \psi_c \sin^2 \psi_c\right) \right.$$
$$\left. \times \left(G_{xxxx} - G_{xxyy} - 2G_{yzyz}\right)\right] M_x^2, \tag{2.82a}$$

$$\chi_{yy}^{(2)} = \left[G_{xxyy} + \frac{3}{2} \cos^2 \psi_c \sin^2 \psi_c \left(G_{xxxx} - G_{xxyy} - 2G_{yzyz}\right) \right] M_x^2, \tag{2.82b}$$

$$\chi_{zz}^{(2)} = \left[G_{xxyy} - \frac{1}{2} \sin^2 \psi_c \left(G_{xxxx} - G_{xxyy} - 2G_{yzyz}\right) \right] M_x^2, \tag{2.82c}$$

$$\chi_{xy}^{(2)} = \chi_{yx}^{(2)} = \frac{1}{2} \cos \psi_c \sin \psi_c \left(1 - 3\cos^2 \psi_c\right) \left(G_{xxxx} - G_{xxyy} - 2G_{yzyz}\right) M_x^2, \tag{2.82d}$$

$$\chi_{yz}^{(1)} = -\chi_{yz}^{(1)} = K_{xyz} M_x, \quad \chi_{zx}^{(1)} = -\chi_{xz}^{(1)} = 0, \quad \chi_{xy}^{(1)} = -\chi_{yx}^{(1)} = 0. \tag{2.82e}$$

The angular dependence of $\chi_{xy}^{(2)} = \chi_{yx}^{(2)}$ displays maxima at $\psi_c = 25.524$ deg and $\psi_c = 71.345$ deg and crosses the zero level at $\psi_c = 0$, $\arccos 3^{-1/2}$, and $\pi/2$.

For the magnetization oriented parallel to the twofold axis, coinciding with the z axis, specified by $\theta_u = 0$, the components of the electric susceptibility tensor are given by

$$\chi_{xx}^{(2)} = \left[G_{xxyy} - \frac{1}{2} \sin^2 \psi_c \left(G_{xxxx} - G_{xxyy} - 2G_{yzyz}\right) \right] M_x^2, \tag{2.83a}$$

$$\chi_{yy}^{(2)} = \left[G_{xxyy} + \frac{1}{2} \cos^2 \psi_c \left(G_{xxxx} - G_{xxyy} - 2G_{yzyz}\right) \right] M_x^2, \tag{2.83b}$$

$$\chi_{zz}^{(2)} = \left[\frac{1}{2} \left(G_{xxxx} - G_{xxyy} - 2G_{yzyz}\right) \right] M_x^2, \tag{2.83c}$$

$$\chi_{xy}^{(2)} = \chi_{yx}^{(2)} = \frac{1}{2} \cos \psi_c \sin \psi_c \left(G_{xxxx} - G_{xxyy} - 2G_{yzyz}\right) M_x^2, \tag{2.83d}$$

$$\chi_{xy}^{(1)} = -\chi_{yx}^{(1)} = K_{xyz} M_z, \quad \chi_{zx}^{(1)} = -\chi_{xz}^{(1)} = 0, \quad \chi_{yz}^{(1)} = -\chi_{zy}^{(1)} = 0. \tag{2.83e}$$

2.4 Frequency Dependence

So far, we have studied the electric susceptibility tensor from the symmetry point of view. This provides the restrictions or consistency test on the form of the tensor. Before entering the discussion of the microscopic models, we discuss consistency tests useful for the analysis of the spectral dependence of the tensor obtained experimentally or deduced from the models.

2.4.1 Time Invariance

We assume that the response of the system is local, *i.e.*, that it can be described in frame of the electric dipole approximation. We also assume that the response, the electric dipole density, $\boldsymbol{P}(t)$, *i.e.*, the polarization, is linear in $\boldsymbol{E}(t)$, the driving electric field of the wave. This means that the properties of the medium are not affected by $\boldsymbol{E}(t)$. The most general relation between $\boldsymbol{P}(t)$ linear in $\boldsymbol{E}(t)$ can be written as [22]

$$\boldsymbol{P}(t) = \varepsilon_{\text{vac}} \int_{-\infty}^{\infty} \boldsymbol{\chi}(t;t') \cdot \boldsymbol{E}(t')\mathrm{d}t'. \tag{2.84}$$

This expresses that the response $\boldsymbol{P}(t)$ in a fixed time t is determined by the action of $\boldsymbol{E}(t')$ in a time interval $-\infty < t' < \infty$. We follow Butcher and Cotter [22] and investigate what restrictions on this relation can be deduced from the assumption on the time invariance of the medium. According to this assumption, the response cannot depend on our choice of the time when the action of the driving field, $\boldsymbol{E}(t')$, starts. In practice, this ignores, *e.g.*, any aging effect on the investigated object. In the Cartesian tensor notation, Eq. (2.84) becomes

$$P_x(t) = \varepsilon_{\text{vac}} \int_{-\infty}^{\infty} \left[\chi_{xx}(t;t')E_x(t') + \chi_{xy}(t;t')E_y(t') + \chi_{xz}(t;t')E_z(t')\right] \mathrm{d}t', \tag{2.85a}$$

$$P_y(t) = \varepsilon_{\text{vac}} \int_{-\infty}^{\infty} \left[\chi_{yx}(t;t')E_x(t') + \chi_{yy}(t;t')E_y(t') + \chi_{yz}(t;t')E_z(t')\right] \mathrm{d}t', \tag{2.85b}$$

$$P_z(t) = \varepsilon_{\text{vac}} \int_{-\infty}^{\infty} \left[\chi_{zx}(t;t')E_x(t') + \chi_{zy}(t;t')E_y(t') + \chi_{zz}(t;t')E_z(t')\right] \mathrm{d}t'. \tag{2.85c}$$

The complete characteristics of the system are contained in the tensor $\boldsymbol{\chi}(t;t')$, which, for our purpose, is assumed to be linear in the driving field amplitude. If the time is changed by a constant (but otherwise arbitrary) time shift, $t_0 \gtrless 0$, Eq. (2.84) becomes

$$\boldsymbol{P}(t+t_0) = \varepsilon_{\text{vac}} \int_{-\infty}^{\infty} \boldsymbol{\chi}(t+t_0;t'+t_0) \cdot \boldsymbol{E}(t'+t_0)\mathrm{d}t'. \tag{2.86}$$

According to the assumption on *time invariance*, the response remains the same, independent of the position on the time scale of the driving field action,

$$\boldsymbol{\chi}(t;t') = \boldsymbol{\chi}(t+t_0;t'+t_0), \tag{2.87}$$

and Eq. (2.86) may be transformed to

$$P(t+t_0) = \varepsilon_{\text{vac}} \int_{-\infty}^{\infty} \chi(t;t') \cdot E(t'+t_0)\mathrm{d}t'. \tag{2.88}$$

On the other hand, the response of the system at the time $t+t_0$ to the same field $E(t')$ is given by a formal substitution $t \to t+t_0$ in Eq. (2.84)

$$P(t+t_0) = \varepsilon_{\text{vac}} \int_{-\infty}^{\infty} \chi(t+t_0;t') \cdot E(t')\mathrm{d}t'. \tag{2.89}$$

In this equation, the origin of time was shifted with respect to that of Eq. (2.84). The substitution $t' = t'' - t_0$ into Eq. (2.88) provides

$$P(t+t_0) = \varepsilon_{\text{vac}} \int_{-\infty}^{\infty} \chi(t;t''-t_0) \cdot E(t'')\mathrm{d}t''. \tag{2.90}$$

For convenience, we replace $t'' \to t'$

$$P(t+t_0) = \varepsilon_{\text{vac}} \int_{-\infty}^{\infty} \chi(t;t'-t_0) \cdot E(t')\mathrm{d}t' \tag{2.91}$$

and compare Eq. (2.89) and Eq. (2.91) to arrive at the expression for the time invariance

$$\chi(t;t'-t_0) = \chi(t+t_0;t'). \tag{2.92}$$

The response of the system is the same with either the driving field advanced by t_0 or the response determined at the time delayed by t_0. In particular, the uniform shift of time in $\chi(t;t')$ either by $-t$ or by $-t'$ has no effect on its value because of the assumption on time invariance

$$\begin{aligned}\chi(t;t') &= \chi(t-t;t'-t) = \chi(t-t';t'-t') \\ &= \chi(0;t'-t) = \chi(t-t';0).\end{aligned} \tag{2.93}$$

The polarization response tensor linear in the driving field amplitude is therefore a function of the time difference $t-t'$, only. Accordingly, we retain the first argument in $\chi(t;t')$, only, and define

$$\chi(t;t') = \chi(t-t'). \tag{2.94}$$

After the substitution of Eq. (2.94) into Eq. (2.84), we arrive at the linear relation between the driving field and the polarization response with restriction to the time invariant media

$$P(t) = \varepsilon_{\text{vac}} \int_{-\infty}^{\infty} \chi(t-t') \cdot E(t')\mathrm{d}t'. \tag{2.95}$$

2.4.2 Causality

The response $\boldsymbol{P}(t)$ always follows the action of the driving field $\boldsymbol{E}(t')$. Consequently, the response tensor $\chi(t-t')$ must be zero for $t-t'<0$. This requirement forms the *causality condition*. As we have seen, in frame of electric dipole (or local) approximation, the time domain response tensor provides a complete material relation for a medium. We mostly deal with the electromagnetic plane waves of a fixed frequency. It is, therefore, more convenient to proceed to the material characterization in terms of frequency domain response tensors, *i.e.*, in terms of *electric susceptibility tensors*.

The time-dependent $\boldsymbol{E}(t)$ is related to the fields $\boldsymbol{E}(\omega)$ at frequency ω by the Fourier integral transform

$$\boldsymbol{E}(t)=\int_{-\infty}^{\infty}\boldsymbol{E}(\omega)\exp(\mathrm{i}\omega t)\mathrm{d}\omega, \tag{2.96}$$

and its inverse

$$\boldsymbol{E}(\omega)=\frac{1}{2\pi}\int_{-\infty}^{\infty}\boldsymbol{E}(t)\exp(-\mathrm{i}\omega t)\mathrm{d}t. \tag{2.97}$$

Using Eq. (2.96) we shift the origin

$$\boldsymbol{E}(t-t')=\int_{-\infty}^{\infty}\boldsymbol{E}(\omega)\exp[\mathrm{i}\omega(t-t')]\mathrm{d}\omega, \tag{2.98}$$

and substitute this result into Eq. (2.84), which gives

$$\begin{aligned}\boldsymbol{P}(t)&=\varepsilon_{\text{vac}}\int_{-\infty}^{\infty}\chi(t')\cdot\boldsymbol{E}(t-t')\mathrm{d}t'\\&=\varepsilon_{\text{vac}}\int_{-\infty}^{\infty}\chi(t')\cdot\left\{\int_{-\infty}^{\infty}\boldsymbol{E}(\omega)\exp[-\mathrm{i}\omega(t-t')]\mathrm{d}\omega\right\}\mathrm{d}t'.\end{aligned} \tag{2.99}$$

The macroscopic quantities entering this equation are represented by reasonably behaving functions. We can, therefore, exchange the order of integration in these infinitesimal double sums as follows

$$\boldsymbol{P}(t)=\varepsilon_{\text{vac}}\int_{-\infty}^{\infty}\left[\int_{-\infty}^{\infty}\chi(t)\exp(\mathrm{i}t')\,\mathrm{d}t'\right]\cdot\boldsymbol{E}(\omega)\exp(-\mathrm{i}\omega t)\,\mathrm{d}\omega, \tag{2.100}$$

where the electric susceptibility tensor is defined as

$$\chi(\omega)=\int_{-\infty}^{\infty}\chi(t')\exp(-\mathrm{i}\omega t')\,\mathrm{d}t'. \tag{2.101}$$

The tensor $\chi(\omega)$ is a complex quantity, in general, which contains both Hermitian and anti-Hermitian parts. In Eq. (2.100), the vector $\boldsymbol{P}$ can also be Fourier transformed

$$\int_{-\infty}^{\infty} d\omega \boldsymbol{P}(\omega) \exp(-i\omega t) = \varepsilon_{\text{vac}} \int_{-\infty}^{\infty} \chi(\omega) \cdot \boldsymbol{E}(\omega) \exp(-i\omega t)\, d\omega. \quad (2.102)$$

Comparing the integrands on both sides at a fixed frequency ω, we arrive at the relation between the Fourier components of the driving field and the polarization vector

$$\boldsymbol{P}(\omega) = \varepsilon_{\text{vac}} \chi(\omega) \cdot \boldsymbol{E}(\omega). \quad (2.103)$$

Using the assumptions on linearity, time invariance, boundedness, and causality, one can show that the real and imaginary parts of any Cartesian complex component of the electric susceptibility tensor, $\chi_{ij}(\omega) = \chi'_{ij}(\omega) + i\chi''_{ij}(\omega)$, where the subscripts i and j are Cartesian x, y, and z, are interdependent through so-called *dispersion relations*.[3] These can be written as

$$\chi'_{ij}(\omega) = \frac{2}{\pi} \mathcal{P} \int_0^{\infty} \omega' \chi''_{ij}(\omega')(\omega'^2 - \omega^2)^{-1} d\omega' \quad (2.104)$$

and

$$\chi''_{ij}(\omega) = \chi''_{ij}(0) - \frac{2\omega}{\pi} \mathcal{P} \int_0^{\infty} \chi'_{ij}(\omega')(\omega'^2 - \omega^2)^{-1} d\omega'. \quad (2.105)$$

Here, $\mathcal{P}$ denotes the Cauchy principal value integral, and $\omega\varepsilon_{\text{vac}}\chi''_{ij}(0)$ is the ij component of the electric susceptibility tensor at zero frequency.

Their validity is not restricted to the special case of material tensors in a homogeneous medium. The assumptions, of linearity, boundedness, and causality are met in many other physical systems. An important example here is the electromagnetic response in multilayer or in compositionally modulated systems.

2.5 Lorentz-Drude Model

2.5.1 Hamiltonian

The Lorentz model represents the simplest microscopic approach to the medium response to an electromagnetic plane wave. Lorentz soon extended his model to incorporate magnetooptic effects. Here, we are interested in media with spontaneous magnetization, which arises from

[3] The dispersion relations were first obtained by Bode in his development of the circuit theory of linear systems. In the treatment of material constants in electromagnetic optics, they are known as Kramers-Kronig relations.

exchange coupling between quantum particles, *i.e.*, electrons with spin. A rigorous account would require the relativistic quantum theory, which is beyond the reach of the classical model. Nevertheless, the Lorentz model provides a useful insight as it properly accounts for the basically different symmetries of two forces acting on the charged particle. These originate from a spherically symmetric (central) Coulomb potential and uniform magnetic flux density field, respectively.

We therefore consider the motion of an elastically coupled electrically charged point particle situated in a vacuum. The particle moves under the action of the Lorentz force produced by a homogeneous time-constant magnetic field and electric and magnetic fields varying in space and time, which constitute an electromagnetic plane wave. The effect of Coulomb potential is replaced by an elastic coupling produced by the potential with a parabolic dependence on the distance from the origin. The action of applied and internal (or exchange) magnetic fields is represented by a uniform time-independent magnetic flux density field.

The explanation of the classical Lorentz model usually starts from the Newton equation. To illustrate the analogy with the semiclassical model, we start from a single-particle classical Hamiltonian of the system [23]

$$\mathcal{H}(\boldsymbol{r}, \boldsymbol{p}, t) = \frac{1}{2m}[\boldsymbol{p} - q\boldsymbol{A}(\boldsymbol{r}, t)]^2 + qU(\boldsymbol{r}, t) + V(\boldsymbol{r}), \tag{2.106}$$

where $V(\boldsymbol{r}) = V(r)$ denotes the central parabolic potential

$$V(r) = \frac{1}{2}m\omega_0^2 r^2. \tag{2.107}$$

Here $m,\ q,\ \boldsymbol{r}$, and $\boldsymbol{p}$ are the particle mass, charge, coordinate, and momentum, respectively. The corresponding electric and magnetic fields are then obtained from the relations [1,23,24]

$$\boldsymbol{E}(\boldsymbol{r}, t) = -\nabla U(\boldsymbol{r}, t) - \frac{\partial}{\partial t}\boldsymbol{A}(\boldsymbol{r}, t) \tag{2.108a}$$

$$\boldsymbol{B}(\boldsymbol{r}, t) = \nabla \times \boldsymbol{A}(\boldsymbol{r}, t). \tag{2.108b}$$

The gauge transformations

$$U'(\boldsymbol{r}, t) = U(\boldsymbol{r}, t) + \frac{\partial \psi(\boldsymbol{r}, t)}{\partial t}, \tag{2.109a}$$

$$\boldsymbol{A}'(\boldsymbol{r}, t) = \boldsymbol{A}(\boldsymbol{r}, t) - \nabla \psi(\boldsymbol{r}, t), \tag{2.109b}$$

where $\psi(\boldsymbol{r}, t)$ is an arbitrary function of the time t and the position vector $\boldsymbol{r}$. In the temporal gauge, the electric fields are completely characterized by $\boldsymbol{A}(\boldsymbol{r}, t)$, the vector potential and the scalar potential, $U(\boldsymbol{r}, t)$, is set to zero, *i.e.*, $U(\boldsymbol{r}, t) = 0$

$$\boldsymbol{E}(\boldsymbol{r}, t) = -\frac{\partial}{\partial t}\boldsymbol{A}(\boldsymbol{r}, t) \tag{2.110a}$$

$$\boldsymbol{B}(\boldsymbol{r}, t) = \nabla \times \boldsymbol{A}(\boldsymbol{r}, t). \tag{2.110b}$$

We take the vector potential as a sum of two components

$$A(r, t) = A_0(r) + A_1(r, t). \tag{2.111}$$

Here,

$$A_0(r) = -\frac{1}{2} r \times B_0 \tag{2.112}$$

denotes the vector potential that represents the uniform time-independent magnetic flux density field, B_0. An elliptically polarized, transverse monochromatic homogenous electromagnetic plane wave, characterized by a real propagation vector, k, is represented by the vector potential[4]

$$A_1(r, t) = \mathcal{A}_1 \exp[i(k \cdot r - \omega t)] + \mathcal{A}_1^* \exp[-i(k \cdot r - \omega t)], \tag{2.113}$$

where ω and k are the angular frequency and the real propagation vector of the wave, respectively.[5] We can express $\mathcal{A}_1$ as

$$\mathcal{A}_1 = \hat{x} |\mathcal{A}_{1x}| e^{i\delta_x} + \hat{y} |\mathcal{A}_{1y}| e^{i\delta_y} + \hat{z} |\mathcal{A}_{1z}| e^{i\delta_z}. \tag{2.114}$$

In order to represent an elliptically polarized plane wave, A_1 should be a complex vector [25,26]. The wave is transverse in the regions free of charges and currents and satisfies the condition

$$\nabla \cdot A_1(r, t) = 0, \tag{2.115}$$

which means that the component of $\mathcal{A}_1$ parallel to k is zero. The electric field of the wave follows from Eq. (2.110a)

$$\begin{aligned} E_1(r, t) &= i\omega \mathcal{A}_1 \exp[i(k \cdot r - \omega t)] - i\omega \mathcal{A}_1^* \exp[-i(k \cdot r - \omega t)] \\ &= \hat{x}\mathcal{E}_{1x} \cos(k \cdot r - \omega t + \delta_x) + \hat{y}\mathcal{E}_{1y} \cos(k \cdot r - \omega t + \delta_y) \\ &\quad + \hat{z}\mathcal{E}_{1z} \cos(k \cdot r - \omega t + \delta_z), \end{aligned} \tag{2.116}$$

where the electric field amplitudes along the Cartesian unit vectors are

$$\mathcal{E}_{1i} = 2\omega |\mathcal{A}_{1i}|, \tag{2.117}$$

and the corresponding phase shifts at $r = 0$ and $t = 0$ are denoted as δ_i, for $i = x, y$, and z.

The magnetic field of the wave follows from Eq. (2.110b)

$$\begin{aligned} B(r, t) &= ik \times \mathcal{A}_1 \exp[i(k \cdot r - \omega t)] - ik \times \mathcal{A}_1^* \exp[-i(k \cdot r - \omega t)] \\ &= \omega^{-1} k \times E_1(r, t). \end{aligned} \tag{2.118}$$

---

[4] In this chapter, we denote the propagation vector (or the wave vector) by k in agreement with the convention applied in the books on quantum theory.

[5] Note that in the treatment of microscopic models, we use the sign convention for the temporal and space dependence of the waves different from that in Eq. 1.15. See footnote 2 in Chapter 2.

2.5.2 Equations of Motion

We now consider Hamilton equations [23]

$$\frac{\mathrm{d}q_i}{\mathrm{d}t} = \frac{\partial \mathcal{H}}{\partial p_i}, \tag{2.119a}$$

$$\frac{\mathrm{d}p_i}{\mathrm{d}t} = -\frac{\partial \mathcal{H}}{\partial q_i}, \tag{2.119b}$$

where $\mathcal{H}$ is given by Eq. (2.106), q_i and p_i are generalized coordinates and momenta. Here, it is sufficient to consider Cartesian components of q_i and p_i with $i = x$, y, and z. Alternatively, the Hamilton equations can be expressed as

$$\frac{\mathrm{d}\boldsymbol{r}}{\mathrm{d}t} = \nabla_p \mathcal{H}, \tag{2.120a}$$

$$\frac{\mathrm{d}\boldsymbol{p}}{\mathrm{d}t} = -\nabla_r \mathcal{H}, \tag{2.120b}$$

where

$$\nabla_p = \hat{\boldsymbol{x}}\frac{\partial}{\partial p_x} + \hat{\boldsymbol{y}}\frac{\partial}{\partial p_y} + \hat{\boldsymbol{z}}\frac{\partial}{\partial p_z}, \tag{2.121a}$$

$$\nabla_r = \hat{\boldsymbol{x}}\frac{\partial}{\partial x} + \hat{\boldsymbol{y}}\frac{\partial}{\partial y} + \hat{\boldsymbol{z}}\frac{\partial}{\partial z}. \tag{2.121b}$$

Starting from the Hamilton equations (2.119), we arrive at the Lorentz force equation. We first compute the x component,

$$\frac{\mathrm{d}x}{\mathrm{d}t} = \frac{\partial \mathcal{H}(t, \boldsymbol{r})}{\partial p_x} = \frac{1}{m} p_x - \frac{q}{m} A_x(t, \boldsymbol{r}), \tag{2.122}$$

$$\begin{aligned}
\frac{\mathrm{d}p_x}{\mathrm{d}t} &= -\frac{\partial \mathcal{H}(t, \boldsymbol{r})}{\partial x} \\
&= -\frac{1}{2m}\frac{\partial}{\partial x}\left\{[p_x - A_x(t, \boldsymbol{r})]^2 + \left[p_y - A_y(t, \boldsymbol{r})\right]^2 + [p_z - A_z(t, \boldsymbol{r})]^2\right\} \\
&\quad -\frac{\partial}{\partial x} V(r) \\
&= -\frac{1}{2m}\frac{\partial}{\partial x}\left[\left(p_x^2 + p_y^2 + p_z^2\right) - 2q\left(p_x A_x + p_y A_y + p_z A_z\right)\right. \\
&\quad \left. + q^2\left(A_x^2 + A_y^2 + A_z^2\right)\right] - \frac{\partial}{\partial x} V(r) \\
&= \frac{q}{m}\left(p_x \frac{\partial A_x}{\partial x} + p_y \frac{\partial A_y}{\partial x} + p_z \frac{\partial A_z}{\partial x}\right) - q^2\left(A_x \frac{\partial A_x}{\partial x} + A_y \frac{\partial A_y}{\partial y} + A_z \frac{\partial A_z}{\partial z}\right) \\
&\quad -\frac{\partial}{\partial x} V(r),
\end{aligned} \tag{2.123}$$

where $\boldsymbol{r} = \boldsymbol{r}(t)$ is the position vector of the moving particle of mass m carrying the charge q. Next, we eliminate p_x from Eq. (2.123) using Eq. (2.122), *i.e.*, $p_x = m\frac{\mathrm{d}x}{\mathrm{d}t} + q\,A_x(t,\boldsymbol{r})$. With analogous expressions, we also eliminate p_y and p_y to get

$$\begin{aligned}
m\frac{\mathrm{d}^2x}{\mathrm{d}t^2} &+ q\frac{\mathrm{d}A_x}{\mathrm{d}t} \\
&= \frac{q}{m}\left[\left(m\frac{\mathrm{d}x}{\mathrm{d}t} + q\,A_x\right)\frac{\partial A_x}{\partial x} + \left(m\frac{\mathrm{d}y}{\mathrm{d}t} + q\,A_y\right)\frac{\partial A_y}{\partial x} + \left(m\frac{\mathrm{d}z}{\mathrm{d}t} + q\,A_z\right)\frac{\partial A_z}{\partial x}\right] \\
&\quad - \frac{q^2}{m}\left(A_x\frac{\partial A_x}{\partial x} + A_y\frac{\partial A_y}{\partial y} + A_z\frac{\partial A_z}{\partial z}\right) - \frac{\partial}{\partial x}V(r) \\
&= q\left(\frac{\mathrm{d}x}{\mathrm{d}t}\frac{\partial A_x}{\partial x} + \frac{\mathrm{d}y}{\mathrm{d}t}\frac{\partial A_y}{\partial x} + \frac{\mathrm{d}z}{\mathrm{d}t}\frac{\partial A_z}{\partial x}\right) - \frac{\partial}{\partial x}V(r)\,. \qquad (2.124)
\end{aligned}$$

We take into account that the total time derivative of $A_x[t,\boldsymbol{r}(t)]$ can be expressed as

$$\frac{\mathrm{d}A_x}{\mathrm{d}t} = \frac{\partial A_x}{\partial t} + \frac{\partial A_x}{\partial x}\frac{\mathrm{d}x}{\mathrm{d}t} + \frac{\partial A_x}{\partial y}\frac{\mathrm{d}y}{\mathrm{d}t} + \frac{\partial A_x}{\partial z}\frac{\mathrm{d}z}{\mathrm{d}t} \qquad (2.125)$$

and transform Eq. (2.124) to

$$\begin{aligned}
m\frac{\mathrm{d}^2x}{\mathrm{d}t^2} &= -q\frac{\partial A_x}{\partial t} - q\left(\frac{\mathrm{d}x}{\mathrm{d}t}\frac{\partial A_x}{\partial x} + \frac{\mathrm{d}y}{\mathrm{d}t}\frac{\partial A_x}{\partial y} + \frac{\mathrm{d}z}{\mathrm{d}t}\frac{\partial A_x}{\partial z}\right) \\
&\quad + q\left(\frac{\mathrm{d}x}{\mathrm{d}t}\frac{\partial A_x}{\partial x} + \frac{\mathrm{d}y}{\mathrm{d}t}\frac{\partial A_y}{\partial x} + \frac{\mathrm{d}z}{\mathrm{d}t}\frac{\partial A_z}{\partial x}\right) - \frac{\partial V(r)}{\partial x}. \qquad (2.126)
\end{aligned}$$

After rearranging, we find for the x component

$$m\frac{\mathrm{d}^2x}{\mathrm{d}t^2} = -q\frac{\partial A_x}{\partial t} + q\left[\frac{\mathrm{d}y}{\mathrm{d}t}\left(\frac{\partial A_y}{\partial x} - \frac{\partial A_x}{\partial y}\right) - \frac{\mathrm{d}z}{\mathrm{d}t}\left(\frac{\partial A_x}{\partial z} - \frac{\partial A_z}{\partial x}\right)\right] - \frac{\partial V(r)}{\partial x}. \qquad (2.127)$$

The substitution for $\boldsymbol{A}$ according to Eq. (2.110) gives

$$m\frac{\mathrm{d}^2x}{\mathrm{d}t^2} = q\left[E_x + \left(\frac{\mathrm{d}y}{\mathrm{d}t}B_z - \frac{\mathrm{d}z}{\mathrm{d}t}B_y\right)\right] - \frac{\partial V(r)}{\partial x}. \qquad (2.128)$$

We could also start directly from Eq. (2.120). Eq. (2.120a) gives

$$m\frac{\mathrm{d}\boldsymbol{r}}{\mathrm{d}t} = \boldsymbol{p} - q\boldsymbol{A}. \qquad (2.129)$$

Taking the time derivative, we obtain

$$\frac{\mathrm{d}\boldsymbol{p}}{\mathrm{d}t} = m\frac{\mathrm{d}^2\boldsymbol{r}}{\mathrm{d}t^2} + q\frac{\partial \boldsymbol{A}}{\partial t} + q\left(\frac{\mathrm{d}\boldsymbol{r}}{\mathrm{d}t}\cdot\nabla\right)\boldsymbol{A}. \tag{2.130}$$

On the other hand, it follows from Eq. (2.120b)

$$\begin{aligned}\frac{\mathrm{d}\boldsymbol{p}}{\mathrm{d}t} &= -\nabla\left[\frac{1}{2m}\left(\boldsymbol{p}-q\boldsymbol{A}\right)^2\right] - \nabla V(r) \\ &= -\frac{1}{2m}\nabla\left[\boldsymbol{p}^2 - 2q\left(\boldsymbol{p}\cdot\boldsymbol{A}\right) + q^2\boldsymbol{A}^2\right] - \nabla V(r) \\ &= -\frac{1}{2m}\left[-2q\nabla\left(\boldsymbol{p}\cdot\boldsymbol{A}\right) + q^2\nabla\boldsymbol{A}^2\right] - \nabla V(r).\end{aligned} \tag{2.131}$$

After the substitution according to Eq. (2.130), we have

$$\begin{aligned}m\frac{\mathrm{d}^2\boldsymbol{r}}{\mathrm{d}t^2} &= -q\frac{\partial \boldsymbol{A}}{\partial t} - q\left(\frac{\mathrm{d}\boldsymbol{r}}{\mathrm{d}t}\cdot\nabla\right)\boldsymbol{A} + \frac{q}{m}\nabla\left(\boldsymbol{p}\cdot\boldsymbol{A}\right) - \frac{q^2}{2m}\nabla\boldsymbol{A}^2 - \nabla V(r) \\ &= -q\frac{\partial \boldsymbol{A}}{\partial t} - q\left(\frac{\mathrm{d}\boldsymbol{r}}{\mathrm{d}t}\cdot\nabla\right)\boldsymbol{A} \\ &\quad + \frac{q}{m}\left[\boldsymbol{p}\times\left(\nabla\times\boldsymbol{A}\right) + \boldsymbol{A}\times\left(\nabla\times\boldsymbol{p}\right) + \left(\boldsymbol{p}\cdot\nabla\right)\boldsymbol{A} + \left(\boldsymbol{A}\cdot\nabla\right)\boldsymbol{p}\right] \\ &\quad - \frac{q^2}{m}\left[\boldsymbol{A}\times\left(\nabla\times\boldsymbol{A}\right) + \left(\boldsymbol{A}\cdot\nabla\right)\boldsymbol{A}\right] - \nabla V(r) \\ &= -q\frac{\partial \boldsymbol{A}}{\partial t} - q\left(\frac{\mathrm{d}\boldsymbol{r}}{\mathrm{d}t}\cdot\nabla\right)\boldsymbol{A} \\ &\quad + \frac{q}{m}\left\{\left(m\frac{\mathrm{d}\boldsymbol{r}}{\mathrm{d}t} + q\boldsymbol{A}\right)\times\left(\nabla\times\boldsymbol{A}\right) + \left[\left(m\frac{\mathrm{d}\boldsymbol{r}}{\mathrm{d}t} + q\boldsymbol{A}\right)\cdot\nabla\right]\boldsymbol{A}\right\} \\ &\quad - \frac{q^2}{m}\left[\boldsymbol{A}\times\left(\nabla\times\boldsymbol{A}\right) + \left(\boldsymbol{A}\cdot\nabla\right)\boldsymbol{A}\right] - \nabla V(r).\end{aligned} \tag{2.132}$$

This finally gives the condition for the equilibrium between the Newtonian inertial force and the forces originating from the elastic coupling, the magnetic flux density field, and the wave electromagnetic field

$$m\frac{\mathrm{d}^2\boldsymbol{r}}{\mathrm{d}t^2} = -q\frac{\partial \boldsymbol{A}}{\partial t} + \frac{q}{m}\left[m\frac{\mathrm{d}\boldsymbol{r}}{\mathrm{d}t}\times\left(\nabla\times\boldsymbol{A}\right)\right] - \nabla V(r). \tag{2.133}$$

Using Eq. (2.110) for the $\boldsymbol{E}$ and $\boldsymbol{B}$ and Eq. (2.107), the expression for the spherically symmetric potential $V(r)$, and adding the phenomenological viscous damping term

$$\Gamma\frac{\mathrm{d}\boldsymbol{r}}{\mathrm{d}t}, \tag{2.134}$$

we arrive at the dynamic equation for the bound charged point particle in homogeneous magnetic flux density field and in the field of a monochromatic electromagnetic plane wave

$$\frac{d^2\boldsymbol{r}}{dt^2} + \Gamma\frac{d\boldsymbol{r}}{dt} + \omega_0^2\boldsymbol{r} = \frac{q}{m}\left\{\boldsymbol{E}_1(\boldsymbol{r},t) + \frac{d\boldsymbol{r}}{dt} \times [\boldsymbol{B}_0 + \boldsymbol{B}_1(\boldsymbol{r},t)]\right\}. \tag{2.135}$$

2.5.2.1 The Propagation Vector Parallel to Magnetization

Let us consider a special case of circularly polarized electromagnetic plane waves propagating parallel to the z axis of a Cartesian system with the propagation vector $\boldsymbol{k} = \hat{z}\omega/c$ parallel to the applied magnetic flux density field $\boldsymbol{B}_0 = \hat{z}B_0$. We employ the real fields

$$E_{1x} = E_{10}\cos(kz - \omega t), \qquad cB_{1y} = E_{10}\cos(kz - \omega t) \tag{2.136a}$$

$$E_{1y} = E_{10}\cos\left(kz - \omega t \mp \frac{\pi}{2}\right), \quad cB_{1x} = -E_{10}\cos\left(kz - \omega t \mp \frac{\pi}{2}\right)$$

$$E_{1y} = \pm E_{10}\sin(kz - \omega t), \qquad cB_{1x} = \mp E_{10}\sin(kz - \omega t) \tag{2.136b}$$

$$E_{1z} = 0, \qquad cB_{1z} = 0. \tag{2.136c}$$

The upper and lower signs correspond to right and left circularly polarized waves, respectively. From Eq. (2.135), we obtain

$$\frac{d^2x}{dt^2} + \Gamma\frac{dx}{dt} + \omega_0^2 x + 2\omega_L\frac{dy}{dt} = \frac{q}{m}\left(1 - c^{-1}\frac{dz}{dt}\right)E_{1x}, \tag{2.137a}$$

$$\frac{d^2y}{dt^2} + \Gamma\frac{dy}{dt} + \omega_0^2 y - 2\omega_L\frac{dx}{dt} = \frac{q}{m}\left(1 - c^{-1}\frac{dz}{dt}\right)E_{1y}, \tag{2.137b}$$

$$\frac{d^2z}{dt^2} + \Gamma\frac{dz}{dt} + \omega_0^2 z = \frac{q}{mc}\left(\frac{dx}{dt}E_{1x} + \frac{dy}{dt}E_{1y}\right). \tag{2.137c}$$

We have expressed the propagation constant for the electromagnetic plane waves in a vacuum as $k = \omega/c$ and have denoted

$$\omega_L = -\frac{q}{2m}B_0. \tag{2.138}$$

To obtain some insight, we consider the following simplified situation. The particle is in the rest position $\boldsymbol{r} = 0$ when $E_{10} = 0$ and $\omega_L = 0$. Under the action of the circularly polarized wave fields and the static magnetic flux density field, the particle is circulating (after reaching the steady state) with the angular frequency ω on a circular orbit of radii $r_0^{\pm}$ dependent on the wave amplitude and handedness of circular polarization. The orbit plane is normal to the magnetic field vector and to the radiation propagation vector.

Accordingly, we look for the solutions of the equation system (2.137) in the form

$$x^{\pm}(t) = r_0^{\pm} \cos\left(kz - \omega t + \varphi_\Gamma^{\pm}\right), \tag{2.139a}$$

$$\begin{aligned} y^{\pm}(t) &= r_0^{\pm} \cos\left(kz - \omega t + \varphi_\Gamma^{\pm} \mp \frac{\pi}{2}\right) \\ &= \pm r_0^{\pm} \sin\left(kz - \omega t + \varphi_\Gamma^{\pm}\right), \end{aligned} \tag{2.139b}$$

where the angles $\varphi_\Gamma^{\pm}$ represent the phase shift of the particle motion with respect to the fields of RCP and LCP waves. We also get the velocity components for the particle

$$\frac{\mathrm{d}x^{\pm}(t)}{\mathrm{d}t} = \omega r_0^{\pm} \sin\left(kz - \omega t + \varphi_\Gamma^{\pm}\right), \tag{2.140a}$$

$$\begin{aligned} \frac{\mathrm{d}y^{\pm}(t)}{\mathrm{d}t} &= \omega r_0^{\pm} \sin\left(kz - \omega t + \varphi_\Gamma^{\pm} \mp \frac{\pi}{2}\right) \\ &= \mp \omega r_0^{\pm} \cos\left(kz - \omega t + \varphi_\Gamma^{\pm}\right). \end{aligned} \tag{2.140b}$$

Substituting into Eq. (2.137c) according to Eq. (2.136a), Eq. (2.136b), and Eq. (2.139), we obtain

$$\frac{q}{mc}\left(\frac{\mathrm{d}x^{\pm}}{\mathrm{d}t} E_{1x} + \frac{\mathrm{d}y^{\pm}}{\mathrm{d}t} E_{1y}\right) = \frac{q\omega E_{10} r_0^{\pm}}{mc} \sin\varphi_\Gamma^{\pm}, \tag{2.141}$$

i.e., the z component of the force originating from the *circularly* polarized wave of defined handedness acting on the particle is time independent. In the steady state, its only effect is to shift the particle from the rest position. As a result, the steady-state z coordinate of the particle is time independent and the z component of velocity is zero. Substituting the assumed solution of Eq. (2.139) into Eq. (2.137a) and Eq. (2.137b), we find

$$\begin{aligned} &\left(\omega_0^2 - \omega^2 \mp 2\omega\omega_L\right)\left[\cos(kz - \omega t)\cos\varphi_\Gamma^{\pm} - \sin(kz - \omega t)\sin\varphi_\Gamma^{\pm}\right] \\ &+ \omega\Gamma\left[\sin(kz - \omega t)\cos\varphi_\Gamma^{\pm} + \cos(kz - \omega t)\sin\varphi_\Gamma^{\pm}\right] = \frac{qE_{10}}{mr_0}\cos(kz - \omega t), \end{aligned} \tag{2.142a}$$

$$\begin{aligned} &\left(\omega_0^2 - \omega^2 \mp 2\omega\omega_L\right)\left[\sin(kz - \omega t)\cos\varphi_\Gamma^{\pm} + \cos(kz - \omega t)\sin\varphi_\Gamma^{\pm}\right] \\ &- \omega\Gamma\left[\cos(kz - \omega t)\cos\varphi_\Gamma^{\pm} - \sin(kz - \omega t)\sin\varphi_\Gamma^{\pm}\right] = \frac{qE_{10}}{mr_0}\sin(kz - \omega t). \end{aligned} \tag{2.142b}$$

We multiply Eq. (2.142a) by $\cos(kz - \omega t)$ and Eq. (2.142b) by $\sin(kz - \omega t)$. From the sum of Eqs. (2.142a) and (2.142b), we get

$$r_0^{\pm} = \frac{|qE_{10}|}{m\left[\left(\omega_0^2 - \omega^2 \mp 2\omega\omega_L\right)^2 + \omega^2\Gamma^2\right]^{1/2}}. \tag{2.143}$$

Next, we multiply Eq. (2.142a) by $\sin(kz - \omega t)$ and Eq. (2.142b) by $\cos(kz - \omega t)$. The resulting equations have their right-hand sides equal to each other. We therefore compare the left-hand sides and obtain

$$\varphi_\Gamma^\pm = \arctan\left(\frac{\omega\Gamma}{\omega_0^2 - \omega^2 \mp 2\omega\omega_L}\right). \tag{2.144}$$

From the trigonometry formulas $\sin\varphi = \tan\varphi\left(1+\tan^2\varphi\right)^{-1/2}$ and $\cos\varphi = \left(1+\tan^2\varphi\right)^{-1/2}$, we obtain

$$\sin\varphi_\Gamma^\pm = \frac{\omega\Gamma}{\left[\left(\omega_0^2 - \omega^2 \mp 2\omega\omega_L\right)^2 + \omega^2\Gamma^2\right]^{1/2}}, \tag{2.145a}$$

$$\cos\varphi_\Gamma^\pm = \frac{\omega_0^2 - \omega^2 \mp 2\omega\omega_L}{\left[\left(\omega_0^2 - \omega^2 \mp 2\omega\omega_L\right)^2 + \omega^2\Gamma^2\right]^{1/2}}. \tag{2.145b}$$

It is convenient to determine the products

$$r_0^\pm \sin\varphi_\Gamma^\pm = \frac{\omega\Gamma\,|q\,E_{10}|}{m\left[\left(\omega_0^2 - \omega^2 \mp 2\omega\omega_L\right)^2 + \omega^2\Gamma^2\right]}, \tag{2.146a}$$

$$r_0^\pm \cos\varphi_\Gamma^\pm = \frac{\left(\omega_0^2 - \omega^2 \mp 2\omega\omega_L\right)|q\,E_{10}|}{m\left[\left(\omega_0^2 - \omega^2 \mp 2\omega\omega_L\right)^2 + \omega^2\Gamma^2\right]^{1/2}}. \tag{2.146b}$$

These quantities characterize the particle displacement from the equilibrium position under the action of RCP (+) and LCP (−) waves. The displacements in phase with the driving field are given by $r_0^\pm \cos\varphi_\Gamma^\pm$. The $\pi/2$ phase shifted displacements are given by $r_0^\pm \sin\varphi_\Gamma^\pm$. The magnetic field reversal $\omega_L \to -\omega_L$ exchanges the roles of $r_0^\pm$ and $\varphi_\Gamma^\pm$, *i.e.*, $r_0^\pm \leftrightarrow r_0^\mp$ and $\varphi_\Gamma^\pm \leftrightarrow \varphi_\Gamma^\mp$. Eq. (2.137c) predicts a small time-independent shift along the z axis, denoted as $z_0^\pm$

$$z_0^\pm = \frac{\omega q\,E_{10}}{mc\omega_0^2} r_0^\pm \sin\varphi_\Gamma^\pm. \tag{2.147}$$

The shift $z_0^\pm$ changes sign when the propagation vector $\boldsymbol{k} = \hat{\boldsymbol{z}}\omega/c$ is reversed. The magnitude of $z_0^\pm$, expressed with help of Eq. (2.143) and Eq. (2.144),

$$\left|z_0^\pm\right| = \frac{\Gamma(q\omega E_{10})^2}{m^2c\omega_0^2\left[\left(\omega_0^2 - \omega^2 \mp 2\omega\omega_L\right)^2 + \omega^2\Gamma^2\right]}, \tag{2.148}$$

reaches its maximum value in the region of resonance.

In order to keep our presentation simple, we have discussed the problem in terms of real quantities only. The use of complex vector quantities, convenient for the treatment of the problem with vector waves [25], allows a more concise presentation. We shall employ this approach to define the RCP and LCP susceptibilities. We start from Eq. (2.137a) and Eq. (2.137b) for a time-independent z coordinate

$$\frac{d^2x}{dt^2} + \Gamma\frac{dx}{dt} + \omega_0^2 x + 2\omega_L\frac{dy}{dt} = \frac{q}{m}E_{1x}, \tag{2.149a}$$

$$\frac{d^2y}{dt^2} + \Gamma\frac{dy}{dt} + \omega_0^2 y - 2\omega_L\frac{dx}{dt} = \frac{q}{m}E_{1y}. \tag{2.149b}$$

Using the CP driving fields in the complex notation

$$E_{1x} = 2^{-1/2}E_{10}\exp[\mathrm{i}\,(kz - \omega t)], \tag{2.150a}$$

$$E_{1y} = \mp \mathrm{i}2^{-1/2}E_{10}\exp[\mathrm{i}\,(kz - \omega t)], \tag{2.150b}$$

and assuming the solution in the form

$$x = 2^{-1/2}r_0^{\pm}\exp\left(\mathrm{i}\varphi_\Gamma^{\pm}\right)\exp[\mathrm{i}\,(kz - \omega t)], \tag{2.151a}$$

$$y = \mp \mathrm{i}2^{-1/2}r_0^{\pm}\exp\left(\mathrm{i}\varphi_\Gamma^{\pm}\right)\exp[\mathrm{i}\,(kz - \omega t)], \tag{2.151b}$$

we get

$$r_0^{\pm}\exp\left(\mathrm{i}\varphi_\Gamma^{\pm}\right)\left(\hat{\boldsymbol{x}} \mp \mathrm{i}\hat{\boldsymbol{y}}\right) = \frac{\frac{q}{m}E_{10}\left(\hat{\boldsymbol{x}} \mp \mathrm{i}\hat{\boldsymbol{y}}\right)}{\omega_0^2 - \omega^2 \mp 2\omega\omega_L - \mathrm{i}\omega\Gamma}. \tag{2.152}$$

The complex electric dipole moment $\hat{\boldsymbol{p}}^{\pm}$ of the system induced by CP waves follows from the product of the charge q with the displacement $r_0^{\pm}\exp\left(\mathrm{i}\varphi_\Gamma^{\pm}\right)$. The complex electric dipole moment, $\hat{\boldsymbol{P}}^{\pm}$, induced with CP waves of N independent particles per unit volume, can be expressed as

$$\hat{\boldsymbol{P}}^{\pm} = N\hat{\boldsymbol{p}}^{\pm} = \left(\frac{Nq^2}{m}\right)\frac{1}{\omega_0^2 - \omega^2 \mp 2\omega\omega_L - \mathrm{i}\omega\Gamma}E_{10}\left(\hat{\boldsymbol{x}} \mp \mathrm{i}\hat{\boldsymbol{y}}\right)\exp\left[\mathrm{i}\,(kz - \omega t)\right]. \tag{2.153}$$

The complex susceptibilities for RCP and LCP waves are then defined as

$$\chi^{\pm}\left(\omega, \omega_L\right) = \left(\frac{Nq^2}{m\varepsilon_{\mathrm{vac}}}\right)\frac{1}{\omega_0^2 - \omega^2 \mp 2\omega\omega_L - \mathrm{i}\omega\Gamma}, \tag{2.154}$$

where $\varepsilon_{\mathrm{vac}}$ denotes the vacuum permittivity. We observe that the real and imaginary part of this expression correspond to Eq. (2.146b) and Eq. (2.146a), respectively. We have expressed the polarization in SI units, using $\boldsymbol{P} = \varepsilon_{\mathrm{vac}}\chi\boldsymbol{E}$. When the magnetic flux density field is reversed, ω_L changes to $-\omega_L$ and

$$\chi^{\pm}(\omega, \omega_L) = \chi^{\mp}(\omega, -\omega_L). \tag{2.155}$$

We observe that under the action of a circularly polarized electromagnetic plane wave propagating along the direction of the magnetic field vector, the particle orbits on a circle in the plane normal to the magnetic flux density field. The radius of the circle $r_0^{\pm}$ and the phase shift $\varphi_\Gamma^{\pm}$ of the particle motion with respect to the electromagnetic plane wave are different for the two opposite circular polarizations.

2.5.2.2 The Propagation Vector Perpendicular to Magnetization

We next consider the response of the system to a linearly polarized electromagnetic plane wave propagating along the x axis normal to the applied magnetic flux density vector $\hat{z}B_{0z}$. Eq. (2.135) gives

$$\frac{\mathrm{d}^2x}{\mathrm{d}t^2} + \Gamma\frac{\mathrm{d}x}{\mathrm{d}t} + \omega_0^2 x + 2\omega_L\frac{\mathrm{d}y}{\mathrm{d}t} = 0, \tag{2.156a}$$

$$\frac{\mathrm{d}^2y}{\mathrm{d}t^2} + \Gamma\frac{\mathrm{d}y}{\mathrm{d}t} + \omega_0^2 y - 2\omega_L\frac{\mathrm{d}x}{\mathrm{d}t} = \frac{q}{m}E_{1y}, \tag{2.156b}$$

$$\frac{\mathrm{d}^2z}{\mathrm{d}t^2} + \Gamma\frac{\mathrm{d}z}{\mathrm{d}t} + \omega_0^2 z = \frac{q}{m}E_{1z}. \tag{2.156c}$$

For the wave polarized in the z direction

$$E_{1z} = E_{10}^{\parallel}\cos(kx - \omega t), \tag{2.157}$$

we obtain from Eq. (2.156c), assuming the particle velocity much smaller than c,

$$z(t) = z_0\cos(kx - \omega t + \varphi_z), \tag{2.158}$$

which describes a time-harmonic motion parallel to the z axis with the amplitude

$$z_0 = \frac{q}{m}\frac{E_{10}^{\parallel}}{\left[(\omega_0^2 - \omega^2)^2 + \omega^2\Gamma^2\right]^{1/2}} \tag{2.159}$$

and the phase shift

$$\varphi_z = \arctan\left(\frac{\omega\Gamma}{\omega_0^2 - \omega^2}\right). \tag{2.160}$$

There is no magnetic force acting on the particle moving parallel to $\hat{z}B_{0z}$, the magnetic flux density vector.

The electric field of the wave polarized parallel to the y axis

$$E_{1y} = E_{10}^{\perp}\cos(kx - \omega t) \tag{2.161}$$

sets the particle into the motion perpendicular to $\hat{z}B_{0z}$. We transform this real field in a phasor form with $E_{10}^{\perp}$, the real amplitude,

$$\tilde{E}_{1y} = E_{10}^{\perp}\exp[\mathrm{i}(kx - \omega t)], \tag{2.162}$$

and choose the complex-steady state solution of Eq. (2.156) given by

$$\tilde{x}(t) = \tilde{x}_0 \exp[\mathrm{i}(kx - \omega t)] \approx \tilde{x}_0 \mathrm{e}^{-\mathrm{i}\omega t}, \tag{2.163a}$$

$$\tilde{y}(t) = \tilde{y}_0 \exp[\mathrm{i}(kx - \omega t)] \approx \tilde{y}_0 \mathrm{e}^{-\mathrm{i}\omega t}, \tag{2.163b}$$

$$\tilde{z}(t) = \tilde{z}_0 \exp[\mathrm{i}(kx - \omega t)] \approx \tilde{z}_0 \mathrm{e}^{-\mathrm{i}\omega t}, \tag{2.163c}$$

where the tildes emphasize that we now deal with complex quantities. We obtain for the complex amplitudes

$$\tilde{x}_0 = \frac{2\mathrm{i}\omega\omega_L \left(\frac{q}{m}\right) E_{10}^{\perp}}{(\omega_0^2 - \omega^2 - \mathrm{i}\omega\Gamma)^2 - (2\omega\omega_L)^2}, \tag{2.164a}$$

$$\tilde{y}_0 = \frac{(\omega_0^2 - \omega^2 - \mathrm{i}\omega\Gamma)\left(\frac{q}{m}\right) E_{10}^{\perp}}{(\omega_0^2 - \omega^2 - \mathrm{i}\omega\Gamma)^2 - (2\omega\omega_L)^2}, \tag{2.164b}$$

$$\tilde{z}_0 = \frac{\left(\frac{q}{m}\right) E_{10}^{\parallel}}{\omega_0^2 - \omega^2 - \mathrm{i}\omega\Gamma}. \tag{2.164c}$$

At zero magnetic field, $\omega_L = 0$, the particle oscillates parallel to the electric field of the incident wave, parallel either to $\hat{y}$ or $\hat{z}$. The combined action of the magnetic flux density field (parallel to $\hat{z}$) oriented perpendicularly to both the propagation vector (parallel to $\hat{x}$) and the wave electric field (parallel to $\hat{y}$) creates a force that acts in the direction of the propagation vector. This gives rise to an oscillatory motion of the particle in the same direction. Its amplitude is, however, small compared to that of the motion along the electric field polarization direction with a possible exception in the resonance region.

Due to the phase shift, the particle moves along an elliptic trajectory. The major axis of the ellipse is parallel to the electric field vector of the wave; the minor axis is parallel to the propagation vector. The handedness of the motion depends on the polarity of the magnetic flux density field vector, $\boldsymbol{B}_0$.

In a way similar to that used in Eq. (2.154), we can also define the susceptibilities[6] for the driving field parallel $\parallel$ and perpendicular $\perp$ to $\boldsymbol{B}_0$

$$\chi^{\parallel}(\omega, \omega_L) = \left(\frac{Nq^2}{m\varepsilon_{\mathrm{vac}}}\right) \frac{1}{\omega_0^2 - \omega^2 - \mathrm{i}\omega\Gamma}, \tag{2.165a}$$

$$\chi^{\perp}(\omega, \omega_L) = \left(\frac{Nq^2}{m\varepsilon_{\mathrm{vac}}}\right) \frac{(\omega_0^2 - \omega^2 - \mathrm{i}\omega\Gamma)}{(\omega_0^2 - \omega^2 - \mathrm{i}\omega\Gamma)^2 - (2\omega\omega_L)^2}. \tag{2.165b}$$

[6] We ignore a small magnetization-induced electric dipole moment parallel to the propagation vector, which follows from Eq. (2.164a).

This concludes the analysis of the two important geometries with magnetic flux density vector, $\boldsymbol{B}_0$, parallel and perpendicular to the radiation propagation vector.

2.5.3 Susceptibility Tensor

Based on the Lorentz model, we now derive the Cartesian electric susceptibility tensor. We limit ourselves to the situation with the particle velocity much smaller than the electromagnetic wave phase velocity in vacuum, c. Consequently, we can neglect the magnetic effect of the wave, $\boldsymbol{B}_1(\boldsymbol{r}, t) = 0$. We further assume that the region of the particle motion is small compared to the wavelength of the electromagnetic radiation. Then, in the local or electric dipole approximation, we can ignore the spatial dependence of $\boldsymbol{E}_1$ and set $\boldsymbol{k} \cdot \boldsymbol{r} = 0$ in Eq. (2.135), *i.e.*,

$$\frac{\mathrm{d}^2\boldsymbol{r}}{\mathrm{d}t^2} + \Gamma\frac{\mathrm{d}\boldsymbol{r}}{\mathrm{d}t} + \omega_0^2\boldsymbol{r} = \frac{q}{m}\left[\boldsymbol{E}_1(t) + \frac{\mathrm{d}\boldsymbol{r}}{\mathrm{d}t} \times \boldsymbol{B}_0\right]. \qquad (2.166)$$

We obtain in the complex phasor notation

$$\frac{\mathrm{d}^2\hat{x}}{\mathrm{d}t^2} + \Gamma\frac{\mathrm{d}\hat{x}}{\mathrm{d}t} + \omega_0^2\hat{x} - \frac{q}{m}\left(\frac{\mathrm{d}\hat{y}}{\mathrm{d}t}B_{0z} - \frac{\mathrm{d}\hat{z}}{\mathrm{d}t}B_{0y}\right) = \frac{q}{m}\hat{E}_{10x}e^{-\mathrm{i}\omega t}, \qquad (2.167\mathrm{a})$$

$$\frac{\mathrm{d}^2\hat{y}}{\mathrm{d}t^2} + \Gamma\frac{\mathrm{d}\hat{y}}{\mathrm{d}t} + \omega_0^2\hat{y} - \frac{q}{m}\left(\frac{\mathrm{d}\hat{z}}{\mathrm{d}t}B_{0x} - \frac{\mathrm{d}\hat{x}}{\mathrm{d}t}B_{0z}\right) = \frac{q}{m}\hat{E}_{10x}e^{-\mathrm{i}\omega t}, \qquad (2.167\mathrm{b})$$

$$\frac{\mathrm{d}^2\hat{z}}{\mathrm{d}t^2} + \Gamma\frac{\mathrm{d}\hat{z}}{\mathrm{d}t} + \omega_0^2\hat{z} - \frac{q}{m}\left(\frac{\mathrm{d}\hat{x}}{\mathrm{d}t}B_{0y} - \frac{\mathrm{d}\hat{y}}{\mathrm{d}t}B_{0x}\right) = \frac{q}{m}\hat{E}_{10z}e^{-\mathrm{i}\omega t}. \qquad (2.167\mathrm{c})$$

Assuming the solution

$$\hat{\boldsymbol{r}} = \hat{\boldsymbol{r}}_0 e^{-\mathrm{i}\omega t}, \qquad (2.168)$$

we find $\hat{x}_0 = D_x/D_0$, $\hat{y}_0 = D_y/D_0$, and $\hat{z}_0 = D_z/D_0$, where

$$D_0 = \left[L^2 - \left(\frac{\omega q}{m}B_{0x}\right)^2 - \left(\frac{\omega q}{m}B_{0y}\right)^2 - \left(\frac{\omega q}{m}B_{0z}\right)^2\right]L, \qquad (2.169\mathrm{a})$$

$$\begin{aligned} D_x = \frac{q}{m}\Bigg\{&\left[L^2 - \left(\frac{\omega q}{m}B_{0x}\right)^2\right]\hat{E}_{10x} - \left[\mathrm{i}L\frac{\omega q}{m}B_{0z} + \left(\frac{\omega q}{m}\right)^2 B_{0x}B_{0y}\right]\hat{E}_{10y} \\ &- \left[\mathrm{i}L\frac{\omega q}{m}B_{0y} - \left(\frac{\omega q}{m}\right)B_{0x}B_{0z}\right]\hat{E}_{10z}\Bigg\} \end{aligned} \qquad (2.169\mathrm{b})$$

with

$$L = \omega_0^2 - \omega^2 - \mathrm{i}\omega\Gamma. \qquad (2.170)$$

Remaining D_y and D_z are obtained from D_x by the cyclic permutation of indices x, y, and z.

We can now define the Cartesian complex electric susceptibility tensor for the applied magnetic flux density field, $\boldsymbol{B}_0$, of an arbitrary orientation as a relation between the complex amplitude of the electric dipole moment $\hat{\boldsymbol{P}}_{10}(\omega)$ of the N-independent charged particles per unit volume and that of the electric field of the wave $\hat{\boldsymbol{E}}_{10}(\omega)$, *i.e.*,

$$\hat{\boldsymbol{P}}_{10}(\omega) = Nq\hat{\boldsymbol{r}}_0(\omega) = \varepsilon_{\text{vac}}\boldsymbol{\chi}(\omega)\hat{\boldsymbol{E}}_{10}(\omega), \tag{2.171}$$

where we have emphasized that the relation pertains to the Fourier components of the fields. The obtained tensor elements satisfy the Onsager relation (2.29), which can now be expressed as

$$\chi_{ij}(B_{0x}, B_{0y}, B_{0z}) = \chi_{ji}(-B_{0x}, -B_{0y}, -B_{0z}). \tag{2.172}$$

For example, the x component of the complex electric dipole moment amplitude

$$\hat{P}_{10x} = \varepsilon_{\text{vac}}\left(\chi_{xx}\hat{E}_{10x} + \chi_{xy}\hat{E}_{10y} + \chi_{xz}\hat{E}_{10z}\right) \tag{2.173}$$

where

$$\chi_{xx} = \frac{Nq^2}{\varepsilon_{\text{vac}}mD_0}\left[L^2 - \left(\frac{\omega q}{m}B_{0x}\right)^2\right], \tag{2.174a}$$

$$\chi_{xy} = \frac{Nq^2}{\varepsilon_{\text{vac}}mD_0}\left[-\mathrm{i}L\frac{\omega q}{m}B_{0z} - \left(\frac{\omega q}{m}\right)^2 B_{0x}B_{0y}\right], \tag{2.174b}$$

$$\chi_{xz} = \frac{Nq^2}{\varepsilon_{\text{vac}}mD_0}\left[\mathrm{i}L\frac{\omega q}{m}B_{0y} - \left(\frac{\omega q}{m}\right)^2 B_{0z}B_{0x}\right]. \tag{2.174c}$$

For isotropic coupling of the particle, characterized by a single natural frequency ω_0, there is no loss in generality, if we choose the homogenous time independent $\boldsymbol{B}_0$ arbitrarily. We therefore choose $\boldsymbol{B}_0 = \hat{\boldsymbol{z}}B_0$, *i.e.*, parallel to the z axis, and obtain the tensor elements in the following simpler form [27,28]

$$\chi_{xx} = \frac{\left[Nq^2/(\varepsilon_{\text{vac}}m)\right](\omega_0^2 - \omega^2 - \mathrm{i}\omega\Gamma)}{(\omega_0^2 - \omega^2 - \mathrm{i}\omega\Gamma)^2 - 4\omega^2\omega_L^2} = \chi_{yy}, \tag{2.175a}$$

$$\chi_{zz} = \frac{\left[Nq^2/(\varepsilon_{\text{vac}}m)\right]}{\omega_0^2 - \omega^2 - \mathrm{i}\omega\Gamma}, \tag{2.175b}$$

$$\chi_{xy} = \frac{-2\mathrm{i}\omega\omega_L\left[Nq^2/(\varepsilon_{\text{vac}}m)\right]}{(\omega_0^2 - \omega^2 - \mathrm{i}\omega\Gamma)^2 - 4\omega^2\omega_L^2} = -\chi_{yx}, \tag{2.175c}$$

with

$$\chi_{xz} = \chi_{zx} = \chi_{yz} = \chi_{zy} = 0. \tag{2.176}$$

Thus, the susceptibility tensor in the approximation $\frac{1}{c}\frac{d\boldsymbol{r}}{dt} = 0$ and $\boldsymbol{k} \cdot \boldsymbol{r} = 0$ takes the form

$$\underline{\chi} = \begin{pmatrix} \chi_{xx} & \chi_{xy} & 0 \\ -\chi_{xy} & \chi_{xx} & 0 \\ 0 & 0 & \chi_{zz} \end{pmatrix}. \tag{2.177}$$

When the magnetic field is reversed, ω_L changes to $-\omega_L$ and $\chi_{xy} = -\chi_{yx}$ changes to $-\chi_{xy} = \chi_{yx}$, in agreement with the Onsager relation (2.29). The Cartesian tensor elements χ_{xx} and χ_{xy} may be expressed in terms of the RCP and LCP susceptibilities from Eq. (2.154) as

$$\chi_{xx} = (\chi^+ + \chi^-)/2, \tag{2.178a}$$

$$-\mathrm{i}\chi_{xy} = (\chi^+ - \chi^-)/2, \tag{2.178b}$$

and vice versa

$$\chi^{\pm} = \chi_{xx} \mp \mathrm{i}\chi_{xy}. \tag{2.179}$$

This agrees with Eq. (1.8) obtained with the convention $\exp(\mathrm{j}\,\omega t)$, whereas we adopt $\exp(-\mathrm{i}\omega t)$ in this chapter.

2.5.4 Real Dielectrics and Metals

The above single Lorentz oscillator model of electric susceptibility tensor provides a satisfactory picture in simple diamagnetic dielectrics [29]. Magnetic materials, dielectrics or metals, are more adequately characterized by the sum of Lorentz oscillators with different resonant frequencies, ω_{0k}

$$\chi_{xx} = \sum_k f_k \frac{\left(\omega_{0k}^2 - \omega^2 - \mathrm{i}\omega\Gamma_k\right)}{\left(\omega_{0k}^2 - \omega^2 - \mathrm{i}\omega\Gamma_k\right)^2 - 4\omega^2\omega_{Lk}^2} = \chi_{yy}, \tag{2.180a}$$

$$\chi_{zz} = \sum_k \frac{f_k}{\omega_{0k}^2 - \omega^2 - \mathrm{i}\omega\Gamma_k}, \tag{2.180b}$$

$$\chi_{xy} = \sum_k \frac{-2\mathrm{i}\omega\omega_{Lk}}{\left(\omega_{0k}^2 - \omega^2 - \mathrm{i}\omega\Gamma_k\right)^2 - 4\omega^2\omega_{Lk}^2} = -\chi_{yx}. \tag{2.180c}$$

In this mostly phenomenological description, we have allowed for the different weight factors f_k, damping constants Γ_k, and Larmor frequency magnetic splitting ω_{Lk} of individual oscillators. Free electrons in metals are modelled by a Lorentz oscillator with zero restoring force, *i.e.*, with zero resonant frequency. The above sums then contain Drude components

$$\chi_{xx}^{(D)} = -f_D \frac{(\omega + \mathrm{i}\Gamma_D)}{\omega\left[(\omega + \mathrm{i}\Gamma_D)^2 - 4\omega_{LD}^2\right]} = \chi_{yy}, \tag{2.181a}$$

$$\chi_{zz}^{(D)} = \frac{-f_D}{\omega\,(\omega + \mathrm{i}\Gamma_D)}, \tag{2.181b}$$

$$\chi_{xy}^{(D)} = -f_D \frac{2\mathrm{i}\omega_{LD}}{\omega\left[(\omega + \mathrm{i}\Gamma_D)^2 - 4\omega_{LD}^2\right]} = -\chi_{yx}. \tag{2.181c}$$

In particular, real ferromagnetic 3d metals (*i.e.*, iron, cobalt, and nickel), noble metals (*e.g.*, copper, gold, and silver), transition metals (*e.g.*, platinum and palladium), *etc.*, display the behavior of both free and bound electrons. For them, the sum in Eq. (2.180a) to (2.180c) include both Drude and Lorentz components. The possibility to decompose the observed spectra into a sum of their Lorentz and Drude components provides a useful check of their consistency. Moreover, the changes induced in a given material, *e.g.*, by patterning, lithography, material incorporation into a multilayer structure, surface treatment, *etc.*, can be traced in the behavior of these components. This aspect presents interest in the optical spectroscopic metrology of magnetic multilayers and periodic nanostructures.

2.6 Semiclassical Susceptibility

2.6.1 Quantum Hamiltonian

The semiclassical model treats the action of a classical electromagnetic plane wave on a quantum system. The purpose of this section is to provide a quantum susceptibility tensor in magnetic media. We limit ourselves to a nonrelativistic model of a single bound electron and include the concepts of electron spin and spin-orbit coupling. The account for the electron spin and spin-orbit coupling is essential for the explanation of the optical response in media with exchange coupled magnetic moments.

The response of an atomic system to electromagnetic plane waves can be found, *e.g.*, in the books by Cohen-Tannoudji *et al.* [23] and by Butcher and Cotter [22]. To explain the Faraday effect in magnetic compounds, the semiclassical model was used, among others, by Shen [30]. He considered the response to circularly polarized waves of a single member in an ensemble of noninteracting atomic systems. The density matrix formalism allowed him to account, in a phenomenological way, for the final linewidth of electron transitions. The semiclassical model was employed by Kahn *et al.* [18], Crossley *et al.* [32], Wittekoek *et al.* [33], and others.

Our treatment focuses on the response of a system expressed in terms of a Cartesian susceptibility tensor. We consider a single electron coupled in a potential well generated by an isotropic [39], *i.e.*, spherically symmetric (central), electrostatic (Coulomb) potential of a particular (magnetic) cation. Superimposed on the central potential, a much weaker but anisotropic electrostatic potential acting on the electron originates from the nearest neighbor anions regularly arranged around the central magnetic cation. This is the crystal field potential. Next, we have to account for the perturbation originating from electron spin and spin-orbit coupling. The combined action of covalency effects (or electron delocalization) and electron–electron interactions results in exchange interactions that, at sufficiently low temperatures, establish magnetic ordering. To avoid difficulties in the microscopic treatment of exchange interactions, we account for their effect by means of an effective exchange magnetic field. This approximate simplified model reflects the symmetry of the quantum system and provides a useful insight into the microscopic origin of magnetooptic effects in crystalline media. In this way, we avoid complicated problems of many correlated electrons.

The quantum Hamiltonian of an electron in the electrostatic potential $\mathcal{V}(\mathbf{R})$ subjected to applied magnetic flux density and exchange fields, in addition to the field of an electromagnetic plane wave, presents a quantum version of Eq. (2.106) and takes the form [23]

$$\mathcal{H} = \frac{1}{2m}\left[\mathbf{P} - qA(\mathbf{R},t)\right]^2 + \mathcal{V}(\mathbf{R}) - \frac{q}{m}\mathbf{S}\cdot\{\nabla\times[A(\mathbf{R},t) + A_{ex}(\mathbf{R})]\} + \left[\left(2m^2c^2\right)^{-1}\mathbf{S}\times\nabla\mathcal{V}(\mathbf{R})\right]\cdot[\mathbf{P} - qA(\mathbf{R},t)], \tag{2.182}$$

where $\mathbf{R}$ and $\mathbf{P}$ are the vector operators of the electron position and momentum, $\mathbf{P} = -i\hbar\nabla$, and $\mathbf{S}$ represents the vector spin operator with Cartesian components in the two-dimensional space of the electron spin states $\{|\alpha\rangle, |\beta\rangle\}$

$$\mathrm{S}_x = \frac{\hbar}{2}\begin{pmatrix}0 & 1\\ 1 & 0\end{pmatrix}, \quad \mathrm{S}_y = \frac{\hbar}{2}\begin{pmatrix}0 & -\mathrm{i}\\ \mathrm{i} & 0\end{pmatrix}, \quad \mathrm{S}_z = \frac{\hbar}{2}\begin{pmatrix}1 & 0\\ 0 & -1\end{pmatrix}, \tag{2.183}$$

where $\hbar = 1.054589(6)\times10^{-34}$ Joule second is the Planck constant. Here, $m = 9.109\,53(5)\times10^{-31}$kg, $q = -1.602\,189\,(5)\times10^{-19}$ Coulomb, and $c = 2.997\,924\,58(1)\times10^{8}$ m/s denote the electron mass, the electron charge, and the electromagnetic light speed in a vacuum, respectively.

The electromagnetic vector potential $A\,(\mathbf{R},t)$ consists of two parts

$$A(\mathbf{R},t) = A_0(\mathbf{R}) + A_1(\mathbf{R},t). \tag{2.184}$$

Here, the first term

$$A_0(\mathbf{R}) = -\frac{1}{2}\mathbf{R}\times B_0 \tag{2.185}$$

represents the effect of an external applied field of magnetic flux density, B_0. The real vector potential, $A_1(\mathbf{R},t)$, of an elliptically polarized monochromatic electromagnetic plane wave follows from Eq. (2.113),

$$A_1(\mathbf{R},t) = \mathcal{A}_1(\omega,k)\exp[\mathrm{i}\,(k\cdot\mathbf{R} - \omega t)] + \mathcal{A}_1^*(\omega,k)\exp[-\mathrm{i}\,(k\cdot\mathbf{R} - \omega t)], \tag{2.186}$$

where $\mathcal{A}_1$ is the complex amplitude vector, $\boldsymbol{k}$ is the vacuum propagation vector, and ω the angular frequency ($|\boldsymbol{k}| = \omega/c$).[7] We choose the temporal gauge [23,24] with the scalar potential of the electromagnetic plane waves set to zero.

According to Eq. (2.116), the electric field of the wave follows from

$$\begin{aligned} \boldsymbol{E}(\mathbf{R}, t) &= -\frac{\partial}{\partial t}\boldsymbol{A}_1(\mathbf{R}, t) \\ &= \mathrm{i}\omega\left\{\mathcal{A}_1(\omega, \boldsymbol{k})\exp[\mathrm{i}(\boldsymbol{k}\cdot\mathbf{R} - \omega t)] - \mathcal{A}_1^*(\omega, \boldsymbol{k})\exp[-\mathrm{i}(\boldsymbol{k}\cdot\mathbf{R} - \omega t)]\right\} \\ &= \mathcal{E}_1(\omega, \boldsymbol{k})\exp[\mathrm{i}(\boldsymbol{k}\cdot\mathbf{R} - \omega t)] + \mathcal{E}_1^*(\omega, \boldsymbol{k})\exp[-\mathrm{i}(\boldsymbol{k}\cdot\mathbf{R} - \omega t)]. \end{aligned} \tag{2.187}$$

For the corresponding magnetic flux density field of the wave, we have

$$\begin{aligned} \boldsymbol{B}(\mathbf{R}, t) &= \nabla \times \boldsymbol{A}_1(\mathbf{R}, t) \\ &= \mathrm{i}\boldsymbol{k} \times \{\mathcal{A}_1(\omega, \boldsymbol{k})\exp[\mathrm{i}(\boldsymbol{k}\cdot\mathbf{R} - \omega t)] + \mathcal{A}_1^*(\omega, \boldsymbol{k})\exp[-\mathrm{i}(\boldsymbol{k}\cdot\mathbf{R} - \omega t)]\} \\ &= \mathcal{B}_1(\omega, \boldsymbol{k})\exp[\mathrm{i}(\boldsymbol{k}\cdot\mathbf{R} - \omega t)] - \mathcal{B}_1^*(\omega, \boldsymbol{k})\exp[-\mathrm{i}(\boldsymbol{k}\cdot\mathbf{R} - \omega t)] \end{aligned} \tag{2.188}$$

The vector potential

$$\boldsymbol{A}_{ex}(\mathbf{R}) = -\frac{1}{2}\mathbf{R} \times \boldsymbol{B}_{ex} \tag{2.189}$$

represents the effect of the exchange field. Its action is restricted to electron spin magnetic moment. The applied magnetic flux density and exchange fields $\boldsymbol{B}_0$ and $\boldsymbol{B}_{ex}$ are assumed independent of $\mathbf{R}$ and time t. From Eq. (2.185) and Eq. (2.189), we observe that $\nabla \cdot \boldsymbol{A}_0 = 0$ and $\nabla \cdot \boldsymbol{A}_{ex} = 0$. For transverse electromagnetic plane waves, it further follows $\nabla \cdot \boldsymbol{A}_1(\mathbf{R}, t) = 0$, which implies that the commutator

$$[\mathbf{P}, \boldsymbol{A}_1(\mathbf{R}, t)] = 0 \tag{2.190}$$

vanishes. The potential $\mathcal{V}(\mathbf{R})$ consists of the dominating spherically symmetric (isotropic) component, the anisotropic part of the cubic crystal field component, and may contain the components of the lower symmetry crystal fields. The last term in Hamiltonian (2.182) represents the energy of spin-orbit coupling. In this term, we account for the dominating effect of the central potential and completely neglect the anisotropy of the potential $\mathcal{V}(\mathbf{R})$. It will be shown below that we can ignore any effect on the electron motion induced by the external $\boldsymbol{B}_0$ ($B_0 \ll B_{ex}$) included in the term

[7] The sign convention in the exponent has been chosen in agreement with the previous section on classical models and according to the common practice in quantum mechanics. In vector wave optics, microwave engineering, as well as in ellipsometry, the opposite sign convention is perhaps more widely established.

$qA(\mathbf{R}, t)$. The isotropic spin-orbit coupling term provides the coupling between the electron orbital and spin spaces essential for the manifestation of the electron spin ordering effect on the optical response. We have

$$\frac{\mathbf{S} \times \nabla \mathcal{V}(R)}{2m^2c^2} \cdot [\mathbf{P} - qA(\mathbf{R}, t)] \approx \frac{1}{2m^2c^2} \frac{1}{R} \frac{\partial \mathcal{V}(R)}{\partial R} \mathbf{S} \cdot (\mathbf{R} \times \mathbf{P})$$
$$= \frac{1}{2m^2c^2} \frac{1}{R} \frac{\partial \mathcal{V}(R)}{\partial R} \mathbf{L} \cdot \mathbf{S}, \tag{2.191}$$

where $\mathbf{L} = \mathbf{R} \cdot \mathbf{P}$, denotes the angular momentum operator.

The total Hamiltonian (2.182) can be written as a sum $\mathcal{H} = \mathcal{H}_0 + \mathcal{H}_1$ of a one-electron quantum system Hamiltonian $\mathcal{H}_0$ and a Hamiltonian $\mathcal{H}_1$ representing the effect of an electromagnetic plane wave on the quantum system. The time-independent Hamiltonian $\mathcal{H}_0$ becomes

$$\mathcal{H}_0 = \frac{1}{2m} (\mathbf{P} - q\mathbf{A}_0)^2 + \mathcal{V}(\mathbf{R}) - \frac{q}{m} \mathbf{S} \cdot [\nabla \times (A_0 + A_{ex})]$$
$$+ \frac{1}{2m^2c^2} [\mathbf{S} \times \nabla \mathcal{V}(R)] \cdot (\mathbf{P} - qA_0) . \tag{2.192}$$

In the absence of electromagnetic wave perturbation, we therefore have

$$\mathcal{H}_0 \left| u_a \right\rangle = E_a \left| u_a \right\rangle , \tag{2.193}$$

where $|u_a\rangle$ and E_a are the proper functions and proper energies of the Hamiltonian $\mathcal{H}_0$. The time-dependent perturbation Hamiltonian $\mathcal{H}_1$, accounting for the effect of electromagnetic plane waves, takes the form

$$\mathcal{H}_1(t) = -\frac{q}{2m} [A_1(\mathbf{R}, t) \cdot (\mathbf{P} - qA_0) + (\mathbf{P} - qA_0) \cdot A_1(\mathbf{R}, t)]$$
$$-\frac{q}{m} \mathbf{S} \cdot [\nabla \times A_1(\mathbf{R}, t)] - \frac{q}{2m^2c^2} [\mathbf{S} \times \nabla \mathcal{V}(R)] \cdot A_1 + \frac{q^2}{2m} A_1^2. \tag{2.194}$$

2.6.2 Density Matrix Method

To include at least phenomenologically relaxation (damping) effects, we shall employ the density matrix method. The equation of motion for the density matrix operator ρ (Liouville equation) is given by [23]

$$\frac{\partial \rho}{\partial t} = \frac{1}{i\hbar} [\mathcal{H}, \rho] + \left(\frac{\partial \rho}{\partial t} \right)_{\text{damping}} . \tag{2.195}$$

The off-diagonal matrix elements of the damping term are phenomenologically described by [23,30]

$$\left[\left(\frac{\partial \rho}{\partial t} \right)_{\text{damping}} \right]_{ba} = -\gamma_{ba} \langle u_b | \rho | u_a \rangle, \tag{2.196}$$

with a frequency-independent parameter, a damping constant $\gamma_{ba} = \gamma_{ab}$. The total density matrix operator is also considered as a sum

$$\rho = \rho_0 + \rho_1 \tag{2.197}$$

where ρ_0 corresponds to the thermodynamic equilibrium. Thus, in the absence of electromagnetic plane wave excitation, the diagonal matrix elements (populations) are independent of time

$$\frac{\partial}{\partial t}\langle u_a|\rho_0|u_a\rangle = 0 \tag{2.198}$$

and the off-diagonal elements (coherences) are zero, *i.e.*,

$$\langle u_a|\rho_0|u_b\rangle = 0, \quad E_a \neq E_b. \tag{2.199}$$

In the zeroth order approximation, Eq. (2.193), Eq. (2.195), Eq. (2.198) and Eq. (2.199) give

$$[\mathcal{H}_0, \rho_0] = 0. \tag{2.200}$$

To first order in ρ_1 and $\mathcal{H}_1$ we obtain

$$\begin{aligned}\frac{\partial}{\partial t}\langle u_b|\rho_1|u_a\rangle &= \frac{1}{i\hbar}\left[\langle u_b|\mathcal{H}_1\rho_0|u_a\rangle - \langle u_b|\rho_0\mathcal{H}_1|u_a\rangle\right] \\ &+ \frac{1}{i\hbar}\left[\langle u_b|\mathcal{H}_0\rho_1|u_a\rangle - \langle u_b|\rho_1\mathcal{H}_0|u_a\rangle\right] - \gamma_{ba}\langle u_b|\rho_1|u_a\rangle.\end{aligned} \tag{2.201}$$

The matrix elements of the operator products can be written as the products of the matrix elements, *i.e.*,

$$\begin{aligned}\frac{\partial}{\partial t}\langle u_b|\rho_1|u_a\rangle &= \frac{1}{i\hbar}\left\{\sum_c\left[\langle u_b|\mathcal{H}_1|u_c\rangle\langle u_c|\rho_0|u_a\rangle\right]\right\} \\ &- \frac{1}{i\hbar}\left\{\sum_c\left[\langle u_b|\rho_0|u_c\rangle\langle u_c|\mathcal{H}_1|u_a\rangle\right] + \sum_c\left[\langle u_b|\mathcal{H}_0|u_c\rangle\langle u_c|\rho_1|u_a\rangle\right]\right\} \\ &- \frac{1}{i\hbar}\left\{\sum_c\left[\langle u_b|\rho_1|u_c\rangle\langle u_c|\mathcal{H}_0|u_a\rangle\right]\right\} - \gamma_{ba}\langle u_b|\rho_1|u_a\rangle.\end{aligned} \tag{2.202}$$

Taking into account that the off-diagonal matrix elements of the operators ρ_0 and $\mathcal{H}_0$ are zero, we have

$$\begin{aligned}\frac{\partial}{\partial t}\langle u_b|\rho_1|u_a\rangle &= \frac{1}{i\hbar}\{\langle u_b|\mathcal{H}_1|u_a\rangle\langle u_a|\rho_0|u_a\rangle\} - \frac{1}{i\hbar}\{\langle u_b|\mathcal{H}_1|u_a\rangle\langle u_b|\rho_0|u_b\rangle \\ &+ (E_b - E_a)\langle u_b|\rho_1|u_a\rangle\} - \gamma_{ba}\langle u_b|\rho_1|u_a\rangle,\end{aligned} \tag{2.203}$$

which may be written as an ordinary differential equation of time

$$\frac{\mathrm{d}}{\mathrm{d}t}\langle u_b|\rho_1|u_a\rangle + \mathrm{i}\left(\omega_{ba} - \mathrm{i}\gamma_{ba}\right)\langle u_b|\rho_1|u_a\rangle = \frac{1}{i\hbar}\left(\rho_a - \rho_b\right)\langle u_b|\mathcal{H}_1|u_a\rangle, \tag{2.204}$$

where the difference in the energies $\hbar\omega_{ba} = E_b - E_a$ and the populations $\rho_a = \langle u_a|\rho_0|u_a\rangle$ and $\rho_b = \langle u_b|\rho_0|u_b\rangle$. The matrix element on the right-hand side of Eq. (2.204) is according to Eq. (2.194)

$$\begin{aligned}\langle u_b|\mathcal{H}_1|u_a\rangle &= -\{\boldsymbol{\mathcal{A}}_1\cdot\langle u_b|\mathbf{Q}^+|u_a\rangle e^{-i\omega t} + \boldsymbol{\mathcal{A}}_1^*\cdot\langle u_b|\mathbf{Q}^-|u_a\rangle e^{i\omega t}\}\\ &\quad + \frac{q^2}{2m}\{2\boldsymbol{\mathcal{A}}_1\cdot\boldsymbol{\mathcal{A}}_1^*\delta_{ba} + \boldsymbol{\mathcal{A}}_1^2 e^{-2i\omega t}\langle u_b|\exp(2i\boldsymbol{k}\cdot\mathbf{R})|u_a\rangle\\ &\quad + \boldsymbol{\mathcal{A}}_1^{*2}e^{2i\omega t}\langle u_b|\exp(-2i\boldsymbol{k}\cdot\mathbf{R})|u_a\rangle\},\end{aligned} \tag{2.205}$$

where the Hermitian vector operators are defined as

$$\begin{aligned}\mathbf{Q}^{\pm} &= \frac{q}{2m}\left\{\exp(\pm i\boldsymbol{k}\cdot\mathbf{R})\left[(\mathbf{P}-qA_0) + \frac{\mathbf{S}\times\nabla\mathcal{V}(R)}{2mc^2} \pm i\mathbf{S}\times\boldsymbol{k}\right]\right\}\\ &\quad + \frac{q}{2m}\left\{\left[(\mathbf{P}-qA_0) + \frac{\mathbf{S}\times\nabla\mathcal{V}(R)}{2mc^2} \pm i\mathbf{S}\times\boldsymbol{k}\right]\exp(\pm i\boldsymbol{k}\cdot\mathbf{R})\right\}.\end{aligned} \tag{2.206}$$

For the evaluation of the linear response of the system to electromagnetic plane waves, only the terms linear in $\boldsymbol{\mathcal{A}}_1$ are relevant, and we therefore simplify Eq. (2.205) to

$$\langle u_b|\mathcal{H}_1|u_a\rangle = -\{\boldsymbol{\mathcal{A}}_1(\omega,\boldsymbol{k})\cdot\langle u_b|\mathbf{Q}^+|u_a\rangle e^{-i\omega t} + \boldsymbol{\mathcal{A}}_1^*(\omega,\boldsymbol{k})\cdot\langle u_b|\mathbf{Q}^-|u_a\rangle e^{i\omega t}\}. \tag{2.207}$$

The solution of Eq. (2.204) can be expressed as

$$\begin{aligned}\langle u_b|\rho_1|u_a\rangle &= \phi(\omega)e^{-a_0 t} + b_0(a_0 - i\omega)^{-1}e^{-i\omega t} + c_0(a_0 + i\omega)^{-1}e^{i\omega t}\\ &\quad + d_0 a_0^{-1} + f_0(a_0 - 2i\omega)^{-1}e^{-2i\omega t} + g_0(a_0 + 2i\omega)^{-1}e^{2i\omega t}\end{aligned} \tag{2.208}$$

Here,

$$a_0 = i(\omega_{ba} - i\gamma_{ba}) \tag{2.209a}$$

$$b_0 = \frac{i}{\hbar}(\rho_a - \rho_b)\boldsymbol{\mathcal{A}}_1\cdot\langle u_b|\mathbf{Q}^+|u_a\rangle \tag{2.209b}$$

$$c_0 = \frac{i}{\hbar}(\rho_a - \rho_b)\boldsymbol{\mathcal{A}}_1^*\cdot\langle u_b|\mathbf{Q}^-|u_a\rangle \tag{2.209c}$$

$$d_0 = \frac{q^2}{i\hbar m}(\rho_a - \rho_b)\boldsymbol{\mathcal{A}}_1\cdot\boldsymbol{\mathcal{A}}_1^*\delta_{ba} \tag{2.209d}$$

$$f_0 = \frac{q^2}{2i\hbar m}(\rho_a - \rho_b)\boldsymbol{\mathcal{A}}_1^2\langle u_b|\exp(2i\boldsymbol{k}\cdot\mathbf{R})|u_a\rangle \tag{2.209e}$$

$$g_0 = \frac{q^2}{2i\hbar m}(\rho_a - \rho_b)\boldsymbol{\mathcal{A}}_1^{*2}\langle u_b|\exp(-2i\boldsymbol{k}\cdot\mathbf{R})|u_a\rangle. \tag{2.209f}$$

We are concerned with the steady-state solutions for $t \gg \gamma_{ba}$ and the response linear in the amplitude $\boldsymbol{\mathcal{A}}_1$. Then, the first term on the right-hand side of Eq. (2.208) proportional to the time-independent function $\phi(\omega)$ disappears. Retaining in Eq. (2.208) the terms linear in $\mathcal{A}_1$, we have

$$\langle u_b|\rho_1|u_a\rangle = \frac{(\rho_a-\rho_b)}{\hbar}\left\{\mathcal{A}_1(\omega,\mathbf{k})\cdot\frac{\langle u_b|\mathbf{Q}^+|u_a\rangle e^{-\mathrm{i}\omega t}}{\omega_{ba}-\omega-\mathrm{i}\gamma_{ba}}\right.$$
$$\left.+\ \mathcal{A}_1^*(\omega,\mathbf{k})\cdot\frac{\langle u_b|\mathbf{Q}^-|u_a\rangle e^{\mathrm{i}\omega t}}{\omega_{ba}+\omega-\mathrm{i}\gamma_{ba}}\right\}. \tag{2.210}$$

2.6.3 Susceptibility Tensor

The quantum operator of the charge current density is defined [23,31] as

$$\mathbf{I} = \frac{q}{2m}\left\{|r,\epsilon\rangle\langle r,\epsilon|\left[\mathbf{P}-q\mathcal{A}_0-q\mathcal{A}_1(\mathbf{R},t)+(2mc^2)^{-1}\mathbf{S}\times\nabla\mathcal{V}(R)+\frac{2\mathrm{i}}{\hbar}(\mathbf{P}\times\mathbf{S})\right]\right\}$$
$$+\frac{q}{2m}\left\{\left[\mathbf{P}-q\mathcal{A}_0-q\mathcal{A}_1(\mathbf{R},t)+\left(2mc^2\right)^{-1}\mathbf{S}\right.\right.$$
$$\left.\left.\times\nabla\mathcal{V}(R)-\frac{2\mathrm{i}}{\hbar}(\mathbf{P}\times\mathbf{S})\right]|r,\epsilon\rangle\langle r,\epsilon|\right\}, \tag{2.211}$$

where ϵ in $|r,\epsilon\rangle$ distinguishes the two possible electron spin states, α and β in ket-vectors $|r,\alpha\rangle$ and $|r,\beta\rangle$ of the electron. The expectation value of the current density is obtained from the trace $\mathrm{Tr}\{\rho\mathbf{I}\}$ of the matrix product [23]

$$\langle\mathbf{I}\rangle = \mathrm{Tr}\{\rho\mathbf{I}\} = \sum_b\langle u_b|\rho\mathbf{I}|u_b\rangle. \tag{2.212}$$

Then, the Fourier component $\mathcal{I}(\omega,\mathbf{k})$ of the expectation value of charge current density vector, $\langle\mathbf{I}\rangle$, linear in $\mathcal{A}_1$ becomes

$$\mathcal{I}(\omega,\mathbf{k}) = -\frac{q^2}{m}\mathcal{A}_1\sum_b\rho_b+\sum_{a,b}(\rho_a-\rho_b)\frac{\langle u_a|\mathbf{Q}^-|u_b\rangle\langle u_b|\mathbf{Q}^+|u_a\rangle}{\hbar(\omega_{ba}-\omega-\mathrm{i}\gamma_{ba})}\cdot\mathcal{A}_1, \tag{2.213}$$

where $\rho_a = \langle u_a|\rho_0|u_a\rangle$ and $\rho_b = \langle u_b|\rho_0|u_b\rangle$ are the populations of the states $|u_a\rangle$ and $|u_b\rangle$. In the next step, we exchange the indices a and b in the terms proportional to ρ_b, employ $\omega_{ba} = -\omega_{ab}$, and assume $\gamma_{ba} = \gamma_{ab}$. This results in

$$\mathcal{I}(\omega,\mathbf{k}) = -\frac{q^2}{m}\mathcal{A}_1(\omega,\mathbf{k})\sum_a\rho_a+\sum_{a,b}\rho_a\left\{\frac{\langle u_a|\mathbf{Q}^-|u_b\rangle\langle u_b|\mathbf{Q}^+|u_a\rangle}{\hbar(\omega_{ba}-\omega-\mathrm{i}\gamma_{ba})}\cdot\mathcal{A}_1(\omega,\mathbf{k})\right.$$
$$\left.+\frac{\langle u_b|\mathbf{Q}^-|u_a\rangle\langle u_a|\mathbf{Q}^+|u_b\rangle}{\hbar(\omega_{ba}+\omega+\mathrm{i}\gamma_{ba})}\cdot\mathcal{A}_1(\omega,\mathbf{k})\right\}. \tag{2.214}$$

We define the generalized susceptibility tensor components, $\chi_{ij}(\omega,\mathbf{k})$, for N systems in the unit volume using the relation

$$N\mathcal{I}(\omega,\mathbf{k}) = \omega^2\varepsilon_{vac}\boldsymbol{\chi}(\omega,\mathbf{k})\cdot\mathcal{A}_1(\omega,\mathbf{k}), \tag{2.215}$$

and obtain

$$\chi_{ij}(\omega, \mathbf{k}) = \frac{N}{\varepsilon_{vac}\omega^2}\left\{-\sum_a \frac{\delta_{ij}\rho_a q^2}{m} + \sum_{a,b}\frac{\rho_a}{\hbar}\left[\frac{\langle u_a|Q_i^-|u_b\rangle\langle u_b|Q_j^+|u_b\rangle}{\omega_{ba}-\omega-\mathrm{i}\gamma_{ba}}\right.\right.$$
$$\left.\left.+\frac{\langle u_b|Q_i^-|u_a\rangle\langle u_a|Q_j^+|u_b\rangle}{\omega_{ba}+\omega+\mathrm{i}\gamma_{ba}}\right]\right\}, \tag{2.216}$$

where δ_{ij} denotes the Kronecker delta, $\delta_{ij} = 1$ for $i = j$ and ϕ otherwise.

Under the assumption that the region where the electron density of an atom or ion has a significant value is much smaller than the wavelength $2\pi/k$ of the electromagnetic plane wave, the exponential function in $\mathbf{Q}^{\pm}$, Eq. (2.206) can be developed in the power series of $\mathrm{i}\boldsymbol{k}\cdot\mathbf{R}$ as follows

$$\begin{aligned}\mathbf{Q}^{\pm} &\approx \frac{q}{m}\left[\mathbf{P} - qA_0 + \left(2mc^2\right)^{-1}\mathbf{S}\times\nabla\mathcal{V}(R)\right]\\ &\pm \frac{q}{2m}\left\{\mathrm{i}\boldsymbol{k}\cdot\mathbf{R}\left[\mathbf{P} - qA_0 + \left(2mc^2\right)^{-1}\mathbf{S}\times\nabla\mathcal{V}(R)\right]\right\}\\ &\pm \frac{q}{2m}\left\{\left[\mathbf{P} - qA_0 + \left(2mc^2\right)^{-1}\mathbf{S}\times\nabla\mathcal{V}(R)\right]\mathrm{i}\boldsymbol{k}\cdot\mathbf{R} + 2\mathrm{i}\mathbf{S}\times\boldsymbol{k}\right\} + \ldots\end{aligned} \tag{2.217}$$

2.6.3.1 Electric Dipole Contribution

The first term on the right-hand side independent of $\boldsymbol{k}$ represents the electric dipole contribution. The second term represents the sum of magnetic dipole and electric quadrupole contributions, *etc.* For the present purpose, we limit ourselves to the terms with $\boldsymbol{k} = 0$ and those linear in $(\boldsymbol{k}\cdot\mathbf{R})$. For $\boldsymbol{k} = 0$ [31] we have from Eq. (2.217)

$$\mathbf{Q}^{\pm} \equiv \mathbf{Q}_0 = \frac{q}{m}\left(\mathbf{P} - qA_0\right) + q\frac{\mathbf{S}\times\nabla\mathcal{V}(R)}{2m^2c^2}. \tag{2.218}$$

The current density vector in Eq. (2.213) becomes

$$\begin{aligned}\mathcal{I}^{(ED)}(\omega) = &-\sum_b \frac{q^2}{m}\rho_b\langle u_b|\mathcal{A}_1(\omega)|u_b\rangle\\ &+\sum_{a,b}(\rho_a - \rho_b)\frac{\langle u_a|\mathbf{Q}_0|u_b\rangle\langle u_b|\mathbf{Q}_0|u_a\rangle}{\hbar\left(\omega_{ba}-\omega-\mathrm{i}\gamma_{ba}\right)}\cdot\mathcal{A}_1(\omega).\end{aligned} \tag{2.219}$$

On the other hand, the commutator

$$[\mathbf{R}, \mathcal{H}_0] = \frac{\mathrm{i}\hbar}{m}(\mathbf{P} - qA_0) + \frac{\mathrm{i}\hbar}{2m^2c^2}\mathbf{S}\times\nabla\mathcal{V}(R). \tag{2.220}$$

For a generic matrix element, we can write

$$\begin{aligned}\langle u_a\left|[\mathbf{R}, \mathcal{H}_0]\right|u_b\rangle &= \sum_c\left[\langle u_a|\mathbf{R}|u_c\rangle\langle u_c|\mathcal{H}_0|u_b\rangle - \langle u_a|\mathcal{H}_0|u_c\rangle\langle u_c|\mathbf{R}|u_b\rangle\right]\\ &= (E_b - E_a)\langle u_a|\mathbf{R}|u_b\rangle.\end{aligned} \tag{2.221}$$

In the electric dipole approximation, we can replace the matrix elements of $\mathbf{Q}_0$ by those of the position operator $\mathbf{R}$

$$\langle u_a|\mathbf{Q}_0|u_b\rangle = -\mathrm{i}q\,\omega_{ba}\langle u_a|\mathbf{R}|u_b\rangle. \tag{2.222}$$

We first write the electric susceptibility tensor as a special case of Eq. (2.216)

$$\chi_{ij}^{(ED)}(\omega) = -\sum_a \frac{Nq^2}{\varepsilon_{vac}m\omega^2}\rho_a\delta_{ij} + \sum_{a,b}\frac{N}{\varepsilon_{vac}\omega^2}\rho_a\left[\frac{\langle u_a|Q_{0i}|u_b\rangle\langle u_b|Q_{0j}|u_a\rangle}{\hbar\,(\omega_{ba}-\omega-\mathrm{i}\gamma_{ba})} + \frac{\langle u_b|Q_{0i}|u_a\rangle\langle u_a|Q_{0j}|u_b\rangle}{\hbar\,(\omega_{ba}+\omega+\mathrm{i}\gamma_{ba})}\right]. \tag{2.223}$$

From the commutation relation

$$\begin{aligned}\delta_{ij} &= \frac{1}{\mathrm{i}\hbar}\langle u_a|\left[X_i, Q_{0j}\right]|u_a\rangle = \frac{1}{\mathrm{i}\hbar}\langle u_a|\left[X_j, Q_{0i}\right]|u_a\rangle \\ &= \frac{1}{\mathrm{i}\hbar}\sum_b\left[\langle u_a|X_i|u_b\rangle\langle u_b|Q_{0j}|u_a\rangle - \langle u_a|Q_{0j}|u_b\rangle\langle u_b|X_i|u_a\rangle\right] \\ &= \frac{1}{\mathrm{i}\hbar}\sum_b\left[\langle u_a|X_j|u_b\rangle\langle u_b|Q_{0i}|u_a\rangle - \langle u_a|Q_{0i}|u_b\rangle\langle u_b|X_j|u_a\rangle\right].\end{aligned} \tag{2.224}$$

The matrix elements of Q_{0i} and Q_{0j} can be eliminated using Eq. (2.222) and we obtain

$$\delta_{ij} = \frac{m}{\hbar}\sum_b\left\{\omega_{ba}\left[\langle u_a|X_i|u_b\rangle\langle u_b|X_j|u_a\rangle + \langle u_a|X_j|u_b\rangle\langle u_b|X_i|u_a\rangle\right]\right\}. \tag{2.225}$$

We now remove the Kronecker's delta in Eq. (2.223) with Eq. (2.225). The position operator is Hermitian, then $\langle u_a|X_i|u_b\rangle = \langle u_b|X_i|u_a\rangle^*$. We finally obtain

$$\chi_{ij}^{(ED)}(\omega) = \sum_{a,b}\frac{Nq^2\rho_a\omega_{ba}(\omega+\mathrm{i}\gamma_{ba})}{\varepsilon_{vac}\hbar\omega^2} \times\left\{\frac{\langle u_a|X_i|u_b\rangle\langle u_b|X_j|u_a\rangle}{\omega_{ba}-\omega-\mathrm{i}\gamma_{ba}} - \frac{\langle u_a|X_i|u_b\rangle^*\langle u_b|X_j|u_a\rangle^*}{\omega_{ba}+\omega+\mathrm{i}\gamma_{ba}}\right\}. \tag{2.226}$$

In particular, under the assumption $\omega \gg \gamma_{ba}$, *i.e.*, $(\omega+\mathrm{i}\gamma_{ba})^2\omega^{-2} \approx 1$, we can write for the diagonal elements

$$\begin{aligned}\chi_{ii}^{(ED)}(\omega) &= \sum_{a,b}\left\{\frac{2Nq^2\rho_a\omega_{ba}(\omega+\mathrm{i}\gamma_{ba})^2}{\varepsilon_{vac}\hbar\omega^2}\frac{|\langle u_a|X_i|u_b\rangle|^2}{\omega_{ba}^2+\gamma_{ba}^2-\omega^2-2\mathrm{i}\gamma_{ba}\omega}\right\} \\ &\approx \sum_{a,b}\frac{2Nq^2\rho_a\omega_{ba}}{\varepsilon_{vac}\hbar}\frac{|\langle u_a|X_i|u_b\rangle|^2}{\omega_{ba}^2+\gamma_{ba}^2-\omega^2-2\mathrm{i}\gamma_{ba}\omega}.\end{aligned} \tag{2.227}$$

Because of the phenomenologically introduced frequency-independent damping, the formula will not work at very low frequencies, *i.e.*, at $\omega \to 0$.

2.6.3.2 Magnetic Dipole and Electric Quadrupole Contributions

In the optical spectral region, the electric dipole contribution is usually dominating. However, in the case of the zero matrix elements of the operator $\mathbf{Q}_0$ (or $\mathbf{R}$) between initial and final states in the spectral interval of concern, it is meaningful to consider the effect of terms linear in $\boldsymbol{k}$. Neglecting the spin-orbit coupling, we can express them as

$$\begin{aligned} Q_1^{\pm} &= \pm\frac{\mathrm{i}}{2}\,[\boldsymbol{k}\cdot\mathbf{R}(\mathbf{P}-q\boldsymbol{A}_0)+(\mathbf{P}-q\boldsymbol{A}_0)\mathbf{R}\cdot\boldsymbol{k}+2\mathbf{S}\times\boldsymbol{k}] \\ &= \pm\frac{\mathrm{i}}{2}\,\{[\mathbf{R}\times(\mathbf{P}-q\boldsymbol{A}_0)+2\mathbf{S}]\times\boldsymbol{k}+(\mathbf{RP}+\mathbf{PR})\cdot\boldsymbol{k}\}, \end{aligned} \tag{2.228}$$

where the magnetic dipole contribution is represented by

$$\pm\frac{\mathrm{i}}{2}\,[\mathbf{R}\times(\mathbf{P}-q\boldsymbol{A}_0)+2\mathbf{S}]\times\boldsymbol{k} \tag{2.229}$$

and that of electric quadrupole by

$$\pm\frac{\mathrm{i}}{2}(\mathbf{RP}+\mathbf{PR})\cdot\boldsymbol{k}. \tag{2.230}$$

The current density vector due to the electronic transitions of magnetic dipole type may be written as

$$\begin{aligned} \mathbf{I}^{(MD)}(\omega,\boldsymbol{k}) &= \sum_{a,b}\frac{q^2(\rho_a-\rho_b)}{2m^2\hbar} \\ &\times\frac{[\mathrm{i}\boldsymbol{k}\times\langle u_a|\mathbf{R}\times(\mathbf{P}-q\boldsymbol{A}_0)+2\mathbf{S}|u_b\rangle][\langle u_b|\mathbf{R}\times(\mathbf{P}-q\boldsymbol{A}_0)+2\mathbf{S}|u_a\rangle\times\mathrm{i}\boldsymbol{k}]}{\omega_{ba}-\omega-\mathrm{i}\gamma_{ba}}\cdot\boldsymbol{\mathcal{A}}_1(\omega,\boldsymbol{k}). \end{aligned} \tag{2.231}$$

We define the macroscopic magnetic susceptibility tensor as a dimensionless quantity

$$\boldsymbol{M}(\omega,\boldsymbol{k}) = \mu_{vac}^{-1}\boldsymbol{\chi}^{(MD)}(\omega)\,[\mathrm{i}\boldsymbol{k}\times\boldsymbol{\mathcal{A}}_1(\omega,\boldsymbol{k})] \tag{2.232}$$

where $\boldsymbol{M}(\omega,\boldsymbol{k})$ is the magnetic dipole moment per unit volume containing N atomic systems and $\mu_{vac}=(\varepsilon_{vac}c^2)^{-1}$ is the magnetic permeability in a vacuum. The current density due to the moment $\boldsymbol{M}(\omega,\boldsymbol{k})$ is assumed to be

$$N\mathbf{I}^{(MD)}(\omega,\boldsymbol{k}) = \mathrm{i}\boldsymbol{k}\times\boldsymbol{M}(\omega,\boldsymbol{k}). \tag{2.233}$$

This allows us to write for the magnetic susceptibility tensor

$$\chi_{ij}^{(MD)}(\omega) = \sum_{a,b} \left\{ \left[\frac{Nq^2(\rho_a - \rho_b)\mu_{vac}}{4m^2\hbar} \right] \times \frac{\langle u_a | [\mathbf{R} \times (\mathbf{P} - qA_0) + 2\mathbf{S}]_i | u_b \rangle \langle u_b | [\mathbf{R} \times (\mathbf{P} - qA_0) + 2\mathbf{S}]_j | u_a \rangle}{\omega_{ba} - \omega - i\gamma_{ba}} \right\}, \quad (2.234)$$

where i, j stand for the Cartesian coordinates x, y, and z. The magnetic susceptibility at optical frequency is a subject of discussions [30,31,32].

2.6.3.3 Remarks

We have given the expression for the complex generalized susceptibility tensor under the presence of magnetic ordering either induced or spontaneous, Eq. (2.216). From the multipole expansion, the electrical susceptibility tensor (in electric dipole approximation), Eq. (2.226), and magnetic susceptibility tensor (in magnetic dipole approximation), Eq. (2.234), were deduced. In the optical region, we are mainly concerned with the contributions of electric dipole transitions, and we therefore focus on Eq. (2.226). Each term on the right-hand side of this equation represents a contribution to the susceptibility tensor due to a single electron transition from an initial state $|u_a\rangle$ to a final state $|u_b\rangle$, which we denote

$$\chi_{ij}^{(ba)}(\omega, u_a \to u_b, \rho_a). \quad (2.235)$$

We find

$$\chi_{ij}^{(ba)}(\omega, u_a \to u_b, 1) = -\chi_{ij}^{(ab)}(\omega, u_b \to u_a, 1). \quad (2.236)$$

A particular term $\chi_{ij}^{(ba)}(\omega, u_a \to u_b, \rho_a)$ can be nonzero, provided the following three conditions are simultaneously fulfilled:

(a) The difference in the energies of the initial and final states is nonzero, *i.e.*,

$$E_b - E_a = \hbar\omega_{ba} \neq 0. \quad (2.237)$$

(b) The occupation probability of the initial state $|u_a\rangle$ is nonzero, *i.e.*,

$$\rho_a \neq 0. \quad (2.238)$$

(c) The matrix element product is nonzero

$$\langle u_a | X_i | u_b \rangle \langle u_b | X_j | u_a \rangle \neq 0. \quad (2.239)$$

Each pair of states $|u_a\rangle$ and $|u_b\rangle$ enters the sum of Eq. (2.226) twice. The total contribution of the pair is zero unless the corresponding occupation probabilities are different, *i.e.*,

$$\rho_a \neq \rho_b. \quad (2.240)$$

The matrix element product for the diagonal tensor elements, Eq. (2.227), cannot be negative

$$|\langle u_a|X_i|u_b\rangle|^2 \geq 0. \tag{2.241}$$

The product for the off-diagonal elements in the absence of magnetic order is in general real. In the crystals of higher symmetries or at special crystallographic orientations, it may vanish. Under the presence of magnetic ordering, the product is in general complex. In the crystals of higher symmetries and for the special orientations, of the magnetization with respect to the crystallographic axes, the Cartesian system can be chosen in a way that the product (2.239) for some off-diagonal elements is imaginary pure while the product for the remaining ones vanishes.

2.6.4 Proper Values and Proper States

In order to determine the susceptibility tensor, we need the information on the proper values of energy and proper states of the Hamiltonian $\mathcal{H}_0$ of Eq. (2.192). This consists of several terms that differ in magnitude and symmetry. The dominating spherically symmetric part takes the form

$$\mathcal{H}_{\text{ctr}} = \frac{1}{2m}\mathbf{P}^2 + \mathcal{V}_{\text{ctr}}(R), \tag{2.242}$$

where $\mathcal{V}_{\text{ctr}}(R)$ is dependent on the scalar operator of distance R from the origin. In the one-electron approximation, we thus deal with a hydrogen-like system, the solution of which is known. The proper states are characterized by the products

$$\begin{pmatrix} \psi_{nl}\alpha \\ \psi_{nl}\beta \end{pmatrix} = \mathcal{R}_{nl}(R)\, Y_l(\theta,\varphi) \begin{pmatrix} \alpha \\ \beta \end{pmatrix}, \tag{2.243}$$

where $\mathcal{R}_{nl}(R)$ is a radial part of the wave function characterized by the radial and azimuthal quantum numbers n and l, respectively; $Y_l(\theta,\varphi)$ denotes spherical harmonics and α and β distinguish the two proper electron spin states. We shall consider the assembly of remaining terms of $\mathcal{H}_0$ as a perturbation to $\mathcal{H}_{\text{ctr}}$ denoted as $\mathcal{H}_p$

$$\mathcal{H}_p = \mathcal{V}_{cf}(\mathbf{R}) - \frac{q}{m}\mathbf{A}_0\cdot\mathbf{P} + \frac{q}{m}\mathbf{A}_0^2 - \frac{q}{m}\mathbf{S}\cdot[\nabla\times(\mathbf{A}_0 + \mathbf{A}_{\text{ex}})] - \frac{1}{2m^2c^2}[\mathbf{S}\times\nabla\mathcal{V}(R)]\cdot(\mathbf{P} - q\mathbf{A}_0). \tag{2.244}$$

Here, $\mathcal{V}_{cf}(\mathbf{R})$ is the potential energy of a charge q in a crystal field created by surrounding ligands as a function of the position vector operator $\boldsymbol{R}$ of the charge. In oxides with 3d magnetic ions, this term dominates the perturbation. For a charge q in the position $\boldsymbol{r}$ near an ion fixed in the center of a ligand octahedron with the point charges Zq fixed in its vertices, we obtain the potential [36,37]

$$\mathcal{V}_{\text{oct}}(\boldsymbol{r}) = \frac{2Zq^2}{4\pi\varepsilon_{\text{vac}}}\left\{\frac{1}{a}+\frac{1}{b}+\frac{1}{c}+\left(\frac{4\pi}{5}\right)^{1/2} r^2\left[\left(\frac{1}{c^3}-\frac{1}{2a^3}-\frac{1}{2b^3}\right)Y_2^{(0)}\right.\right.$$
$$+\left.\left(\frac{3}{8}\right)^{1/2}\left(\frac{1}{a^3}-\frac{1}{b^3}\right)\left(Y_2^{(2)}+Y_2^{(-2)}\right)\right]$$
$$+\left(\frac{4\pi}{9}\right)^{1/2}\frac{7}{8}r^4\left(\frac{1}{a^5}+\frac{1}{b^5}\right)\left[Y_4^{(0)}+\left(\frac{5}{14}\right)^{1/2}\left(Y_4^{(4)}+Y_4^{(-4)}\right)\right]$$
$$+\left(\frac{4\pi}{9}\right)^{1/2} r^4\left[\left(\frac{1}{c^5}-\frac{1}{2a^5}-\frac{1}{2b^5}\right)Y_4^{(0)}\right.$$
$$\left.\left.-\left(\frac{5}{32}\right)^{1/2}\left(Y_4^{(2)}+Y_4^{(-2)}\right)\right]+\cdots\right\}. \tag{2.245}$$

Here, the octahedron center is in the origin of the Cartesian system and the six point charges Zq are in the positions $\boldsymbol{r}_{1,2} = \pm a\hat{\boldsymbol{x}}$, $\boldsymbol{r}_{3,4} = \pm b\hat{\boldsymbol{y}}$, and $\boldsymbol{r}_{5,6} = \pm c\hat{\boldsymbol{z}}$. We can write $\mathcal{V}_{cf}$ as a sum of the crystal field of a regular octahedron with $a = b = c$, which displays the cubic symmetry

$$\mathcal{V}_{\text{cub}}(\boldsymbol{r}) = \frac{6Zq^2}{4\pi\varepsilon_{\text{vac}}a}$$
$$+\left(\frac{4\pi}{9}\right)^{1/2}\frac{7Zq^2}{8\pi\varepsilon_{\text{vac}}a^5}r^4\left[Y_4^{(0)}+\left(\frac{5}{14}\right)^{1/2}\left(Y_4^{(4)}+Y_4^{(-4)}\right)\right]$$
$$+\left(\frac{4\pi}{13}\right)^{1/2}\frac{3Zq^2}{16\pi\varepsilon_{\text{vac}}a^7}r^6\left[Y_6^0-\left(\frac{7}{2}\right)^{1/2}\left(Y_6^4+Y_6^{-4}\right)\right]+\cdots, \tag{2.246}$$

and the orthorhombic crystal field component (up to an additive constant)

$$\mathcal{V}_{\text{orth}}(\boldsymbol{r}) = \frac{2Zq^2}{4\pi\varepsilon_{\text{vac}}}\left\{\left(\frac{4\pi}{5}\right)^{1/2} r^2\left[\left(\frac{1}{c^3}-\frac{1}{2a^3}-\frac{1}{2b^3}\right)Y_2^{(0)}\right.\right.$$
$$+\left.\left(\frac{3}{8}\right)^{1/2}\left(\frac{1}{a^3}-\frac{1}{b^3}\right)\left(Y_2^{(2)}+Y_2^{(-2)}\right)\right]$$
$$+\left(\frac{4\pi}{9}\right)^{1/2} r^4\left[\left(\frac{1}{c^5}-\frac{1}{2a^5}-\frac{1}{2b^5}\right)Y_4^{(0)}\right.$$
$$\left.\left.-\left(\frac{5}{32}\right)^{1/2}\left(\frac{1}{a^5}-\frac{1}{b^5}\right)\left(Y_4^2+Y_4^{-2}\right)\right]+\cdots\right\}. \tag{2.247}$$

The expression for the crystal field potential $\mathcal{V}_{\text{cub}}$ in Eq. (2.246) can include the orbital states up to $l = 3$, corresponding to f states. The expressions for

$\mathcal{V}_{\text{oct}}$ and $\mathcal{V}_{\text{orth}}$ are restricted to the terms in the developments of the crystal field potential that allow the investigation of the effect of the crystal field on the states with $l \leq 2$ (*i.e.*, on the p and d states) of the central ion.

In practice, we evaluate the splitting of the energy levels in a system from the symmetry of the perturbation only. The reason is that the factors with q^2 in the equation for the crystal field potential are difficult to determine. The same is valid for angularly independent terms. In practical calculations, these factors and terms are replaced by experimentally determined quantities.

Eq. (2.246) indicates that the p states ($l = 1$) are not split in cubic crystal field. The splitting starts for $l \geq 2$. For the evaluation of the octahedral cubic crystal field on d states ($l = 2$), only the second term in Eq. (2.246), proportional to $[Y_4^{(0)} + (5/14)^{1/2}(Y_4^{(4)} + Y_4^{(-4)})]$, is relevant.

We introduce the parameter $\lambda^{(nl)}$ dependent on the quantum numbers n and l and defined as

$$\lambda^{(nl)} = \frac{1}{2m^2c^2}\frac{1}{R}\frac{\mathrm{d}\mathcal{V}(R)}{\mathrm{d}R}. \tag{2.248}$$

This allows us to express the spin-orbit coupling contribution to $\mathcal{H}_p$ in Eq. (2.191) and Eq. (2.244) more concisely as

$$\lambda^{(nl)}\mathbf{L}\cdot\mathbf{S} = \lambda^{(nl)}\left[\mathrm{L}_z\mathrm{S}_z + \frac{1}{2}\left(\mathrm{L}_+\mathrm{S}_- + \mathrm{L}_-\mathrm{S}_+\right)\right], \tag{2.249}$$

where $\mathrm{L}_\pm = \mathrm{L}_x \pm \mathrm{i}\mathrm{L}_y$ and $\mathrm{S}_\pm = \mathrm{S}_x \pm \mathrm{i}\mathrm{S}_y$. In the calculation of the operator matrix elements, we shall make use of the following relations between the proper functions of the orbital and spin angular momentum operators $\mathbf{L}$ and $\mathbf{S}$ valid for the total angular momentum operator $\mathbf{J}$

$$\mathrm{J}^2\,|k, j, m\rangle = j\,(j+1)\hbar^2\,|k, j, m\rangle\,, \tag{2.250a}$$

$$\mathrm{J}_z\,|k, j, m\rangle = m\hbar\,|k, j, m\rangle\,, \tag{2.250b}$$

$$\mathrm{J}_+\,|k, j, m\rangle = \hbar\,[j\,(j+1) - m\,(m+1)]^{1/2}\,|k, j, m+1\rangle\,, \tag{2.250c}$$

$$\mathrm{J}_-\,|k, j, m\rangle = \hbar\,[j\,(j+1) - m\,(m-1)]^{1/2}\,|k, j, m-1\rangle\,. \tag{2.250d}$$

Here, the index k distinguishes the different proper vectors with equal values of $j(j+1)\hbar^2$ and $m\hbar$. We limit ourselves to the so-called Zeeman approximation. This ignores the diamagnetic term proportional to A_0^2. We transform the perturbation to

$$\begin{aligned}\mathcal{H}_p &= -\frac{q}{m}\boldsymbol{A}_0\cdot\mathbf{P} + \mathcal{V}_{cf}\,(\mathbf{R}) - \frac{q}{m}\mathbf{S}\cdot[\nabla\times(\boldsymbol{A}_0 + \boldsymbol{A}_{\text{ex}})] \\ &\quad - \frac{1}{2m^2c^2}\,[\mathbf{S}\times\nabla\mathcal{V}\,(R)]\cdot(\mathbf{P} - q\boldsymbol{A}_0) \\ &= -\frac{\mu_B}{\hbar}\mathbf{L}\cdot\boldsymbol{B}_0 + \mathcal{V}_{\text{cf}}\,(\mathbf{R}) - \frac{2\mu_B}{\hbar}\mathbf{S}\cdot(\boldsymbol{B}_0 + \boldsymbol{B}_{\text{ex}}) + \lambda^{(nl)}\mathbf{L}\cdot\mathbf{S},\end{aligned} \tag{2.251}$$

where the external field of magnetic flux density and exchange field are related to the corresponding vector potentials by $\boldsymbol{B}_0 = \nabla \times \boldsymbol{A}_0$, and $\boldsymbol{B}_{\text{ex}} = \nabla \times \boldsymbol{A}_{\text{ex}}$. Here,

$$\mu_B = \frac{q\hbar}{2m} = -9.27408(4) \times 10^{-24}\text{JouleTesla}^{-1} \tag{2.252}$$

denotes the Bohr magneton. The effect of the external and exchange fields can be expressed, as in the case of Eq. (2.138), with help of the angular frequencies

$$\omega_L = \frac{-q\,B_0}{2m}, \qquad \omega_{\text{ex}}^{(nl)} = \frac{-q\,B_{\text{ex}}}{2m}, \tag{2.253}$$

characterizing the strength of applied external and exchange fields, respectively.

The perturbation Hamiltonian takes a more compact form

$$\mathcal{H}_p = \mathcal{V}_{\text{cf}}\,(\mathbf{R}) + +\lambda^{(nl)}\mathbf{L}\cdot\text{S} + \omega_L \text{L}_u + 2\left(\omega_L + \omega_{\text{ex}}^{(nl)}\right)\text{S}_u, \tag{2.254}$$

where L_u and S_u are the components of the orbital and spin angular momenta parallel to the unit vector $\hat{\boldsymbol{u}}$, which determines the orientation of $\boldsymbol{B}_0$ and $\boldsymbol{B}_{\text{ex}}$,

$$\hat{\boldsymbol{u}} = \hat{\boldsymbol{x}}\sin\theta_u\cos\varphi_u + \hat{\boldsymbol{y}}\sin\theta_u\sin\varphi_u + \hat{\boldsymbol{z}}\cos\theta_u. \tag{2.255}$$

Then, the component L_u of the orbital angular momentum operator in the direction $\hat{\boldsymbol{u}}$ becomes

$$\text{L}_u = \mathbf{L}\cdot\hat{\boldsymbol{u}} = \text{L}_x\sin\theta_u\cos\varphi_u + \text{L}_y\sin\theta_u\sin\varphi_u + \text{L}_z\cos\theta_u, \tag{2.256}$$

where the polar spherical angles θ_u and φ_u specify the direction of $\hat{\boldsymbol{u}}$, and

$$\text{L}_x = \frac{1}{2}\left(\text{L}_+ + \text{L}_-\right), \qquad \text{L}_x = -\text{i}\frac{1}{2}\left(\text{L}_+ - \text{L}_-\right). \tag{2.257}$$

For $l = 1$, corresponding to the space of the p states, the operator L_u may be represented by the matrix

$$\text{L}_\text{u} = \hbar\begin{pmatrix} \cos\theta_u & \frac{1}{\sqrt{2}}\sin\theta_u \text{e}^{-\text{i}\varphi_u} & 0 \\ \frac{1}{\sqrt{2}}\sin\theta_u \text{e}^{\text{i}\varphi_u} & 0 & \frac{1}{\sqrt{2}}\sin\theta_u \text{e}^{-\text{i}\varphi_u} \\ 0 & \frac{1}{\sqrt{2}}\sin\theta_u \text{e}^{\text{i}\varphi_u} & \cos\theta_u \end{pmatrix} \tag{2.258}$$

for the order in the sequence of the states with $m_l = 1, 0 - 1$, where m_l denotes the magnetic quantum number. The component S_u of the electron spin angular momentum in the direction $\hat{\boldsymbol{u}}$ becomes

$$\text{S}_u = \mathbf{S}\cdot\hat{\boldsymbol{u}} = \text{S}_x\sin\theta_u\cos\varphi_u + \text{S}_y\sin\theta_u\sin\varphi_u + \text{S}_z\cos\theta_u, \tag{2.259}$$

and the matrix representation follows from Eq. (2.183)

$$\mathsf{S}_u = \frac{\hbar}{2}\begin{pmatrix} \cos\theta_u & \sin\theta_u \exp(-\mathrm{i}\varphi_u) \\ \sin\theta_u \exp(\mathrm{i}\varphi_u) & -\cos\theta_u \end{pmatrix}, \tag{2.260}$$

for the order in the sequence of the states $\langle\alpha| = \langle 1/2, +1/2|$ and $\langle\beta| = \langle 1/2, -1/2|$ with the electron spin projections $s_z = \pm 1/2$.

The energy of the exchange field can be evaluated with the help of the Curie temperature T_C. For $T_C \approx 1400\,\mathrm{K}$, we have $kT_C \lesssim 0.12\,\mathrm{eV}$. This energy is still small compared to the level of splitting by the cubic crystal field, which is of the order of 1 eV, *i.e.*, of the same order in magnitude as photon energies $\hbar\omega$ in the optical region. For the external magnetic flux densities of the order of 1 Tesla, the effect of $\boldsymbol{A}_0$ on the proper values of $\mathcal{H}_{\mathrm{ctr}}$ is smaller by three orders of magnitude than the effect of the exchange field $\boldsymbol{A}_{\mathrm{ex}}$ and can usually be neglected. Well below T_C, the main role of the external field $\boldsymbol{B}_0$ consists of the ordering of exchange coupled magnetic moments. In most cases, its effect on the susceptibility tensor may be ignored (see p. 78).

Both the radial and angular parts of the wave functions are normalized to unity. The spherical harmonics are given by

$$Y_0^{(0)} = (4\pi)^{-1/2}, \tag{2.261a}$$

$$Y_1^{(0)} = \left(\frac{3}{4\pi}\right)^{1/2} \cos\theta, \tag{2.261b}$$

$$Y_1^{(\pm 1)} = \mp\left(\frac{3}{8\pi}\right)^{1/2} \sin\theta \exp(\pm\mathrm{i}\varphi), \tag{2.261c}$$

$$Y_2^{(0)} = \left(\frac{5}{16\pi}\right)^{1/2} \left(3\cos^2\theta - 1\right), \tag{2.261d}$$

$$Y_2^{(\pm 1)} = \mp\left(\frac{15}{8\pi}\right)^{1/2} \sin\theta\cos\theta \exp(\pm\mathrm{i}\varphi), \tag{2.261e}$$

$$Y_2^{(\pm 2)} = \mp\left(\frac{15}{32\pi}\right)^{1/2} \sin^2\theta \exp(\pm 2\mathrm{i}\varphi). \tag{2.261f}$$

The classification of the orbital states in the cubic and orthorhombic crystal field is summarized in Table 2.1 [36]. To find the susceptibility tensor, we evaluate the proper states constructed from s and p states.

2.6.4.1 *s* States

The crystal field produces a uniform shift of energy levels; the spin-orbit coupling does not affect these states. The exchange field removes the spin degeneracy. In Eq. (2.251) it is sufficient to consider the term

$$-\frac{q}{m}\mathbf{S}\cdot[\nabla\times(\boldsymbol{A}_0 + \boldsymbol{A}_{\mathrm{ex}})] = 2\left(\omega_L + \omega_{\mathrm{ex}}^{(s)}\right)\mathsf{S}_u. \tag{2.262}$$

TABLE 2.1

Classification of the orbital states in central field and in crystal field of cubic octahedral and orthorhombic symmetry

O(3)	O_h	D_{2h}			
s	A_{1g}	A	$\|s\rangle$	$Y_0^{(0)}$	$(4\pi)^{-1/2}$
		B_3'	$\|x\rangle$	$2^{-1/2}\left(Y_1^{(-1)} - Y_1^{(1)}\right)$	$\left(\frac{3}{4\pi}\right)^{1/2} \sin\theta\cos\varphi$
p	T_{1u}	B_2'	$\|y\rangle$	$\mathrm{i}2^{-1/2}\left(Y_1^{(-1)} + Y_1^{(1)}\right)$	$\left(\frac{3}{4\pi}\right)^{1/2} \sin\theta\sin\varphi$
		B_1'	$\|z\rangle$	$Y_1^{(0)}$	$\left(\frac{3}{4\pi}\right)^{1/2} \cos\theta$
	E_g	A	$\left\|3z^2 - r^2\right\rangle$	$Y_2^{(0)}$	$\left(\frac{5}{4\pi}\right)^{1/2}\left(\frac{1}{4}\right)^{1/2}\left(3\cos^2 - 1\right)$
		A	$\left\|x^2 - y^2\right\rangle$	$2^{-1/2}\left(Y_2^{(-2)} + Y_2^{(2)}\right)$	$\left(\frac{5}{4\pi}\right)^{1/2}\left(\frac{3}{4}\right)^{1/2} \sin^2\theta\left(\cos^2\varphi - \sin^2\varphi\right)$
d		B_3	$\|yz\rangle$	$\mathrm{i}2^{-1/2}\left(Y_2^{(-1)} + Y_2^{(1)}\right)$	$\left(\frac{5}{4\pi}\right)^{1/2} 3^{1/2}\cos\theta\sin\theta\sin\varphi$
	T_{2g}	B_2	$\|zx\rangle$	$2^{-1/2}\left(Y_2^{(-1)} - Y_2^{(1)}\right)$	$\left(\frac{5}{4\pi}\right)^{1/2} 3^{1/2}\cos\theta\sin\theta\cos\varphi$
		B_1	$\|xy\rangle$	$\mathrm{i}2^{-1/2}\left(Y_2^{(-2)} + Y_2^{(2)}\right)$	$\left(\frac{5}{4\pi}\right)^{1/2} 3^{1/2}\sin^2\theta\cos\varphi\sin\varphi$

The proper values of energy are $E = E^{(s)} \pm \left(\omega_L + \omega_{\text{ex}}^{(s)}\right)$ where E_s is the proper value of energy at zero external and exchange fields. The proper states for a general orientation of the exchange field can be written as

$$|\psi_+\rangle = \cos(\theta_u/2)\exp(-\mathrm{i}\varphi_u/2)\,|s\alpha\rangle + \sin(\theta_u/2)\exp(\mathrm{i}\varphi_u/2)\,|s\beta\rangle\,, \quad (2.263a)$$

$$|\psi_-\rangle = -\sin(\theta_u/2)\exp(-\mathrm{i}\varphi_u/2)\,|s\alpha\rangle + \cos(\theta_u/2)\exp(\mathrm{i}\varphi_u/2)\,|s\beta\rangle. \quad (2.263b)$$

2.6.4.2 *p States*

As already mentioned, the crystal field of cubic symmetry cannot remove the orbital degeneracy of the p states. Its effect is reduced to a uniform shift in the proper energy levels. It is, therefore, sufficient to consider lower symmetry components of the crystal field. For the orthorhombic component, we get from Eq. (2.244)

$$\mathcal{H}_p = \mathcal{V}_{\text{orth}}(\mathbf{R}) + \lambda^{(np)}\mathbf{L}\cdot\mathbf{S} + 2\omega_{\text{ex}}^{(np)}S_u, \quad (2.264)$$

where we have neglected the contribution from the external field by setting $\omega_L = 0$. The matrix representation of this perturbation constructed with Table 2.1 is displayed in Table 2.2.

Because of spin-orbit coupling, the proper values of energy depend on the orientation of the exchange field with respect to the orthorhombic axes

TABLE 2.2

Matrix representation of a perturbation acting on p states. The perturbation includes the orthorhombic component of the crystal field, spin-orbit coupling, and exchange field oriented in the direction determined by the angles θ and φ of spherical coordinates. The two electron spin states are distinguished by α and β.

	$\|x\alpha\rangle$	$\|y\alpha\rangle$	$\|z\beta\rangle$	$\|x\beta\rangle$	$\|y\beta\rangle$	$\|z\alpha\rangle$
$\langle x\alpha\|$	$A_p + \hbar\omega_{\text{ex}}^{(p)} \cos\theta$	$-\text{i}\lambda^{(p)}\left(\frac{\hbar^2}{2}\right)$	$\lambda^{(p)}\left(\frac{\hbar^2}{2}\right)$	$\hbar\omega_{\text{ex}}^{(p)} \sin\theta \text{e}^{-\text{i}\varphi}$	0	0
$\langle y\alpha\|$	$\text{i}\lambda^{(p)}\left(\frac{\hbar^2}{2}\right)$	$B_p + \hbar\omega_{\text{ex}}^{(p)} \cos\theta$	$-\text{i}\lambda^{(p)}\left(\frac{\hbar^2}{2}\right)$	0	$\hbar\omega_{\text{ex}}^{(p)} \sin\theta \text{e}^{-\text{i}\varphi}$	0
$\langle z\beta\|$	$\lambda^{(p)}\left(\frac{\hbar^2}{2}\right)$	$\text{i}\lambda^{(p)}\left(\frac{\hbar^2}{2}\right)$	$C_p - \hbar\omega_{\text{ex}}^{(p)} \cos\theta$	0	0	$\hbar\omega_{\text{ex}}^{(p)} \sin\theta \text{e}^{\text{i}\varphi}$
$\langle x\beta\|$	$\hbar\omega_{\text{ex}}^{(p)} \sin\theta \text{e}^{\text{i}\varphi}$	0	0	$A_p - \hbar\omega_{\text{ex}}^{(p)} \cos\theta$	$\text{i}\lambda^{(p)}\left(\frac{\hbar^2}{2}\right)$	$-\lambda^{(p)}\left(\frac{\hbar^2}{2}\right)$
$\langle y\beta\|$	0	$\hbar\omega_{\text{ex}}^{(p)} \sin\theta \text{e}^{\text{i}\varphi}$	0	$-\text{i}\lambda^{(p)}\left(\frac{\hbar^2}{2}\right)$	$B_p - \hbar\omega_{\text{ex}}^{(p)} \cos\theta$	$-\text{i}\lambda^{(p)}\left(\frac{\hbar^2}{2}\right)$
$\langle z\alpha\|$	0	0	$\hbar\omega_{\text{ex}}^{(p)} \sin\theta \text{e}^{-\text{i}\varphi}$	$-\lambda^{(p)}\left(\frac{\hbar^2}{2}\right)$	$\text{i}\lambda^{(p)}\left(\frac{\hbar^2}{2}\right)$	$C_p + \hbar\omega_{\text{ex}}^{(p)} \cos\theta$

[38]. Ignoring the radial dependence, we have for the matrix elements of the orthorhombic crystal field

$$\langle x\,|\mathcal{V}_{\text{orth}}|\,x\rangle = a_p - b_p = A_p, \tag{2.265a}$$

$$\langle y\,|\mathcal{V}_{\text{orth}}|\,y\rangle = a_p + b_p = B_p, \tag{2.265b}$$

$$\langle z\,|\mathcal{V}_{\text{orth}}|\,z\rangle = -2a_p = C_p, \tag{2.265c}$$

where

$$a_p = 2q^2\left(\frac{1}{c^3} - \frac{1}{2a^3} - \frac{1}{2b^3}\right) r^2 \left\langle Y_1^{(1)} \middle| Y_2^{(0)} \middle| Y_1^{(1)} \right\rangle, \tag{2.266a}$$

$$b_p = 2q^2\left(\frac{1}{a^3} - \frac{1}{b^3}\right) r^2 \left(\frac{3}{8}\right)^{1/2} r^2 \left\langle Y_1^{(1)} \middle| Y_2^{(2)} \middle| Y_1^{(-1)} \right\rangle, \tag{2.266b}$$

with

$$\left\langle Y_1^{(0)} \middle| Y_2^{(0)} \middle| Y_1^{(0)} \right\rangle = -2\left\langle Y_1^{(1)} \middle| Y_2^{(0)} \middle| Y_1^{(1)} \right\rangle = (5\pi)^{-1/2}, \tag{2.267a}$$

$$\left\langle Y_1^{(1)} \middle| Y_2^{(2)} \middle| Y_1^{(-1)} \right\rangle = -\left(\frac{3}{160\pi}\right)^{1/2}. \tag{2.267b}$$

The remaining matrix elements are zero. At the exchange field parallel to the z axis, corresponding to the $\theta = 0$, the representation in Table 2.2 reduces to a 3×3 block diagonal form and the characteristic equation of sixth degree required for the representation of Table 2.2 splits to two independent equations of third degree

$$E^3 \mp \hbar\omega_{\text{ex}}^{(p)} E^2 - \left[\left(C_p \mp \hbar\omega_{\text{ex}}^{(p)}\right)^2 - A_p B_p + 3\left(\frac{\hbar^2}{2}\lambda^{(p)}\right)^2\right] E + 2\left(\frac{\hbar^2}{2}\lambda^{(p)}\right)^3$$
$$- A_p B_p C_p \pm \hbar\omega_{\text{ex}}^{(p)}\left[\left(\frac{\hbar^2}{2}\lambda^{(p)}\right)^2 + A_p B_p + C_p^2\right] - 2\left(\hbar\omega_{\text{ex}}^{(p)}\right)^2 C_p \pm \left(\hbar\omega_{\text{ex}}^{(p)}\right)^3 = 0. \tag{2.268}$$

By the permutation of the parameters A_p, B_p, and C_p, we obtain the characteristic equations for the remaining special orientations, *i.e.*, parallel to the x and y axes, of the exchange field. We have set $A_p + B_p + C_p = 0$. In case two of the parameters A_p, B_p, and C_p are equal, the problem reduces to that of tetragonal crystal field. In the octahedral cubic crystal field, we have $A_p = B_p = C_p = 0$. Then, from Table 2.2 it can be shown that the proper values of energy do not depend on the orientation of the exchange field.

The proper states associated with the proper values of energy $E_p^{(i)}$ $i = 1, \ldots, 3$ solving the characteristic equation (2.268) for the upper signs, are given by the linear superpositions

$$c_1^{(i)}\,|x\alpha\rangle + c_2^{(i)}\,|y\alpha\rangle + c_3^{(i)}\,|z\beta\rangle\,. \tag{2.269}$$

Similarly, the proper states associated with the proper values of energy $E_p^{(i)}$, where $i = 4, \ldots, 6$, solving the characteristic equation (2.268) for the lower signs, are given by the linear superpositions with the reversed electron spins

$$d_1^{(i)}\,|x\beta\rangle + d_2^{(i)}\,|y\beta\rangle + d_3^{(i)}\,|z\alpha\rangle\,. \tag{2.270}$$

After the substitution of the proper values of energy $E_p^{(i)}$ corresponding to the upper signs in Eq. (2.268) into Eq. (2.264), it follows

$$\left(A_p + \hbar\omega_{\text{ex}}^{(p)} - E_p^{(i)}\right)c_1^{(i)} - \mathrm{i}\frac{\hbar^2}{2}\lambda^{(p)}c_2^{(i)} = -\frac{\hbar^2}{2}\lambda^{(p)}c_3^{(i)}, \tag{2.271a}$$

$$\mathrm{i}\frac{\hbar^2}{2}\lambda^{(p)}c_1^{(i)} + \left(B_p + \hbar\omega_{\text{ex}}^{(p)} - E_p^{(i)}\right)c_2^{(i)} - = \mathrm{i}\frac{\hbar^2}{2}\lambda^{(p)}c_3^{(i)}. \tag{2.271b}$$

The coefficients are related by

$$c_1^{(i)} = \frac{\left[E_p^{(i)} - \left(B_p + \hbar\omega_{\text{ex}}^{(p)} + \dfrac{\hbar^2}{2}\lambda^{(p)}\right)\right]\dfrac{\hbar^2}{2}\lambda^{(p)}}{\left(E_p^{(i)} - A_p - \hbar\omega_{\text{ex}}^{(p)}\right)\left(E_p^{(i)} - B_p - \hbar\omega_{\text{ex}}^{(p)}\right) - \dfrac{\hbar^4}{4}\lambda^{(p)2}}c_3^{(i)}, \tag{2.272a}$$

$$\mathrm{i}c_2^{(i)} = \frac{\left[E_p^{(i)} - \left(A_p + \hbar\omega_{\text{ex}}^{(p)} + \dfrac{\hbar^2}{2}\lambda^{(p)}\right)\right]\dfrac{\hbar^2}{2}\lambda^{(p)}}{\left(E_p^{(i)} - A_p - \hbar\omega_{\text{ex}}^{(p)}\right)\left(E_p^{(i)} - B_p - \hbar\omega_{\text{ex}}^{(p)}\right) - \dfrac{\hbar^4}{4}\lambda^{(p)2}}c_3^{(i)}. \tag{2.272b}$$

Similarly, for the set with the lower signs in Eq. (2.268)

$$\left(A_p - \hbar\omega_{\text{ex}}^{(p)} - E_p^{(i)}\right)d_1^{(i)} - \mathrm{i}\frac{\hbar^2}{2}\lambda^{(p)}d_2^{(i)} = \frac{\hbar^2}{2}\lambda^{(p)}d_3^{(i)}, \tag{2.273a}$$

$$-\mathrm{i}\frac{\hbar^2}{2}\lambda^{(p)}d_1^{(i)} + \left(B_p - \hbar\omega_{\text{ex}}^{(p)} - E_p^{(i)}\right)d_2^{(i)} - = \mathrm{i}\frac{\hbar^2}{2}\lambda^{(p)}d_3^{(i)} \tag{2.273b}$$

and the coefficients are now related by

$$d_1^{(i)} = \frac{-\left[E_p^{(i)} - \left(B_p - \hbar\omega_{\text{ex}}^{(p)} + \dfrac{\hbar^2}{2}\lambda^{(p)}\right)\right]\dfrac{\hbar^2}{2}\lambda^{(p)}}{\left(E_p^{(i)} - A_p + \hbar\omega_{\text{ex}}^{(p)}\right)\left(E_p^{(i)} - B_p + \hbar\omega_{\text{ex}}^{(p)}\right) - \dfrac{\hbar^4}{4}\lambda^{(p)2}}d_3^{(i)}, \tag{2.274a}$$

$$\mathrm{i}d_2^{(i)} = \frac{\left[E_p^{(i)} - \left(A_p - \hbar\omega_{\text{ex}}^{(p)} + \dfrac{\hbar^2}{2}\lambda^{(p)}\right)\right]\dfrac{\hbar^2}{2}\lambda^{(p)}}{\left(E_p^{(i)} - A_p + \hbar\omega_{\text{ex}}^{(p)}\right)\left(E_p^{(i)} - B_p + \hbar\omega_{\text{ex}}^{(p)}\right) - \dfrac{\hbar^4}{4}\lambda^{(p)2}}d_3^{(i)}. \tag{2.274b}$$

The normalized proper states can be expressed as

$$\left|\psi_p^{(i)}\right\rangle = \left(\left|c_1^{(i)}\right|^2 + \left|c_2^{(i)}\right|^2 + \left|c_3^{(i)}\right|^2\right)^{-1/2} \left(c_1^{(i)}\left|x\alpha\right\rangle + c_2^{(i)}\left|y\alpha\right\rangle + c_3^{(i)}\left|z\beta\right\rangle\right), \tag{2.275a}$$

for $i = 1, \ldots, 3$, and

$$\left|\psi_p^{(i)}\right\rangle = \left(\left|d_1^{(i)}\right|^2 + \left|d_2^{(i)}\right|^2 + \left|d_3^{(i)}\right|^2\right)^{-1/2} \left(d_1^{(i)}\left|x\beta\right\rangle + d_2^{(i)}\left|y\beta\right\rangle + d_3^{(i)}\left|z\alpha\right\rangle\right) \tag{2.275b}$$

for the states with the reversed electron spins, where $i = 4, \ldots, 6$.

In the cubic crystal field, the characteristic equation (2.268) simplifies to

$$\left(E \mp \hbar\omega_{\mathrm{ex}}^{(p)} - \frac{\hbar^2}{2}\lambda^{(p)}\right)\left[E^2 + \left(\frac{\hbar^2}{2}\lambda^{(p)}\right)E \pm \left(\frac{\hbar^2}{2}\lambda^{(p)}\right)\left(\hbar\omega_{\mathrm{ex}}^{(p)}\right) - \left(\hbar\omega_{\mathrm{ex}}^{(p)}\right)^2 - 2\left(\frac{\hbar^2}{2}\lambda^{(p)}\right)^2\right] = 0. \tag{2.276}$$

The proper values of energy become

$$E_{1,2}^{(p)} = \pm\hbar\omega_{\mathrm{ex}}^{(p)} + \frac{\hbar^2}{2}\lambda^{(p)}, \tag{2.277a}$$

$$E_{3,4}^{(p)} = \frac{1}{2}\left\{-\left(\frac{\hbar^2}{2}\lambda^{(p)}\right) \pm \left[9\left(\frac{\hbar^2}{2}\lambda^{(p)}\right)^2 + 4\left(\frac{\hbar^2}{2}\lambda^{(p)}\right)\left(\hbar\omega_{\mathrm{ex}}^{(p)}\right) + 4\left(\hbar\omega_{\mathrm{ex}}^{(p)}\right)^2\right]^{1/2}\right\}, \tag{2.277b}$$

$$E_{5,6}^{(p)} = \frac{1}{2}\left\{-\left(\frac{\hbar^2}{2}\lambda^{(p)}\right) \pm \left[9\left(\frac{\hbar^2}{2}\lambda^{(p)}\right)^2 + 4\left(\frac{\hbar^2}{2}\lambda^{(p)}\right)\left(\hbar\omega_{\mathrm{ex}}^{(p)}\right) - 4\left(\hbar\omega_{\mathrm{ex}}^{(p)}\right)^2\right]^{1/2}\right\}. \tag{2.277c}$$

where the uniform shift in energy is ignored. For this special case of the cubic crystal field with the exchange field parallel to the z axis and specified by $\theta_u = 0$, we construct the proper states and the matrix representing the operator $\lambda^{(p)}\mathbf{L}\cdot\mathbf{S} + 2\omega_{\text{ex}}^{(p)}S_z$ from the basis of spherical harmonics

$$\left|\psi_p^{(1)}\right\rangle = \left|Y_1^{(1)}\alpha\right\rangle, \tag{2.278a}$$

$$\left|\psi_p^{(2)}\right\rangle = \left|Y_1^{(-1)}\beta\right\rangle, \tag{2.278b}$$

$$\left|\psi_p^{(3)}\right\rangle = \cos\left(\frac{\Theta_{34}}{2}\right)\left|Y_1^{(1)}\beta\right\rangle + \sin\left(\frac{\Theta_{34}}{2}\right)\left|Y_1^{(0)}\alpha\right\rangle, \tag{2.278c}$$

$$\left|\psi_p^{(4)}\right\rangle = -\sin\left(\frac{\Theta_{34}}{2}\right)\left|Y_1^{(1)}\beta\right\rangle + \cos\left(\frac{\Theta_{34}}{2}\right)\left|Y_1^{(0)}\alpha\right\rangle, \tag{2.278d}$$

$$\left|\psi_p^{(5)}\right\rangle = \cos\left(\frac{\Theta_{56}}{2}\right)\left|Y_1^{(-1)}\alpha\right\rangle + \sin\left(\frac{\Theta_{56}}{2}\right)\left|Y_1^{(0)}\beta\right\rangle, \tag{2.278e}$$

$$\left|\psi_p^{(6)}\right\rangle = -\sin\left(\frac{\Theta_{56}}{2}\right)\left|Y_1^{(-1)}\alpha\right\rangle + \cos\left(\frac{\Theta_{56}}{2}\right)\left|Y_1^{(0)}\beta\right\rangle, \tag{2.278f}$$

where the angles Θ_{34} and Θ_{56} are defined by

$$\tan\Theta_{34} = \frac{-2^{1/2}\lambda^{(p)}\hbar^2}{\frac{\hbar^2}{2}\lambda^{(p)} + 2\hbar\omega_{\text{ex}}^{(p)}}, \tag{2.279a}$$

$$\tan\Theta_{56} = \frac{-2^{1/2}\lambda^{(p)}\hbar^2}{\frac{\hbar^2}{2}\lambda^{(p)} - 2\hbar\omega_{\text{ex}}^{(p)}}. \tag{2.279b}$$

There may exist situations where our assumption on the relative strength of the applied field of magnetic flux density, $\boldsymbol{B}_0$, and the effective exchange field, $\boldsymbol{B}_{ex}$, expressed by $B_0 \ll B_{ex}$, should be modified to account for $\boldsymbol{B}_0$. Table 2.3 represents the perturbation resulting from the simultaneous action of cubic crystal field, spin-orbit coupling, exchange field, $\boldsymbol{B}_{ex}$, and external field of magnetic flux density, $\boldsymbol{B}_0$. Both $\boldsymbol{B}_{ex}$ and $\boldsymbol{B}_0$ are oriented parallel to the quantization z axis.

2.6.4.3 d States

In magnetic compounds, the most interesting cases include the cubic ligand fields of a regular octahedron and that of a regular tetrahedron. The center of inversion belongs to the symmetry group of the former, while it is absent in the latter. The parity selection rule affects the electron transitions allowed in the electric dipole approximation. The electron transitions strictly forbidden in the regular octahedron may be partially allowed in the regular tetrahedron.

The cubic crystal field of a regular ligand octahedron splits the d states of fivefold orbital degeneracy into two groups of the t_{2g} and e_g symmetry (Table 2.1). The orbital triplet t_{2g} states and the orbital doublet e_g states

TABLE 2.3

Matrix representation of a perturbation acting on p states in the space specified by $m_l = 0, \pm 1$ and $m_s = \pm\frac{1}{2}$ (the two electron spin states are distinguished by α and β). The perturbation includes the spin-orbit coupling, the exchange field, and applied field of external magnetic flux density. Both exchange and applied fields are oriented parallel to the z axis.

	$\lvert Y_1^{(1)}\alpha\rangle$	$\lvert Y_1^{(-1)}\beta\rangle$	$\lvert Y_1^{(1)}\beta\rangle$	$\lvert Y_1^{(0)}\alpha\rangle$	$\lvert Y_1^{(-1)}\alpha\rangle$	$\lvert Y_1^{(0)}\beta\rangle$
$\langle Y_1^{(1)}\alpha\rvert$	$\frac{\hbar^2}{2}\lambda^{(p)} + \hbar\omega_{ex}^{(p)} + 2\hbar\omega_L$	0	0	0	0	0
$\langle Y_1^{(-1)}\beta\rvert$	0	$\frac{\hbar^2}{2}\lambda^{(p)} - \hbar\omega_{ex}^{(p)} - 2\hbar\omega_L$	0	0	0	0
$\langle Y_1^{(1)}\beta\rvert$	0	0	$-\frac{\hbar^2}{2}\lambda^{(p)} - \hbar\omega_{ex}^{(p)}$	$2^{-1/2}\lambda^{(p)}\hbar^2$	0	0
$\langle Y_1^{(0)}\alpha\rvert$	0	0	$2^{-1/2}\lambda^{(p)}\hbar^2$	$\hbar\omega_{ex}^{(p)} + \hbar\omega_L$	0	0
$\langle Y_1^{(-1)}\alpha\rvert$	0	0	0	0	$-\frac{\hbar^2}{2}\lambda^{(p)} + \hbar\omega_{ex}^{(p)}$	$2^{-1/2}\lambda^{(p)}\hbar^2$
$\langle Y_1^{(0)}\beta\rvert$	0	0	0	0	$2^{-1/2}\lambda^{(p)}\hbar^2$	$-\hbar\omega_{ex}^{(p)} - \hbar\omega_L$

are coupled via spin-orbit coupling, which makes the dependence of the proper values of energy on the orientation of the exchange field with respect to the cubic crystallographic axes possible. This circumstance provides the microscopic basis for the anisotropy of the even magnetooptic effects in cubic crystals found from the symmetry arguments in Section 2.3.

2.6.5 Spectra

In Section 2.6.1, we obtained the expression for the complex generalized susceptibility tensor under the presence of magnetic ordering summarized in Eq. (2.216). From the multipole expansion, the electrical susceptibility tensor (in electric dipole approximation), Eq. (2.226), and magnetic susceptibility tensor (in magnetic dipole approximation), Eq. (2.234), were deduced. The expression for the generalized susceptibility tensor in Eq. (2.216) combined with the information on the proper values of energy and on the proper states allow us to compute the dependence of the susceptibility tensor components on the photon energy. In the optical region, we are mainly concerned with the contributions from electric dipole transitions, and we therefore focus on Eq. (2.226). We ignore the common factor $\frac{Nq^2}{\varepsilon_{vac}\hbar}$ and write

$$\chi_{ij}(\omega) \propto \sum_{a,b} \frac{\rho_a \omega_{ba}\,(\omega + \mathrm{i}\gamma_{ba})}{\omega^2} \left\{ \frac{\langle u_a|X_i|u_b\rangle\langle u_b|X_j|u_a\rangle}{\omega_{ba} - \omega - \mathrm{i}\gamma_{ba}} - \frac{\langle u_a|X_i|u_b\rangle^*\langle u_b|X_j|u_a\rangle^*}{\omega_{ba} + \omega + \mathrm{i}\gamma_{ba}} \right\}. \tag{2.280}$$

We give the examples of the spectra computed from Eq. (2.280) for the electron transitions from the s states ($l = 0$) to p states ($l = 1$). The off-diagonal sp submatrix representing the operators Z, X, and Y are schematically shown in Table 2.4. The only nonzero matrix elements are designated as c_{11}, c_{12}, and c_{13}. Their angular parts are summarized in Tables 2.5, Table 2.6, and Table 2.7. We shall be concerned with the situations where the products $\langle u_a|X_i|u_b\rangle\langle u_b|X_j|u_a\rangle$ are real and positive as in the case of the diagonal elements of susceptibility tensor, χ_{ii} *i.e.*, $|\langle u_a|X_i|u_b\rangle|^2 \geq 0$. Then, a single

TABLE 2.4

Off-diagonal sp submatrix representing the operators Z/R, X/R, and Y/R used in the evaluation of the electric dipole transitions contributing to the susceptibility tensor. The two electron spin states are distinguished by α and β.

	$\left\|Y_1^{(1)}\alpha\right\rangle$	$\left\|Y_1^{(1)}\beta\right\rangle$	$\left\|Y_1^{(0)}\alpha\right\rangle$	$\left\|Y_1^{(0)}\beta\right\rangle$	$\left\|Y_1^{(-1)}\alpha\right\rangle$	$\left\|Y_1^{(-1)}\beta\right\rangle$
$\left\langle Y_0^{(0)}\alpha\right\|$	c_{11}	0	c_{10}	0	c_{12}	0
$\left\langle Y_0^{(0)}\beta\right\|$	0	c_{11}	0	c_{10}	0	c_{12}

TABLE 2.5

Angular part for the off-diagonal *sp* submatrix representing the operator $\frac{Z}{R} = \frac{4\pi}{3} Y_1^{(0)}$. The two electron spin states are distinguished by α and β.

	$\lvert Y_1^{(1)}\alpha\rangle$	$\lvert Y_1^{(1)}\beta\rangle$	$\lvert Y_1^{(0)}\alpha\rangle$	$\lvert Y_1^{(0)}\beta\rangle$	$\lvert Y_1^{(-1)}\alpha\rangle$	$\lvert Y_1^{(-1)}\beta\rangle$
$\langle Y_0^{(0)}\alpha\rvert$	0	0	$3^{-1/2}$	0	0	0
$\langle Y_0^{(0)}\beta\rvert$	0	0	0	$3^{-1/2}$	0	0

TABLE 2.6

Angular part of the off-diagonal *sp* submatrix representing the electric dipole operator $\frac{X}{R} = \frac{2\pi}{3}\left(Y_1^{(-1)} - Y_1^{(1)}\right)$. The two electron spin states are distinguished by α and β.

	$\lvert Y_1^{(1)}\alpha\rangle$	$\lvert Y_1^{(1)}\beta\rangle$	$\lvert Y_1^{(0)}\alpha\rangle$	$\lvert Y_1^{(0)}\beta\rangle$	$\lvert Y_1^{(-1)}\alpha\rangle$	$\lvert Y_1^{(-1)}\beta\rangle$
$\langle Y_0^{(0)}\alpha\rvert$	$-6^{1/2}$	0	0	0	$6^{1/2}$	0
$\langle Y_0^{(0)}\beta\rvert$	0	$-6^{1/2}$	0	0	0	$6^{1/2}$

TABLE 2.7

Angular part for the off-diagonal *sp* submatrix representing the electric dipole operator $\frac{Y}{R} = i\frac{2\pi}{3}\left(Y_1^{(-1)} + Y_1^{(1)}\right)$. The two electron spin states are distinguished by α and β.

	$\lvert Y_1^{(1)}\alpha\rangle$	$\lvert Y_1^{(1)}\beta\rangle$	$\lvert Y_1^{(0)}\alpha\rangle$	$\lvert Y_1^{(0)}\beta\rangle$	$\lvert Y_1^{(-1)}\alpha\rangle$	$\lvert Y_1^{(-1)}\beta\rangle$
$\langle Y_0^{(0)}\alpha\rvert$	$-i6^{1/2}$	0	0	0	$-i6^{1/2}$	0
$\langle Y_0^{(0)}\beta\rvert$	0	$-i6^{1/2}$	0	0	0	$-i6^{1/2}$

term in the sum of Eq. (2.280) becomes

$$\chi_{ii}^{(ba)}(\omega) = \rho_a\, |\langle u_a|X_i|u_b\rangle|^2 \left\{ \frac{2\omega_{ba}\left[\omega_{ba}^2\left(\omega^2-\gamma_{ba}^2\right)-\left(\omega^2+\gamma_{ba}^2\right)^2\right]}{\omega^2\left[\left(\omega_{ba}^2-\omega^2+\gamma_{ba}^2\right)^2+4\omega^2\gamma_{ba}^2\right]} + \frac{4\mathrm{i}\omega_{ba}^3\gamma_{ba}}{\omega\left[\left(\omega_{ba}^2-\omega^2+\gamma_{ba}^2\right)^2+4\omega^2\gamma_{ba}^2\right]} \right\}, \tag{2.281}$$

where $\chi_{ii}^{(ba)}(\omega) = -\chi_{ii}^{(ab)}(\omega)$ because of $\omega_{ba} = -\omega_{ab}$ in agreement with Eq. (2.236).

We shall be further concerned with the situations where the product of matrix elements is imaginary pure, *i.e*, $\langle u_a|X_i|u_b\rangle\langle u_b|X_j|u_a\rangle = \pm\mathrm{i}|\langle u_a|X_i|u_b\rangle \langle u_b|X_j|u_a\rangle|$. Now, a single term in the sum of Eq. (2.280) becomes

$$\chi_{ij}^{(ba)}(\omega) = \pm\mathrm{i}\rho_a \left|\langle u_a|X_i|u_b\rangle\langle u_b|X_j|u_a\rangle\right| \left\{ \frac{2\omega_{ba}^2\left(\omega_{ba}^2-\omega^2-\gamma_{ba}^2\right)}{\omega\left[\left(\omega_{ba}^2-\omega^2+\gamma_{ba}^2\right)^2+4\omega^2\gamma_{ba}^2\right]} + \frac{2\mathrm{i}\gamma_{ba}\omega_{ba}^2\left(\omega_{ba}^2+\omega^2+\gamma_{ba}^2\right)}{\omega^2\left[\left(\omega_{ba}^2-\omega^2+\gamma_{ba}^2\right)^2+4\omega^2\gamma_{ba}^2\right]} \right\}, \tag{2.282}$$

where $\chi_{ij}^{(ba)}(\omega) = -\chi_{ij}^{(ab)}(\omega)$ because of $\langle u_a|X_i|u_b\rangle\langle u_b|X_j|u_a\rangle = -\langle u_b|X_i|u_a\rangle \langle u_a|X_j|u_b\rangle$ in agreement with a more general Eq. (2.236). Using Eq. (2.179), we can express the susceptibility tensor in terms of its CP components.

Scalar susceptibility. The simplest situation takes place for the electron transitions from the initial s state to the final p states in the absence of the magnetic ordering when the effect of the electron spin is ignored and only the initial state is occupied, $\rho_a = 1$. This is an electron transition from an orbital singlet ground state to an orbital triplet of excited states. We choose the value of the energy difference, $\hbar\omega_{ba} = 1$, and that of the relaxation (damping) constant $\gamma_{ba} = 0.05$. For simplicity, we have chosen the same value of the relaxation constant for all pairs of states. The response of this system to an electromagnetic plane wave is independent of the wave polarization. For a wave with its electric field polarized parallel to the z axis, it is sufficient to consider $\langle u_a|Z|u_b\rangle = 3^{-1/2}$ as the only nonzero matrix element. Here, $|u_a\rangle$ is the initial proper state s $(l = 0)$ and $|u_b\rangle$ is the final proper state p $(l = 1)$. The magnetic quantum number distinguishing the spherical harmonics, $m_l = 0$, in the matrix element $\langle u_a|Z|u_b\rangle$ connecting the initial state with $l = 0$ and the final state with $l = 1$. By substituting these values into Eq. (2.280) we obtain the spectrum of the complex electric susceptibility. The susceptibility is a scalar quantity, and the response is

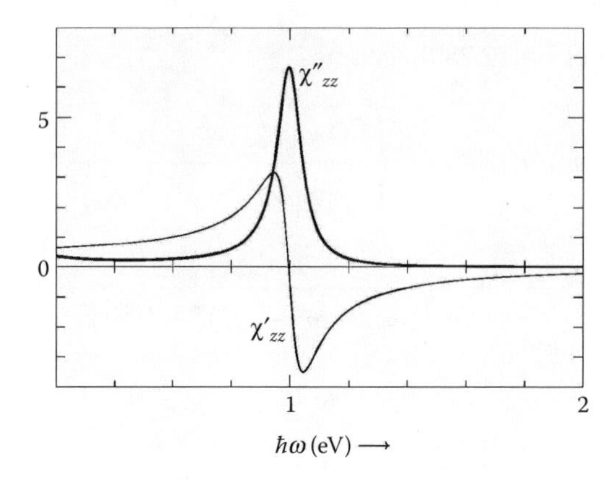

FIGURE 2.1
Complex susceptibility, $\chi_{zz} = \chi'_{zz} + \mathrm{i}\chi''_{zz}$, spectrum (in arbitrary units) for an electron transition from an orbital singlet s state to orbital triplet p states. The exchange field and the field of the external magnetic flux density are zero. The effect of electron spin is ignored.

independent of the wave polarization, as expected. Figure 2.1 shows the spectrum.

The line corresponds to a single term in the sum of Eq. (2.280) and represents the contribution to the diagonal component of the susceptibility tensor given by Eq. (2.281). We should get the same spectra for the waves polarized parallel to the x or y axis. However, in the spherical harmonics representation, this would require two terms in Eq. (2.280) with $m_l = \pm 1$ in the final state.

Figure 2.2 shows a single term in the sum of Eq. (2.280) as represented in Eq. (2.282) contributing to the off-diagonal element of the susceptibility tensor. These off-diagonal contributions appear in pairs with opposite signs and cancel at the zero magnetization. At magnetic ordering, these pairs have slightly different resonant frequencies and may also have different populations of the initial states. As a result, a small contribution to the off-diagonal element of the susceptibility tensor is generated as shown below. Note that imaginary pure $\mathrm{i}\chi''_{xy}$ and real χ'_{xy} represent the Hermitian and anti-Hermitian parts of the off-diagonal element.

External field. The external field of magnetic flux density oriented parallel to the quantization z axis, $\boldsymbol{B}_0 = \hat{z} B_0$ of the magnitude specified by $\hbar\omega_L = 0.05$ splits the p energy level with degeneracy into three levels (effect of electron spin is ignored) with the energies $\hbar\omega = 0.95, 1.00, 1.05$. Using Eq. (2.261a) and Table 2.1, we find for the matrix elements (always ignoring the radial dependence)

$$\langle u_a | \mathrm{Z} | u_b \rangle = 3^{-1/2}, \tag{2.283a}$$

$$\langle u_a | \mathrm{X} | u_b \rangle = \mathrm{i} \langle u_a | \mathrm{Y} | u_b \rangle = 6^{-1/2}. \tag{2.283b}$$

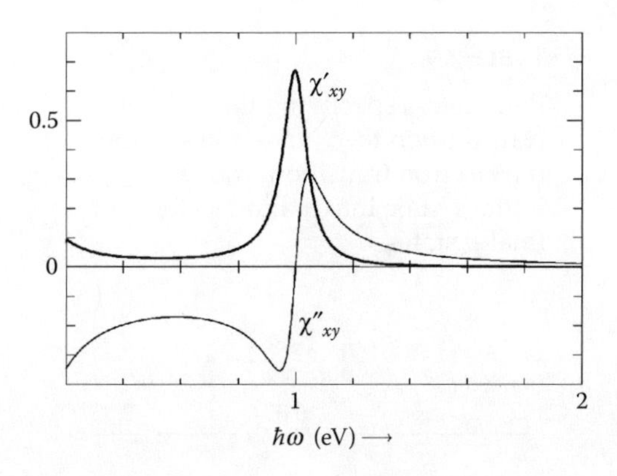

FIGURE 2.2
Elementary contribution to the off-diagonal susceptibility tensor element $\chi_{xy} = \chi'_{xy} + i\chi''_{xy}$ spectrum (in arbitrary units) for an electron transition from an orbital singlet s state to an orbital triplet of p states. The exchange field and the field of the external magnetic flux density are zero. The effect of electron spin is ignored. Evaluated for $\langle u_a | X | u_b \rangle \langle u_b | Y | u_a \rangle = -i/6, \rho_a = 1$ and $\hbar\omega_{ba} = 1$ and $\hbar\Gamma_{ba} = 0.05$.

The calculation of χ_{zz} requires a single term of Eq. (2.280) with the parameters $\rho_a = 1, \hbar\omega_{ba} = 1$, and $\langle u_a \langle | \, Z \, | u_b \rangle = 3^{-1/2}$. The calculation of χ_{xx} requires two terms of Eq. (2.280), with the final states $|u_b\rangle = |l = 1, m_l = \pm 1\rangle$, specified by the set of parameters in Table 2.8 including the matrix element given in Eq. (2.283b). The calculation of χ_{yy} gives the same result. The calculation of the off-diagonal element χ_{xy} requires two terms of Eq. (2.280) specified by the set of parameters in Table 2.9. The parameters corresponding to the reversed magnetic moment are shown in Table 2.10. The evaluation of χ_{yx} shows

$$\chi_{yx}(\hbar\omega) = -\chi_{xy}(\hbar\omega) . \quad (2.284)$$

Upon the reversal of the external field, $\mathbf{B}_0 \rightarrow -\mathbf{B}_0$, we have $\hbar\omega_L \rightarrow -\hbar\omega_L$. Now, the off-diagonal element χ_{xy} follows simply from the set with a reversed order. We observe

$$\chi_{xy}(\hbar\omega_L) = -\chi_{xy}(-\hbar\omega_L) . \quad (2.285)$$

TABLE 2.8

Parameters specifying the contribution to χ_{xx} from an electron transition from an initial s state into magnetically split final p states.

ρ_a	1	1
$\hbar\omega_{ba}$	0.95	1.05
$\langle u_a | \mathbf{X} | u_b \rangle$	$6^{-1/2}$	$-6^{-1/2}$
$\hbar\gamma_{ba}$	0.05	0.05

TABLE 2.9

Parameters specifying the contribution to $\chi_{xy}(\hbar\omega_L)$ assuming the electron transition from an initial s state into magnetically split final p states.

ρ_a	1	1
$\hbar\omega_{ba}$	0.95	1.05
$\langle u_a \vert \mathrm{X} \vert u_b \rangle$	$6^{-1/2}$	$-6^{-1/2}$
$\langle u_a \vert \mathrm{Y} \vert u_b \rangle$	$-\mathrm{i}6^{-1/2}$	$-\mathrm{i}6^{-1/2}$
$\hbar\gamma_{ba}$	0.05	0.05

The magnetic ordering transformed the originally isotropic medium with the scalar susceptibility into an anisotropic one with the tensorial susceptibility. The spectra of the complex tensor elements of the susceptibility tensor are summarized in Figure 2.3 to Figure 2.5. We include the plot of $\chi_{xx} - \chi_{zz}$ illustrating the magnetooptic effects even in magnetization. Despite the fundamentally different starting points, the spectra deduced from the semiclassical model for the electric dipole transitions between the orbital s and p states show a close resemblance with those obtained from the classical Lorentz oscillator model as expressed by Eq. (2.175).

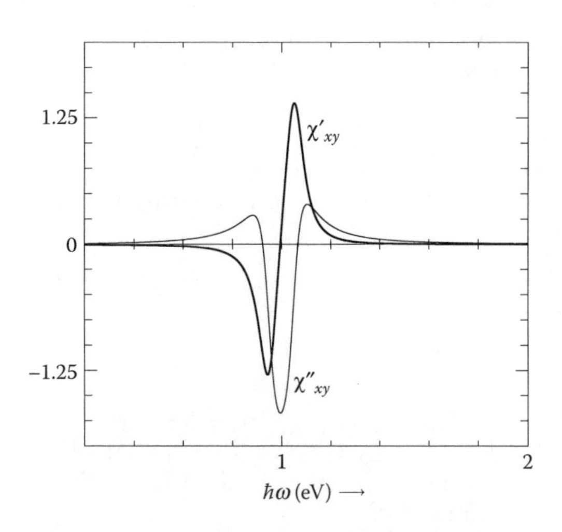

FIGURE 2.3
Spectrum of the complex off-diagonal susceptibility tensor element $\chi_{xy} = \chi'_{xy} + \mathrm{i}\chi''_{xy}$ (in arbitrary units) for the electron transition from an orbital singlet s state to orbital triplet p states split by the field of the external magnetic flux density parallel to the z axis. The effect of electron spin is ignored.

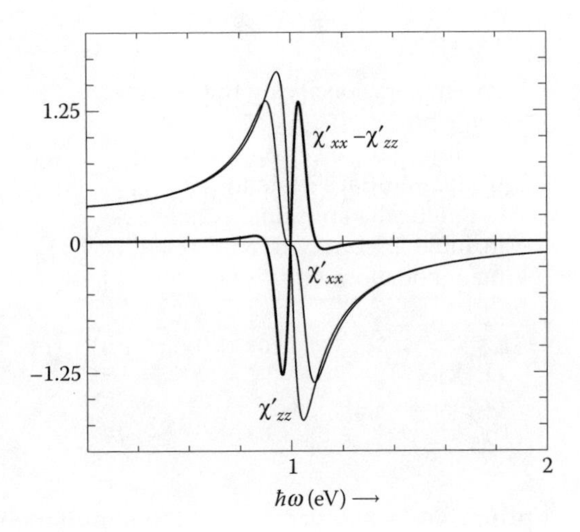

FIGURE 2.4
Spectrum of the real part of the diagonal susceptibility tensor elements χ'_{zz} and χ'_{xx} and their difference $\chi'_{xx} - \chi'_{zz}$ (in arbitrary units) for an electron transition from an orbital singlet s state to orbital triplet p states split by the field of the external magnetic flux density parallel to the z axis. The effect of electron spin is ignored.

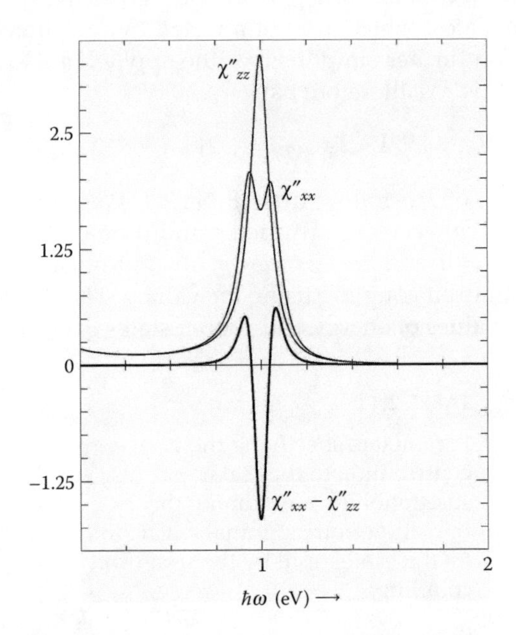

FIGURE 2.5
Spectrum of the imaginary part of the diagonal susceptibility tensor elements χ''_{zz} and χ''_{xx} and their difference $\chi''_{xx} - \chi''_{zz}$ for an electron transition from an orbital singlet s state to orbital triplet p states split by the field of the external magnetic flux density parallel to the z axis. The effect of electron spin is ignored.

TABLE 2.10

Parameters specifying the contribution to $\chi_{xy}(-\hbar\omega_L)$ assuming the electron transition from an initial s state into magnetically split final p states. As in Table 2.9 but with a reversed magnetic moment.

ρ_a	1	1
$\hbar\omega_{ba}$	1.05	0.95
$\langle u_a\| X \|u_b\rangle$	$6^{-1/2}$	$-6^{-1/2}$
$\langle u_a\| Y \|u_b\rangle$	$-i6^{-1/2}$	$-i6^{-1/2}$
$\hbar\gamma_{ba}$	0.05	0.05

Spin-orbit coupling. We now account for the simultaneous action of electron spin and spin-orbit coupling. The effect of spin-orbit coupling on spectra deserves special attention. Strong magnetooptic effects in the bismuth doped iron garnets are attributed to the enhancement of the spin-orbit coupling in the final states of the electric dipole transitions transferred from Bi^{3+} ion excited states. The perturbation given in Eq. (2.254) can be simplified for p states, which are not affected by the cubic crystal field. The representation further simplifies for the applied and exchange fields parallel to the z axis. We therefore have

$$\mathcal{H}_p = \lambda^{(p)}\mathbf{L}\cdot\mathrm{S} + \omega_L \mathrm{L_z} + 2\left(\omega_L + \omega_{\mathrm{ex}}^{(nl)}\right)\mathrm{S_z}. \tag{2.286}$$

The proper values of $\mathcal{H}_p$ follow from Table 2.3. We continue to take the energy separation between unperturbed s and p states equal to unity, *i.e.*, $\hbar\omega_{ba} = 1$ (at $\lambda^{(p)} = 0$ and $\hbar\omega_{\mathrm{ex}} = \hbar\omega_L = 0$), and illustrate the effect for the energy of spin-orbit coupling using the value $\lambda^{(p)}\hbar^2/2 = 0.06$. Table 2.3 provides proper values of energy and proper states of electrons.

TABLE 2.11

Parameters specifying the contribution to the scalar susceptibility χ assuming the transition from an initial s state into final p states split by the spin-orbit coupling.

ρ_a	0.5	0.5
$\hbar\omega_{ba}$	0.88	1.06
$\langle u_a\| X_i \|u_b\rangle$	$\frac{1}{3}$	$\frac{\sqrt{2}}{3}$
$\hbar\gamma_{ba}$	0.05	0.05
degeneracy	doublet	quartet

TABLE 2.12
Matrix elements and photon energies for the electron transitions between E_s and E_p energy levels split by the combined effect of the spin-orbit coupling and the external field of magnetic flux density. The perturbation energies are characterized by $\lambda_p\hbar^2/2 = 0.06$ eV and $\hbar\omega_L = 0.05$ eV. The energy separation between unperturbed E_s and E_p was chosen to be 1 eV, and the transition photon energy follow as $\hbar\omega_{ba} = 1 - E_s + E_p$.

#	$\hbar\omega_{ba}$	E_s	E_p	$\langle u_a \|Z\| u_b\rangle$	$\langle u_a \|X\| u_b\rangle$	$\langle u_a \|Y\| u_b\rangle$
1	0.910	−0.050	−0.140	−0.3961		
2	1.080	−0.050	0.030	0.4201		
3	0.944	−0.050	−0.106		−0.3587	−0.3587 i
4	1.010	−0.050	−0.040		$6^{-1/2}$	$-6^{-1/2}$ i
5	1.146	−0.050	0.0961		−0.1950	−0.1950 i
6	0.844	0.050	−0.106	−0.2757		
7	1.046	0.050	0.096	0.5073		
8	0.810	0.050	−0.140		$(3/34)^{1/2}$	$-(3/34)^{1/2}$ i
9	0.980	0.050	0.030		$(4/51)^{1/2}$	$-(4/51)^{1/2}$ i
10	1.110	0.050	0.160		$-6^{-1/2}$	$-6^{-1/2}$ i

In the absence of magnetic order, $\hbar\omega_{\text{ex}} = \hbar\omega_L = 0$, the energy level of the sextet of p states splits into a doublet and quartet due to the spin-orbit coupling. The scalar susceptibility is given by two terms of the sum in Eq. (2.280) with the parameters of Table 2.11. Note that at $\hbar\omega_L = \hbar\omega_{\text{ex}}^{(s)} = 0$, the s state displays electron spin degeneracy, and the two initial s states are occupied with the same occupation probability, ρ_a, equal to 1/2. Neither the energy level splitting nor the values of the matrix elements depend on the polarization of electromagnetic plane waves. The susceptibility displayed in Figure 2.6 is a scalar quantity.

Magnetic order. We now add the effect of the external applied field of magnetic flux density, which completely removes both the orbital and spin degeneracy. Figure 2.7 displays the allowed electric dipole transitions in the system. We denote by π and $\sigma_\pm$ the electron transitions associated with the absorption of the waves with the electric field polarized parallel to $\boldsymbol{B}_0$ (or $\boldsymbol{B}_{ex}^{(p)}$)[8] and right and left circularly polarized ones in the plane perpendicular to $\boldsymbol{B}_0$ (or $\boldsymbol{B}_{ex}^{(p)}$), respectively. The situation is illustrated using the parameters collected in Table 2.12. Figure 2.8 and Figure 2.9 show the spectra of the off-diagonal element, χ_{xy}, and the difference in the diagonal elements, $\chi_{zz} - \chi_{xx}$, of the corresponding susceptibility tensor, respectively.

Effect of temperature. We recall that the effective exchange field, $\boldsymbol{B}_{ex}$, acting on the electron spin magnetic moments, provides the simplest

[8] We restrict ourselves to the discussion of the collinear spin configurations with spin and orbital magnetic moments either parallel or antiparallel to the external field.

way for the account of the interactions among magnetic ions resulting in spontaneous magnetic order. In crystals with collinear ferrimagnetic order, this field can be either parallel or antiparallel in a given crystallographic site with respect to the external fields $\boldsymbol{B}_0$. In strong external fields, $\boldsymbol{B}_0$, or at low $\boldsymbol{B}_{ex}^{(p)}$ (*e.g.*, close to the Curie temperature), a spin-flip can be induced by $\boldsymbol{B}_0$. The temperature affects the population of the spin split initial s states. The lower energy s state is occupied by an electron with the probability

$$\varrho_{0\beta} = \frac{\exp[\hbar\left(\omega_L + \omega_{ex}\right)/kT]}{\exp[\hbar\left(\omega_L + \omega_{ex}\right)/kT] + \exp[-\hbar\left(\omega_L + \omega_{ex}\right)/kT]}, \tag{2.287}$$

where $k = 1.380\ 66(4) \times 10^{-23}$. Joule Kelvin^{-1} denotes the Boltzmann's constant and T is the absolute temperature in Kelvin. The occupation probability of the upper s state is given by $\varrho_{0\alpha} = 1 - \varrho_{0\beta}$. The energy of the exchange field may be estimated from the Curie temperature T_C of a magnetic crystal using the relation $\hbar\omega_{ex} \approx kT_C$. Note that the phenomenological relaxation (damping) constant γ_{ab} should increase with temperature. The effect of temperature on the line shape manifests itself in paramagnetic materials where an additional structure appears in the magnetooptic spectra as the upper states of the ground state manifold become populated. In materials with spontaneous magnetization, an increase of temperature or that of diamagnetic substitution level decreases the number of exchange

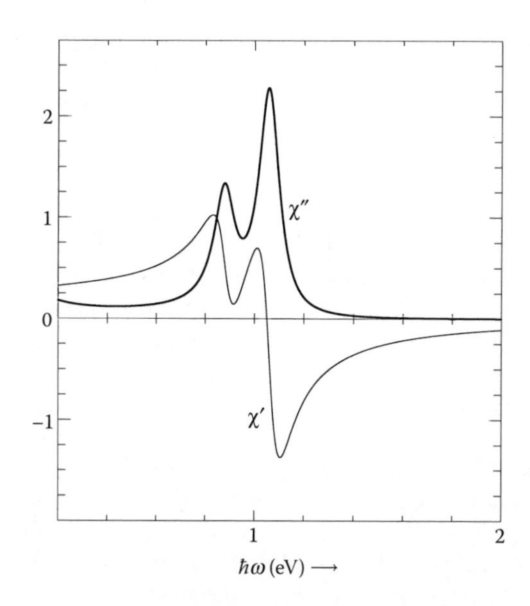

FIGURE 2.6
Spectrum of the scalar complex electric susceptibility $\chi = \chi' + i\chi''$ (in arbitrary units) for the electron transition from an orbital singlet (spin doublet) s states to an orbital triplet (spin sextet) of p states. Spin-orbit coupling splits the p states to a lower doublet and an upper quartet.

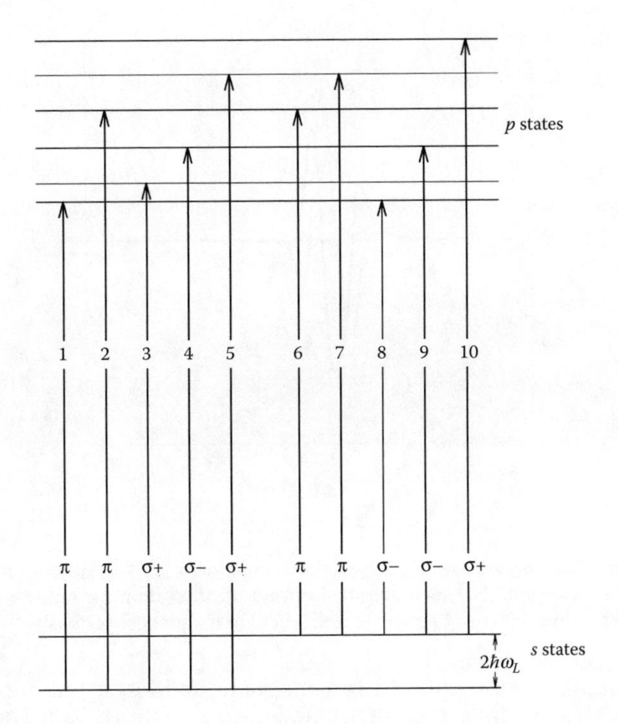

FIGURE 2.7
Electron transitions from the s to p energy levels split by the spin-orbit coupling. The external field of magnetic flux density completely removes the electron orbital and spin degeneracy. The circular ($\sigma_{\pm}$) and linear (π) polarizations are indicated. The photon energies at resonances and matrix elements are given in Table 2.12.

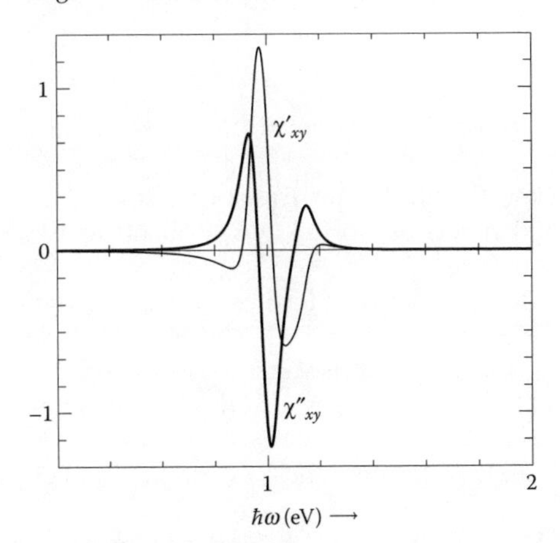

FIGURE 2.8
Spectrum of the complex off-diagonal element of the susceptibility tensor, $\chi_{xy} = \chi'_{xy} + i\chi''_{xy}$ for an electron transition from the lower s state to the p states split by spin-orbit coupling and external field of magnetic flux density.

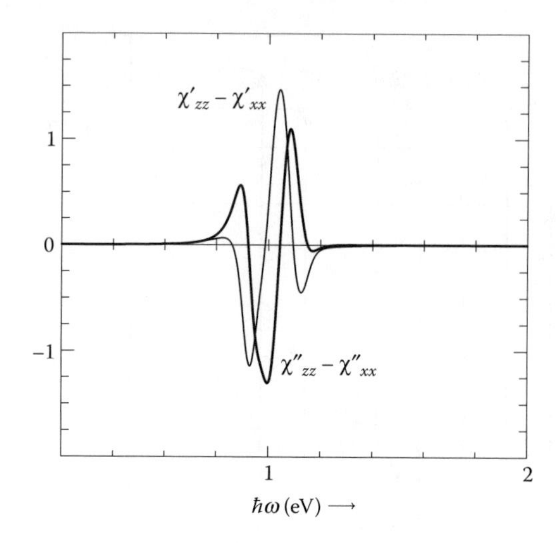

FIGURE 2.9
Spectrum of the differences of the real, $\chi'_{zz} - \chi'_{xx}$, and imaginary, $\chi''_{zz} - \chi''_{xx}$, parts of the diagonal elements of the susceptibility tensor, for an electron transition from the lower s state to the p states split by spin-orbit coupling and external field of magnetic flux density.

coupled magnetic moments. In ferrimagnetic oxides, a temperature increase manifests itself by line broadening and amplitude reduction in the spectra.

References

1. J. D. Jackson, *Classical Electrodynamics, Third Edition* (John Wiley & Sons, Inc. 1999).
2. J. H. Heinbockel, *Introduction to Tensor Calculus and Continuum Mechanics* (Old Dominion University, Norfoik, VA, 1996, http://www.math.odu.edu/jhh/counter2.html).
3. W. Wettling, "Magneto-optics of ferrites," J. Magn. Magn. Mater. **3**, 147–160, 1976.
4. V. V. Eremenko, *Vvedenie v opticheskuyu spektroskopiyu magnetikov* (in Russian, *Introduction to Optical Spectroscopy of Magnetic Substances*, Naukova Dumka, Kiev, 1975).
5. G. A. Smolenskii, R. V. Pisarev, and I. G. Simii, Usp. Fiz. Nauk (in Russian) **116**, 231, 1975.
6. J. F. Dillon, Jr., J. P. Remeika, and C. R. Staton, "Linear magnetic birefringence in the ferrimagnetic garnets," J. Appl. Phys. **41**, 4613–4619, 1970.
7. A. M. Prokhorov, G. A. Smolenskii, and A. N. Ageev, Usp. Fiz. Nauk (in Russian) **143**, 33, 1984.
8. J. Ferré and G. A. Gehring, Rep. Prog. Phys. **47**, 513, 1984.

9. I. P. Kaminoff, *An Introduction to Electrooptic Devices* (Academic Press, New York, 1974).
10. A. Yariv, *Quantum Electronics* (Wiley, New York 1975).
11. R. M. White, *Quantum Theory of Magnetism* (Springer Verlag, Berlin, 1983), Chapter 1.
12. R. R. Birss, *Symmetry and Magnetism* (North-Holland, Amsterdam, 1964).
13. D. B. Melrose and R. C. McPhedran, *Electromagnetic Processes in Dispersive Media: A Treatment Based on Dielectric Tensor* (Cambridge University Press, Cambridge, 1991), pp. 83–86.
14. F. Bassani and G. Pastori Parravicini, *Electronic States and Optical Transitions in Solids* (Pergamon Press, Oxford, 1975), Chapter 1.
15. J. F. Nye, *Physical Properties of Crystals* (Clarendon Press, Oxford, 1964).
16. B. A. Auld and D. A. Wilson, "Bragg scattering of infrared radiation from coherent spin waves," J. Appl. Phys. **38**, 3331, 1967.
17. W. J. Tabor and F. S. Chen, "Electromagnetic propagation through materials possessing both Faraday rotation and birefingence: Experiments with ytterbium orthoferrite," J. Appl. Phys. **40**, 2760–2765, 1969.
18. F. J. Kahn, P. S. Pershan, and J. P. Remeika, "Ultraviolet magneto-optical properties of single-crystal orthoferrites, garnets, and other ferric oxide compounds," Phys. Rev. **186**, 891–918, 1969.
19. Š. Višňovský, "Magneto–optical permittivity tensor in crystals," Czech. J. Phys. B **36**, 1424–1433, 1986.
20. A. Yariv and P. Yeh, *Optical Waves in Crystals* (John Wiley & Sons, Inc., New York, 1983) Chapter 7.
21. Š. Višňovský, "Anisotropy of magneto optic interaction in magnetic crystals," IEEE Transactions on Magnetics **26**, 2786–2788, 1990.
22. P. N. Butcher, and D. Cotter, *The Elements of Nonlinear Optics* (Cambridge Studies in Modern Optics 9, Cambridge University Press 1990).
23. C. Cohen-Tannoudji, B. Diu, and F. Laloe, *Mécanique Quantique* (Hermann, Paris, 1977) Chapter 13.
24. D. B. Melrose and R. C. McPhedran, *Electromagnetic Processes in Dispersive Media. A Treatment Based on Dielectric Tensor* (Cambridge University Press, Cambridge, 1991), pp. 6–9.
25. I. V. Lindell, *Methods of Electromagnetic Field Analysis* (Clarendon Press, Oxford, 1992), Chapter 1.
26. R. M. A. Azzam, and N. M. Bashara, *Ellipsometry and Polarized Light* (Elsevier, Amsterdam, 1987).
27. T. S. Moss, *Optical Properties of Semi-Conductors* (Butterworths, London, 1959), Chapter 5.
28. M. J. Freiser, "A survey of magnetooptic effects," IEEE Trans. Magn. **4**, 152–161, 1968.
29. S. H. Wemple and M. DiDomenico, Jr., "Behavior of the electronic dielectric constant in covalent and ionic materials," Phys. Rev. B **3**, 1338–51, 1971.
30. Y. R. Shen, "Farady rotation of rare-earth ions. I. Theory," Phys. Rev. **133**, A511–A515, 1964.
31. P. S. Pershan, "Magneto-optical effects," J. Appl. Phys. **38**, 1482–1490, 1967.
32. W. A. Crossley, R. W. Cooper, J. L. Page, and R. P. van Stapele, "Faraday rotation in rare-earth iron garnets," Phys. Rev. **181**, 896–904, 1969.

33. S. Wittekoek, T. J. A. Popma, J. M. A. Robertson, and P. F. Bongers, "Magneto-optic spectra and dielectric tensor elements of bismuth-substituted iron garnets at photon enegies between 2.2–5.2 eV," Phys. Rev. B **12**, 2777–2788, 1975.
34. G. A. Allen and G. F. Dionne, "Application of permittivity tensor for accurate interpretation of magneto-optical spectra," J. Appl. Phys. **73**, 6130–6132, 1993.
35. L. E. Helseth, R. W. Hansen, E. I. Il'yashenko, M. Baziljevich, and T. H. Johanson, "Faraday rotation spectra of bismuth-substituted ferrite garnet films with in-plane magnetization," Phys. Rev. B **64**, 174406(6), 2001.
36. C. J. Ballhausen, *Introduction to Ligand Field Theory* (McGraw-Hill Book Company, New York, 1962).
37. S. Sugano, Y. Tanabe, and H. Kamimura, *Multiplets of Transition-Metal Ions in Crystals* (Academic Press, New York and London 1970).
38. Š. Višňovský, "Magnetooptical effects in orthoferrites: A simple model," Czech. J. Phys. B **34**, 1344–1348, 1984.
39. Š. Višňovský, DrSc thesis, Charles University, Prague, 1989.

3

Anisotropic Multilayers

3.1 Introduction

The response to electromagnetic vector waves in anisotropic multilayers in general and in thin film magnetooptic media and multilayers in particular received attention in recent years. The main motivation comes from the practical applications of nanostructures in optoelectronics, microelectronics, and magnetic or magnetooptic recording. In magnetooptics, the systematic interest started in the early 60s when the concept of magnetooptic memory and that of perpendicular recording was introduced. In 1967, Hunt published the response analysis of a metallic magnetooptic film sandwiched between dielectric layered structures [1]. Hunt rederived and completed the previous results of Voigt [2] and Robinson [3]. A similar problem was also treated by Smith [4]. References to earlier works can be found in the bibliography by Palik and Henvis [5].

An analysis of Faraday rotation in multilayers confined to the normal incidence and nonabsorbing media was performed by Sansalone [6]. The work was an extension of the papers dealing with the effect of multiple reflections on Faraday rotation in a plate published by Donovan and Medcalf [7] and Piller [8]. The effect of incoherent multiple reflections on the complex Faraday effect in multilayers was considered in References 9 and 10. Reflection and transmission in gyroelectromagnetic biaxial layered media at the special orientations of optical axes with respect to interfaces were the subject of the paper by Damaskos *et al.* [11]. Schwelb treated stratified lossy anisotropic media with polar and longitudinal magnetizations [12]. Gamble and Lissberger analyzed, among others, electromagnetic field distributions in multilayers [13].

This chapter deals with the generalization of the Yeh's formalism [14] to absorbing layered magnetically ordered media [15,16]. This represents a unified approach to the problem of electromagnetic response in multilayer anisotropic planar structures. Several simpler situations are included as special cases, *e.g.*, the complex Faraday and magnetooptic Kerr effects, the enhancement of the magnetooptic response with dielectric or metallic films, and the effect of interfaces between magnetic films in multilayers, *etc*. Note that there are no restrictions on the anisotropy and on magnetization profiles across the structure, except they should be homogeneous within a particular layer. In other words, there are no restrictions on the permittivity tensor as a function of the position on the axis normal to the multilayer interface planes.

The problem geometry is as follows: The planar structure consisting of $\mathcal{N}$ layers separated by the parallel interface planes normal to the common axis is sandwiched between two half spaces. Each layer is homogeneous and characterized by a general complex relative permittivity tensor, $\varepsilon^{(n)}$ $(n = 1, \ldots, \mathcal{N})$. Snell law requires the invariance of the propagation vector components parallel to the interface planes. Therefore, these components remain the same in all layers of the stack. We look for the plane wave solutions of $\mathcal{N}$ wave equations in infinite media characterized by these permittivity tensors. For each layer treated as an infinite medium, this is a proper value problem that provides four propagation vector components normal to the interface planes and the corresponding four proper polarizations or proper polarization modes. We assume that the fields in an individual layer, which may be of arbitrary thickness, can be expressed as a linear superposition of monochromatic plane waves with these four proper polarizations (Fraunhofer approximation). The four relations between the proper polarization amplitudes are provided by the boundary conditions at the interfaces. These require the continuity at interfaces of the total electric and magnetic field components parallel to the interface planes.

The boundary conditions can be expressed in a 4×4 matrix form as a relation between the four field amplitudes in one layer and the corresponding four field amplitudes in the adjacent layer. The corresponding matrix designated as *transfer matrix* represents the product of three matrices: inverse dynamical matrix in the $(n-1)$-th layer, dynamical matrix in the n-th layer, and the diagonal propagation matrix in the n-th layer.

The multilayer structure is then represented as a product of $\mathcal{N}+1$ transfer matrices. This product relates the field amplitudes in the isotropic half spaces that sandwich the multilayer. The 4×1 electric field amplitude vectors in the isotropic half spaces represent the complex amplitudes of two waves with two different (preferably) orthogonal polarizations propagating toward the multilayer structure and the corresponding complex amplitudes of two outgoing waves.

We usually set the field amplitudes of waves propagating towards the multilayer structure in one of the sandwiching half spaces to zero. This allows us to define the global transmission and reflection coefficients of the multilayer. With this restriction, the 4×4 matrix formalism can be reduced to a (2×2) matrix formalism [17]. The global transmission and reflection coefficients represent the information required for the complete characterization of the multilayer optical response. They are conventionally expressed as Jones reflection and transmission matrices of the multilayer structure.

3.2 Proper Modes

We shall consider the structure shown in Figure 3.1 consisting of $\mathcal{N}$ homogeneous layers separated by the interface planes $z = z^{(n)}$ $(n = 0, \ldots, \mathcal{N})$ in a Cartesian coordinate system, $z^{(n-1)} < z^{(n)}$. The structure is sandwiched between a homogeneous half space (0) with $z < z^{(0)}$ and a homogeneous half space ($\mathcal{N}+1$) with $z > z^{(\mathcal{N})}$, which for simplicity may be nonmagnetic, nonabsorbing, and isotropic. The n-th homogeneous layer confined by the

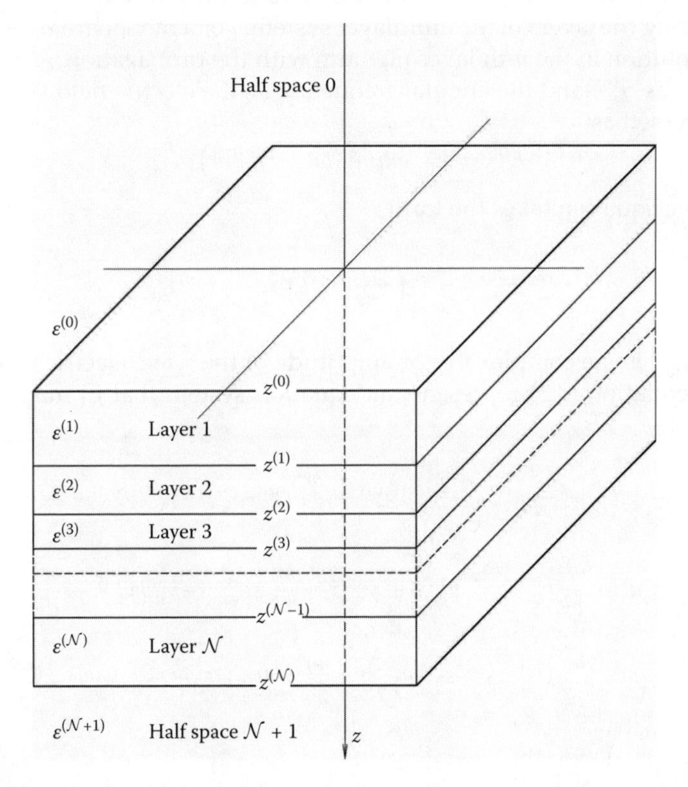

FIGURE 3.1
Anisotropic multilayer structure consisting of $\mathcal{N}$ anisotropic layers characterized by permittivity tensors $\varepsilon^{(n)}$, with $n = 1, \cdots, \mathcal{N}$. The interface planes are normal to a common axis parallel to the z axis of the Cartesian coordinate system.

interface planes $z = z^{(n-1)}$ and $z = z^{(n)}$ is characterized by the relative permittivity tensor

$$\varepsilon^{(n)} = \begin{pmatrix} \varepsilon_{xx}^{(n)} & \varepsilon_{xy}^{(n)} & \varepsilon_{xz}^{(n)} \\ \varepsilon_{yx}^{(n)} & \varepsilon_{yy}^{(n)} & \varepsilon_{yz}^{(n)} \\ \varepsilon_{zx}^{(n)} & \varepsilon_{zy}^{(n)} & \varepsilon_{zz}^{(n)} \end{pmatrix}, \tag{3.1}$$

which can account for arbitrary crystalline or induced anisotropies and for an arbitrary orientation of the magnetization vector. According to Chapter 2, in anisotropic absorbing magnetic media, the tensor elements $\varepsilon_{ij}^{(n)}$ are in general complex and the tensor itself is nonsymmetric, *i.e.*, $\varepsilon_{ij}^{(n)} \neq \varepsilon_{ji}^{(n)}$. The magnetic permeability may be assumed a scalar constant in a given layer. This corresponds to the electric dipole approximation justified at optical frequencies, as mentioned in Section 1.2. We shall assume that the magnetic permeability in all layers is the same and equal to its vacuum value, μ_{vac}. The solution to the wave equation in anisotropic media was already considered in Section 1.5. We follow similar lines and introduce the superscripts numbering the layers of the multilayer system. For a monochromatic plane wave solution in the n-th layer medium with the propagation vector now denoted as $\boldsymbol{\gamma}^{(n)}$ and the angular frequency, ω, the electric field vector can be expressed as

$$\boldsymbol{E}^{(n)} = \boldsymbol{E}_0^{(n)} \exp\left[\mathrm{j}\left(\omega t - \boldsymbol{\gamma}^{(n)} \cdot \boldsymbol{r}\right)\right], \tag{3.2}$$

the wave equation takes the form

$$\boldsymbol{\gamma}^{(n)2} \boldsymbol{E}_0^{(n)} - \boldsymbol{\gamma}^{(n)} \left(\boldsymbol{\gamma}^{(n)} \cdot \boldsymbol{E}_0^{(n)}\right) = \frac{\omega^2}{c^2} \boldsymbol{\varepsilon}^{(n)} \boldsymbol{E}_0^{(n)}, \tag{3.3}$$

where $\boldsymbol{E}_0^{(n)}$ is the complex vector amplitude of the wave electric field. The vector equation (3.3) represents an equation system that in matrix form becomes

$$\begin{pmatrix} \gamma_y^{(n)2} + \gamma_z^{(n)2} - \frac{\omega^2}{c^2}\varepsilon_{xx}^{(n)} & -\gamma_x^{(n)}\gamma_y^{(n)} - \frac{\omega^2}{c^2}\varepsilon_{xy}^{(n)} & -\gamma_x^{(n)}\gamma_z^{(n)} - \frac{\omega^2}{c^2}\varepsilon_{xz}^{(n)} \\ -\gamma_y^{(n)}\gamma_x^{(n)} - \frac{\omega^2}{c^2}\varepsilon_{yx}^{(n)} & \gamma_z^{(n)2} + \gamma_x^{(n)2} - \frac{\omega^2}{c^2}\varepsilon_{yy}^{(n)} & -\gamma_y^{(n)}\gamma_z^{(n)} - \frac{\omega^2}{c^2}\varepsilon_{yz}^{(n)} \\ -\gamma_z^{(n)}\gamma_x^{(n)} - \frac{\omega^2}{c^2}\varepsilon_{zx}^{(n)} & -\gamma_z^{(n)}\gamma_y^{(n)} - \frac{\omega^2}{c^2}\varepsilon_{zy}^{(n)} & \gamma_x^{(n)2} + \gamma_y^{(n)2} - \frac{\omega^2}{c^2}\varepsilon_{zz}^{(n)} \end{pmatrix}$$

$$\times \begin{pmatrix} E_{0x}^{(n)} \\ E_{0y}^{(n)} \\ E_{0z}^{(n)} \end{pmatrix} = 0. \tag{3.4}$$

We write the propagation vector in the n-th layer medium in Cartesian components as

$$\gamma^{(n)} = \frac{\omega}{c}\left(N_x^{(n)}\hat{x} + N_y^{(n)}\hat{y} + N_z^{(n)}\hat{z}\right). \tag{3.5}$$

The Snell law requirement that plane waves propagate across the structure with parallel to planar interface components of their propagation vector invariant can be expressed as

$$\gamma^{(n)} \cdot \hat{x} = \frac{\omega}{c} N_x^{(0)}, \tag{3.6a}$$

$$\gamma^{(n)} \cdot \hat{y} = \frac{\omega}{c} N_y^{(0)}, \tag{3.6b}$$

for $n = 1$ to $\mathcal{N}+1$. For given values of N_x and N_y, the equation system (3.4) provides the proper values of $N_{zj}^{(n)}$, $j = 1, ..., 4$, which are the solutions of the proper value equation expressed as a determinant

$$\begin{vmatrix} \varepsilon_{xx}^{(n)} - N_y^2 - N_z^{(n)2} & \varepsilon_{xy}^{(n)} + N_x N_y & \varepsilon_{xz}^{(n)} + N_x N_z^{(n)} \\ \varepsilon_{yx}^{(n)} + N_x N_y & \varepsilon_{yy}^{(n)} - N_x^2 - N_z^{(n)2} & \varepsilon_{yz}^{(n)} + N_y N_z^{(n)} \\ \varepsilon_{zx}^{(n)} + N_x N_z^{(n)} & \varepsilon_{zy}^{(n)} + N_y N_z^{(n)} & \varepsilon_{zz}^{(n)} - N_x^2 - N_y^2 \end{vmatrix} = 0. \tag{3.7}$$

We arrive at a quartic equation of the following form

$$\begin{aligned}
&\varepsilon_{zz}^{(n)} N_z^{(n)4} + \left[N_x\left(\varepsilon_{zx}^{(n)} + \varepsilon_{xz}^{(n)}\right) + N_y\left(\varepsilon_{yz}^{(n)} + \varepsilon_{zy}^{(n)}\right)\right] N_z^{(n)3} \\
&- \left[\varepsilon_{zz}^{(n)}\left(\varepsilon_{xx}^{(n)} - N_x^2\right) + \varepsilon_{zz}^{(n)}\left(\varepsilon_{yy}^{(n)} - N_y^2\right) - \varepsilon_{xx}^{(n)} N_x^2 - \varepsilon_{yy}^{(n)} N_y^2\right. \\
&\left. - N_x N_y\left(\varepsilon_{xy}^{(n)} + \varepsilon_{yx}^{(n)}\right) - \varepsilon_{zx}^{(n)}\varepsilon_{xz}^{(n)} - \varepsilon_{yz}^{(n)}\varepsilon_{zy}^{(n)}\right] N_z^{(n)2} \\
&- \left[\left(\varepsilon_{xx}^{(n)} - N_y^2\right) N_y\left(\varepsilon_{yz}^{(n)} + \varepsilon_{zy}^{(n)}\right) + \left(\varepsilon_{yy}^{(n)} - N_x^2\right) N_x\left(\varepsilon_{zx}^{(n)} + \varepsilon_{xz}^{(n)}\right)\right. \\
&\left. - \left(\varepsilon_{xy}^{(n)} + N_x N_y\right)\left(N_x\varepsilon_{yz}^{(n)} + N_y\varepsilon_{zx}^{(n)}\right) - \left(\varepsilon_{yx}^{(n)} + N_x N_y\right)\left(N_x\varepsilon_{zy}^{(n)} + N_y\varepsilon_{xz}^{(n)}\right)\right] N_z^{(n)} \\
&+ \left(\varepsilon_{zz}^{(n)} - N_x^2 - N_y^2\right)\left[\left(\varepsilon_{xx}^{(n)} - N_y^2\right)\left(\varepsilon_{yy}^{(n)} - N_x^2\right) - \left(\varepsilon_{xy}^{(n)} + N_x N_y\right)\left(\varepsilon_{yx}^{(n)} + N_x N_y\right)\right] \\
&- \left(\varepsilon_{xx}^{(n)} - N_y^2\right)\varepsilon_{yz}^{(n)}\varepsilon_{zy}^{(n)} - \left(\varepsilon_{yy}^{(n)} - N_x^2\right)\varepsilon_{zx}^{(n)}\varepsilon_{xz}^{(n)} \\
&+ N_x N_y\left(\varepsilon_{yz}^{(n)}\varepsilon_{zx}^{(n)} + \varepsilon_{zy}^{(n)}\varepsilon_{xz}^{(n)}\right) + \varepsilon_{xy}^{(n)}\varepsilon_{zx}^{(n)}\varepsilon_{yz}^{(n)} + \varepsilon_{yx}^{(n)}\varepsilon_{xz}^{(n)}\varepsilon_{zy}^{(n)} = 0.
\end{aligned} \tag{3.8}$$

Under special circumstances, this equation reduces to a biquadratic one. One common case is represented by an originally isotropic medium uniformly magnetized along an arbitrary axis, for which $\varepsilon_{ij} = -\varepsilon_{ji}$, $i, j = x, y$, and $z (i \neq j)$. In magnetic crystals or in media displaying induced anisotropy, the form of the permittivity tensor is more complicated and the situation requires the general approach outlined below. The electric field proper polarizations (or the proper polarization modes) follow, *e.g.*, from the first and third equations of the system (3.4) and can be written as the Cartesian vectors

$$e_j^{(n)} = \begin{pmatrix} \left(\varepsilon_{zz}^{(n)} - N_x^2 - N_y^2\right)\left(N_x N_y + \varepsilon_{xy}^{(n)}\right) - \left(N_x N_{zj}^{(n)} + \varepsilon_{xz}^{(n)}\right)\left(N_y N_{zj}^{(n)} + \varepsilon_{zy}^{(n)}\right) \\ -\left(\varepsilon_{zz}^{(n)} - N_x^2 - N_y^2\right)\left(\varepsilon_{xx}^{(n)} - N_y^2 - N_{zj}^{(n)2}\right) + \left(N_x N_{zj}^{(n)} + \varepsilon_{zx}^{(n)}\right)\left(N_x N_{zj}^{(n)} + \varepsilon_{xz}^{(n)}\right) \\ \left(\varepsilon_{xx}^{(n)} - N_y^2 - N_{zj}^{(n)2}\right)\left(N_y N_{zj}^{(n)} + \varepsilon_{zy}^{(n)}\right) - \left(N_x N_{zj}^{(n)} + \varepsilon_{zx}^{(n)}\right)\left(N_x N_y + \varepsilon_{xy}^{(n)}\right) \end{pmatrix}, \tag{3.9}$$

where $j = 1, \ldots, 4$, distinguishes the proper values of the z component of the reduced propagation vector, $N_{zj}^{(n)}$, and the associated proper polarizations $e_j^{(n)}$. The way we number the proper values, $N_{zj}^{(n)}$, and the associated proper polarizations in a given layer medium is a matter of convention.[1]

Let us assume that the complex propagation vectors in one of the isotropic half spaces are restricted to a plane perpendicular to the interfaces, which defines the plane of incidence. The problem can then be simplified by eliminating the Cartesian component of the complex propagation vector parallel to the interfaces and normal to the plane of incidence. This is achieved by an appropriate rotation of the coordinate system. There is no loss in generality. Any rotation transforms the original general permittivity tensor to another tensor with the same number of independent components. Choosing the plane of incidence parallel to the yz plane and normal to the x axis, we have $N_x = 0$ and the propagation vector is reduced to

$$\gamma^{(n)} = \frac{\omega}{c}\left(N_y^{(n)}\hat{y} + N_z^{(n)}\hat{z}\right). \tag{3.10}$$

Snell law (3.6) simplifies to

$$\gamma^{(n)} \cdot \hat{y} = \frac{\omega}{c} N_y. \tag{3.11}$$

If the entrance medium is a vacuum, N_y simply represents the sine of the angle of incidence. The wave equation (3.4) reduces to the form

$$\begin{pmatrix} \left(N_y^2 + N_z^{(n)2} - \varepsilon_{xx}^{(n)}\right) & -\varepsilon_{xy}^{(n)} & -\varepsilon_{xz}^{(n)} \\ -\varepsilon_{yx}^{(n)} & \left(N_z^{(n)2} - \varepsilon_{yy}^{(n)}\right) & -\left(N_y N_z^{(n)} + \varepsilon_{yz}^{(n)}\right) \\ -\varepsilon_{zx}^{(n)} & -\left(N_y N_z^{(n)} + \varepsilon_{zy}^{(n)}\right) & \left(N_y^2 - \varepsilon_{zz}^{(n)}\right) \end{pmatrix} \begin{pmatrix} E_{0x}^{(n)} \\ E_{0y}^{(n)} \\ E_{0z}^{(n)} \end{pmatrix} = 0. \tag{3.12}$$

[1] In higher symmetry media, the proper polarization degeneracy may take place leading to indefinite expressions for the proper polarization vectors. In most cases, these can be removed by inspection. The proper polarizations in high symmetry media can always be found from the wave equation with a particular form of the permittivity tensor.

The proper value equation to the wave equation system (3.12) follows from Eq. (3.8) by setting $N_x = 0$

$$\begin{aligned}
&\varepsilon_{zz}^{(n)} N_z^{(n)4} + \left(\varepsilon_{yz}^{(n)} + \varepsilon_{zy}^{(n)}\right) N_y N_z^{(n)3} \\
&\quad - \left[\varepsilon_{yy}^{(n)}\left(\varepsilon_{zz}^{(n)} - N_y^2\right) + \varepsilon_{zz}^{(n)}\left(\varepsilon_{xx}^{(n)} - N_y^2\right) - \varepsilon_{xz}^{(n)}\varepsilon_{zx}^{(n)} - \varepsilon_{yz}^{(n)}\varepsilon_{zy}^{(n)}\right] N_z^{(n)2} \\
&\quad - \left[\left(\varepsilon_{xx}^{(n)} - N_y^2\right)\left(\varepsilon_{yz}^{(n)} + \varepsilon_{zy}^{(n)}\right) - \varepsilon_{xy}^{(n)}\varepsilon_{zx}^{(n)} - \varepsilon_{yx}^{(n)}\varepsilon_{xz}^{(n)}\right] N_y N_z^{(n)} \\
&\quad + \varepsilon_{yy}^{(n)}\left[\left(\varepsilon_{xx}^{(n)} - N_y^2\right)\left(\varepsilon_{zz}^{(n)} - N_y^2\right) - \varepsilon_{xz}^{(n)}\varepsilon_{zx}^{(n)}\right] - \varepsilon_{xy}^{(n)}\varepsilon_{yx}^{(n)}\left(\varepsilon_{zz}^{(n)} - N_y^2\right) \\
&\quad - \varepsilon_{yz}^{(n)}\varepsilon_{zy}^{(n)}\left(\varepsilon_{xx}^{(n)} - N_y^2\right) + \varepsilon_{xy}^{(n)}\varepsilon_{zx}^{(n)}\varepsilon_{yz}^{(n)} + \varepsilon_{yx}^{(n)}\varepsilon_{xz}^{(n)}\varepsilon_{zy}^{(n)} = 0.
\end{aligned} \tag{3.13}$$

The electric field in the n-th layer can be written as

$$\boldsymbol{E}^{(n)} = \sum_{j=1}^{4} \mathrm{E}_{0j}^{(n)} \boldsymbol{e}_j^{(n)} \exp\left\{\mathrm{j}\omega t - \mathrm{j}\frac{\omega}{c}\left[N_y y + N_{zj}^{(n)}\left(z - z^{(n)}\right)\right]\right\}, \tag{3.14}$$

where $\mathrm{E}_{0j}^{(n)}$ is the complex amplitude of the j-th proper polarization mode and $\boldsymbol{e}_j^{(n)}$ is the complex vector that specifies the proper polarization mode polarization. Note that $\mathrm{E}_{0j}^{(n)}$ represents the amplitude of the modal field in the plane $z = z^{(n)}$ inside the n-th layer at the interface shared with the $(n+1)$-st layer. It is sometimes advantageous to work with normalized proper polarizations and to impose

$$\left[\boldsymbol{e}_j^{(n)}\right]^{\dagger} \boldsymbol{e}_j^{(n)} = 1, \tag{3.15}$$

where the dagger † indicates Hermitian adjoint. The total phase of the product $\mathrm{E}_{0j}^{(n)} \boldsymbol{e}_j^{(n)}$ can be arbitrarily distributed between $\mathrm{E}_{0j}^{(n)}$ and $\boldsymbol{e}_j^{(n)}$. In the classical ellipsometry of isotropic surfaces, the distribution is a matter of convention [18]. There, the positive sense of the unit vectors characterizing linear polarizations, either perpendicular or parallel with respect to the plane of incidence, is determined with respect to the propagation vectors of the incident, reflected, and refracted waves. Here, we determine the positive sense of $\boldsymbol{e}_j^{(n)}$ in a coordinate system fixed to the multilayer. From the first and third row of the matrix wave equation (3.13), we have

$$\boldsymbol{e}_j^{(n)} = C_j^{(n)} \begin{pmatrix} -\varepsilon_{xy}^{(n)}\left(\varepsilon_{zz}^{(n)} - N_y^2\right) + \varepsilon_{xz}^{(n)}\left(\varepsilon_{zy}^{(n)} + N_y N_{zj}^{(n)}\right) \\ \left(\varepsilon_{zz}^{(n)} - N_y^2\right)\left(\varepsilon_{xx}^{(n)} - N_y^2 - N_{zj}^{(n)2}\right) - \varepsilon_{xz}^{(n)}\varepsilon_{zx}^{(n)} \\ -\left(\varepsilon_{xx}^{(n)} - N_y^2 - N_{zj}^{(n)2}\right)\left(\varepsilon_{zy}^{(n)} + N_y N_{zj}^{(n)}\right) + \varepsilon_{zx}^{(n)}\varepsilon_{xy}^{(n)} \end{pmatrix} \tag{3.16}$$

or

$$\begin{aligned}\boldsymbol{e}_j^{(n)} = C_j^{(n)} \Big\{ \hat{\boldsymbol{x}} \Big[-\varepsilon_{xy}^{(n)} \left(\varepsilon_{zz}^{(n)} - N_y^2\right) + \varepsilon_{xz}^{(n)} \left(\varepsilon_{zy}^{(n)} + N_y N_{zj}^{(n)}\right) \Big] \\ + \hat{\boldsymbol{y}} \Big[\left(\varepsilon_{zz}^{(n)} - N_y^2\right) \left(\varepsilon_{xx}^{(n)} - N_y^2 - N_{zj}^{(n)2}\right) - \varepsilon_{xz}^{(n)} \varepsilon_{zx}^{(n)} \Big] \\ + \hat{\boldsymbol{z}} \Big[- \left(\varepsilon_{xx}^{(n)} - N_y^2 - N_{zj}^{(n)2}\right) \left(\varepsilon_{zy}^{(n)} + N_y N_{zj}^{(n)}\right) + \varepsilon_{zx}^{(n)} \varepsilon_{xy}^{(n)} \Big] \Big\}, \end{aligned} \tag{3.17}$$

where $C_j^{(n)}$ is the corresponding normalizing coefficient. As in Eq. (3.9), the determination of the proper polarizations from Eq. (3.16) or Eq. (3.17) may lead to singularities when the general permittivity tensor is reduced to a special form in higher symmetry media. These can be eliminated analytically or by appropriate numerical procedures. The magnetic fields of the proper polarization modes are obtained from the Faraday law, $\nabla \times \boldsymbol{E} = -\partial \boldsymbol{B}/\partial t$

$$c\boldsymbol{B}^{(n)} = \sum_{j=1}^{4} E_{0j}^{(n)} \boldsymbol{b}_j^{(n)} \exp\left\{ \mathrm{j}\omega t - \mathrm{j}\frac{\omega}{c} \left[N_y y + N_{zj}^{(n)} (z - z^{(n)}) \right] \right\}, \tag{3.18}$$

here,

$$\boldsymbol{b}_j^{(n)} = C_j^{(n)} \begin{pmatrix} -\left(\varepsilon_{xx}^{(n)} - N_y^2 - N_{zj}^{(n)2}\right)\left(N_{zj}^{(n)} \varepsilon_{zz}^{(n)} + N_y \varepsilon_{zy}^{(n)}\right) + \varepsilon_{zx}^{(n)} \left(N_y \varepsilon_{xy}^{(n)} + N_{zj}^{(n)} \varepsilon_{xz}^{(n)}\right) \\ N_{zj}^{(n)} \left[-\varepsilon_{xy}^{(n)} \left(\varepsilon_{zz}^{(n)} - N_y^2\right) + \varepsilon_{xz}^{(n)} \left(\varepsilon_{zy}^{(n)} + N_y N_{zj}^{(n)}\right) \right] \\ -N_y \left[-\varepsilon_{xy}^{(n)} \left(\varepsilon_{zz}^{(n)} - N_y^2\right) + \varepsilon_{xz}^{(n)} \left(\varepsilon_{zy}^{(n)} + N_y N_{zj}^{(n)}\right) \right] \end{pmatrix}, \tag{3.19}$$

which was obtained from

$$\begin{aligned} \boldsymbol{b}_j^{(n)} &= \left(N_y \hat{\boldsymbol{y}} + N_{zj}^{(n)} \hat{\boldsymbol{z}}\right) \times \boldsymbol{e}_j^{(n)} \\ &= C_j^{(n)} \Big\{ \hat{\boldsymbol{x}} \Big[- \left(\varepsilon_{xx}^{(n)} - N_y^2 - N_{zj}^{(n)2}\right) \left(N_{zj}^{(n)} \varepsilon_{zz}^{(n)} + N_y \varepsilon_{zy}^{(n)}\right) \\ &\quad + \varepsilon_{zx}^{(n)} \left(N_y \varepsilon_{xy}^{(n)} + N_{zj}^{(n)} \varepsilon_{xz}^{(n)}\right) \Big] \\ &\quad + \hat{\boldsymbol{y}} N_{zj}^{(n)} \Big[-\varepsilon_{xy}^{(n)} \left(\varepsilon_{zz}^{(n)} - N_y^2\right) + \varepsilon_{xz}^{(n)} \left(\varepsilon_{zy}^{(n)} + N_y N_{zj}^{(n)}\right) \Big] \\ &\quad - \hat{\boldsymbol{z}} N_y \Big[-\varepsilon_{xy}^{(n)} \left(\varepsilon_{zz}^{(n)} - N_y^2\right) + \varepsilon_{xz}^{(n)} \left(\varepsilon_{zy}^{(n)} + N_y N_{zj}^{(n)}\right) \Big] \Big\}. \end{aligned} \tag{3.20}$$

The mean value of the mode Poynting vector for the j-th mode is given by the real part (denoted by $\Re$) of the vector product

$$\boldsymbol{S}_j^{(n)} = \frac{1}{2\mu_{\text{vac}}} \Re \left[\boldsymbol{E}_j^{(n)*} \times \boldsymbol{B}_j^{(n)} \right], \tag{3.21}$$

in watts/m^2, where asterisk * denotes the complex conjugate and $\mu_{\text{vac}} = 4\pi \times 10^{-7} \text{mkgs}^{-2}\text{A}^{-2}$ is the magnetic permeability in a vacuum. Using

Eq. (3.14) and Eqs. (3.16) to (3.19), we obtain

$$\mathbf{S}_j^{(n)}(z) = \frac{|\mathrm{E}_{0j}^{(n)}|^2}{2Z_{\mathrm{vac}}} \mathrm{Re}\left\{ \left(\boldsymbol{e}_j^{(n)}\right)^* \times \left[\left(N_y \hat{\boldsymbol{y}} + N_{zj}^{(n)} \hat{\boldsymbol{z}} \right) \times \boldsymbol{e}_j^{(n)} \right] \right\} \times \exp\left[-\frac{2\omega}{c} k_{zj}^{(n)} \left(z - z^{(n)}\right) \right], \tag{3.22}$$

where $Z_{\mathrm{vac}} = 376.63$ ohms is the vacuum impedance. We have to write $N_{zj}^{(n)} = n_{zj}^{(n)} - \mathrm{j}k_{zj}^{(n)}$ because of our sign choice in the complex exponential of Eq. (3.2). Then, $(2\omega/c)|k_{zj}^{(n)}|$ can be understood as the absorption coefficient of the j-th mode in the n-th layer.

3.3 Matrix Representation of Planar Structures

The requirement of the continuity of the tangential field components at the interface $z = z^{(n-1)}$ separating the $(n-1)$-th and n-th layers provides the conditions

$$\sum_{j=1}^{4} \mathrm{E}_{0j}^{(n-1)} \boldsymbol{e}_j^{(n-1)} \cdot \hat{\boldsymbol{x}} = \sum_{j=1}^{4} \mathrm{E}_{0j}^{(n)} \boldsymbol{e}_j^{(n)} \cdot \hat{\boldsymbol{x}} \exp\left(\mathrm{j}\frac{\omega}{c} N_{zj}^{(n)} d^{(n)} \right), \tag{3.23a}$$

$$\sum_{j=1}^{4} \mathrm{E}_{0j}^{(n-1)} \boldsymbol{e}_j^{(n-1)} \cdot \hat{\boldsymbol{y}} = \sum_{j=1}^{4} \mathrm{E}_{0j}^{(n)} \boldsymbol{e}_j^{(n)} \cdot \hat{\boldsymbol{y}} \exp\left(\mathrm{j}\frac{\omega}{c} N_{zj}^{(n)} d^{(n)} \right), \tag{3.23b}$$

$$\sum_{j=1}^{4} \mathrm{E}_{0j}^{(n-1)} \boldsymbol{b}_j^{(n-1)} \cdot \hat{\boldsymbol{x}} = \sum_{j=1}^{4} \mathrm{E}_{0j}^{(n)} \boldsymbol{b}_j^{(n)} \cdot \hat{\boldsymbol{x}} \exp\left(\mathrm{j}\frac{\omega}{c} N_{zj}^{(n)} d^{(n)} \right), \tag{3.23c}$$

$$\sum_{j=1}^{4} \mathrm{E}_{0j}^{(n-1)} \boldsymbol{b}_j^{(n-1)} \cdot \hat{\boldsymbol{y}} = \sum_{j=1}^{4} \mathrm{E}_{0j}^{(n)} \boldsymbol{b}_j^{(n)} \cdot \hat{\boldsymbol{y}} \exp\left(\mathrm{j}\frac{\omega}{c} N_{zj}^{(n)} d^{(n)} \right). \tag{3.23d}$$

Here, $d^{(n)}$ is the thickness of the n-th layer. The introduction of the layer thickness removes the dependence of the equations on the position of the interface on the z axis. This step is important for the construction of the matrix formalism to be explained below. The modal electric field amplitudes at the interface $z^{(n-1)}$ in the n-th layer are given by the modal amplitudes $\mathrm{E}_{0j}^{(n)}$ in the same layer at the opposite interface $z^{(n)}$ multiplied by the factor $\exp(\mathrm{j}\frac{\omega}{c} N_{zj}^{(n)} d^{(n)})$. Note that the energy dissipated per unit area in the n-th layer of the structure is represented by the sum over the proper polarization modes

$$W_{\mathrm{dis}}^{(n)} = \sum_{j=1}^{4} \left| \boldsymbol{S}_j^{(n)}(0) - \boldsymbol{S}_j^{(n)}(d^{(n)}) \right| . \tag{3.24}$$

Eqs. (3.23) can be expressed in a matrix form, with a rearranged row order, as

$$\begin{pmatrix} \boldsymbol{e}_1^{(n-1)}\cdot\hat{x} & \boldsymbol{e}_2^{(n-1)}\cdot\hat{x} & \boldsymbol{e}_3^{(n-1)}\cdot\hat{x} & \boldsymbol{e}_4^{(n-1)}\cdot\hat{x} \\ \boldsymbol{b}_1^{(n-1)}\cdot\hat{y} & \boldsymbol{b}_2^{(n-1)}\cdot\hat{y} & \boldsymbol{b}_3^{(n-1)}\cdot\hat{y} & \boldsymbol{b}_4^{(n-1)}\cdot\hat{y} \\ \boldsymbol{e}_1^{(n-1)}\cdot\hat{y} & \boldsymbol{e}_2^{(n-1)}\cdot\hat{y} & \boldsymbol{e}_3^{(n-1)}\cdot\hat{y} & \boldsymbol{e}_4^{(n-1)}\cdot\hat{y} \\ \boldsymbol{b}_1^{(n-1)}\cdot\hat{x} & \boldsymbol{b}_2^{(n-1)}\cdot\hat{x} & \boldsymbol{b}_3^{(n-1)}\cdot\hat{x} & \boldsymbol{b}_4^{(n-1)}\cdot\hat{x} \end{pmatrix} \begin{pmatrix} \mathrm{E}_{01}^{(n-1)} \\ \mathrm{E}_{02}^{(n-1)} \\ \mathrm{E}_{03}^{(n-1)} \\ \mathrm{E}_{04}^{(n-1)} \end{pmatrix}$$

$$= \begin{pmatrix} \boldsymbol{e}_1^{(n)}\cdot\hat{x} & \boldsymbol{e}_2^{(n)}\cdot\hat{x} & \boldsymbol{e}_3^{(n)}\cdot\hat{x} & \boldsymbol{e}_4^{(n)}\cdot\hat{x} \\ \boldsymbol{b}_1^{(n)}\cdot\hat{y} & \boldsymbol{b}_2^{(n)}\cdot\hat{y} & \boldsymbol{b}_3^{(n)}\cdot\hat{y} & \boldsymbol{b}_4^{(n)}\cdot\hat{y} \\ \boldsymbol{e}_1^{(n)}\cdot\hat{y} & \boldsymbol{e}_2^{(n)}\cdot\hat{y} & \boldsymbol{e}_3^{(n)}\cdot\hat{y} & \boldsymbol{e}_4^{(n)}\cdot\hat{y} \\ \boldsymbol{b}_1^{(n)}\cdot\hat{x} & \boldsymbol{b}_2^{(n)}\cdot\hat{x} & \boldsymbol{b}_3^{(n)}\cdot\hat{x} & \boldsymbol{b}_4^{(n)}\cdot\hat{x} \end{pmatrix}$$

$$\times \begin{pmatrix} \exp\left(\mathrm{j}\frac{\omega}{c}N_{z1}^{(n)}d^{(n)}\right) & 0 & 0 & 0 \\ 0 & \exp\left(\mathrm{j}\frac{\omega}{c}N_{z2}^{(n)}d^{(n)}\right) & 0 & 0 \\ 0 & 0 & \exp\left(\mathrm{j}\frac{\omega}{c}N_{z3}^{(n)}d^{(n)}\right) & 0 \\ 0 & 0 & 0 & \exp\left(\mathrm{j}\frac{\omega}{c}N_{z4}^{(n)}d^{(n)}\right) \end{pmatrix}$$

$$\times \begin{pmatrix} \mathrm{E}_{01}^{n} \\ \mathrm{E}_{02}^{n} \\ \mathrm{E}_{03}^{n} \\ \mathrm{E}_{04}^{n} \end{pmatrix} . \tag{3.25}$$

This can be concisely rewritten as

$$\mathbf{D}^{(n-1)}\mathbf{E}_0^{(n-1)} = \mathbf{D}^{(n)}\mathbf{P}^{(n)}\mathbf{E}_0^{(n)} , \tag{3.26}$$

or

$$\mathbf{E}_0^{(n-1)} = \left(\mathbf{D}^{(n-1)}\right)^{-1}\mathbf{D}^{(n)}\mathbf{P}^{(n)}\mathbf{E}_0^{(n)} , \tag{3.27}$$

where $\mathbf{E}_0^{(n)}$ is a four-component vector of the complex proper polarization amplitudes

$$\mathbf{E}_0^{(n)} = \begin{pmatrix} E_{01}^{(n)} \\ E_{02}^{(n)} \\ E_{03}^{(n)} \\ E_{04}^{(n)} \end{pmatrix} \tag{3.28}$$

and $\mathbf{D}^{(n)}$ is a 4×4 matrix

$$\mathbf{D}^{(n)} = \begin{pmatrix} \boldsymbol{e}_1^{(n)} \cdot \hat{x} & \boldsymbol{e}_2^{(n)} \cdot \hat{x} & \boldsymbol{e}_3^{(n)} \cdot \hat{x} & \boldsymbol{e}_4^{(n)} \cdot \hat{x} \\ \boldsymbol{b}_1^{(n)} \cdot \hat{y} & \boldsymbol{b}_2^{(n)} \cdot \hat{y} & \boldsymbol{b}_3^{(n)} \cdot \hat{y} & \boldsymbol{b}_4^{(n)} \cdot \hat{y} \\ \boldsymbol{e}_1^{(n)} \cdot \hat{y} & \boldsymbol{e}_2^{(n)} \cdot \hat{y} & \boldsymbol{e}_3^{(n)} \cdot \hat{y} & \boldsymbol{e}_4^{(n)} \cdot \hat{y} \\ \boldsymbol{b}_1^{(n)} \cdot \hat{x} & \boldsymbol{b}_2^{(n)} \cdot \hat{x} & \boldsymbol{b}_3^{(n)} \cdot \hat{x} & \boldsymbol{b}_4^{(n)} \cdot \hat{x} \end{pmatrix}, \tag{3.29}$$

the elements $D_{ij}^{(n)}$ of which are obtained from Eq. (3.16) or Eq. (3.17) and Eq. (3.18)

$$D_{1j}^{(n)} = \boldsymbol{e}_j^{(n)} \cdot \hat{x} = C_j^{(n)} \left[-\varepsilon_{xy}^{(n)} \left(\varepsilon_{zz}^{(n)} - N_y^2 \right) + \varepsilon_{xz}^{(n)} \left(\varepsilon_{zy}^{(n)} + N_y N_{zj}^{(n)} \right) \right] \tag{3.30a}$$

$$D_{2j}^{(n)} = \boldsymbol{b}_j^{(n)} \cdot \hat{y} = C_j^{(n)} N_{zj}^{(n)} \left[-\varepsilon_{xy}^{(n)} \left(\varepsilon_{zz}^{(n)} - N_y^2 \right) + \varepsilon_{xz}^{(n)} \left(\varepsilon_{zy}^{(n)} + N_y N_{zj}^{(n)} \right) \right] \tag{3.30b}$$

$$D_{3j}^{(n)} = \boldsymbol{e}_j^{(n)} \cdot \hat{y} = C_j^{(n)} \left[\left(\varepsilon_{zz}^{(n)} - N_y^2 \right) \left(\varepsilon_{xx}^{(n)} - N_y^2 - N_{zj}^{(n)2} \right) - \varepsilon_{xz}^{(n)} \varepsilon_{zx}^{(n)} \right] \tag{3.30c}$$

$$D_{4j}^{(n)} = \boldsymbol{b}_j^{(n)} \cdot \hat{x} = C_j^{(n)} \left[-\left(\varepsilon_{xx}^{(n)} - N_y^2 - N_{zj}^{(n)2} \right) \left(N_y \varepsilon_{zy}^{(n)} + N_{zj}^{(n)} \varepsilon_{zz}^{(n)} \right) \right.$$
$$\left. + N_{zj}^{(n)} \varepsilon_{xz}^{(n)} \varepsilon_{zx}^{(n)} + N_y \varepsilon_{zx}^{(n)} \varepsilon_{xy}^{(n)} \right]. \tag{3.30d}$$

We recall the note already made in connection with Eq. (3.16) on singularities in higher symmetry media, which should be accounted for in the design of numerical procedures [19]. Here, $\mathbf{P}^{(n)}$ is a diagonal matrix

$$\mathbf{P}^{(n)} = \begin{pmatrix} \exp\left(j\frac{\omega}{c} N_{z1}^{(n)} d^{(n)}\right) & 0 & 0 & 0 \\ 0 & \exp\left(j\frac{\omega}{c} N_{z2}^{(n)} d^{(n)}\right) & 0 & 0 \\ 0 & 0 & \exp\left(j\frac{\omega}{c} N_{z3}^{(n)} d^{(n)}\right) & 0 \\ 0 & 0 & 0 & \exp\left(j\frac{\omega}{c} N_{z4}^{(n)} d^{(n)}\right) \end{pmatrix}. \tag{3.31}$$

In thick absorbing layers, the $\mathbf{P}^{(n)}$ matrix elements may cause the overflow in the numerical evaluation. In such a situation, there is no transmitted

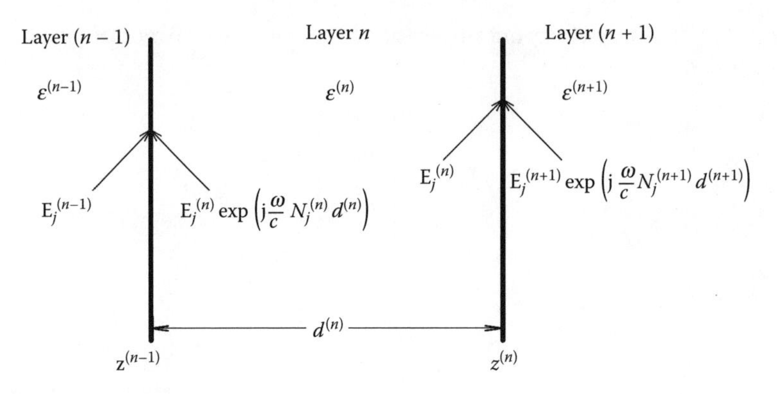

FIGURE 3.2
Transfer matrix relates the proper polarization fields of amplitudes $E_j^{(n-1)}$ in the layer $(n-1)$ at its right interface with the corresponding fields $E_j^{(n)}$ at the right interface of the layer n.

wave, and the problem can be easily solved by a division of all elements of $\mathbf{P}^{(n)}$ by an appropriate reduction factor.

The matrix $\mathbf{D}^{(n)}$, so-called *dynamical matrix*, the matrix $\mathbf{P}^{(n)}$, so-called *propagation matrix*, and the four-component vector $\mathbf{E}^{(n)}$ have been introduced by Yeh [14,15]. The crucial relation between the field $\mathbf{E}_0^{(n-1)}$ in the $(n-1)$-th layer at the interface $z = z^{(n-1)}$ and the field $\mathbf{E}_0^{(n)}$ in the n-th layer at the interface $z = z^{(n)}$ is given by Eq. (3.27). The relation may be expressed with the so-called *transfer matrix* defined by Yeh as

$$\mathbf{T}^{(n-1,n)} = \left(\mathbf{D}^{(n-1)}\right)^{-1} \mathbf{D}^{(n)}\mathbf{P}^{(n)}. \tag{3.32}$$

The transfer matrix represents a building block of the 4×4 matrix formalism (Figure 3.2). At the zero thickness, $d^{(n)} = 0$, $\mathbf{P}^{(n)} = \mathbf{1}$. Then, the transfer matrix simplifies to $\mathbf{T}^{(n-1,n)} = (\mathbf{D}^{(n-1)})^{-1}\mathbf{D}^{(n)}$. It relates the field amplitudes in the planes on either side of the interface $z = z_{n-1}$ between the $(n-1)$-th and the n-th layers. In this special case, these planes are situated at a distance from the interface plane and from each other, which tends to zero ($z_n \to z_{n-1}$). The two adjacent media may be characterized by general permittivity tensors. The special form of the transfer matrix with $\mathbf{P}^{(n)} = \mathbf{1}$ represents an interface between two half spaces, which may be characterized by general permittivity tensors. More importantly, the transfer matrix with $\mathbf{P}^{(n)} = \mathbf{1}$ is employed as a last segment in the matrix product representing the multilayer.

Indeed, the product of transfer matrices relates the field amplitudes in the isotropic half spaces in the sandwiching planes, which in the limit coincide with the multilayer boundaries at $z = z^{(0)}$ and $z = z^{(\mathcal{N})}$

$$\mathbf{E}_0^{(0)} = \prod_{n=1}^{\mathcal{N}+1} \mathbf{T}^{(n-1,n)} \mathbf{E}_0^{(\mathcal{N}+1)} = \mathbf{M}\mathbf{E}_0^{(\mathcal{N}+1)}. \tag{3.33}$$

According to what has been just said above, the last transfer matrix of the product is given by

$$\mathbf{T}^{(\mathcal{N},\mathcal{N}+1)} = \left(\mathbf{D}^{(\mathcal{N})}\right)^{-1} \mathbf{D}^{(\mathcal{N}+1)}. \tag{3.34}$$

The field distribution $\mathbf{E}^{(m)}$ $(1 \leq m \leq \mathcal{N})$ in a particular layer of the structure (occupying the region $z^{(m-1)} < z < z^{(m)}$) can be obtained from

$$\mathbf{E}_0^{(m)} = \left(\prod_{n=1}^{m} \mathbf{T}^{(n-1,n)}\right)^{-1} \mathbf{E}_0^{(0)} \tag{3.35}$$

or

$$\mathbf{E}_0^{(m)} = \prod_{n=m+1}^{\mathcal{N}+1} \mathbf{T}^{(n-1,n)} \mathbf{E}_0^{(\mathcal{N}+1)}, \tag{3.36}$$

provided either $\mathbf{E}_0^{(0)}$ or $\mathbf{E}_0^{(\mathcal{N}+1)}$ are already known from the solution of Eq. (3.33). The field distribution inside the m-th layer follows from

$$\begin{pmatrix} \mathrm{E}_{01}^{(m)}(z) \\ \mathrm{E}_{02}^{(m)}(z) \\ \mathrm{E}_{03}^{(m)}(z) \\ \mathrm{E}_{04}^{(m)}(z) \end{pmatrix} = \begin{pmatrix} e^{j\beta_1^{(m)}} & 0 & 0 & 0 \\ 0 & e^{j\beta_2^{(m)}} & 0 & 0 \\ 0 & 0 & e^{j\beta_3^{(m)}} & 0 \\ 0 & 0 & 0 & e^{j\beta_4^{(m)}} \end{pmatrix} \begin{pmatrix} \mathrm{E}_{01}^{(m)}(z^{(m)}) \\ \mathrm{E}_{02}^{(m)}(z^{(m)}) \\ \mathrm{E}_{03}^{(m)}(z^{(m)}) \\ \mathrm{E}_{04}^{(m)}(z^{(m)}) \end{pmatrix}, \tag{3.37}$$

where $\beta_j^{(m)} = \frac{\omega}{c} N_{zj}^{(m)} (z^{(m)} - z)$ for $j = 1 \cdots 4$.

Note that the mode normalization of $\boldsymbol{e}_j^{(n)}$ is not required except for $n = 0$ and $\mathcal{N}+1$ as far as we are concerned with the fields outside the structure. To see this, let us suppose that the polarization vectors $\boldsymbol{e}_j^{(n)}$ of the n-th layer inside the structure are not normalized. If the normalizing coefficients are all set to the unity (*i.e.*, effectively removed), the dynamical matrix $\mathbf{D}^{(n)}$ built from normalized $\boldsymbol{e}_j^{(n)}$ becomes a non-normalized one, $[\mathbf{D}^{(n)}(\mathbf{C}^{(n)})^{-1}]$. Inserting the unit matrix $(\mathbf{C}^{(n)})^{-1}\mathbf{C}^{(n)} = \mathbf{1}$ into the right-hand side of Eq. (3.27), we have

$$\mathbf{E}_0^{(n-1)} = \left(\mathbf{D}^{(n-1)}\right)^{-1} \left[\mathbf{D}^{(n)}(\mathbf{C}^{(n)})^{-1}\right] \mathbf{P}^{(n)}\mathbf{C}^{(n)}\mathbf{E}_0^{(n)}, \tag{3.38}$$

where $\mathbf{C}^{(n)}$ is a 4×4 diagonal matrix of the normalizing coefficients from Eq. (3.16)

$$\mathbf{C}^{(n)} = \begin{pmatrix} C_1^{(n)} & 0 & 0 & 0 \\ 0 & C_2^{(n)} & 0 & 0 \\ 0 & 0 & C_3^{(n)} & 0 \\ 0 & 0 & 0 & C_4^{(n)} \end{pmatrix}. \tag{3.39}$$

We have used the fact that the diagonal matrices $\mathbf{P}^{(n)}$ and $\mathbf{C}^{(n)}$ commute. We now express $\mathbf{E}_0^{(n)}$ in terms of the product of the transfer matrix $\mathbf{T}^{(n,n+1)}$ with $\mathbf{E}_0^{(n+1)}$, as in Eq. (3.27), *i.e.*,

$$\mathbf{E}_0^{(n)} = \mathbf{T}^{(n,n+1)}\mathbf{E}_0^{(n+1)} \tag{3.40}$$

with Eq. (3.32) and obtain

$$\mathbf{E}_0^{(n-1)} = \left(\mathbf{D}^{(n-1)}\right)^{-1} \left[\mathbf{D}^{(n)} \left(\mathbf{C}^{(n)}\right)^{-1}\right] \mathbf{P}^{(n)} \left[\mathbf{C}^{(n)} \left(\mathbf{D}^{(n)}\right)^{-1}\right] \mathbf{D}^{(n+1)}\mathbf{P}^{(n+1)}\mathbf{E}_0^{(n+1)}, \tag{3.41}$$

where

$$\left[\mathbf{D}^{(n)} \left(\mathbf{C}^{(n)}\right)^{-1}\right]^{-1} = \left[\mathbf{C}^{(n)} \left(\mathbf{D}^{(n)}\right)^{-1}\right]. \tag{3.42}$$

We have thus shown that the relation between $\mathbf{E}_0^{(n-1)}$ and $\mathbf{E}_0^{(n+1)}$ is independent of $\mathbf{C}^{(n)}$.

The eight components $\mathbf{E}^{(0)}$ and $\mathbf{E}_0^{(\mathcal{N}+1)}$ are related by four equations (3.33). In a more general way, the $(\mathcal{N}+1)$ transfer matrices relate the four $(\mathcal{N}+2)$ components of $\mathbf{E}_0^{(m)}$. The fields in the structure will then be completely determined if the four components of any vector $\mathbf{E}_0^{(m)}$ are given, provided the corresponding proper polarizations are normalized. Moreover, it is sufficient to specify the modal amplitudes $E_{0j}^{(m)}(z)$ of normalized proper polarization modes in an arbitrary plane $z^{(m-1)} < z < z^{(m)}$ perpendicular to the z axis inside the m-th layer. They are simply related to the amplitudes $E_{0j}^{(m)} = E_{0j}^{(m)}\left(z^{(m)}\right)$ at the interface $z = z^{(m)}$ through Eq. (3.37).

3.4 Waves in Isotropic Regions

In this section, we obtain the dynamical and propagation matrices for the simplest special case of isotropic half spaces and isotropic layers. In particular, the isotropic half spaces $z < z^{(0)}$ and $z > z^{(\mathcal{N})}$ are specified by the scalar relative permittivities $(N^{(s)})^2$, where $s = 0$ or $\mathcal{N}+1$. The propagation vectors take two possible values

$$\gamma_{\pm}^{(s)} = \frac{\omega}{c}\left[N_y\hat{\mathbf{y}} \pm \left(N^{(s)2} - N_y^2\right)^{1/2}\hat{\mathbf{z}}\right]. \tag{3.43}$$

In the nonabsorbing isotropic half spaces, as well as in nonabsorbing layers, the extinction coefficients are zero, *i.e.*, $k^{(s)} = 0$ for $s = 0, s = \mathcal{N}+1$, and for $1 \leq s \leq \mathcal{N}$ corresponding to isotropic nonabsorbing layers. Then, in the regions with $k^{(s)} = 0$, real angles of incidence, reflection, and refraction can be defined (provided the total reflection does not occur). The propagation vectors with the positive and negative normal components become

$$\gamma_{\pm}^{(s)} = \frac{\omega}{c} N^{(s)} \left[\hat{\boldsymbol{y}} \sin \varphi^{(s)} \pm \hat{\boldsymbol{z}} \cos \varphi^{(s)} \right]. \tag{3.44}$$

The substitution into the wave equation (3.12), shows that the proper polarization modes may be always chosen as two pairs of orthogonal elliptically polarized waves.

The orthogonal elliptically polarized transverse electromagnetic plane waves propagating, with the *positive* projection of the propagation vector to the z axis and the proper polarizations $\boldsymbol{e}_1^{(s)}$ and $\boldsymbol{e}_3^{(s)}$, and the orthogonal elliptically polarized transverse waves propagating with the *negative* projection of the propagation vector to the z axis and the proper polarizations $\boldsymbol{e}_2^{(s)}$ and $\boldsymbol{e}_4^{(s)}$, are characterized (apart from complex amplitudes) by the fields

$$\boldsymbol{e}_{1,2}^{(s)} \exp\left(\mathrm{j}\omega t \mp \mathrm{j}\frac{\omega}{c} N^{(s)} z\right) = \left(\hat{\boldsymbol{x}} p^{(s)} + \hat{\boldsymbol{y}} q^{(s)}\right) \exp\left(\mathrm{j}\omega t \mp \mathrm{j}\frac{\omega}{c} N^{(s)} z\right), \tag{3.45a}$$

$$\boldsymbol{e}_{3,4}^{(s)} \exp\left(\mathrm{j}\omega t \mp \mathrm{j}\frac{\omega}{c} N^{(s)} z\right) = \left(-\hat{\boldsymbol{x}} q^{(s)*} + \hat{\boldsymbol{y}} p^{(s)*}\right) \exp\left(\mathrm{j}\omega t \mp \mathrm{j}\frac{\omega}{c} N^{(s)} z\right). \tag{3.45b}$$

Here, the normalized orthogonal proper polarizations may be expressed as the Cartesian column vectors [15]

$$\boldsymbol{e}_{1,2}^{(s)} = \begin{pmatrix} p^{(s)} \\ q^{(s)} \\ 0 \end{pmatrix} \tag{3.46}$$

and

$$\boldsymbol{e}_{3,4}^{(s)} = \begin{pmatrix} -q^{(s)*} \\ p^{(s)*} \\ 0 \end{pmatrix}, \tag{3.47}$$

where the complex numbers $p^{(s)}$ and $q^{(s)}$ satisfy the normalization condition

$$p^{(s)} p^{(s)*} + q^{(s)} q^{(s)*} = 1. \tag{3.48}$$

In nonabsorbing media [18]

$$p^{(s)} = \cos\theta^{(s)} \cos\epsilon^{(s)} - \mathrm{j} \sin\theta^{(s)} \sin\epsilon^{(s)}, \tag{3.49a}$$

$$q^{(s)} = \sin\theta^{(s)} \cos\epsilon^{(s)} + \mathrm{j} \cos\theta^{(s)} \sin\epsilon^{(s)}, \tag{3.49b}$$

where $\theta^{(s)}$ and $\tan\epsilon^{(s)}$ are the azimuth and ellipticity of the polarized wave. An orthogonally polarized wave propagating in the same sense is specified by the azimuth $\theta + \pi/2$ and by the ellipticity of the opposite sign, $-\tan\epsilon$. For example, we get a set of normalized waves orthogonally linearly polarized parallel to $\hat{x}$ and $\hat{y}$ by choosing $\theta = 0$ and $\pi/2$ with $\tan\epsilon = 0$. This corresponds to $p = 1$ and $q = 0$

$$\boldsymbol{e}_{1,2}^{(s)} = \begin{pmatrix} 1 \\ 0 \\ 0 \end{pmatrix} \tag{3.50}$$

and

$$\boldsymbol{e}_{3,4}^{(s)} = \begin{pmatrix} 0 \\ 1 \\ 0 \end{pmatrix}. \tag{3.51}$$

One possible normalized orthogonal set of RCP and LCP waves is obtained with $\theta = \pi/4$ and $\tan\epsilon = \pi/4$ for RCP and $\theta = -\pi/4$ and $\tan\epsilon = -\pi/4$ for LCP. This corresponds to $p = (1 - \mathrm{j})/2$ and $q = (1 + \mathrm{j})/2$

$$\boldsymbol{e}_{1,2}^{(s)} = \frac{1}{2}\begin{pmatrix} 1-\mathrm{j} \\ 1+\mathrm{j} \\ 0 \end{pmatrix} = \frac{1+\mathrm{j}}{4}\begin{pmatrix} 1 \\ \mathrm{j} \\ 0 \end{pmatrix}, \tag{3.52}$$

and

$$\boldsymbol{e}_{3,4}^{(s)} = \frac{1}{2}\begin{pmatrix} 1-\mathrm{j} \\ -1-\mathrm{j} \\ 0 \end{pmatrix} = \frac{1+\mathrm{j}}{4}\begin{pmatrix} 1 \\ -\mathrm{j} \\ 0 \end{pmatrix} = \frac{1+\mathrm{j}}{4}\begin{pmatrix} \mathrm{j} \\ 1 \\ 0 \end{pmatrix}. \tag{3.53}$$

To get a general orientation of the propagation vector in the plane of incidence $x = 0$, we apply the rotation about the x axis by an angle $\pm\varphi^{(s)}$. For the modes propagating with the propagation vector $\gamma_+^{(s)}$, we obtain

$$\boldsymbol{e}_1^{(s)} = \begin{pmatrix} 1 & 0 & 0 \\ 0 & \cos\varphi^{(s)} & \sin\varphi^{(s)} \\ 0 & -\sin\varphi^{(s)} & \cos\varphi^{(s)} \end{pmatrix}\begin{pmatrix} p^{(s)} \\ q^{(s)} \\ 0 \end{pmatrix} = \begin{pmatrix} p^{(s)} \\ q^{(s)}\cos\varphi^{(s)} \\ -q^{(s)}\sin\varphi^{(s)} \end{pmatrix}, \tag{3.54}$$

and in the same way

$$\boldsymbol{e}_3^{(s)} = \begin{pmatrix} -q^{(s)*} \\ p^{(s)*}\cos\varphi^{(s)} \\ -p^{(s)*}\sin\varphi^{(s)} \end{pmatrix}. \tag{3.55}$$

For the propagation vector $\gamma_-^{(s)}$ with the opposite sense of its normal component, we have

$$e_2^{(s)} = \begin{pmatrix} 1 & 0 & 0 \\ 0 & \cos\varphi^{(s)} & -\sin\varphi^{(s)} \\ 0 & \sin\varphi^{(s)} & \cos\varphi^{(s)} \end{pmatrix} \begin{pmatrix} p^{(s)} \\ q^{(s)} \\ 0 \end{pmatrix} = \begin{pmatrix} p^{(s)} \\ q^{(s)}\cos\varphi^{(s)} \\ q^{(s)}\sin\varphi^{(s)} \end{pmatrix}, \tag{3.56}$$

and in the same way

$$e_4^{(s)} = \begin{pmatrix} -q^{(s)*} \\ p^{(s)*}\cos\varphi^{(s)} \\ p^{(s)*}\sin\varphi^{(s)} \end{pmatrix}. \tag{3.57}$$

In nonabsorbing media, $N_y = N^{(s)}\sin\varphi^{(s)}$ with $N^{(s)}$ and $\varphi^{(s)}$ real. For the magnetic fields of the waves with $\gamma_+^{(s)}$, we obtain using Eq. (3.19)

$$b_1^{(s)} = N^{(s)} \begin{pmatrix} -q^{(s)} \\ p^{(s)}\cos\varphi^{(s)} \\ -p^{(s)}\sin\varphi^{(s)} \end{pmatrix}, \tag{3.58a}$$

$$b_3^{(s)} = N^{(s)} \begin{pmatrix} -p^{(s)*} \\ -q^{(s)}\cos\varphi^{(s)} \\ q^{(s)*}\sin\varphi^{(s)} \end{pmatrix}, \tag{3.58b}$$

and for the waves propagating with $\gamma_-^{(s)}$

$$b_2^{(s)} = N^{(s)} \begin{pmatrix} q^{(s)} \\ -p^{(s)}\cos\varphi^{(s)} \\ -p^{(s)}\sin\varphi^{(s)} \end{pmatrix}, \tag{3.58c}$$

$$b_4^{(s)} = N^{(s)} \begin{pmatrix} p^{(s)*} \\ q^{(s)*}\cos\varphi^{(s)} \\ q^{(s)*}\sin\varphi^{(s)} \end{pmatrix}. \tag{3.58d}$$

We have made our choice in numbering the proper values of the propagation vectors,

$$\gamma_1^{(s)} = \gamma_3^{(s)} = \gamma_+^{(s)} \tag{3.59a}$$

$$\gamma_2^{(s)} = \gamma_4^{(s)} = \gamma_-^{(s)}, \tag{3.59b}$$

for the forward ($\gamma_+^{(s)}$) and retrograde ($\gamma_-^{(s)}$) waves and associated proper polarizations. This choice affects the structure of the dynamical and propagation matrices as well as that of the transfer matrices.

In nonabsorbing media $\hat{\boldsymbol{e}}_j^{(s)} \cdot \hat{\boldsymbol{b}}_j^{(s)} = 0, \hat{\boldsymbol{e}}_j^{(s)} \cdot \boldsymbol{\gamma}_j^{(s)} = 0$, and $\hat{\boldsymbol{b}}_j^{(s)} \cdot \boldsymbol{\gamma}_j^{(s)} = 0$, for $j = 1, \ldots, 4$. The associated (*i.e.*, with the same j) propagation and Poynting vectors are parallel. The forward proper polarizations are orthogonal, *i.e.*, $\hat{\boldsymbol{e}}_1^{(s)} \cdot \hat{\boldsymbol{e}}_3^{(s)}=0$. Similarly, the retrograde proper polarizations also obey the orthogonality condition, $\hat{\boldsymbol{e}}_2^{(s)} \cdot \hat{\boldsymbol{e}}_4^{(s)} = 0$. Note that neither of these relations is automatically fulfilled in anisotropic absorbing media displaying magnetic order.

Using Eqs. (3.30), we can now write the dynamical matrix $\mathbf{D}^{(s)}$ in isotropic media

$$\mathbf{D}^{(s)} = \begin{pmatrix} p^{(s)} & p^{(s)} & -q^{(s)*} & -q^{(s)*} \\ N^{(s)} p^{(s)} \cos\varphi^{(s)} & -N^{(s)} p^{(s)} \cos\varphi^{(s)} & -N^{(s)} q^{(s)*} \cos\varphi^{(s)} & N^{(s)} q^{(s)*} \cos\varphi^{(s)} \\ q^{(s)} \cos\varphi^{(s)} & q^{(s)} \cos\varphi^{(s)} & p^{(s)*} \cos\varphi^{(s)} & p^{(s)*} \cos\varphi^{(s)} \\ -N^{(s)} q^{(s)} & N^{(s)} q^{(s)} & -N^{(s)} p^{(s)*} & N^{(s)} p^{(s)*} \end{pmatrix}, \tag{3.60}$$

and its inverse

$$\left(\mathbf{D}^{(s)}\right)^{-1} = \left(2N^{(s)} \cos\varphi^{(s)}\right)^{-1} \begin{pmatrix} N^{(s)} p^{(s)*} \cos\varphi^{(s)} & p^{(s)*} & N^{(s)} q^{(s)*} & -q^{(s)*} \cos\varphi^{(s)} \\ N^{(s)} p^{(s)*} \cos\varphi^{(s)} & -p^{(s)*} & N^{(s)} q^{(s)*} & q^{(s)*} \cos\varphi^{(s)} \\ -N^{(s)} q^{(s)} \cos\varphi^{(s)} & -q^{(s)} & N^{(s)} p^{(s)} & -p^{(s)} \cos\varphi^{(s)} \\ -N^{(s)} q^{(s)} \cos\varphi^{(s)} & q^{(s)} & N^{(s)} p^{(s)} & p^{(s)} \cos\varphi^{(s)} \end{pmatrix}. \tag{3.61}$$

The propagation matrix in isotropic regions becomes

$$\boldsymbol{P}^{(s)} = \begin{pmatrix} e^{j\beta^{(s)}} & 0 & 0 & 0 \\ 0 & e^{-j\beta^{(s)}} & 0 & 0 \\ 0 & 0 & e^{j\beta^{(s)}} & 0 \\ 0 & 0 & 0 & e^{-j\beta^{(s)}} \end{pmatrix}, \tag{3.62}$$

where $\beta^{(s)} = (\omega/c)\left(N^{(s)2} - N_y^2\right)^{1/2} d^{(s)}$.

3.5 Reflection and Transmission

Eq. (3.30) and Eq. (3.31) for anisotropic regions and Eqs. (3.60) to (3.62) for isotropic regions allow us to obtain the transfer matrices defined in Eq. (3.32) for n equal to 1 through $(\mathcal{N} + 1)$. According to Eq. (3.33), the product of the transfer matrices specifies the total matrix **M** of the structure

$$\begin{pmatrix} E_{01}^{(0)} \\ E_{02}^{(0)} \\ E_{03}^{(0)} \\ E_{04}^{(0)} \end{pmatrix} = \begin{pmatrix} M_{11} & M_{12} & M_{13} & M_{14} \\ M_{21} & M_{22} & M_{23} & M_{24} \\ M_{31} & M_{32} & M_{33} & M_{34} \\ M_{41} & M_{42} & M_{43} & M_{44} \end{pmatrix} \begin{pmatrix} E_{01}^{(\mathcal{N}+1)} \\ E_{02}^{(\mathcal{N}+1)} \\ E_{03}^{(\mathcal{N}+1)} \\ E_{04}^{(\mathcal{N}+1)} \end{pmatrix}. \tag{3.63}$$

Eq. (3.63) provides the required information on the electromagnetic response of the multilayer structure to the fields incident from the sandwiching half spaces. To simplify the situation, we set the amplitudes of the waves propagating towards the structure from one of the isotropic half spaces equal to ϕ. Choosing, *e.g.*, the half space $(\mathcal{N}+1)$, we have for $z > z^{(\mathcal{N})}$

$$E_{02}^{(\mathcal{N}+1)} = E_{04}^{(\mathcal{N}+1)} = 0, \tag{3.64}$$

which simplifies Eq. (3.63) to

$$\begin{pmatrix} E_{01}^{(0)} \\ E_{02}^{(0)} \\ E_{03}^{(0)} \\ E_{04}^{(0)} \end{pmatrix} = \begin{pmatrix} M_{11} & M_{12} & M_{13} & M_{14} \\ M_{21} & M_{22} & M_{23} & M_{24} \\ M_{31} & M_{32} & M_{33} & M_{34} \\ M_{41} & M_{42} & M_{43} & M_{44} \end{pmatrix} \begin{pmatrix} E_{01}^{(\mathcal{N}+1)} \\ 0 \\ E_{03}^{(\mathcal{N}+1)} \\ 0 \end{pmatrix}. \tag{3.65}$$

This allows us to conveniently describe the response of the multilayer structure to the waves incident from the half space 0 situated at $z < z^{(0)}$. Assuming that no waves arrive towards the structure from the half space (0), we have for $z < z^{(0)}$

$$E_{01}^{(0)} = E_{03}^{(0)} = 0, \tag{3.66}$$

which simplifies Eq. (3.63) to

$$\begin{pmatrix} 0 \\ E_{02}^{(0)} \\ 0 \\ E_{04}^{(0)} \end{pmatrix} = \begin{pmatrix} M_{11} & M_{12} & M_{13} & M_{14} \\ M_{21} & M_{22} & M_{23} & M_{24} \\ M_{31} & M_{32} & M_{33} & M_{34} \\ M_{41} & M_{42} & M_{43} & M_{44} \end{pmatrix} \begin{pmatrix} E_{01}^{(\mathcal{N}+1)} \\ E_{02}^{(\mathcal{N}+1)} \\ E_{03}^{(\mathcal{N}+1)} \\ E_{04}^{(\mathcal{N}+1)} \end{pmatrix}. \tag{3.67}$$

Let us return to the case of Eq. (3.65) and assume that two orthogonally polarized *incident* waves propagate towards the structure from the

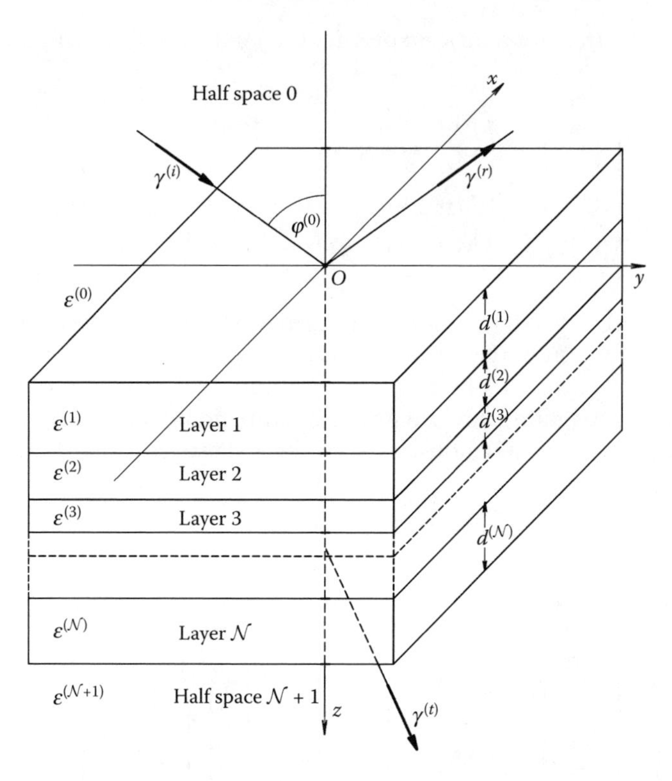

FIGURE 3.3
The cross section of an anisotropic multilayer structure consisting of $\mathcal{N}$ anisotropic layers. The incident, reflected, and transmitted propagation vectors in sandwiching isotropic non-absorbing half spaces (0) and ($\mathcal{N}$ + 1) characterized by the permittivity tensors $\varepsilon^{(n)}$, with $n = 1, \cdots, \mathcal{N}$, are denoted as $\gamma^{(i)}$, $\gamma^{(r)}$, and $\gamma^{(t)}$, respectively, and $\varphi^{(0)}$ denotes the angle of incidence.

isotropic medium $z < z^{(0)}$ (Figure 3.3). Their wave amplitudes are denoted as $E_{01}^{(0)}$ and $E_{03}^{(0)}$, and their proper polarization vectors are given, according to Eq. (3.54) and Eq. (3.55), by

$$e_1^{(0)} = \begin{pmatrix} p^{(0)} \\ q^{(0)} \cos\varphi^{(0)} \\ -q^{(0)} \sin\varphi^{(0)} \end{pmatrix}, \tag{3.68a}$$

$$e_3^{(0)} = \begin{pmatrix} -q^{(0)*} \\ p^{(0)*} \cos\varphi^{(0)} \\ -p^{(0)*} \sin\varphi^{(0)} \end{pmatrix}, \tag{3.68b}$$

where $\varphi^{(0)}$ denotes the angle of incidence.

Then, there are two orthogonally polarized *reflected* waves propagating away from the structure into the isotropic medium of incidence at $z < z^{(0)}$.

Their amplitudes are $E_{02}^{(0)}$ and $E_{04}^{(0)}$, and their polarization vectors are given, according to Eq. (3.54) and Eq. (3.55), by

$$e_2^{(0)} = \begin{pmatrix} p^{(0)} \\ q^{(0)} \cos\varphi^{(0)} \\ q^{(0)} \sin\varphi^{(0)} \end{pmatrix}, \tag{3.68c}$$

$$e_4^{(0)} = \begin{pmatrix} -q^{(0)*} \\ p^{(0)*} \cos\varphi^{(0)} \\ p^{(0)*} \sin\varphi^{(0)} \end{pmatrix}, \tag{3.68d}$$

where $\varphi^{(0)}$ denotes the angle of incidence equal to the angle of reflection.

There are two orthogonally polarized waves *transmitted* across the multilayer structure into the isotropic half space $(\mathcal{N}+1)$ with $z > z^{(\mathcal{N})}$. Their amplitudes are $E_{01}^{(\mathcal{N}+1)}$ and $E_{03}^{(\mathcal{N}+1)}$, and their polarization vectors are given, according to Eq. (3.54) and Eq. (3.55), by

$$e_1^{(\mathcal{N}+1)} = \begin{pmatrix} p^{(\mathcal{N}+1)} \\ q^{(\mathcal{N}+1)} \cos\varphi^{(\mathcal{N}+1)} \\ -q^{(\mathcal{N}+1)} \sin\varphi^{(\mathcal{N}+1)} \end{pmatrix}, \tag{3.69a}$$

$$e_3^{(\mathcal{N}+1)} = \begin{pmatrix} -q^{(\mathcal{N}+1)*} \\ p^{(\mathcal{N}+1)*} \cos\varphi^{(\mathcal{N}+1)} \\ -p^{(\mathcal{N}+1)*} \sin\varphi^{(\mathcal{N}+1)} \end{pmatrix}, \tag{3.69b}$$

where $\varphi^{(\mathcal{N}+1)}$ denotes the angle of refraction.

For orthogonally polarized incident, transmitted, and reflected modes in the sandwiching half spaces, we can define the *global reflection and transmission coefficients* (there are no waves arriving at the structure from the half space $(\mathcal{N}+1)$ situated at $z > z^{(\mathcal{N})}$)

$$r_{21}^{(0,\mathcal{N}+1)} = \left(\frac{E_{02}^{(0)}}{E_{01}^{(0)}}\right)_{E_{03}^{(0)}=0} = \frac{M_{21}M_{33} - M_{23}M_{31}}{M_{11}M_{33} - M_{13}M_{31}}, \tag{3.70a}$$

$$r_{41}^{(0,\mathcal{N}+1)} = \left(\frac{E_{04}^{(0)}}{E_{01}^{(0)}}\right)_{E_{03}^{(0)}=0} = \frac{M_{41}M_{33} - M_{43}M_{31}}{M_{11}M_{33} - M_{13}M_{31}}, \tag{3.70b}$$

$$r_{43}^{(0,\mathcal{N}+1)} = \left(\frac{E_{04}^{(0)}}{E_{03}^{(0)}}\right)_{E_{01}^{(0)}=0} = \frac{M_{11}M_{43} - M_{41}M_{13}}{M_{11}M_{33} - M_{13}M_{31}}, \tag{3.70c}$$

$$r_{23}^{(0,\mathcal{N}+1)} = \left(\frac{E_{02}^{(0)}}{E_{03}^{(0)}}\right)_{E_{01}^{(0)}=0} = \frac{M_{11}M_{23} - M_{21}M_{13}}{M_{11}M_{33} - M_{13}M_{31}}, \tag{3.70d}$$

$$t_{11}^{(0,\mathcal{N}+1)} = \left(\frac{E_{01}^{(\mathcal{N}+1)}}{E_{01}^{(0)}} \right)_{E_{03}^{(0)}=0} = \frac{M_{33}}{M_{11}M_{33} - M_{13}M_{31}}, \tag{3.70e}$$

$$t_{31}^{(0,\mathcal{N}+1)} = \left(\frac{E_{03}^{(\mathcal{N}+1)}}{E_{01}^{(0)}} \right)_{E_{03}^{(0)}=0} = \frac{-M_{31}}{M_{11}M_{33} - M_{13}M_{31}}, \tag{3.70f}$$

$$t_{33}^{(0,\mathcal{N}+1)} = \left(\frac{E_{03}^{(\mathcal{N}+1)}}{E_{03}^{(0)}} \right)_{E_{01}^{(0)}=0} = \frac{M_{11}}{M_{11}M_{33} - M_{13}M_{31}}, \tag{3.70g}$$

$$t_{13}^{(0,\mathcal{N}+1)} = \left(\frac{E_{01}^{(\mathcal{N}+1)}}{E_{03}^{(0)}} \right)_{E_{01}^{(0)}=0} = \frac{-M_{13}}{M_{11}M_{33} - M_{13}M_{31}}. \tag{3.70h}$$

We observe

$$t_{11}^{(0,\mathcal{N}+1)} t_{33}^{(0,\mathcal{N}+1)} - t_{31}^{(0,\mathcal{N}+1)} t_{13}^{(0,\mathcal{N}+1)} = \frac{1}{M_{11}M_{33} - M_{13}M_{31}}, \tag{3.71a}$$

$$r_{21}^{(0,\mathcal{N}+1)} r_{43}^{(0,\mathcal{N}+1)} - r_{41}^{(0,\mathcal{N}+1)} r_{23}^{(0,\mathcal{N}+1)} = \frac{M_{21}M_{43} - M_{23}M_{41}}{M_{11}M_{33} - M_{13}M_{31}}. \tag{3.71b}$$

Similarly, we can define the global reflection and transmission coefficients for the case where there are no waves arriving at the structure from the half space (0) situated at $z < z^{(0)}$ characterized by Eq. (3.66) and Eq. (3.67)

$$r_{12}^{(\mathcal{N}+1,\,0)} = \left(\frac{E_{01}^{(\mathcal{N}+1)}}{E_{02}^{(\mathcal{N}+1)}} \right)_{E_{04}^{(\mathcal{N}+1)}=0} = \frac{M_{13}M_{32} - M_{12}M_{33}}{M_{11}M_{33} - M_{13}M_{31}}, \tag{3.72a}$$

$$r_{32}^{(\mathcal{N}+1,\,0)} = \left(\frac{E_{03}^{(\mathcal{N}+1)}}{E_{02}^{(\mathcal{N}+1)}} \right)_{E_{04}^{(\mathcal{N}+1)}=0} = \frac{M_{12}M_{31} - M_{11}M_{32}}{M_{11}M_{33} - M_{13}M_{31}}, \tag{3.72b}$$

$$r_{34}^{(\mathcal{N}+1,\,0)} = \left(\frac{E_{03}^{(\mathcal{N}+1)}}{E_{04}^{(\mathcal{N})}} \right)_{E_{02}^{(\mathcal{N}+1)}=0} = \frac{M_{14}M_{31} - M_{11}M_{34}}{M_{11}M_{33} - M_{13}M_{31}}, \tag{3.72c}$$

$$r_{14}^{(\mathcal{N}+1,\,0)} = \left(\frac{E_{01}^{(\mathcal{N}+1)}}{E_{04}^{(\mathcal{N})}} \right)_{E_{02}^{(\mathcal{N}+1)}=0} = \frac{M_{13}M_{34} - M_{14}M_{33}}{M_{11}M_{33} - M_{13}M_{31}}, \tag{3.72d}$$

$$t_{22}^{(\mathcal{N}+1,\,0)} = \left(\frac{E_{02}^{(0)}}{E_{02}^{(\mathcal{N}+1)}} \right)_{E_{04}^{(\mathcal{N}+1)}=0}$$
$$= M_{22} + \frac{M_{21}\left(M_{13}M_{32} - M_{12}M_{33}\right) + M_{23}\left(M_{12}M_{31} - M_{11}M_{32}\right)}{M_{11}M_{33} - M_{13}M_{31}}, \tag{3.72e}$$

$$t_{42}^{(\mathcal{N}+1,\,0)} = \left(\frac{E_{04}^{(0)}}{E_{02}^{(\mathcal{N})}}\right)_{E_{04}^{(\mathcal{N}+1)}=0}$$
$$= M_{42} + \frac{M_{41}\left(M_{13}M_{32} - M_{12}M_{33}\right) + M_{43}\left(M_{12}M_{31} - M_{11}M_{32}\right)}{M_{11}M_{33} - M_{13}M_{31}}, \tag{3.72f}$$

$$t_{44}^{(\mathcal{N}+1,\,0)} = \left(\frac{E_{04}^{(0)}}{E_{04}^{(\mathcal{N})}}\right)_{E_{02}^{(\mathcal{N}+1)}=s}$$
$$= M_{44} + \frac{M_{41}\left(M_{13}M_{34} - M_{14}M_{33}\right) + M_{43}\left(M_{14}M_{31} - M_{11}M_{34}\right)}{M_{11}M_{33} - M_{13}M_{31}}, \tag{3.72g}$$

$$t_{24}^{(\mathcal{N}+1,\,0)} = \left(\frac{E_{02}^{(\mathcal{N}+1)}}{E_{04}^{(\mathcal{N})}}\right)_{E_{02}^{(\mathcal{N}+1)}=0}$$
$$= M_{24} + \frac{M_{21}\left(M_{13}M_{34} - M_{14}M_{33}\right) + M_{23}\left(M_{14}M_{31} - M_{11}M_{34}\right)}{M_{11}M_{33} - M_{13}M_{31}}. \tag{3.72h}$$

We observe

$$r_{12}^{(\mathcal{N}+1,\,0)} r_{34}^{(\mathcal{N}+1,\,0)} - r_{14}^{(\mathcal{N}+1,\,0)} r_{32}^{(\mathcal{N}+1,\,0)} = \frac{M_{12}M_{34} - M_{14}M_{32}}{M_{11}M_{33} - M_{13}M_{31}}. \tag{3.73}$$

The reflection and transmission coefficients defined in Eqs. (3.70) can be arranged to form (2 × 2) Jones reflection and transmission matrices, respectively, in a proper polarization mode representation chosen for the half spaces (0) and $(\mathcal{N}+1)$

$$\begin{pmatrix} E_{02}^{(0)} \\ E_{04}^{(0)} \end{pmatrix} = \begin{pmatrix} r_{21}^{(0,\mathcal{N}+1)} & r_{23}^{(0,\mathcal{N}+1)} \\ r_{41}^{(0,\mathcal{N}+1)} & r_{43}^{(0,\mathcal{N}+1)} \end{pmatrix} \begin{pmatrix} E_{01}^{(0)} \\ E_{03}^{(0)} \end{pmatrix} \tag{3.74a}$$

$$\begin{pmatrix} E_{01}^{(\mathcal{N}+1)} \\ E_{03}^{(\mathcal{N}+1)} \end{pmatrix} = \begin{pmatrix} t_{11}^{(0,\mathcal{N}+1)} & t_{13}^{(0,\mathcal{N}+1)} \\ t_{31}^{(0,\mathcal{N}+1)} & t_{33}^{(0,\mathcal{N}+1)} \end{pmatrix} \begin{pmatrix} E_{01}^{(0)} \\ E_{03}^{(0)} \end{pmatrix}. \tag{3.74b}$$

For nonzero angles of incidence, the common choice $p^{(s)} = 1$ and $q^{(s)} = 0$ in Eq. (3.49) would be the most convenient for the modes in the sandwiching isotropic (nonabsorbing) half spaces. Then, the waves are polarized linearly perpendicular and parallel to the plane of incidence, and the reflection and transmission coefficients reduce to those defined by Yeh [14]. The ratios of the reflection and transmission coefficients in Eq. (3.70) represent generalized complex ellipsometric parameters.

On the right-hand side of Eq. (3.70) and Eq. (3.72), the denominator

$$M_{11}M_{33} - M_{13}M_{31}$$

remains the same, representing, after a suitable normalization, the effect of multiple reflections of the propagating modes. Then, the numerators of Eq. (3.70e) to Eq. (3.70h) and Eq. (3.72e) to Eq. (3.72h) for the transmission coefficients represent a single pass across the structure, while those of Eqs. (3.70a) to (3.70d) and Eqs. (3.72a) to (3.72d) for the reflection coefficients represent a double pass across the structure. The analysis can be extended to include evanescent waves. For $E_{01}^{(0)} = 0$ and $E_{03}^{(0)} = 0$, Eq. (3.65) provides

$$0 = M_{11}E_{01}^{(\mathcal{N}+1)} + M_{13}E_{03}^{(\mathcal{N}+1)} \quad (3.75a)$$

$$0 = M_{13}E_{01}^{(\mathcal{N}+1)} + M_{33}E_{03}^{(\mathcal{N}+1)}. \quad (3.75b)$$

As shown by Yeh, the condition for nontrivial solution of this system expressed by the zero determinant

$$M_{11}M_{33} - M_{13}M_{31} = 0 \quad (3.76)$$

represents the dispersion relation for the propagation of guided waves in the structure, *i.e.*, the waveguiding condition. The fields $E_{01}^{(0)}$ $E_{03}^{(0)}$, $E_{01}^{(\mathcal{N}+1)}$, and $E_{03}^{(\mathcal{N}+1)}$ must vanish at infinity and correspond therefore to the evanescent waves. The condition expressed in Eq. (3.76) proved to be useful in the design of planar multilayer magnetooptic waveguides.

3.6 Single Interface

In this section, we consider the general case of an interface between isotropic and anisotropic regions. The situation covers, among others, the magnetooptic Kerr effects at arbitrary orientations of magnetization, crystalline axes, interface, and incident radiation propagation vector. The results can be employed in the MO magnetometry of magnetic vicinal surfaces. The **M** matrix of the single interface is

$$\mathbf{M} = \left(\mathbf{D}^{(0)}\right)^{-1} \mathbf{D}^{(1)}, \quad (3.77)$$

where the matrices $\left(\mathbf{D}^{(0)}\right)^{-1}$ and $\mathbf{D}^{(1)}$ are given by Eq. (3.61) and Eq. (3.30). We assume the anisotropic medium sufficiently thick and absorbing so that the transmitted waves as well as the waves incident at the interface from the magnetic medium need not be considered. The propagation vectors

$\gamma^{(i)}$ and $\gamma^{(r)}$ of the incident and reflected waves, respectively, are related by the transformation of rotation by an angle of $(\pi - 2\varphi_0)$

$$\gamma^{(r)} = \begin{pmatrix} 1 & 0 & 0 \\ 0 & \cos\left(\pi - \varphi^{(0)}\right) & \sin\left(\pi - 2\varphi^{(0)}\right) \\ 0 & -\sin\left(\pi - 2\varphi^{(0)}\right) & \cos\left(\pi - 2\varphi^{(0)}\right) \end{pmatrix} \gamma^{(i)} \tag{3.78}$$

or expressed more explicitly

$$\frac{\omega}{c} N^{(0)} \begin{pmatrix} 0 \\ \sin\varphi^{(0)} \\ -\cos\varphi^{(0)} \end{pmatrix} = \begin{pmatrix} 1 & 0 & 0 \\ 0 & \cos\left(\pi - 2\varphi^{(0)}\right) & \sin\left(\pi - 2\varphi^{(0)}\right) \\ 0 & -\sin\left(\pi - 2\varphi^{(0)}\right) & \cos\left(\pi - 2\varphi^{(0)}\right) \end{pmatrix} \times \frac{\omega}{c} N^{(0)} \begin{pmatrix} 0 \\ \sin\varphi^{(0)} \\ \cos\varphi^{(0)} \end{pmatrix}. \tag{3.79}$$

At oblique incidence, the most convenient choice of the proper polarization modes in the isotropic medium are the waves linearly polarized normal and parallel with respect to the plane of incidence $x = 0$ with $p^{(0)} = 1$ and $q^{(0)} = 0$. Then the inspection of Eq. (3.60) and Eq. (3.61) shows that the matrices $\mathbf{D}^{(s)}$ and $(\mathbf{D}^{(s)})^{-1}$ become (2×2) block diagonal. In special cases, this would also lead to a (2×2) block-diagonal global $\mathbf{M}$ matrix, *e.g.*, in the limit of the interface between isotropic media. However, in general, the off-diagonal (2×2) blocks are nonzero where an anisotropic medium is involved.

The electric fields of the four linearly polarized proper polarization modes in the isotropic region $z < z^{(0)}$ are represented by

$$\boldsymbol{E}_1^{(0)} = \mathrm{E}_{01}^{(0)} \hat{\boldsymbol{x}} \exp\left[\mathrm{j}\omega t - \mathrm{j}N^{(0)}\frac{\omega}{c}\left(y\sin\varphi^{(0)} + z\cos\varphi^{(0)}\right)\right], \tag{3.80a}$$

$$\boldsymbol{E}_2^{(0)} = \mathrm{E}_{02}^{(0)} \hat{\boldsymbol{x}} \exp\left[\mathrm{j}\omega t - \mathrm{j}N^{(0)}\frac{\omega}{c}\left(y\sin\varphi^{(0)} - z\cos\varphi^{(0)}\right)\right], \tag{3.80b}$$

$$\boldsymbol{E}_3^{(0)} = \mathrm{E}_{03}^{(0)} \left(\cos\varphi^{(0)}\hat{\boldsymbol{y}} - \sin\varphi^{(0)}\hat{\boldsymbol{z}}\right) \exp\left[\mathrm{j}\omega t - \mathrm{j}N^{(0)}\frac{\omega}{c}\left(y\sin\varphi^{(0)} + z\cos\varphi^{(0)}\right)\right], \tag{3.80c}$$

$$\boldsymbol{E}_4^{(0)} = \mathrm{E}_{04}^{(0)} \left(\cos\varphi^{(0)}\hat{\boldsymbol{y}} + \sin\varphi^{(0)}\hat{\boldsymbol{z}}\right) \exp\left[\mathrm{j}\omega t - \mathrm{j}N^{(0)}\frac{\omega}{c}\left(y\sin\varphi^{(0)} - z\cos\varphi^{(0)}\right)\right], \tag{3.80d}$$

where $\varphi^{(0)}$ is the angle of incidence. The corresponding magnetic fields follow from the Faraday law $\nabla \times \boldsymbol{E} = -(\partial\boldsymbol{B}/\partial t)$ and may be expressed as

$$c\boldsymbol{B}_1^{(0)} = N^{(0)}\mathrm{E}_{01}^{(0)}\left(\cos\varphi^{(0)}\hat{\boldsymbol{y}} - \sin\varphi^{(0)}\hat{\boldsymbol{z}}\right) \times \exp\left[\mathrm{j}\omega t - \mathrm{j}N^{(0)}\frac{\omega}{c}\left(y\sin\varphi^{(0)} + z\cos\varphi^{(0)}\right)\right], \tag{3.81a}$$

$$
\begin{aligned}
c\boldsymbol{B}_2^{(0)} &= N^{(0)}\mathrm{E}_{02}^{(0)}\left(-\cos\varphi^{(0)}\hat{\boldsymbol{y}} - \sin\varphi^{(0)}\hat{\boldsymbol{z}}\right) \\
&\quad \times \exp\left[\mathrm{j}\omega t - \mathrm{j}N^{(0)}\frac{\omega}{c}\left(y\sin\varphi^{(0)} - z\cos\varphi^{(0)}\right)\right], \qquad (3.81b) \\
c\boldsymbol{B}_3^{(0)} &= N^{(0)}\mathrm{E}_{03}^{(0)}\left(-\hat{\boldsymbol{x}}\right)\exp\left[\mathrm{j}\omega t - \mathrm{j}N^{(0)}\frac{\omega}{c}\left(y\sin\varphi^{(0)} + z\cos\varphi^{(0)}\right)\right], \qquad (3.81c) \\
c\boldsymbol{B}_4^{(0)} &= N^{(0)}\mathrm{E}_{04}^{(0)}\hat{\boldsymbol{x}}\exp\left[\mathrm{j}\omega t - \mathrm{j}N^{(0)}\frac{\omega}{c}\left(y\sin\varphi^{(0)} - z\cos\varphi^{(0)}\right)\right]. \qquad (3.81d)
\end{aligned}
$$

We can compute the scalar products of the four proper polarizations with the Cartesian unit vectors parallel to the interface and construct the dynamical matrix defined in Eq. (3.29). Alternatively, we can apply Eq. (3.60) and Eq. (3.61) for $\mathbf{D}^{(s)}$ and $\left(\mathbf{D}^{(s)}\right)^{-1}$. Setting $p^{(s)} = 1$ and $q^{(s)} = 0$, we obtain

$$
\mathbf{D}^{(s)} = \begin{pmatrix} 1 & 1 & 0 & 0 \\ N^{(s)}\cos\varphi^{(s)} & -N^{(s)}\cos\varphi^{(s)} & 0 & 0 \\ 0 & 0 & \cos\varphi^{(s)} & \cos\varphi^{(s)} \\ 0 & 0 & -N^{(s)} & N^{(s)} \end{pmatrix}, \qquad (3.82a)
$$

$$
\left(\mathbf{D}^{(s)}\right)^{-1} = \left(2N^{(s)}\cos\varphi^{(s)}\right)^{-1}\begin{pmatrix} N^{(s)}\cos\varphi^{(s)} & 1 & 0 & 0 \\ N^{(s)}\cos\varphi^{(s)} & -1 & 0 & 0 \\ 0 & 0 & N^{(s)} & -\cos\varphi^{(s)} \\ 0 & 0 & N^{(s)} & \cos\varphi^{(s)} \end{pmatrix}. \qquad (3.82b)
$$

In the matrix product of Eq. (3.77), we make use of Eq. (3.82b) with $s = 0$. Using the general form of the $\mathbf{D}^{(n)}$ matrix given in Eq. (3.29) and Eq. (3.30), we obtain the $\mathbf{M}$ matrix elements relevant for the reflection characteristics from the side of the isotropic medium

$$
\mathrm{M}_{\substack{11\\21}} = \left[\left(\mathbf{D}^{(0)}\right)^{-1}\mathbf{D}^{(1)}\right]_{\substack{11\\21}} = \frac{1}{2}\left[D_{11}^{(1)} \pm \left(N^{(0)}\cos\varphi^{(0)}\right)^{-1}D_{21}^{(1)}\right], \qquad (3.83a)
$$

$$
\mathrm{M}_{\substack{31\\41}} = \left[\left(\mathbf{D}^{(0)}\right)^{-1}\mathbf{D}^{(1)}\right]_{\substack{31\\41}} = \frac{1}{2}\left[\left(\cos\varphi^{(0)}\right)^{-1}D_{31}^{(1)} \mp \left(N^{(0)}\right)^{-1}D_{41}^{(1)}\right], \qquad (3.83b)
$$

$$
\mathrm{M}_{\substack{13\\23}} = \left[\left(\mathbf{D}^{(0)}\right)^{-1}\mathbf{D}^{(1)}\right]_{\substack{13\\23}} = \frac{1}{2}\left[D_{13}^{(1)} \pm \left(N^{(0)}\cos\varphi^{(0)}\right)^{-1}D_{23}^{(1)}\right], \qquad (3.83c)
$$

$$
\mathrm{M}_{\substack{33\\43}} = \left[\left(\mathbf{D}^{(0)}\right)^{-1}\mathbf{D}^{(1)}\right]_{\substack{33\\43}} = \frac{1}{2}\left[\left(\cos\varphi^{(0)}\right)^{-1}D_{33}^{(1)} \mp \left(N^{(0)}\right)^{-1}D_{43}^{(1)}\right]. \qquad (3.83d)
$$

It was already shown that the normalization of the polarization vector $\boldsymbol{e}_j^{(1)}$ in the anisotropic medium is not required as far as we are concerned with the reflection characteristics and the fields in the isotropic region only.

We define the (2×2) Cartesian Jones reflection matrix by the relation between the proper polarization modes polarized perpendicular and parallel to the plane of incidence as a special case of Eq. (3.74a)

$$\begin{pmatrix} \mathrm{E}_{02}^{(0)} \\ \mathrm{E}_{04}^{(0)} \end{pmatrix} = \begin{pmatrix} r_{21}^{(01)} & r_{23}^{(01)} \\ r_{41}^{(01)} & r_{43}^{(01)} \end{pmatrix} \begin{pmatrix} \mathrm{E}_{01}^{(0)} \\ \mathrm{E}_{03}^{(0)} \end{pmatrix}. \tag{3.84}$$

Here, $\mathrm{E}_{01}^{(0)}$ and $\mathrm{E}_{02}^{(0)}$ represent the complex proper polarization amplitudes of the incident and reflected waves, respectively, polarized parallel to the x axis of our Cartesian coordinate system, *i.e.*, perpendicular to the plane of incidence. Similarly, $\mathrm{E}_{03}^{(0)}$ and $\mathrm{E}_{04}^{(0)}$ denote the corresponding amplitudes of the polarization modes polarized in the yz plane, *i.e.*, parallel to the plane of incidence.

To correctly evaluate the phase relationship between the incident and reflected fields and consequently apply the sign conventions chosen, it is useful to have these relations in a vectorial form. We thus express the electric field vectors of the reflected proper polarization modes as functions of the electric fields of the incident proper polarization modes in the plane $z = 0$

$$\boldsymbol{E}_2^{(0)} = \begin{pmatrix} 1 & 0 & 0 \\ 0 & \cos 2\varphi^{(0)} & -\sin 2\varphi^{(0)} \\ 0 & \sin 2\varphi^{(0)} & \cos 2\varphi^{(0)} \end{pmatrix} r_{21}^{(01)} \boldsymbol{E}_1^{(0)}, \tag{3.85}$$

for $\mathrm{E}_{03}^{(0)} = 0$,

$$\boldsymbol{E}_4^{(0)} = \begin{pmatrix} 1 & 0 & 0 \\ 0 & \cos 2\varphi^{(0)} & -\sin 2\varphi^{(0)} \\ 0 & \sin 2\varphi^{(0)} & \cos 2\varphi^{(0)} \end{pmatrix} r_{43}^{(01)} \boldsymbol{E}_3^{(0)}, \tag{3.86}$$

for $\mathrm{E}_{01}^{(0)} = 0$. The following two relations express mode coupling

$$\begin{aligned} \boldsymbol{E}_4^{(0)} &= \begin{pmatrix} 1 & 0 & 0 \\ 0 & \cos 2\varphi(0) & \sin 2\varphi^{(0)} \\ 0 & -\sin 2\varphi^{(0)} & \cos 2\varphi^{(0)} \end{pmatrix} \begin{pmatrix} 1 & 0 & 0 \\ 0 & \cos \varphi^{(0)} & \sin \varphi^{(0)} \\ 0 & -\sin \varphi^{(0)} & \cos \varphi^{(0)} \end{pmatrix} \\ &\quad \times \begin{pmatrix} 0 & -1 & 0 \\ 1 & 0 & 0 \\ 0 & 0 & 1 \end{pmatrix} \begin{pmatrix} 1 & 0 & 0 \\ 0 & \cos \varphi^{(0)} & -\sin \varphi^{(0)} \\ 0 & \sin \varphi^{(0)} & \cos \varphi^{(0)} \end{pmatrix} r_{41}^{(01)} \boldsymbol{E}_1^{(0)} \\ &= \begin{pmatrix} 1 & 0 & 0 \\ 0 & \cos \varphi^{(0)} & -\sin \varphi^{(0)} \\ 0 & \sin \varphi^{(0)} & \cos \varphi^{(0)} \end{pmatrix} \begin{pmatrix} 0 & -1 & 0 \\ 1 & 0 & 0 \\ 0 & 0 & 1 \end{pmatrix} \\ &\quad \times \begin{pmatrix} 1 & 0 & 0 \\ 0 & \cos \varphi^{(0)} & -\sin \varphi^{(0)} \\ 0 & \sin \varphi^{(0)} & \cos \varphi^{(0)} \end{pmatrix} r_{41}^{(01)} \boldsymbol{E}_1^{(0)} \end{aligned} \tag{3.87}$$

for $E_{03}^{(0)} = 0$. We have first rotated the incident field about the axis parallel to γ_i by a positive angle $\pi/2$ from $\hat{x}$ into the plane of incidence. Subsequently, we have rigidly turned the resulting vector by an angle $2\varphi^{(0)}$ from the direction of γ_i to that of γ_r. Alternatively, we perform the transformation in three steps. We start with the rotation of the incident field by $\varphi^{(0)}$ followed by the rotation from $\hat{x}$ to $\hat{y}$ about the $\hat{z}$. In the third step, we rotate the resulting vector about $\hat{x}$ again by $\varphi^{(0)}$, the angle between $\hat{z}$ and $\boldsymbol{k}_r$. Next, we transform the field $\boldsymbol{E}_3^{(0)}$ into $\boldsymbol{E}_2^{(0)}$, which has the positive $\hat{x}$ direction

$$\boldsymbol{E}_2^{(0)} = \begin{pmatrix} 1 & 0 & 0 \\ 0 & \cos\varphi^{(0)} & -\sin\varphi^{(0)} \\ 0 & \sin\varphi^{(0)} & \cos\varphi^{(0)} \end{pmatrix} \begin{pmatrix} 0 & 1 & 0 \\ -1 & 0 & 0 \\ 0 & 0 & 1 \end{pmatrix} \begin{pmatrix} 1 & 0 & 0 \\ 0 & \cos\varphi^{(0)} & -\sin\varphi^{(0)} \\ 0 & \sin\varphi^{(0)} & \cos\varphi^{(0)} \end{pmatrix} r_{23}^{(0)} \boldsymbol{E}_3^{(0)}, \tag{3.88}$$

for $E_{01}^{(0)} = 0$. We observe that at $\varphi^{(0)} = 0$, the incident $\boldsymbol{E}$ fields transform into the reflected $\boldsymbol{E}$ fields according to the following rules

$$r_{21}^{(01)} \hat{x} E_{01}^{(0)} \rightarrow \hat{x} E_{02}^{(0)}, \tag{3.89a}$$

$$r_{43}^{(01)} \hat{y} E_{03}^{(0)} \rightarrow \hat{y} E_{04}^{(0)}, \tag{3.89b}$$

$$r_{41}^{(01)} \hat{x} E_{01}^{(0)} \rightarrow \hat{y} E_{04}^{(0)}, \tag{3.89c}$$

$$r_{23}^{(01)} \hat{y} E_{03}^{(0)} \rightarrow \hat{x} E_{02}^{(0)}. \tag{3.89d}$$

Note that in the classical ellipsometry of isotropic surfaces, the opposite sign is conventionally established on the right-hand side of Eq. (3.89b). This naturally follows from the duality transformation between $\boldsymbol{E}$ (polar vector) and $\boldsymbol{B}$ (axial vector or pseudotensor) plane wave fields in the derivation of the Fresnel equations. This is not very convenient in the present context, as we often deal with fundamental circularly polarized fields. For example, our convention naturally describes the helicity change upon the normal reflection where an RCP (LCP) incident wave transforms into an LCP (RCP) reflected wave.

The magnetic fields of the incident and reflected proper polarization modes in the plane $z = 0$ are related in the following way

$$\boldsymbol{B}_2^{(0)} = \begin{pmatrix} -1 & 0 & 0 \\ 0 & -\cos 2\varphi(0) & \sin 2\varphi^{(0)} \\ 0 & -\sin 2\varphi^{(0)} & -\cos 2\varphi^{(0)} \end{pmatrix} r_{21}^{(01)} \boldsymbol{B}_1^{(0)} \tag{3.90}$$

for $E_{03}^{(0)} = 0$,

$$\boldsymbol{B}_4^{(0)} = \begin{pmatrix} -1 & 0 & 0 \\ 0 & -\cos 2\varphi(0) & \sin 2\varphi^{(0)} \\ 0 & -\sin 2\varphi^{(0)} & -\cos 2\varphi^{(0)} \end{pmatrix} r_{43}^{(01)} \boldsymbol{B}_3^{(0)} \tag{3.91}$$

for $E_{01}^{(0)} = 0$. The following two relations express mode coupling

$$B_4^{(0)} = \begin{pmatrix} 1 & 0 & 0 \\ 0 & \cos 2\varphi(0) & -\sin 2\varphi^{(0)} \\ 0 & \sin 2\varphi^{(0)} & \cos 2\varphi^{(0)} \end{pmatrix} \begin{pmatrix} 1 & 0 & 0 \\ 0 & \cos\varphi^{(0)} & \sin\varphi^{(0)} \\ 0 & -\sin\varphi^{(0)} & \cos\varphi^{(0)} \end{pmatrix}$$
$$\times \begin{pmatrix} 0 & 1 & 0 \\ -1 & 0 & 0 \\ 0 & 0 & 1 \end{pmatrix} \begin{pmatrix} 1 & 0 & 0 \\ 0 & \cos\varphi^{(0)} & -\sin\varphi^{(0)} \\ 0 & \sin\varphi^{(0)} & \cos\varphi^{(0)} \end{pmatrix} r_{41}^{(0)} B_1^{(0)}$$
$$= \begin{pmatrix} 1 & 0 & 0 \\ 0 & \cos\varphi^{(0)} & -\sin\varphi^{(0)} \\ 0 & \sin\varphi^{(0)} & \cos\varphi^{(0)} \end{pmatrix} \begin{pmatrix} 0 & 1 & 0 \\ -1 & 0 & 0 \\ 0 & 0 & 1 \end{pmatrix}$$
$$\times \begin{pmatrix} 1 & 0 & 0 \\ 0 & \cos\varphi^{(0)} & -\sin\varphi^{(0)} \\ 0 & \sin\varphi^{(0)} & \cos\varphi^{(0)} \end{pmatrix} r_{41}^{(0)} B_1^{(0)}, \tag{3.92}$$

for $E_{03}^{(0)} = 0$ and

$$B_2^{(0)} = \begin{pmatrix} 1 & 0 & 0 \\ 0 & \cos\varphi^{(0)} & -\sin\varphi^{(0)} \\ 0 & \sin\varphi^{(0)} & \cos\varphi^{(0)} \end{pmatrix} \begin{pmatrix} 0 & -1 & 0 \\ 1 & 0 & 0 \\ 0 & 0 & 1 \end{pmatrix}$$
$$\times \begin{pmatrix} 1 & 0 & 0 \\ 0 & \cos\varphi^{(0)} & -\sin\varphi^{(0)} \\ 0 & \sin\varphi^{(0)} & \cos\varphi^{(0)} \end{pmatrix} r_{23}^{(0)} B_3^{(0)} \tag{3.93}$$

for $E_{01}^{(0)} = 0$.

We observe that at $\varphi^{(0)} = 0$, the incident $\boldsymbol{B}$ fields transform into the reflected $\boldsymbol{B}$ fields according to the following rules consistent with Eq. (3.89)

$$r_{21}^{(01)} \hat{\boldsymbol{y}} B_{01}^{(0)} \rightarrow -\hat{\boldsymbol{y}} B_{02}^{(0)}, \tag{3.94a}$$

$$-r_{43}^{(01)} \hat{\boldsymbol{x}} B_{03}^{(0)} \rightarrow \hat{\boldsymbol{x}} B_{04}^{(0)}, \tag{3.94b}$$

$$r_{41}^{(01)} \hat{\boldsymbol{y}} B_{01}^{(0)} \rightarrow \hat{\boldsymbol{x}} B_{04}^{(0)}, \tag{3.94c}$$

$$-r_{23}^{(01)} \hat{\boldsymbol{x}} B_{03}^{(0)} \rightarrow -\hat{\boldsymbol{y}} B_{02}^{(0)}. \tag{3.94d}$$

We notice that the relation between the Cartesian x and y components of the vector field amplitudes of the modes linearly polarized perpendicular and parallel to the plane of incidence can be expressed in the matrix form

$$\begin{pmatrix} E_{02x}^{(0)} \\ E_{04y}^{(0)} \end{pmatrix} = \begin{pmatrix} r_{21}^{(01)} & r_{23}^{(01)} \cos^{-1} \varphi^{(0)} \\ r_{41}^{(01)} \cos \varphi^{(0)} & r_{43}^{(01)} \end{pmatrix} \begin{pmatrix} E_{01x}^{(0)} \\ E_{03y}^{(0)} \end{pmatrix}, \tag{3.95}$$

where $E_{01x}^{(0)} = E_{01}^{(0)}$, $E_{02x}^{(0)} = E_{02}^{(0)}$, $E_{03y}^{(0)} = E_{03}^{(0)} \cos \varphi^{(0)}$, and $E_{04y}^{(0)} = E_{04}^{(0)} \cos \varphi^{(0)}$.

The substitution of the matrix elements (3.83) into Eqs. (3.70a) to (3.70d) provides the reflection coefficients at the interface between an isotropic medium and a medium characterized by a general permittivity tensor defined in Eq. (3.1). We have for the incident wave polarized perpendicular to the plane of incidence, conventionally indexed as an s polarized wave (from German *senkrecht*)

$$r_{21}^{(01)} = r_{ss}^{(01)} = \frac{N^{(0)} \cos^2 \varphi^{(0)} \mathcal{P}^{(1)} + N^{(0)2} \cos \varphi^{(0)} \mathcal{Q}^{(1)} - \cos \varphi^{(0)} \mathcal{S}^{(1)} - N^{(0)} \mathcal{T}^{(1)}}{N^{(0)} \cos^2 \varphi^{(0)} \mathcal{P}^{(1)} + N^{(0)2} \cos \varphi^{(0)} \mathcal{Q}^{(1)} + \cos \varphi^{(0)} \mathcal{S}^{(1)} + N^{(0)} \mathcal{T}^{(1)}}, \tag{3.96a}$$

$$r_{41}^{(01)} = r_{ps}^{(01)} = \frac{2N^{(0)} \cos \varphi^{(0)} \left(D_{41}^{(1)} D_{33}^{(1)} - D_{43}^{(1)} D_{31}^{(1)} \right)}{N^{(0)} \cos^2 \varphi^{(0)} \mathcal{P}^{(1)} + N^{(0)2} \cos \varphi^{(0)} \mathcal{Q}^{(1)} + \cos \varphi^{(0)} \mathcal{S}^{(1)} + N^{(0)} \mathcal{T}^{(1)}}. \tag{3.96b}$$

When the incident wave is polarized parallel to the plane of incidence, conventionally indexed as a p polarized wave, we obtain

$$r_{43}^{(01)} = r_{pp}^{(01)} = \frac{-N^{(0)} \cos^2 \varphi^{(0)} \mathcal{P}^{(1)} + N^{(0)2} \cos \varphi^{(0)} \mathcal{Q}^{(1)} - \cos \varphi^{(0)} \mathcal{S}^{(1)} + N^{(0)} \mathcal{T}^{(1)}}{N^{(0)} \cos^2 \varphi^{(0)} \mathcal{P}^{(1)} + N^{(0)2} \cos \varphi^{(0)} \mathcal{Q}^{(1)} + \cos \varphi^{(0)} \mathcal{S}^{(1)} + N^{(0)} \mathcal{T}^{(1)}}, \tag{3.97a}$$

$$r_{23}^{(01)} = r_{sp}^{(01)} = \frac{2N^{(0)} \cos \varphi^{(0)} \left(D_{21}^{(1)} D_{13}^{(1)} - D_{11}^{(1)} D_{23}^{(1)} \right)}{N^{(0)} \cos^2 \varphi^{(0)} \mathcal{P}^{(1)} + N^{(0)2} \cos \varphi^{(0)} \mathcal{Q}^{(1)} + \cos \varphi^{(0)} \mathcal{S}^{(1)} + N^{(0)} \mathcal{T}^{(1)}}, \tag{3.97b}$$

where the reflection coefficients may be organized into a Cartesian Jones reflection matrix equivalent to that defined in Eq. (1.99) and Eq. (3.84)

$$\begin{pmatrix} E_{0s}^{(0)} \\ E_{0p}^{(0)} \end{pmatrix} = \begin{pmatrix} r_{ss}^{(01)} & r_{sp}^{(01)} \\ r_{ps}^{(01)} & r_{pp}^{(01)} \end{pmatrix} \begin{pmatrix} E_{0s}^{(0)} \\ E_{0p}^{(0)} \end{pmatrix}. \tag{3.98}$$

We recall that in the classical ellipsometry the sign convention is established [18] according to which the positive sense for the incident and reflected electric field polarized in the plane of incidence are antiparallel to each other in the limit of the normal light incidence. In the present choice, the positive senses of the p polarized wave at $\varphi^{(0)}$ are congruent. Consequently, the sign of the matrix elements involving reflected p polarized waves, *i.e.*, the elements $r_{ps}^{(01)}$ and $r_{pp}^{(01)}$, calculated with the classical ellipsometry convention [18] would be opposite.

We obtain the equivalents (up to a common factor) of Eqs. (1.89)

$$\mathcal{P}^{(1)} = \left(D_{13}^{(1)} D_{41}^{(1)} - D_{11}^{(1)} D_{43}^{(1)}\right), \tag{3.99a}$$

$$\mathcal{Q}^{(1)} = \left(D_{11}^{(1)} D_{33}^{(1)} - D_{13}^{(1)} D_{31}^{(1)}\right), \tag{3.99b}$$

$$\mathcal{S}^{(1)} = \left(D_{23}^{(1)} D_{41}^{(1)} - D_{21}^{(1)} D_{43}^{(1)}\right), \tag{3.99c}$$

$$\mathcal{T}^{(1)} = \left(D_{21}^{(1)} D_{33}^{(1)} - D_{23}^{(1)} D_{31}^{(1)}\right). \tag{3.99d}$$

It will be useful to abbreviate the expressions entering the off-diagonal elements of the Cartesian Jones reflection matrix as

$$\mathcal{U}^{(1)} = \left(D_{13}^{(1)} D_{21}^{(1)} - D_{11}^{(1)} D_{23}^{(1)}\right), \tag{3.99e}$$

$$\mathcal{W}^{(1)} = \left(D_{41}^{(1)} D_{33}^{(1)} - D_{43}^{(1)} D_{31}^{(1)}\right). \tag{3.99f}$$

We list the explicit form of these formulae below [20]. They contain the complete information on the reflection at the interface between an isotropic ambient and a medium of arbitrary anisotropy and are useful as a starting point from which all special cases can be derived. We substitute into Eq. (3.99) for the $\mathbf{D}^{(1)}$ matrix elements according to Eq. (3.30). It is useful to realize that the difference of two products of the elements of the matrix $\mathbf{D}^{(1)}$ is antisymmetric with respect to the exchange $N_{z1}^{(1)} \leftrightarrow N_{z3}^{(1)}$. Up to a common factor not relevant for the present purpose, we arrive at the following expressions consistent with Eq. (1.94)

$$\begin{aligned}
\mathcal{P}^{(1)} = \Big\{ & \varepsilon_{xz}^{(1)} \varepsilon_{zz}^{(1)} N_y \left(N_{z1}^{(1)} N_{z3}^{(1)}\right) \left(N_{z1}^{(1)} + N_{z3}^{(1)}\right) \\
& - \varepsilon_{zz}^{(1)} \left[\varepsilon_{xy}^{(1)} \left(\varepsilon_{zz}^{(1)} - N_y^2\right) - \varepsilon_{xz}^{(1)} \varepsilon_{zy}^{(1)}\right] \left(N_{z1}^{(1)2} + N_{z3}^{(1)2}\right) \\
& - \left[\varepsilon_{zz}^{(1)} \varepsilon_{xy}^{(1)} \left(\varepsilon_{zz}^{(1)} - N_y^2\right) - \varepsilon_{xz}^{(1)} \varepsilon_{zy}^{(1)} \left(\varepsilon_{zz}^{(1)} + N_y^2\right)\right] \left(N_{z1}^{(1)} N_{z3}^{(1)}\right) \\
& - \varepsilon_{zy}^{(1)} N_y \left[\varepsilon_{xy}^{(1)} \left(\varepsilon_{zz}^{(1)} - N_y^2\right) - \varepsilon_{xz}^{(1)} \varepsilon_{zy}^{(1)}\right] \left(N_{z1}^{(1)} + N_{z3}^{(1)}\right) \\
& + \left(\varepsilon_{xy}^{(1)} \varepsilon_{zz}^{(1)} - \varepsilon_{xz}^{(1)} \varepsilon_{zy}^{(1)}\right) \left[\left(\varepsilon_{xx}^{(1)} - N_y^2\right) \left(\varepsilon_{zz}^{(1)} - N_y^2\right) - \varepsilon_{zx}^{(1)} \varepsilon_{xz}^{(1)}\right] \Big\},
\end{aligned} \tag{3.100a}$$

$$\begin{aligned}
\mathcal{Q}^{(1)} = \Big\{ & \varepsilon_{xz}^{(1)} N_y \left(\varepsilon_{zz}^{(1)} - N_y^2\right) \left(N_{z1}^{(1)} N_{z3}^{(1)}\right) \\
& - \left(\varepsilon_{zz}^{(1)} - N_y^2\right) \left[\varepsilon_{xy}^{(1)} \left(\varepsilon_{zz}^{(1)} - N_y^2\right) - \varepsilon_{xz}^{(1)} \varepsilon_{zy}^{(1)}\right] \left(N_{z1}^{(1)} + N_{z3}^{(1)}\right) \\
& + \varepsilon_{xz}^{(1)} N_y \left[\left(\varepsilon_{xx}^{(1)} - N_y^2\right) \left(\varepsilon_{zz}^{(1)} - N_y^2\right) - \varepsilon_{xz}^{(1)} \varepsilon_{zx}^{(1)}\right] \Big\},
\end{aligned} \tag{3.100b}$$

$$\begin{aligned}\mathcal{S}^{(1)} = \Big\{ & \varepsilon_{xz}^{(1)}\varepsilon_{zz}^{(1)} N_y \left(N_{z1}^{(1)} N_{z3}^{(1)}\right)^2 \\ & -\varepsilon_{zz}^{(1)}\left[\varepsilon_{xy}^{(1)}\left(\varepsilon_{zz}^{(1)} - N_y^2\right) - \varepsilon_{xz}^{(1)}\varepsilon_{zy}^{(1)}\right]\left(N_{z1}^{(1)} N_{z3}^{(1)}\right)\left(N_{z1}^{(1)} + N_{z3}^{(1)}\right) \\ & +\varepsilon_{xz}^{(1)} N_y \left[\varepsilon_{zz}^{(1)}\left(\varepsilon_{xx}^{(1)} - N_y^2\right) - \varepsilon_{xz}^{(1)}\varepsilon_{zx}^{(1)}\right]\left(N_{z1}^{(1)} N_{z3}^{(1)}\right) \\ & -\varepsilon_{zy}^{(1)} N_y \left[\varepsilon_{xy}^{(1)}\left(\varepsilon_{zz}^{(1)} - N_y^2\right) - \varepsilon_{xz}^{(1)}\varepsilon_{zy}^{(1)}\right]\left(N_{z1}^{(1)} N_{z3}^{(1)}\right) \\ & +\varepsilon_{xz}^{(1)} N_y^2 \left[\varepsilon_{zy}^{(1)}\left(\varepsilon_{xx}^{(1)} - N_y^2\right) - \varepsilon_{xy}^{(1)}\varepsilon_{zx}^{(1)}\right]\left(N_{z1}^{(1)} + N_{z3}^{(1)}\right) \\ & -N_y \left[\varepsilon_{xy}^{(1)}\left(\varepsilon_{zz}^{(1)} - N_y^2\right) - \varepsilon_{xz}^{(1)}\varepsilon_{zy}^{(1)}\right]\left[\varepsilon_{zy}^{(1)}\left(\varepsilon_{xx}^{(1)} - N_y^2\right) - \varepsilon_{xy}^{(1)}\varepsilon_{zx}^{(1)}\right]\Big\}, \end{aligned} \tag{3.100c}$$

$$\begin{aligned}\mathcal{T}^{(1)} = \Big\{ & -\left(\varepsilon_{zz}^{(1)} - N_y^2\right)\left[\varepsilon_{xy}^{(1)}\left(\varepsilon_{zz}^{(1)} - N_y^2\right) - \varepsilon_{xz}^{(1)}\varepsilon_{zy}^{(1)}\right]\left(N_{z1}^{(1)} N_{z3}^{(1)}\right) \\ & +\varepsilon_{xz}^{(1)} N_y \left[\left(\varepsilon_{xx}^{(1)} - N_y^2\right)\left(\varepsilon_{zz}^{(1)} - N_y^2\right) - \varepsilon_{xz}^{(1)}\varepsilon_{zx}^{(1)}\right]\left(N_{z1}^{(1)} + N_{z3}^{(1)}\right) \\ & -\left[\varepsilon_{xy}^{(1)}\left(\varepsilon_{zz}^{(1)} - N_y^2\right) - \varepsilon_{xz}^{(1)}\varepsilon_{zy}^{(1)}\right]\left[\left(\varepsilon_{xx}^{(1)} - N_y^2\right)\left(\varepsilon_{zz}^{(1)} - N_y^2\right) - \varepsilon_{xz}^{(1)}\varepsilon_{zx}^{(1)}\right]\Big\}, \end{aligned} \tag{3.100d}$$

$$\begin{aligned}\mathcal{U}^{(1)} = {} & \varepsilon_{xz}^2 N_y^2 \left(N_{z1}^{(1)} N_{z3}^{(1)}\right) - \varepsilon_{xz}^{(1)} N_y \left[\varepsilon_{xy}^{(1)}\left(\varepsilon_{zz}^{(1)} - N_y^2\right) - \varepsilon_{xz}^{(1)}\varepsilon_{zy}^{(1)}\right]\left(N_{z1}^{(1)} + N_{z3}^{(1)}\right) \\ & + \left[\varepsilon_{xy}\left(\varepsilon_{zz} - N_y^2\right) - \varepsilon_{xz}\varepsilon_{zy}\right]^2, \end{aligned} \tag{3.101}$$

$$\begin{aligned}\mathcal{W}^{(1)} = {} & -\varepsilon_{zz}^{(1)}\left(\varepsilon_{zz}^{(1)} - N_y^2\right)\left(N_{z1}^{(1)} N_{z3}^{(1)}\right)^2 \\ & +\varepsilon_{zz}^{(1)}\left[\left(\varepsilon_{xx}^{(1)} - N_y^2\right)\left(\varepsilon_{zz} - N_y^2\right) - \varepsilon_{xz}^{(1)}\varepsilon_{zx}^{(1)}\right]\left(N_{z1}^{(1)2} + N_{z3}^{(1)2}\right) \\ & -\varepsilon_{xz}^{(1)}\varepsilon_{zx}^{(1)} N_y^2 \left(N_{z1}^{(1)} N_{z3}^{(1)}\right) \\ & -\varepsilon_{zx}^{(1)} N_y \left[\varepsilon_{xy}^{(1)}\left(\varepsilon_{zz} - N_y^2\right) - \varepsilon_{xz}^{(1)}\varepsilon_{zy}^{(1)}\right]\left(N_{z1}^{(1)} + N_{z3}^{(1)}\right) \\ & -\left[\left(\varepsilon_{zz}^{(1)} - N_y^2\right)\left(\varepsilon_{xx}^{(1)} - N_y^2\right) - \varepsilon_{xz}^{(1)}\varepsilon_{zx}^{(1)}\right]\left[\varepsilon_{zz}^{(1)}\left(\varepsilon_{xx}^{(1)} - N_y^2\right) - \varepsilon_{xz}^{(1)}\varepsilon_{zx}^{(1)}\right]. \end{aligned} \tag{3.102}$$

In the analysis of more complicated structures, the single interface **M** matrices enter as partial results into more general expressions. Then, the complete knowledge of **M** matrix is required. The remaining matrix elements are given by

$$\mathrm{M}_{\substack{12\\22}} = \left[\left(\mathbf{D}^{(0)}\right)^{-1}\mathbf{D}^{(1)}\right]_{\substack{12\\22}} = \frac{1}{2}\left[D_{12}^{(1)} \pm \left(N^{(0)}\cos\varphi^{(0)}\right)^{-1} D_{22}^{(1)}\right], \tag{3.103a}$$

$$\mathrm{M}_{\substack{32\\42}} = \left[\left(\mathbf{D}^{(0)}\right)^{-1}\mathbf{D}^{(1)}\right]_{\substack{32\\42}} = \frac{1}{2}\left[\left(\cos\varphi^{(0)}\right)^{-1} D_{32}^{(1)} \mp \left(N^{(0)}\right)^{-1} D_{42}^{(1)}\right], \tag{3.103b}$$

$$\mathrm{M}_{\substack{14\\24}} = \left[\left(\mathbf{D}^{(0)}\right)^{-1}\mathbf{D}^{(1)}\right]_{\substack{14\\24}} = \frac{1}{2}\left[D_{14}^{(1)} \pm \left(N^{(0)}\cos\varphi^{(0)}\right)^{-1} D_{24}^{(1)}\right], \tag{3.103c}$$

$$\mathrm{M}_{\substack{34\\44}} = \left[\left(\mathbf{D}^{(0)}\right)^{-1}\mathbf{D}^{(1)}\right]_{\substack{34\\44}} = \frac{1}{2}\left[\left(\cos\varphi^{(0)}\right)^{-1} D_{34}^{(1)} \mp \left(N^{(0)}\right)^{-1} D_{44}\right]. \tag{3.103d}$$

In this section, we have illustrated the application of Yeh's formalism to an ambient-magnetic crystal interface. The usefulness of the formalism increases with the problem complexity. The explicit expressions for the multilayer structures are in most cases not practical, and the problems are to be handled by a computer. Nevertheless, it is instructive to evaluate the transfer matrices in some special cases, as this gives some insight to the problem and helps us in obtaining simpler approximate expressions. This will be the subject of the following chapters.

References

1. R. P. Hunt, "Magneto-optic scattering from thin solid films," J. Appl. Phys. **37**, 1652–1671, 1967.
2. W. Voigt, *Magneto- und Elektro-Optik* (Teubner, Leipzig, 1908).
3. C. C. Robinson, J. Opt. Soc. Am. **54**, 1220 (1964).
4. D. O. Smith, "Magneto-optical scattering from multilayer magnetic and dielectric films," Opt. Acta **12**, 13, 1965.
5. E. D. Palik and B. W. Henvis, "A bibliography of magnetooptics of solids," Appl. Opt. **6**, 603–630, 1967.
6. C. J. Sansalone, "Faraday rotation of a system of thin layers containing a thick layer," Appl. Opt. **10**, 2332–2335, 1971.
7. B. Donovan and T. Medcalf, "The inclusion of multiple reflections in the theory of the Faraday effect in semiconductors," Brit. J. Appl. Phys. **15**, 1139–1151, 1964.
8. H. Piller, "Effect of internal reflection on optical faraday rotation," J. Appl. Phys. **37**, 763–767, 1966.
9. Š. Višňovský, V. Prosser, and R. Krishnan, "Effect of multiple internal reflections on Faraday rotation in multilayer structures," J. Appl. Phys. **49**, 403–408, 1978.
10. Š. Višňovský and R. Krishnan, "Complex Faraday effect in multilayer structures," J. Opt. Soc. Am. **71**, 315–320, 1981.
11. N. J. Damaskos, A. L. Maffet and P. L. E. Uslenghi, "Reflection and transmission for gyroelectromagnetic biaxial layered media," J. Soc. Am. A **2**, 454–461, 1985.
12. O. Schwelb, "Stratified lossy anisotropic media: General characteristics," J. Opt. Soc. Am. A **3**, 188–193, 1986.
13. R. Gamble and P. H. Lissberger, "Electromagnetic field distributions in multilayer thin films for magneto-optical recording," J. Opt. Soc. Am. A **5**, 1533–1542, 1988.

14. P. Yeh, "Optics of anisotropic layered media: A new 4 × 4 matrix algebra," Surf. Sci. **96**, 41–53, 1980; P. Yeh, *Optical Waves in Layered Media* (John Wiley & Sons, New York 1988), Chapter 9.
15. Š. Višňovský, "Magnetooptical ellipsometry," Czech. J. Phys. B **36**, 625–650, 1986.
16. Š. Višňovský, "Optics of magnetic multilayers," Czech. J. Phys. **41**, 663–694, 1991.
17. M. Mansuripur, *The Physical Principles of Magneto-optical Recording* (Cambridge University Press, London, 1996).
18. R. M. A. Azzam and N. M. Bashara, *Ellipsometry and Polarized Light* (Elsevier, Amsterdam, 1987).
19. W. Xu, L. T. Wood, and T. D. Godling, "Optical degeneracies in anisotropic layered media: Treatment of singularities in 4 × 4 matrix formalism," Phys. Rev. B **61**, 1740–1743, 2000.
20. Š. Višňovský, "Magnetooptical longitudinal and transverse Kerr and birefringence effect in orthorhombic crystals," Czech. J. Phys. B **34**, 969–980, 1984.

4

Polar Magnetization

4.1 Introduction

This chapter deals with multilayers at polar magnetization. Both the interface and propagation effects contribute to the global response. The cases of the single interface either at normal or oblique light incidence considered in Section 1.3.2 and Section 1.6.1, respectively, and that of the Faraday effect of Section 1.3 represent just the simplest special cases. In multilayers with polar magnetization, the magnetization vectors are perpendicular to the plane parallel interfaces in all layers. In this configuration, the strongest linear magnetooptic effects, useful for practical applications, take place.

We first discuss the situation where the propagation vectors of the incident, transmitted, and reflected radiation are all parallel to the common axis of the layer magnetization vectors. The configuration includes Faraday effect and normal incidence MO polar Kerr effects. It is distinguished by its practical importance and ultimate simplicity. The configuration is exploited in MO sensing of the thermomagnetically recorded information using focused polarized laser beams. The MO disks employ the perpendicular recording concept and the normal incidence. This allows us to focus the read/write beam to a minimum area. The beam focusing involves obliquely incident rays, but fortunately the polar magnetooptic response depends only weakly on the angle of incidence in a rather broad range around the zero angle of incidence.

At polar magnetization and normal incidence, the use of the 4×4 matrix formalism in modelling the MO effects linear in magnetization can be demonstrated with a minimum effort. The simplicity of the problem geometry allows us to write the explicit formula for transmission and reflection, magnetic circular birefringence, and dichroism, and construct corresponding Jones and Mueller matrices even for multilayers consisting of several layers. The procedure can be extended to multilayers with periodic composition profile (magnetic superlattices).

Then, we shall treat the case of an arbitrary orientation of the propagation vector in the plane of incidence. At the beginning, we assume the permittivity tensor of a medium of orthorhombic symmetry with one of the orthorhombic axes parallel to the magnetization and perpendicular to the interfaces. This enables us to account for both perpendicular and in-plane anisotropies originating from crystallographic symmetry of the layer medium or induced during the deposition process. In particular, the effect of substrate birefringence, which must be eliminated from the reading process in magnetooptic memories, can be analyzed in this way.

We construct dynamical and propagation matrices and express the global **M** matrix as a product of transfer matrices. As an example, the reflection on a three-layer system consisting of an isotropic ambient and magnetic film–magnetic substrate system is considered [1]. This is the simplest system in which we can evaluate the effect of a buried interface between two magnetic media at oblique angles of incidence. The analysis requires a considerable algebra but allows us to identify several effects contributing to the response, which cannot be directly appreciated in numerical procedures. The importance of their account in the device performance will increase with continuing reduction of the information bit area to nanometer range.

The analysis of the magnetic film–magnetic substrate system covers several special cases, *e.g.*, the magnetooptic Kerr effect in a single plate,[1] enhancement of the MO Kerr effects by dielectric overcoats, ellipsometry of a film-substrate system, *etc*. Three simple cases illustrate the practical application of the general results, *i.e.*, the normal incidence on an ambient–film–substrate system, oblique incidence on an optically uniaxial crystal, and an oblique incidence on magnetic film–magnetic substrate system at polar magnetization. In this last case, we shall limit ourselves to an approximate solution for the simplest form of the permittivity tensor. We will check the consistency of the results with those obtained for these configurations by other procedures.

4.2 Normal Incidence

We start with the case of multilayers at the polar magnetization induced in originally isotropic layer media at normal light incidence (Figure 4.1). The pertinent permittivity tensor for the magnetization parallel to the z axis follows from Eq. (2.23) and Eq. (2.24) using Eq. (1.10). Its form is consistent with that already employed in Eq. (1.22) and appropriate for the present Cartesian coordinate system with the z axis normal to the interfaces and parallel to both the magnetization and the radiation propagation vectors.

[1] In a plate, we are concerned with the combination of the interface (polar Kerr) and propagation (Faraday) effects.

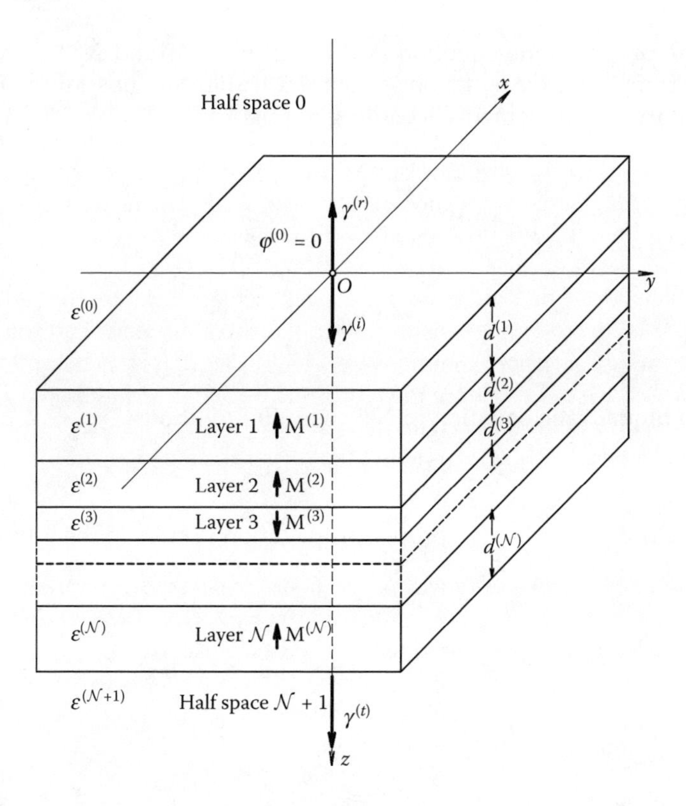

FIGURE 4.1
Normal light incidence ($\varphi^{(0)} = 0$) at a multilayer consisting of $\mathcal{N}$ anisotropic layers with polar magnetizations, $\boldsymbol{M}^{(n)}$ ($n = 1, \ldots, \mathcal{N}$). The layers are characterized by permittivity tensors $\varepsilon^{(n)}$. The incident, reflected, and transmitted propagation vectors in sandwiching isotropic nonabsorbing half spaces (0) and ($\mathcal{N} + 1$) are denoted as $\gamma^{(i)}$, $\gamma^{(r)}$, and $\gamma^{(t)}$, respectively.

Although the symmetry restricts the magnetizations, $\boldsymbol{M}$, in all layers to the direction of the z axis, we still have the freedom to choose their magnitudes and signs. Moreover, at strictly normal incidence, any in-plane magnetization components would not affect the response to first order in $\boldsymbol{M}$. In the experiment, the response is determined from a difference between the signals at $\boldsymbol{M}$ and $-\boldsymbol{M}$. This procedure automatically eliminates the contributions proportional to $\boldsymbol{M}^2$ originating from the in-plane components of $\boldsymbol{M}$. The relative permittivity tensor in the n-th layer magnetized along the z axis can be expressed as

$$\varepsilon^{(n)} = \begin{pmatrix} \varepsilon_0^{(n)} & \mathrm{j}\,\varepsilon_1^{(n)} & 0 \\ -\mathrm{j}\,\varepsilon_1^{(n)} & \varepsilon_0^{(n)} & 0 \\ 0 & 0 & \varepsilon_3^{(n)} \end{pmatrix}. \tag{4.1}$$

From the proper value equation (3.13) with $N_y = 0$ and $N_z^{(n)} = N^{(n)}$ or directly from the wave equation system (1.23) after the substitution for the permittivity tensor elements according to Eq. (4.1),

$$\left(N^{(n)2} - \varepsilon_0^{(n)}\right) E_{0x}^{(n)} - \mathrm{j}\,\varepsilon_1^{(n)} E_{0y}^{(n)} = 0 \tag{4.2a}$$

$$\mathrm{j}\,\varepsilon_1^{(n)} E_{0x}^{(n)} + \left(N^{(n)2} - \varepsilon_0^{(n)}\right) E_{0y}^{(n)} = 0 \tag{4.2b}$$

$$\varepsilon_3^{(n)} E_{0z}^{(n)} = 0, \tag{4.2c}$$

we have

$$N^{(n)4} - 2\varepsilon_0^{(n)} N^{(n)2} + \varepsilon_0^{(n)2} - \varepsilon_1^{(n)2} = 0. \tag{4.3}$$

This is a biquadratic equation in $N^{(n)}$ with the solutions[2]

$$N_{\pm}^{(n)2} = \varepsilon_0^{(n)} \mp \varepsilon_1^{(n)}. \tag{4.4}$$

There are thus four allowed proper propagation vectors in the n-th layer

$$\frac{\omega}{c} N_1^{(n)} = \frac{\omega}{c} N_{+}^{(n)} = \frac{\omega}{c}\left(\varepsilon_0^{(n)} - \varepsilon_1^{(n)}\right)^{1/2}, \tag{4.5a}$$

$$\frac{\omega}{c} N_2^{(n)} = -\frac{\omega}{c} N_{+}^{(n)} = -\frac{\omega}{c}\left(\varepsilon_0^{(n)} - \varepsilon_1^{(n)}\right)^{1/2}, \tag{4.5b}$$

$$\frac{\omega}{c} N_3^{(n)} = \frac{\omega}{c} N_{-}^{(n)} = \frac{\omega}{c}\left(\varepsilon_0^{(n)} + \varepsilon_1^{(n)}\right)^{1/2}, \tag{4.5c}$$

$$\frac{\omega}{c} N_4^{(n)} = -\frac{\omega}{c} N_{-}^{(n)} = -\frac{\omega}{c}\left(\varepsilon_0^{(n)} + \varepsilon_1^{(n)}\right)^{1/2}. \tag{4.5d}$$

The associated proper polarizations follow from the substitution for $N_{\pm}^{(n)2}$ according to Eq. (4.4) into the wave equation system (4.2)

$$\mp\,\varepsilon_1^{(n)} E_{0x}^{(n)} - \mathrm{j}\,\varepsilon_1^{(n)} E_{0y}^{(n)} = 0, \tag{4.6a}$$

$$\mathrm{j}\,\varepsilon_1^{(n)} E_{0x}^{(n)} \mp \varepsilon_1^{(n)} E_{0y}^{(n)} = 0, \tag{4.6b}$$

$$\varepsilon_3^{(n)} E_{0z}^{(n)} = 0. \tag{4.6c}$$

The proper polarizations (or the proper polarization modes) are purely transverse ($E_{0z}^{(n)} = 0$) circularly polarized waves with

$$E_{0y} = \pm \mathrm{j} E_{0x}. \tag{4.7}$$

For the waves propagating in the positive sense of the z axis, the solution with the upper and lower sign represents, according to the convention

[2] The assignment of the solutions $N_{\pm}^{(n)2}$ of Eq. (4.3) is a matter of convention, and a particular choice affects the signs in the final formulas. The present choice is consistent with Eq. (1.25) for $\varepsilon_{xy}^{(n)} = \mathrm{j}\,\varepsilon_1^{(n)}$.

[2], the right (RCP) and left (LCP) circularly polarized waves, respectively. Thus, the complex indices of refraction $N_+^{(n)}$ and $N_-^{(n)}$ belong to the RCP and LCP waves propagating in the positive sense of the z axis, respectively. On the other hand, the RCP and LCP waves propagating in the opposite, *i.e.*, negative, sense of the z axis have the indices $N_-^{(n)}$ and $N_+^{(n)}$, respectively. This feature explains the already mentioned (see Section 1.1) phase accumulation in the Faraday effect and forms the physical basis for the nonreciprocal microwave devices like isolators or circulators and their optical counterparts. The magnetization reversal $\boldsymbol{M} \rightarrow -\boldsymbol{M}$ produces $\varepsilon_1^{(n)} \rightarrow -\varepsilon_1^{(n)}$ and results in the exchange of RCP and LCP as seen from Eq. (4.2). In the n-th layer of the multilayer structure, the proper propagation vectors and the associated proper polarizations are grouped together as follows

$$\pm\frac{\omega}{c}N_+^{(n)}, \qquad \boldsymbol{e}_{1,2} = 2^{-1/2}\begin{pmatrix}1\\ \mathrm{j}\\ 0\end{pmatrix} = 2^{-1/2}(\hat{x} + \mathrm{j}\hat{y}), \tag{4.8a}$$

$$\pm\frac{\omega}{c}N_-^{(n)}, \qquad \boldsymbol{e}_{3,4} = 2^{-1/2}\begin{pmatrix}1\\ -\mathrm{j}\\ 0\end{pmatrix} = 2^{-1/2}(\hat{x} - \mathrm{j}\hat{y}). \tag{4.8b}$$

For a given sense of propagation, the proper CP polarizations are normalized and orthogonal to each other according to the definition of Eq. (3.15). They are consistent with Eq. (3.16) and Eq. (3.17). Eq. (3.14), Eq. (3.16), and Eq. (4.7) allow us to write the total electric wave field in the n-th layer ($z^{(n-1)} < z < z^{(n)}$) as a sum of the CP proper polarization modes[3]

$$\begin{aligned}\mathbf{E}^{(n)} = \frac{\sqrt{2}}{2}\Big\{&\mathrm{E}_{01}^{(n)}(\hat{x} + \mathrm{j}\hat{y})\exp\left[\mathrm{j}\omega t - \mathrm{j}\frac{\omega}{c}N_+^{(n)}(z - z^{(n)})\right]\\ &+ \mathrm{E}_{02}^{(n)}(\hat{x} + \mathrm{j}\hat{y})\exp\left[\mathrm{j}\omega t + \mathrm{j}\frac{\omega}{c}N_+^{(n)}(z - z^{(n)})\right]\\ &+ \mathrm{E}_{03}^{(n)}(\hat{x} - \mathrm{j}\hat{y})\exp\left[\mathrm{j}\omega t - \mathrm{j}\frac{\omega}{c}N_-^{(n)}(z - z^{(n)})\right]\\ &+ \mathrm{E}_{04}^{(n)}(\hat{x} - \mathrm{j}\hat{y})\exp\left[\mathrm{j}\omega t + \mathrm{j}\frac{\omega}{c}N_-^{(n)}(z - z^{(n)})\right]\Big\}.\end{aligned} \tag{4.9}$$

Here, $\mathrm{E}_{01}^{(n)}$ and $\mathrm{E}_{03}^{(n)}$ are the complex amplitudes of the RCP and LCP waves, respectively, propagating in the direction of positive sense of the z axis (forward modes); $\mathrm{E}_{02}^{(n)}$ and $\mathrm{E}_{04}^{(n)}$ denote the amplitudes of the LCP and RCP waves, respectively, propagating in the direction of negative sense of

[3] We recall that the classical ellipsometry [2] classifies as RCP the waves proportional to $\Re\left\{\left(\hat{x} \pm \mathrm{j}\hat{y}\right)\exp\left[\mathrm{j}\omega t \mp \mathrm{j}\frac{\omega}{c}N_+^{(n)}\left(z - z^{(n)}\right)\right]\right\}$ and as LCP those proportional to $\Re\left\{\left(\hat{x} \mp \mathrm{j}\hat{y}\right)\exp\left[\mathrm{j}\omega t \mp \mathrm{j}\frac{\omega}{c}N_+^{(n)}\left(z - z^{(n)}\right)\right]\right\}$.

the z axis (backward modes). According to Eq. (3.18) and Eq. (3.19), the corresponding magnetic flux density field becomes

$$\begin{aligned} c\mathbf{B}^{(n)} = \frac{\sqrt{2}}{2} \Big\{ & -\mathrm{j}N_+^{(n)} E_{01}^{(n)} (\hat{x} + \mathrm{j}\hat{y}) \exp\left[\mathrm{j}\omega t - \mathrm{j}\frac{\omega}{c} N_+^{(n)} (z - z^{(n)})\right] \\ & + \mathrm{j}N_+^{(n)} E_{02}^{(n)} (\hat{x} + \mathrm{j}\hat{y}) \exp\left[\mathrm{j}\omega t + \mathrm{j}\frac{\omega}{c} N_+^{(n)} (z - z^{(n)})\right] \\ & + \mathrm{j}N_-^{(n)} E_{03}^{(n)} (\hat{x} - \mathrm{j}\hat{y}) \exp\left[\mathrm{j}\omega t - \mathrm{j}\frac{\omega}{c} N_-^{(n)} (z - z^{(n)})\right] \\ & - \mathrm{j}N_-^{(n)} E_{04}^{(n)} (\hat{x} - \mathrm{j}\hat{y}) \exp\left[\mathrm{j}\omega t + \mathrm{j}\frac{\omega}{c} N_-^{(n)} (z - z^{(n)})\right] \Big\}. \end{aligned} \tag{4.10}$$

4.2.1 Circular Representation

The dynamical matrix (3.29), its inverse and propagation matrix (3.31) in the circular (CP) representation entering the transfer matrix (3.32) for $n = 1$ to $\mathcal{N} + 1$, are obtained with help of Eqs. (4.5), Eq. (4.9), and Eq. (4.10)

$$(\mathbf{D}^{(n-1)})^{-1} = \frac{\sqrt{2}}{4} \begin{pmatrix} 1 & \left(N_+^{(n-1)}\right)^{-1} & -\mathrm{j} & \mathrm{j}\left(N_+^{(n-1)}\right)^{-1} \\ 1 & -\left(N_+^{(n-1)}\right)^{-1} & -\mathrm{j} & -\mathrm{j}\left(N_+^{(n-1)}\right)^{-1} \\ 1 & \left(N_-^{(n-1)}\right)^{-1} & \mathrm{j} & -\mathrm{j}\left(N_-^{(n-1)}\right)^{-1} \\ 1 & -\left(N_-^{(n-1)}\right)^{-1} & \mathrm{j} & \mathrm{j}\left(N_-^{(n-1)}\right)^{-1} \end{pmatrix}, \tag{4.11}$$

$$\mathbf{D}^{(n)} = \frac{\sqrt{2}}{2} \begin{pmatrix} 1 & 1 & 1 & 1 \\ N_+^{(n)} & -N_+^{(n)} & N_-^{(n)} & -N_-^{(n)} \\ \mathrm{j} & \mathrm{j} & -\mathrm{j} & -\mathrm{j} \\ -\mathrm{j}N_+^{(n)} & \mathrm{j}N_+^{(n)} & \mathrm{j}N_-^{(n)} & -\mathrm{j}N_-^{(n)} \end{pmatrix}, \tag{4.12}$$

$$\mathbf{P}^{(n)} = \begin{pmatrix} \exp\left(\mathrm{j}\frac{\omega}{c} N_+^{(n)} d^{(n)}\right) & 0 & 0 & 0 \\ 0 & \exp\left(-\mathrm{j}\frac{\omega}{c} N_+^{(n)} d^{(n)}\right) & 0 & 0 \\ 0 & 0 & \exp\left(\mathrm{j}\frac{\omega}{c} N_-^{(n)} d^{(n)}\right) & 0 \\ 0 & 0 & 0 & \exp\left(-\mathrm{j}\frac{\omega}{c} N_-^{(n)} d^{(n)}\right) \end{pmatrix}. \tag{4.13}$$

According to Eq. (3.32), the product of the matrices in Eqs. (4.11) to (4.13) defines the transfer matrix $\mathbf{T}_{n-1,n}$ between the $(n-1)$-th and n-th layers.

The absence of coupling between RCP and LCP modes makes all transfer matrices (2 × 2) block diagonal

$$\mathbf{T}_{n-1,n} = \begin{pmatrix} (T_{n-1,n})_{11} & (T_{n-1,n})_{12} & 0 & 0 \\ (T_{n-1,n})_{21} & (T_{n-1,n})_{22} & 0 & 0 \\ 0 & 0 & (T_{n-1,n})_{33} & (T_{n-1,n})_{34} \\ 0 & 0 & (T_{n-1,n})_{43} & (T_{n-1,n})_{44} \end{pmatrix}, \tag{4.14}$$

where

$$(T_{n-1,n})_{11} = \frac{1}{2}\left(1 + u_{+}^{(n-1,n)}\right)\exp\left(\mathrm{j}\beta_{+}^{(n)}\right) \tag{4.15a}$$

$$(T_{n-1,n})_{33} = \frac{1}{2}\left(1 + u_{-}^{(n-1,n)}\right)\exp\left(\mathrm{j}\beta_{-}^{(n)}\right) \tag{4.15b}$$

$$(T_{n-1,n})_{12} = \frac{1}{2}\left(1 - u_{+}^{(n-1,n)}\right)\exp\left(-\mathrm{j}\beta_{+}^{(n)}\right) \tag{4.15c}$$

$$(T_{n-1,n})_{34} = \frac{1}{2}\left(1 - u_{-}^{(n-1,n)}\right)\exp\left(-\mathrm{j}\beta_{-}^{(n)}\right) \tag{4.15d}$$

$$(T_{n-1,n})_{21} = \frac{1}{2}\left(1 - u_{+}^{(n-1,n)}\right)\exp\left(\mathrm{j}\beta_{-}^{(n)}\right) \tag{4.15e}$$

$$(T_{n-1,n})_{43} = \frac{1}{2}\left(1 - u_{-}^{(n-1,n)}\right)\exp\left(\mathrm{j}\beta_{-}^{(n)}\right) \tag{4.15f}$$

$$(T_{n-1,n})_{22} = \frac{1}{2}\left(1 + u_{+}^{(n-1,n)}\right)\exp\left(-\mathrm{j}\beta_{+}^{(n)}\right) \tag{4.15g}$$

$$(T_{n-1,n})_{44} = \frac{1}{2}\left(1 + u_{-}^{(n-1,n)}\right)\exp\left(-\mathrm{j}\beta_{-}^{(n)}\right) \tag{4.15h}$$

with

$$u_{\pm}^{(n-1,n)} = \frac{N_{\pm}^{(n)}}{N_{\pm}^{(n-1)}} \tag{4.16}$$

and

$$\beta_{\pm}^{(n)} = \frac{\omega}{c} N_{\pm}^{(n)} d^{\,(n)}. \tag{4.17}$$

In isotropic regions, including the sandwiching half spaces $n = 0$ $(z < 0)$ and $\mathcal{N} + 1$ $(z > z_{\mathcal{N}})$, any pair of orthogonally polarized proper polarizations can be chosen. The Cartesian dynamical matrix for an isotropic region at normal incidence and its inverse follow from Eq. (3.82a) and Eq. (3.82b) with $\varphi^{(0)}$ replaced by a general $\varphi^{(s)}$ and set to zero, $\varphi^{(s)} = 0$. It is, however, advantageous to choose CP polarization modes in the sandwiching isotropic half spaces as well as in isotropic layers. The circular (CP) dynamical matrix and its inverse in isotropic regions follows from Eq. (4.12) and Eq. (4.11), respectively, by setting $N_{+}^{(s)} = N_{-}^{(s)} = N^{(s)}$, with s indexing isotropic regions. For example, in the isotropic half space of incidence, we obtain the first matrix forming the product **M** with help of Eq. (4.11)

$$(\mathbf{D}^{(0)})^{-1} = \frac{\sqrt{2}}{4}\begin{pmatrix} 1 & (N^{(0)})^{-1} & -\mathrm{j} & \mathrm{j}(N^{(0)})^{-1} \\ 1 & -(N^{(0)})^{-1} & -\mathrm{j} & -\mathrm{j}\left(N^{(0)}\right)^{-1} \\ 1 & (N^{(0)})^{-1} & \mathrm{j} & -\mathrm{j}(N^{(0)})^{-1} \\ 1 & -(N^{(0)})^{-1} & \mathrm{j} & \mathrm{j}(N^{(0)})^{-1} \end{pmatrix}. \tag{4.18}$$

Similarly, the last matrix in the product **M** is obtained from Eq. (4.12)

$$\mathbf{D}^{(\mathcal{N}+1)} = \frac{\sqrt{2}}{2}\begin{pmatrix} 1 & 1 & 1 & 1 \\ N^{(\mathcal{N}+1)} & -N^{(\mathcal{N}+1)} & N^{(\mathcal{N}+1)} & -N^{(\mathcal{N}+1)} \\ \mathrm{j} & \mathrm{j} & -\mathrm{j} & -\mathrm{j} \\ -\mathrm{j}N^{(\mathcal{N}+1)} & \mathrm{j}N^{(\mathcal{N}+1)} & \mathrm{j}N^{(\mathcal{N}+1)} & -\mathrm{j}N^{(\mathcal{N}+1)} \end{pmatrix}. \tag{4.19}$$

The global **M** matrix of the structure consisting of $\mathcal{N}$ layers in the CP representation follows from Eq. (3.33) and, as a product of block diagonal matrices, should also be a block diagonal one. Assuming, as before, that there are no incident waves from the region $z > z^{(\mathcal{N})}$, the CP wave fields are related by

$$\begin{pmatrix} E_{01}^{(0)} \\ E_{02}^{(0)} \\ E_{03}^{(0)} \\ E_{04}^{(0)} \end{pmatrix} = \begin{pmatrix} M_{11} & M_{12} & 0 & 0 \\ M_{21} & M_{22} & 0 & 0 \\ 0 & 0 & M_{33} & M_{34} \\ 0 & 0 & M_{43} & M_{44} \end{pmatrix}\begin{pmatrix} E_{01}^{(\mathcal{N}+1)} \\ 0 \\ E_{03}^{(\mathcal{N}+1)} \\ 0 \end{pmatrix}, \tag{4.20}$$

from which we obtain, *e.g.*, relation between the incident (0) and transmitted $(\mathcal{N}+1)$ RCP and LCP waves

$$E_{01}^{(\mathcal{N}+1)} = M_{11}^{-1}E_{01}^{(0)}, \tag{4.21a}$$

$$E_{03}^{(\mathcal{N}+1)} = M_{33}^{-1}E_{03}^{(0)}. \tag{4.21b}$$

We express Eq. (4.20) with a more explicit identification of the proper polarizations

$$\begin{pmatrix} E_{+}^{(i)} \\ E_{+}^{(r)} \\ E_{-}^{(i)} \\ E_{-}^{(r)} \end{pmatrix} = \begin{pmatrix} M_{11} & M_{12} & 0 & 0 \\ M_{21} & M_{22} & 0 & 0 \\ 0 & 0 & M_{33} & M_{34} \\ 0 & 0 & M_{43} & M_{44} \end{pmatrix}\begin{pmatrix} E_{+}^{(t)} \\ 0 \\ E_{-}^{(t)} \\ 0 \end{pmatrix}. \tag{4.22}$$

Here, we have replaced the mode amplitudes of a pair of CP waves, $E_{01}^{(0)}$ and $E_{03}^{(0)}$, incident (i) on the structure as follows

$$E_{01}^{(0)} = E_{+}^{(i)}, \quad E_{03}^{(0)} = E_{-}^{(i)}. \tag{4.23}$$

There are also pairs of CP transmitted (t) and reflected (r) waves labelled respectively

$$E_{01}^{(\mathcal{N}+1)} = E_{+}^{(t)}, \quad E_{03}^{(\mathcal{N}+1)} = E_{-}^{(t)}, \tag{4.24a}$$

$$E_{02}^{(0)} = E_{+}^{(r)}, \quad E_{04}^{(0)} = E_{-}^{(r)}. \tag{4.24b}$$

Note that for isotropic layers as well as for the sandwiching isotropic semi-infinite media, $N_{+}^{(s)} = N_{-}^{(s)} = N^{(s)}$. In isotropic multilayers, the two sub-blocks in the **M** are identical and the response at normal incidence can be treated with a 2×2 **M** matrix of reduced dimension.

In the CP representation the polarizations states of the incident (i), reflected (r), and transmitted (t) waves are specified by the complex ratios determined at the front and back interfaces of the multilayer [2]

$$\chi_i^{(CP)} = \frac{E_{03}^{(0)}}{E_{01}^{(0)}}, \qquad \text{at } z = 0, \tag{4.25a}$$

$$\chi_r^{(CP)} = \frac{E_{04}^{(0)}}{E_{02}^{(0)}}, \qquad \text{at } z = 0, \tag{4.25b}$$

$$\chi_t^{(CP)} = \frac{E_{03}^{(\mathcal{N}+1)}}{E_{01}^{(\mathcal{N}+1)}}, \qquad \text{at } z = \sum_{n=1}^{\mathcal{N}} d^{(n)}. \tag{4.25c}$$

In order to obtain the wave azimuths, θ_l, and ellipticities, $\tan\epsilon_l$ ($l = i, t$ or r), we employ CP column 2×1 Jones vectors [2]

$$\begin{pmatrix} E_{+}^{(l)} \\ E_{-}^{(l)} \end{pmatrix} = \frac{\left|E_0^{(l)}\right|}{\sqrt{2}} e^{j\phi_0^{(l)}} \begin{pmatrix} (\cos\epsilon_l + \sin\epsilon_l)\, e^{(-j\theta_l)} \\ (\cos\epsilon_l - \sin\epsilon_l)\, e^{(j\theta_l)} \end{pmatrix}, \tag{4.26}$$

where $|E_0^{(l)}|$ and $\varphi_0^{(l)}$ are the common amplitude and the initial phase. The ellipsometric parameters, the complex plane representation of the polarization state, $\chi_l^{(CP)}$, the azimuth, θ_l, and the ellipticity, $\tan\epsilon_l$, all in the CP representation of polarization states, are obtained from the following relations consistent with Eqs. (1.58) to (1.61) employed above in the single-interface problem. We have

$$\chi_l^{(\mathrm{CP})} = \frac{E_-^{(l)}}{E_+^{(l)}} = \tan\left(\frac{\pi}{4} - \epsilon_l\right) e^{2j\theta_l}, \tag{4.27}$$

$$\theta_l = \frac{1}{2} \arg \chi_l^{(\mathrm{CP})}, \tag{4.28}$$

$$\tan \epsilon_l = \frac{1 - \left|\chi_l^{(\mathrm{CP})}\right|}{1 + \left|\chi_l^{(\mathrm{CP})}\right|}. \tag{4.29}$$

Note that the electric field amplitudes are defined with respect to our Cartesian coordinate system fixed to the multilayer structure. Consequently, $\chi_i^{(\mathrm{CP})}$ and $\chi_t^{(\mathrm{CP})}$ are defined by the amplitude ratio of LCP to RCP waves, while $\chi_r^{(\mathrm{CP})}$ is defined by the amplitude ratio of RCP to LCP waves. The field vectors of waves with positive ellipticity rotate from the positive y axis to the positive x axis. The response in reflection of the multilayer system is given by the **M** matrix elements according to Eq. (3.70a) and Eq. (3.70c)

$$\chi_r^{(\mathrm{CP})} = \frac{M_{11} M_{43}}{M_{21} M_{33}} \chi_i^{(\mathrm{CP})}, \tag{4.30}$$

and the response in transmission follows from Eq. (3.70e) and Eq. (3.70g)

$$\chi_t^{(\mathrm{CP})} = \frac{M_{11}}{M_{33}} \chi_i^{(\mathrm{CP})}. \tag{4.31}$$

The global complex transmission and amplitude reflection coefficients for the incident RCP (+) and LCP (−) waves are defined by

$$t_+ = |t_+|\, e^{j\tau_+} = \frac{E_{01}^{(\mathcal{N}+1)}}{E_{01}^{(0)}} = M_{11}^{-1}, \tag{4.32a}$$

$$t_- = |t_-|\, e^{j\tau_-} = \frac{E_{03}^{(\mathcal{N}+1)}}{E_{03}^{(0)}} = M_{33}^{-1}, \tag{4.32b}$$

$$r_+ = |r_+|\, e^{j\varrho_+} = \frac{E_{02}^{(0)}}{E_{01}^{(0)}} = M_{21} M_{11}^{-1}, \tag{4.32c}$$

$$r_- = |r_-|\, e^{j\varrho_-} = \frac{E_{04}^{(0)}}{E_{03}^{(0)}} = M_{43} M_{33}^{-1}. \tag{4.32d}$$

Let us first consider the case of an incident wave linearly polarized parallel to the x axis, *i.e.*, with the zero azimuth and $\chi_i^{(\mathrm{CP})} = 1$. The expression for the multilayer complex Faraday effect follows from Eq. (4.25c), Eq. (4.27), Eq. (4.31), and Eq. (4.32)

$$\chi_t^{(CP)} = \frac{t_-}{t_+} = \frac{M_{11}}{M_{33}}. \tag{4.33}$$

The azimuth rotation in transmission follows with Eq. (4.28)

$$\theta_t = \frac{1}{2}\arg\left(\chi_t^{(CP)}\right) = \frac{1}{2}\arg\left(\frac{M_{11}}{M_{33}}\right) = \frac{1}{2}\left(\tau_- - \tau_+\right), \tag{4.34}$$

and the ellipticity in transmission follows with Eq. (4.29)

$$\tan \epsilon_t = \frac{1 - \left|\chi_t^{(CP)}\right|}{1 + \left|\chi_t^{(CP)}\right|} = \frac{|M_{33}| - |M_{11}|}{|M_{33}| + |M_{11}|} = \frac{|t_+| - |t_-|}{|t_+| + |t_-|}. \tag{4.35}$$

Assuming again $\chi_i^{(CP)} = 1$ in Eq. (4.30), we have for the normal incidence multilayer MO Kerr effect with Eq. (4.25b), Eq. (4.27), Eq. (4.30), Eq. (4.32c), and Eq. (4.32d)

$$\chi_r^{(CP)} = \frac{r_-}{r_+} = \frac{M_{11}M_{43}}{M_{21}M_{33}}. \tag{4.36}$$

The azimuth rotation in reflection follows with Eq. (4.28)

$$\theta_r = \frac{1}{2}\arg\left(\chi_r^{(CP)}\right) = \frac{1}{2}\arg\left(\frac{M_{11}M_{43}}{M_{21}M_{33}}\right) = \frac{1}{2}\left(\varrho_- - \varrho_+\right), \tag{4.37}$$

and the corresponding ellipticity in reflection follows with Eq. (4.29)

$$\tan \varepsilon_r = \frac{1 - \left|\chi_r^{(CP)}\right|}{1 + \left|\chi_r^{(CP)}\right|} = \frac{|M_{21}M_{33}| - |M_{11}M_{43}|}{|M_{21}M_{33}| + |M_{11}M_{43}|} = \frac{|r_+| - |r^-|}{|r_+| + |r_-|}. \tag{4.38}$$

The situation remains simple even when the incident wave has a nonzero azimuth θ_i but a zero ellipticity, $\tan \epsilon_i = 0$. Then, $\chi_i^{(CP)} = \exp\left(2j\theta_i\right)$. Then, the azimuth rotations in transmission and reflection follow from the differences $\theta_t - \theta_i$ and $\theta_r - \theta_i$, respectively, and the corresponding ellipticities remain unchanged. The case of a nonzero incident azimuth and ellipticity requires the use of the transformation rules given in Eq. (4.30) and Eq. (4.31). The transmission ($\mathcal{I}_\mathcal{T}$) and reflection ($\mathcal{I}_\mathcal{R}$) coefficient for intensities are obtained from

$$\mathcal{I}_\mathcal{T} = \frac{1}{2}\left(|M_{11}|^{-2} + |M_{33}|^{-2}\right) \tag{4.39}$$

and

$$\mathcal{I}_\mathcal{R} = \frac{1}{2}\left(\left|\frac{M_{21}}{M_{11}}\right|^2 + \left|\frac{M_{43}}{M_{33}}\right|^2\right). \tag{4.40}$$

4.2.1.1 Single Interface

The interface between an isotropic medium and a medium at polar magnetization is characterized by the matrix product

$$\mathbf{M} = (\mathbf{D}^{(0)})^{-1}\mathbf{D}^{(1)}. \tag{4.41}$$

In the CP representation, we have

$$\mathbf{M} = (\mathbf{D}^{(0)})^{-1}\mathbf{D}^{(1)}$$
$$= 2^{-3/2}\begin{pmatrix} 1 & N^{(0)-1} & -\mathrm{j} & \mathrm{j}N^{(0)-1} \\ 1 & -N^{(0)-1} & -\mathrm{j} & -\mathrm{j}N^{(0)-1} \\ 1 & N^{(0)-1} & \mathrm{j} & -\mathrm{j}N^{(0)-1} \\ 1 & -N^{(0)-1} & \mathrm{j} & \mathrm{j}N^{(0)-1} \end{pmatrix} 2^{-1/2}\begin{pmatrix} 1 & 1 & 1 & 1 \\ N_+^{(1)} & -N_+^{(1)} & N_-^{(1)} & -N_-^{(1)} \\ \mathrm{j} & \mathrm{j} & -\mathrm{j} & -\mathrm{j} \\ -\mathrm{j}N_+^{(1)} & \mathrm{j}N_+^{(1)} & \mathrm{j}N_-^{(1)} & -\mathrm{j}N_-^{(1)} \end{pmatrix}, \tag{4.42}$$

i.e.,

$$\mathbf{M} = \frac{1}{2}\begin{pmatrix} \left(1+\frac{N_+^{(1)}}{N^{(0)}}\right) & \left(1-\frac{N_+^{(1)}}{N^{(0)}}\right) & 0 & 0 \\ \left(1-\frac{N_+^{(1)}}{N^{(0)}}\right) & \left(1+\frac{N_+^{(1)}}{N^{(0)}}\right) & 0 & 0 \\ 0 & 0 & \left(1+\frac{N_-^{(1)}}{N^{(0)}}\right) & \left(1-\frac{N_-^{(1)}}{N^{(0)}}\right) \\ 0 & 0 & \left(1-\frac{N_-^{(1)}}{N^{(0)}}\right) & \left(1+\frac{N_-^{(1)}}{N^{(0)}}\right) \end{pmatrix}, \tag{4.43}$$

and using Eq. (3.70a), Eq. (3.70c), Eq. (3.70e), and Eq. (3.70g), we obtain the relations formally identical with the Fresnel (single-interface) reflection and transmission formulae reduced to the normal incidence

$$r_{\pm}^{(01)} = \frac{N^{(0)} - N_{\pm}^{(1)}}{N^{(0)} + N_{\pm}^{(1)}}, \tag{4.44a}$$

and

$$t_{\pm}^{(01)} = \frac{2N^{(0)}}{N^{(0)} + N_{\pm}^{(1)}}. \tag{4.44b}$$

4.2.1.2 Thin Plate

The structure consisting of a single magnetic layer between two isotropic half spaces has the following $\mathbf{M}$ matrix

$$\mathbf{M} = (\mathbf{D}^{(0)})^{-1}\mathbf{D}^{(1)}\mathbf{P}^{(1)}(\mathbf{D}^{(1)})^{-1}\mathbf{D}^{(2)}, \tag{4.45}$$

where $\mathbf{D}^{(2)}$ follows from Eq. (4.19). The reflection and transmission coefficients are obtained from Eq. (4.32) using the matrix elements in the CP representation

$$\mathrm{M}_{11} = \left(t_+^{(01)}\right)^{-1}\left(t_+^{(12)}\right)^{-1}\mathrm{e}^{\mathrm{j}\beta_+^{(1)}}\left[1 + r_+^{(01)}r_+^{(12)}\mathrm{e}^{-2\mathrm{j}\beta_+^{(1)}}\right], \tag{4.46a}$$

$$\mathrm{M}_{33} = \left(t_-^{(01)}\right)^{-1}\left(t_-^{(12)}\right)^{-1}\mathrm{e}^{\mathrm{j}\beta_-^{(1)}}\left[1 + r_-^{(01)}r_-^{(12)}\mathrm{e}^{-2\mathrm{j}\beta_-^{(1)}}\right], \tag{4.46b}$$

$$\mathrm{M}_{21} = \left(t_+^{(01)}\right)^{-1}\left(t_+^{(12)}\right)^{-1}\mathrm{e}^{\mathrm{j}\beta_+^{(1)}}\left[r_+^{(01)} + r_+^{(12)}\mathrm{e}^{-2\mathrm{j}\beta_+^{(1)}}\right], \tag{4.46c}$$

$$\mathrm{M}_{43} = \left(t_-^{(01)}\right)^{-1}\left(t_-^{(12)}\right)^{-1}\mathrm{e}^{\mathrm{j}\beta_-^{(1)}}\left[r_-^{(01)} + r_-^{(12)}\mathrm{e}^{-2\mathrm{j}\beta_-^{(1)}}\right], \tag{4.46d}$$

where

$$\beta_\pm^{(1)} = (\omega/c)\, N_\pm^{(1)} d^{\,(1)}, \tag{4.47}$$

and the single-interface reflection and transmission coefficients represent expressions analogous to the Fresnel equations

$$r_\pm^{(01)} = \frac{N^{(0)} - N_\pm^{(1)}}{N^{(0)} + N_\pm^{(1)}}, \tag{4.48a}$$

$$t_\pm^{(01)} = \frac{2N^{(0)}}{N^{(0)} + N_\pm^{(1)}}, \tag{4.48b}$$

$$r_\pm^{(12)} = \frac{N_\pm^{(1)} - N^{(2)}}{N_\pm^{(1)} + N^{(2)}}, \tag{4.48c}$$

$$t_\pm^{(12)} = \frac{2N_\pm^{(1)}}{N_\pm^{(1)} + N^{(2)}}. \tag{4.48d}$$

This situation was considered by Donovan and Medcalf [3], Piller [4], and Sansalone [5] in their treatment of the multiple reflections effect in a thin plate on the observed Faraday rotation.

For the purpose of illustration, let us consider in some detail the case of a magnetic plate in a vacuum where $N^{(0)} = N^{(2)} = 1$. Then from Eqs. (4.48), it follows $r_\pm^{(01)} = -r_\pm^{(12)} = -r_\pm^{(10)}$ and $t_\pm^{(12)} = t_\pm^{(10)}$. The system is characterized by the matrix product given in Eq. (4.45) with $\mathbf{D}^{(2)} = \mathbf{D}^{(0)}$. We observe that

$$(\mathbf{D}^{(0)})^{-1}\mathbf{D}^{(1)} = \frac{1}{2}\begin{pmatrix} 1 + N_+^{(1)} & 1 - N_+^{(1)} & 0 & 0 \\ 1 - N_+^{(1)} & 1 + N_+^{(1)} & 0 & 0 \\ 0 & 0 & 1 + N_-^{(1)} & 1 - N_-^{(1)} \\ 0 & 0 & 1 - N_-^{(1)} & 1 + N_-^{(1)} \end{pmatrix}.$$

The matrix product characterizing the response in this simplified system is given by

$$\mathbf{M} = (\mathbf{D}^{(0)})^{-1}\mathbf{D}^{(1)}\mathbf{P}^{(1)}(\mathbf{D}^{(1)})^{-1}\mathbf{D}^{(0)} = 2^{-3/2}\begin{pmatrix} 1 & 1 & -j & j \\ 1 & -1 & -j & -j \\ 1 & 1 & j & -j \\ 1 & -1 & j & j \end{pmatrix}$$

$$\times\, 2^{-1/2}\begin{pmatrix} 1 & 1 & 1 & 1 \\ N_+^{(1)} & -N_+^{(1)} & N_-^{(1)} & -N_-^{(1)} \\ j & j & -j & -j \\ -jN_+^{(1)} & jN_+^{(1)} & jN_-^{(1)} & -jN_-^{(1)} \end{pmatrix}\begin{pmatrix} e^{j\beta_+^{(1)}} & 0 & 0 & 0 \\ 0 & e^{-j\beta_+^{(1)}} & 0 & 0 \\ 0 & 0 & e^{j\beta_-^{(1)}} & 0 \\ 0 & 0 & 0 & e^{-j\beta_-^{(1)}} \end{pmatrix}$$

$$\times\, 2^{-3/2}\left(N_+^{(1)}N_-^{(1)}\right)^{-1}\begin{pmatrix} N_+^{(1)}N_-^{(1)} & N_-^{(1)} & -jN_+^{(1)}N_-^{(1)} & jN_-^{(1)} \\ N_+^{(1)}N_-^{(1)} & -N_-^{(1)} & -jN_+^{(1)}N_-^{(1)} & -jN_-^{(1)} \\ N_+^{(1)}N_-^{(1)} & N_+^{(1)} & jN_+^{(1)}N_-^{(1)} & -jN_+^{(1)} \\ N_+^{(1)}N_-^{(1)} & -N_+^{(1)} & jN_+^{(1)}N_-^{(1)} & jN_+^{(1)} \end{pmatrix}$$

$$\times\, 2^{-1/2}\begin{pmatrix} 1 & 1 & 1 & 1 \\ 1 & -1 & 1 & -1 \\ j & j & -j & -j \\ -j & j & j & -j \end{pmatrix}. \qquad (4.49)$$

The resulting matrix is block diagonal with the following nonzero M matrix elements

$$\mathrm{M}_{\substack{11\\33}} = \left(4N_\pm^{(1)}\right)^{-1}\left[\left(1 + N_\pm^{(1)}\right)^2 e^{j\beta_\pm^{(1)}} - \left(1 - N_\pm^{(1)}\right)^2 e^{-j\beta_\pm^{(1)}}\right], \qquad (4.50a)$$

$$\mathrm{M}_{\substack{21\\43}} = -\mathrm{M}_{\substack{12\\34}} = \left(4N_\pm^{(1)}\right)^{-1}\left(1 - N_\pm^{(1)2}\right)\left(e^{j\beta_\pm^{(1)}} - e^{-j\beta_\pm^{(1)}}\right), \qquad (4.50b)$$

$$\mathrm{M}_{\substack{22\\44}} = \left(4N_\pm^{(1)}\right)^{-1}\left[\left(1 + N_\pm^{(1)}\right)^2 e^{-j\beta_\pm^{(1)}} - \left(1 - N_\pm^{(1)}\right)^2 e^{j\beta_\pm^{(1)}}\right]. \qquad (4.50c)$$

consistent with Eq. (4.46). This information is sufficient for computing the reflection and transmission coefficients in circular representation from Eq. (4.32).

In some cases, the information on the fields inside the magnetic layer may be of interest. This can be obtained, *e.g.*, from the matrix relation

$$\mathbf{E}_0^{(1)} = (\mathbf{D}^{(1)})^{-1}\mathbf{D}^{(0)}\mathbf{E}_0^{(2)} \qquad (4.51)$$

or

$$\begin{pmatrix} E_{01}^{(1)} \\ E_{02}^{(1)} \\ E_{03}^{(1)} \\ E_{04}^{(1)} \end{pmatrix} = \left(2N_+^{(1)}N_-^{(1)}\right)^{-1}$$
$$\times \begin{pmatrix} N_-^{(1)}\left(N_+^{(1)}+1\right) & N_-^{(1)}\left(N_+^{(1)}-1\right) & 0 & 0 \\ N_-^{(1)}\left(N_+^{(1)}-1\right) & N_-^{(1)}\left(N_+^{(1)}+1\right) & 0 & 0 \\ 0 & 0 & N_+^{(1)}\left(N_-^{(1)}+1\right) & N_+^{(1)}\left(N_-^{(1)}-1\right) \\ 0 & 0 & N_+^{(1)}\left(N_-^{(1)}-1\right) & N_+^{(1)}\left(N_-^{(1)}+1\right) \end{pmatrix}$$
$$\times \begin{pmatrix} E_{01}^{(2)} \\ 0 \\ E_{03}^{(2)} \\ 0 \end{pmatrix}. \tag{4.52}$$

To express the fields in the layer in terms of the incident amplitudes, we eliminate the transmitted amplitudes of $\mathbf{E}_0^{(2)}$ using Eq. (4.50a)

$$\begin{aligned} E_{\substack{01\\03}}^{(2)} &= M_{\substack{11\\33}}^{-1} E_{\substack{01\\03}}^{(0)} \\ &= \frac{t_\pm^{(01)} t_\pm^{(10)} e^{-j\beta_\pm^{(1)}}}{1 - r^{(01)2} e^{-2j\beta_\pm^{(1)}}} E_{\substack{01\\03}}^{(0)} \end{aligned} \tag{4.53}$$

and obtain the relation between the CP proper polarization wave amplitudes in the plane $z^{(1)}$ at the exit boundary inside the layer (1) and those of the incident CP proper polarizations in the plane $z^{(0)}$ ($d^{(1)} = z^{(1)} - z^{(0)}$ is the thickness of the magnetic layer) situated in the medium (0) at the entrance boundary outside the layer (1), *i.e.*,

$$E_{01}^{(1)}\left(z^{(1)}\right) = t_+^{(01)} e^{-j\beta_+^{(1)}} \left[1 - \left(r_+^{(01)}\right)^2 e^{-j2\beta_+^{(1)}}\right]^{-1} E_{01}^{(0)}\left(z^{(0)}\right), \tag{4.54a}$$

$$E_{02}^{(1)}\left(z^{(1)}\right) = t_+^{(01)} r_+^{(01)} e^{-j\beta_+^{(1)}} \left[1 - \left(r_+^{(01)}\right)^2 e^{-j2\beta_+^{(1)}}\right]^{-1} E_{01}^{(0)}\left(z^{(0)}\right), \tag{4.54b}$$

$$E_{03}^{(1)}\left(z^{(1)}\right) = t_-^{(01)} e^{-j\beta_-^{(1)}} \left[1 - \left(r_-^{(01)}\right)^2 e^{-j2\beta_-^{(1)}}\right]^{-1} E_{03}^{(0)}\left(z^{(0)}\right), \tag{4.54c}$$

$$E_{04}^{(1)}\left(z^{(1)}\right) = t_-^{(01)} r_-^{(01)} e^{-j\beta_-^{(1)}} \left[1 - \left(r_-^{(01)}\right)^2 e^{-j2\beta_-^{(1)}}\right]^{-1} E_{03}^{(0)}\left(z^{(0)}\right). \tag{4.54d}$$

The field profile inside the layer (1) follows from the linear superposition of the proper polarizations

$$\begin{pmatrix} E_{01}^{(1)}(z) \\ E_{02}^{(1)}(z) \\ E_{03}^{(1)}(z) \\ E_{04}^{(1)}(z) \end{pmatrix} = \begin{pmatrix} e^{j\omega N_+^{(1)}(z^{(1)}-z)/c} & 0 & 0 & 0 \\ 0 & e^{-j\omega N_+^{(1)}(z^{(1)}-z)/c} & 0 & 0 \\ 0 & 0 & e^{j\omega N_-^{(1)}(z^{(1)}-z)/c} & 0 \\ 0 & 0 & 0 & e^{j\omega N_-^{(1)}(z^{(1)}-z)/c} \end{pmatrix} \times \begin{pmatrix} E_{01}^{(1)}(z^{(1)}) \\ E_{02}^{(1)}(z^{(1)}) \\ E_{03}^{(1)}(z^{(1)}) \\ E_{04}^{(1)}(z^{(1)}) \end{pmatrix}. \quad (4.55)$$

Note that we have written the solution to the proper value equation (4.3) in terms of $N_j^{(1)} = n_j^{(1)} - jk_j^{(1)}$. The proper polarizations are classified as follows

$$N_1^{(1)} = n_1^{(1)} - jk_1^{(1)} = N_+^{(1)} = n_+^{(1)} - jk_+^{(1)}, \quad (4.56a)$$
$$N_2^{(1)} = n_2^{(1)} - jk_2^{(1)} = -N_+^{(1)} = -n_+^{(1)} + jk_+^{(1)}, \quad (4.56b)$$
$$N_3^{(3)} = n_3^{(1)} - jk_3^{(1)} = N_-^{(1)} = n_-^{(1)} - jk_-^{(1)}, \quad (4.56c)$$
$$N_4^{(4)} = n_4^{(1)} - jk_4^{(1)} = -N_-^{(1)} = -n_-^{(1)} + jk_-^{(1)}. \quad (4.56d)$$

Here, $n_\pm^{(1)} > 0$ and $k_\pm^{(1)} > 0$ are the real index of refraction and the extinction coefficient for CP wave polarization modes in the region (1), respectively. The Poynting vectors of these proper polarizations take the form

$$S_j^{(1)}(z) = \hat{z} n_j^{(1)} (2Z_{vac})^{-1} \left|E_{0j}^{(1)}\right|^2 \exp\left[-\frac{2\omega}{c} k_j^{(1)} (z - z^{(1)})\right], \quad (4.57)$$

where $Z_{vac} = (\mu_{vac}/\varepsilon_{vac})$ is the vacuum impedance. From the dissipated energy, the temperature increase during the thermomagnetic writing process in magnetooptic memory disk can be evaluated.

The procedure was demonstrated on a single-layer system where the analytical illustration is easy. Other systems, *e.g.*, those with dielectric overlayers and back reflectors, can be analyzed in a similar way. The system consisting of a single magnetic layer on top of a thick magnetic substrate (at $z > z^{(1)}$) and an isotropic ambient (at $z < z^{(0)}$) is described by the $\mathbf{M}$ matrix of the same form as in Eq. (4.45). The last matrix in the product sequence, $\mathbf{D}^{(2)}$, represents magnetic region and has, therefore, the form given by Eq. (4.12). The relevant matrix elements follow from Eq. (4.46) where the coefficients $r_{\pm}^{(01)}$ and $t_{\pm}^{(01)}$ are given by Eq. (4.48a) and Eq. (4.48b). The coefficients $r_{\pm}^{(12)}$ and $t_{\pm}^{(12)}$ become

$$r_{\pm}^{(12)} = \frac{N_{\pm}^{(1)} - N_{\pm}^{(2)}}{N_{\pm}^{(1)} + N_{\pm}^{(2)}}, \tag{4.58a}$$

$$t_{\pm}^{(12)} = \frac{2N_{\pm}^{(1)}}{N_{\pm}^{(1)} + N_{\pm}^{(2)}}. \tag{4.58b}$$

The global CP complex reflection coefficients in the circular (CP) representation of a system consisting of isotropic ambient, magnetic film, and magnetic substrate are given by

$$r_{\pm}^{(02)} = \frac{r_{\pm}^{(01)} + r_{\pm}^{(12)} \mathrm{e}^{-2\mathrm{j}\beta_{\pm}^{(1)}}}{1 + r_{\pm}^{(01)} r_{\pm}^{(12)} \mathrm{e}^{-2\mathrm{j}\beta_{\pm}^{(1)}}}. \tag{4.59}$$

The analysis can be extended to multilayer systems. The transformation of the polarization state upon transmission and reflection can be studied by means of Eq. (4.30) and Eq. (4.31).

4.2.2 LP Representation

So far, we have characterized the multilayer response at polar magnetization and normal incidence in terms of the global reflection and transmission coefficients and related complex number parameters $\chi_i^{(CP)}$, $\chi_t^{(CP)}$, and $\chi_r^{(CP)}$ in the circular (CP) representation. The latter parameters convey the most concise information on the wave limited to the description of its polarization state, eliminating the redundant information on the common amplitude and the initial phase. In practice, it is often convenient to characterize the response in terms of Jones reflection and Jones transmission matrices. The multilayer response is experimentally evaluated in a sequence of optical elements that can be also represented by their Jones matrices. The sequence can then be represented by a product of corresponding Jones matrices from which the required information on the multilayer parameters can be extracted. In this context, the multilayer at the polar magnetization and normal incidence can be regarded as a single element in an optical system. Its circular (CP) Jones transmission and reflection matrices are diagonal and follow from Eqs. (4.32)

$$\begin{pmatrix} E_+^{(t)} \\ E_-^{(t)} \end{pmatrix} = \begin{pmatrix} t_+^{(0,\mathcal{N}+1)} & 0 \\ 0 & t_-^{(0,\mathcal{N}+1)} \end{pmatrix} \begin{pmatrix} E_+^{(i)} \\ E_-^{(i)} \end{pmatrix} = \begin{pmatrix} M_{11}^{-1} & 0 \\ 0 & M_{33}^{-1} \end{pmatrix} \begin{pmatrix} E_+^{(i)} \\ E_-^{(i)} \end{pmatrix} \tag{4.60a}$$

$$\begin{pmatrix} E_+^{(r)} \\ E_-^{(r)} \end{pmatrix} = \begin{pmatrix} r_+^{(0,\mathcal{N}+1)} & 0 \\ 0 & r_-^{(0,\mathcal{N}+1)} \end{pmatrix} \begin{pmatrix} E_+^{(i)} \\ E_-^{(i)} \end{pmatrix} = \begin{pmatrix} M_{21}M_{11}^{-1} & 0 \\ 0 & M_{43}M_{33}^{-1} \end{pmatrix} \begin{pmatrix} E_+^{(i)} \\ E_-^{(i)} \end{pmatrix}, \tag{4.60b}$$

where

$$\begin{pmatrix} E_+^{(t)} \\ E_-^{(t)} \end{pmatrix}, \quad \begin{pmatrix} E_+^{(r)} \\ E_-^{(r)} \end{pmatrix}, \quad \text{and} \quad \begin{pmatrix} E_+^{(i)} \\ E_-^{(i)} \end{pmatrix} \tag{4.61}$$

are the Jones vectors in circular (CP) representation of the transmitted, reflected, and incident waves, respectively. Usually, the common amplitude of the incident wave is normalized to the unity, and its initial phase is set to zero. Our choice of the CP proper polarizations in the isotropic media on both sides of the multilayer structure (at $z < 0$ and $z > z^{(n)}$) has allowed us to calculate the multilayer response in a convenient way. In the experiment, the multilayer is studied as a part of a sequence of optical elements; most of them are naturally represented in a linearly polarized (or Cartesian) basis. There are two aspects that make the Cartesian representation interesting. First, it allows a direct relation of the Cartesian parameters to the experiment, and second, it represents a convenient starting point for approximations used in the evaluation of the multilayer response. We therefore look for the Cartesian representation of the multilayer. In order to distinguish the circular and Cartesian representations, we add the suffixes CP and xy, respectively, *e.g.*, $\mathbf{M}^{(\mathrm{CP})}$ and $\mathbf{M}^{(xy)}$. To this purpose, we employ the proper polarizations linearly polarized along the Cartesian x and y axes corresponding to $p^{(s)} = 1$ and $q^{(s)} = 0$ in Eq. (3.49).

To transform a (2×2) block diagonal circular matrix $\mathbf{M}^{(\mathrm{CP})}$, obtained as a product of the (2×2) block diagonal transfer matrices given in

Eq. (3.18), to the Cartesian matrix $\mathbf{M}^{(xy)}$, we have to perform the matrix transformation

$$\mathbf{M}^{(xy)} = \left(\mathbf{D}^{(0)}_{(xy)}\right)^{-1}\mathbf{D}^{(0)}_{(\mathrm{CP})}\mathbf{M}^{(\mathrm{CP})}\left(\mathbf{D}^{(\mathcal{N}+1)}_{(\mathrm{CP})}\right)^{-1}\mathbf{D}^{(\mathcal{N}+1)}_{(xy)}, \tag{4.62}$$

where

$$\begin{aligned}
&\left(\mathbf{D}^{(0)}_{(xy)}\right)^{-1}\mathbf{D}^{(0)}_{(\mathrm{CP})} \\
&= \frac{\sqrt{2}}{4N^{(0)}}\begin{pmatrix} N^{(0)} & 1 & 0 & 0 \\ N^{(0)} & -1 & 0 & 0 \\ 0 & 0 & N^{(0)} & -1 \\ 0 & 0 & N^{(0)} & 1 \end{pmatrix}\begin{pmatrix} 1 & 1 & 1 & 1 \\ N^{(0)} & -N^{(0)} & N^{(0)} & -N^{(0)} \\ \mathrm{j} & \mathrm{j} & -\mathrm{j} & -\mathrm{j} \\ -\mathrm{j}N^{(0)} & \mathrm{j}N^{(0)} & \mathrm{j}N^{(0)} & -\mathrm{j}N^{(0)} \end{pmatrix} \\
&= 2^{-1/2}\begin{pmatrix} 1 & 0 & 1 & 0 \\ 0 & 1 & 0 & 1 \\ \mathrm{j} & 0 & -\mathrm{j} & 0 \\ 0 & \mathrm{j} & 0 & -\mathrm{j} \end{pmatrix}
\end{aligned} \tag{4.63}$$

and

$$\begin{aligned}
\left(\mathbf{D}^{(\mathcal{N}+1)}_{(\mathrm{CP})}\right)^{-1}\mathbf{D}^{(\mathcal{N}+1)}_{(xy)} &= \frac{\sqrt{2}}{4}\begin{pmatrix} 1 & \left(N^{(\mathcal{N}+1)}\right)^{-1} & -\mathrm{j} & \mathrm{j}\left(N^{(\mathcal{N}+1)}\right)^{-1} \\ 1 & -\left(N^{(\mathcal{N}+1)}\right)^{-1} & -\mathrm{j} & -\mathrm{j}\left(N^{(\mathcal{N}+1)}\right)^{-1} \\ 1 & \left(N^{(\mathcal{N}+1)}\right)^{-1} & \mathrm{j} & -\mathrm{j}\left(N^{(\mathcal{N}+1)}\right)^{-1} \\ 1 & -\left(N^{(\mathcal{N}+1)}\right)^{-1} & \mathrm{j} & \mathrm{j}\left(N^{(\mathcal{N}+1)}\right)^{-1} \end{pmatrix} \\
&\quad\times\begin{pmatrix} 1 & 1 & 0 & 0 \\ N^{(\mathcal{N}+1)} & -N^{(\mathcal{N}+1)} & 0 & 0 \\ 0 & 0 & 1 & 1 \\ 0 & 0 & -N^{(\mathcal{N}+1)} & N^{(\mathcal{N}+1)} \end{pmatrix} \\
&= 2^{-1/2}\begin{pmatrix} 1 & 0 & -\mathrm{j} & 0 \\ 0 & 1 & 0 & -\mathrm{j} \\ 1 & 0 & \mathrm{j} & 0 \\ 0 & 1 & 0 & \mathrm{j} \end{pmatrix}.
\end{aligned} \tag{4.64}$$

The use was made of Eq. (3.82), Eq. (4.18), and Eq. (4.19). In terms of the $\mathbf{M}^{(\mathrm{CP})}$ matrix elements, the $\mathbf{M}^{(xy)}$ matrix can be expressed by a suitable choice of the initial phases for the waves polarized parallel to $\hat{x}$ and $\hat{y}$, as

$$\mathbf{M}^{(xy)} = \begin{pmatrix} \mathbf{M}^{(xy)}_{11} & -\mathbf{M}^{(xy)}_{21} \\ \mathbf{M}^{(xy)}_{21} & \mathbf{M}^{(xy)}_{11} \end{pmatrix}, \tag{4.65}$$

where

$$\mathbf{M}_{11}^{(xy)} = \frac{1}{2}\begin{pmatrix} \mathrm{M}_{11}^{(\mathrm{CP})} + \mathrm{M}_{33}^{(\mathrm{CP})} & \mathrm{M}_{12}^{(\mathrm{CP})} + \mathrm{M}_{34}^{(\mathrm{CP})} \\ \mathrm{M}_{21}^{(\mathrm{CP})} + \mathrm{M}_{43}^{(\mathrm{CP})} & \mathrm{M}_{22}^{(\mathrm{CP})} + \mathrm{M}_{44}^{(\mathrm{CP})} \end{pmatrix}, \tag{4.66}$$

$$\mathbf{M}_{21}^{(xy)} = \frac{1}{2}\begin{pmatrix} \mathrm{j}(\mathrm{M}_{11}^{(\mathrm{CP})} - \mathrm{M}_{33}^{(\mathrm{CP})}) & \mathrm{j}(\mathrm{M}_{12}^{(\mathrm{CP})} - \mathrm{M}_{34}^{(\mathrm{CP})}) \\ \mathrm{j}(\mathrm{M}_{21}^{(\mathrm{CP})} - \mathrm{M}_{43}^{(\mathrm{CP})}) & \mathrm{j}(\mathrm{M}_{22}^{(\mathrm{CP})} - \mathrm{M}_{44}^{(\mathrm{CP})}) \end{pmatrix}. \tag{4.67}$$

Note that the diagonal (2×2) blocks are equal to each other, whereas the off-diagonal ones differ in sign. We form the product of **M** matrix elements of Eqs. (3.70a) to (3.70d) and obtain, up to an irrelevant common factor,

$$\mathrm{M}_{21}^{(xy)}\mathrm{M}_{33}^{(xy)} - \mathrm{M}_{23}^{(xy)}\mathrm{M}_{31}^{(xy)} = 2(\mathrm{M}_{43}^{(\mathrm{CP})}\mathrm{M}_{11}^{(\mathrm{CP})} + \mathrm{M}_{21}^{(\mathrm{CP})}\mathrm{M}_{33}^{(\mathrm{CP})}), \tag{4.68a}$$

$$\mathrm{M}_{41}^{(xy)}\mathrm{M}_{33}^{(xy)} - \mathrm{M}_{43}^{(xy)}\mathrm{M}_{31}^{(xy)} = 2\mathrm{j}(\mathrm{M}_{21}^{(\mathrm{CP})}\mathrm{M}_{33}^{(\mathrm{CP})} - \mathrm{M}_{11}^{(\mathrm{CP})}\mathrm{M}_{43}^{(\mathrm{CP})}), \tag{4.68b}$$

$$\mathrm{M}_{11}^{(xy)}\mathrm{M}_{43}^{(xy)} - \mathrm{M}_{41}^{(xy)}\mathrm{M}_{13}^{(xy)} = 2(\mathrm{M}_{21}^{(\mathrm{CP})}\mathrm{M}_{33}^{(\mathrm{CP})} + \mathrm{M}_{11}^{(\mathrm{CP})}\mathrm{M}_{43}^{(\mathrm{CP})}), \tag{4.68c}$$

$$\mathrm{M}_{11}^{(xy)}\mathrm{M}_{23}^{(xy)} - \mathrm{M}_{21}^{(xy)}\mathrm{M}_{13}^{(xy)} = -2\mathrm{j}(\mathrm{M}_{21}^{(\mathrm{CP})}\mathrm{M}_{33}^{(\mathrm{CP})} - \mathrm{M}_{11}^{(\mathrm{CP})}\mathrm{M}_{43}^{(\mathrm{CP})}), \tag{4.68d}$$

$$\mathrm{M}_{11}^{(xy)}\mathrm{M}_{33}^{(xy)} - \mathrm{M}_{13}^{(xy)}\mathrm{M}_{31}^{(xy)} = 4\mathrm{M}_{11}^{(\mathrm{CP})}\mathrm{M}_{33}^{(\mathrm{CP})}. \tag{4.68e}$$

According to Eqs. (3.70a) to (3.70d), the reflection coefficients of the interface are given by

$$r_{21}^{(0,\mathcal{N}+1)} = r_{xx}^{(0,\mathcal{N}+1)} = \frac{1}{2}\left(\frac{\mathrm{M}_{21}^{(\mathrm{CP})}}{\mathrm{M}_{11}^{(\mathrm{CP})}} + \frac{\mathrm{M}_{43}^{(\mathrm{CP})}}{\mathrm{M}_{33}^{(\mathrm{CP})}}\right), \tag{4.69a}$$

$$r_{41}^{(0,\mathcal{N}+1)} = r_{yx}^{(0,\mathcal{N}+1)} = \frac{\mathrm{j}}{2}\left(\frac{\mathrm{M}_{21}^{(\mathrm{CP})}}{\mathrm{M}_{11}^{(\mathrm{CP})}} - \frac{\mathrm{M}_{43}^{(\mathrm{CP})}}{\mathrm{M}_{33}^{(\mathrm{CP})}}\right), \tag{4.69b}$$

$$r_{43}^{(0,\mathcal{N}+1)} = r_{yy}^{(0,\mathcal{N}+1)} = \frac{1}{2}\left(\frac{\mathrm{M}_{21}^{(\mathrm{CP})}}{\mathrm{M}_{11}^{(\mathrm{CP})}} + \frac{\mathrm{M}_{43}^{(\mathrm{CP})}}{\mathrm{M}_{33}^{(\mathrm{CP})}}\right), \tag{4.69c}$$

$$r_{23}^{(0,\mathcal{N}+1)} = r_{xy}^{(0,\mathcal{N}+1)} = -\frac{\mathrm{j}}{2}\left(\frac{\mathrm{M}_{21}^{(\mathrm{CP})}}{\mathrm{M}_{11}^{(\mathrm{CP})}} - \frac{\mathrm{M}_{43}^{(\mathrm{CP})}}{\mathrm{M}_{33}^{(\mathrm{CP})}}\right). \tag{4.69d}$$

The special forms of both $\mathbf{M}^{(xy)}$ and $\mathbf{M}^{(\mathrm{CP})}$ reflect the symmetry of the system at the polar magnetization and normal incidence. In particular, the off-diagonal blocks of $\mathbf{M}^{(xy)}$ provide a measure of the mode coupling between two linearly polarized modes traversing the multilayer.

The Cartesian Jones reflection and transmission matrix elements have already been obtained in Eq. (1.162) and Eq. (1.163) from the corresponding circular representations. The polarization states of the incident, reflected, and transmitted waves in the Cartesian representation are characterized by the complex ellipsometric parameters $\chi_i^{(xy)}$, $\chi_r^{(xy)}$, and $\chi_t^{(xy)}$. The corresponding parameters in the circular (CP) representation were given in Eq. (4.25). The amplitudes $E_{01}^{(0)} = E_{0x}^{(i)}$ and $E_{03}^{(0)} = E_{0y}^{(i)}$ now correspond to the incident modes linearly polarized parallel to x and y axes, respectively. Then, $E_{02}^{(0)} = E_{0x}^{(r)}$ and $E_{04}^{(0)} = E_{0y}^{(r)}$ correspond to the reflected x- and y-polarized LP modes and $E_{01}^{(\mathcal{N}+1)} = E_{0x}^{(t)}$ and $E_{03}^{(\mathcal{N}+1)} = E_{0y}^{(t)}$ correspond to the transmitted x- and y-polarized LP modes.

The Cartesian complex plane representation of the polarization states $\chi_l^{(xy)}$ follows from the Cartesian Jones vector given in Eq. (1.38)

$$\begin{pmatrix} E_{0x}^{(l)} \\ E_{0y}^{(l)} \end{pmatrix} = E_{0l}^{(0)} \exp\left(j\phi_{0l}^{(0)}\right) \begin{pmatrix} \cos\theta_l \cos\epsilon_l - j\sin\theta_l \sin\epsilon_l \\ \sin\theta_l \cos\epsilon_l + j\cos\theta_l \sin\epsilon_l \end{pmatrix}, \tag{4.70}$$

where $l = i, r,$ and t. Here, $E_{0l}^{(0)}$ and $\phi_{0l}^{(0)}$ are the common amplitude and the initial phase [2].

The complex ratios $\chi_l^{(xy)}$ are related to the azimuths and ellipticities through the relation [2]

$$\chi_l^{(xy)} = \frac{E_{0y}^{(l)}}{E_{0x}^{(l)}} = \frac{\tan\theta_l + j\tan\epsilon_l}{1 - j\tan\theta_l \tan\epsilon_l}, \tag{4.71}$$

which leads to

$$\tan 2\theta_l = \frac{2\Re\left(\chi_l^{(xy)}\right)}{1 - \left|\chi_l^{(xy)}\right|^2} \tag{4.72}$$

and

$$\sin 2\epsilon_l = \frac{2\Im\left(\chi_l^{(xy)}\right)}{1 + \left|\chi_l^{(xy)}\right|^2} \tag{4.73}$$

for $l = i, r,$ or t.

To establish the relations between the polarization states of incident and reflected or transmitted waves, *i.e.*, to determine the multilayer response, we start from the Cartesian Jones reflection and transmission matrices expressed in terms of the $\mathbf{M}^{(xy)}$ matrix elements, *i.e.*,

$$\begin{pmatrix} r_{xx}^{(0,\mathcal{N}+1)} & r_{xy}^{(0,\mathcal{N}+1)} \\ r_{yx}^{(0,\mathcal{N}+1)} & r_{xx}^{(0,\mathcal{N}+1)} \end{pmatrix} = \left(M_{11}^{(xy)2} + M_{13}^{(xy)2}\right)^{-1}$$
$$\times \begin{pmatrix} \left(M_{11}^{(xy)}M_{21}^{(xy)} + M_{23}^{(xy)}M_{13}^{(xy)}\right) & \left(M_{11}^{(xy)}M_{23}^{(xy)} - M_{21}^{(xy)}M_{13}^{(xy)}\right) \\ -\left(M_{11}^{(xy)}M_{23}^{(xy)} - M_{21}^{(xy)}M_{13}^{(xy)}\right) & \left(M_{11}^{(xy)}M_{21}^{(xy)} + M_{23}^{(xy)}M_{13}^{(xy)}\right) \end{pmatrix} \tag{4.74}$$

and

$$\begin{pmatrix} t_{xx}^{(0,\mathcal{N}+1)} & t_{xy}^{(0,\mathcal{N}+1)} \\ t_{yx}^{(0,\mathcal{N}+1)} & t_{xx}^{(0,\mathcal{N}+1)} \end{pmatrix} = \left(\mathrm{M}_{11}^{(xy)2} + \mathrm{M}_{13}^{(xy)2}\right)^{-1} \begin{pmatrix} \mathrm{M}_{11}^{(xy)} & -\mathrm{M}_{13}^{(xy)} \\ \mathrm{M}_{13}^{(xy)} & \mathrm{M}_{11}^{(xy)} \end{pmatrix}. \tag{4.75}$$

Upon the reflection and transmission, the general polarization state of the incident wave, characterized by the complex number $\chi_i^{(xy)}$, transforms according to

$$\chi_r^{(xy)} = \frac{r_{xx}^{(0,\mathcal{N}+1)} \chi_i^{(xy)} + r_{yx}^{(0,\mathcal{N}+1)}}{r_{xx}^{(0,\mathcal{N}+1)} + r_{xy}^{(0,\mathcal{N}+1)} \chi_i^{(xy)}}$$

$$= \frac{\left(\mathrm{M}_{11}^{(xy)}\mathrm{M}_{21}^{(xy)} + \mathrm{M}_{23}^{(xy)}\mathrm{M}_{13}^{(xy)}\right)\chi_i^{(xy)} - \left(\mathrm{M}_{11}^{(xy)}\mathrm{M}_{23}^{(xy)} - \mathrm{M}_{21}^{(xy)}\mathrm{M}_{13}^{(xy)}\right)}{\left(\mathrm{M}_{11}^{(xy)}\mathrm{M}_{21}^{(xy)} + \mathrm{M}_{23}^{(xy)}\mathrm{M}_{13}^{(xy)}\right) + \left(\mathrm{M}_{11}^{(xy)}\mathrm{M}_{23}^{(xy)} - \mathrm{M}_{21}^{(xy)}\mathrm{M}_{13}^{(xy)}\right)\chi_i^{(xy)}}, \tag{4.76a}$$

$$\chi_t^{(xy)} = \frac{t_{xx}^{(0,\mathcal{N}+1)} \chi_i^{(xy)} + t_{yx}^{(0,\mathcal{N}+1)}}{t_{xx}^{(0,\mathcal{N}+1)} + t_{xy}^{(0,\mathcal{N}+1)} \chi_i^{(xy)}} = \frac{\mathrm{M}_{11}^{(xy)}\chi_i^{(xy)} + \mathrm{M}_{13}^{(xy)}}{\mathrm{M}_{11}^{(xy)} - \mathrm{M}_{13}^{(xy)}\chi_i^{(xy)}}. \tag{4.76b}$$

Here, the response in reflection and transmission is characterized with the Cartesian $\mathbf{M}^{(xy)}$ matrix elements given by Eqs. (3.70). We have accounted for the symmetry of the $\mathbf{M}$ matrix at the polar configuration and normal incidence expressed in Eq. (4.65).

In the special case of the incident wave linearly polarized parallel to the x axis ($\chi_i^{(xy)} = 0$), we obtain (by assigning $\theta_r = \theta_K$, $\epsilon_r = \epsilon_K$, $\theta_t = \theta_F$, and $\epsilon_t = \epsilon_F$) for Faraday rotation, θ_F,

$$\tan 2\theta_F = \frac{2\Re\left(\dfrac{\mathrm{M}_{13}^{(xy)}}{\mathrm{M}_{11}^{(xy)}}\right)}{\left(1 - \left|\dfrac{\mathrm{M}_{13}^{(xy)}}{\mathrm{M}_{11}^{(xy)}}\right|^2\right)}, \tag{4.77}$$

Faraday ellipticity, $\tan\epsilon_F$,

$$\sin 2\epsilon_F = \frac{2\Im\left(\dfrac{\mathrm{M}_{13}^{(xy)}}{\mathrm{M}_{11}^{(xy)}}\right)}{\left(1 + \left|\dfrac{\mathrm{M}_{13}^{(xy)}}{\mathrm{M}_{11}^{(xy)}}\right|^2\right)}, \tag{4.78}$$

MO Kerr azimuth rotation, θ_K,

$$\tan 2\theta_K = \frac{2\Re\left[-\dfrac{\left(M_{11}^{(xy)}M_{23}^{(xy)} - M_{21}^{(xy)}M_{13}^{(xy)}\right)}{\left(M_{11}^{(xy)}M_{21}^{(xy)} + M_{23}^{(xy)}M_{13}^{(xy)}\right)}\right]}{1 - \left|\dfrac{\left(M_{11}^{(xy)}M_{23}^{(xy)} - M_{21}^{(xy)}M_{13}^{(xy)}\right)}{\left(M_{11}^{(xy)}M_{21}^{(xy)} + M_{23}^{(xy)}M_{13}^{(xy)}\right)}\right|^2}, \tag{4.79}$$

and MO Kerr ellipticity, $\tan \epsilon_K$,

$$\sin 2\epsilon_K = \frac{2\Im\left[-\dfrac{\left(M_{11}^{(xy)}M_{23}^{(xy)} - M_{21}^{(xy)}M_{13}^{(xy)}\right)}{\left(M_{11}^{(xy)}M_{21}^{(xy)} + M_{23}^{(xy)}M_{13}^{(xy)}\right)}\right]}{1 + \left|\dfrac{\left(M_{11}^{(xy)}M_{23}^{(xy)} - M_{21}^{(xy)}M_{13}^{(xy)}\right)}{\left(M_{11}^{(xy)}M_{21}^{(xy)} + M_{23}^{(xy)}M_{13}^{(xy)}\right)}\right|^2}. \tag{4.80}$$

These formula allow us to evaluate the complex Faraday and polar magnetooptic Kerr effects of the multilayer structure. They may be written more concisely in terms of the global Cartesian reflection and transmission coefficients as

$$\tan 2\theta_F = 2\Re(t_{yx}/t_{xx})/\left(1 - |t_{yx}/t_{xx}|^2\right), \tag{4.81a}$$

$$\sin 2\epsilon_F = 2\Im(t_{yx}/t_{xx})/\left(1 + |t_{yx}/t_{xx}|^2\right), \tag{4.81b}$$

$$\tan 2\theta_K = 2\Re(r_{yx}/r_{xx})/\left(1 - |r_{yx}/r_{xx}|^2\right), \tag{4.81c}$$

$$\sin 2\epsilon_K = 2\Im(r_{yx}/r_{xx})/\left(1 + |r_{yx}/r_{xx}|^2\right), \tag{4.81d}$$

where the superscripts $(0, \mathcal{N}+1)$ were suppressed.

The Jones matrix method assumes ideal optical components and perfectly defined polarization states. In case the multilayer structure enters the depolarizing optical system, Mueller's approach may be more suitable. The corresponding Mueller transmission ($\mathbf{T}_M$) and reflection matrices ($\mathbf{R}_M$) are obtained by the procedure described in Reference 2 from Eq. (4.75) and Eq. (4.74),

$$\mathbf{T}_M = \begin{pmatrix} |t_{xx}|^2 + |t_{xy}|^2 & 0 & 0 & 2\Im\left(t_{xy}^* t_{xx}\right) \\ 0 & |t_{xx}|^2 - |t_{xy}|^2 & 2\Re\left(t_{xy}^* t_{xx}\right) & 0 \\ 0 & -2\Re\left(t_{xy}^* t_{xx}\right) & |t_{xx}|^2 - |t_{xy}|^2 & 0 \\ 2\Im\left(t_{xy}^* t_{xx}\right) & 0 & 0 & |t_{xx}|^2 + |t_{xy}|^2 \end{pmatrix} \tag{4.82a}$$

and

$$\mathbf{R}_M = \begin{pmatrix} |r_{xx}|^2 + |r_{xy}|^2 & 0 & 0 & 2\Im\left(r_{xy}^* r_{xx}\right) \\ 0 & |r_{xx}|^2 - |r_{xy}|^2 & 2\Re\left(r_{xy}^* r_{xx}\right) & 0 \\ 0 & -2\Re\left(r_{xy}^* r_{xx}\right) & |r_{xx}|^2 - |r_{xy}|^2 & 0 \\ 2\Im\left(r_{xy}^* r_{xx}\right) & 0 & 0 & |r_{xx}|^2 + |r_{xy}|^2 \end{pmatrix}. \tag{4.82b}$$

The Mueller matrices relate the transmitted and reflected Stokes vectors to the incident Stokes vector. For the coherent waves, the Stokes vectors of the transmitted and reflected waves can be written respectively as

$$S^{(t)} \begin{pmatrix} 1 \\ \cos 2\theta_t \cos 2\epsilon_t \\ \sin 2\theta_t \cos 2\epsilon_t \\ \sin 2\epsilon_t \end{pmatrix} = \mathbf{T}_M \begin{pmatrix} 1 \\ \cos 2\theta_i \cos 2\epsilon_i \\ \sin 2\theta_i \cos 2\epsilon_i \\ \sin 2\epsilon_i \end{pmatrix} \tag{4.83a}$$

$$S^{(r)} \begin{pmatrix} 1 \\ \cos 2\theta_r \cos 2\epsilon_r \\ \sin 2\theta_r \cos 2\epsilon_r \\ \sin 2\epsilon_r \end{pmatrix} = \mathbf{R}_M \begin{pmatrix} 1 \\ \cos 2\theta_i \cos 2\epsilon_i \\ \sin 2\theta_i \cos 2\epsilon_i \\ \sin 2\epsilon_i \end{pmatrix}, \tag{4.83b}$$

where for simplicity the intensity of the incident wave of the azimuth θ_i and the ellipticity $\tan \epsilon_i$ was normalized to the unity. The magnitudes of the Stokes vectors of the transmitted and reflected waves are denoted $S^{(t)}$ and $S^{(r)}$, respectively.

4.3 Analytical Formulae

We now consider the response of the multilayer at polar magnetization and normal incidence from a slightly different point of view. The magnetic order in originally isotropic materials induces only a small perturbation to their optical permittivities. We shall, therefore, assume from the beginning that the off-diagonal elements of the permittivity tensor are much smaller than the diagonal ones and restrict ourselves to the terms linear in the off-diagonal elements in all layers. This procedure splits the MO response to the sum of contributions from individual layers. Each contribution consists of two components that could be approximatively assigned to the interface and propagation effects, respectively. Off-diagonal elements of the Jones matrices $r_{xy}^{(0,\mathcal{N}+1)}$ and $r_{xy}^{(0,\mathcal{N}+1)}$ along with the complex plane representations $\chi_r^{(0,\mathcal{N}+1)}$ and $\chi_t^{(0,\mathcal{N}+1)}$ for multilayers containing up to seven layers will be given. They are useful both in a qualitative comprehension and in quantitative simulations of the experimental MO Kerr spectra of a wide variety of ultrathin magnetic structures. They can be used to separate the different contributions arising from the interfaces due to intermixing or spin polarization (including interface roughness) as well as to simulate the optical response in magnetically coupled multilayers.

The approximations are summarized in Section 4.3.1. In Section 4.3.2, the analytical expression for the Jones transmission and reflection matrices linear in the off-diagonal tensor element are derived. Section 4.3.3 lists the formulae for the reflection and transmission in a stack consisting of up to seven layers sandwiched between two semi-infinite media. Section 4.3.4 deals with the approximation employed in the ultrathin layers. The procedure is illustrated in Section 4.3.5 on a system consisting of a magnetic film separated from the magnetic substrate by a nonmagnetic film.

4.3.1 Approximations

The simplifications result from the two following approximations:

1. Because $\varepsilon_{xy}^{(n)} \ll \varepsilon_{xx}^{(n)}$, the MO effects are developed to first order, in the off-diagonal permittivity tensor elements $\varepsilon_{xy}^{(n)}$ (*i.e.*, in $\varepsilon_{xy}^{(n)}$).
2. In most practical cases, the azimuth rotations and ellipticities induced by magnetization are small. This justifies the approximation of small ellipsometric angles. Eq. (4.71) then simplifies to

$$\chi_l \approx \theta_l + \mathrm{j}\epsilon_l \qquad \text{with } l = t, r. \tag{4.84}$$

3. When the n-th layer thickness becomes much smaller than the radiation wavelength (in the n-th layer medium), the exponentials may be developed in the power series up to terms linear in $(\omega/c)N^{(n)}d^{\,(n)}$

$$\exp\left(-2\mathrm{j}\frac{\omega}{c}N^{(n)}d^{(n)}\right) \approx 1 - 2\mathrm{j}\frac{\omega}{c}N^{(n)}d^{\,(n)}. \tag{4.85}$$

This forms the basis of the *ultrathin film approximation* [6].

In practice, the first two approximations applied to absorbing magnetic multilayers made of ultrathin films have little effect on the accuracy with respect to the 4×4 matrix approach. On the other hand, the third one requires some care. A reasonable agreement in the region of small $(\omega/c)N^{(n)}d^{\,(n)}$ with the rigorous 4×4 matrix formalism is achieved for the off-diagonal elements of Cartesian Jones reflection and transmission matrices. The diagonal elements, which are perturbed by terms of second order in the off-diagonal permittivity tensor elements, can be replaced by their value corresponding to the absence of magnetization, *i.e.*, by those for isotropic multilayers, provided the MO ellipsometric angles are small. The ultrathin film approximation applied to the MO ellipsometric angles, defined as ratios of the off-diagonal to diagonal elements of the corresponding Jones matrices, does not provide satisfactory results, in general. The formula are, however, simple. With some care, they may be used as a guide to expected trends in the MO effects when the parameters of individual layers change.

4.3.1.1 Single Interface

Let us first consider the case already discussed in Section 1.3.2 of the normal incidence to a single interface between isotropic region (0) and the region (1) at the polar magnetization. The response is represented by $\mathbf{M} = (\mathbf{D}^{(0)})^{-1}\mathbf{D}^{(1)}$. The CP representation of the $\mathbf{M}$ matrix, here denoted as $\mathbf{M}^{(\mathrm{CP})}$, was obtained in Eq. (4.43). The mixed matrix $\mathbf{M}$ relating the Cartesian fields in the region (0) and CP fields in the region (1) follows from the product of the inverse to the dynamical matrix, $(\mathbf{D}^{(0)}_{(xy)})^{-1}$ of Eq. (3.82b) with $\varphi^{(0)} = 0$ and the circular dynamical matrix, $\mathbf{D}^{(1)}_{\mathrm{CP}}$, of Eq. (4.12) for $n = 1$,

$$\mathbf{M} = \left(\mathbf{D}^{(0)}_{(xy)}\right)^{-1}\mathbf{D}^{(1)}_{(\mathrm{CP})}$$

$$= \frac{\sqrt{2}}{4N^{(0)}}\begin{pmatrix} N^{(0)} & 1 & 0 & 0 \\ N^{(0)} & -1 & 0 & 0 \\ 0 & 0 & N^{(0)} & -1 \\ 0 & 0 & N^{(0)} & 1 \end{pmatrix}\begin{pmatrix} 1 & 1 & 1 & 1 \\ N_+^{(1)} & -N_+^{(1)} & N_-^{(1)} & -N_-^{(1)} \\ \mathrm{j} & \mathrm{j} & -\mathrm{j} & -\mathrm{j} \\ -\mathrm{j}N_+^{(1)} & \mathrm{j}N_+^{(1)} & \mathrm{j}N_-^{(1)} & -\mathrm{j}N_-^{(1)} \end{pmatrix}$$

$$= \frac{\sqrt{2}}{4N^{(0)}}\begin{pmatrix} N^{(0)} + N_+^{(1)} & N^{(0)} - N_+^{(1)} & N^{(0)} + N_-^{(1)} & N^{(0)} - N_-^{(1)} \\ N^{(0)} - N_+^{(1)} & N^{(0)} + N_+^{(1)} & N^{(0)} - N_-^{(1)} & N^{(0)} + N_-^{(1)} \\ \mathrm{j}(N^{(0)} + N_+^{(1)}) & \mathrm{j}(N^{(0)} - N_+^{(1)}) & -\mathrm{j}(N^{(0)} + N_-^{(1)}) & -\mathrm{j}(N^{(0)} - N_-^{(1)}) \\ \mathrm{j}(N^{(0)} - N_+^{(1)}) & \mathrm{j}(N^{(0)} + N_+^{(1)}) & -\mathrm{j}(N^{(0)} - N_-^{(1)}) & -\mathrm{j}(N^{(0)} + N_-^{(1)}) \end{pmatrix}. \tag{4.86}$$

We form the product of $\mathbf{M}$ matrix elements of Eqs. (3.70a) to (3.70d) and obtain, up to an irrelevant common factor,

$$\mathrm{M}_{21}^{(xy)}\mathrm{M}_{33}^{(xy)} - \mathrm{M}_{23}^{(xy)}\mathrm{M}_{31}^{(xy)} = N^{(0)2} - N_+^{(1)}N_-^{(1)}, \tag{4.87a}$$

$$\mathrm{M}_{41}^{(xy)}\mathrm{M}_{33}^{(xy)} - \mathrm{M}_{43}^{(xy)}\mathrm{M}_{31}^{(xy)} = -\mathrm{j}(N_+^{(1)} - N_-^{(1)}), \tag{4.87b}$$

$$\mathrm{M}_{11}^{(xy)}\mathrm{M}_{43}^{(xy)} - \mathrm{M}_{41}^{(xy)}\mathrm{M}_{13}^{(xy)} = N^{(0)2} - N_+^{(1)}N_-^{(1)}, \tag{4.87c}$$

$$\mathrm{M}_{11}^{(xy)}\mathrm{M}_{23}^{(xy)} - \mathrm{M}_{21}^{(xy)}\mathrm{M}_{13}^{(xy)} = \mathrm{j}(N_+^{(1)} - N_-^{(1)}), \tag{4.87d}$$

$$\mathrm{M}_{11}^{(xy)}\mathrm{M}_{33}^{(xy)} - \mathrm{M}_{13}^{(xy)}\mathrm{M}_{31}^{(xy)} = (N^{(0)} + N_+^{(1)})(N^{(0)} + N_-^{(1)}). \tag{4.87e}$$

According to Eqs. (3.70a) to (3.70d), the Cartesian reflection coefficients of the interface are given by

$$r_{21}^{(01)} = r_{xx}^{(01)} = \frac{N^{(0)2} - N_+^{(1)} N_-^{(1)}}{\left(N^{(0)} + N_+^{(1)}\right)\left(N^{(0)} + N_-^{(1)}\right)}, \tag{4.88a}$$

$$r_{41}^{(01)} = r_{yx}^{(01)} = \frac{-\mathrm{j}\left(N_+^{(1)} - N_-^{(1)}\right)}{\left(N^{(0)} + N_+^{(1)}\right)\left(N^{(0)} + N_-^{(1)}\right)}, \tag{4.88b}$$

$$r_{43}^{(01)} = r_{yy}^{(01)} = \frac{N^{(0)2} - N_+^{(1)} N_-^{(1)}}{\left(N^{(0)} + N_+^{(1)}\right)\left(N^{(0)} + N_-^{(1)}\right)}, \tag{4.88c}$$

$$r_{23}^{(01)} = r_{xy}^{(01)} = \frac{\mathrm{j}\left(N_+^{(1)} - N_-^{(1)}\right)}{\left(N^{(0)} + N_+^{(1)}\right)\left(N^{(0)} + N_-^{(1)}\right)}. \tag{4.88d}$$

We observe that $r_{xx}^{(01)} = r_{yy}^{(01)}$ and $r_{xy}^{(01)} = -r_{yx}^{(01)}$. The Cartesian complex ellipsometric parameter

$$\chi_r^{(01)} = \frac{r_{yx}^{(01)}}{r_{yy}^{(01)}} = -\frac{r_{xy}^{(01)}}{r_{xx}^{(01)}} = \frac{\mathrm{j}\left(N_+^{(1)} - N_-^{(1)}\right)}{N_+^{(1)} N_-^{(1)} - N^{(0)2}} \tag{4.89}$$

characterizes single-interface normal incidence complex MO polar Kerr effect. This result corresponds to Eq. (1.49) derived for $N^{(0)} = 1$. We now make use of Eq. (4.4) and write

$$N_+^{(1)2} - N_-^{(1)2} = -2\varepsilon_1^{(1)}$$

$$\left(N_+^{(1)} - N_-^{(1)}\right)\left(N_+^{(1)} + N_-^{(1)}\right) = -2\varepsilon_1^{(1)}$$

$$2\varepsilon_0^{(1)1/2}\left(N_+^{(1)} - N_-^{(1)}\right) \approx -2\varepsilon_1^{(1)}$$

$$N_+^{(1)} - N_-^{(1)} \approx -\frac{\varepsilon_1^{(1)}}{\varepsilon_0^{(1)1/2}}, \tag{4.90}$$

where, to first order in $\varepsilon_1^{(1)}$, we have set $N_+^{(1)} + N_-^{(1)} \approx 2\varepsilon_0^{(1)1/2}$. We substitute this result into Eq. (4.89) and use another approximation following from Eq. (4.4), *i.e.*, $N_+^{(1)} N_-^{(1)} \approx \varepsilon_0^{(1)}$, and obtain the often-used formula for the normal incidence MO polar Kerr effect at a single interface

$$\chi_r^{(01)} = \frac{\mathrm{j}\,\varepsilon_1^{(1)}}{\varepsilon_0^{(1)1/2}\left(\varepsilon^{(0)} - \varepsilon_0^{(1)}\right)}, \tag{4.91}$$

which reduces to Eq. (1.54) for the interface with vacuum. Here, $N^{(0)2} = \varepsilon^{(0)}$.

4.3.2 Linear Expressions

We now provide the Jones matrix elements for the multilayer with polar magnetization at normal light incidence restricted to the terms of zero and first order in the off-diagonal element of the permittivity tensors characterizing individual layers. In a given layer of the multilayer system, the relative permittivity tensor, $\varepsilon^{(n)}$, of an originally isotropic medium uniformly magnetized along the z axis perpendicular to interfaces (polar configuration) writes

$$\varepsilon^{(n)} = \begin{pmatrix} \varepsilon_{xx}^{(n)} & \varepsilon_{xy}^{(n)} & 0 \\ -\varepsilon_{xy}^{(n)} & \varepsilon_{xx}^{(n)} & 0 \\ 0 & 0 & \varepsilon_{zz}^{(n)} \end{pmatrix}. \tag{4.92}$$

Here, we have employed the form deduced from the general tensor given in Eq. (3.1). Alternatively, we could use Eq. (4.1). The substitution of the tensor (4.92) into the wave equation provides the complex indices of refraction $N_{\pm}^{(n)}$ associated with the circular polarizations ($N_{\pm}^{(n)} = n_{\pm}^{(n)} - \mathrm{j}k_{\pm}^{(n)}$ with $n_{\pm}^{(n)}$ and $k_{\pm}^{(n)}$ corresponding respectively to the real index of refraction and extinction coefficient), according to Eq. (1.26),

$$N_{\pm}^{(n)2} = \varepsilon_{xx}^{(n)} \pm \mathrm{j}\varepsilon_{xy}^{(n)}. \tag{4.93}$$

From this, we can write the off-diagonal element as

$$\begin{aligned} \varepsilon_{xy}^{(n)} &= \frac{N_{+}^{(n)2} - N_{-}^{(n)2}}{2\mathrm{j}} \\ &= \frac{\left(N_{+}^{(n)} - N_{-}^{(n)}\right)\left(N_{+}^{(n)} + N_{-}^{(n)}\right)}{2\mathrm{j}} \\ &\approx -\mathrm{j}\,\varepsilon_{xx}^{(n)1/2}\left(N_{+}^{(n)} - N_{-}^{(n)}\right), \end{aligned} \tag{4.94}$$

where $\varepsilon_{xx}^{(n)1/2} = N^{(n)} \approx (N_{+}^{(n)} + N_{-}^{(n)})/2$. We have assumed that $\varepsilon_{xy}^{(n)}/\varepsilon_{xx}^{(n)} \ll 1$, which justifies the development of Eq. (4.93)

$$N_{\pm}^{(n)} = \varepsilon_{xx}^{(n)1/2}\left(1 + \mathrm{j}\frac{\varepsilon_{xy}^{(n)1/2}}{\varepsilon_{xx}^{(n)1/2}}\right)^{1/2} \approx \varepsilon_{xx}^{(n)1/2} \pm \mathrm{j}\frac{\varepsilon_{xy}^{(n)}}{2\varepsilon_{xx}^{(n)1/2}}. \tag{4.95}$$

Note that the sum $N_{+}^{(n)} + N_{-}^{(n)}$ is an even function of $\varepsilon_{xy}^{(n)}$. We denote as $\Delta N^{(n)}$ the difference

$$\Delta N^{(n)} = \frac{\left(N_{+}^{(n)} - N_{-}^{(n)}\right)}{2} = \frac{\mathrm{j}\,\varepsilon_{xy}^{(n)}}{2\varepsilon_{xx}^{(n)1/2}}. \tag{4.96}$$

To obtain the formula precise up to first order in $\varepsilon_{xy}^{(n)}$, we start from the expressions for the elements of the Jones transmission and reflection matrices in Cartesian representation expressed in terms of the corresponding matrix elements in circular representation according to Eq. (4.74) and Eq. (4.75) as

$$t_{xy}^{(0,\mathcal{N}+1)} = -\frac{j}{2}\left(t_{+}^{(0,\mathcal{N}+1)} - t_{-}^{(0,\mathcal{N}+1)}\right) = j\frac{M_{11}^{(CP)} - M_{33}^{(CP)}}{2M_{11}^{(CP)}M_{33}^{(CP)}}, \tag{4.97a}$$

$$r_{xy}^{(0,\mathcal{N}+1)} = -\frac{j}{2}\left(r_{+}^{(0,\mathcal{N}+1)} - r_{-}^{(0,\mathcal{N}+1)}\right) = j\frac{M_{11}^{(CP)}M_{43}^{(CP)} - M_{21}^{(CP)}M_{33}^{(CP)}}{2M_{11}^{(CP)}M_{33}^{(CP)}}, \tag{4.97b}$$

$$t_{xx}^{(0,\mathcal{N}+1)} = \frac{1}{2}\left(t_{+}^{(0,\mathcal{N}+1)} + t_{-}^{(0,\mathcal{N}+1)}\right) = \frac{M_{11}^{(CP)} + M_{33}^{(CP)}}{2M_{11}^{(CP)}M_{33}^{(CP)}}, \tag{4.97c}$$

$$r_{xx}^{(0,\mathcal{N}+1)} = \frac{1}{2}\left(r_{+}^{(0,\mathcal{N}+1)} + r_{-}^{(0,\mathcal{N}+1)}\right) = \frac{M_{11}^{(CP)}M_{43}^{(CP)} + M_{21}^{(CP)}M_{33}^{(CP)}}{2M_{11}^{(CP)}M_{33}^{(CP)}}. \tag{4.97d}$$

Upon the magnetization reversal, $\varepsilon_{xy}^{(n)}(\boldsymbol{M}) \rightarrow -\varepsilon_{xy}^{(n)}(-\boldsymbol{M})$, $\Delta N^{(n)} \rightarrow -\Delta N^{(n)}$ because of Eq. (4.96). Then, the CP polarizations exchange their roles. The diagonal 2×2 blocks of the $\mathbf{M}^{(CP)}$ matrix in Eq. (4.22) are consequently also exchanged. The inspection of Eqs. (4.97) shows that $t_{xy}^{(0,\mathcal{N}+1)}$ and $r_{xy}^{(0,\mathcal{N}+1)}$ are the odd functions of $\varepsilon_{xy}^{(n)}$ (they change the sign upon the magnetization, $\boldsymbol{M}$, reversal). On the other hand, $t_{xx}^{(0,\mathcal{N}+1)}$ and $r_{xx}^{(0,\mathcal{N}+1)}$ are even functions of $\varepsilon_{xy}^{(n)}$ (they do not change the sign upon $\boldsymbol{M}$ reversal). At zero magnetization, $\boldsymbol{M} = 0$, we have $M_{11}^{(CP)} = M_{33}^{(CP)}$ and $M_{21}^{(CP)} = M_{43}^{(CP)}$. Then, to first order in $\Delta N^{(n)}$, the elements of the Cartesian Jones transmission and reflection matrices given by Eqs. (4.97) can be obtained from $M_{11}^{(CP)}$ and $M_{21}^{(CP)}$ corresponding to $\boldsymbol{M} = 0$ [13]

$$t_{xx}^{(0,\mathcal{N}+1)} = \frac{1}{M_{11}^{(CP)}}, \tag{4.98a}$$

$$\begin{aligned} t_{yx}^{(0,\mathcal{N}+1)} &= j\Delta t_{xx}^{(0,\mathcal{N}+1)} = j\Delta\left(\frac{1}{M_{11}^{(CP)}}\right) \\ &= \frac{-j}{M_{11}^{(CP)2}}\sum_{n=1}^{\mathcal{N}+1}\frac{\partial M_{11}^{(CP)}}{\partial N^{(n)}}\Delta N^{(n)} \\ &= \frac{-j}{M_{11}^{(CP)}}\sum_{n=1}^{\mathcal{N}+1}\left\{\frac{\partial}{\partial N^{(n)}}\left[\ln\left(M_{11}^{(CP)}\right)\right]\right\}\Delta N^{(n)} \\ &= j\sum_{n=1}^{\mathcal{N}+1}\left[\frac{\partial}{\partial N^{(n)}}\left(t_{xx}^{(0,\mathcal{N}+1)}\right)\right]\Delta N^{(n)}, \end{aligned} \tag{4.98b}$$

$$r_{xx}^{(0,\mathcal{N}+1)} = \frac{\mathrm{M}_{21}^{(\mathrm{CP})}}{\mathrm{M}_{11}^{(\mathrm{CP})}}, \tag{4.98c}$$

$$\begin{aligned}
r_{yx}^{(0,\mathcal{N}+1)} &= \mathrm{j}\Delta r_{xx}^{(0,\mathcal{N}+1)} = \mathrm{j}\Delta\left(\frac{\mathrm{M}_{21}^{(\mathrm{CP})}}{\mathrm{M}_{11}^{(\mathrm{CP})}}\right) \\
&= \mathrm{j}\frac{\mathrm{M}_{11}^{(\mathrm{CP})}\Delta\left(\mathrm{M}_{21}^{(\mathrm{CP})}\right) - \mathrm{M}_{21}^{(\mathrm{CP})}\Delta\left(\mathrm{M}_{11}^{(\mathrm{CP})}\right)}{\mathrm{M}_{11}^{(\mathrm{CP})2}} \\
&= \frac{\mathrm{j}}{\mathrm{M}_{11}^{(\mathrm{CP})2}}\left(\mathrm{M}_{11}^{(\mathrm{CP})}\sum_{n=1}^{\mathcal{N}+1}\frac{\partial \mathrm{M}_{21}^{(\mathrm{CP})}}{\partial N^{(n)}}\Delta N^{(n)} - \mathrm{M}_{21}^{(\mathrm{CP})}\sum_{n=1}^{\mathcal{N}+1}\frac{\partial \mathrm{M}_{11}^{(\mathrm{CP})}}{\partial N^{(n)}}\Delta N^{(n)}\right) \\
&= \mathrm{j}\frac{\mathrm{M}_{21}^{(\mathrm{CP})}}{\mathrm{M}_{11}^{(\mathrm{CP})}}\sum_{n=1}^{\mathcal{N}+1}\left\{\frac{\partial}{\partial N^{(n)}}\left[\ln\left(\frac{\mathrm{M}_{21}^{(\mathrm{CP})}}{\mathrm{M}_{11}^{(\mathrm{CP})}}\right)\right]\right\}\Delta N^{(n)} \\
&= \mathrm{j}\sum_{n=1}^{\mathcal{N}+1}\left[\frac{\partial}{\partial N^{(n)}}\left(r_{xx}^{(0,\mathcal{N}+1)}\right)\right]\Delta N^{(n)}.
\end{aligned} \tag{4.98d}$$

The off-diagonal transmission or reflection matrix elements are expressed as a sum of contributions from individual layers proportional to ΔN_n (in nonmagnetic layers $\Delta N_n = 0$). They can be obtained from the Cartesian transmission and reflection coefficients $t_{xx}^{(0,\mathcal{N}+1)}$ and $r_{xx}^{(0,\mathcal{N}+1)}$ or from the **M** matrix of the structure corresponding to the isotropic case (*i.e.*, to the absence of magnetic order) where $N_+^{(n)} = N_-^{(n)} = N^{(n)}$ for $n = 0, \ldots, \mathcal{N}+1$, and

$$\mathrm{M}_{11}^{(\mathrm{CP})} = \mathrm{M}_{33}^{(\mathrm{CP})}, \quad \mathrm{M}_{21}^{(\mathrm{CP})} = \mathrm{M}_{43}^{(\mathrm{CP})}. \tag{4.99}$$

Appendix D illustrates the procedure summarized in Eq. (4.98) for the case of a single magnetic film sandwiched between two isotropic half spaces. The isotropic multilayer can be treated by the procedure described in Section 1.7 or equivalently with the help of simplified transfer matrices forming the product in Eq. (3.33). At the polar magnetization and normal incidence, the transfer matrix (4.14) consists of two slightly different diagonal 2×2 blocks

$$\mathbf{T}^{(n-1,n)} = \begin{pmatrix} \left(\mathbf{T}^{(n-1,n)}\right)_+ & 0 \\ 0 & \left(\mathbf{T}^{(n-1,n)}\right)_- \end{pmatrix}, \tag{4.100}$$

antisymmetrically displaced from their values at zero magnetization. In the absence of magnetization, the two blocks are equal to each other and take the following form

$$\left(\mathbf{T}^{(n-1,n)}\right)_A = \frac{\mathrm{e}^{\mathrm{j}\beta^{(n)}}}{t^{(n-1,n)}}\begin{pmatrix} 1 & r^{(n-1,n)}\mathrm{e}^{-2\mathrm{j}\beta^{(n)}} \\ r^{(n-1,n)} & \mathrm{e}^{-2\mathrm{j}\beta^{(n)}} \end{pmatrix}, \tag{4.101}$$

where the single-interface Fresnel reflection ($r^{(n-1,n)}$) and transmission ($t^{(n-1,n)}$) coefficients are given by

$$r^{(n-1,n)} = \frac{N^{(n-1)} - N^{(n)}}{N^{(n-1)} + N^{(n)}} \tag{4.102a}$$

$$t^{(n-1,n)} = 1 + r^{(n-1,n)} = \frac{2N^{(n-1)}}{N^{(n-1)} + N^{(n)}}, \tag{4.102b}$$

for $n = 1, \ldots, \mathcal{N} + 1$, and

$$\beta^{(n)} = \frac{\omega}{c} N^{(n)} d^{(n)}, \tag{4.102c}$$

for $n = 1, \ldots, \mathcal{N}$.

Consequently, in order to obtain the **M** matrix of the structure, it is sufficient to consider the product of 2×2 matrices representing an isotropic multilayer at normal incidence. Then, the diagonal elements of the Cartesian Jones transmission and reflection matrices correspond to those for the isotropic multilayer, and the off-diagonal ones follow from the expressions given above and containing differentiation with respect to $N^{(n)}$. In Eq. (4.98), we set $\Delta N^{(0)} = 0$, which corresponds to a nonmagnetic semi-infinite incident medium (*e.g.*, vacuum or air). For practical reasons, we set $\Delta N^{(\mathcal{N}+1)} = 0$ in the case of transmission (nonmagnetic exit medium) as opposed to that of reflection, where we may easily have $\Delta N^{(\mathcal{N}+1)} \neq 0$ (semi-infinite magnetic substrate).

For an incident wave of zero azimuth and ellipticity ($\chi_i^{(xy)} = 0$), *i.e.*, for an incident wave linearly polarized parallel to the x axis in our coordinate system, the complex number representation of the polarization state of the transmitted and reflected waves in the linear approximation can be expressed according to Eq. (3.70) or Eq. (4.97) as

$$\begin{aligned}
\chi_t^{(0,\mathcal{N}+1)} &= \frac{t_{31}^{(0,\mathcal{N}+1)}}{t_{11}^{(0,\mathcal{N}+1)}} = -\frac{t_{13}^{(0,\mathcal{N}+1)}}{t_{33}^{(0,\mathcal{N}+1)}} = \frac{t_{yx}^{(0,\mathcal{N}+1)}}{t_{xx}^{(0,\mathcal{N}+1)}} = -\frac{t_{xy}^{(0,\mathcal{N}+1)}}{t_{yy}^{(0,\mathcal{N}+1)}} \\
&= -\mathrm{j}\frac{\Delta\left(\mathrm{M}_{11}^{(\mathrm{CP})}\right)}{\mathrm{M}_{11}^{(\mathrm{CP})}} = -\mathrm{j}\sum_{n=1}^{\mathcal{N}+1}\left\{\frac{\partial}{\partial N^{(n)}}\left[\ln\left(\mathrm{M}_{11}^{(\mathrm{CP})}\right)\right]\right\}\Delta N^{(n)} \\
&= \mathrm{j}\frac{\Delta\left(t_{xx}^{(0,\mathcal{N}+1)}\right)}{t_{xx}^{(0,\mathcal{N}+1)}} = \mathrm{j}\sum_{n=1}^{\mathcal{N}+1}\left\{\frac{\partial}{\partial N^{(n)}}\left[\ln\left(t_{xx}^{(0,\mathcal{N}+1)}\right)\right]\right\}\Delta N^{(n)},
\end{aligned} \tag{4.103a}$$

and

$$\chi_r^{(0,\mathcal{N}+1)} = \frac{r_{41}^{(0,\mathcal{N}+1)}}{r_{21}^{(0,\mathcal{N}+1)}} = -\frac{r_{23}^{(0,\mathcal{N}+1)}}{r_{43}^{(0,\mathcal{N}+1)}} = \frac{r_{yx}^{(0,\mathcal{N}+1)}}{r_{xx}^{(0,\mathcal{N}+1)}} = -\frac{r_{xy}^{(0,\mathcal{N}+1)}}{r_{yy}^{(0,\mathcal{N}+1)}}$$
$$= \mathrm{j}\frac{\mathrm{M}_{11}^{(\mathrm{CP})}\Delta\left(\mathrm{M}_{21}^{(\mathrm{CP})}\right) - \mathrm{M}_{21}^{(\mathrm{CP})}\Delta\left(\mathrm{M}_{11}^{(\mathrm{CP})}\right)}{\mathrm{M}_{11}^{(\mathrm{CP})}\mathrm{M}_{21}^{(\mathrm{CP})}}$$
$$= \mathrm{j}\sum_{n=1}^{\mathcal{N}+1}\left\{\frac{\partial}{\partial N^{(n)}}\left[\ln\left(\frac{\mathrm{M}_{21}^{(\mathrm{CP})}}{\mathrm{M}_{11}^{(\mathrm{CP})}}\right)\right]\right\}\Delta N^{(n)}$$
$$= \mathrm{j}\frac{\Delta\left(r_{xx}^{(0,\mathcal{N}+1)}\right)}{r_{xx}^{(0,\mathcal{N}+1)}} = \mathrm{j}\sum_{n=1}^{\mathcal{N}+1}\left\{\frac{\partial}{\partial N^{(n)}}\left[\ln\left(r_{xx}^{(0,\mathcal{N}+1)}\right)\right]\right\}\Delta N^{(n)}, \quad (4.103b)$$

where $\chi_t^{(0,\mathcal{N}+1)}$ and $\chi_r^{(0,\mathcal{N}+1)}$ are related to the elements of the Jones transmission and reflection matrices in the circular (CP) representation by the expressions following from Eq. (4.74) and Eq. (4.75)

$$\chi_t^{(xy)} = \frac{t_{yx}}{t_{xx}} = -\frac{t_{xy}}{t_{xx}} = \mathrm{j}\frac{t_+ - t_-}{t_+ + t_-}, \quad (4.104)$$

$$\chi_r^{(xy)} = \frac{r_{yx}}{r_{xx}} = -\frac{r_{xy}}{r_{xx}} = \mathrm{j}\frac{r_+ - r_-}{r_+ + r_-}. \quad (4.105)$$

The complex numbers $\chi_t^{(0,\mathcal{N}+1)}$ and $\chi_r^{(0,\mathcal{N}+1)}$ for an incident wave with $\chi_i^{(xy)} = 0$ characterize the multilayer complex Faraday effect (or complex magnetic circular birefringence in transmission), and MO complex polar Kerr effect (or complex magnetic circular birefringence in reflection), respectively. Eq. (4.98b), Eq. (4.98d), and Eq. (4.103) can be transformed as follows

$$r_{yx}^{(0,\mathcal{N}+1)}$$
$$= \frac{1}{\mathrm{M}_0^{(\mathcal{N})2}}\sum_{n=1}^{\mathcal{N}}\frac{\Delta N^{(n)}}{2N^{(n)}}\left(1 - r^{(n-1,n)2}\right)\left[4\beta^{(n)}\mathrm{e}^{-2\mathrm{j}\beta^{(n)}}H_n^{(\mathcal{N})} - \mathrm{j}\left(1-\mathrm{e}^{-2\mathrm{j}\beta^{(n)}}\right)K_n^{(\mathcal{N})}\right]$$
$$-\mathrm{j}\frac{\Delta N^{(\mathcal{N}+1)}}{2N^{(\mathcal{N}+1)}}\frac{K_{\mathcal{N}+1}^{(\mathcal{N})}}{\mathrm{M}_0^{(\mathcal{N})2}}, \quad (4.106a)$$

$$\chi_r^{(0,\mathcal{N}+1)} = \frac{1}{\mathrm{M}_0^{(\mathcal{N})}\mathrm{M}_1^{(\mathcal{N})}}\sum_{n=1}^{\mathcal{N}}\frac{\Delta N^{(n)}}{2N^{(n)}}\left(1 - r^{(n-1,n)2}\right)$$
$$\times\left[4\beta^{(n)}\mathrm{e}^{-2\mathrm{j}\beta^{(n)}}H_n^{(\mathcal{N})} - \mathrm{j}\left(1-\mathrm{e}^{-2\mathrm{j}\beta^{(n)}}\right)K_n^{(\mathcal{N})}\right]$$
$$-\mathrm{j}\frac{\Delta N^{(\mathcal{N}+1)}}{2N^{(\mathcal{N}+1)}}\frac{K_{\mathcal{N}+1}^{(\mathcal{N})}}{\mathrm{M}_0^{(\mathcal{N})}\mathrm{M}_1^{(\mathcal{N})}}, \quad (4.106b)$$

$$t_{yx}^{(0,\mathcal{N}+1)} = \frac{1}{Q^{(\mathcal{N})}\mathrm{M}_0^{(\mathcal{N})2}} \sum_{n=1}^{\mathcal{N}} \frac{\Delta N^{(n)}}{2N^{(n)}} \left[2\beta^{(n)} J_n^{(\mathcal{N})} + \mathrm{j}(1 - \mathrm{e}^{-2\mathrm{j}\beta^{(n)}}) I_n^{(\mathcal{N})}\right], \quad (4.106c)$$

$$\chi_t^{(0,\mathcal{N}+1)} = \frac{1}{\mathrm{M}_0^{(\mathcal{N})}} \sum_{n=1}^{\mathcal{N}} \frac{\Delta N^{(n)}}{2N^{(n)}} \left[2\beta^{(n)} J_n^{(\mathcal{N})} + \mathrm{j}(1 - \mathrm{e}^{-2\mathrm{j}\beta^{(n)}}) I_n^{(\mathcal{N})}\right]. \quad (4.106d)$$

Here, $\mathrm{M}_0^{(\mathcal{N})}$ and $\mathrm{M}_1^{(\mathcal{N})}$ are the reduced matrix elements given by $\mathrm{M}_{11}^{(\mathrm{CP})} = Q^{(\mathcal{N})}\mathrm{M}_0^{(\mathcal{N})}$ and $\mathrm{M}_{21}^{(\mathrm{CP})} = Q^{(\mathcal{N})}\mathrm{M}_0^{(\mathcal{N})}$, respectively,

$$Q^{(\mathcal{N})} = \prod_{n=0}^{\mathcal{N}} \frac{\mathrm{e}^{\mathrm{j}\beta^{(n)}}}{t^{(n,n+1)}}. \quad (4.107)$$

We set $\beta^{(0)} = 0$, which corresponds to the incident amplitudes determined in the region (0) in the reference plane $z^{(0)}$ adjacent to the first layer of the multilayer system. The complex parameters $H_n^{(\mathcal{N})}$, $K_n^{(\mathcal{N})}$, $J_n^{(\mathcal{N})}$, and $I_n^{(\mathcal{N})}$ are functions of the optical constants and thicknesses of the layers forming the multilayer system and may be expressed in terms of the Fresnel coefficients and $\beta^{(n)}$ given in Eq. (4.102). The parameters $H_n^{(\mathcal{N})}$, $K_n^{(\mathcal{N})}$, $J_n^{(\mathcal{N})}$, and $I_n^{(\mathcal{N})}$ depend on the position in the stack of the n-th magnetic layer. As a result, $\chi_t^{(0,\mathcal{N}+1)}$ displays the symmetry with respect to the reversal in the layer order, whereas in $\chi_r^{(0,\mathcal{N}+1)}$ the contribution from the layers situated closer to the surface is enhanced with respect to those buried deeper in the structure. The explicit representation of $\chi_r^{(0,\mathcal{N}+1)}$ and $\chi_t^{(0,\mathcal{N}+1)}$ given in Eq. (4.106) allows one to better appreciate the simplifications in MO response for multilayers formed by identical magnetic layers.

According to Eq. (4.98) and Eq. (4.106), the contribution of each magnetic layer appears as a sum of two terms. The first one is proportional to $4\beta^{(n)}\mathrm{e}^{-2\mathrm{j}\beta^{(n)}}$ for χ_r (and to $2\beta^{(n)}$ for χ_t). This term represents the "propagation" effects. The second one, proportional to $(1 - \mathrm{e}^{-2\mathrm{j}\beta^{(n)}})$, represents the "interface" effects. In absorbing layers, this term saturates at the thickness that far exceeds the penetration depth.

4.3.3 Seven-Layer System

In this section, we provide expressions for χ_r and χ_t, from which analytical formulae for any multilayer system of up to seven layers ($\mathcal{N} \leq 7$) sandwiched between a semi-infinite isotropic medium and a semi-infinite magnetic substrate can be obtained. Analytical expressions for the response in a stack of seven layers is rather complicated. Nevertheless, this seven-layer system is useful to describe magnetic structures grown on a seed metallic layer, such as two magnetic layers coupled through a polarized nonmagnetic spacer or a simple magnetic layer with interface intermixing. Moreover, from them the formulae for simpler structures can be easily deduced. Thus, the main purpose of the complete expression for a seven-layer system is to provide sufficient freedom in the choice for the position of the magnetic layers, magnetic interfaces, and nonmagnetic spacers in the stack.

The analytical expressions are given in terms of Fresnel single-interface reflection coefficients, as this makes the formulas more concise. Furthermore, this form is more convenient if one wishes to account for the effect of roughness through an adequate reduction of the amplitude of the Fresnel reflection coefficients at rough interfaces [7].

In order to carry out easily the differentiation required by Eq. (4.98) and Eq. (4.103), the relevant elements of the **M** matrix are represented in an explicit form as

$$\begin{aligned} \mathrm{M}_{11} &= Q^{(7)}\left[(A_1A_3+B_1C_3)(A_5A_7+B_5C_7)+(A_1B_3+B_1D_3)(C_5A_7+D_5C_7)\right] \\ &= Q^{(7)}\mathrm{M}_0^{(7)}, \qquad (4.108a) \\ \mathrm{M}_{21} &= Q^{(7)}\left[(C_1A_3+D_1C_3)(A_5A_7+B_5C_7)+(C_1B_3+D_1D_3)(C_5A_7+D_5C_7)\right] \\ &= Q^{(7)}\mathrm{M}_1^{(7)}. \qquad (4.108b) \end{aligned}$$

Here, $Q^{(7)}$ is given by Eq. (4.107) and

$$\begin{aligned} A_n &= 1+r^{(n-1,n)}r^{(n,n+1)}\mathrm{e}^{-2\mathrm{j}\beta^{(n)}}, \qquad (4.109a) \\ B_n &= \mathrm{e}^{-2\mathrm{j}\beta^{(n+1)}}\left(r^{(n-1,n)}\mathrm{e}^{-2\mathrm{j}\beta^{(n)}}+r^{(n,n+1)}\right), \qquad (4.109b) \\ C_n &= r^{(n-1,n)}+r^{(n,n+1)}\mathrm{e}^{-2\mathrm{j}\beta^{(n)}}, \qquad (4.109c) \\ D_n &= \mathrm{e}^{-2\mathrm{j}\beta^{(n+1)}}\left(r^{(n-1,n)}r^{(n,n+1)}+\mathrm{e}^{-2\mathrm{j}\beta^{(n)}}\right). \qquad (4.109d) \end{aligned}$$

The interface reflection coefficients $r^{(n-1,n)}$ are given by Eq. (4.102a). Note that the sense of the polar magnetization of each layer can be chosen arbitrarily through the sign of $\Delta N^{(n)}$. This is useful in the MO studies of magnetically coupled multilayers. We assume $\beta^{(0)} = 0$, but $\Delta N^{(8)} \neq 0$ (corresponding to a magnetic exit medium). The reduced matrix elements $\mathrm{M}_0^{(7)}$ and $\mathrm{M}_1^{(7)}$ follow from Eq. (4.108).

4.3.3.1 Reflection

We obtain from Eq. (4.106a) and Eq. (4.106b)

$$\begin{aligned} r_{yx}^{(0,8)} &= \frac{1}{\mathrm{M}_0^{(7)2}}\left\{\sum_{n=1}^{7}\frac{\Delta N^{(n)}}{2N^{(n)}}\left(1-r^{(n-1,n)2}\right)\left[4\beta^{(n)}\mathrm{e}^{-2\mathrm{j}\beta^{(n)}}H_n^{(7)}\right.\right. \\ &\left.\left.\quad -\,\mathrm{j}\left(1-\mathrm{e}^{-2\mathrm{j}\beta^{(n)}}\right)K_n^{(7)}\right]-\mathrm{j}\frac{\Delta N^{(8)}}{2N^{(8)}}\left(1-r^{(78)2}\right)K_8^{(7)}\right\}, \qquad (4.110) \end{aligned}$$

$$\chi_r^{(0,8)} = \frac{1}{M_0^{(7)}M_1^{(7)}} \left\{ \sum_{n=1}^{7} \frac{\Delta N^{(n)}}{2N^{(n)}} (1 - r^{(n-1,n)2}) \left[4\beta^{(n)} e^{-2j\beta^{(n)}} H_n^{(7)} \right.\right.$$
$$\left.\left. - j(1 - e^{-2j\beta^{(n)}}) K_n^{(7)} \right] - j\frac{\Delta N^{(8)}}{2N^{(8)}} (1 - r^{(78)2}) K_8^{(7)} \right\}. \quad (4.111)$$

The coefficients $H_n^{(7)}$ and $K_n^{(7)}$, expressed in terms of $r^{(n-1,n)}$ and $\beta^{(n)}$, given by Eq. (4.102c) using the abbreviations A_n, B_n, C_n, and D_n summarized in Eq. (4.109), become

$$H_1^{(7)} = \left\{ [A_3 (A_5 A_7 + B_5 C_7) + B_3 (C_5 A_7 + D_5 C_7)] \right.$$
$$\left. + r^{(12)} e^{-2j\beta^{(2)}} [C_3 (A_5 A_7 + B_5 C_7) + D_3 (C_5 A_7 + D_5 C_7)] \right\}$$
$$\times \left\{ r^{(12)} [A_3 (A_5 A_7 + B_5 C_7) + B_3 (C_5 A_7 + D_5 C_7)] \right.$$
$$\left. + e^{-2j\beta^{(2)}} [C_3 (A_5 A_7 + B_5 C_7) + D_3 (C_5 A_7 + D_5 C_7)] \right\}, \quad (4.112a)$$

$$H_2^{(7)} = (1 - r^{(01)2}) e^{-2j\beta^{(1)}} [A_3 (A_5 A_7 + B_5 C_7) + B_3 (C_5 A_7 + D_5 C_7)]$$
$$\times [C_3 (A_5 A_7 + B_5 C_7) + D_3 (C_5 A_7 + D_5 C_7)], \quad (4.112b)$$

$$H_3^{(7)} = (A_1 D_1 - B_1 C_1) \left[(A_5 A_7 + B_5 C_7) + r^{(34)} e^{-2j\beta^{(4)}} (C_5 A_7 + D_5 C_7) \right]$$
$$\times \left[r^{(34)} (A_5 A_7 + B_5 C_7) + e^{-2j\beta^{(4)}} (C_5 A_7 + D_5 C_7) \right], \quad (4.112c)$$

$$H_4^{(7)} = (A_1 D_1 - B_1 C_1) (1 - r^{(23)2}) e^{-2j\beta^{(3)}} (A_5 A_7 + B_5 C_7)$$
$$\times (C_5 A_7 + D_5 C_7), \quad (4.112d)$$

$$H_5^{(7)} = (A_1 D_1 - B_1 C_1)(A_3 D_3 - B_3 C_3) \left(A_7 + r^{(56)} e^{-2j\beta^{(6)}} C_7 \right)$$
$$\times \left(r^{(56)} A_7 + e^{-2j\beta^{(6)}} C_7 \right), \quad (4.112e)$$

$$H_6^{(7)} = (A_1 D_1 - B_1 C_1)(A_3 D_3 - B_3 C_3)(1 - r^{(45)2}) e^{-2j\beta^{(5)}} A_7 C_7, \quad (4.112f)$$

$$H_7^{(7)} = (A_1 D_1 - B_1 C_1)(A_3 D_3 - B_3 C_3)(A_5 D_5 - B_5 C_5) r^{(78)}, \quad (4.112g)$$

and

$$K_1^{(7)} = \left\{ [A_3 (A_5 A_7 + B_5 C_7) + B_3 (C_5 A_7 + D_5 C_7)] \right.$$
$$\left. + r^{(12)} e^{-2j\beta^{(2)}} \left[C_3 (A_5 A_7 + B_5 C_7) + D_3 (C_5 A_7 + D_5 C_7) \right] \right\}^2$$
$$+ e^{-2j\beta^{(1)}} \left\{ r^{(12)} [A_3 (A_5 A_7 + B_5 C_7) + B_3 (C_5 A_7 + D_5 C_7)] \right.$$
$$\left. + e^{-2j\beta^{(2)}} \left[C_3 (A_5 A_7 + B_5 C_7) + D_3 (C_5 A_7 + D_5 C_7) \right] \right\}^2, \quad (4.113a)$$

$$K_2^{(7)} = (1 - r^{(01)2}) e^{-2j\beta^{(1)}} \left\{ [A_3 (A_5 A_7 + B_5 C_7) + B_3 (C_5 A_7 + D_5 C_7)]^2 \right.$$
$$\left. + e^{-2j\beta^{(2)}} [C_3 (A_5 A_7 + B_5 C_7) + D_3 (C_5 A_7 + D_5 C_7)]^2 \right\}, \quad (4.113b)$$

$$K_3^{(7)} = (A_1 D_1 - B_1 C_1) \Big\{ \big[(A_5 A_7 + B_5 C_7) + r^{(34)} e^{-2j\beta^{(4)}} (C_5 A_7 + D_5 C_7) \big]^2 + e^{-2j\beta^{(3)}} \big[r^{(34)} (A_5 A_7 + B_5 C_7) + e^{-2j\beta^{(4)}} (C_5 A_7 + D_5 C_7) \big]^2 \Big\}, \quad (4.113c)$$

$$K_4^{(7)} = (A_1 D_1 - B_1 C_1) \left(1 - r^{(23)2}\right) e^{-2j\beta^{(3)}} \big[(A_5 A_7 + B_5 C_7)^2 + e^{-2j\beta^{(4)}} \times (C_5 A_7 + D_5 C_7)^2 \big], \quad (4.113d)$$

$$K_5^{(7)} = (A_1 D_1 - B_1 C_1)(A_3 D_3 - B_3 C_3) \times \left[\left(A_7 + r^{(56)} e^{-2j\beta^{(6)}} C_7 \right)^2 + e^{-2j\beta^{(5)}} \left(r^{(56)} A_7 + e^{-2j\beta^{(6)}} C_7 \right)^2 \right] \quad (4.113e)$$

$$K_6^{(7)} = (A_1 D_1 - B_1 C_1)(A_3 D_3 - B_3 C_3) \left(1 - r^{(45)2}\right) \times e^{-2j\beta^{(5)}} \left(A_7^2 + e^{-2j\beta^{(6)}} C_7^2 \right), \quad (4.113f)$$

$$K_7^{(7)} = (A_1 D_1 - B_1 C_1)(A_3 D_3 - B_3 C_3)(A_5 D_5 - B_5 C_5) \times \left(1 + r^{(78)2} e^{-2j\beta^{(7)}} \right), \quad (4.113g)$$

$$K_8^{(7)} = (A_1 D_1 - B_1 C_1)(A_3 D_3 - B_3 C_3)(A_5 D_5 - B_5 C_5) \times (1 - r^{(67)2}) e^{-2j\beta^{(7)}}. \quad (4.113h)$$

Note that

$$A_n D_n - B_n C_n = \left(1 - r^{(n-1,n)2}\right)\left(1 - r^{(n,n+1)2}\right) e^{-2j\beta^{(n)}} e^{-2j\beta^{(n+1)}} \quad (4.114)$$

for $n = 1, 3, 5,$ and 7.

The number of terms forming the sum in Eq. (4.111) is reduced to that of magnetic layers, *i.e.*, the only nonzero $H_n^{(7)}$ and $K_n^{(7)}$ are those for which n pertains to a magnetic layer. When the number of layers involved in the system is $\mathcal{N} < 7$, the corresponding response is obtained from the above formulae by setting $\beta^{(n)} = 0$ or $d^{(n)} = 0$ in suitably chosen $(7 - \mathcal{N})$ layers. In this way, the excessive $(7 - \mathcal{N})$ layers are removed from the stack.

The formulae may be considerably simplified when the system displays symmetry elements. This results, for example, when two or more layers have identical permittivities or when the composition profile displays periodicity. The expressions may be extended to periodic multilayers, with the seven-layer system serving as a basic building block [8] as explained in the next section.

Note the symmetry of $H_n^{(7)}$ and $K_n^{(7)}$, which allows the extension to arbitrary $\mathcal{N}$. They can be viewed as a product of two factors. The first one pertains to the layers situated before the n-th magnetic layer. The second one is a function of the reflection coefficient, denoted as $r^{(n,8)}$, of the structure consisting of layers behind the n-th layer. Both $H_n^{(7)}$ and $K_n^{(7)}$ contain a common factor formed by the product of the Fresnel transmission coefficients and exponentials

$$\prod_{j=1}^{n-1}\left(1-r^{(j-1,j)2}\right)e^{2j\beta^{(j)}}=\prod_{j=1}^{n-1}t^{(j-1,j)}t^{(j,j-1)}e^{2j\beta^{(j)}}. \tag{4.115}$$

The second factors entering $H_n^{(7)}$ and $K_n^{(7)}$ may be expressed as $\mathrm{M}_1^{(7-n)}\mathrm{M}_0^{(7-n)}$ and $[\mathrm{M}_0^{(7-n)}]^2+[\mathrm{M}_1^{(7-n)}]^2e^{2j\beta^{(n)}}$, respectively. The coefficients $\mathrm{M}_1^{(7-n)}$ and $\mathrm{M}_0^{(7-n)}$ have a similar meaning as the reduced matrix elements $\mathrm{M}_1^{(\mathcal{N})}$ and $\mathrm{M}_0^{(\mathcal{N})}$ in Eq. (4.108). They give the reflection coefficient of the truncated multilayer structure with the first n layers removed (the entrance medium is characterized by $N^{(n)}$), *i.e.*,

$$r^{(n,8)}=\frac{\mathrm{M}_1^{(7-n)}}{\mathrm{M}_0^{(7-n)}}. \tag{4.116}$$

Using Eq. (4.112) and Eq. (4.113), this can be illustrated for the last three parameters $H_n^{(7)}$ and $K_n^{(7)}$ as

$$H_7^{(7)}\propto r^{(78)}, \tag{4.117a}$$
$$H_6^{(7)}\propto r^{(68)}\left(1+r^{(67)}r^{(78)}e^{-2j\beta^{(7)}}\right)^2, \tag{4.117b}$$
$$H_5^{(7)}\propto r^{(58)}\left(A_7+r^{(56)}e^{-2j\beta^{(6)}}C_7\right), \tag{4.117c}$$

and

$$K_8^{(7)}\propto 1, \tag{4.118a}$$
$$K_7^{(7)}\propto 1+r^{(78)2}e^{-2j\beta^{(7)}}, \tag{4.118b}$$
$$K_6^{(7)}\propto A_7^2\left(1+r^{(68)2}e^{-2j\beta^{(6)}}\right). \tag{4.118c}$$

4.3.3.2 Transmission

From Eq. (4.106c) and Eq. (4.106d), the off-diagonal element of the Jones transmission matrix and the transformation of the incident linear polarization (Faraday effect) in the stack of seven layers are determined respectively by

$$t_{yx}^{(0,8)}=\frac{1}{Q^{(7)}\mathrm{M}_0^{(7)2}}\sum_{n=1}^{7}\frac{\Delta N^{(n)}}{2N^{(n)}}\left[j\left(1-e^{-2j\beta^{(n)}}\right)I_n^{(7)}+2\beta^{(n)}J_n^{(7)}\right], \tag{4.119}$$

$$\chi_t^{(0,8)}=\frac{1}{\mathrm{M}_0^{(7)}}\sum_{n=1}^{7}\frac{\Delta N^{(n)}}{2N^{(n)}}\left[j\left(1-e^{-2j\beta^{(n)}}\right)I_n^{(7)}+2\beta^{(n)}J_n^{(7)}\right]. \tag{4.120}$$

The reduced matrix element $\mathrm{M}_0^{(7)}$ is given by Eq. (4.108a), and $I_n^{(7)}$ and $J_n^{(7)}$ are expressed in terms of $r^{(n-1,n)}$ and $\beta^{(n)}$ given by Eq. (4.102c) using the abbreviations A_n, B_n, C_n, and D_n, summarized in Eq. (4.109). They are given by

$$
\begin{aligned}
I_1^{(7)} &= \left[\left(r^{(01)} - r^{(12)}\right) A_3 - \left(1 - r^{(01)} r^{(12)}\right) e^{-2j\beta^{(2)}} C_3\right] (A_5 A_7 + B_5 C_7) \\
&\quad + \left[\left(r^{(01)} - r^{(12)}\right) B_3 - \left(1 - r^{(01)} r^{(12)}\right) e^{-2j\beta^{(2)}} D_3\right] \\
&\quad \times (C_5 A_7 + D_5 C_7), \qquad (4.121a) \\
I_2^{(7)} &= \left[\left(r^{(12)} - r^{(23)}\right)\left(1 - r^{(01)} r^{(34)} e^{-2j\beta^{(1)}} e^{-2j\beta^{(3)}}\right)\right. \\
&\quad \left. + \left(1 - r^{(12)} r^{(23)}\right)\left(r^{(01)} e^{-2j\beta^{(1)}} - r^{(34)} e^{-2j\beta^{(3)}}\right)\right] (A_5 A_7 + B_5 C_7) \\
&\quad + \left[\left(r^{(12)} - r^{(23)}\right)\left(r^{(34)} - r^{(01)} e^{-2j\beta^{(1)}} e^{-2j\beta^{(3)}}\right)\right. \\
&\quad \left. + \left(1 - r^{(12)} r^{(23)}\right)\left(r^{(01)} r^{(34)} e^{-2j\beta^{(1)}} - e^{-2j\beta^{(3)}}\right)\right] \\
&\quad \times e^{-2j\beta^{(4)}} (C_5 A_7 + D_5 C_7), \qquad (4.121b) \\
I_3^{(7)} &= \left[\left(r^{(23)} - r^{(34)}\right) A_1 + \left(1 - r^{(23)} r^{(34)}\right) B_1\right] (A_5 A_7 + B_5 C_7) \\
&\quad - \left[\left(1 - r^{(23)} r^{(34)}\right) A_1 + \left(r^{(23)} - r^{(34)}\right) B_1\right] e^{-2j\beta^{(4)}} \\
&\quad \times (C_5 A_7 + D_5 C_7), \qquad (4.121c) \\
I_4^{(7)} &= \left[\left(r^{(23)} e^{-2j\beta^{(3)}} + r^{(34)}\right) A_1 + \left(r^{(23)} r^{(34)} + e^{-2j\beta^{(3)}}\right) B_1\right] (A_5 A_7 + B_5 C_7) \\
&\quad - (A_1 A_3 + B_1 C_3)(C_5 A_7 + D_5 C_7), \qquad (4.121d) \\
I_5^{(7)} &= \left[\left(r^{(45)} - r^{(56)}\right) A_7 - \left(1 - r^{(45)} r^{(56)}\right) e^{-2j\beta^{(6)}} C_7\right] (A_1 A_3 + B_1 C_3) \\
&\quad + \left[\left(1 - r^{(45)} r^{(56)}\right) A_7 - \left(r^{(45)} - r^{(56)}\right) e^{-2j\beta^{(6)}} C_7\right] \\
&\quad \times (A_1 B_3 + B_1 D_3), \qquad (4.121e) \\
I_6^{(7)} &= \left[\left(r^{(45)} e^{-2j\beta^{(5)}} + r^{(56)}\right) A_7 - A_5 C_7\right] (A_1 A_3 + B_1 C_3) \\
&\quad + \left[\left(r^{(45)} r^{(56)} + e^{-2j\beta^{(5)}}\right) A_7 - C_5 C_7\right] (A_1 B_3 + B_1 D_3), \qquad (4.121f) \\
I_7^{(7)} &= \left[\left(r^{(67)} - r^{(78)}\right) A_5 + \left(1 - r^{(67)} r^{(78)}\right) B_5\right] (A_1 A_3 + B_1 C_3) \\
&\quad + \left[\left(r^{(67)} - r^{(78)}\right) C_5 + \left(1 - r^{(67)} r^{(78)}\right) D_5\right] (A_1 B_3 + B_1 D_3), \qquad (4.121g)
\end{aligned}
$$

and

$$
\begin{aligned}
J_1^{(7)} &= \left[\left(1 - r^{(01)} r^{(12)} e^{-2j\beta^{(1)}}\right) A_3 - \left(r^{(01)} e^{-2j\beta^{(1)}} - r^{(12)}\right) e^{-2j\beta^{(2)}} C_3\right] \\
&\quad \times (A_5 A_7 + B_5 C_7) + \left[\left(1 - r^{(01)} r^{(12)} e^{-2j\beta^{(1)}}\right) B_3\right. \\
&\quad \left. - \left(r^{(01)} e^{-2j\beta^{(1)}} - r^{(12)}\right) e^{-2j\beta^{(2)}} D_3\right] (C_5 A_7 + D_5 C_7), \qquad (4.122a)
\end{aligned}
$$

$$J_2^{(7)} = (A_1 A_3 - B_1 C_3)(A_5 A_7 + B_5 C_7) + (A_1 B_3 - B_1 D_3) \times (C_5 A_7 + D_5 C_7), \tag{4.122b}$$

$$J_3^{(7)} = \left[\left(1 - r^{(23)} r^{(34)} e^{-2j\beta^{(3)}}\right) A_1 + \left(r^{(23)} - r^{(34)} e^{-2j\beta^{(3)}}\right) B_1\right] (A_5 A_7 + B_5 C_7) + \left[\left(-r^{(23)} e^{-2j\beta^{(3)}} + r^{(34)}\right) A_1 + \left(r^{(23)} r^{(34)} - e^{-2j\beta^{(3)}}\right) B_1\right] \times e^{-2j\beta^{(4)}} (C_5 A_7 + D_5 C_7), \tag{4.122c}$$

$$J_4^{(7)} = (A_1 A_3 + B_1 C_3)(A_5 A_7 + B_5 C_7) - (A_1 B_3 + B_1 D_3) \times (C_5 A_7 + D_5 C_7), \tag{4.122d}$$

$$J_5^{(7)} = \left[\left(1 - r^{(45)} r^{(56)} e^{-2j\beta^{(5)}}\right) A_7 - \left(r^{(45)} e^{-2j\beta^{(5)}} - r^{(56)}\right) e^{-2j\beta^{(6)}} C_7\right] \times (A_1 A_3 + B_1 C_3) + \left[\left(r^{(45)} - r^{(56)} e^{-2j\beta^{(5)}}\right) A_7 + \left(r^{(45)} r^{(56)} - e^{-2j\beta^{(5)}}\right) e^{-2j\beta^{(6)}} C_7\right] \times (A_1 B_3 + B_1 D_3), \tag{4.122e}$$

$$J_6^{(7)} = (A_5 A_7 - B_5 C_7)(A_1 A_3 + B_1 C_3) + (C_5 A_7 - D_5 C_7) \times (A_1 B_3 + B_1 D_3), \tag{4.122f}$$

$$J_7^{(7)} = \left[\left(1 - r^{(67)} r^{(78)} e^{-2j\beta^{(7)}}\right) A_5 + \left(r^{(67)} - r^{(78)} e^{-2j\beta^{(7)}}\right) B_5\right] (A_1 A_3 + B_1 C_3) + \left[\left(1 - r^{(67)} r^{(78)} e^{-2j\beta^{(7)}}\right) C_5 + \left(r^{(67)} - r^{(78)} e^{-2j\beta^{(7)}}\right) D_5\right] \times (A_1 B_3 + B_1 D_3). \tag{4.122g}$$

Note that the expressions for χ_t are invariant with respect to the reversal in the order of the layers.

4.3.4 Ultrathin Film Approximation

In the ultrathin film approximation, Eq. (4.106) is reduced with the help of Eq. (4.85). The multilayer response becomes the sum of contributions from individual layers. The contribution of the n-th layer becomes proportional to $d^{(n)}$, the layer thickness, but may still be a function of other $\beta^{(m)}$ $(m \neq n)$. The exponentials $e^{-j\beta^{(n)}}$ occur in $M_0^{(\mathcal{N})2}$, $M_1^{(\mathcal{N})2}$, $K_n^{(\mathcal{N})}$, and $J_n^{(\mathcal{N})}$. The reduced matrix elements $M_0^{(\mathcal{N})}$ and $M_1^{(\mathcal{N})}$ are easy to determine, as they pertain to isotropic multilayers without ultrathin film approximation. If we wish to apply the approximation to them, they must enter the sums in Eq. (4.106). We have

$$
\begin{aligned}
r_{yx}^{(0,\mathcal{N}+1)} &= \frac{1}{\mathrm{M}_0^{(\mathcal{N})2}} \sum_{n=1}^{\mathcal{N}} \Delta \mathrm{N}^{(n)} \frac{\omega}{c} d^{(n)} (1 - r^{(n-1,n)2}) \left(2H_n^{(\mathcal{N})} + K_n^{(\mathcal{N},n)}\right) \\
&\quad -\mathrm{j} \frac{\Delta \mathrm{N}^{(\mathcal{N}+1)}}{2\mathrm{N}^{(\mathcal{N}+1)}} \frac{K_{\mathcal{N}+1}^{(\mathcal{N})}}{\mathrm{M}_0^{(\mathcal{N})2}} \\
&= \sum_{n=1}^{\mathcal{N}} \frac{1}{\mathrm{M}_0^{(\mathcal{N},n)2}} \Delta \mathrm{N}^{(n)} \frac{\omega}{c} d^{(n)} (1 - r^{(n-1,n)2}) \left(2H_n^{(\mathcal{N})} + K_n^{(\mathcal{N},n)}\right) \\
&\quad -\mathrm{j} \frac{\Delta \mathrm{N}^{(\mathcal{N}+1)}}{2\mathrm{N}^{(\mathcal{N}+1)}} \frac{K_{\mathcal{N}+1}^{(\mathcal{N})}}{\mathrm{M}_0^{(\mathcal{N})2}},
\end{aligned} \tag{4.123a}
$$

$$
\begin{aligned}
\chi_r &= \sum_{n=1}^{\mathcal{N}} \frac{1}{\mathrm{M}_0^{(\mathcal{N},n)} \mathrm{M}_1^{(\mathcal{N},n)}} \Delta \mathrm{N}^{(n)} \frac{\omega}{c} d^{(n)} \left(1 - r_{n-1,n}^2\right) \left(2H_n^{(\mathcal{N})} + K_n^{(\mathcal{N},n)}\right) \\
&\quad -\mathrm{j} \left(1 - r_{\mathcal{N},\mathcal{N}+1}^2\right) \frac{\Delta \mathrm{N}_{\mathcal{N}+1}}{2\mathrm{N}_{\mathcal{N}+1}} \frac{K_{\mathcal{N}+1}^{(\mathcal{N})}}{\mathrm{M}_0^{(\mathcal{N})} \mathrm{M}_1^{(\mathcal{N})}},
\end{aligned} \tag{4.123b}
$$

$$
\begin{aligned}
t_{yx}^{(0,\mathcal{N}+1)} &= \frac{1}{\mathrm{Q}^{(\mathcal{N})} \mathrm{M}_0^{(\mathcal{N})2}} \sum_{n=1}^{\mathcal{N}} \Delta \mathrm{N}^{(n)} \frac{\omega}{c} d^{(n)} \left(J_n^{(\mathcal{N},n)} - I_n^{(\mathcal{N})}\right) \\
&= \sum_{n=1}^{\mathcal{N}} \frac{1}{\mathrm{Q}^{(\mathcal{N},n)} \mathrm{M}_0^{(\mathcal{N},n)2}} \Delta \mathrm{N}^{(n)} \frac{\omega}{c} d^{(n)} \left(J_n^{(\mathcal{N},n)} - I_n^{(\mathcal{N})}\right),
\end{aligned} \tag{4.123c}
$$

$$
\chi_t = \sum_{n=1}^{\mathcal{N}} \frac{1}{\mathrm{M}_0^{(\mathcal{N},n)}} \Delta \mathrm{N}^{(n)} \frac{\omega}{c} d^{(n)} \left(J_n^{(\mathcal{N},n)} - I_n^{(\mathcal{N})}\right), \tag{4.123d}
$$

where $\mathrm{M}_0^{(\mathcal{N},n)}$, $\mathrm{M}_1^{(\mathcal{N},n)}$, $\mathrm{Q}_1^{(\mathcal{N},n)}$, $K_n^{(\mathcal{N},n)}$, and $J_n^{(\mathcal{N},n)}$ correspond to $\mathrm{M}_0^{(\mathcal{N})}$, $\mathrm{M}_1^{(\mathcal{N})}$, $K_n^{(\mathcal{N})}$, and $J_n^{(\mathcal{N}),n}$ with $\beta^{(n)} = 0$. With restriction to $\mathcal{N} = 7$, we obtain for $2H_n^{(7)} + K_n^{(7,n)}$ and $J_n^{(\mathcal{N})} - I_n^{(\mathcal{N})}$

$$
\begin{aligned}
2H_1^{(7)} + K_1^{(7,1)} &= \left(1 + r^{(12)}\right)^2 \Big[\left(A_3 + \mathrm{e}^{-2\mathrm{j}\beta^{(2)}} C_3\right) (A_5 A_7 + B_5 C_7) \\
&\quad + \left(B_3 + \mathrm{e}^{-2\mathrm{j}\beta^{(2)}} D_3\right) (C_5 A_7 + D_5 C_7) \Big]^2,
\end{aligned} \tag{4.124a}
$$

$$2H_2^{(7)} + K_2^{(7,2)} = \left(1 - r^{(01)2}\right) e^{-2j\beta^{(1)}} \left(1 + r^{(23)}\right)^2 \times \left[\left(1 + r^{(34)} e^{-2j\beta^{(3)}}\right) \times (A_5 A_7 + B_5 C_7) + e^{-2j\beta^{(4)}} \times \left(e^{-2j\beta^{(3)}} + r^{(34)}\right)(C_5 A_7 + D_5 C_7)\right]^2, \quad (4.124b)$$

$$2H_3^{(7)} + K_3^{(7,3)} = \left(1 - r^{(01)2}\right) e^{-2j\beta^{(1)}} \left(1 - r^{(12)2}\right) e^{-2j\beta^{(2)}} \left(1 + r^{(34)}\right)^2 \times \left[\left(A_5 + e^{-2j\beta^{(4)}} C_5\right) A_7 \left(B_5 + e^{-2j\beta^{(4)}} D_5\right) C_7\right]^2, \quad (4.124c)$$

$$2H_4^{(7)} + K_4^{(7,4)} = \left(1 - r^{(01)2}\right) e^{-2j\beta^{(1)}} \left(1 - r^{(12)2}\right) e^{-2j\beta^{(2)}} \left(1 - r^{(23)2}\right) \times e^{-2j\beta^{(3)}} \left(1 + r^{(45)}\right)^2 \left[\left(1 + r^{(56)} e^{-2j\beta^{(5)}}\right) A_7 + e^{-2j\beta^{(6)}} \left(e^{-2j\beta^{(5)}} + r^{(56)}\right) C_7\right]^2, \quad (4.124d)$$

$$2H_5^{(7)} + K_5^{(7,5)} = \left(1 - r^{(01)2}\right) e^{-2j\beta^{(1)}} \left(1 - r^{(12)2}\right) e^{-2j\beta^{(2)}} \left(1 - r^{(23)2}\right) \times e^{-2j\beta^{(3)}} \left(1 - r^{(34)2}\right) e^{-2j\beta^{(4)}} \left(1 + r^{(56)}\right)^2 \times \left(A_7 + e^{-2j\beta^{(6)}} C_7\right)^2, \quad (4.124e)$$

$$2H_6^{(7)} + K_6^{(7,6)} = \left(1 - r^{(01)2}\right) e^{-2j\beta^{(1)}} \left(1 - r^{(12)2}\right) e^{-2j\beta^{(2)}} \left(1 - r^{(23)2}\right) \times e^{-2j\beta^{(3)}} \left(1 - r^{(34)2}\right) e^{-2j\beta^{(4)}} \left(1 - r^{(45)2}\right) \times e^{-2j\beta^{(5)}} \left(1 + r^{(67)}\right)^2 \left(1 + e^{-2j\beta^{(7)}} r^{(78)}\right)^2, \quad (4.124f)$$

$$2H_7^{(7)} + K_7^{(7,7)} = \left(1 - r^{(01)2}\right) e^{-2j\beta^{(1)}} \left(1 - r^{(12)2}\right) e^{-2j\beta^{(2)}} \left(1 - r^{(23)2}\right) \times e^{-2j\beta^{(3)}} \left(1 - r^{(34)2}\right) e^{-2j\beta^{(4)}} \left(1 - r^{(45)2}\right) \times e^{-2j\beta^{(5)}} \left(1 - r^{(56)2}\right) e^{-2j\beta^{(6)}} \left(1 + r^{(78)}\right)^2. \quad (4.124g)$$

$$J_1^{(7,1)} - I_1^{(7)} = \left(1 - r^{(01)}\right)\left(1 + r^{(12)}\right)\left[\left(A_3 + e^{-2j\beta^{(2)}} C_3\right)(A_5 A_7 + B_5 C_7) + \left(B_3 + e^{-2j\beta^{(2)}} D_3\right)(C_5 A_7 + D_5 C_7)\right], \quad (4.125a)$$

$$J_2^{(7,2)} - I_2^{(7)} = \left(1 - r^{(01)} e^{-2j\beta^{(1)}}\right)\left(1 - r^{(12)}\right)\left(1 + r^{(23)}\right)\left[\left(1 + r^{(34)} e^{-2j\beta^{(3)}}\right) \times (A_5 A_7 + B_5 C_7) + e^{-2j\beta^{(4)}} \left(e^{-2j\beta^{(3)}} + r^{(34)}\right) \times (C_5 A_7 + D_5 C_7)\right], \quad (4.125b)$$

$$J_3^{(7,3)} - I_3^{(7)} = (A_1 - B_1)\left(1 - r^{(23)}\right)\left(1 + r^{(34)}\right) \times \left[\left(A_5 + e^{-2j\beta^{(4)}} C_5\right) A_7 + \left(B_5 + e^{-2j\beta^{(4)}} D_5\right) C_7\right], \quad (4.125c)$$

$$J_4^{(7,4)} - I_4^{(7)} = \left[A_1\left(1 - r^{(23)}e^{-2j\beta^{(3)}}\right) - B_1\left(e^{-2j\beta^{(3)}} - r^{(23)}\right)\right] \times \left(1 - r^{(34)}\right)\left(1 + r^{(45)}\right)\left[\left(1 + r^{(56)}e^{-2j\beta^{(5)}}\right)A_7 + e^{-2j\beta^{(6)}}\left(r^{(56)} + e^{-2j\beta^{(5)}}\right)C_7\right], \tag{4.125d}$$

$$J_5^{(7,5)} - I_5^{(7)} = \left[A_1\left(A_3 - B_3\right) + B_1\left(C_3 - D_3\right)\right] \times \left(1 - r^{(45)}\right)\left(1 + r^{(56)}\right)\left(A_7 + e^{-2j\beta^{(6)}}C_7\right), \tag{4.125e}$$

$$J_6^{(7,6)} - I_6^{(7)} = \left[\left(1 - r^{(45)}e^{-2j\beta^{(5)}}\right)\left(A_1A_3 + B_1C_3\right) - \left(e^{-2j\beta^{(5)}} - r^{(45)}\right) \times \left(A_1B_3 + B_1D_3\right)\right]\left(1 - r^{(56)}\right)\left(1 + r^{(67)}\right) \times \left(1 + r^{(78)}e^{-2j\beta^{(7)}}\right), \tag{4.125f}$$

$$J_7^{(7,7)} - I_7^{(7)} = \left[\left(A_1A_3 + B_1C_3\right)\left(A_5 - B_5\right) + \left(A_1B_3 + B_1D_3\right)\left(C_5 - D_5\right)\right] \times \left(1 - r^{(67)}\right)\left(1 + r^{(78)}\right). \tag{4.125g}$$

4.3.5 Exchange Coupled Film

The procedure can be best illustrated with a simple case. We choose a system with $\mathcal{N} = 2$ consisting of a magnetic film (1), nonmagnetic spacer (2), and magnetic substrate (3). Under favorable conditions, the magnetic film may be exchange coupled to the magnetic substrate across a nonmagnetic spacer. The system response is represented by

$$\mathbf{M}^{(CP)} = \left[\mathbf{D}^{(0)}\right]^{-1}\mathbf{D}^{(1)}\mathbf{P}^{(1)}\left[\mathbf{D}^{(1)}\right]^{-1}\mathbf{D}^{(2)}\mathbf{P}^{(2)}\left[\mathbf{D}^{(2)}\right]^{-1}\mathbf{D}^{(3)}. \tag{4.126}$$

This configuration is often used in the studies of exchange coupling of a magnetic layer across a nonmagnetic spacer. To first order in $\varepsilon_{xy}^{(n)}/\varepsilon_{xx}^{(n)}$, the elements of the Jones reflection matrix may be expressed as

$$r_{xx}^{(03)} = M_1^{(2)}M_0^{(2)-1}, \tag{4.127a}$$

$$r_{xy}^{(03)} = jM_0^{(2)-2}\left\{\frac{\Delta N^{(1)}}{2N^{(1)}}\left(1 - r^{(01)2}\right)\left[4j\beta^{(1)}H_1^{(2)}e^{-2j\beta^{(1)}} + \left(1 - e^{-2j\beta^{(1)}}\right)K_1^{(2)}\right] + \frac{\Delta N^{(3)}}{2N^{(3)}}\left(1 - r^{(23)2}\right)K_3^{(2)}\right\}, \tag{4.127b}$$

where

$$M_0^{(2)} = 1 + r^{(12)}r^{(23)}e^{-2j\beta^{(2)}} + r^{(01)}e^{-2j\beta^{(1)}}\left(r^{(12)} + r^{(23)}e^{-2j\beta^{(2)}}\right), \tag{4.128a}$$

$$M_1^{(2)} = r^{(01)}\left(1 + r^{(12)}r^{(23)}e^{-2j\beta^{(2)}}\right) + e^{-2j\beta^{(1)}}\left(r^{(12)} + r^{(23)}e^{-2j\beta^{(2)}}\right), \tag{4.128b}$$

$$H_1^{(2)} = \left(1 + r^{(12)} r^{(23)} e^{-2j\beta^{(2)}}\right) \left(r^{(12)} + r^{(23)} e^{-2j\beta^{(2)}}\right), \tag{4.128c}$$

$$K_1^{(2)} = \left(1 + r^{(12)} r^{(23)} e^{-2j\beta^{(2)}}\right)^2 + \left(r^{(12)} + r^{(23)} e^{-2j\beta^{(2)}}\right)^2 e^{-2j\beta^{(1)}}, \tag{4.128d}$$

$$K_3^{(2)} = (1 - r^{(01)2}) e^{-2j\beta^{(1)}} (1 - r^{(12)2})\, e^{-2j\beta^{(2)}}. \tag{4.128e}$$

In the nonmagnetic spacer (2), $H_2^{(2)} = 0$ and $K_2^{(2)} = 0$.

To first order in magnetization, the polar magnetooptic Kerr effect (MOKE) at normal light incidence for a linearly polarized incident radiation with zero azimuth is defined as a complex ellipsometric angle

$$\begin{aligned} \chi_r^{(03)} &= -\frac{r_{xy}^{(03)}}{r_{xx}^{(03)}} \\ &= -j(M_0^{(2)} M_1^{(2)})^{-1} \left\{ \frac{\Delta N^{(1)}}{2N^{(1)}} (1 - r^{(01)2}) \left[4j\beta^{(1)} H_1^{(2)} e^{-2j\beta^{(1)}} \right.\right. \\ &\quad \left.\left. + \left(1 - e^{-2j\beta^{(1)}}\right) K_1^{(2)} \right] + \frac{\Delta N^{(3)}}{2N^{(3)}} (1 - r^{(23)2}) K_3^{(2)} \right\}. \end{aligned} \tag{4.129}$$

Thus, in the approximation limited to the terms linear in $\varepsilon_{xy}^{(n)}$ or $\Delta N^{(n)}$, the MOKE response can be considered as a sum of the contributions arising separately from the magnetic film and the magnetic substrate and proportional to $\Delta N^{(1)}$ and $\Delta N^{(3)}$, respectively [6]. In the reflection configuration, a selectivity is possible as the contribution from the upper ultrathin magnetic film, which might be only a few atomic monolayers thick and giving only a low response, is favored whereas that from the thick magnetic substrate is reduced and phase shifted. In the ultrathin film approximation applied to the magnetic film and specified by Eq. (4.85), ($e^{-2j\beta^{(1)}} \approx 1 - 2j\beta^{(1)}$), Eq. (4.129) simplifies to

$$\begin{aligned} \chi_r^{(03)} &\approx (M_0^{(2,1)} M_1^{(2,1)})^{-1} \left[\frac{\Delta N^{(1)}}{N^{(1)}} \beta^{(1)} (1 - r^{(01)2}) (2H_1^{(2)} + K_1^{(2,1)}) \right] \\ &\quad - j \frac{\Delta N^{(3)}}{2N^{(3)}} (1 - r^{(23)2}) K_3^{(2)} (M_0^{(2)} M_1^{(2)})^{-1}. \end{aligned} \tag{4.130}$$

This can be written in a more explicit form as

$$\begin{aligned} \chi_r^{(03)} &\approx (1 - r^{(01)2}) \\ &\quad \times \left[\Delta N^{(1)} \frac{\omega}{c} d^{(1)} (1 + r^{(12)})^2 \left(1 + r^{(23)} e^{-2j\beta^{(2)}}\right)^2 \right] \\ &\quad \times \left[(1 + r^{(01)} r^{(12)}) + \left(r^{(01)} + r^{(12)}\right) r^{(23)} e^{-2j\beta^{(2)}} \right]^{-1} \\ &\quad \times \left[(r^{(01)} + r^{(12)}) + (1 + r^{(01)} r^{(12)}) r^{(23)} e^{-2j\beta^{(2)}} \right]^{-1} \\ &\quad - j \frac{\Delta N^{(3)}}{2N^{(3)}} (1 - r^{(01)2}) (1 - r^{(12)2}) (1 - r^{(23)2}) e^{-2j\beta^{(2)}} \end{aligned}$$

$$\times\left[1+r^{(12)}r^{(23)}e^{-2j\beta^{(2)}}+r^{(01)}e^{-2j\beta^{(1)}}\left(r^{(12)}+r^{(23)}e^{-2j\beta^{(2)}}\right)\right]^{-1}$$
$$\times\left[r^{(01)}\left(1+r^{(12)}r^{(23)}e^{-2j\beta^{(2)}}\right)+e^{-2j\beta^{(1)}}\left(r^{(12)}+r^{(23)}e^{-2j\beta^{(2)}}\right)\right]^{-1}. \tag{4.131}$$

We note that

$$\frac{r^{(01)}+r^{(12)}}{1+r^{(01)}r^{(12)}}=r^{(02)} \tag{4.132}$$

is the ambient–nonmagnetic spacer Fresnel reflection coefficient and obtain

$$1-r^{(02)2}=\frac{\left(1-r^{(01)2}\right)\left(1-r^{(12)2}\right)}{\left(1+r^{(01)}r^{(12)}\right)^2}, \tag{4.133}$$

$$\frac{N^{(1)}}{N^{(2)}}\left(1-r^{(02)2}\right)=\frac{\left(1-r^{(01)2}\right)\left(1+r^{(12)}\right)^2}{\left(1+r^{(01)}r^{(12)}\right)^2}. \tag{4.134}$$

The observed polar MO Kerr effect can be understood as a sum of two components from magnetic regions ($\underline{1}$) and ($\underline{3}$)

$$\chi_r^{(03)}=\chi_r^{(0\underline{1}23)}+\chi_r^{(012\underline{3})}. \tag{4.135}$$

The first component is due to the ultrathin magnetic film ($\underline{1}$)

$$\chi_r^{(0\underline{1}23)}\approx\Delta N^{(1)}\frac{\omega}{c}d^{(1)}\frac{N_1}{N_2}\frac{\left(1-r^{(02)2}\right)\left(1+r^{(23)}e^{-2j\beta^{(2)}}\right)^2}{\left(1+r^{(02)}r^{(23)}e^{-2j\beta^{(2)}}\right)\left(r^{(02)}+r^{(23)}e^{-2j\beta^{(2)}}\right)}. \tag{4.136}$$

The second one originates from the magnetic substrate ($\underline{3}$)

$$\chi_r^{(012\underline{3})}\approx-j\frac{\Delta N^{(3)}}{2N^{(3)}}\frac{\left(1-r^{(02)2}\right)\left(1-r^{(23)2}\right)e^{-2j\beta^{(2)}}}{\left(1+r^{(02)}r^{(23)}e^{-2j\beta^{(2)}}\right)\left(r^{(02)}+r^{(23)}e^{-2j\beta^{(2)}}\right)}. \tag{4.137}$$

In this expression, we have effectively removed the optical effect of the ultrathin magnetic film (1) by setting $\beta^{(1)}=0$. The $\chi_r^{(012\underline{3})}$ component models the MOKE enhancement by dielectric films [9]. The phase shift between the components $\chi_r^{(0\underline{1}23)}$ and $\chi_r^{(012\underline{3})}$ is determined by the complex factors $\Delta N^{(1)}(\omega/c)d^{(1)}(N^{(1)}/N^{(2)})(1+r^{(23)}e^{-2j\beta^{(2)}})^2$ and $-j(\Delta N^{(3)}/2N^{(3)})\,(1-r^{(23)2})$ $e^{-2j\beta^{(2)}}$. The phase shift difference can be controlled by the parameters of the nonmagnetic spacer $N^{(2)}$ and $d^{(2)}$. When the two magnetic media are identical $N^{(1)} = N^{(3)}$, $\Delta N^{(1)} = \Delta N^{(3)}$, and $r^{(23)} = -r^{(12)}$ and we get

$$\chi_r^{(03)} \approx \Delta N^{(1)}\left(1 - r^{(02)2}\right)\frac{\left[\frac{\omega}{c} d^{(1)} \frac{N^{(1)}}{N^{(\mathcal{N}+1)}} \left(1 - r^{(12)} e^{-2j\beta^{(2)}}\right)^2 - j\frac{e^{-2j\beta^{(2)}}}{2N^{(1)}}\left(1 - r^{(12)2}\right)\right]}{\left(1 - r^{(02)} r^{(12)} e^{-2j\beta^{(2)}}\right)\left(r^{(02)} - r^{(12)} e^{-2j\beta^{(2)}}\right)}. \tag{4.138}$$

When the spacer is thick and absorbing, then $e^{-2j\beta^{(2)}} \approx 0$. The contribution of the magnetic substrate is effectively removed from the response, and we get from Eq. (4.136)

$$\chi_r^{(0123)} \approx \Delta N^{(1)} \frac{\omega}{c} d^{(1)} \frac{N^{(1)}}{N^{(2)}} \left(\frac{1 - r^{(02)2}}{r^{(02)}}\right), \tag{4.139}$$

or

$$\chi_r^{(0123)} \approx \Delta N^{(1)} \frac{\omega}{c} d^{(1)} \frac{4N^{(0)} N^{(1)}}{N^{(0)2} - N^{(2)2}} = \frac{2j\left(\varepsilon_{xx}^{(0)}\right)^{1/2} \varepsilon_{xy}^{(1)}}{\varepsilon_{xx}^{(0)} - \varepsilon_{xx}^{(2)}} \frac{\omega}{c} d^{(1)}. \tag{4.140}$$

In the ultrathin film limit, the MO response of the magnetic film becomes simply proportional to its thickness [6]. If the film and spacer are removed, we arrive at the classical result (4.91) for normal incidence polar MO Kerr effect

$$\chi_r \approx \frac{-j\frac{\Delta N^{(3)}}{2N^{(3)}}\left(1 - r^{(03)2}\right)}{r^{(03)}} = -2j\frac{\Delta N^{(3)} N^{(0)}}{N^{(0)2} - N^{(3)2}}$$
$$= \frac{\varepsilon_{xy}^{(3)}}{\left(\varepsilon_{xx}^{(3)}\right)^{1/2}\left(\varepsilon_{xx}^{(0)} - \varepsilon_{xx}^{(3)}\right)}, \tag{4.141}$$

where

$$r^{(03)} = \frac{N^{(0)} - N^{(3)}}{N^{(0)} + N^{(3)}}. \tag{4.142}$$

The same result follows from Eq. (4.129) by setting a zero spacer thickness ($\beta^{(2)} = 0$) and by assuming that the two magnetic media are identical, which implies $r^{(12)} = r^{(23)} = 0$.

The simulations of the response at normal incidence and polar magnetization in magnetic multilayers with ultrathin films of ferromagnetic

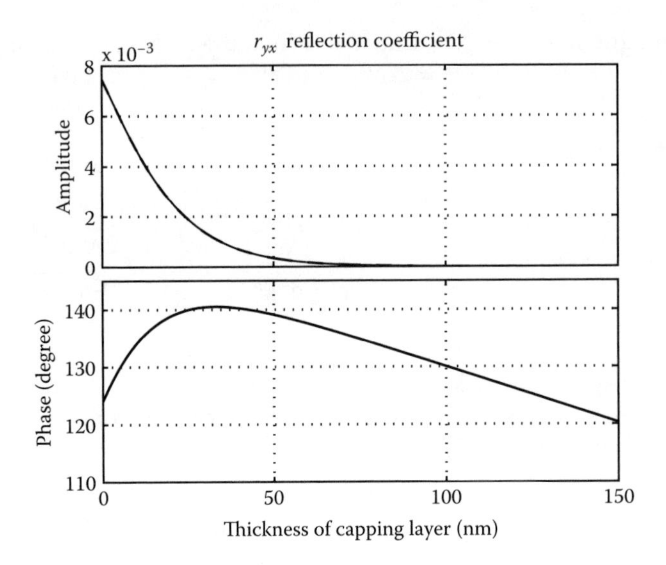

FIGURE 4.2
Off-diagonal reflection coefficient r_{yx} at polar magnetization and normal incidence in Au(d)/Fe(50 nm)/Au(buffer) sandwich as a function of the Au capping layer thickness, d. The simulation pertains to the wavelength of 632.8 nm and employs the diagonal and off-diagonal permittivity tensor elements for Fe given by $\varepsilon_{xx}^{(\mathrm{Fe})} = \varepsilon_0^{(\mathrm{Fe})} = -0.8845 - \mathrm{j}\,17.938$ and $\varepsilon_{xy}^{(\mathrm{Fe})} = \mathrm{j}\,\varepsilon_1^{(\mathrm{Fe})} = -0.6676 - \mathrm{j}\,0.08988$, respectively [10]. The complex index of refraction for Au is given by $N^{(\mathrm{Au})} = 0.174 - \mathrm{j}\,3.65$ [11].

metals have been the subject of papers [12–14].To enable the *ex situ* studies, the ferromagnetic films must be protected against contamination from ambient. One of the best protections is achieved with noble metal capping layers. These are however absorbing and reduce the amplitude of the MO response. Figure 4.2 demonstrates the effect of the thickness of Au capping layer on the off-diagonal reflection coefficient in an Au–Fe–Au sandwich. The same situation for the diagonal reflection coefficient is shown in Figure 4.3. Figure 4.4 shows the polar off-diagonal element, r_{yx}, of the Jones reflection matrix in a system consisting of a Fe layer deposited on a thick Au buffer and covered by an Au layer 4 nm thick. The thickness of the Fe layer varies from 0 to 150 nm. At the beginning, the amplitude, $|r_{yx}|$, increases linearly from zero with the Fe layer thickness. Then comes an overshoot region, and finally the amplitude reaches the saturation value at the Fe thickness exceeding the penetration depth. Figure 4.5 shows the detail of Figure 4.4 in the low Fe layer thickness region. The linear dependence of r_{yx} on the Fe layer thickness, corresponding to the ultrathin film approximation, reasonably follows the rigorously simulated plot. It is instructive to follow the behavior of the diagonal reflection coefficient r_{xx},

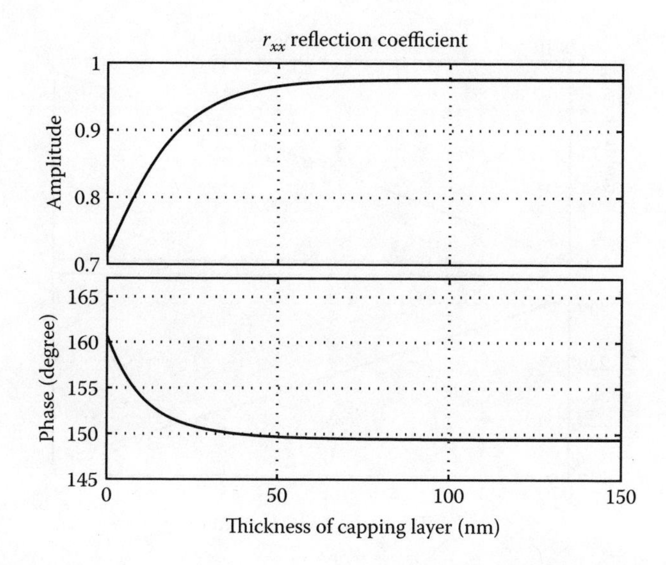

FIGURE 4.3
Diagonal reflection coefficient r_{xx} at polar magnetization and normal incidence in Au(d)/Fe(50 nm)/Au(buffer) sandwich as a function of the Au capping layer thickness, d, computed with the data of Figure 4.2.

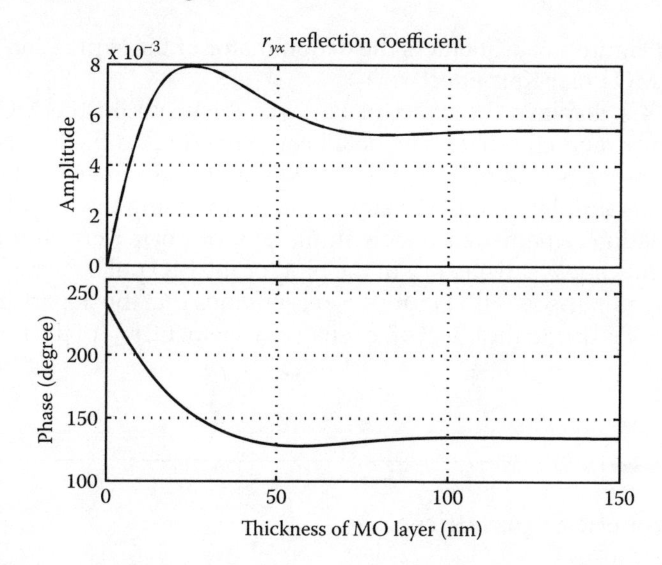

FIGURE 4.4
The effect of the ferromagnetic layer thickness, d, on the normal incidence polar off-diagonal reflection coefficients r_{yx} in a structure consisting of a Fe layer sandwiched between optically thick Au buffer and the Au capping layer 4 nm thick. The protective Au capping layer separates the Fe(d) layer from the ambient medium of incidence (air). Computed with the data of Figure 4.2.

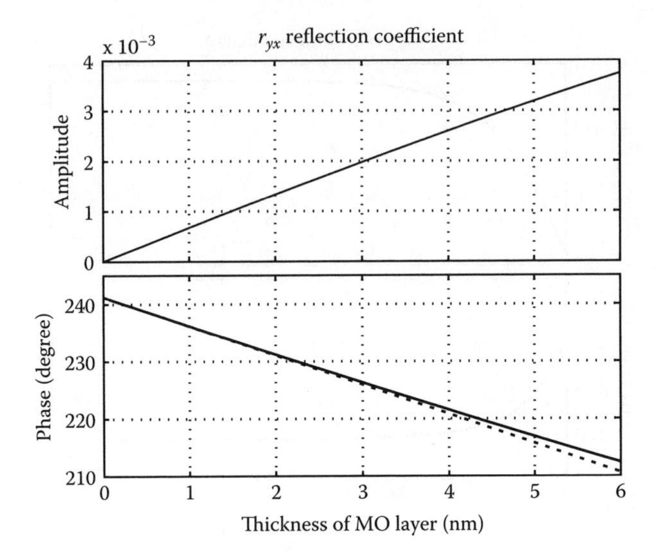

FIGURE 4.5
Detail of Figure 4.4. The off-diagonal element, r_{yx}, of the Jones reflection matrix for small thickness of the MO layer. The dotted line represents the ultrathin film approximation.

shown in Figure 4.6, as it enters the denominator of the expression for the complex MO polar Kerr effect.

Figure 4.7 displays the effect of Fe layer thickness on the MO polar Kerr rotation and ellipticity. The detail corresponding to the low Fe layer thickness range is shown in Figure 4.8. Although the agreement between the rigorous simulation and the straight line expressing the ultrathin film approximation remains reasonable in the plot of polar Kerr ellipticity, it becomes much less satisfactory in the plot for the MO polar Kerr rotation. From the comparison with Figure 4.5, it is obvious that the departure from the linearity is due to the effect of the diagonal element, r_{xx}, of the reflection matrix.

4.4 Magnetic Superlattices

Magnetic superlattices represent magnetic multilayers with periodic composition profile. Their optical response can, therefore, be analyzed with the 4×4 matrix formalism. However, in the absence of the mode coupling, *i.e.*, when the **M** matrix of the structure in Eq. (4.20) consists of two diagonal 2×2 blocks, the analysis can be simplified using an approach originally developed by Abelès [15] for isotropic periodic multilayers at

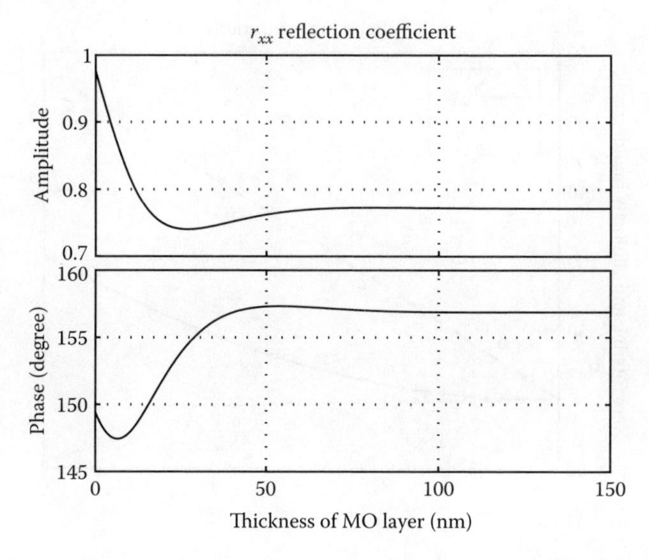

FIGURE 4.6
Effect of the ferromagnetic Fe layer thickness, d, on the normal incidence diagonal reflection coefficient, r_{xx}. The Fe layer is sandwiched between optically thick Au buffer and the Au protective capping layer 4 nm thick, which separates the Fe layer from the ambient medium of incidence (air). Computed with the data of Figure 4.2.

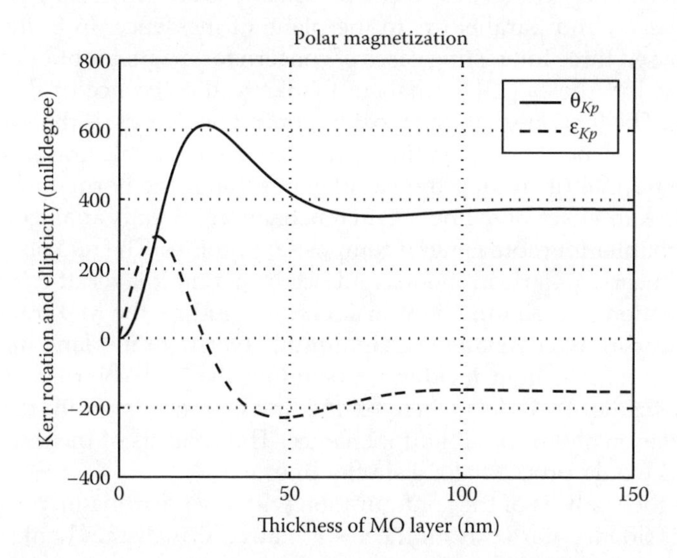

FIGURE 4.7
Effect of Fe layer thickness on the normal incidence MO polar Kerr rotation θ_{Kp} and ellipticity ϵ_{Kp} in the system consisting of a Fe layer sandwiched between optically thick Au buffer and the Au capping layer 4 nm thick, which separates the Fe layer from the ambient medium of incidence (air). Computed with the data of Figure 4.2.

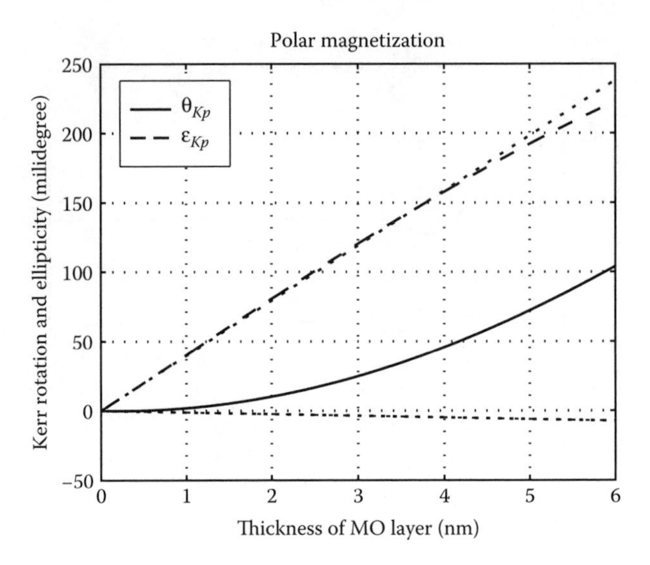

FIGURE 4.8
Detail of Figure 4.7 near the zero thickness of the Fe layer. The dotted straight lines indicate the ultrathin film approximation.

oblique incidence. There, the proper polarizations are linear (LP), *i.e.*, perpendicular (s) and parallel (p) to the plane of incidence. In Section 1.7, we discussed three interesting cases of concern for magnetooptics that can be treated by Abelès [15] formalism thanks to the absence of the mode coupling. The first one, and the most important one for us, is that of circularly polarized modes propagating parallel to the magnetization and to the interface normal (including the Faraday effect and the normal incidence polar MO Kerr effect, both linear in magnetization), already analyzed in the present chapter for more general composition profiles. The remaining two involve linearly polarized modes and occur in a multilayer at transverse magnetization and oblique light incidence (including the MO transverse effect linear in magnetization) or in a multilayer with in-plane magnetization and normal light incidence (including the MO Voigt effect, even in magnetization). We focus here on the first case dealing with the polar magnetization and normal light incidence. The analysis of the remaining two cases would proceed along similar lines.

So far, the analysis of the configuration with the polar magnetization at normal incidence started from the 4×4 matrix formalism. Thanks to the restriction to terms linear in the off-diagonal permittivity tensor, simplified analytical expressions could be derived. These represented the global MO response of the structure expressed as weighted sums of contributions from individual layers. The extension of this approach to periodic profiles leads to analytical formulas expressed in terms of the Chebyshev polynomials and their derivatives. The analysis can indicate the depth to which deeply buried layers of the periodic system contribute to the MO response in reflection. In order to simplify the problem, we restrict ourselves to periodic multilayers built of symmetric blocks [8,16,17]. The analysis makes use of Sections 4.2 and 4.3.

4.4.1 Characteristic Matrix

The matrix **M** was most often expressed as a product of transfer matrices. Here, we employ representation using characteristic matrices introduced by Abelès [15]

$$\mathbf{M} = [\mathbf{D}^{(0)}]^{-1} \prod_{n=1}^{\mathcal{N}} \mathbf{S}^{(n)} \mathbf{D}^{(\mathcal{N}+1)}, \tag{4.143}$$

where the block diagonal characteristic matrix, $\mathbf{S}^{(n)}$, for the case of normal incidence, is given by

$$\begin{aligned}\mathbf{S}^{(n)} &= \mathbf{D}^{(n)}\mathbf{P}^{(n)}[\mathbf{D}^{(n)}]^{-1}\\ &= \begin{pmatrix} \cos\beta_+^{(n)} & \dfrac{j}{N_+^{(n)}}\sin\beta_+^{(n)} & 0 & 0 \\ jN_+^{(n)}\sin\beta_+^{(n)} & \cos\beta_+^{(n)} & 0 & 0 \\ 0 & 0 & \cos\beta_-^{(n)} & \dfrac{j}{N_-^{(n)}}\sin\beta_-^{(n)} \\ 0 & 0 & jN_-^{(n)}\sin\beta_-^{(n)} & \cos\beta_-^{(n)} \end{pmatrix}\end{aligned} \tag{4.144}$$

for $n = 1, 2, \ldots, \mathcal{N}$; here

$$\beta_\pm^{(n)} = \frac{\omega}{c} N_\pm^{(n)} d^{(n)}. \tag{4.145}$$

The characteristic matrix consists of two diagonal 2×2 blocks approximately linearly dependent on the magnetization. The changes induced by the magnetization in the blocks are small and of the opposite sign. Note that $\mathbf{S}^{(n)}$ becomes a unit matrix when $d^{(n)} = 0$ and the n-th layer is then effectively removed from the structure. The inverse matrix $(\mathbf{S}^{(n)})^{-1}$ is obtained from $\mathbf{S}^{(n)}$ by $d^{(n)} \to -d^{(n)}$ and $\det \mathbf{S}^{(n)} = 1$. For a nonmagnetic layer, $N_+^{(n)} = N_-^{(n)}$, the characteristic matrix consists of identical 2×2 blocks. If the layer is also nonabsorbing, $N^{(n)}$ becomes real and the blocks correspond to the characteristic matrix introduced by Abelès [15] restricted to the case of normal incidence. In ultrathin films, where the thickness is much smaller than radiation wavelength, the characteristic matrix can be approximated by

$$\mathbf{S}^{(n)} \approx \begin{pmatrix} 1 & \frac{\mathrm{j}}{N_+^{(n)}}\beta_+^{(n)} & 0 & 0 \\ \mathrm{j}N_+^{(n)}\beta_+^{(n)} & 1 & 0 & 0 \\ 0 & 0 & 1 & \frac{\mathrm{j}}{N_-^{(n)}}\beta_-^{(n)} \\ 0 & 0 & \mathrm{j}N_-^{(n)}\beta_-^{(n)} & 1 \end{pmatrix}, \tag{4.146}$$

and $|\det \mathbf{S}^{(n)}| \approx 1$ to the first order in $\beta_\pm^{(n)}$.

4.4.2 M Matrix in Periodic Structures

The representation of $\mathbf{M}$ in terms of a product of $\mathbf{S}^{(n)}$ in Eq. (4.143) is suitable for the treatment of multilayers containing a periodically stratified region. To this purpose, we write

$$\mathbf{M} = \mathbf{C}\mathbf{L}^q\mathbf{W}, \tag{4.147}$$

where $\mathbf{C}$ represents the incident medium (the semi-infinite medium on the left) and cover layer(s) of the total number $\mathcal{C}$

$$\mathbf{C} = \left[\mathbf{D}^{(0)}\right]^{-1}\prod_{n=1}^{\mathcal{C}}\mathbf{S}^{(n)}, \tag{4.148}$$

$\mathbf{L}^q$ is the periodic region with the unit $\mathbf{L}$ repeated q times, and $\mathbf{W}$ the buffer layer(s), substrate, and the exit medium (the semi-infinite medium on the right). The $\mathbf{L}$ matrix itself can be considered as a product of $\mathcal{L}$ matrices representing homogenous layers

$$\mathbf{L} = \prod_{n=1}^{\mathcal{L}}\mathbf{S}^{(n)}, \tag{4.149}$$

and can be expressed as

$$\mathbf{L} = \begin{pmatrix} m_{11}^+ & m_{12}^+ & 0 & 0 \\ m_{21}^+ & m_{22}^+ & 0 & 0 \\ 0 & 0 & m_{11}^- & m_{12}^- \\ 0 & 0 & m_{21}^- & m_{22}^- \end{pmatrix}. \tag{4.150}$$

We shall restrict ourselves to symmetric sandwiches with $m_{11}^{\pm} = m_{22}^{\pm}$ [8]. The q-th power of $\mathbf{L}$ is then given by [8,15]

$$\mathbf{L}^q = \begin{pmatrix} m_{11}^{+}p_q^{+} - p_{q-1}^{+} & m_{12}^{+}p_q^{+} & 0 & 0 \\ m_{21}^{+}p_q^{+} & m_{11}^{+}p_q^{+} - p_{q-1}^{+} & 0 & 0 \\ 0 & 0 & m_{11}^{-}p_q^{-} - p_{q-1}^{-} & m_{12}^{-}p_q^{-} \\ 0 & 0 & m_{21}^{-}p_q^{-} & m_{11}^{-}p_q^{-} - p_{q-1}^{-} \end{pmatrix}. \tag{4.151}$$

Here, $p_q^{\pm}(m_{11}^{\pm})$ denote the *Chebyshev polynomials of second kind* with the argument $m_{11}^{\pm}$ (see Appendix E). The $\mathbf{W}$ matrix, representing $\mathcal{W}$ layers and the exit medium, can be written as

$$\mathbf{W} = \prod_{n=1}^{\mathcal{W}} \mathbf{S}^{(n)}\mathbf{D}^{(\mathcal{N}+1)}. \tag{4.152}$$

The total number of layers is $\mathcal{N} = \mathcal{C} + q\mathcal{L} + \mathcal{W}$.

4.4.3 Approximate Treatment

In order to obtain the MO characteristics of a magnetic multilayer with periodic regions, it is sufficient to apply Eq. (4.98) and Eq. (4.103) to the 2×2 isotropic equivalent representation of a given periodic multilayer with $m_{11}^{\pm}$, $m_{12}^{\pm}$, and $m_{21}^{\pm}$ replaced by m_{11}, m_{12}, and m_{21}, respectively. We start from Eq. (4.147) assuming $N_{+}^{(n)} = N_{-}^{(n)} = N^{(n)}$ and consider a multilayer represented by a 2×2 matrix that contains a region represented by $\mathbf{L}$ repeated q times

$$\mathbf{M} = \mathbf{C}\mathbf{L}^q\mathbf{W} = \begin{pmatrix} C_1 & C_2 \\ C_3 & C_4 \end{pmatrix} \begin{pmatrix} L_1^{(q)} & L_2^{(q)} \\ L_3^{(q)} & L_4^{(q)} \end{pmatrix} \begin{pmatrix} W_1 & W_2 \\ W_3 & W_4 \end{pmatrix}. \tag{4.153}$$

With restriction to symmetric $\mathbf{L}$, the relevant matrix elements are

$$M_{11} = (C_1W_1 + C_2W_3)L_1^{(q)} + C_1W_3L_2^{(q)} + C_2W_1L_3^{(q)}, \tag{4.154a}$$

$$M_{21} = (C_3W_1 + C_4W_3)L_1^{(q)} + C_3W_3L_2^{(q)} + C_4W_1L_3^{(q)}, \tag{4.154b}$$

where, according to Eq. (4.151),

$$L_1^{(q)} = m_{11}p_q - p_{q-1}, \tag{4.155a}$$

$$L_2^{(q)} = m_{12}p_q, \tag{4.155b}$$

$$L_3^{(q)} = m_{21}p_q. \tag{4.155c}$$

We now assume that the periodic structure represented by $\mathbf{L}^q$ is the only source of MO effects in the structure represented by Eq. (4.147). Let the complex index of refraction $N_n^{(L)}$ characterize the n-th magnetic layer in the symmetric building block $\mathbf{L}$ (repeated q times) of the periodic structure. Then, the elements of the Jones transmission and reflection matrices for a system represented by the matrix product in Eq. (4.147) are given by

$$t_{xx}^{(C,q\underline{L},W)} = \frac{1}{M_{11}} = \left[(C_1W_1 + C_2W_3)L_1^{(q)} + C_1W_3L_2^{(q)} + C_2W_1L_3^{(q)}\right]^{-1}, \quad (4.156a)$$

$$t_{xy}^{(C,q\underline{L},W)} = j\sum_{n=1}^{\mathcal{L}} \frac{\partial M_{11}}{\partial N_i^{(L)}} \Delta N_n^{(L)}$$

$$= j[(C_1W_1 + C_2W_3)L_1^{(q)} + C_1W_3L_2^{(q)} + C_2W_1L_3^{(q)}]^{-2}$$

$$\times \sum_{n=1}^{\mathcal{L}} \left[(C_1W_1 + C_2W_3)\frac{\partial L_1^{(q)}}{\partial N_n^{(L)}} + C_1W_3\frac{\partial L_2^{(q)}}{\partial N_n^{(L)}} + C_2W_1\frac{\partial L_3^{(q)}}{\partial N_n^{(L)}}\right] \Delta N_n^{(L)},$$

$$r_{xx}^{(C,q\underline{L},W)} = \frac{M_{21}}{M_{11}} = \frac{(C_3W_1 + C_4W_3)L_1^{(q)} + C_3W_3L_2^{(q)} + C_4W_1L_3^{(q)}}{(C_1W_1 + C_2W_3)L_1^{(q)} + C_1W_3L_2^{(q)} + C_2W_1L_3^{(q)}}, \quad (4.156b)$$

$$r_{xy}^{(C,q\underline{L},W)} = \frac{j}{M_{11}^2}\left(M_{21}\sum_{n=1}^{\mathcal{L}} \frac{\partial M_{11}}{\partial N_n^{(L)}} - M_{11}\sum_{n=1}^{\mathcal{L}} \frac{\partial M_{21}}{\partial NV_n}\right) \Delta N_n^{(L)}$$

$$= -j[(C_3W_1 + C_4W_3)L_1^{(q)} + C_3W_3L_2^{(q)} + C_4W_1L_3^{(q)}]^{-1}$$

$$\times \sum_{n=1}^{\mathcal{L}} \left[(C_3W_1 + C_4W_3)\frac{\partial L_1^{(q)}}{\partial N_n^{(L)}} + C_3W_3\frac{\partial L_2^{(q)}}{\partial N_n^{(L)}} + C_4W_1\frac{\partial L_3^{(q)}}{\partial N_n^{(L)}}\right] \Delta N_n^{(L)}$$

$$- j[(C_1W_1 + C_2W_3)L_1^{(q)} + C_1W_3L_2^{(q)} + C_2W_1L_3^{(q)}]^{-1}$$

$$\times \sum_{n=1}^{\mathcal{L}} \left[(C_1W_1 + C_2W_3)\frac{\partial L_1^{(q)}}{\partial N_n^{(L)}} + C_1W_3\frac{\partial L_2^{(q)}}{\partial N_n^{(L)}} + C_2W_1\frac{\partial L_3^{(q)}}{\partial N_n^{(L)}}\right] \Delta N_n^{(L)}. \quad (4.156c)$$

The underlined $\underline{L}$ pertains to magnetic blocks. Using Eq. (4.155), this can be alternatively expressed as

$$t_{xx}^{(C,q\underline{L},W)} = \frac{1}{M_{11}}$$

$$= \left[(C_1W_1 + C_2W_3)(m_{11}p_q - p_{q-1}) + C_1W_3m_{12}p_q + C_2W_1m_{21}p_q\right]^{-1}, \quad (4.157a)$$

$$
\begin{aligned}
t_{xy}^{(C,q\underline{L},W)} &= \mathrm{j}\frac{\Delta\left(\mathrm{M}_{11}\right)}{\mathrm{M}_{11}^{2}} \\
&= \mathrm{j}\left[\left(C_1 W_1 + C_2 W_3\right)\left(m_{11} p_q - p_{q-1}\right) + C_1 W_3 m_{12} p_q \right. \\
&\quad \left. + C_2 W_1 m_{21} p_q\right]^{-2}\left[\left(C_1 W_1 + C_2 W_3\right)\left(p_q + m_{11} p_q' - p_{q-1}'\right)\right. \\
&\quad + C_1 W_3\left(p_q' m_{12} \Delta m_{11} + p_q \Delta m_{12}\right) \\
&\quad \left. + C_2 W_1\left(p_q' m_{21} \Delta m_{11} + p_q \Delta m_{21}\right)\right], \qquad (4.157b)
\end{aligned}
$$

$$
\begin{aligned}
r_{xx}^{(C,q\underline{L},W)} &= \frac{\mathrm{M}_{21}}{\mathrm{M}_{11}} \\
&= \left[\left(C_1 W_1 + C_2 W_3\right)\left(m_{11} p_q - p_{q-1}\right) + C_1 W_3 m_{12} p_q + C_2 W_1 m_{21} p_q\right]^{-1} \\
&\quad \times\left[\left(C_3 W_1 + C_4 W_3\right)\left(m_{11} p_q - p_{q-1}\right) + C_3 W_3 m_{12} p_q + C_4 W_1 m_{21} p_q\right], \qquad (4.157c)
\end{aligned}
$$

$$
\begin{aligned}
r_{xy}^{(C,q\underline{L},W)} &= -\mathrm{j}\frac{\mathrm{M}_{11}\Delta\left(\mathrm{M}_{21}\right) - \mathrm{M}_{21}\Delta\left(\mathrm{M}_{11}\right)}{\mathrm{M}_{11}^{2}} \\
&= -\mathrm{j}\left(C_1 C_4 - C_2 C_3\right)\left[\left(C_1 W_1 + C_2 W_3\right)\left(m_{11} p_q - p_{q-1}\right)\right. \\
&\quad \left. + C_1 W_3 m_{12} p_q + C_2 W_1 m_{21} p_q\right]^{-2} \\
&\quad \times\left[\left(W_3^2 m_{12} - W_1^2 m_{21}\right)\left(p_q^2 + p_{q-1} p_q' - p_q p_{q-1}'\right)\Delta m_{11}\right. \\
&\quad - \left(W_3^2 \Delta m_{12} - W_1^2 \Delta m_{21}\right)\left(m_{11} p_q - p_{q-1}\right) p_q \\
&\quad \left. + W_1 W_3 p_q^2\left(m_{12}\Delta m_{21} - m_{21}\Delta m_{12}\right)\right]. \qquad (4.157d)
\end{aligned}
$$

The prime indicates the differentiation of Chebyshev polynomials with respect to their argument m_{11}. The MO response is expressed as a sum of contributions from individual layers.

Let us consider a periodic structure of alternating A and B layers of the thickness $d^{(A)}$ and $d^{(B)}$, respectively. For simplicity and without any loss in generality, we may focus on a multilayer built of symmetric trilayer units $(A/B/A)$

$$
\mathbf{L} = \mathbf{S}^{(A)}\mathbf{S}^{(B)}\mathbf{S}^{(A)}, \qquad (4.158)
$$

where $\mathbf{S}^{(A)}$ and $\mathbf{S}^{(B)}$ are given by Eq. (4.144) for $n = A$ and B, respectively. In the symmetric $A/B/A$ block, the sandwiching layers A have the thickness $d^{(A)}/2$. Then, $\beta_{\pm}^{(n)} = \beta_{\pm}^{(A)}/2$ for $\mathbf{S}^{(A)}$ and $\beta_{\pm}^{(n)} = \beta_{\pm}^{(B)}$ for $\mathbf{S}^{(B)}$. Here, $\beta_{\pm}^{(n)}$ is given by Eq. (4.145). The matrix elements of the symmetric unit $\mathbf{L}$ in Eq. (4.150) become

$$
m_{11}^{\pm} = \cos\beta_{\pm}^{(A)}\cos\beta_{\pm}^{(B)} - \frac{1}{2}\left(\frac{N_{\pm}^{(A)}}{N_{\pm}^{(B)}} + \frac{N_{\pm}^{(B)}}{N_{\pm}^{(A)}}\right)\sin\beta_{\pm}^{(A)}\sin\beta_{\pm}^{(B)}, \qquad (4.159a)
$$

$$m_{12}^{\pm} = \frac{\mathrm{j}}{N_{\pm}^{(A)}}\left[\sin\beta_{\pm}^{(A)}\cos\beta_{\pm}^{(B)} + \frac{1}{2}\left(\frac{N_{\pm}^{(A)}}{N_{\pm}^{(B)}} + \frac{N_{\pm}^{(B)}}{N_{\pm}^{(A)}}\right)\cos\beta_{\pm}^{(A)}\sin\beta_{\pm}^{(B)} + \frac{1}{2}\left(\frac{N_{\pm}^{(A)}}{N_{\pm}^{(B)}} - \frac{N_{\pm}^{(B)}}{N_{\pm}^{(A)}}\right)\sin\beta_{\pm}^{(B)}\right], \tag{4.159b}$$

$$m_{21}^{\pm} = \mathrm{j}N_{\pm}^{(A)}\left[\sin\beta_{\pm}^{(A)}\cos\beta_{\pm}^{(B)} + \frac{1}{2}\left(\frac{N_{\pm}^{(A)}}{N_{\pm}^{(B)}} + \frac{N_{\pm}^{(B)}}{N_{\pm}^{(A)}}\right)\cos\beta_{\pm}^{(A)}\sin\beta_{\pm}^{(B)} - \frac{1}{2}\left(\frac{N_{\pm}^{(A)}}{N_{\pm}^{(B)}} - \frac{N_{\pm}^{(B)}}{N_{\pm}^{(A)}}\right)\sin\beta_{\pm}^{(B)}\right]. \tag{4.159c}$$

For illustration, we consider the periodic structure with a symmetric unit represented by **L** repeated q times and sandwiched between semi-infinite isotropic media (0) and $(\mathcal{N}+1)$. The extension to more complicated sandwiching structures (including cover, buffer, and substrate layers) can be done by the insertion of the corresponding matrices to the product. As before, in order to find $\chi_t^{(0,q\underline{L},\mathcal{N}+1)}$ and $\chi_r^{(0,q\underline{L},\mathcal{N}+1)}$, we start from Eq. (4.147) for $N_+^{(n)} = N_-^{(n)} = N^{(n)}$ (corresponding to the absence of magnetic order) and consider 2×2 matrix product representing an isotropic multilayer at normal incidence

$$\mathbf{M} = \mathbf{C}\mathbf{L}^q\mathbf{W} = \frac{1}{2N^{(0)}}\begin{pmatrix} N^{(0)} & 1 \\ N^{(0)} & -1 \end{pmatrix}\begin{pmatrix} \mathrm{L}_{11}^{(q)} & \mathrm{L}_{12}^{(q)} \\ \mathrm{L}_{21}^{(q)} & \mathrm{L}_{11}^{(q)} \end{pmatrix}\begin{pmatrix} 1 & 1 \\ N^{(\mathcal{N}+1)} & -N^{(\mathcal{N}+1)} \end{pmatrix}$$
$$= \frac{1}{2N^{(0)}}\begin{pmatrix} N^{(0)} & 1 \\ N^{(0)} & -1 \end{pmatrix}\begin{pmatrix} m_{11} & m_{12} \\ m_{21} & m_{11} \end{pmatrix}^q\begin{pmatrix} 1 & 1 \\ N^{(\mathcal{N}+1)} & -N^{(\mathcal{N}+1)} \end{pmatrix}. \tag{4.160}$$

The relevant **M** matrix elements are

$$M_{11} = \frac{1}{2N^{(0)}}\left[(N^{(0)} + N^{(\mathcal{N}+1)})L_{11}^{(q)} + N^{(0)}N^{(\mathcal{N}+1)}L_{12}^{(q)} + L_{21}^{(q)}\right]$$
$$= \frac{1}{2N^{(0)}}\left\{p_q\left[N^{(0)}\left(m_{11} + N^{(\mathcal{N}+1)}m_{12}\right) + \left(N^{(\mathcal{N}+1)}m_{11} + m_{21}\right)\right] - p_{q-1}\left(N^{(0)} + N^{(\mathcal{N}+1)}\right)\right\}, \tag{4.161a}$$

$$M_{21} = \frac{1}{2N^{(0)}}\left[(N^{(0)} + N^{(\mathcal{N}+1)})L_{11}^{(q)} + N^{(0)}N^{(\mathcal{N}+1)}L_{12}^{(q)} + L_{21}^{(q)}\right]$$
$$= \frac{1}{2N^{(0)}}\left\{p_q\left[N^{(0)}\left(m_{11} + N^{(\mathcal{N}+1)}m_{12}\right) - \left(N^{(\mathcal{N}+1)}m_{11} + m_{21}\right)\right] - p_{q-1}\left(N^{(0)} - N^{(\mathcal{N}+1)}\right)\right\}. \tag{4.161b}$$

With Eq. (4.103a) and Eq. (4.103b), the MO Faraday and Kerr effects of this periodic structure are given by

$$\begin{aligned}\chi_t^{(0,q\underline{L},\mathcal{N}+1)} &= -\mathrm{j}\left\{p_q\left[N^{(0)}\left(m_{11}+N^{(\mathcal{N}+1)}m_{12}\right)+\left(N^{(\mathcal{N}+1)}m_{11}+m_{21}\right)\right]\right.\\ &\quad \left.-\,p_{q-1}\left(N^{(0)}+N^{(\mathcal{N}+1)}\right)\right\}^{-1}\left\{\left[\left(p_q-p'_{q-1}\right)\left(N^{(0)}+N^{(\mathcal{N}+1)}\right)\right.\right.\\ &\quad \left.+\,p'_qN^{(0)}\left(m_{11}+N^{(\mathcal{N}+1)}m_{12}\right)+p'_q\left(N^{(\mathcal{N}+1)}m_{11}+m_{21}\right)\right]\Delta m_{11}\\ &\quad \left.+\,p_q\left(N^{(0)}N^{(\mathcal{N}+1)}\Delta m_{12}+\Delta m_{21}\right)\right\}, \qquad (4.162a)\end{aligned}$$

$$\begin{aligned}\chi_r^{(0,q\underline{L},\mathcal{N}+1)} &= -2\mathrm{j}N^{(0)}\left\{p_q\left[N^{(0)}\left(m_{11}+N^{(\mathcal{N}+1)}m_{12}\right)+\left(N^{(\mathcal{N}+1)}m_{11}+m_{21}\right)\right]\right.\\ &\quad \left.-\,p_{q-1}\left(N^{(0)}+N^{(\mathcal{N}+1)}\right)\right\}^{-1}\\ &\quad \times\left\{p_q\left[N^{(0)}\left(m_{11}+N^{(\mathcal{N}+1)}m_{12}\right)-\left(N^{(\mathcal{N}+1)}m_{11}+m_{21}\right)\right]\right.\\ &\quad \left.-\,-p_{q-1}\left(N^{(0)}-N^{(\mathcal{N}+1)}\right)\right\}^{-1}\\ &\quad \times\left\{\Delta m_{11}(N^{(\mathcal{N}+1)2}m_{12}-m_{21})(p_q^2+p'_qp_{q-1}-p_qp'_{q-1})\right.\\ &\quad -\Delta m_{12}N^{(\mathcal{N}+1)}p_q\left[\left(N^{(\mathcal{N}+1)}m_{11}+m_{21}\right)p_q-N^{(\mathcal{N}+1)}p_{q-1}\right]\\ &\quad \left.+\,\Delta m_{21}p_q\left[p_q\left(m_{11}+N^{(\mathcal{N}+1)}m_{12}\right)-p_{q-1}\right]\right\}. \qquad (4.162b)\end{aligned}$$

We have to find Δm_{11}, Δm_{12}, and Δm_{21}. For an isotropic symmetric unit $A/B/A$, characterized by the refractive indices $N^{(A)}$ and $N^{(B)}$, the **L** matrix elements are given by Eq. (4.159), with $m_{11}^{\pm}$, $m_{12}^{\pm}$, and $m_{21}^{\pm}$ replaced by m_{11}, m_{12}, and m_{21}, respectively. The trigonometric functions may be replaced by exponentials, and Eq. (4.159) in the absence of magnetization becomes

$$m_{11} = U_{AB}\left[\left(1-r^{(AB)2}\mathrm{e}^{-2\mathrm{j}\beta^{(A)}}\right)+\mathrm{e}^{-2\mathrm{j}\beta^{(B)}}\left(\mathrm{e}^{-2\mathrm{j}\beta^{(A)}}-r^{(AB)2}\right)\right], \qquad (4.163a)$$

$$m_{12} = \frac{1}{N^{(A)}}U_{AB}\left[\left(1+r^{(AB)}\mathrm{e}^{-\mathrm{j}\beta^{(A)}}\right)^2-\mathrm{e}^{-2\mathrm{j}\beta^{(B)}}\left(r^{(AB)}+\mathrm{e}^{-\mathrm{j}\beta^{(A)}}\right)^2\right], \qquad (4.163b)$$

$$m_{21} = N^{(A)}U_{AB}\left[\left(1-r^{(AB)}\mathrm{e}^{-\mathrm{j}\beta^{(A)}}\right)^2-\mathrm{e}^{-2\mathrm{j}\beta^{(B)}}\left(r^{(AB)}-\mathrm{e}^{-\mathrm{j}\beta^{(A)}}\right)^2\right], \qquad (4.163c)$$

where

$$U_{AB} = \frac{1}{2}\frac{\exp[\mathrm{j}\beta^{(A)}]\exp[\mathrm{j}\beta^{(B)}]}{t^{(AB)}t^{(BA)}} \qquad (4.164)$$

(with $\beta_{\pm}^{(A)}$ and $\beta_{\pm}^{(B)}$ replaced by $\beta^{(A)}$ and $\beta^{(B)}$, respectively). Here, $r^{(AB)}$, $t^{(AB)}$, and $t^{(BA)}$ are the Fresnel interface coefficients defined in Eq. (4.102a) and Eq. (4.102b). We now assume that only the central layer B is magnetic and displays MO effects characterized by $\Delta N^{(B)} \neq 0$. Then, $\Delta m_{ij} = (dm_{ij}/dN^{(B)})\Delta N^{(B)}$. Using Eqs. (4.159), we obtain

$$\begin{aligned}\Delta m_{11} &= \mathrm{j}\frac{\Delta N^{(B)}}{N^{(B)}}U_{AB}\left\{\beta^{(B)}\left[1-r^{(AB)2}\mathrm{e}^{-2\mathrm{j}\beta^{(A)}}-\mathrm{e}^{-2\mathrm{j}\beta^{(B)}}\left(\mathrm{e}^{-2\mathrm{j}\beta^{(A)}}-r^{(AB)2}\right)\right]\right.\\ &\quad \left.+\,\mathrm{j}r^{(AB)}\left(1-\mathrm{e}^{-2\mathrm{j}\beta^{(A)}}\right)\left(1-\mathrm{e}^{-2\mathrm{j}\beta^{(B)}}\right)\right\}, \qquad (4.165a)\end{aligned}$$

$$\Delta m_{12} = \mathrm{j}\frac{\Delta N^{(B)}}{N^{(A)}N^{(B)}}U_{AB}\left\{\beta^{(B)}\left[\left(1+r^{(AB)}\mathrm{e}^{-\mathrm{j}\beta^{(A)}}\right)^2+\mathrm{e}^{-2\mathrm{j}\beta^{(B)}}\left(r^{(AB)}+\mathrm{e}^{-\mathrm{j}\beta^{(A)}}\right)^2\right]\right.$$
$$\left.+\mathrm{j}\left(1+r^{(AB)}\mathrm{e}^{-\mathrm{j}\beta^{(A)}}\right)\left(r^{(AB)}+\mathrm{e}^{-\mathrm{j}\beta^{(A)}}\right)\left(1-\mathrm{e}^{-2\mathrm{j}\beta^{(B)}}\right)\right\}, \qquad (4.165\mathrm{b})$$

$$\Delta m_{21} = \mathrm{j}N^{(A)}\frac{\Delta N^{(B)}}{N^{(B)}}U_{AB}\left\{\beta^{(B)}\left[\left(1-r^{(AB)}\mathrm{e}^{-\mathrm{j}\beta^{(A)}}\right)^2+\mathrm{e}^{-2\mathrm{j}\beta^{(B)}}\left(r^{(AB)}-\mathrm{e}^{-\mathrm{j}\beta^{(A)}}\right)^2\right]\right.$$
$$\left.+\mathrm{j}\left(1-r^{(AB)}\mathrm{e}^{-\mathrm{j}\beta^{(A)}}\right)\left(r^{(AB)}-\mathrm{e}^{-\mathrm{j}\beta^{(A)}}\right)\left(1-\mathrm{e}^{-2\mathrm{j}\beta^{(B)}}\right)\right\}. \qquad (4.165\mathrm{c})$$

In these expressions, we recognize the "propagation" and "interface" contributions proportional to $\beta^{(B)}$ and $(1-\mathrm{e}^{-2\mathrm{j}\beta^{(B)}})$, respectively. The remaining factors at the Chebyshev polynomials in Eq. (4.162) may be written down as

$$m_{11} + N^{(\mathcal{N}+1)}m_{12} = \frac{1}{N^{(A)}}U_{AB}\left(N^{(A)}+N^{(\mathcal{N}+1)}\right)$$
$$\times\left[1-r^{(AB)2}\mathrm{e}^{-2\mathrm{j}\beta^{(B)}}-t^{(A,\mathcal{N}+1)}\mathrm{e}^{-2\mathrm{j}\beta^{(B)}}\left(r^{(AB)2}-\mathrm{e}^{-2\mathrm{j}\beta^{(A)}}\right)\right.$$
$$\left.+t^{(\mathcal{N}+1,A)}r^{(AB)}\mathrm{e}^{-\mathrm{j}\beta^{(A)}}\left(1-\mathrm{e}^{-2\mathrm{j}\beta^{(B)}}\right)\right], \qquad (4.166\mathrm{a})$$

$$N^{(\mathcal{N}+1)}m_{11} + m_{21} = U_{AB}\left(N^{(A)}+N^{(\mathcal{N}+1)}\right)$$
$$\times\left[1-r^{(AB)2}\mathrm{e}^{-2\mathrm{j}\beta^{(B)}}+t^{(A,\mathcal{N}+1)}\mathrm{e}^{-2\mathrm{j}\beta^{(B)}}\left(r^{(AB)2}-\mathrm{e}^{-2\mathrm{j}\beta^{(A)}}\right)\right.$$
$$\left.-t^{(\mathcal{N}+1,A)}r^{(AB)}\mathrm{e}^{-\mathrm{j}\beta^{(A)}}\left(1-\mathrm{e}^{-2\mathrm{j}\beta^{(B)}}\right)\right], \qquad (4.166\mathrm{b})$$

$$N^{(\mathcal{N}+1)2}m_{12} - m_{21} = \frac{1}{N^{(A)}}U_{AB}\left(N^{(A)}+N^{(\mathcal{N}+1)}\right)^2$$
$$\times\left\{-t^{(A,\mathcal{N}+1)}\left[1+r^{(AB)2}\mathrm{e}^{-2\mathrm{j}\beta^{(A)}}-\mathrm{e}^{-2\mathrm{j}\beta^{(B)}}\left(\mathrm{e}^{-2\mathrm{j}\beta^{(A)}}+r^{(AB)2}\right)\right]\right.$$
$$\left.+\left(1+t^{(A,\mathcal{N}+1)2}\right)r^{(AB)}\mathrm{e}^{-\mathrm{j}\beta^{(A)}}\left(1-\mathrm{e}^{-2\mathrm{j}\beta^{(B)}}\right)\right\}. \qquad (4.166\mathrm{c})$$

They are considerably simplified for $N^{(\mathcal{N}+1)} = N^{(A)}$ when the spacer and substrate layers are the same materials or in situations where the penetration depth for the radiation is much smaller than the thickness of periodic region. The limiting case corresponds to the periodic region reduced to a single unit **L**, then p_q for $q = 1$ becomes unity and other Chebyshev polynomials and their derivatives vanish. We obtain from Eq. (4.162)

$$\chi_t^{(0,\underline{L},0)} = \frac{\Delta N^{(B)}}{N^{(B)}}\left[\left(1+r^{(0A)}r^{(AB)}\mathrm{e}^{-\mathrm{j}\beta^{(A)}}\right)^2-\mathrm{e}^{-2\mathrm{j}\beta^{(B)}}\left(r^{(0A)}\mathrm{e}^{-\mathrm{j}\beta^{(A)}}+r^{(AB)}\right)^2\right]^{-1}$$
$$\times\left\{\beta^{(B)}\left[\left(1+r^{(0A)}r^{(AB)}\mathrm{e}^{-\mathrm{j}\beta^{(A)}}\right)^2+\mathrm{e}^{-2\mathrm{j}\beta^{(B)}}\left(r^{(0A)}\mathrm{e}^{-\mathrm{j}\beta^{(A)}}+r^{(AB)}\right)^2\right]\right.$$
$$+\mathrm{j}\left(1-\mathrm{e}^{-2\mathrm{j}\beta^{(B)}}\right)\left[\left(1+r^{(0A)2}\mathrm{e}^{-2\mathrm{j}\beta^{(A)}}\right)r^{(AB)}$$
$$\left.+r^{(0A)}\mathrm{e}^{-\mathrm{j}\beta^{(A)}}\left(1+r^{(AB)2}\right)\right]\right\}, \qquad (4.167\mathrm{a})$$

for the same incident and exit media, *i.e.*, $N^{(\mathcal{N}+1)} = N^{(0)}$, and

$$\chi_r^{(0,L,A)} = \frac{\Delta N^{(B)}}{2N^{(B)}} \left(1 - r^{(0A)2}\right) e^{-j\beta^{(A)}} \left(1 - r^{(AB)2}\right)$$
$$\times \left[-4\beta^{(B)} r^{(AB)} e^{-2j\beta^{(B)}} - j\left(1 - e^{-2j\beta^{(B)}}\right)\left(1 + r^{(AB)2} e^{-2j\beta^{(B)}}\right)\right]$$
$$\times \left[\left(1 + r^{(0A)} r^{(AB)} e^{-j\beta^{(A)}}\right) - e^{-2j\beta^{(B)}} r^{(AB)} \left(r^{(0A)} e^{-j\beta^{(A)}} + r^{(AB)}\right)\right]^{-1}$$
$$\times \left[\left(r^{(0A)} + r^{(AB)} e^{-j\beta^{(A)}}\right) - e^{-2j\beta^{(B)}} r^{(AB)} \left(r^{(0A)} r^{(AB)} + e^{-j\beta^{(A)}}\right)\right]^{-1}, \tag{4.167b}$$

with the exit medium $N^{(\mathcal{N}+1)} = N^{(A)}$. In ultrathin film approximation expressed in Eq. (4.85) and applied to the central magnetic layer B, the elements of the characteristic matrix in the symmetric unit become

$$m_{11} \approx \cos\beta^{(A)}, \tag{4.168a}$$
$$m_{12} \approx \frac{j}{N^{(A)}} \sin\beta^{(A)}, \tag{4.168b}$$
$$m_{21} \approx jN^{(A)} \sin\beta^{(A)}. \tag{4.168c}$$

and

$$\Delta m_{11} \approx -\beta^{(B)} \sin\beta^{(A)} \frac{\Delta N^{(B)}}{N^{(A)}}, \tag{4.169a}$$
$$\Delta m_{12} \approx -j\beta^{(B)} \left(1 - \cos\beta^{(A)}\right) \frac{\Delta N^{(B)}}{N^{(A)2}}, \tag{4.169b}$$
$$\Delta m_{21} \approx j\beta^{(B)} \left(1 + \cos\beta^{(A)}\right) \Delta N^{(B)}. \tag{4.169c}$$

In order to account for the optical effect of B layers, we retain the argument of the Chebyshev polynomials in the form given in Eq. (4.163a). Then, the MO response in reflection of the periodic structure becomes

$$\chi_r^{(0,q\underline{L},A)} = \frac{4N^{(0)} \beta^{(B)} \Delta N^{(B)} p_q}{\left(N^{(0)2} - N^{(A)2}\right)\left(p_q e^{i\beta^{(A)}} - p_{q-1}\right)}. \tag{4.170}$$

The optical effect of B layers is completely ignored when we replace the argument of the Chebyshev polynomials by $\cos\beta^{(A)}$. Then, making use of the properties of Chebyshev polynomials

$$e^{iq\beta^{(A)}} = p_q e^{i\beta^{(A)}} - p_{q-1} \tag{4.171a}$$
$$p_q = e^{i(q-1)\beta^{(A)}} \left(1 + e^{-2i\beta^{(A)}} + e^{-4i\beta^{(A)}} + \cdots + e^{-2i(q-1)\beta^{(A)}}\right), \tag{4.171b}$$

we arrive at a simplified representation of $\chi_r^{(0,q\underline{L},A)}$

$$\chi_r^{(0,q\underline{L},A)} = \frac{4N^{(0)} \beta^{(B)} \Delta N^{(B)}}{\left(N^{(0)2} - N^{(A)2}\right)} e^{-i\beta^{(A)}}$$
$$\times \left(1 + e^{-2i\beta^{(A)}} + e^{-4i\beta^{(A)}} + \cdots + e^{-2i(q-1)\beta^{(A)}}\right). \tag{4.172}$$

The formula may be used for a qualitative evaluation of trends in $\chi_r^{(0,q\underline{L},A)}$ when the multilayer parameters change. It is consistent with the saturation of the MO polar Kerr effect in absorbing multilayers for a large number of periods.

When simultaneously $|\beta^{(A)}| \ll 1$ and $|\beta^{(B)}| \ll 1$, we can put in Eq. (4.162), to first order in $\beta^{(A)}$ and $\beta^{(B)}$, $m_{11} \approx 1$, $m_{12} \approx 0$, $m_{21} \approx 0$, $\Delta m_{11} \approx 0$, $\Delta m_{12} \approx 0$, and $\Delta m_{21} \approx 2j\beta^{(B)}\Delta N^{(B)}$. We obtain

$$\chi_t^{(0,q\underline{L},\mathcal{N}+1)} \approx \frac{2p_q\beta^{(B)}\Delta N^{(B)}}{(p_q - p_{q-1})(N^{(0)} + N^{(\mathcal{N}+1)})}, \tag{4.173a}$$

$$\chi_r^{(0,q\underline{L},\mathcal{N}+1)} \approx \frac{4p_q\beta^{(B)}N^{(0)}\Delta N^{(B)}}{(p_q - p_{q-1})(N^{(0)2} - N^{(\mathcal{N}+1)2})}. \tag{4.173b}$$

Then, $p_q - p_{q-1} \approx 1$ and $p_q \approx q$. The MO effects can be expressed as a q times contribution of an individual block

$$\chi_t^{(0,q\underline{L},\mathcal{N}+1)} \approx \frac{2q\beta^{(B)}\Delta N^{(B)}}{(N^{(0)} + N^{(\mathcal{N}+1)})}, \tag{4.174a}$$

$$\chi_r^{(0,q\underline{L},\mathcal{N}+1)} \approx \frac{4q\beta^{(B)}N^{(0)}\Delta N^{(B)}}{(N^{(0)2} - N^{(\mathcal{N}+1)2})}. \tag{4.174b}$$

At $q = 1$ ($p_1 = 1$), these relations reduce to the contribution of a single ultrathin magnetic layer B sandwiched in an $A/B/A$ block. In this approximation, the multilayer Faraday effect is proportional to the product of specific Faraday effect in the magnetic medium times the total magnetic film thickness $qd^{(B)}$ multiplied by a factor of $2N^{(B)}/(N^{(0)} + N^{(\mathcal{N}+1)})$. Note that the specific Faraday effect takes place when the distance travelled by polarized plane wave parallel to the magnetization in medium (n) has a unit length. We have, according to Eq. (1.37) and Eq. (4.96), for the specific Faraday effect

$$\frac{\chi_F^{(n)}}{d^{(n)}} = j\frac{\omega}{2c}\frac{\varepsilon_{xy}^{(n)}}{\sqrt{\varepsilon_{xx}^{(n)}}} = \frac{\omega}{c}\Delta N^{(n)}. \tag{4.175}$$

Similarly, the normal incidence polar Kerr effect in multilayers is proportional to double pass Faraday effect in the magnetic film times q and a factor of $2N^{(0)}N^{(B)}/(N^{(0)2} - N^{(\mathcal{N}+1)2})$.

We now return to the case of periodic structures with cover layers, represented by Eq. (4.153). The corresponding Jones reflection matrix is given by Eq. (4.156b) and Eq. (4.156c). We consider the multilayer with cover layers situated above q $(A/B/A)$ sandwiches with central magnetic layer B and the exit medium characterized by $N^{(\mathcal{N}+1)} = N^{(A)}$. A case with arbitrary $\mathbf{W}$ where the total thickness of q $(A/B/A)$ sandwiches is much higher than the radiation penetration depth is also included. Using Eq. (4.157c) and Eq. (4.157d), we obtain for the elements of the Jones reflection matrix

$$r_{xx}^{(C,q\underline{L},A)} = \frac{M_{21}}{M_{11}}$$
$$= \left[(C_1 + C_2 N^{(A)})\left(m_{11} p_q - p_{q-1}\right) + C_1 N^{(A)} m_{12} p_q + C_2 m_{21} p_q\right]^{-1}$$
$$\times \left[(C_3 + C_4 N^{(A)})\left(m_{11} p_q - p_{q-1}\right) + C_3 N^{(A)} m_{12} p_q + C_4 m_{21} p_q\right], \tag{4.176a}$$

$$r_{xy}^{(C,q\underline{L},A)} = -\mathrm{j}\frac{M_{11}\Delta\left(M_{21}\right) - M_{21}\Delta\left(M_{11}\right)}{M_{11}^2}$$
$$= -\mathrm{j}\left(C_1 C_4 - C_2 C_3\right)\left[(C_1 + C_2 N^{(A)})\left(m_{11} p_q - p_{q-1}\right) + C_1 N^{(A)} m_{12} p_q\right.$$
$$\left. + C_2 m_{21} p_q\right]^{-2}\left[\left(N^{(A)2} m_{12} - m_{21}\right)\left(p_q^2 + p_{q-1} p_q' - p_q p_{q-1}'\right)\Delta m_{11}\right.$$
$$+ \left(m_{11} p_q - p_{q-1}\right) p_q \left(\Delta m_{21} - N^{(A)2} \Delta m_{12}\right)$$
$$\left. + N^{(A)} p_q^2 \left(m_{12}\Delta m_{21} - m_{21}\Delta m_{12}\right)\right]. \tag{4.176b}$$

4.4.4 Practical Aspects

We have considered the analytical representation of the optical response in the most common and also most important arrangement of the normal incidence to the planar interfaces giving rise to the Faraday and normal incidence magnetooptic polar Kerr effects. To simplify the algebra, we have focused on the periodic regions formed by symmetrical sandwiches of magnetic and nonmagnetic films. The analysis of the periodic structures consisting of nonsymmetrical blocks can be carried out along the same lines. The formulae represent the magnetooptic effects with an accuracy that is practically identical to that obtained with the 4×4 matrix formalism, provided the ellipsometric angles are small, which is indeed the case for absorbing multilayers with the layer thickness of nanometer range. This is not always the case of the formulae simplified with the ultrathin film approximation of Eq. (4.85).

In Figure 4.9, we illustrate the effect of the number of periods on the periodic multilayer response expressed in terms of the off-diagonal element of the Jones reflection matrix. Figure 4.10 displays the effect of the nonmagnetic ultrathin Au spacer on the off-diagonal elements of the reflection matrix. The situation where the thickness of the magnetic layer varies is displayed in Figure 4.11. Before showing the complex polar Kerr effect in these periodic structures, we present in Figure 4.12, Figure 4.13, and Figure 4.14 the diagonal elements of the Jones reflection matrix. The plots of the normal incidence polar Kerr rotation and ellipticity are given in Figure 4.15, Figure 4.16, and Figure 4.17.

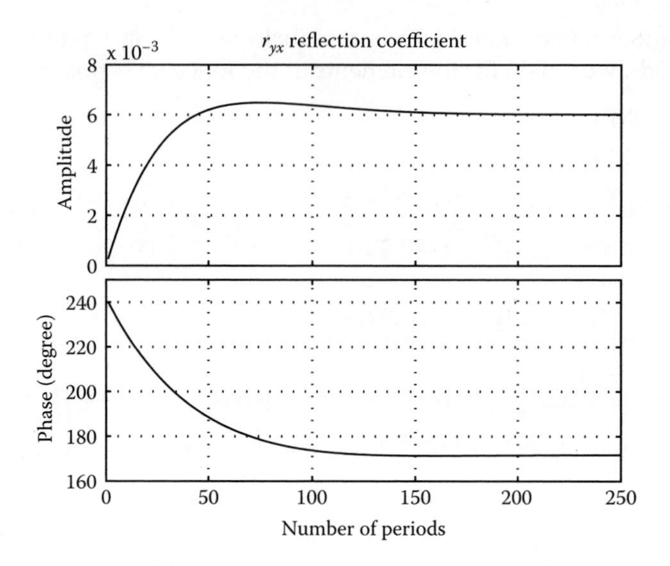

FIGURE 4.9
Effect of the number of periods on the normal incidence polar off-diagonal reflection coefficients r_{yx} in a periodic structure consisting of the blocks Au(0.3 nm)/Fe(0.3 nm) and deposited on an optically thick Au buffer. The first layer exposed to the ambient is Au. Computed with the data of Figure 4.2.

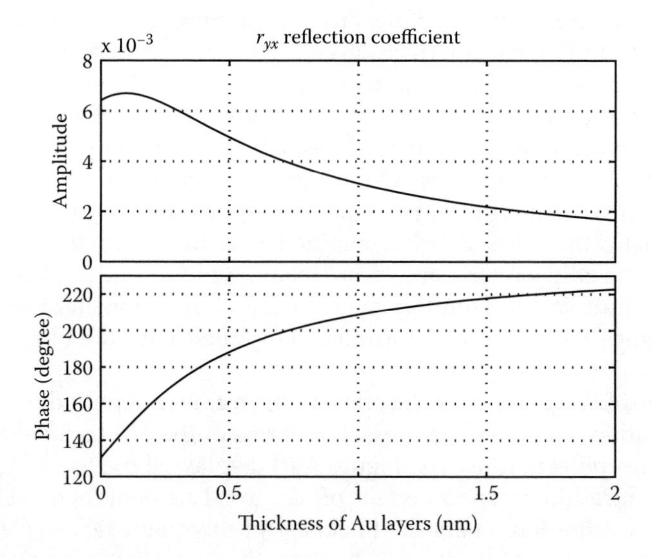

FIGURE 4.10
Effect of the Au layer thickness, d, on the normal incidence polar off-diagonal reflection coefficient, r_{yx}, in a periodic structure consisting of the blocks Au(d)/Fe(0.3 nm) repeated 500 times and deposited on an optically thick Au buffer. The first layer exposed to the ambient is Au. Computed with the data of Figure 4.2.

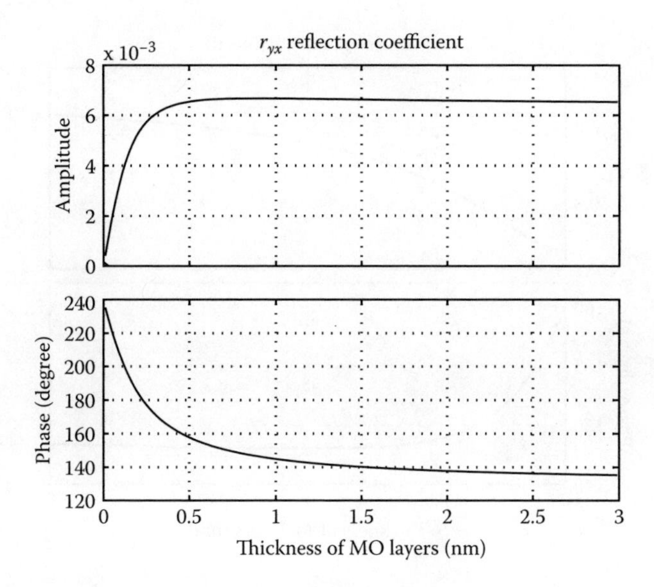

FIGURE 4.11
Effect of the Fe layer thickness, d, on the normal incidence polar off-diagonal reflection coefficient, r_{yx}, in a periodic structure consisting of the blocks Au(0.3 nm)/Fe(d) repeated 500 times and deposited on an optically thick Au buffer. The first layer exposed to the ambient is Au. Computed with the data of Figure 4.2.

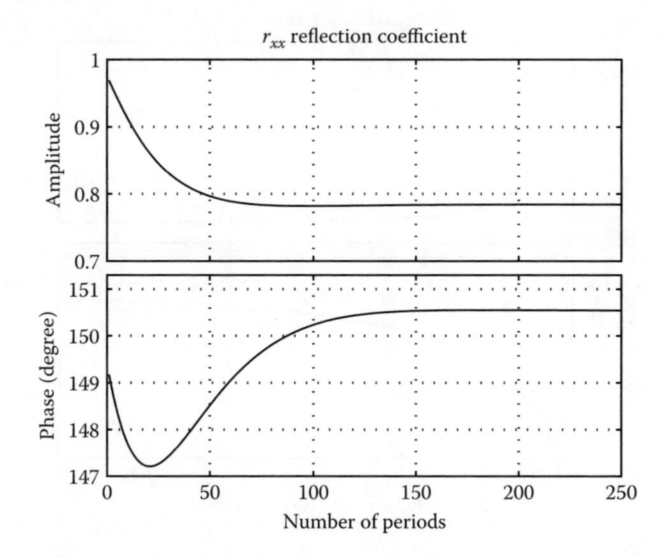

FIGURE 4.12
Effect of the number of periods on the normal incidence diagonal reflection coefficient, r_{xx}, in a periodic structure consisting of the blocks Au(0.3 nm)/Fe(0.3 nm) and deposited on an optically thick Au buffer. The first layer exposed to the ambient is Au. Computed with the data of Figure 4.2.

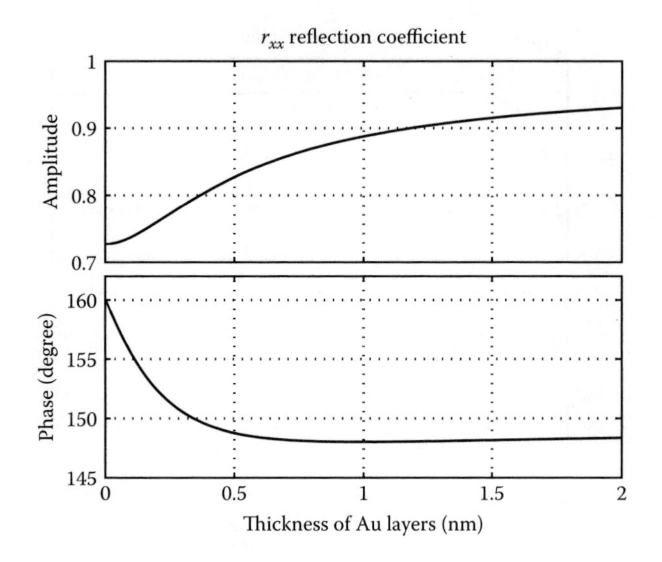

FIGURE 4.13
Effect of the Au layer thickness, d, on the diagonal reflection coefficient, r_{xx}, in a periodic structure consisting of the blocks Au(d)/Fe(0.3 nm) repeated 500 times and deposited on an optically thick Au buffer. The first layer exposed to the ambient is Au. Computed with the data of Figure 4.2.

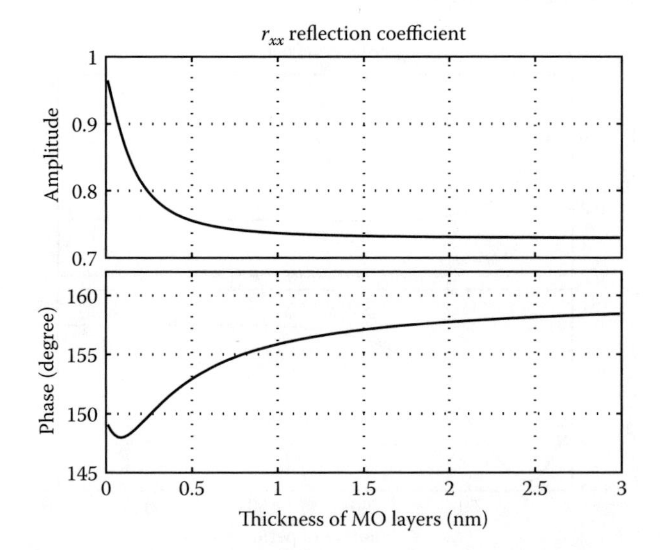

FIGURE 4.14
Effect of the Fe layer thickness, d, on the normal incidence diagonal reflection coefficients r_{xx} in a periodic structure consisting of the blocks Au(0.3 nm)/Fe(d) repeated 500 times and deposited on an optically thick Au buffer. The first layer exposed to the ambient is Au. The simulation employs the data of Figure 4.4.

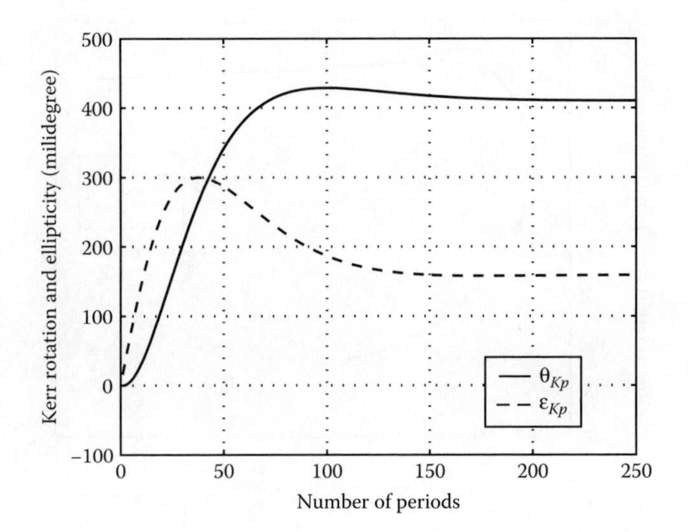

FIGURE 4.15
Effect of the number of periods on the normal incidence polar Kerr rotation, θ_{Kp}, and ellipticity, ϵ_{Kp}, in the periodic structure of Figure 4.9 consisting of the blocks Au(0.3 nm)/Fe(0.3 nm) and deposited on an optically thick Au buffer. The first layer exposed to the ambient is Au. Computed with the data of Figure 4.2.

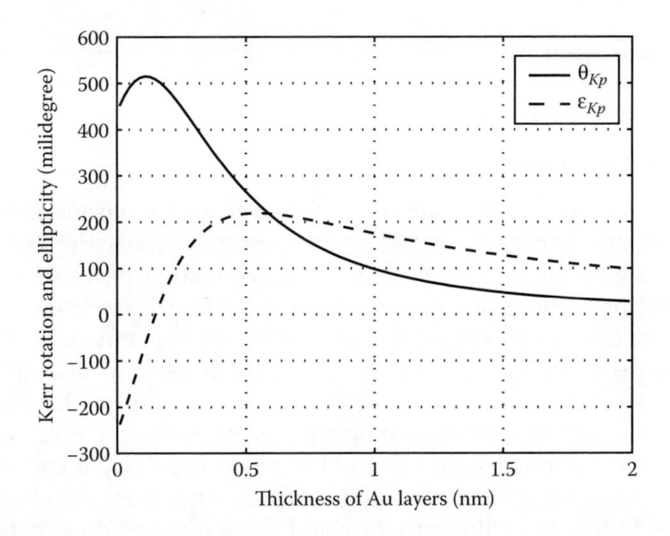

FIGURE 4.16
The effect of the Au layer thickness, d, on the normal incidence polar Kerr rotation, θ_{Kp}, and ellipticity, ϵ_{Kp}, in the periodic structure of Figure 4.10 consisting of the blocks Au(d)/Fe(0.3 nm) repeated 500 times and deposited on an optically thick Au buffer. The first layer exposed to the ambient is Au. Computed with the data of Figure 4.2.

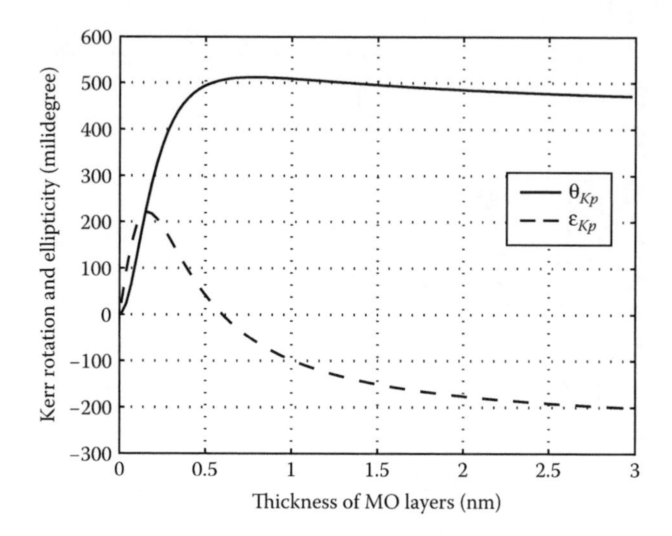

FIGURE 4.17
Effect of the Fe layer thickness, d, on the normal incidence polar Kerr rotation, θ_{Kp}, and ellipticity, ϵ_{Kp}, in a periodic structure of Figure 4.11 consisting of the blocks Au(0.3 nm)/Fe(d) repeated 500 times and deposited on an optically thick Au buffer. The first layer exposed to the ambient is Au. Computed with the data of Figure 4.2.

4.5 Oblique Incidence

4.5.1 In-Plane Anisotropy

Figure 4.18 shows the diagram of a multilayer with polar magnetization along with the incident, reflected, and transmitted propagation vectors in sandwiching half spaces for the case of nonzero angle of incidence. As it has already been shown in Eq. (3.33), the transfer matrix represents a building block in the matrix representation of the multilayer structure. We shall now give the Cartesian transfer matrix for the important case of the polar magnetization at an *arbitrary angle of incidence* with the LP modes in sandwiching isotropic nonabsorbing regions (0) and $\mathcal{N}+1$ polarized perpendicular and parallel to the plane of incidence. In principle, each layer of the structure may also display the anisotropies of nonmagnetic origin, *e.g.*, those due to the crystalline structure and those induced during the layer deposition, by interfacial stresses, *etc*. Special magnetic and crystalline anisotropies are induced with the layer growth on single crystal substrates with vicinal surfaces. The corresponding permittivity tensor would have nearly as many independent components as the general tensor of Eq. (3.1). We simplify the problem to most common situations by assuming that the

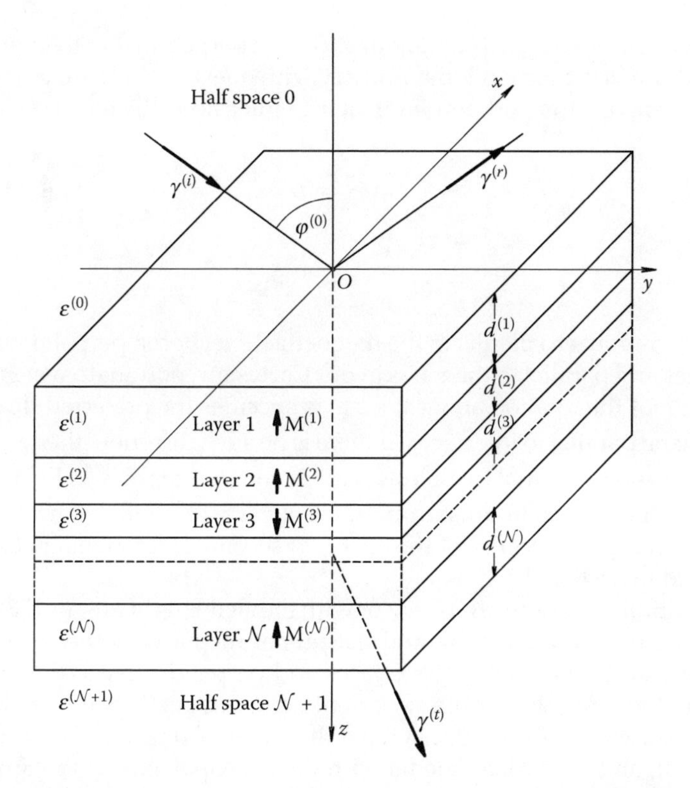

FIGURE 4.18
Oblique light incidence ($\varphi^{(0)} \neq 0$) at a multilayer consisting of $\mathcal{N}$ anisotropic layers with polar magnetizations, $\boldsymbol{M}^{(n)}$ ($n = 1, \ldots, \mathcal{N}$). The layers are characterized by permittivity tensors $\varepsilon^{(n)}$. The incident, reflected, and transmitted propagation vectors in sandwiching isotropic nonabsorbing half spaces (0) and ($\mathcal{N} + 1$) are denoted as $\gamma^{(i)}$, $\gamma^{(r)}$, and $\gamma^{(t)}$, respectively.

anisotropies manifest themselves by preferred directions either normal or parallel to the interface planes. The relative permittivity tensor in the n-th layer then takes the form

$$\varepsilon^{(n)} = \begin{pmatrix} \varepsilon_{xx}^{(n)} & \varepsilon_{xy}^{(n)} & 0 \\ \varepsilon_{yx}^{(n)} & \varepsilon_{yy}^{(n)} & 0 \\ 0 & 0 & \varepsilon_{zz}^{(n)} \end{pmatrix}. \tag{4.177}$$

The rotation about the z axis transforms the elements ε_{xx}, ε_{xy}, ε_{yx}, and ε_{yy} among themselves. The symmetry of tensor remains unchanged for any orientation of the plane of incidence, and our choice of the plane of incidence normal to the x axis, *i.e.*, with $N_x = 0$, brings no loss in generality. It covers most of the symmetries in atomic ordering we meet in two-dimensional ultrathin films and multilayers with magnetization perpendicular to interfaces. The lowest included symmetry is that of a monoclinic medium, the axes of which, as well as its magnetization, are perpendicular to interfaces. In a medium of orthorhombic symmetry with one of the three axes parallel to the magnetization and perpendicular to interfaces, the tensor elements become

$$\varepsilon^{(n)}_{\substack{xx\\yy}} = \varepsilon^{(n)}_0 \pm \varepsilon^{(n)}_2 \cos 2\alpha^{(n)}, \tag{4.178a}$$

$$\varepsilon^{(n)}_{zz} = \varepsilon^{(n)}_3, \tag{4.178b}$$

$$\varepsilon^{(n)}_{\substack{xy\\yx}} = \pm \mathrm{j}\,\varepsilon^{(n)}_1 + \varepsilon^{(n)}_2 \sin 2\alpha^{(n)}. \tag{4.178c}$$

Here, $\varepsilon_1^{(n)}$ is nonzero thanks to the magnetization chosen perpendicular to interfaces and parallel to the z axis in our Cartesian coordinate system. The angle $\alpha^{(n)}$ of the rotation about the z axis specifies the preferred direction in the plane parallel to interfaces. If there is no such direction, then $\varepsilon_2^{(n)} = 0$. The anisotropies normal to interfaces are accounted for by $\varepsilon_3^{(n)}$. Even in the absence of nonmagnetic anisotropies, the element $\varepsilon_3^{(n)}$ will depart from the scalar permittivity because of the effects of second order in magnetization discussed in Section 2.2.4.

In addition, the permittivity tensor with the elements of Eqs. (4.178) characterizes trigonal, tetragonal, and hexagonal media with their principal axis confined to the plane of interfaces and perpendicular to the magnetization at an otherwise arbitrary orientation, specified by the angle $\alpha^{(n)}$. At the restriction to $\epsilon_2 = 0$, we have the cases of uniaxial, *i.e.*, trigonal, tetragonal, and hexagonal, media with the principal axis and magnetization perpendicular to interfaces and those of cubic media with the magnetization parallel to either three- or fourfold cubic axes and perpendicular to the interface planes.

Without any loss in generality, we thus choose the plane of incidence perpendicular to the x axis. Note that the only effect of the rotation about the z axis is the displacement of the origin for counting the angle $\alpha^{(n)}$. By substituting the tensor of Eq. (4.177) into Eq. (3.12), we arrive at the proper value biquadratic equation

$$\begin{aligned}\varepsilon^{(n)}_{zz} N^{(n)4}_z - \left[\varepsilon^{(n)}_{yy}\left(\varepsilon^{(n)}_{zz} - N_y^2\right) + \varepsilon^{(n)}_{zz}\left(\varepsilon^{(n)}_{xx} - N_y^2\right)\right] N^{(n)2}_z \\ + \varepsilon^{(n)}_{yy}\left(\varepsilon^{(n)}_{xx} - N_y^2\right)\left(\varepsilon^{(n)}_{zz} - N_y^2\right) - \varepsilon^{(n)}_{xy}\varepsilon^{(n)}_{yx}\left(\varepsilon^{(n)}_{zz} - N_y^2\right) = 0,\end{aligned} \tag{4.179}$$

which provides four solutions to the normal components of the proper propagation vectors. Because of the reduction of the proper value equation to a biquadratic form, the proper values of the normal propagation vector components can be grouped in two pairs. The components in each pair are of the same absolute values, *i.e.*,

$$\frac{\omega}{c} N^{(n)}_{z1}, \qquad \frac{\omega}{c} N^{(n)}_{z2} = -\frac{\omega}{c} N^{(n)}_{z1}, \tag{4.180a}$$

$$\frac{\omega}{c} N^{(n)}_{z3}, \qquad \frac{\omega}{c} N^{(n)}_{z4} = -\frac{\omega}{c} N^{(n)}_{z3}. \tag{4.180b}$$

We can, therefore, express the proper propagation vectors as

$$\boldsymbol{\gamma}_1^{(n)} = \frac{\omega}{c}\left(\hat{\boldsymbol{y}}N_y + \hat{\boldsymbol{z}}N_{z1}^{(n)}\right), \tag{4.181a}$$

$$\boldsymbol{\gamma}_2^{(n)} = \frac{\omega}{c}\left(\hat{\boldsymbol{y}}N_y - \hat{\boldsymbol{z}}N_{z1}^{(n)}\right), \tag{4.181b}$$

$$\boldsymbol{\gamma}_3^{(n)} = \frac{\omega}{c}\left(\hat{\boldsymbol{y}}N_y + \hat{\boldsymbol{z}}N_{z3}^{(n)}\right), \tag{4.181c}$$

$$\boldsymbol{\gamma}_4^{(n)} = \frac{\omega}{c}\left(\hat{\boldsymbol{y}}N_y - \hat{\boldsymbol{z}}N_{z3}^{(n)}\right). \tag{4.181d}$$

The solution of Eq. (4.179) provides

$$N_{z\pm}^{(n)2} = \frac{\varepsilon_{yy}^{(n)}\left(\varepsilon_{zz}^{(n)} - N_y^2\right) + \varepsilon_{zz}^{(n)}\left(\varepsilon_{xx}^{(n)} - N_y^2\right)}{2\varepsilon_{zz}^{(n)}} \pm \frac{\sqrt{\left[\varepsilon_{yy}^{(n)}\left(\varepsilon_{zz}^{(n)} - N_y^2\right) - \varepsilon_{zz}^{(n)}\left(\varepsilon_{xx}^{(n)} - N_y^2\right)\right]^2 + 4\varepsilon_{zz}^{(n)}\varepsilon_{xy}^{(n)}\varepsilon_{yx}^{(n)}\left(\varepsilon_{zz}^{(n)} - N_y^2\right)}}{2\varepsilon_{zz}^{(n)}}. \tag{4.182}$$

Here, the subscripts +/− distinguish the different absolute values of $N_{zj}^{(n)}$, *i.e.*, $N_{z1}^{(n)} = N_{z+}^{(n)}$, $N_{z2}^{(n)} = -N_{z+}^{(n)}$, $N_{z3}^{(n)} = N_{z-}^{(n)}$, and $N_{z4}^{(n)} = -N_{z-}^{(n)}$. The following relations are often useful

$$N_{z+}^{(n)2} + N_{z-}^{(n)2} = \frac{\varepsilon_{yy}^{(n)}\left(\varepsilon_{zz}^{(n)} - N_y^2\right) + \varepsilon_{zz}^{(n)}\left(\varepsilon_{xx}^{(n)} - N_y^2\right)}{\varepsilon_{zz}^{(n)}}, \tag{4.183a}$$

$$N_{z+}^{(n)2}N_{z-}^{(n)2} = \frac{\left(\varepsilon_{zz}^{(n)} - N_y^2\right)\left[\varepsilon_{yy}^{(n)}\left(\varepsilon_{xx}^{(n)} - N_y^2\right) - \varepsilon_{xy}^{(n)}\varepsilon_{yx}^{(n)}\right]}{\varepsilon_{zz}^{(n)}}. \tag{4.183b}$$

The proper polarizations follow from Eq. (3.16) and Eq. (3.19)

$$\mathbf{e}_j^{(n)} = C_j^{(n)}\begin{pmatrix} -\varepsilon_{xy}^{(n)}\left(\varepsilon_{zz}^{(n)} - N_y^2\right) \\ \left(\varepsilon_{zz}^{(n)} - N_y^2\right)\left(\varepsilon_{xx}^{(n)} - N_y^2 - N_{zj}^{(n)2}\right) \\ -N_yN_{zj}^{(n)}\left(\varepsilon_{xx}^{(n)} - N_y^2 - N_{zj}^{(n)2}\right) \end{pmatrix} \tag{4.184a}$$

$$\mathbf{b}_j^{(n)} = C_j^{(n)}\begin{pmatrix} -\varepsilon_{zz}^{(n)}N_{zj}^{(n)}\left(\varepsilon_{xx}^{(n)} - N_y^2 - N_{zj}^{(n)2}\right) \\ -\varepsilon_{xy}^{(n)}N_{zj}^{(n)}\left(\varepsilon_{zz}^{(n)} - N_y^2\right) \\ \varepsilon_{xy}^{(n)}N_y\left(\varepsilon_{zz}^{(n)} - N_y^2\right) \end{pmatrix}. \tag{4.184b}$$

In general, the proper polarizations are elliptically polarized and nonorthogonal. Only at the zero angle of incidence, they reduce to orthogonal RCP and LCP waves provided $\varepsilon_{xx}^{(n)} = \varepsilon_{yy}^{(n)}$, *i.e.*, $\varepsilon_2^{(n)} = 0$. At nonzero angles of incidence, the solutions depend also on the anisotropy component in $\varepsilon_{zz}^{(n)}$. We observe that the normalization coefficients are related and can be expressed as

$$C_{+}^{(n)} = C_{1}^{(n)} = C_{2}^{(n)}, \tag{4.185a}$$

$$C_{-}^{(n)} = C_{3}^{(n)} = C_{4}^{(n)}. \tag{4.185b}$$

Making use of Eqs. (3.30), we obtain

$$\mathbf{D}^{n} = \begin{pmatrix} D_{11}^{(n)} & D_{11}^{(n)} & D_{11}^{(n)} & D_{11}^{(n)} \\ D_{21}^{(n)} & -D_{21}^{(n)} & D_{23}^{(n)} & -D_{23}^{(n)} \\ D_{31}^{(n)} & D_{31}^{(n)} & D_{33}^{(n)} & D_{33}^{(n)} \\ D_{41}^{(n)} & -D_{41}^{(n)} & D_{43}^{(n)} & -D_{43}^{(n)} \end{pmatrix}. \tag{4.186}$$

We have to evaluate six elements only. Omitting the normalization coefficient $C_{j}^{(n)}$, we get

$$D_{1j}^{(n)} = D_{11}^{(n)} = -\varepsilon_{xy}^{(n)} \left(\varepsilon_{zz}^{(n)} - N_{y}^{2}\right), \quad j = 1, \ldots, 4, \tag{4.187a}$$

$$D_{21}^{(n)} = -D_{22}^{(n)} = N_{z}^{(n)+} \left[-\varepsilon_{xy}^{(n)} \left(\varepsilon_{zz}^{(n)} - N_{y}^{2}\right)\right], \tag{4.187b}$$

$$D_{23}^{(n)} = -D_{24}^{(n)} = N_{z-}^{(n)} \left[-\varepsilon_{xy}^{(n)} \left(\varepsilon_{zz}^{(n)} - N_{y}^{2}\right)\right], \tag{4.187c}$$

$$D_{31}^{(n)} = D_{32}^{(n)} = \left(\varepsilon_{zz}^{(n)} - N_{y}^{2}\right)\left[\varepsilon_{xx}^{(n)} - N_{y}^{2} - \left(N_{z+}^{(n)}\right)^{2}\right], \tag{4.187d}$$

$$D_{33}^{(n)} = D_{34}^{(n)} = \left(\varepsilon_{zz}^{(n)} - N_{y}^{2}\right)\left[\varepsilon_{xx}^{(n)} - N_{y}^{2} - \left(N_{z-}^{(n)}\right)^{2}\right], \tag{4.187e}$$

$$D_{41}^{(n)} = -D_{42}^{(n)} = -N_{z+}^{(n)} \varepsilon_{zz}^{(n)} \left[\varepsilon_{xx}^{(n)} - N_{y}^{2} - \left(N_{z+}^{(n)}\right)^{2}\right], \tag{4.187f}$$

$$D_{43}^{(n)} = -D_{44}^{(n)} = -N_{z-}^{(n)} \varepsilon_{zz}^{(n)} \left[\varepsilon_{xx}^{(n)} - N_{y}^{2} - \left(N_{z-}^{(n)}\right)^{2}\right]. \tag{4.187g}$$

If needed, the normalized dynamical matrix can be obtained from the matrix product

$$\mathbf{D}_{\text{norm}}^{(n)} = \begin{pmatrix} D_{11}^{(n)} & D_{11}^{(n)} & D_{11}^{(n)} & D_{11}^{(n)} \\ D_{21}^{(n)} & -D_{21}^{(n)} & D_{23}^{(n)} & -D_{23}^{(n)} \\ D_{31}^{(n)} & D_{31}^{(n)} & D_{33}^{(n)} & D_{33}^{(n)} \\ D_{41}^{(n)} & -D_{41}^{(n)} & D_{43}^{(n)} & -D_{43}^{(n)} \end{pmatrix} \begin{pmatrix} C_{+}^{(n)} & 0 & 0 & 0 \\ 0 & C_{+}^{(n)} & 0 & 0 \\ 0 & 0 & C_{-}^{(n)} & 0 \\ 0 & 0 & 0 & C_{-}^{(n)} \end{pmatrix}. \tag{4.188}$$

It is useful to establish expressions

$$\begin{aligned}\mathcal{P}^{(n)} &= D_{11}^{(n)}\left(D_{41}^{(n)} - D_{43}^{(n)}\right)\\ &= -\varepsilon_{xy}^{(n)}\left(\varepsilon_{zz}^{(n)} - N_y^2\right)\left(N_{z+}^{(n)} - N_{z-}^{(n)}\right)\left[\varepsilon_{yy}^{(n)}\left(\varepsilon_{zz}^{(n)} - N_y^2\right) + \varepsilon_{zz}^{(n)}\left(N_{z+}^{(n)}N_{z-}^{(n)}\right)\right],\end{aligned} \tag{4.189a}$$

$$\begin{aligned}\mathcal{Q}^{(n)} &= D_{11}^{(n)}\left(D_{33}^{(n)} - D_{31}^{(n)}\right)\\ &= -\varepsilon_{xy}^{(n)}\left(\varepsilon_{zz}^{(n)} - N_y^2\right)\left(N_{z+}^{(n)} - N_{z-}^{(n)}\right)\left[\left(\varepsilon_{zz}^{(n)} - N_y^2\right)\left(N_{z+}^{(n)} + N_{z-}^{(n)}\right)\right],\end{aligned} \tag{4.189b}$$

$$\begin{aligned}\mathcal{S}^{(n)} &= \left(D_{23}^{(n)}D_{41}^{(n)} - D_{21}^{(n)}D_{43}^{(n)}\right)\\ &= -\varepsilon_{xy}^{(n)}\left(\varepsilon_{zz}^{(n)} - N_y^2\right)\left(N_{z+}^{(n)} - N_{z-}^{(n)}\right)\left[\varepsilon_{zz}^{(n)}\left(N_{z+}^{(n)}N_{z-}^{(n)}\right)\left(N_{z+}^{(n)} + N_{z-}^{(n)}\right)\right],\end{aligned} \tag{4.189c}$$

$$\begin{aligned}\mathcal{T}^{(n)} &= \left(D_{21}^{(n)}D_{33}^{(n)} - D_{23}^{(n)}D_{31}^{(n)}\right)\\ &= -\varepsilon_{xy}^{(n)}\left(\varepsilon_{zz}^{(n)} - N_y^2\right)^2\left(N_{z+}^{(n)} - N_{z-}^{(n)}\right)\left[\left(\varepsilon_{xx}^{(n)} - N_y^2\right) + \left(N_{z+}^{(n)}N_{z-}^{(n)}\right)\right],\end{aligned} \tag{4.189d}$$

$$\mathcal{U}^{(n)} = D_{11}^{(n)}\left(D_{21}^{(n)} - D_{23}^{(n)}\right) = \varepsilon_{xy}^{(n)2}\left(N_{z+}^{(n)} - N_{z-}^{(n)}\right)\left(\varepsilon_{zz}^{(n)} - N_y^2\right)^2, \tag{4.189e}$$

$$\mathcal{W}^{(n)} = \left(D_{33}^{(n)}D_{41}^{(n)} - D_{31}^{(n)}D_{43}^{(n)}\right) = \varepsilon_{xy}^{(n)}\varepsilon_{yx}^{(n)}\left(N_{z+}^{(n)} - N_{z-}^{(n)}\right)\left(\varepsilon_{zz}^{(n)} - N_y^2\right)^2. \tag{4.189f}$$

Up to a common factor, they represent a special case of Eq. (3.100) pertinent to the polar magnetization. We further need

$$\begin{aligned}&D_{11}^{(n)}\left(D_{41}^{(n)} + D_{43}^{(n)}\right)\\ &= -\varepsilon_{xy}^{(n)}\left(\varepsilon_{zz}^{(n)} - N_y^2\right)\left(N_{z+}^{(n)} + N_{z-}^{(n)}\right)\left[\varepsilon_{yy}^{(n)}\left(\varepsilon_{zz}^{(n)} - N_y^2\right) - \varepsilon_{zz}^{(n)}\left(N_{z+}^{(n)}N_{z-}^{(n)}\right)\right],\end{aligned} \tag{4.190a}$$

$$\begin{aligned}&\left(D_{21}^{(n)}D_{33}^{(n)} + D_{23}^{(n)}D_{31}^{(n)}\right)\\ &= -\varepsilon_{xy}^{(n)}\left(\varepsilon_{zz}^{(n)} - N_y^2\right)^2\left(N_{z+}^{(n)} + N_{z-}^{(n)}\right)\left[\left(\varepsilon_{xx}^{(n)} - N_y^2\right) - \left(N_{z+}^{(n)}N_{z-}^{(n)}\right)\right],\end{aligned} \tag{4.190b}$$

$$D_{11}^{(n)}\left(D_{21}^{(n)} + D_{23}^{(n)}\right) = \varepsilon_{xy}^{(n)2}\left(N_{z+}^{(n)} + N_{z-}^{(n)}\right)\left(\varepsilon_{zz}^{(n)} - N_y^2\right)^2, \tag{4.190c}$$

$$\left(D_{33}^{(n)}D_{41}^{(n)} + D_{31}^{(n)}D_{43}^{(n)}\right) = \varepsilon_{xy}^{(n)}\varepsilon_{yx}^{(n)}\left(N_{z+}^{(n)} + N_{z-}^{(n)}\right)\left(\varepsilon_{zz}^{(n)} - N_y^2\right)^2, \tag{4.190d}$$

$$\begin{aligned}&D_{11}^{(n)}\left(D_{33}^{(n)} + D_{31}^{(n)}\right)\\ &= -\varepsilon_{xy}^{(n)}\left(\varepsilon_{zz}^{(n)} - N_y^2\right)^2\varepsilon_{zz}^{(n)-1}\left[\varepsilon_{zz}^{(n)}\left(\varepsilon_{xx}^{(n)} - N_y^2\right) - \varepsilon_{yy}^{(n)}\left(\varepsilon_{zz}^{(n)} - N_y^2\right)\right],\end{aligned} \tag{4.190e}$$

$$\begin{aligned}&\left(D_{21}^{(n)}D_{43}^{(n)} + D_{23}^{(n)}D_{41}^{(n)}\right)\\ &= \varepsilon_{xy}^{(n)}\left(\varepsilon_{zz}^{(n)} - N_y^2\right)\left(N_{z+}^{(n)}N_{z-}^{(n)}\right)\left[\varepsilon_{zz}^{(n)}\left(\varepsilon_{xx}^{(n)} - N_y^2\right) - \varepsilon_{yy}^{(n)}\left(\varepsilon_{zz}^{(n)} - N_y^2\right)\right].\end{aligned} \tag{4.190f}$$

The inverse matrix $(\mathbf{D}^{(n)})^{-1}$ is given by

$$(\mathbf{D}^{(n)})^{-1} = \left[\det(\mathbf{D}^{(n)})\right]^{-1} \begin{pmatrix} L_{11}^{(n)} & L_{12}^{(n)} & L_{13}^{(n)} & L_{14}^{(n)} \\ L_{11}^{(n)} & -L_{12}^{(n)} & L_{13}^{(n)} & -L_{14}^{(n)} \\ L_{31}^{(n)} & L_{32}^{(n)} & -L_{13}^{(n)} & L_{34}^{(n)} \\ L_{31}^{(n)} & -L_{32}^{(n)} & -L_{13}^{(n)} & -L_{34}^{(n)} \end{pmatrix}, \tag{4.191}$$

where the determinant of $\mathbf{D}^{(n)}$,

$$\mathcal{D}_{\mathrm{pol}}^{(n)} = \det(\mathbf{D}^{(n)}) = -4D_{11}^{(n)}\left(D_{33}^{(n)} - D_{31}^{(n)}\right)\left(D_{23}^{(n)}D_{41}^{(n)} - D_{21}^{(n)}D_{43}^{(n)}\right), \tag{4.192}$$

and

$$L_{11}^{(n)} = -2D_{33}^{(n)}\left(D_{23}^{(n)}D_{41}^{(n)} - D_{21}^{(n)}D_{43}^{(n)}\right), \tag{4.193a}$$
$$L_{31}^{(n)} = 2D_{31}^{(n)}\left(D_{23}^{(n)}D_{41}^{(n)} - D_{21}^{(n)}D_{43}^{(n)}\right), \tag{4.193b}$$
$$L_{12}^{(n)} = 2D_{11}^{(n)}D_{43}^{(n)}\left(D_{33}^{(n)} - D_{31}^{(n)}\right), \tag{4.193c}$$
$$L_{32}^{(n)} = -2D_{11}^{(n)}D_{41}^{(n)}\left(D_{33}^{(n)} - D_{31}^{(n)}\right), \tag{4.193d}$$
$$L_{13}^{(n)} = 2D_{11}^{(n)}\left(D_{23}^{(n)}D_{41}^{(n)} - D_{21}^{(n)}D_{43}^{(n)}\right), \tag{4.193e}$$
$$L_{14}^{(n)} = -2D_{11}^{(n)}D_{23}^{(n)}\left(D_{33}^{(n)} - D_{31}^{(n)}\right), \tag{4.193f}$$
$$L_{34}^{(n)} = 2D_{11}^{(n)}D_{21}^{(n)}\left(D_{33}^{(n)} - D_{31}^{(n)}\right). \tag{4.193g}$$

We form the matrix $\mathbf{F}$ as a product

$$\mathbf{F} = \begin{pmatrix} F_{11} & F_{12} & F_{13} & F_{14} \\ F_{12} & F_{11} & F_{14} & F_{13} \\ F_{31} & F_{32} & F_{33} & F_{34} \\ F_{32} & F_{31} & F_{34} & F_{33} \end{pmatrix}$$
$$= \begin{pmatrix} L_{11}^{(n-1)} & L_{12}^{(n-1)} & L_{13}^{(n-1)} & L_{14}^{(n-1)} \\ L_{11}^{(n-1)} & -L_{12}^{(n-1)} & L_{13}^{(n-1)} & -L_{14}^{(n-1)} \\ L_{31}^{(n-1)} & L_{32}^{(n-1)} & -L_{13}^{(n-1)} & L_{34}^{(n-1)} \\ L_{31}^{(n-1)} & -L_{32}^{(n-1)} & -L_{13}^{(n-1)} & -L_{34}^{(n-1)} \end{pmatrix} \begin{pmatrix} D_{11}^{(n)} & D_{11}^{(n)} & D_{11}^{(n)} & D_{11}^{(n)} \\ D_{21}^{(n)} & -D_{21}^{(n)} & D_{23}^{(n)} & -D_{23}^{(n)} \\ D_{31}^{(n)} & D_{31}^{(n)} & D_{33}^{(n)} & D_{33}^{(n)} \\ D_{41}^{(n)} & -D_{41}^{(n)} & D_{43}^{(n)} & -D_{43}^{(n)} \end{pmatrix}. \tag{4.194}$$

From Eq. (4.194) we find

$$F_{\substack{11\\12}} = L_{11}^{(n-1)}D_{11}^{(n)} \pm L_{12}^{(n-1)}D_{21}^{(n)} + L_{13}^{(n-1)}D_{31}^{(n)} \pm L_{14}^{(n-1)}D_{41}^{(n)}, \tag{4.195a}$$
$$F_{\substack{13\\14}} = L_{11}^{(n-1)}D_{11}^{(n)} \pm L_{12}^{(n-1)}D_{23}^{(n)} + L_{13}^{(n-1)}D_{33}^{(n)} \pm L_{14}^{(n-1)}D_{43}^{(n)}, \tag{4.195b}$$
$$F_{\substack{31\\32}} = L_{31}^{(n-1)}D_{11}^{(n)} \pm L_{32}^{(n-1)}D_{21}^{(n)} - L_{13}^{(n-1)}D_{31}^{(n)} \pm L_{34}^{(n-1)}D_{41}^{(n)}, \tag{4.195c}$$
$$F_{\substack{33\\34}} = L_{31}^{(n-1)}D_{11}^{(n)} \pm L_{32}^{(n-1)}D_{23}^{(n)} - L_{13}^{(n-1)}D_{33}^{(n)} \pm L_{34}^{(n-1)}D_{43}^{(n)}. \tag{4.195d}$$

We may ignore normalizing matrices $\mathbf{C}^{(n-1)}$ and $\mathbf{C}^{(n)}$ unless we are interested in the fields inside the structure. Moreover, the common factors, *e.g.*, $[\det(\mathbf{D}^{(n-1)})]^{-1}$, are not important in determining both the multilayer reflection coefficients of Eqs. (3.70a) to (3.70d) and the waveguiding condition from Eq. (3.76).

The transfer matrix between two layers displaying magnetic order parallel to the z axis becomes

$$\mathbf{T}^{(n-1,n)} = [\det(\mathbf{D}^{(n-1)})]^{-1}(\mathbf{C}^{(n-1)})^{-1}\mathbf{F}\mathbf{C}^{(n)}\mathbf{P}^{(n)}, \tag{4.196}$$

where $\mathbf{C}^{(n-1)}$ and $\mathbf{C}^{(n)}$ are the matrices of normalizing coefficients $C_j^{(n-1)}$, $C_j^{(n)}$, and

$$\mathbf{P}^{(n)} = \begin{pmatrix} e^{j\beta_+^{(n)}} & 0 & 0 & 0 \\ 0 & e^{-j\beta_+^{(n)}} & 0 & 0 \\ 0 & 0 & e^{j\beta_-^{(n)}} & 0 \\ 0 & 0 & 0 & e^{-j\beta_-^{(n)}} \end{pmatrix}, \tag{4.197}$$

with $\beta_\pm^{(n)} = (\omega/c)N_{z\pm}^{(n)}d^{\,(n)}$, $d^{\,(n)}$ is the thickness of the n-th layer.

The transfer matrices $\mathbf{T}_{0,1}$ and $\mathbf{T}_{\mathcal{N},\mathcal{N}+1}$, which pertain to the multilayer left and right boundaries with the isotropic half spaces, are special cases of transfer matrices $\mathbf{T}^{(n-1,n)}$ given in Eq. (4.196). For convenience, we shall write them down. To describe the transfer matrix $\mathbf{T}_{0,1}$ at the left (front) interface $z = z_0$, we need the product

$$\mathbf{A} = (\mathbf{D}^{(0)})^{-1}\mathbf{D}^{(1)}, \tag{4.198}$$

which has already been obtained for the general form of $\mathbf{D}^{(1)}$ in Eq. (3.83) and Eq. (3.103). The use was made of the matrix $(\mathbf{D}^{(0)})^{-1}$ for the isotropic half space given in Eq. (3.82b) and corresponding to the proper polarizations linearly polarized perpendicular and parallel to the plane of incidence.

Taking into account Eq. (4.188) with $\mathbf{C}^{(1)} = \mathbf{1}$, we find that the $\mathbf{A}$ matrix has the structure of the more general matrix $\mathbf{F}$ given in Eq. (4.194)

$$\mathbf{A} = \begin{pmatrix} A_{11} & A_{12} & A_{13} & A_{14} \\ A_{12} & A_{11} & A_{14} & A_{13} \\ A_{31} & A_{32} & A_{33} & A_{34} \\ A_{32} & A_{31} & A_{34} & A_{33} \end{pmatrix}. \tag{4.199}$$

Its elements are

$$A_{\substack{11\\12}} = \left(2N^{(0)}\cos\varphi^{(0)}\right)^{-1}\left(D_{11}^{(1)}N^{(0)}\cos\varphi^{(0)} \pm D_{21}^{(1)}\right), \tag{4.200a}$$

$$A_{\substack{13\\14}} = \left(2N^{(0)}\cos\varphi^{(0)}\right)^{-1}\left(D_{11}^{(1)}N^{(0)}\cos\varphi^{(0)} \pm D_{23}^{(1)}\right), \tag{4.200b}$$

$$A_{\substack{31\\32}} = \left(2N^{(0)}\cos\varphi^{(0)}\right)^{-1}\left(D_{31}^{(1)}N^{(0)} \mp D_{41}^{(1)}\cos\varphi^{(0)}\right), \tag{4.200c}$$

$$A_{\substack{33\\34}} = \left(2N^{(0)}\cos\varphi^{(0)}\right)^{-1}\left(D_{33}^{(1)}N^{(0)} \mp D_{43}^{(1)}\cos\varphi^{(0)}\right). \tag{4.200d}$$

The transfer matrix at the right interface $z = z_N$ is given by the product

$$\mathbf{T}_{\mathcal{N},\mathcal{N}+1} = \left[\det\left(\mathbf{D}^{(\mathcal{N})}\right)\right]^{-1}\left(\mathbf{C}^{(\mathcal{N})}\right)^{-1}\mathbf{Z} \tag{4.201}$$

with

$$\mathbf{Z} = \begin{pmatrix} L_{11}^{(\mathcal{N})} & L_{12}^{(\mathcal{N})} & L_{13}^{(\mathcal{N})} & L_{14}^{(\mathcal{N})} \\ L_{11}^{(\mathcal{N})} & -L_{12}^{(\mathcal{N})} & L_{13}^{(\mathcal{N})} & -L_{14}^{(\mathcal{N})} \\ L_{31}^{(\mathcal{N})} & L_{32}^{(\mathcal{N})} & -L_{13}^{(\mathcal{N})} & L_{34}^{(\mathcal{N})} \\ L_{31}^{(\mathcal{N})} & -L_{32}^{(\mathcal{N})} & -L_{13}^{(\mathcal{N})} & -L_{34}^{(\mathcal{N})} \end{pmatrix}\mathbf{D}^{(\mathcal{N}+1)}. \tag{4.202}$$

The dynamical matrix $\mathbf{D}^{(\mathcal{N}+1)}$ of the other isotropic half space for the proper polarizations linearly polarized perpendicular and parallel to the plane of incidence follows from Eq. (3.60) with $s = \mathcal{N}+1$, $p_{(\mathcal{N}+1)} = 1$, and $q_{(\mathcal{N}+1)} = 0$,

$$\mathbf{D}^{(\mathcal{N}+1)} = \begin{pmatrix} 1 & 1 & 0 & 0 \\ N^{(\mathcal{N}+1)}\cos\varphi^{(\mathcal{N}+1)} & -N^{(\mathcal{N}+1)}\cos\varphi^{(\mathcal{N}+1)} & 0 & 0 \\ 0 & 0 & \cos\varphi^{(\mathcal{N}+1)} & \cos\varphi^{(\mathcal{N}+1)} \\ 0 & 0 & -N^{(\mathcal{N}+1)} & N^{(\mathcal{N}+1)} \end{pmatrix}. \tag{4.203}$$

Performing the matrix multiplication in Eq. (4.202) with Eq. (4.203) we find that **Z** has the same structure as **F** and **A** given in Eq. (4.194) and Eq. (4.199). The elements of **Z** are given (up to a common factor) by

$$Z_{\substack{11\\12}} = L_{11}^{(\mathcal{N})} \pm L_{12}^{(\mathcal{N})}N^{(\mathcal{N}+1)}\cos\varphi^{(\mathcal{N}+1)}, \tag{4.204a}$$

$$Z_{\substack{13\\14}} = L_{13}^{(\mathcal{N})}\cos\varphi^{(\mathcal{N}+1)} \mp L_{14}^{(\mathcal{N})}N^{(\mathcal{N}+1)}, \tag{4.204b}$$

$$Z_{\substack{31\\32}} = L_{31}^{(\mathcal{N})} \pm L_{32}^{(\mathcal{N})}N^{(\mathcal{N}+1)}\cos\varphi^{(\mathcal{N}+1)}, \tag{4.204c}$$

$$Z_{\substack{33\\34}} = -L_{13}^{(\mathcal{N})}\cos\varphi^{(\mathcal{N}+1)} \mp L_{34}^{(\mathcal{N})}N^{(\mathcal{N}+1)}. \tag{4.204d}$$

In this way, we have collected all the information necessary for the construction of the **M** matrix of the multilayer structure as a product of matrices of specific symmetry typical for the regions characterized by the permittivity tensor of Eq. (4.177).

4.5.2 Film–Substrate System

4.5.2.1 M *Matrix*

We shall now apply the information on the dynamical, propagation, and transfer matrices at oblique incidence with polar magnetization to the analysis of the reflection in a system consisting of an isotropic ambient, magnetic film, and magnetic substrate. This is the simplest situation that includes the interface between two magnetic regions. It is of both fundamental and practical importance. From the analysis, we can identify various contributions to the response. Of particular interest are the effects of the in-plane and perpendicular anisotropy, parallel or antiparallel magnetizations in the film and the substrate, those of the film thickness, angle of incidence, and radiation wavelength. At oblique angles of incidence, a rather complicated mode coupling takes place upon the internal reflection at film–ambient and film–substrate interfaces. The solution of the electromagnetic problem provides the basis for the microscopic models on electron spin arrangement, magnetic coupling between film and substrate, and interface-induced changes in the electron level structure.

In our analysis, we therefore assume an arbitrary angle of incidence $\varphi^{(0)}$ of the incident plane wave in the ambient of the refractive index $N^{(0)}$. As before, the condition $x = 0$ specifies the plane of incidence. The film and substrate media are characterized by the permittivity tensor of Eq. (4.177) including the effect of in-plane anisotropy. We situate the ambient–magnetic film interface into the plane $z = 0$ and the magnetic film–magnetic surface interface into the plane $z = d^{(1)}$ with their magnetization parallel to the z axis. The substrate is assumed thick and absorbing so that we can ignore the waves travelling from the substrate medium, corresponding to the region of $z > d^{(1)}$. We thus limit ourselves to the reflection characteristics. Then, the normalizing matrices $\mathbf{C}^{(n)}$ can be replaced by the unit matrices, and the common factors can be ignored. They should only be restored in case we wish to determine the fields inside the structure.

The **M** matrix of this system takes the form

$$\mathbf{M} = (\mathbf{D}^{(0)})^{-1}\mathbf{D}^{(1)}\mathbf{P}^{(1)}(\mathbf{D}^{(1)})^{-1}\mathbf{D}^{(2)}. \tag{4.205}$$

Making use of the matrix **F** defined in Eq. (4.194)

$$\mathbf{F} = \det(\mathbf{D}^{(n-1)})(\mathbf{D}^{(n-1)})^{-1}\mathbf{D}^{(n)}, \tag{4.206}$$

the elements of which are given in Eq. (4.194) and Eq. (4.195), we can represent the system by the product

$$\mathbf{M} = \mathbf{A}\mathbf{P}^{(1)}\mathbf{F}. \tag{4.207}$$

Here, the matrix $\mathbf{A} = (\mathbf{D}^{(0)})^{-1}\mathbf{D}^{(1)}$ defined in Eq. (4.198) represents a special case of the matrix $[\det(\mathbf{D}^{(n-1)})]^{-1}\mathbf{F}$. The elements of **A** were obtained in

Eqs. (4.200). The propagation matrix $\mathbf{P}^{(1)}$ follows from Eq. (4.197) with $n = 1$.

The complete set of the **M** matrix elements, up to a common factor, is given by

$$\mathrm{M}_{\substack{11\\22}} = A_{11}F_{11}\mathrm{e}^{\pm \mathrm{j}\beta_+} + A_{12}F_{12}\mathrm{e}^{\mp \mathrm{j}\beta_+} + A_{13}F_{31}\mathrm{e}^{\pm \mathrm{j}\beta_-} + A_{14}F_{32}\mathrm{e}^{\mp \mathrm{j}\beta_-}, \quad (4.208a)$$

$$\mathrm{M}_{\substack{12\\21}} = A_{11}F_{12}\mathrm{e}^{\pm \mathrm{j}\beta_+} + A_{12}F_{11}\mathrm{e}^{\mp \mathrm{j}\beta_+} + A_{13}F_{32}\mathrm{e}^{\pm \mathrm{j}\beta_-} + A_{14}F_{31}\mathrm{e}^{\mp \mathrm{j}\beta_-}, \quad (4.208b)$$

$$\mathrm{M}_{\substack{13\\24}} = A_{11}F_{13}\mathrm{e}^{\pm \mathrm{j}\beta_+} + A_{12}F_{14}\mathrm{e}^{\mp \mathrm{j}\beta_+} + A_{13}F_{33}\mathrm{e}^{\pm \mathrm{j}\beta_-} + A_{14}F_{34}\mathrm{e}^{\mp \mathrm{j}\beta_-}, \quad (4.208c)$$

$$\mathrm{M}_{\substack{14\\23}} = A_{11}F_{14}\mathrm{e}^{\pm \mathrm{j}\beta_+} + A_{12}F_{13}\mathrm{e}^{\mp \mathrm{j}\beta_+} + A_{13}F_{34}\mathrm{e}^{\pm \mathrm{j}\beta_-} + A_{14}F_{33}\mathrm{e}^{\mp \mathrm{j}\beta_-}, \quad (4.208d)$$

$$\mathrm{M}_{\substack{31\\42}} = A_{31}F_{11}\mathrm{e}^{\pm \mathrm{j}\beta_+} + A_{32}F_{12}\mathrm{e}^{\mp \mathrm{j}\beta_+} + A_{33}F_{31}\mathrm{e}^{\pm \mathrm{j}\beta_-} + A_{34}F_{32}\mathrm{e}^{\mp \mathrm{j}\beta_-}, \quad (4.208e)$$

$$\mathrm{M}_{\substack{32\\41}} = A_{31}F_{12}\mathrm{e}^{\pm \mathrm{j}\beta_+} + A_{32}F_{11}\mathrm{e}^{\mp \mathrm{j}\beta_+} + A_{33}F_{32}\mathrm{e}^{\pm \mathrm{j}\beta_-} + A_{34}F_{31}\mathrm{e}^{\mp \mathrm{j}\beta_-}, \quad (4.208f)$$

$$\mathrm{M}_{\substack{33\\44}} = A_{31}F_{13}\mathrm{e}^{\pm \mathrm{j}\beta_+} + A_{32}F_{14}\mathrm{e}^{\mp \mathrm{j}\beta_+} + A_{33}F_{33}\mathrm{e}^{\pm \mathrm{j}\beta_-} + A_{34}F_{34}\mathrm{e}^{\mp \mathrm{j}\beta_-}, \quad (4.208g)$$

$$\mathrm{M}_{\substack{34\\43}} = A_{31}F_{14}\mathrm{e}^{\pm \mathrm{j}\beta_+} + A_{32}F_{13}\mathrm{e}^{\mp \mathrm{j}\beta_+} + A_{33}F_{34}\mathrm{e}^{\pm \mathrm{j}\beta_-} + A_{34}F_{33}\mathrm{e}^{\mp \mathrm{j}\beta_-}. \quad (4.208h)$$

Here, $\beta^{\pm}$ stands for $(\omega/c)\, N_z^{(1)\pm} d^{\,(1)}$. Each element is made of four terms corresponding to the four proper polarizations.

4.5.2.2 Expressions Entering the Reflection Matrix

The expressions entering the Cartesian Jones reflection matrix of the multilayer system can be written as

$$\begin{aligned}
&\mathrm{M}_{11}\mathrm{M}_{33} - \mathrm{M}_{13}\mathrm{M}_{31} \\
&= -\left(A_{12}A_{31} - A_{11}A_{32}\right)\left(F_{11}F_{14} - F_{12}F_{13}\right) \\
&\quad - \left(A_{13}A_{34} - A_{14}A_{33}\right)\left(F_{32}F_{33} - F_{31}F_{34}\right) \\
&\quad + \mathrm{e}^{\mathrm{j}(\beta_+ + \beta_-)}\left(A_{11}A_{33} - A_{13}A_{31}\right)\left(F_{11}F_{33} - F_{13}F_{31}\right) \\
&\quad + \mathrm{e}^{-\mathrm{j}(\beta_+ + \beta_-)}\left(A_{12}A_{34} - A_{14}A_{32}\right)\left(F_{12}F_{34} - F_{14}F_{32}\right) \\
&\quad + \mathrm{e}^{\mathrm{j}(\beta_+ - \beta_-)}\left(A_{11}A_{34} - A_{14}A_{31}\right)\left(F_{11}F_{34} - F_{13}F_{32}\right) \\
&\quad + \mathrm{e}^{-\mathrm{j}(\beta_+ - \beta_-)}\left(A_{12}A_{33} - A_{13}A_{32}\right)\left(F_{12}F_{33} - F_{14}F_{31}\right),
\end{aligned} \quad (4.209a)$$

$$\begin{aligned}
&\mathrm{M}_{21}\mathrm{M}_{33} - \mathrm{M}_{23}\mathrm{M}_{31} \\
&= \left(A_{12}A_{32} - A_{11}A_{31}\right)\left(F_{11}F_{14} - F_{12}F_{13}\right) \\
&\quad + \left(A_{13}A_{33} - A_{14}A_{34}\right)\left(F_{32}F_{33} - F_{31}F_{34}\right) \\
&\quad + \mathrm{e}^{\mathrm{j}(\beta_+ + \beta_-)}\left(A_{12}A_{33} - A_{14}A_{31}\right)\left(F_{11}F_{33} - F_{13}F_{31}\right) \\
&\quad + \mathrm{e}^{-\mathrm{j}(\beta_+ + \beta_-)}\left(A_{11}A_{34} - A_{13}A_{32}\right)\left(F_{12}F_{34} - F_{14}F_{32}\right) \\
&\quad + \mathrm{e}^{\mathrm{j}(\beta_+ - \beta_-)}\left(A_{12}A_{34} - A_{13}A_{31}\right)\left(F_{11}F_{34} - F_{13}F_{32}\right) \\
&\quad + \mathrm{e}^{-\mathrm{j}(\beta_+ - \beta_-)}\left(A_{11}A_{33} - A_{14}A_{32}\right)\left(F_{12}F_{33} - F_{14}F_{31}\right),
\end{aligned} \quad (4.209b)$$

$$\begin{aligned}
&M_{11}M_{43} - M_{13}M_{41} \\
&= (A_{11}A_{31} - A_{12}A_{32})(F_{11}F_{14} - F_{12}F_{13}) \\
&\quad + (A_{14}A_{34} - A_{13}A_{33})(F_{32}F_{33} - F_{31}F_{34}) \\
&\quad + e^{j(\beta_+ + \beta_-)}(A_{11}A_{34} - A_{13}A_{32})(F_{11}F_{33} - F_{13}F_{31}) \\
&\quad + e^{-j(\beta_+ + \beta_-)}(A_{12}A_{33} - A_{14}A_{31})(F_{12}F_{34} - F_{14}F_{32}) \\
&\quad + e^{j(\beta_+ - \beta_-)}(A_{11}A_{33} - A_{32}A_{14})(F_{11}F_{34} - F_{13}F_{32}) \\
&\quad + e^{-j(\beta_+ - \beta_-)}(A_{12}A_{34} - A_{13}A_{31})(F_{12}F_{33} - F_{14}F_{31}),
\end{aligned} \tag{4.209c}$$

$$\begin{aligned}
&M_{11}M_{23} - M_{21}M_{13} \\
&= \left(A_{11}^2 - A_{12}^2\right)(F_{11}F_{14} - F_{12}F_{13}) - \left(A_{13}^2 - A_{14}^2\right)(F_{32}F_{33} - F_{31}F_{34}) \\
&\quad + (A_{11}A_{14} - A_{12}A_{13})\left[e^{j(\beta_+ + \beta_-)}(F_{11}F_{33} - F_{13}F_{31})\right. \\
&\quad \left. - e^{-j(\beta_+ + \beta_-)}(F_{12}F_{34} - F_{14}F_{32})\right] + (A_{11}A_{13} - A_{12}A_{14}) \\
&\quad \times \left[e^{j(\beta_+ - \beta_-)}(F_{11}F_{34} - F_{13}F_{32}) - e^{-j(\beta_+ - \beta_-)}(F_{12}F_{33} - F_{14}F_{31})\right],
\end{aligned} \tag{4.209d}$$

and

$$\begin{aligned}
&M_{41}M_{33} - M_{43}M_{31} \\
&= -\left(A_{31}^2 - A_{32}^2\right)(F_{11}F_{14} - F_{12}F_{13}) + \left(A_{33}^2 - A_{34}^2\right)(F_{32}F_{33} - F_{31}F_{34}) \\
&\quad + (A_{32}A_{33} - A_{31}A_{34})\left[e^{j(\beta_+ + \beta_-)}(F_{11}F_{33} - F_{13}F_{31})\right. \\
&\quad \left. - e^{-j(\beta_+ + \beta_-)}(F_{12}F_{34} - F_{14}F_{32})\right] + (A_{32}A_{34} - A_{31}A_{33}) \\
&\quad \times \left[e^{j(\beta_+ - \beta_-)}(F_{11}F_{34} - F_{13}F_{32}) - e^{-j(\beta_+ - \beta_-)}(F_{12}F_{33} - F_{14}F_{31})\right].
\end{aligned} \tag{4.209e}$$

The terms independent of $\exp\left(j\beta^{\pm}\right)$ indicate the interface contribution originating from polar magnetization and other anisotropies characterized by the tensor in Eq. (4.177). The terms proportional to $e^{\pm j(\beta_+ - \beta_-)}$ include MO effects originating from the wave propagation across the film. Setting $e^{j(\beta_+ \pm \beta_-)} = 1$, which corresponds to $d^{(1)} = 0$, we effectively remove the film and return to the case of an interface between isotropic half space and anisotropic half space discussed in Section 3.6 for a general permittivity tensor (3.1) replaced by the special one (4.177) restricted to polar magnetization.

4.5.2.3 *Interface Reflection Coefficients*

The inspection of Eq. (3.70a) to Eq. (3.70d), Eq. (3.77), and Eq. (4.198) shows that

$$r_{21}^{(01)} = \left(\frac{E_{02}^{(0)}}{E_{01}^{(0)}}\right)_{E_{03}^{(0)}=0} = (A_{12}A_{33} - A_{14}A_{31})\,A_0^{-1}, \tag{4.210a}$$

$$r_{41}^{(01)} = \left(\frac{E_{04}^{(0)}}{E_{01}^{(0)}}\right)_{E_{03}^{(0)}=0} = (A_{32}A_{33} - A_{31}A_{34})\,A_0^{-1}, \tag{4.210b}$$

$$r_{43}^{(01)} = \left(\frac{E_{02}^{(0)}}{E_{03}^{(0)}}\right)_{E_{01}^{(0)}=0} = (A_{11}A_{34} - A_{13}A_{32})\,A_0^{-1}, \tag{4.210c}$$

$$r_{23}^{(01)} = \left(\frac{E_{02}^{(0)}}{E_{03}^{(0)}}\right)_{E_{01}^{(0)}=0} = (A_{11}A_{14} - A_{12}A_{13})\,A_0^{-1}, \tag{4.210d}$$

are the Cartesian reflection coefficients at the interface $z = 0$ of two semi-infinite media: isotropic one, indexed (0), and magnetized one in the direction normal to the interface, indexed (1). They correspond to the reflection coefficients of Eq. (3.96) simplified from the general tensor (3.1) to the special one (4.177) for higher symmetry and polar magnetization. It is convenient to put the single-interface reflection coefficients into the form that relates the linearly polarized incident and reflected proper polarizations in the isotropic medium polarized perpendicular and parallel to the plane of incidence. Here, $A_0 = A_{11}A_{33} - A_{13}A_{31}$. With help of Eq. (4.200) and taking into account Eq. (3.99) or Eq. (4.189), we obtain, after skipping the common factor $\left(2N^{(0)}\cos\varphi^{(0)}\right)^2$ on the left-hand side of the following equations, the following useful relations

$$\begin{aligned}
&(A_{11}A_{33} - A_{13}A_{31})\\
&\quad = N^{(0)}\cos^2\varphi^{(0)}\mathcal{P}^{(1)} + N^{(0)2}\cos\varphi^{(0)}\mathcal{Q}^{(1)} + \cos\varphi^{(0)}\mathcal{S}^{(1)} + N^{(0)}\mathcal{T}^{(1)}\\
&\quad = N^{(0)}\cos^2\varphi^{(0)}D_{11}^{(1)}\left(D_{41}^{(1)} - D_{43}^{(1)}\right) + N^{(0)2}\cos\varphi^{(0)}D_{11}^{(1)}\left(D_{33}^{(1)} - D_{31}^{(1)}\right)\\
&\qquad + \cos\varphi^{(0)}\left(D_{41}^{(1)}D_{23}^{(1)} - D_{43}^{(1)}D_{21}^{(1)}\right) + N^{(0)}\left(D_{21}^{(1)}D_{33}^{(1)} - D_{23}^{(1)}D_{31}^{(1)}\right),
\end{aligned} \tag{4.211a}$$

$$\begin{aligned}
&(A_{12}A_{33} - A_{14}A_{31})\\
&\quad = N^{(0)}\cos^2\varphi^{(0)}\mathcal{P}^{(1)} + N^{(0)2}\cos\varphi^{(0)}\mathcal{Q}^{(1)} - \cos\varphi^{(0)}\mathcal{S}^{(1)} - N^{(0)}\mathcal{T}^{(1)}\\
&\quad = N^{(0)}\cos^2\varphi^{(0)}D_{11}^{(1)}\left(D_{41}^{(1)} - D_{43}^{(1)}\right) + N^{(0)2}\cos\varphi^{(0)}D_{11}^{(1)}\left(D_{33}^{(1)} - D_{31}^{(1)}\right)\\
&\quad - \cos\varphi^{(0)}\left(D_{41}^{(1)}D_{23}^{(1)} - D_{43}^{(1)}D_{21}^{(1)}\right) - N^{(0)}\left(D_{21}^{(1)}D_{33}^{(1)} - D_{23}^{(1)}D_{31}^{(1)}\right),
\end{aligned} \tag{4.211b}$$

$$\begin{aligned}
&(A_{11}A_{34} - A_{13}A_{32})\\
&\quad = -N^{(0)}\cos^2\varphi^{(0)}\mathcal{P}^{(1)} + N^{(0)2}\cos\varphi^{(0)}\mathcal{Q}^{(1)} - \cos\varphi^{(0)}\mathcal{S}^{(1)} + N^{(0)}\mathcal{T}^{(1)}\\
&\quad = -N^{(0)}\cos^2\varphi^{(0)}D_{11}^{(1)}\left(D_{41}^{(1)} - D_{43}^{(1)}\right) + N^{(0)2}\cos\varphi^{(0)}D_{11}^{(1)}\left(D_{33}^{(1)} - D_{31}^{(1)}\right)\\
&\qquad - \cos\varphi^{(0)}\left(D_{41}^{(1)}D_{23}^{(1)} - D_{43}^{(1)}D_{21}^{(1)}\right) + N^{(0)}\left(D_{21}^{(1)}D_{33}^{(1)} - D_{23}^{(1)}D_{31}^{(1)}\right),
\end{aligned} \tag{4.211c}$$

$$
\begin{aligned}
&(A_{12}A_{34} - A_{14}A_{32}) \\
&\quad = -N^{(0)}\cos^2\varphi^{(0)}\mathcal{P}^{(1)} + N^{(0)2}\cos\varphi^{(0)}\mathcal{Q}^{(1)} + \cos\varphi^{(0)}\mathcal{S}^{(1)} - N^{(0)}\mathcal{T}^{(1)} \\
&\quad = -N^{(0)}\cos^2\varphi^{(0)}D_{11}^{(1)}\left(D_{41}^{(1)} - D_{43}^{(1)}\right) + N^{(0)2}\cos\varphi^{(0)}D_{11}^{(1)}\left(D_{33}^{(1)} - D_{31}^{(1)}\right) \\
&\qquad + \cos\varphi^{(0)}\left(D_{41}^{(1)}D_{23}^{(1)} - D_{43}^{(1)}D_{21}^{(1)}\right) - N^{(0)}\left(D_{21}^{(1)}D_{33}^{(1)} - D_{23}^{(1)}D_{31}^{(1)}\right).
\end{aligned} \tag{4.211d}
$$

After substituting for the dynamical matrix elements according to Eq. (4.187), we may further write

$$
\begin{aligned}
(A_{11}A_{14} - A_{12}A_{13}) &= 2N^{(0)}\cos\varphi^{(0)}D_{11}^{(1)}\left(D_{21}^{(1)} - D_{23}^{(1)}\right) \\
&= 2N^{(0)}\cos\varphi^{(0)}\varepsilon_{xy}^{(1)2}\left(\varepsilon_{zz}^{(1)} - N_y^2\right)^2\left(N_{z+}^{(1)} - N_{z-}^{(1)}\right),
\end{aligned} \tag{4.211e}
$$

$$
\begin{aligned}
(A_{32}A_{33} - A_{31}A_{34}) &= 2N^{(0)}\cos\varphi^{(0)}\left(D_{41}^{(1)}D_{33}^{(1)} - D_{43}^{(1)}D_{31}^{(1)}\right) \\
&= -\left(2N^{(0)}\cos\varphi^{(0)}\right)\left(N_{z+}^{(1)} - N_{z-}^{(1)}\right)\left(\varepsilon_{zz}^{(1)} - N_y^2\right)\varepsilon_{zz} \\
&\quad \times \left[\left(\varepsilon_{xx}^{(1)} - N_y^2\right)^2 - \left(\varepsilon_{xx}^{(1)} - N_y^2\right)\left(N_{z+}^{(1)2} + N_{z-}^{(1)2}\right) + N_{z+}^{(1)2}N_{z-}^{(1)2}\right] \\
&= \left(2N^{(0)}\cos\varphi^{(0)}\right)\left(N_{z+}^{(1)} - N_{z-}^{(1)}\right)\varepsilon_{xy}^{(1)}\varepsilon_{yx}^{(1)}\left(\varepsilon_{zz}^{(1)} - N_y^2\right)^2,
\end{aligned} \tag{4.211f}
$$

$$
\begin{aligned}
(A_{12}A_{32} - A_{11}A_{31}) &= 2N^{(0)}\left(D_{11}^{(1)}D_{41}^{(1)}\cos^2\varphi^{(0)} - D_{21}^{(1)}D_{31}^{(1)}\right) \\
&= 2N^{(0)}\varepsilon_{xy}^{(1)}\left(\varepsilon_{zz}^{(1)} - N_y^2\right)N_{z+}^{(1)}\left[\left(\varepsilon_{xx}^{(1)} - N_y^2\right) - N_{z+}^{(1)2}\right] \\
&\quad \times \left[\varepsilon_{zz}^{(1)}\cos^2\varphi^{(0)} + \left(\varepsilon_{zz}^{(1)} - N_y^2\right)\right],
\end{aligned} \tag{4.211g}
$$

$$
\begin{aligned}
(A_{13}A_{33} - A_{14}A_{34}) &= 2N^{(0)}\left(D_{11}^{(1)}D_{43}^{(1)}\cos^2\varphi^{(0)} - D_{23}^{(1)}D_{33}^{(1)}\right) \\
&= 2N^{(0)}\varepsilon_{xy}^{(1)}\left(\varepsilon_{zz}^{(1)} - N_y^2\right)N_{z-}^{(1)}\left[\left(\varepsilon_{xx}^{(1)} - N_y^2\right) - N_{z-}^{(1)2}\right] \\
&\quad \times \left[\varepsilon_{zz}^{(1)}\cos^2\varphi^{(0)} + \left(\varepsilon_{zz}^{(1)} - N_y^2\right)\right].
\end{aligned} \tag{4.211h}
$$

The use was made of Eq. (4.183), providing

$$
\begin{aligned}
&\varepsilon_{zz}^{(1)}\left[\varepsilon_{xx}^{(n)} - N_y^2 - \left(N_{z+}^{(n)}\right)^2\right]\left[\varepsilon_{xx}^{(n)} - N_y^2 - \left(N_{z-}^{(n)}\right)^2\right] \\
&\quad = -\varepsilon_{xy}^{(1)}\varepsilon_{yx}^{(1)}\left(\varepsilon_{zz}^{(1)} - N_y^2\right).
\end{aligned} \tag{4.211i}
$$

The differences of the squared **A** matrix elements can be expressed explicitly as

$$
\left(A_{11}^2 - A_{12}^2\right) = 4N^{(0)}\cos\varphi^{(0)}D_{11}^{(1)}D_{21}^{(1)} = 4N^{(0)}\cos\varphi^{(0)}\varepsilon_{xy}^{(1)2}\left(\varepsilon_{zz}^{(1)} - N_y^2\right)^2 N_{z+}^{(1)}, \tag{4.211j}
$$

$$
\left(A_{13}^2 - A_{14}^2\right) = 4N^{(0)}\cos\varphi^{(0)}D_{11}^{(1)}D_{23}^{(1)} = 4N^{(0)}\cos\varphi^{(0)}\varepsilon_{xy}^{(1)2}\left(\varepsilon_{zz}^{(1)} - N_y^2\right)^2 N_{z-}^{(1)}, \tag{4.211k}
$$

$$\left(A_{31}^2 - A_{32}^2\right) = -4N^{(0)}\cos\varphi^{(0)} D_{31}^{(1)} D_{41}^{(1)}$$
$$= 4N^{(0)}\cos\varphi^{(0)} N_{z+}^{(1)} \varepsilon_{zz}^{(1)} \left(\varepsilon_{zz}^{(1)} - N_y^2\right)\left[\varepsilon_{xx}^{(1)} - N_y^2 - \left(N_{z+}^{(1)}\right)^2\right]^2, \qquad (4.211l)$$
$$\left(A_{33}^2 - A_{34}^2\right) = -4N^{(0)}\cos\varphi^{(0)} D_{33}^{(1)} D_{43}^{(1)}$$
$$= 4N^{(0)}\cos\varphi^{(0)} N_{z-}^{(1)} \varepsilon_{zz}^{(1)} \left(\varepsilon_{zz}^{(1)} - N_y^2\right)\left[\varepsilon_{xx}^{(1)} - N_y^2 - \left(N_{z-}^{(1)}\right)^2\right]^2. \qquad (4.211m)$$

The reflection coefficients for the proper polarizations incident on this interface from the side of the medium (1) are

$$r_{12}^{(10)} = \left(\frac{E_{01}^{(1)}}{E_{02}^{(1)}}\right)_{E_{04}^{(1)}=0} = \left(A_{13}A_{32} - A_{12}A_{33}\right) A_0^{-1}, \qquad (4.212a)$$

$$r_{32}^{(10)} = \left(\frac{E_{03}^{(1)}}{E_{02}^{(1)}}\right)_{E_{04}^{(1)}=0} = \left(A_{12}A_{31} - A_{11}A_{32}\right) A_0^{-1}, \qquad (4.212b)$$

$$r_{34}^{(10)} = \left(\frac{E_{03}^{(1)}}{E_{04}^{(1)}}\right)_{E_{02}^{(1)}=0} = \left(A_{14}A_{31} - A_{11}A_{34}\right) A_0^{-1}, \qquad (4.212c)$$

$$r_{14}^{(10)} = \left(\frac{E_{01}^{(1)}}{E_{04}^{(1)}}\right)_{E_{02}^{(1)}=0} = \left(A_{13}A_{34} - A_{14}A_{33}\right) A_0^{-1}. \qquad (4.212d)$$

Here, $E_{02}^{(1)}$ and $E_{04}^{(1)}$ are the amplitudes at the interface $z = 0$ of the proper polarizations in the region (1) corresponding to the magnetic layer. The proper polarization modes propagate in the magnetized medium (1) with $(\omega/c)N_{z2}^{(1)} = -(\omega/c)N_{z1}^{(1)}$ and $(\omega/c)N_{z4}^{(1)} = -(\omega/c)N_{z3}^{(1)}$, respectively, towards the interface $z = 0$. With Eq. (4.200), we obtain, again after skipping the common factor $(2N^{(0)}\cos\varphi^{(0)})^2$ on the left-hand side of the following equations,

$$\left(A_{13}A_{32} - A_{12}A_{33}\right)$$
$$= -N^{(0)2}\cos\varphi^{(0)} D_{11}^{(1)}\left(D_{33}^{(1)} - D_{31}^{(1)}\right) + N^{(0)}\cos^2\varphi^{(0)} D_{11}^{(1)}\left(D_{41}^{(1)} + D_{43}^{(1)}\right)$$
$$+N^{(0)}\left(D_{21}^{(1)}D_{33}^{(1)} + D_{23}^{(1)}D_{31}^{(1)}\right) + \cos\varphi^{(0)}\left(D_{41}^{(1)}D_{23}^{(1)} - D_{43}^{(1)}D_{21}^{(1)}\right), \qquad (4.213a)$$
$$\left(A_{14}A_{31} - A_{11}A_{34}\right)$$
$$= -N^{(0)2}\cos\varphi^{(0)} D_{11}^{(1)}\left(D_{33}^{(1)} - D_{31}^{(1)}\right) - N^{(0)}\cos^2\varphi^{(0)} D_{11}^{(1)}\left(D_{41}^{(1)} + D_{43}^{(1)}\right)$$
$$-N^{(0)}\left(D_{21}^{(1)}D_{33}^{(1)} + D_{23}^{(1)}D_{31}^{(1)}\right) + \cos\varphi^{(0)}\left(D_{41}^{(1)}D_{23}^{(1)} - D_{43}^{(1)}D_{21}^{(1)}\right). \qquad (4.213b)$$

After substituting for the dynamical matrix elements according to Eq. (4.187), we may write

$$\left(A_{11}A_{13} - A_{12}A_{14}\right) = 2N^{(0)}\cos\varphi^{(0)} D_{11}^{(1)}\left(D_{21}^{(1)} + D_{23}^{(1)}\right)$$
$$= 2N^{(0)}\cos\varphi^{(0)} \varepsilon_{xy}^{(1)2}\left(\varepsilon_{zz}^{(1)} - N_y^2\right)^2\left(N_{z+}^{(1)} + N_{z-}^{(1)}\right), \qquad (4.213c)$$

$$(A_{32}A_{34} - A_{31}A_{33}) = 2N^{(0)}\cos\varphi^{(0)}\left(D_{41}^{(1)}D_{33}^{(1)} + D_{43}^{(1)}D_{31}^{(1)}\right)$$
$$= -\left(2N^{(0)}\cos\varphi^{(0)}\right)\left(N_{z+}^{(1)} + N_{z-}^{(1)}\right)\left(\varepsilon_{zz}^{(1)} - N_y^2\right)\varepsilon_{zz}$$
$$\times\left[\left(\varepsilon_{xx}^{(1)} - N_y^2\right)^2 - \left(\varepsilon_{xx}^{(1)} - N_y^2\right)\left(N_{z+}^{(1)2} + N_{z-}^{(1)2}\right) + N_{z+}^{(1)2}N_{z-}^{(1)2}\right]$$
$$= \left(2N^{(0)}\cos\varphi^{(0)}\right)\left(N_{z+}^{(1)} + N_{z-}^{(1)}\right)\varepsilon_{xy}^{(1)}\varepsilon_{yx}^{(1)}\left(\varepsilon_{zz}^{(1)} - N_y^2\right)^2, \tag{4.213d}$$
$$(A_{12}A_{31} - A_{11}A_{32}) = 2N^{(0)}\left(D_{11}^{(1)}D_{41}^{(1)}\cos^2\varphi^{(0)} + D_{21}^{(1)}D_{31}^{(1)}\right)$$
$$= 2N^{(0)}\varepsilon_{xy}^{(1)}\left(\varepsilon_{zz}^{(1)} - N_y^2\right)N_{z+}^{(1)}\left[\left(\varepsilon_{xx}^{(1)} - N_y^2\right) - N_{z+}^{(n)2}\right]$$
$$\times\left[\varepsilon_{zz}\cos^2\varphi^{(0)} - \left(\varepsilon_{zz}^{(1)} - N_y^2\right)\right], \tag{4.213e}$$
$$(A_{13}A_{34} - A_{14}A_{33}) = 2N^{(0)}\left(D_{11}^{(1)}D_{43}^{(1)}\cos^2\varphi^{(0)} + D_{23}^{(1)}D_{33}^{(1)}\right)$$
$$= 2N^{(0)}\varepsilon_{xy}^{(1)}\left(\varepsilon_{zz}^{(1)} - N_y^2\right)N_{z-}^{(1)}\left[\left(\varepsilon_{xx}^{(1)} - N_y^2\right) - N_{z-}^{(n)2}\right]$$
$$\times\left[\varepsilon_{zz}\cos^2\varphi^{(0)} - \left(\varepsilon_{zz}^{(1)} - N_y^2\right)\right]. \tag{4.213f}$$

We further have

$$\tag{4.213g}$$

$$(A_{12}A_{34} - A_{13}A_{31})$$
$$= N^{(0)}\cos^2\varphi^{(0)}D_{11}^{(1)}\left(D_{41}^{(1)} + D_{43}^{(1)}\right) + N^{(0)2}\cos\varphi^{(0)}D_{11}^{(1)}\left(D_{33}^{(1)} - D_{31}^{(1)}\right)$$
$$+\cos\varphi^{(0)}\left(D_{23}^{(1)}D_{41}^{(1)} - D_{21}^{(1)}D_{43}^{(1)}\right) - N^{(0)}\left(D_{21}^{(1)}D_{33}^{(1)} + D_{23}^{(1)}D_{31}^{(1)}\right), \tag{4.213h}$$
$$(A_{11}A_{33} - A_{14}A_{32})$$
$$= -N^{(0)}\cos^2\varphi^{(0)}D_{11}^{(1)}\left(D_{41}^{(1)} + D_{43}^{(1)}\right) + N^{(0)2}\cos\varphi^{(0)}D_{11}^{(1)}\left(D_{33}^{(1)} - D_{31}^{(1)}\right)$$
$$+\cos\varphi^{(0)}\left(D_{23}^{(1)}D_{41}^{(1)} - D_{21}^{(1)}D_{43}^{(1)}\right) + N^{(0)}\left(D_{21}^{(1)}D_{33}^{(1)} + D_{23}^{(1)}D_{31}^{(1)}\right). \tag{4.213i}$$

We observe that the following expression is invariant with respect to the exchange of the region indices (0) and (1)

$$r_{21}^{(01)}r_{43}^{(01)} - r_{41}^{(01)}r_{23}^{(01)} = r_{12}^{(10)}r_{34}^{(10)} - r_{14}^{(10)}r_{32}^{(10)} = \left(A_{12}A_{34} - A_{14}A_{32}\right)A_0^{-1}. \tag{4.214}$$

Note that $r_{32}^{(10)}$ and $r_{14}^{(10)}$, which characterize the mode coupling, have the meaning of the reflection coefficients in the film medium (1) if the proper polarizations in the region (1) are normalized. In Eq. (4.210), Eq. (4.212), and Eq. (4.214), we have identified most of the expressions built from the elements of **A** entering Eq. (4.209).

Denoting $F_0 = F_{11}F_{33} - F_{13}F_{31}$, we can write, in a similar way, for the film substrate interface at $z = d^{(1)}$

$$r_{21}^{(12)} = \left(\frac{E_{02}^{(1)}}{E_{01}^{(1)}}\right)_{E_{03}^{(1)}=0} = \left(F_{12}F_{33} - F_{14}F_{31}\right)F_0^{-1}, \tag{4.215a}$$

$$r_{41}^{(12)} = \left(\frac{E_{04}^{(1)}}{E_{01}^{(1)}}\right)_{E_{03}^{(1)}=0} = \left(F_{32}F_{33} - F_{31}F_{34}\right)F_0^{-1}, \tag{4.215b}$$

$$r_{43}^{(12)} = \left(\frac{E_{04}^{(1)}}{E_{03}^{(1)}}\right)_{E_{01}^{(1)}=0} = \left(F_{11}F_{34} - F_{13}F_{32}\right)F_0^{-1}, \tag{4.215c}$$

$$r_{23}^{(12)} = \left(\frac{E_{02}^{(1)}}{E_{03}^{(1)}}\right)_{E_{01}^{(1)}=0} = \left(F_{11}F_{14} - F_{12}F_{13}\right)F_0^{-1}, \tag{4.215d}$$

where $E_{01}^{(1)}$ and $E_{03}^{(1)}$ are the amplitudes of the proper polarization modes propagating in the medium (1) with $+(\omega/c)N_{z1}^{(1)}$ and $+(\omega/c)N_{z3}^{(1)}$ towards the interface $z = d^{(1)}$ between the magnetic film (1) and the magnetic substrate (2). In view of Eq. (4.185), the expressions for $r_{41}^{(12)}$ and $r_{23}^{(12)}$ have the meaning of the reflection coefficients if the proper polarization modes in the film medium (1) are normalized. They indicate rather delicate polarization mode coupling between the proper polarization modes of the film medium upon the reflection at the film–substrate interface, when both the media are characterized by the permittivity tensor of the same symmetry and given in Eq. (4.177). With Eq. (4.195), we obtain for the expression formed by the elements of the $\mathbf{F}$ matrix

$$\begin{aligned}(F_{11}F_{14} - F_{12}F_{13}) &= 2L_{11}^{(n-1)}L_{12}^{(n-1)}D_{11}^{(n)}\left(D_{21}^{(n)} - D_{23}^{(n)}\right)\\ &+ 2L_{11}^{(n-1)}L_{14}^{(n-1)}D_{11}^{(n)}\left(D_{41}^{(n)} - D_{43}^{(n)}\right)\\ &+ 2L_{12}^{(n-1)}L_{13}^{(n-1)}\left(D_{21}^{(n)}D_{33}^{(n)} - D_{23}^{(n)}D_{31}^{(n)}\right)\\ &+ 2L_{13}^{(n-1)}L_{14}^{(n-1)}\left(D_{33}^{(n)}D_{41}^{(n)} - D_{31}^{(n)}D_{43}^{(n)}\right)\end{aligned} \tag{4.216a}$$

$$\begin{aligned}&(F_{31}F_{34} - F_{32}F_{33})\\ &= 2L_{31}^{(n-1)}L_{32}^{(n-1)}D_{11}^{(n)}\left(D_{21}^{(n)} - D_{23}^{(n)}\right) + 2L_{31}^{(n-1)}L_{34}^{(n-1)}D_{11}^{(n)}\left(D_{41}^{(n)} - D_{43}^{(n)}\right)\\ &- 2L_{13}^{(n-1)}L_{32}^{(n-1)}\left(D_{21}^{(n)}D_{33}^{(n)} - D_{23}^{(n)}D_{31}^{(n)}\right)\\ &- 2L_{13}^{(n-1)}L_{34}^{(n-1)}\left(D_{33}^{(n)}D_{41}^{(n)} - D_{31}^{(n)}D_{43}^{(n)}\right)\end{aligned} \tag{4.216b}$$

$$\begin{aligned}&(F_{11}F_{33} - F_{13}F_{31})\\ &= -\left(L_{11}^{(n-1)}L_{32}^{(n-1)} - L_{12}^{(n-1)}L_{31}^{(n-1)}\right)D_{11}^{(n)}\left(D_{21}^{(n)} - D_{23}^{(n)}\right)\\ &- L_{13}^{(n-1)}\left(L_{11}^{(n-1)} + L_{31}^{(n-1)}\right)D_{11}^{(n)}\left(D_{33}^{(n)} - D_{31}^{(n)}\right)\\ &- \left(L_{11}^{(n-1)}L_{34}^{(n-1)} - L_{14}^{(n-1)}L_{31}^{(n-1)}\right)D_{11}^{(n)}\left(D_{41}^{(n)} - D_{43}^{(n)}\right)\\ &- L_{13}^{(n-1)}\left(L_{12}^{(n-1)} + L_{32}^{(n-1)}\right)\left(D_{21}^{(n)}D_{33}^{(n)} - D_{23}^{(n)}D_{31}^{(n)}\right)\\ &- \left(L_{12}^{(n-1)}L_{34}^{(n-1)} - L_{14}^{(n-1)}L_{32}^{(n-1)}\right)\left(D_{23}^{(n)}D_{41}^{(n)} - D_{21}^{(n)}D_{43}^{(n)}\right)\\ &- L_{13}^{(n-1)}\left(L_{34}^{(n-1)} + L_{14}^{(n-1)}\right)\left(D_{33}^{(n)}D_{41}^{(n)} - D_{31}^{(n)}D_{43}^{(n)}\right)\end{aligned} \tag{4.216c}$$

$$
\begin{aligned}
&\left(F_{12}F_{34} - F_{14}F_{32}\right) \\
&\quad = \left(L_{11}^{(n-1)}L_{32}^{(n-1)} - L_{12}^{(n-1)}L_{31}^{(n-1)}\right)D_{11}^{(n)}\left(D_{21}^{(n)} - D_{23}^{(n)}\right) \\
&\quad - L_{13}^{(n-1)}\left(L_{11}^{(n-1)} + L_{31}^{(n-1)}\right)D_{11}^{(n)}\left(D_{33}^{(n)} - D_{31}^{(n)}\right) \\
&\quad + \left(L_{11}^{(n-1)}L_{34}^{(n-1)} - L_{14}^{(n-1)}L_{31}^{(n-1)}\right)D_{11}^{(n)}\left(D_{41}^{(n)} - D_{43}^{(n)}\right) \\
&\quad + L_{13}^{(n-1)}\left(L_{12}^{(n-1)} + L_{32}^{(n-1)}\right)\left(D_{21}^{(n)}D_{33}^{(n)} - D_{23}^{(n)}D_{31}^{(n)}\right) \\
&\quad - \left(L_{12}^{(n-1)}L_{34}^{(n-1)} - L_{14}^{(n-1)}L_{32}^{(n-1)}\right)\left(D_{23}^{(n)}D_{41}^{(n)} - D_{21}^{(n)}D_{43}^{(n)}\right) \\
&\quad + L_{13}^{(n-1)}\left(L_{34}^{(n-1)} + L_{14}^{(n-1)}\right)\left(D_{33}^{(n)}D_{41}^{(n)} - D_{31}^{(n)}D_{43}^{(n)}\right)
\end{aligned}
\tag{4.216d}
$$

$$
\begin{aligned}
&\left(F_{11}F_{34} - F_{13}F_{32}\right) \\
&\quad = \left(L_{11}^{(n-1)}L_{32}^{(n-1)} + L_{12}^{(n-1)}L_{31}^{(n-1)}\right)D_{11}^{(n)}\left(D_{21}^{(n)} - D_{23}^{(n)}\right) \\
&\quad - L_{13}^{(n-1)}\left(L_{11}^{(n-1)} + L_{31}^{(n-1)}\right)D_{11}^{(n)}\left(D_{33}^{(n)} - D_{31}^{(n)}\right) \\
&\quad + \left(L_{11}^{(n-1)}L_{34}^{(n-1)} + L_{14}^{(n-1)}L_{31}^{(n-1)}\right)D_{11}^{(n)}\left(D_{41}^{(n)} - D_{43}^{(n)}\right) \\
&\quad + L_{13}^{(n-1)}\left(L_{32}^{(n-1)} - L_{12}^{(n-1)}\right)\left(D_{21}^{(n)}D_{33}^{(n)} - D_{23}^{(n)}D_{31}^{(n)}\right) \\
&\quad + \left(L_{12}^{(n-1)}L_{34}^{(n-1)} - L_{14}^{(n-1)}L_{32}^{(n-1)}\right)\left(D_{23}^{(n)}D_{41}^{(n)} - D_{21}^{(n)}D_{43}^{(n)}\right) \\
&\quad + L_{13}^{(n-1)}\left(L_{34}^{(n-1)} - L_{14}^{(n-1)}\right)\left(D_{33}^{(n)}D_{41}^{(n)} - D_{31}^{(n)}D_{43}^{(n)}\right)
\end{aligned}
\tag{4.216e}
$$

$$
\begin{aligned}
&\left(F_{12}F_{33} - F_{14}F_{31}\right) \\
&\quad = -\left(L_{11}^{(n-1)}L_{32}^{(n-1)} + L_{12}^{(n-1)}L_{31}^{(n-1)}\right)D_{11}^{(n)}\left(D_{21}^{(n)} - D_{23}^{(n)}\right) \\
&\quad - L_{13}^{(n-1)}\left(L_{11}^{(n-1)} + L_{31}^{(n-1)}\right)D_{11}^{(n)}\left(D_{33}^{(n)} - D_{31}^{(n)}\right) \\
&\quad - \left(L_{11}^{(n-1)}L_{34}^{(n-1)} + L_{14}^{(n-1)}L_{31}^{(n-1)}\right)D_{11}^{(n)}\left(D_{41}^{(n)} - D_{43}^{(n)}\right) \\
&\quad + L_{13}^{(n-1)}\left(L_{12}^{(n-1)} - L_{32}^{(n-1)}\right)\left(D_{21}^{(n)}D_{33}^{(n)} - D_{23}^{(n)}D_{31}^{(n)}\right) \\
&\quad + \left(L_{12}^{(n-1)}L_{34}^{(n-1)} - L_{14}^{(n-1)}L_{32}^{(n-1)}\right)\left(D_{23}^{(n)}D_{41}^{(n)} - D_{21}^{(n)}D_{43}^{(n)}\right) \\
&\quad - L_{13}^{(n-1)}\left(L_{34}^{(n-1)} - L_{14}^{(n-1)}\right)\left(D_{33}^{(n)}D_{41}^{(n)} - D_{31}^{(n)}D_{43}^{(n)}\right)
\end{aligned}
\tag{4.216f}
$$

with $n = 2$. We can replace L_{ij} by the elements of the dynamical matrix using Eq. (4.193)

$$L_{11}^{(n-1)}L_{12}^{(n-1)} = \mathcal{D}_{\mathrm{pol}}^{(n-1)}D_{33}^{(n-1)}D_{43}^{(n-1)} \tag{4.217a}$$

$$L_{11}^{(n-1)}L_{14}^{(n-1)} = -\mathcal{D}_{\mathrm{pol}}^{(n-1)}D_{23}^{(n-1)}D_{33}^{(n-1)} \tag{4.217b}$$

$$L_{12}^{(n-1)}L_{13}^{(n-1)} = -\mathcal{D}_{\mathrm{pol}}^{(n-1)}D_{11}^{(n-1)}D_{43}^{(n-1)} \tag{4.217c}$$

$$L_{13}^{(n-1)}L_{14}^{(n-1)} = \mathcal{D}_{\mathrm{pol}}^{(n-1)}D_{11}^{(n-1)}D_{23}^{(n-1)} \tag{4.217d}$$

$$L_{31}^{(n-1)}L_{32}^{(n-1)} = \mathcal{D}_{\text{pol}}^{(n-1)}D_{31}^{(n-1)}D_{41}^{(n-1)} \tag{4.217e}$$
$$L_{31}^{(n-1)}L_{34}^{(n-1)} = -\mathcal{D}_{\text{pol}}^{(n-1)}D_{21}^{(n-1)}D_{31}^{(n-1)} \tag{4.217f}$$
$$L_{13}^{(n-1)}L_{32}^{(n-1)} = \mathcal{D}_{\text{pol}}^{(n-1)}D_{11}^{(n-1)}D_{41}^{(n-1)} \tag{4.217g}$$
$$L_{13}^{(n-1)}L_{34}^{(n-1)} = -\mathcal{D}_{\text{pol}}^{(n-1)}D_{11}^{(n-1)}D_{21}^{(n-1)} \tag{4.217h}$$

$$L_{12}^{(n-1)}L_{31}^{(n-1)} - L_{11}^{(n-1)}L_{32}^{(n-1)}$$
$$= \mathcal{D}_{\text{pol}}^{(n-1)}\left(D_{33}^{(n-1)}D_{41}^{(n-1)} - D_{31}^{(n-1)}D_{43}^{(n-1)}\right) \tag{4.217i}$$
$$L_{13}^{(n-1)}\left(L_{11}^{(n-1)} + L_{31}^{(n-1)}\right) = \mathcal{D}_{\text{pol}}^{(n-1)}\left(D_{23}^{(n-1)}D_{41}^{(n-1)} - D_{21}^{(n-1)}D_{43}^{(n-1)}\right) \tag{4.217j}$$
$$L_{11}^{(n-1)}L_{34}^{(n-1)} - L_{14}^{(n-1)}L_{31}^{(n-1)}$$
$$= \mathcal{D}_{\text{pol}}^{(n-1)}\left(D_{21}^{(n-1)}D_{33}^{(n-1)} - D_{23}^{(n-1)}D_{31}^{(n-1)}\right) \tag{4.217k}$$
$$L_{13}^{(n-1)}\left(L_{12}^{(n-1)} + L_{32}^{(n-1)}\right) = \mathcal{D}_{\text{pol}}^{(n-1)}D_{11}^{(n-1)}\left(D_{41}^{(n-1)} - D_{43}^{(n-1)}\right) \tag{4.217l}$$

$$L_{12}^{(n-1)}L_{34}^{(n-1)} - L_{14}^{(n-1)}L_{32}^{(n-1)} = \mathcal{D}_{\text{pol}}^{(n-1)}D_{11}^{(n-1)}\left(D_{33}^{(n-1)} - D_{31}^{(n-1)}\right) \tag{4.217m}$$
$$L_{13}^{(n-1)}\left(L_{14}^{(n-1)} + L_{34}^{(n-1)}\right) = -\mathcal{D}_{\text{pol}}^{(n-1)}D_{11}^{(n-1)}\left(D_{21}^{(n-1)} - D_{23}^{(n-1)}\right) \tag{4.217n}$$

$$L_{11}^{(n-1)}L_{32}^{(n-1)} + L_{12}^{(n-1)}L_{31}^{(n-1)}$$
$$= -\mathcal{D}_{\text{pol}}^{(n-1)}\left(D_{33}^{(n-1)}D_{41}^{(n-1)} + D_{31}^{(n-1)}D_{43}^{(n-1)}\right) \tag{4.217o}$$
$$L_{11}^{(n-1)}L_{34}^{(n-1)} + L_{14}^{(n-1)}L_{31}^{(n-1)}$$
$$= \mathcal{D}_{\text{pol}}^{(n-1)}\left(D_{21}^{(n-1)}D_{33}^{(n-1)} + D_{23}^{(n-1)}D_{31}^{(n-1)}\right) \tag{4.217p}$$
$$L_{13}^{(n-1)}\left(L_{32}^{(n-1)} - L_{12}^{(n-1)}\right) = \mathcal{D}_{\text{pol}}^{(n-1)}D_{11}^{(n-1)}\left(D_{41}^{(n-1)} + D_{43}^{(n-1)}\right) \tag{4.217q}$$
$$L_{13}^{(n-1)}\left(L_{34}^{(n-1)} - L_{14}^{(n-1)}\right) = -\mathcal{D}_{\text{pol}}^{(n-1)}D_{11}^{(n-1)}\left(D_{21}^{(n-1)} + D_{23}^{(n-1)}\right). \tag{4.217r}$$

The sum of the first and the last term in Eq. (4.216c) and Eq. (4.216d) can be evaluated from

$$\begin{aligned}
&-\left(L_{11}^{(n-1)}L_{32}^{(n-1)} - L_{12}^{(n-1)}L_{31}^{(n-1)}\right)D_{11}^{(n)}\left(D_{21}^{(n)} - D_{23}^{(n)}\right)\\
&\quad -L_{13}^{(n-1)}\left(L_{34}^{(n-1)} + L_{14}^{(n-1)}\right)\left(D_{41}^{(n)}D_{33}^{(n)} - D_{43}^{(n)}D_{31}^{(n)}\right)\\
&= \varepsilon_{xy}^{(n-1)}\varepsilon_{xy}^{(n)}\left(\varepsilon_{xy}^{(n-1)}\varepsilon_{yx}^{(n)} + \varepsilon_{yx}^{(n-1)}\varepsilon_{xy}^{(n)}\right)\\
&\quad \times\left(N_{z+}^{(n-1)} - N_{z-}^{(n-1)}\right)\left(N_{z+}^{(n)} - N_{z-}^{(n)}\right)\left(\varepsilon_{zz}^{(n-1)} - N_y^2\right)^2\left(\varepsilon_{zz}^{(n)} - N_y^2\right)^2\\
&= \mathcal{U}^{(n-1)}\mathcal{W}^{(n)} + \mathcal{W}^{(n-1)}\mathcal{U}^{(n)}.
\end{aligned} \tag{4.218}$$

Similarly, the sum of the first and the last term in Eq. (4.216e) and Eq. (4.216f) can be evaluated from

$$\begin{aligned}
&\left(L_{11}^{(n-1)}L_{32}^{(n-1)} + L_{12}^{(n-1)}L_{31}^{(n-1)}\right)D_{11}^{(n)}\left(D_{21}^{(n)} - D_{23}^{(n)}\right) \\
&\quad + L_{13}^{(n-1)}\left(L_{34}^{(n-1)} - L_{14}^{(n-1)}\right)\left(D_{41}^{(n)}D_{33}^{(n)} - D_{43}^{(n)}D_{31}^{(n)}\right) \\
&= -\varepsilon_{xy}^{(n-1)}\varepsilon_{xy}^{(n)}\left(\varepsilon_{xy}^{(n-1)}\varepsilon_{yx}^{(n)} + \varepsilon_{yx}^{(n-1)}\varepsilon_{xy}^{(n)}\right) \\
&\quad \times\left(N_{z+}^{(n-1)} + N_{z-}^{(n-1)}\right)\left(N_{z+}^{(n)} - N_{z-}^{(n)}\right)\left(\varepsilon_{zz}^{(n-1)} - N_y^2\right)^2\left(\varepsilon_{zz}^{(n)} - N_y^2\right)^2 \\
&= -\mathcal{U}^{(n-1)}\mathcal{W}^{(n)} - \mathcal{W}^{(n-1)}\mathcal{U}^{(n)}.
\end{aligned} \tag{4.219}$$

We note the relation analogous to Eq. (4.214)

$$r_{21}^{(12)}r_{43}^{(12)} - r_{23}^{(12)}r_{41}^{(12)} = \left(F_{14}F_{32} - F_{12}F_{34}\right)F_0^{-1}. \tag{4.220}$$

The characteristic equation for the guided wave propagation (waveguiding condition) in the multilayer system (3.76)

$$\mathrm{M}_{11}\mathrm{M}_{33} - \mathrm{M}_{13}\mathrm{M}_{31} = 0 \tag{4.221}$$

for the polar magnetization follows from Eq. (4.209a) with Eq. (4.210), Eq. (4.214), and Eq. (4.215). It can be written in a compact form as

$$\begin{aligned}
&1 - \mathrm{e}^{-\mathrm{j}(\beta_+ + \beta_-)}\left(r_{14}^{(10)}r_{41}^{(12)} + r_{32}^{(10)}r_{23}^{(12)}\right) \\
&+ \mathrm{e}^{-2\mathrm{j}(\beta_+ + \beta_-)}\left[\left(r_{12}^{(10)}r_{34}^{(10)} - r_{14}^{(10)}r_{32}^{(10)}\right)\left(r_{21}^{(12)}r_{43}^{(12)} - r_{23}^{(12)}r_{41}^{(12)}\right)\right] \\
&- \mathrm{e}^{-2\mathrm{j}\beta_+}r_{12}^{(10)}r_{21}^{(12)} - \mathrm{e}^{-2\mathrm{j}\beta_-}r_{34}^{(10)}r_{43}^{(12)} = 0.
\end{aligned} \tag{4.222}$$

We can order the terms according their magnitudes to obtain

$$\begin{aligned}
&\left(1 - r_{12}^{(10)}r_{21}^{(12)}\mathrm{e}^{-2\mathrm{j}\beta_+}\right)\left(1 - r_{34}^{(10)}r_{43}^{(12)}\mathrm{e}^{-2\mathrm{j}\beta_-}\right) \\
&- \mathrm{e}^{-\mathrm{j}(\beta_+ + \beta_-)}\left[\left(r_{14}^{(10)}r_{41}^{(12)} + r_{32}^{(10)}r_{23}^{(12)}\right)\right. \\
&+ \left.\mathrm{e}^{-\mathrm{j}(\beta_+ + \beta_-)}\left(r_{21}^{(12)}r_{43}^{(12)}r_{14}^{(10)}r_{32}^{(10)} + r_{12}^{(10)}r_{34}^{(10)}r_{23}^{(12)}r_{41}^{(12)}\right)\right] \\
&+ \mathrm{e}^{-2\mathrm{j}(\beta_+ + \beta_-)}r_{14}^{(10)}r_{32}^{(10)}r_{23}^{(12)}r_{41}^{(12)} = 0.
\end{aligned} \tag{4.223}$$

We look for the solutions for the longitudinal propagation constant $(\omega/c)N_y$. The second, third, and fourth terms on the left-hand side are responsible for the coupling between transverse electric and transverse magnetic modes. In the limits of $\varepsilon_{xx}^{(n)} = \varepsilon_{yy}^{(n)} = \varepsilon_{zz}^{(n)}$ and $\varepsilon_{xy}^{(n)} = \varepsilon_{yx}^{(n)} = 0$ in Eq. (4.177), corresponding to the isotropic regions $n = 0, 1,$ and 2, only the first term is nonzero, and we obtain the separate characteristic equations (waveguiding condition) for the TE and TM guided modes.

4.5.2.4 Explicit Form of the Reflection Matrix

We now return to Eq. (4.189) and Eq. (4.190) and define

$$\mathcal{P}_{\pm}^{(n)} = \varepsilon_{yy}^{(n)}\left(\varepsilon_{zz}^{(n)} - N_y^2\right) \pm \varepsilon_{zz}^{(n)}\left(N_{z+}^{(n)} N_{z-}^{(n)}\right) \tag{4.224a}$$

$$\mathcal{Q}_{\pm}^{(n)} = \left(\varepsilon_{zz}^{(n)} - N_y^2\right)\left(N_{z+}^{(n)} \pm N_{z-}^{(n)}\right) \tag{4.224b}$$

$$\mathcal{S}_{\pm}^{(n)} = \varepsilon_{zz}^{(n)}\left(N_{z+}^{(n)} N_{z-}^{(n)}\right)\left(N_{z+}^{(n)} \pm N_{z-}^{(n)}\right) \tag{4.224c}$$

$$\mathcal{T}_{\pm}^{(n)} = \left(\varepsilon_{zz}^{(n)} - N_y^2\right)\left[\left(N_{z+}^{(n)} N_{z-}^{(n)}\right) \pm \left(\varepsilon_{xx}^{(n)} - N_y^2\right)\right] \tag{4.224d}$$

$$\mathcal{U}_{\pm}^{(n)} = -\varepsilon_{xy}^{(n)}\left(\varepsilon_{zz}^{(n)} - N_y^2\right) \tag{4.224e}$$

$$\mathcal{W}_{\pm}^{(n)} = -\varepsilon_{yx}^{(n)}\left(\varepsilon_{zz}^{(n)} - N_y^2\right), \tag{4.224f}$$

where

$$D_{11}^{(n)}\left(D_{41}^{(n)} \mp D_{43}^{(n)}\right) = -\varepsilon_{xy}^{(n)}\left(\varepsilon_{zz}^{(n)} - N_y^2\right)\left(N_{z+}^{(n)} \mp N_{z-}^{(n)}\right)\mathcal{P}_{\pm}^{(n)}, \tag{4.225a}$$

$$D_{11}^{(n)}\left(D_{33}^{(n)} - D_{31}^{(n)}\right) = -\varepsilon_{xy}^{(n)}\left(\varepsilon_{zz}^{(n)} - N_y^2\right)\left(N_{z+}^{(n)} \mp N_{z-}^{(n)}\right)\mathcal{Q}_{\pm}^{(n)}, \tag{4.225b}$$

$$D_{41}^{(n)} D_{23}^{(n)} - D_{43}^{(n)} D_{21}^{(n)} = -\varepsilon_{xy}^{(n)}\left(\varepsilon_{zz}^{(n)} - N_y^2\right)\left(N_{z+}^{(n)} \mp N_{z-}^{(n)}\right)\mathcal{S}_{\pm}^{(n)}, \tag{4.225c}$$

$$\pm D_{21}^{(n)} D_{33}^{(n)} - D_{23}^{(n)} D_{31}^{(n)} = -\varepsilon_{xy}^{(n)}\left(\varepsilon_{zz}^{(n)} - N_y^2\right)\left(N_{z+}^{(n)} \mp N_{z-}^{(n)}\right)\mathcal{T}_{\pm}^{(n)}, \tag{4.225d}$$

$$D_{11}^{(n)}\left(D_{21}^{(n)} \mp D_{23}^{(n)}\right) = -\varepsilon_{xy}^{(n)}\left(\varepsilon_{zz}^{(n)} - N_y^2\right)\left(N_{z+}^{(n)} \mp N_{z-}^{(n)}\right)\mathcal{U}_{\pm}^{(n)} \tag{4.225e}$$

$$D_{33}^{(n)} D_{41}^{(n)} \mp D_{31}^{(n)} D_{43}^{(n)} = -\varepsilon_{yx}^{(n)}\left(\varepsilon_{zz}^{(n)} - N_y^2\right)\left(N_{z+}^{(n)} \mp N_{z-}^{(n)}\right)\mathcal{W}_{\pm}^{(n)}, \tag{4.225f}$$

with $\mathcal{U}_{+}^{(n)} = \mathcal{U}_{-}^{(n)}$ and $\mathcal{W}_{+}^{(n)} = \mathcal{W}_{-}^{(n)}$. For reasons of symmetry, we may further define in the isotropic medium

$$\mathcal{P}^{(0)} = N^{(0)} \tag{4.226a}$$

$$\mathcal{Q}^{(0)} = \cos\varphi^{(0)} \tag{4.226b}$$

$$\mathcal{S}^{(0)} = N^{(0)2}\cos\varphi^{(0)} \tag{4.226c}$$

$$\mathcal{T}^{(0)} = N^{(0)}\cos^2\varphi^{(0)}. \tag{4.226d}$$

The expressions formed by the **A** matrix elements can be written more concisely as

$$\left(A_{11}A_{33} - A_{13}A_{31}\right) = \left(N_{z+}^{(n)} - N_{z-}^{(n)}\right)\left(\mathcal{T}^{(0)}\mathcal{P}_{+}^{(1)} + \mathcal{S}^{(0)}\mathcal{Q}_{+}^{(1)} + \mathcal{Q}^{(0)}\mathcal{S}_{+}^{(1)} + \mathcal{P}^{(0)}\mathcal{T}_{+}^{(1)}\right) \tag{4.227a}$$

$$\left(A_{12}A_{33} - A_{14}A_{31}\right) = \left(N_{z+}^{(n)} - N_{z-}^{(n)}\right)\left(\mathcal{T}^{(0)}\mathcal{P}_{+}^{(1)} + \mathcal{S}^{(0)}\mathcal{Q}_{+}^{(1)} - \mathcal{Q}^{(0)}\mathcal{S}_{+}^{(1)} - \mathcal{P}^{(0)}\mathcal{T}_{+}^{(1)}\right) \tag{4.227b}$$

$$\left(A_{11}A_{34} - A_{13}A_{32}\right) = \left(N_{z+}^{(n)} - N_{z-}^{(n)}\right)\left(-\mathcal{T}^{(0)}\mathcal{P}_{+}^{(1)} + \mathcal{S}^{(0)}\mathcal{Q}_{+}^{(1)} - \mathcal{Q}^{(0)}\mathcal{S}_{+}^{(1)} + \mathcal{P}^{(0)}\mathcal{T}_{+}^{(1)}\right) \tag{4.227c}$$

$$\left(A_{12}A_{34} - A_{14}A_{32}\right) = \left(N_{z+}^{(n)} - N_{z-}^{(n)}\right)\left(-\mathcal{T}^{(0)}\mathcal{P}_{+}^{(1)} + \mathcal{S}^{(0)}\mathcal{Q}_{+}^{(1)} + \mathcal{Q}^{(0)}\mathcal{S}_{+}^{(1)} - \mathcal{P}^{(0)}\mathcal{T}_{+}^{(1)}\right) \tag{4.227d}$$

$$
\begin{aligned}
&(A_{12}A_{34} - A_{13}A_{31}) \\
&\quad = (N_{z+}^{(n)} + N_{z-}^{(n)})(\mathcal{T}^{(0)}\mathcal{P}_-^{(1)} + \mathcal{S}^{(0)}\mathcal{Q}_-^{(1)} + \mathcal{Q}^{(0)}\mathcal{S}_-^{(1)} + \mathcal{P}^{(0)}\mathcal{T}_-^{(1)}) \qquad (4.228a)\\
&(A_{13}A_{32} - A_{12}A_{33}) \\
&\quad = (N_{z+}^{(n)} + N_{z-}^{(n)})(\mathcal{T}^{(0)}\mathcal{P}_-^{(1)} - \mathcal{S}^{(0)}\mathcal{Q}_-^{(1)} + \mathcal{Q}^{(0)}\mathcal{S}_-^{(1)} - \mathcal{P}^{(0)}\mathcal{T}_-^{(1)}) \qquad (4.228b)\\
&(A_{14}A_{31} - A_{11}A_{34}) \\
&\quad = (N_{z+}^{(n)} + N_{z-}^{(n)})(-\mathcal{T}^{(0)}\mathcal{P}_-^{(1)} - \mathcal{S}^{(0)}\mathcal{Q}_-^{(1)} + \mathcal{Q}^{(0)}\mathcal{S}_-^{(1)} + \mathcal{P}^{(0)}\mathcal{T}_-^{(1)}) \qquad (4.228c)\\
&(A_{11}A_{33} - A_{14}A_{32}) \\
&\quad = (N_{z+}^{(n)} + N_{z-}^{(n)})(-\mathcal{T}^{(0)}\mathcal{P}_-^{(1)} + \mathcal{S}^{(0)}\mathcal{Q}_-^{(1)} + \mathcal{Q}^{(0)}\mathcal{S}_-^{(1)} - \mathcal{P}^{(0)}\mathcal{T}_-^{(1)}). \qquad (4.228d)
\end{aligned}
$$

The equations in (4.216) characterizing the reflection at the film–substrate interface can be expressed as

$$
\begin{aligned}
&(F_{11}F_{14} - F_{12}F_{13}) \\
&\quad = 2D_{33}^{(1)}D_{43}^{(1)}\mathcal{U}^{(2)} + 2D_{11}^{(1)}D_{23}^{(1)}\mathcal{W}^{(2)} - 2D_{23}^{(1)}D_{33}^{(1)}\mathcal{P}^{(2)} - 2D_{11}^{(1)}D_{43}^{(1)}\mathcal{T}^{(2)} \qquad (4.229a)
\end{aligned}
$$

$$
\begin{aligned}
&(F_{32}F_{33} - F_{31}F_{34}) \\
&\quad = -2D_{31}^{(1)}D_{41}^{(1)}\mathcal{U}^{(2)} - 2D_{11}^{(1)}D_{21}^{(1)}\mathcal{W}^{(2)} + 2D_{21}^{(1)}D_{31}^{(1)}\mathcal{P}^{(2)} + 2D_{11}^{(1)}D_{41}^{(1)}\mathcal{T}^{(2)} \qquad (4.229b)
\end{aligned}
$$

$$
\begin{aligned}
&(F_{11}F_{33} - F_{13}F_{31}) \\
&\quad = (D_{33}^{(1)}D_{41}^{(1)} - D_{31}^{(1)}D_{43}^{(1)})\mathcal{U}^{(2)} + D_{11}^{(1)}(D_{21}^{(1)} - D_{23}^{(1)})\mathcal{W}^{(2)} \\
&\quad -(D_{23}^{(1)}D_{41}^{(1)} - D_{21}^{(1)}D_{43}^{(1)})\mathcal{Q}^{(2)} - (D_{21}^{(1)}D_{33}^{(1)} - D_{23}^{(1)}D_{31}^{(1)})\mathcal{P}^{(2)} \\
&\quad -D_{11}^{(1)}(D_{41}^{(1)} - D_{43}^{(1)})\mathcal{T}^{(2)} - D_{11}^{(1)}(D_{33}^{(1)} - D_{31}^{(1)})\mathcal{S}^{(2)}, \qquad (4.229c)
\end{aligned}
$$

where $n = 2$.

$$
\begin{aligned}
&(F_{12}F_{34} - F_{14}F_{32}) \\
&\quad = -(D_{33}^{(1)}D_{41}^{(1)} - D_{31}^{(1)}D_{43}^{(1)})\mathcal{U}^{(2)} - D_{11}^{(1)}(D_{21}^{(1)} - D_{23}^{(1)})\mathcal{W}^{(2)} \\
&\quad -(D_{23}^{(1)}D_{41}^{(1)} - D_{21}^{(1)}D_{43}^{(1)})\mathcal{Q}^{(2)} + (D_{21}^{(1)}D_{33}^{(1)} - D_{23}^{(1)}D_{31}^{(1)})\mathcal{P}^{(2)} \\
&\quad +D_{11}^{(1)}(D_{41}^{(1)} - D_{43}^{(1)})\mathcal{T}^{(2)} - D_{11}^{(1)}(D_{33}^{(1)} - D_{31}^{(1)})\mathcal{S}^{(2)} \qquad (4.229d)
\end{aligned}
$$

$$
\begin{aligned}
&(F_{11}F_{34} - F_{13}F_{32}) \\
&\quad = -(D_{33}^{(1)}D_{41}^{(1)} + D_{31}^{(1)}D_{43}^{(1)})\mathcal{U}^{(2)} - D_{11}^{(1)}(D_{21}^{(1)} + D_{23}^{(1)})\mathcal{W}^{(2)} \\
&\quad -(D_{23}^{(1)}D_{41}^{(1)} - D_{21}^{(1)}D_{43}^{(1)})\mathcal{Q}^{(2)} + (D_{21}^{(1)}D_{33}^{(1)} + D_{23}^{(1)}D_{31}^{(1)})\mathcal{P}^{(2)} \\
&\quad + D_{11}^{(1)}(D_{41}^{(1)} + D_{43}^{(1)})\mathcal{T}^{(2)} + D_{11}^{(1)}(D_{33}^{(1)} - D_{31}^{(1)})\mathcal{S}^{(2)} \qquad (4.229e)
\end{aligned}
$$

$$
\begin{aligned}
&(F_{12}F_{33} - F_{14}F_{31}) \\
&\quad = (D_{33}^{(1)}D_{41}^{(1)} + D_{31}^{(1)}D_{43}^{(1)})\mathcal{U}^{(2)} \\
&\quad + D_{11}^{(1)}(D_{21}^{(1)} + D_{23}^{(1)})\mathcal{W}^{(2)} \\
&\quad - (D_{23}^{(1)}D_{41}^{(1)} - D_{21}^{(1)}D_{43}^{(1)})\mathcal{Q}^{(2)} - (D_{21}^{(1)}D_{33}^{(1)} + D_{23}^{(1)}D_{31}^{(1)})\mathcal{P}^{(2)} \\
&\quad - D_{11}^{(1)}(D_{41}^{(1)} + D_{43}^{(1)})\mathcal{T}^{(2)} + D_{11}^{(1)}(D_{33}^{(1)} - D_{31}^{(1)})\mathcal{S}^{(2)}.
\end{aligned}
\tag{4.229f}
$$

We now make use of the abbreviations defined in Eq. (4.224) and Eq. (4.226) and obtain (up to a common factor)

$$
\begin{aligned}
&(F_{11}F_{33} - F_{13}F_{31}) \\
&\quad = (N_{z+}^{(1)} - N_{z-}^{(1)})(\mathcal{W}_{+}^{(1)}\mathcal{U}_{+}^{(2)} + \mathcal{U}_{+}^{(1)}\mathcal{W}_{+}^{(2)} \\
&\quad - \mathcal{S}_{+}^{(1)}\mathcal{Q}_{+}^{(2)} - \mathcal{T}_{+}^{(1)}\mathcal{P}_{+}^{(2)} - \mathcal{P}_{+}^{(1)}\mathcal{T}_{+}^{(2)} - \mathcal{Q}_{+}^{(1)}\mathcal{S}_{+}^{(2)})
\end{aligned}
\tag{4.230a}
$$

$$
\begin{aligned}
&(F_{12}F_{34} - F_{14}F_{32}) \\
&\quad = (N_{z+}^{(1)} - N_{z-}^{(1)})(-\mathcal{W}_{+}^{(1)}\mathcal{U}_{+}^{(2)} - \mathcal{U}_{+}^{(1)}\mathcal{W}_{+}^{(2)} \\
&\quad - \mathcal{S}_{+}^{(1)}\mathcal{Q}_{+}^{(2)} + \mathcal{T}_{+}^{(1)}\mathcal{P}_{+}^{(2)} + \mathcal{P}_{+}^{(1)}\mathcal{T}_{+}^{(2)} - \mathcal{Q}_{+}^{(1)}\mathcal{S}_{+}^{(2)})
\end{aligned}
\tag{4.230b}
$$

$$
\begin{aligned}
&(F_{11}F_{34} - F_{13}F_{32}) \\
&\quad = (N_{z+}^{(1)} + N_{z-}^{(1)})(-\mathcal{W}_{-}^{(1)}\mathcal{U}_{+}^{(2)} - \mathcal{U}_{-}^{(1)}\mathcal{W}_{+}^{(2)} \\
&\quad - \mathcal{S}_{-}^{(1)}\mathcal{Q}_{+}^{(2)} - \mathcal{T}_{-}^{(1)}\mathcal{P}_{+}^{(2)} + \mathcal{P}_{-}^{(1)}\mathcal{T}_{+}^{(2)} + \mathcal{Q}_{-}^{(1)}\mathcal{S}_{+}^{(2)})
\end{aligned}
\tag{4.230c}
$$

$$
\begin{aligned}
&(F_{12}F_{33} - F_{14}F_{31}) \\
&\quad = (N_{z+}^{(1)} + N_{z-}^{(1)})(\mathcal{W}_{-}^{(1)}\mathcal{U}_{+}^{(2)} + \mathcal{U}_{-}^{(1)}\mathcal{W}_{+}^{(2)} \\
&\quad - \mathcal{S}_{-}^{(1)}\mathcal{Q}_{+}^{(2)} + \mathcal{T}_{-}^{(1)}\mathcal{P}_{+}^{(2)} - \mathcal{P}_{-}^{(1)}\mathcal{T}_{+}^{(2)} + \mathcal{Q}_{-}^{(1)}\mathcal{S}_{+}^{(2)}).
\end{aligned}
\tag{4.230d}
$$

The first two terms in Eq. (4.209a), proportional to the expression $(r_{14}^{(10)}r_{41}^{(12)} + r_{32}^{(10)}r_{23}^{(12)})$, formed by the product of the internal reflection coefficients, are given, up to the common factor $\mathcal{D}_{\text{pol}}^{(n)}(2N^{(0)}\cos\varphi^{(0)})^{-2}$,

$$
\begin{aligned}
&-(A_{12}A_{31} - A_{11}A_{32})(F_{11}F_{14} - F_{12}F_{13}) - (A_{13}A_{34} - A_{14}A_{33}) \\
&\quad \times (F_{32}F_{33} - F_{31}F_{34}) = 4N^{(0)}(\cos^2\varphi^{(0)}D_{11}^{(1)}D_{41}^{(1)} + D_{21}^{(1)}D_{31}^{(1)}) \\
&\quad \times (D_{33}^{(1)}D_{43}^{(1)}\mathcal{U}^{(2)} + D_{11}^{(1)}D_{23}^{(1)}\mathcal{W}^{(2)} - D_{23}^{(1)}D_{33}^{(1)}\mathcal{P}^{(2)} - D_{11}^{(1)}D_{43}^{(1)}\mathcal{T}^{(2)}) \\
&\quad + 4N^{(0)}(\cos^2\varphi^{(0)}D_{11}^{(1)}D_{43}^{(1)} + D_{23}^{(1)}D_{33}^{(1)}) \\
&\quad \times (D_{31}^{(1)}D_{41}^{(1)}\mathcal{U}^{(2)} + D_{11}^{(1)}D_{21}^{(1)}\mathcal{W}^{(2)} - D_{21}^{(1)}D_{31}^{(1)}\mathcal{P}^{(2)} - D_{11}^{(1)}D_{41}^{(1)}\mathcal{T}^{(2)}) \\
&\quad = 4N^{(0)}\{[\cos^2\varphi^{(0)}D_{11}^{(1)}D_{41}^{(1)}D_{43}^{(1)}(D_{31}^{(1)} + D_{33}^{(1)})
\end{aligned}
$$

$$+ D_{31}^{(1)} D_{33}^{(1)} \left(D_{21}^{(1)} D_{43}^{(1)} + D_{23}^{(1)} D_{41}^{(1)} \right)] \mathcal{U}^{(2)}$$
$$+ \left[\cos^2 \varphi^{(0)} D_{11}^{(1)2} \left(D_{21}^{(1)} D_{43}^{(1)} + D_{23}^{(1)} D_{41}^{(1)} \right) + D_{11}^{(1)} D_{21}^{(1)} D_{23}^{(1)} \left(D_{31}^{(1)} + D_{33}^{(1)} \right) \right] \mathcal{W}^{(2)}$$
$$- \left[\cos^2 \varphi^{(0)} D_{11}^{(1)} \left(D_{21}^{(1)} D_{31}^{(1)} D_{43}^{(1)} + D_{23}^{(1)} D_{33}^{(1)} D_{41}^{(1)} \right) + 2 D_{21}^{(1)} D_{23}^{(1)} D_{31}^{(1)} D_{33}^{(1)} \right] \mathcal{P}^{(2)}$$
$$- \left[2 \cos^2 \varphi^{(0)} D_{11}^{(1)2} D_{41}^{(1)} D_{43}^{(1)} + D_{11}^{(1)} \left(D_{21}^{(1)} D_{31}^{(1)} D_{43}^{(1)} + D_{23}^{(1)} D_{33}^{(1)} D_{41}^{(1)} \right) \right] \mathcal{T}^{(2)} \Big\}, \tag{4.231}$$

where

$$D_{11}^{(1)} D_{41}^{(1)} D_{43}^{(1)} \left(D_{33}^{(1)} + D_{31}^{(1)} \right) = \varepsilon_{xy}^{(1)2} \varepsilon_{yx}^{(1)} \left(\varepsilon_{zz}^{(1)} - N_y^2 \right)^3 N_{z+}^{(1)} N_{z-}^{(1)} \left[\varepsilon_{zz}^{(1)} \left(\varepsilon_{xx}^{(1)} - N_y^2 \right) - \varepsilon_{yy}^{(1)} \left(\varepsilon_{zz}^{(1)} - N_y^2 \right) \right] \tag{4.232a}$$

$$D_{11}^{(1)2} \left(D_{23}^{(1)} D_{41}^{(1)} + D_{21}^{(1)} D_{43}^{(1)} \right) = \varepsilon_{xy}^{(1)3} \left(\varepsilon_{zz}^{(1)} - N_y^2 \right)^3 N_{z+}^{(1)} N_{z-}^{(1)} \left[\varepsilon_{zz}^{(1)} \left(\varepsilon_{xx}^{(1)} - N_y^2 \right) - \varepsilon_{yy}^{(1)} \left(\varepsilon_{zz}^{(1)} - N_y^2 \right) \right] \tag{4.232b}$$

$$D_{31}^{(1)} D_{33}^{(1)} \left(D_{23}^{(1)} D_{41}^{(1)} + D_{21}^{(1)} D_{43}^{(1)} \right) = -\varepsilon_{xy}^{(1)2} \varepsilon_{yx}^{(1)} \varepsilon_{zz}^{(1)-1} \left(\varepsilon_{zz}^{(1)} - N_y^2 \right)^4 N_{z+}^{(1)} N_{z-}^{(1)} \left[\varepsilon_{zz}^{(1)} \left(\varepsilon_{xx}^{(1)} - N_y^2 \right) - \varepsilon_{yy}^{(1)} \left(\varepsilon_{zz}^{(1)} - N_y^2 \right) \right] \tag{4.232c}$$

$$D_{11}^{(1)} D_{21}^{(1)} D_{23}^{(1)} \left(D_{31}^{(1)} + D_{33}^{(1)} \right) = -\varepsilon_{xy}^{(1)3} \varepsilon_{zz}^{(1)-1} \left(\varepsilon_{zz}^{(1)} - N_y^2 \right)^4 N_{z+}^{(1)} N_{z-}^{(1)} \left[\varepsilon_{zz}^{(1)} \left(\varepsilon_{xx}^{(1)} - N_y^2 \right) - \varepsilon_{yy}^{(1)} \left(\varepsilon_{zz}^{(1)} - N_y^2 \right) \right] \tag{4.232d}$$

$$D_{11}^{(1)} \left(D_{23}^{(1)} D_{33}^{(1)} D_{41}^{(1)} + D_{21}^{(1)} D_{31}^{(1)} D_{43}^{(1)} \right) = 2 \varepsilon_{xy}^{(1)3} \varepsilon_{yx}^{(1)} \left(\varepsilon_{zz}^{(1)} - N_y^2 \right)^4 N_{z+}^{(1)} N_{z-}^{(1)} \tag{4.232e}$$

$$2 D_{11}^{(1)2} D_{41}^{(1)} D_{43}^{(1)} = -2 \varepsilon_{xy}^{(1)3} \varepsilon_{yx}^{(1)} \left(\varepsilon_{zz}^{(1)} - N_y^2 \right)^3 \varepsilon_{zz}^{(1)} N_{z+}^{(1)} N_{z-}^{(1)} \tag{4.232f}$$

$$2 D_{21}^{(1)} D_{23}^{(1)} D_{31}^{(1)} D_{33}^{(1)} = -2 \varepsilon_{xy}^{(1)3} \varepsilon_{yx}^{(1)} \varepsilon_{zz}^{(1)-1} \left(\varepsilon_{zz}^{(1)} - N_y^2 \right)^5 N_{z+}^{(1)} N_{z-}^{(1)}. \tag{4.232g}$$

The use was made of Eqs. (4.183), providing

$$2 \varepsilon_{zz}^{(1)} \left(\varepsilon_{xx}^{(1)} - N_y^2 \right) - \varepsilon_{zz}^{(1)} \left(N_{z+}^{(1)2} + N_{z-}^{(1)2} \right) = \varepsilon_{zz}^{(1)} \left(\varepsilon_{xx}^{(1)} - N_y^2 \right) - \varepsilon_{yy}^{(1)} \left(\varepsilon_{zz}^{(1)} - N_y^2 \right). \tag{4.233}$$

After the substitution and removing the appropriate common factor $\varepsilon_{xy}^{(1)2} \left(\varepsilon_{zz}^{(1)} - N_y^2 \right)^2$ deduced from Eq. (4.225), we obtain

$$-\left(A_{12} A_{31} - A_{11} A_{32} \right) \left(F_{11} F_{14} - F_{12} F_{13} \right) - \left(A_{13} A_{34} - A_{14} A_{33} \right) \times \left(F_{32} F_{33} - F_{31} F_{34} \right) = 4 N^{(0)} \left(\varepsilon_{zz}^{(1)} - N_y^2 \right) \varepsilon_{zz}^{(1)-1} N_{z+}^{(1)} N_{z-}^{(1)}$$

$$\times\left[\varepsilon_{zz}^{(1)}\cos^2\varphi^{(0)}-\left(\varepsilon_{zz}^{(1)}-N_y^2\right)\right]\left\{\left[\varepsilon_{zz}^{(1)}\left(\varepsilon_{xx}^{(1)}-N_y^2\right)-\varepsilon_{yy}^{(1)}\left(\varepsilon_{zz}^{(1)}-N_y^2\right)\right]\right.$$
$$\left.\times\left(\varepsilon_{yx}^{(1)}\mathcal{U}_+^{(2)}+\varepsilon_{xy}^{(1)}\mathcal{W}_+^{(2)}\right)-2\varepsilon_{xy}^{(1)}\varepsilon_{yx}^{(1)}\left[\left(\varepsilon_{zz}^{(1)}-N_y^2\right)\mathcal{P}_+^{(2)}-\varepsilon_{zz}^{(1)}\mathcal{T}_+^{(2)}\right]\right\}. \quad (4.234)$$

Note that this expression is of second order in the off-diagonal elements of the permittivity tensor and vanishes at the normal incidence when $\cos^2\varphi^{(0)}=1$. Eq. (4.209a) finally becomes

$$\begin{aligned}
&\mathrm{M}_{11}\mathrm{M}_{33}-\mathrm{M}_{13}\mathrm{M}_{31}\\
&\quad=4N^{(0)}\left(\varepsilon_{zz}^{(1)}-N_y^2\right)\varepsilon_{zz}^{(1)-1}N_{z+}^{(1)}N_{z-}^{(1)}\left[\varepsilon_{zz}^{(1)}\cos^2\varphi^{(0)}-\left(\varepsilon_{zz}^{(1)}-N_y^2\right)\right]\\
&\quad\times\left\{\left[\varepsilon_{zz}^{(1)}\left(\varepsilon_{xx}^{(1)}-N_y^2\right)-\varepsilon_{yy}^{(1)}\left(\varepsilon_{zz}^{(1)}-N_y^2\right)\right]\left(\varepsilon_{yx}^{(1)}\mathcal{U}_+^{(2)}+\varepsilon_{xy}^{(1)}\mathcal{W}_+^{(2)}\right)\right.\\
&\quad\left.-2\varepsilon_{xy}^{(1)}\varepsilon_{yx}^{(1)}\left[\left(\varepsilon_{zz}^{(1)}-N_y^2\right)\mathcal{P}_+^{(2)}-\varepsilon_{zz}^{(1)}\mathcal{T}_+^{(2)}\right]\right\}\\
&\quad+\mathrm{e}^{\mathrm{j}(\beta_++\beta_-)}\left(N_{z+}^{(1)}-N_{z-}^{(1)}\right)^2\left(\mathcal{T}^{(0)}\mathcal{P}_+^{(1)}+\mathcal{S}^{(0)}\mathcal{Q}_+^{(1)}+\mathcal{Q}^{(0)}\mathcal{S}_+^{(1)}+\mathcal{P}^{(0)}\mathcal{T}_+^{(1)}\right)\\
&\quad\times\left(\mathcal{W}_+^{(1)}\mathcal{U}_+^{(2)}+\mathcal{U}_+^{(1)}\mathcal{W}_+^{(2)}-\mathcal{T}_+^{(1)}\mathcal{P}_+^{(2)}-\mathcal{S}_+^{(1)}\mathcal{Q}_+^{(2)}-\mathcal{Q}_+^{(1)}\mathcal{S}_+^{(2)}-\mathcal{P}_+^{(1)}\mathcal{T}_+^{(2)}\right)\\
&\quad+\mathrm{e}^{-\mathrm{j}(\beta_++\beta_-)}\left(N_{z+}^{(1)}-N_{z-}^{(1)}\right)^2\left(-\mathcal{T}^{(0)}\mathcal{P}_+^{(1)}+\mathcal{S}^{(0)}\mathcal{Q}_+^{(1)}+\mathcal{Q}^{(0)}\mathcal{S}_+^{(1)}-\mathcal{P}^{(0)}\mathcal{T}_+^{(1)}\right)\\
&\quad\times\left(-\mathcal{W}_+^{(1)}\mathcal{U}_+^{(2)}-\mathcal{U}_+^{(1)}\mathcal{W}_+^{(2)}+\mathcal{T}_+^{(1)}\mathcal{P}_+^{(2)}-\mathcal{S}_+^{(1)}\mathcal{Q}_+^{(2)}-\mathcal{Q}_+^{(1)}\mathcal{S}_+^{(2)}+\mathcal{P}_+^{(1)}\mathcal{T}_+^{(2)}\right)\\
&\quad+\mathrm{e}^{\mathrm{j}(\beta_+-\beta_-)}\left(N_{z+}^{(1)}+N_{z-}^{(1)}\right)^2\left(\mathcal{T}^{(0)}\mathcal{P}_-^{(1)}+\mathcal{S}^{(0)}\mathcal{Q}_-^{(1)}-\mathcal{Q}^{(0)}\mathcal{S}_-^{(1)}-\mathcal{P}^{(0)}\mathcal{T}_-^{(1)}\right)\\
&\quad\times\left(-\mathcal{W}_-^{(1)}\mathcal{U}_+^{(2)}-\mathcal{U}_-^{(1)}\mathcal{W}_+^{(2)}-\mathcal{T}_-^{(1)}\mathcal{P}_+^{(2)}-\mathcal{S}_-^{(1)}\mathcal{Q}_+^{(2)}+\mathcal{Q}_-^{(1)}\mathcal{S}_+^{(2)}+\mathcal{P}_-^{(1)}\mathcal{T}_+^{(2)}\right)\\
&\quad+\mathrm{e}^{-\mathrm{j}(\beta_+-\beta_-)}\left(N_{z+}^{(1)}+N_{z-}^{(1)}\right)^2\left(-\mathcal{T}^{(0)}\mathcal{P}_-^{(1)}+\mathcal{S}^{(0)}\mathcal{Q}_-^{(1)}-\mathcal{Q}^{(0)}\mathcal{S}_-^{(1)}+\mathcal{P}^{(0)}\mathcal{T}_-^{(1)}\right)\\
&\quad\times\left(\mathcal{W}_-^{(1)}\mathcal{U}_+^{(2)}+\mathcal{U}_-^{(1)}\mathcal{W}_+^{(2)}+\mathcal{T}_-^{(1)}\mathcal{P}_+^{(2)}-\mathcal{S}_-^{(1)}\mathcal{Q}_+^{(2)}+\mathcal{Q}_-^{(1)}\mathcal{S}_+^{(2)}-\mathcal{P}_-^{(1)}\mathcal{T}_+^{(2)}\right).
\end{aligned} \quad (4.235)$$

The first two terms in Eq. (4.209b), and the first two terms in Eq. (4.209c) with the inverted sign, are given, up to the common factor $\mathcal{D}_{\mathrm{pol}}^{(n)}\left(2N^{(0)}\cos\varphi^{(0)}\right)^{-2}$,

$$\begin{aligned}
&\left(A_{12}A_{32}-A_{11}A_{31}\right)\left(F_{11}F_{14}-F_{12}F_{13}\right)+\left(A_{13}A_{33}-A_{14}A_{34}\right)\left(F_{32}F_{33}-F_{31}F_{34}\right)\\
&\quad=4N^{(0)}\left(\cos^2\varphi^{(0)}D_{11}^{(1)}D_{41}^{(1)}-D_{21}^{(1)}D_{31}^{(1)}\right)\\
&\quad\times\left(D_{33}^{(1)}D_{43}^{(1)}\mathcal{U}^{(2)}+D_{11}^{(1)}D_{23}^{(1)}\mathcal{W}^{(2)}-D_{23}^{(1)}D_{33}^{(1)}\mathcal{P}^{(2)}-D_{11}^{(1)}D_{43}^{(1)}\mathcal{T}^{(2)}\right)\\
&\quad+4N^{(0)}\left(\cos^2\varphi^{(0)}D_{11}^{(1)}D_{43}^{(1)}-D_{23}^{(1)}D_{33}^{(1)}\right)\\
&\quad\times\left(D_{31}^{(1)}D_{41}^{(1)}\mathcal{U}^{(2)}+D_{11}^{(1)}D_{21}^{(1)}\mathcal{W}^{(2)}-D_{21}^{(1)}D_{31}^{(1)}\mathcal{P}^{(2)}-D_{11}^{(1)}D_{41}^{(1)}\mathcal{T}^{(2)}\right).
\end{aligned} \quad (4.236)$$

After the substitution and removing the appropriate common factor $\varepsilon_{xy}^{(1)2}\left(\varepsilon_{zz}^{(1)} - N_y^2\right)^2$,

$$\begin{aligned}
&\left(A_{12}A_{32} - A_{11}A_{31}\right)\left(F_{11}F_{14} - F_{12}F_{13}\right) + \left(A_{13}A_{33} - A_{14}A_{34}\right)\left(F_{32}F_{33} - F_{31}F_{34}\right)\\
&\quad = 4N^{(0)}\left(\varepsilon_{zz}^{(1)} - N_y^2\right)\varepsilon_{zz}^{(1)-1}N_{z+}^{(1)}N_{z-}^{(1)}\left[\varepsilon_{zz}^{(1)}\cos^2\varphi^{(0)} + \left(\varepsilon_{zz}^{(1)} - N_y^2\right)\right]\\
&\quad \times\left\{\left[\varepsilon_{zz}^{(1)}\left(\varepsilon_{xx}^{(1)} - N_y^2\right) - \varepsilon_{yy}^{(1)}\left(\varepsilon_{zz}^{(1)} - N_y^2\right)\right]\left(\varepsilon_{yx}^{(1)}\mathcal{U}_+^{(2)} + \varepsilon_{xy}^{(1)}\mathcal{W}_+^{(2)}\right)\right.\\
&\quad \left. - 2\varepsilon_{xy}^{(1)}\varepsilon_{yx}^{(1)}\left[\left(\varepsilon_{zz}^{(1)} - N_y^2\right)\mathcal{P}_+^{(2)} - \varepsilon_{zz}^{(1)}\mathcal{T}_+^{(2)}\right]\right\},
\end{aligned} \tag{4.237}$$

which differs from Eq. (4.234) by the sign change in the factor to $\left[\varepsilon_{zz}^{(1)}\cos^2\varphi^{(0)} + \left(\varepsilon_{zz}^{(1)} - N_y^2\right)\right]$. Eq. (4.209b) and Eq. (4.209c) finally become

$$\begin{aligned}
&M_{21}M_{33} - M_{23}M_{31}\\
&\quad = 4N^{(0)}\left(\varepsilon_{zz}^{(1)} - N_y^2\right)\varepsilon_{zz}^{(1)-1}N_{z+}^{(1)}N_{z-}^{(1)}\left[\varepsilon_{zz}^{(1)}\cos^2\varphi^{(0)} + \left(\varepsilon_{zz}^{(1)} - N_y^2\right)\right]\\
&\quad \times\left\{\left[\varepsilon_{zz}^{(1)}\left(\varepsilon_{xx}^{(1)} - N_y^2\right) - \varepsilon_{yy}^{(1)}\left(\varepsilon_{zz}^{(1)} - N_y^2\right)\right]\left(\varepsilon_{yx}^{(1)}\mathcal{U}_+^{(2)} + \varepsilon_{xy}^{(1)}\mathcal{W}_+^{(2)}\right)\right.\\
&\quad \left. - 2\varepsilon_{xy}^{(1)}\varepsilon_{yx}^{(1)}\left[\left(\varepsilon_{zz}^{(1)} - N_y^2\right)\mathcal{P}_+^{(2)} - \varepsilon_{zz}^{(1)}\mathcal{T}_+^{(2)}\right]\right\}\\
&\quad + e^{j(\beta_+ + \beta_-)}\left(N_{z+}^{(1)} - N_{z-}^{(1)}\right)^2\left(\mathcal{T}^{(0)}\mathcal{P}_+^{(1)} + \mathcal{S}^{(0)}\mathcal{Q}_+^{(1)} - \mathcal{Q}^{(0)}\mathcal{S}_+^{(1)} - \mathcal{P}^{(0)}\mathcal{T}_+^{(1)}\right)\\
&\quad \times\left(\mathcal{W}_+^{(1)}\mathcal{U}_+^{(2)} + \mathcal{U}_+^{(1)}\mathcal{W}_+^{(2)} - \mathcal{T}_+^{(1)}\mathcal{P}_+^{(2)} - \mathcal{S}_+^{(1)}\mathcal{Q}_+^{(2)} - \mathcal{Q}_+^{(1)}\mathcal{S}_+^{(2)} - \mathcal{P}_+^{(1)}\mathcal{T}_+^{(2)}\right)\\
&\quad + e^{-j(\beta_+ + \beta_-)}\left(N_{z+}^{(1)} - N_{z-}^{(1)}\right)^2\left(-\mathcal{T}^{(0)}\mathcal{P}_+^{(1)} + \mathcal{S}^{(0)}\mathcal{Q}_+^{(1)} - \mathcal{Q}^{(0)}\mathcal{S}_+^{(1)} + \mathcal{P}^{(0)}\mathcal{T}_+^{(1)}\right)\\
&\quad \times\left(-\mathcal{W}_+^{(1)}\mathcal{U}_+^{(2)} - \mathcal{U}_+^{(1)}\mathcal{W}_+^{(2)} + \mathcal{T}_+^{(1)}\mathcal{P}_+^{(2)} - \mathcal{S}_+^{(1)}\mathcal{Q}_+^{(2)} - \mathcal{Q}_+^{(1)}\mathcal{S}_+^{(2)} + \mathcal{P}_+^{(1)}\mathcal{T}_+^{(2)}\right)\\
&\quad + e^{j(\beta_+ - \beta_-)}\left(N_{z+}^{(1)} + N_{z-}^{(1)}\right)^2\left(\mathcal{T}^{(0)}\mathcal{P}_-^{(1)} + \mathcal{S}^{(0)}\mathcal{Q}_-^{(1)} + \mathcal{Q}^{(0)}\mathcal{S}_-^{(1)} + \mathcal{P}^{(0)}\mathcal{T}_-^{(1)}\right)\\
&\quad \times\left(-\mathcal{W}_-^{(1)}\mathcal{U}_+^{(2)} - \mathcal{U}_-^{(1)}\mathcal{W}_+^{(2)} - \mathcal{T}_-^{(1)}\mathcal{P}_+^{(2)} - \mathcal{S}_-^{(1)}\mathcal{Q}_+^{(2)} + \mathcal{Q}_-^{(1)}\mathcal{S}_+^{(2)} + \mathcal{P}_-^{(1)}\mathcal{T}_+^{(2)}\right)\\
&\quad + e^{-j(\beta_+ - \beta_-)}\left(N_{z+}^{(1)} + N_{z-}^{(1)}\right)^2\left(-\mathcal{T}^{(0)}\mathcal{P}_-^{(1)} + \mathcal{S}^{(0)}\mathcal{Q}_-^{(1)} + \mathcal{Q}^{(0)}\mathcal{S}_-^{(1)} - \mathcal{P}^{(0)}\mathcal{T}_-^{(1)}\right)\\
&\quad \times\left(\mathcal{W}_-^{(1)}\mathcal{U}_+^{(2)} + \mathcal{U}_-^{(1)}\mathcal{W}_+^{(2)} + \mathcal{T}_-^{(1)}\mathcal{P}_+^{(2)} - \mathcal{S}_-^{(1)}\mathcal{Q}_+^{(2)} + \mathcal{Q}_-^{(1)}\mathcal{S}_+^{(2)} - \mathcal{P}_-^{(1)}\mathcal{T}_+^{(2)}\right)
\end{aligned} \tag{4.238}$$

$$\begin{aligned}
&M_{11}M_{43} - M_{13}M_{41}\\
&\quad = -4N^{(0)}\left(\varepsilon_{zz}^{(1)} - N_y^2\right)\varepsilon_{zz}^{(1)-1}N_{z+}^{(1)}N_{z-}^{(1)}\left[\varepsilon_{zz}^{(1)}\cos^2\varphi^{(0)} + \left(\varepsilon_{zz}^{(1)} - N_y^2\right)\right]\\
&\quad \times\left\{\left[\varepsilon_{zz}^{(1)}\left(\varepsilon_{xx}^{(1)} - N_y^2\right) - \varepsilon_{yy}^{(1)}\left(\varepsilon_{zz}^{(1)} - N_y^2\right)\right]\left(\varepsilon_{yx}^{(1)}\mathcal{U}_+^{(2)} + \varepsilon_{xy}^{(1)}\mathcal{W}_+^{(2)}\right)\right.\\
&\quad \left. - 2\varepsilon_{xy}^{(1)}\varepsilon_{yx}^{(1)}\left[\left(\varepsilon_{zz}^{(1)} - N_y^2\right)\mathcal{P}_+^{(2)} - \varepsilon_{zz}^{(1)}\mathcal{T}_+^{(2)}\right]\right\}\\
&\quad + e^{j(\beta_+ + \beta_-)}\left(N_{z+}^{(1)} - N_{z-}^{(1)}\right)^2\left(-\mathcal{T}^{(0)}\mathcal{P}_+^{(1)} + \mathcal{S}^{(0)}\mathcal{Q}_+^{(1)} - \mathcal{Q}^{(0)}\mathcal{S}_+^{(1)} + \mathcal{P}^{(0)}\mathcal{T}_+^{(1)}\right)\\
&\quad \times\left(\mathcal{W}_+^{(1)}\mathcal{U}_+^{(2)} + \mathcal{U}_+^{(1)}\mathcal{W}_+^{(2)} - \mathcal{T}_+^{(1)}\mathcal{P}_+^{(2)} - \mathcal{S}_+^{(1)}\mathcal{Q}_+^{(2)} - \mathcal{Q}_+^{(1)}\mathcal{S}_+^{(2)} - \mathcal{P}_+^{(1)}\mathcal{T}_+^{(2)}\right)\\
&\quad + e^{-j(\beta_+ + \beta_-)}\left(N_{z+}^{(1)} - N_{z-}^{(1)}\right)^2\left(\mathcal{T}^{(0)}\mathcal{P}_+^{(1)} + \mathcal{S}^{(0)}\mathcal{Q}_+^{(1)} - \mathcal{Q}^{(0)}\mathcal{S}_+^{(1)} - \mathcal{P}^{(0)}\mathcal{T}_+^{(1)}\right)\\
&\quad \times\left(-\mathcal{W}_+^{(1)}\mathcal{U}_+^{(2)} - \mathcal{U}_+^{(1)}\mathcal{W}_+^{(2)} + \mathcal{T}_+^{(1)}\mathcal{P}_+^{(2)} - \mathcal{S}_+^{(1)}\mathcal{Q}_+^{(2)} - \mathcal{Q}_+^{(1)}\mathcal{S}_+^{(2)} + \mathcal{P}_+^{(1)}\mathcal{T}_+^{(2)}\right)
\end{aligned}$$

$$
\begin{aligned}
&+\mathrm{e}^{\mathrm{j}(\beta_+-\beta_-)}\left(N_{z+}^{(1)}+N_{z-}^{(1)}\right)^2\left(-\mathcal{T}^{(0)}\mathcal{P}_-^{(1)}+\mathcal{S}^{(0)}\mathcal{Q}_-^{(1)}+\mathcal{Q}^{(0)}\mathcal{S}_-^{(1)}-\mathcal{P}^{(0)}\mathcal{T}_-^{(1)}\right)\\
&\times\left(-\mathcal{W}_-^{(1)}\mathcal{U}_+^{(2)}-\mathcal{U}_-^{(1)}\mathcal{W}_+^{(2)}-\mathcal{T}_-^{(1)}\mathcal{P}_+^{(2)}-\mathcal{S}_-^{(1)}\mathcal{Q}_+^{(2)}+\mathcal{Q}_-^{(1)}\mathcal{S}_+^{(2)}+\mathcal{P}_-^{(1)}\mathcal{T}_+^{(2)}\right)\\
&+\mathrm{e}^{-\mathrm{j}(\beta_+-\beta_-)}\left(N_{z+}^{(1)}+N_{z-}^{(1)}\right)^2\left(\mathcal{T}^{(0)}\mathcal{P}_-^{(1)}+\mathcal{S}^{(0)}\mathcal{Q}_-^{(1)}+\mathcal{Q}^{(0)}\mathcal{S}_-^{(1)}+\mathcal{P}^{(0)}\mathcal{T}_-^{(1)}\right)\\
&\times\left(\mathcal{W}_-^{(1)}\mathcal{U}_+^{(2)}+\mathcal{U}_-^{(1)}\mathcal{W}_+^{(2)}+\mathcal{T}_-^{(1)}\mathcal{P}_+^{(2)}-\mathcal{S}_-^{(1)}\mathcal{Q}_+^{(2)}+\mathcal{Q}_-^{(1)}\mathcal{S}_+^{(2)}-\mathcal{P}_-^{(1)}\mathcal{T}_+^{(2)}\right).
\end{aligned}
\tag{4.239}
$$

The expression entering the off-diagonal element of Cartesian Jones reflection matrix contains the off-diagonal reflection coefficients for the film–substrate interface

$$
\begin{aligned}
&-\left(A_{11}^2-A_{12}^2\right)\left(F_{11}F_{14}-F_{12}F_{13}\right)-\left(A_{13}^2-A_{14}^2\right)\left(F_{32}F_{33}-F_{31}F_{34}\right)\\
&=8N^{(0)}\cos\varphi^{(0)}\left[D_{11}^{(1)}\left(D_{21}^{(1)}D_{33}^{(1)}D_{43}^{(1)}+D_{23}^{(1)}D_{31}^{(1)}D_{41}^{(1)}\right)\mathcal{U}^{(2)}+2D_{11}^{(1)2}D_{21}^{(1)}D_{23}^{(1)}\right.\\
&\left.\mathcal{W}^{(2)}-D_{11}^{(1)}D_{21}^{(1)}D_{23}^{(1)}\left(D_{33}^{(1)}+D_{31}^{(1)}\right)\mathcal{P}^{(2)}-D_{11}^{(1)2}\left(D_{21}^{(1)}D_{43}^{(1)}+D_{23}^{(1)}D_{41}^{(1)}\right)\mathcal{T}^{(2)}\right]\\
&=8\left[\varepsilon_{xy}^{(1)2}\left(\varepsilon_{zz}^{(1)}-N_y^2\right)^2\right]N^{(0)}\cos\varphi^{(0)}\varepsilon_{xy}^{(1)2}\left(\varepsilon_{zz}^{(1)}-N_y^2\right)^2\varepsilon_{zz}^{(1)-1}\left(\varepsilon_{zz}^{(1)}-N_y^2\right)\\
&\times N_{z+}^{(1)}N_{z-}^{(1)}\left\{\left[\varepsilon_{zz}^{(1)}\left(\varepsilon_{xx}^{(1)}-N_y^2\right)-\varepsilon_{yy}^{(1)}\left(\varepsilon_{zz}^{(1)}-N_y^2\right)\right]^2\mathcal{U}^{(2)}\right.\\
&+2\varepsilon_{xy}^{(1)}\varepsilon_{yx}^{(1)}\varepsilon_{zz}^{(1)}\left(\varepsilon_{zz}^{(1)}-N_y^2\right)\mathcal{U}^{(2)}+2\varepsilon_{xy}^{(1)2}\varepsilon_{zz}^{(1)}\left(\varepsilon_{zz}^{(1)}-N_y^2\right)\mathcal{W}^{(2)}\\
&\left.\varepsilon_{xy}^{(1)}\left[\varepsilon_{zz}^{(1)}\left(\varepsilon_{xx}^{(1)}-N_y^2\right)-\varepsilon_{yy}^{(1)}\left(\varepsilon_{zz}^{(1)}-N_y^2\right)\right]\left[\left(\varepsilon_{zz}^{(1)}-N_y^2\right)\mathcal{P}^{(2)}-\varepsilon_{zz}^{(1)}\mathcal{T}^{(2)}\right]\right\}.
\end{aligned}
\tag{4.240}
$$

Here, with Eq. (4.183),

$$
\begin{aligned}
&D_{11}^{(1)}\left(D_{21}^{(1)}D_{33}^{(1)}D_{43}^{(1)}+D_{23}^{(1)}D_{31}^{(1)}D_{41}^{(1)}\right)\\
&=-\varepsilon_{xy}^{(1)2}\left(\varepsilon_{zz}^{(1)}-N_y^2\right)^3\varepsilon_{zz}^{(1)}N_{z+}^{(1)}N_{z-}^{(1)}\\
&\qquad\times\left\{\left[\left(\varepsilon_{xx}^{(1)}-N_y^2\right)-N_{z+}^{(1)2}\right]^2\left[\left(\varepsilon_{xx}^{(1)}-N_y^2\right)-N_{z-}^{(1)2}\right]^2\right\}\\
&=-\varepsilon_{xy}^{(1)2}\left(\varepsilon_{zz}^{(1)}-N_y^2\right)^3N_{z+}^{(1)}N_{z-}^{(1)}\\
&\qquad\times\left[2\varepsilon_{zz}^{(1)}\left(\varepsilon_{xx}^{(1)}-N_y^2\right)^2-2\varepsilon_{zz}^{(1)}\left(\varepsilon_{xx}^{(1)}-N_y^2\right)\left(N_{z+}^{(1)2}+N_{z-}^{(1)2}\right)\right.\\
&\quad+\varepsilon_{zz}^{(1)}\left(N_{z+}^{(1)4}+N_{z-}^{(1)4}\right)-2\varepsilon_{zz}^{(1)}N_{z+}^{(1)2}N_{z-}^{(1)2}\\
&\quad\left.+2\varepsilon_{yy}^{(1)}\left(\varepsilon_{xx}^{(1)}-N_y^2\right)\left(\varepsilon_{zz}^{(1)}-N_y^2\right)-2\varepsilon_{xy}^{(1)}\varepsilon_{yx}^{(1)}\left(\varepsilon_{zz}^{(1)}-N_y^2\right)\right]\\
&=-\varepsilon_{xy}^{(1)2}\left(\varepsilon_{zz}^{(1)}-N_y^2\right)^3N_{z+}^{(1)}N_{z-}^{(1)}\left[\varepsilon_{zz}^{(1)}\left(N_{z+}^{(1)2}-N_{z-}^{(1)2}\right)^2-2\varepsilon_{xy}^{(1)}\varepsilon_{yx}^{(1)}\left(\varepsilon_{zz}^{(1)}-N_y^2\right)\right]\\
&=-\left[\varepsilon_{xy}^{(1)2}\left(\varepsilon_{zz}^{(1)}-N_y^2\right)^2\right]N_{z+}^{(1)}N_{z-}^{(1)}\left(\varepsilon_{zz}^{(1)}-N_y^2\right)\varepsilon_{zz}^{(1)-1}\\
&\quad\times\left\{\left[\varepsilon_{zz}^{(1)}\left(\varepsilon_{xx}^{(1)}-N_y^2\right)-\varepsilon_{yy}^{(1)}\left(\varepsilon_{zz}^{(1)}-N_y^2\right)\right]^2+2\varepsilon_{xy}^{(1)}\varepsilon_{yx}^{(1)}\varepsilon_{zz}^{(1)}\left(\varepsilon_{zz}^{(1)}-N_y^2\right)\right\}.
\end{aligned}
\tag{4.241}
$$

We get for the expression in Eq. (4.209d) with Eq. (4.240) divided by the common factor in the brackets $\left[\varepsilon_{xy}^{(1)2}\left(\varepsilon_{zz}^{(1)} - N_y^2\right)^2\right]$

$$
\begin{aligned}
&\mathrm{M}_{11}\mathrm{M}_{23} - \mathrm{M}_{21}\mathrm{M}_{13} \\
&\quad = 8N^{(0)} \cos\varphi^{(0)} \varepsilon_{zz}^{(1)-1}\left(\varepsilon_{zz}^{(1)} - N_y^2\right) N_{z+}^{(1)} N_{z-}^{(1)} \\
&\quad \times \left\{ -\left[\varepsilon_{zz}^{(1)}\left(\varepsilon_{xx}^{(1)} - N_y^2\right) - \varepsilon_{yy}^{(1)}\left(\varepsilon_{zz}^{(1)} - N_y^2\right)\right]^2 \mathcal{U}_+^{(2)} \right. \\
&\quad - 2\varepsilon_{xy}^{(1)}\varepsilon_{yx}^{(1)}\varepsilon_{zz}^{(1)}\left(\varepsilon_{zz}^{(1)} - N_y^2\right)\mathcal{U}_+^{(2)} + 2\varepsilon_{xy}^{(1)2}\varepsilon_{zz}^{(1)}\left(\varepsilon_{zz}^{(1)} - N_y^2\right)\mathcal{W}_+^{(2)} \\
&\quad \left. + \varepsilon_{xy}^{(1)}\left[\varepsilon_{zz}^{(1)}\left(\varepsilon_{xx}^{(1)} - N_y^2\right) - \varepsilon_{yy}^{(1)}\left(\varepsilon_{zz}^{(1)} - N_y^2\right)\right]\left[\left(\varepsilon_{zz}^{(1)} - N_y^2\right)\mathcal{P}_+^{(2)} - \varepsilon_{zz}^{(1)}\mathcal{T}_+^{(2)}\right]\right\} \\
&\quad + 2N^{(0)} \cos\varphi^{(0)} \mathrm{e}^{\mathrm{j}(\beta_+ + \beta_-)}\left(N_{z+}^{(1)} - N_{z-}^{(1)}\right)^2 \mathcal{U}_+^{(1)} \\
&\quad \times \left(\mathcal{W}_+^{(1)}\mathcal{U}_+^{(2)} + \mathcal{U}_+^{(1)}\mathcal{W}_+^{(2)} - \mathcal{T}_+^{(1)}\mathcal{P}_+^{(2)} - \mathcal{S}_+^{(1)}\mathcal{Q}_+^{(2)} - \mathcal{Q}_+^{(1)}\mathcal{S}_+^{(2)} - \mathcal{P}_+^{(1)}\mathcal{T}_+^{(2)}\right) \\
&\quad - 2N^{(0)} \cos\varphi^{(0)} \mathrm{e}^{-\mathrm{j}(\beta_+ + \beta_-)}\left(N_{z+}^{(1)} - N_{z-}^{(1)}\right)^2 \mathcal{U}_+^{(1)} \\
&\quad \times \left(-\mathcal{W}_+^{(1)}\mathcal{U}_+^{(2)} - \mathcal{U}_+^{(1)}\mathcal{W}_+^{(2)} + \mathcal{T}_+^{(1)}\mathcal{P}_+^{(2)} - \mathcal{S}_+^{(1)}\mathcal{Q}_+^{(2)} - \mathcal{Q}_+^{(1)}\mathcal{S}_+^{(2)} + \mathcal{P}_+^{(1)}\mathcal{T}_+^{(2)}\right) \\
&\quad + 2N^{(0)} \cos\varphi^{(0)} \mathrm{e}^{\mathrm{j}(\beta_+ - \beta_-)}\left(N_{z+}^{(1)} + N_{z-}^{(1)}\right)^2 \mathcal{U}_-^{(1)} \\
&\quad \times \left(-\mathcal{W}_-^{(1)}\mathcal{U}_+^{(2)} - \mathcal{U}_-^{(1)}\mathcal{W}_+^{(2)} - \mathcal{T}_-^{(1)}\mathcal{P}_+^{(2)} - \mathcal{S}_-^{(1)}\mathcal{Q}_+^{(2)} + \mathcal{Q}_-^{(1)}\mathcal{S}_+^{(2)} + \mathcal{P}_-^{(1)}\mathcal{T}_+^{(2)}\right) \\
&\quad - 2N^{(0)} \cos\varphi^{(0)} \mathrm{e}^{-\mathrm{j}(\beta_+ - \beta_-)}\left(N_{z+}^{(1)} + N_{z-}^{(1)}\right)^2 \mathcal{U}_-^{(1)} \\
&\quad \times \left(\mathcal{W}_-^{(1)}\mathcal{U}_+^{(2)} + \mathcal{U}_-^{(1)}\mathcal{W}_+^{(2)} + \mathcal{T}_-^{(1)}\mathcal{P}_+^{(2)} - \mathcal{S}_-^{(1)}\mathcal{Q}_+^{(2)} + \mathcal{Q}_-^{(1)}\mathcal{S}_+^{(2)} - \mathcal{P}_-^{(1)}\mathcal{T}_+^{(2)}\right).
\end{aligned}
\tag{4.242}
$$

The expression entering the second off-diagonal element of Cartesian Jones reflection matrix contains the off-diagonal reflection coefficients for the film–substrate interface

$$
\begin{aligned}
&-\left(A_{31}^2 - A_{32}^2\right)\left(F_{11}F_{14} - F_{12}F_{13}\right) - \left(A_{33}^2 - A_{34}^2\right)\left(F_{32}F_{33} - F_{31}F_{34}\right) \\
&\quad = 8N^{(0)} \cos\varphi^{(0)}\left[2D_{31}^{(1)}D_{33}^{(1)}D_{41}^{(1)}D_{43}^{(1)}\mathcal{U}^{(2)}\right. \\
&\quad + D_{11}^{(1)}\left(D_{23}^{(1)}D_{31}^{(1)}D_{41}^{(1)} + D_{21}^{(1)}D_{33}^{(1)}D_{43}^{(1)}\right)\mathcal{W}^{(2)} \\
&\quad \left. - D_{31}^{(1)}D_{33}^{(1)}\left(D_{23}^{(1)}D_{41}^{(1)} + D_{21}^{(1)}D_{43}^{(1)}\right)\mathcal{P}^{(2)} - D_{11}^{(1)}D_{41}^{(1)}D_{43}^{(1)}\left(D_{31}^{(1)} + D_{33}^{(1)}\right)\mathcal{T}^{(2)}\right] \\
&\quad = 8\left[\varepsilon_{xy}^{(1)2}\left(\varepsilon_{zz}^{(1)} - N_y^2\right)^2\right] N^{(0)} \cos\varphi^{(0)} \varepsilon_{xy}^{(1)2}\left(\varepsilon_{zz}^{(1)} - N_y^2\right)^2 \varepsilon_{zz}^{(1)-1}\left(\varepsilon_{zz}^{(1)} - N_y^2\right) \\
&\quad \times N_{z+}^{(1)}N_{z-}^{(1)}\left\{-2\varepsilon_{yx}^{(1)2}\varepsilon_{zz}^{(1)}\left(\varepsilon_{zz}^{(1)} - N_y^2\right)\mathcal{U}^{(2)}\right. \\
&\quad + \left[\varepsilon_{zz}^{(1)}\left(\varepsilon_{xx}^{(1)} - N_y^2\right) - \varepsilon_{yy}^{(1)}\left(\varepsilon_{zz}^{(1)} - N_y^2\right)\right]^2\mathcal{W}^{(2)} + 2\varepsilon_{xy}^{(1)}\varepsilon_{yx}^{(1)}\varepsilon_{zz}^{(1)}\left(\varepsilon_{zz}^{(1)} - N_y^2\right)\mathcal{W}^{(2)} \\
&\quad \left. + \varepsilon_{yx}^{(1)}\left[\varepsilon_{zz}^{(1)}\left(\varepsilon_{xx}^{(1)} - N_y^2\right) - \varepsilon_{yy}^{(1)}\left(\varepsilon_{zz}^{(1)} - N_y^2\right)\right]\left[\left(\varepsilon_{zz}^{(1)} - N_y^2\right)\mathcal{P}^{(2)} - \varepsilon_{zz}^{(1)}\mathcal{T}^{(2)}\right]\right\}.
\end{aligned}
\tag{4.243}
$$

We get for the expression in Eq. (4.209e) with Eq. (4.243) divided by the common factor in the brackets $[\varepsilon_{xy}^{(1)2}(\varepsilon_{zz}^{(1)} - N_y^2)^2]$

$$\begin{aligned}
&M_{41}M_{33} - M_{43}M_{31}\\
&\quad = 8N^{(0)}\cos\varphi^{(0)}\varepsilon_{zz}^{(1)-1}\left(\varepsilon_{zz}^{(1)} - N_y^2\right)N_{z+}^{(1)}N_{z-}^{(1)}\\
&\quad \times\Big\{2\varepsilon_{yx}^{(1)2}\varepsilon_{zz}^{(1)}\left(\varepsilon_{zz}^{(1)} - N_y^2\right)\mathcal{U}_+^{(2)} - \left[\varepsilon_{zz}^{(1)}\left(\varepsilon_{xx}^{(1)} - N_y^2\right) - \varepsilon_{yy}^{(1)}\left(\varepsilon_{zz}^{(1)} - N_y^2\right)\right]^2\mathcal{W}_+^{(2)}\\
&\quad - 2\varepsilon_{xy}^{(1)}\varepsilon_{yx}^{(1)}\varepsilon_{zz}^{(1)}\left(\varepsilon_{zz}^{(1)} - N_y^2\right)\mathcal{W}_+^{(2)}\\
&\quad + \varepsilon_{yx}^{(1)}\left[\varepsilon_{zz}^{(1)}\left(\varepsilon_{xx}^{(1)} - N_y^2\right) - \varepsilon_{yy}^{(1)}\left(\varepsilon_{zz}^{(1)} - N_y^2\right)\right]\left[\left(\varepsilon_{zz}^{(1)} - N_y^2\right)\mathcal{P}_+^{(2)} - \varepsilon_{zz}^{(1)}\mathcal{T}_+^{(2)}\right]\Big\}\\
&\quad + 2N^{(0)}\cos\varphi^{(0)}e^{j(\beta_+ + \beta_-)}\left(N_{z+}^{(1)} - N_{z-}^{(1)}\right)^2\mathcal{W}_+^{(1)}\\
&\quad \times\left(\mathcal{W}_+^{(1)}\mathcal{U}_+^{(2)} + \mathcal{U}_+^{(1)}\mathcal{W}_+^{(2)} - \mathcal{T}_+^{(1)}\mathcal{P}_+^{(2)} - \mathcal{S}_+^{(1)}\mathcal{Q}_+^{(2)} - \mathcal{Q}_+^{(1)}\mathcal{S}_+^{(2)} - \mathcal{P}_+^{(1)}\mathcal{T}_+^{(2)}\right)\\
&\quad - 2N^{(0)}\cos\varphi^{(0)}e^{-j(\beta_+ + \beta_-)}\left(N_{z+}^{(1)} - N_{z-}^{(1)}\right)^2\mathcal{W}_+^{(1)}\\
&\quad \times\left(-\mathcal{W}_+^{(1)}\mathcal{U}_+^{(2)} - \mathcal{U}_+^{(1)}\mathcal{W}_+^{(2)} + \mathcal{T}_+^{(1)}\mathcal{P}_+^{(2)} - \mathcal{S}_+^{(1)}\mathcal{Q}_+^{(2)} - \mathcal{Q}_+^{(1)}\mathcal{S}_+^{(2)} + \mathcal{P}_+^{(1)}\mathcal{T}_+^{(2)}\right)\\
&\quad + 2N^{(0)}\cos\varphi^{(0)}e^{j(\beta_+ - \beta_-)}\left(N_{z+}^{(1)} + N_{z-}^{(1)}\right)^2\mathcal{W}_-^{(1)}\\
&\quad \times\left(-\mathcal{W}_-^{(1)}\mathcal{U}_+^{(2)} - \mathcal{U}_-^{(1)}\mathcal{W}_+^{(2)} - \mathcal{T}_-^{(1)}\mathcal{P}_+^{(2)} - \mathcal{S}_-^{(1)}\mathcal{Q}_+^{(2)} + \mathcal{Q}_-^{(1)}\mathcal{S}_+^{(2)} + \mathcal{P}_-^{(1)}\mathcal{T}_+^{(2)}\right)\\
&\quad - 2N^{(0)}\cos\varphi^{(0)}e^{-j(\beta_+ - \beta_-)}\left(N_{z+}^{(1)} + N_{z-}^{(1)}\right)^2\mathcal{W}_-^{(1)}\\
&\quad \times\left(\mathcal{W}_-^{(1)}\mathcal{U}_+^{(2)} + \mathcal{U}_-^{(1)}\mathcal{W}_+^{(2)} + \mathcal{T}_-^{(1)}\mathcal{P}_+^{(2)} - \mathcal{S}_-^{(1)}\mathcal{Q}_+^{(2)} + \mathcal{Q}_-^{(1)}\mathcal{S}_+^{(2)} - \mathcal{P}_-^{(1)}\mathcal{T}_+^{(2)}\right).
\end{aligned} \tag{4.244}$$

We have set $\beta_\pm = \frac{\omega}{c}N_{z\pm}^{(n)}d^{(1)}$. Eq. (4.235), Eq. (4.238), Eq. (4.239), Eq. (4.242), and Eq. (4.244) allow us to construct the Jones reflection matrix, as given by Eq. (3.72), for a magnetic film on a magnetic substrate system. They are rather complicated for a direct numerical evaluation. Nevertheless, they convey useful information. They allow us to appreciate various effects contributing to the response, develop approximations, and trace the numerical procedures including complicated phase relations. We observe that each of Eq. (4.235), Eq. (4.238), Eq. (4.239), Eq. (4.242), and Eq. (4.244) contains a term independent of the factors $\exp\left(\pm j\beta^{(1)\pm}\right)$, which can be assigned to the mode coupling upon the reflection on the film–substrate interface.

We further observe from Eq. (4.235), Eq. (4.238), Eq. (4.239), Eq. (4.242), and Eq. (4.244) that these expressions are perturbed by the magnetizations in both the film and substrate in such a way that the diagonal elements of the Cartesian Jones reflection matrix contain only even powers of the off-diagonal tensor elements, *i.e.*, $\varepsilon_{xy}^{(1)2}$, $\varepsilon_{xy}^{(2)2}$, and $\varepsilon_{xy}^{(1)}\varepsilon_{xy}^{(2)}$ or $\varepsilon_{xy}^{(1)}\varepsilon_{yx}^{(1)}$, *etc.* As a result, the diagonal Cartesian Jones reflection matrix elements are not affected by a simultaneous reversal of the magnetization in the film and the substrate. However, they should be sensitive to the transitions from a parallel electron spin arrangement (ferromagnetic coupling) to an antiparallel one (antiferromagnetic coupling) between the film and the substrate. The off-diagonal elements of the Jones reflection matrix are odd functions of the off-diagonal elements of the permittivity tensor. In particular, they have the components proportional separately to either $\varepsilon_1^{(1)}$ or $\varepsilon_1^{(2)}$. They change the sign when the magnetization is reversed in both the film and the substrate. Note the presence of the expressions characterizing the perpendicular anisotropy, *i.e.*, $\left[\varepsilon_{zz}^{(1)}\left(\varepsilon_{xx}^{(1)} - N_y^2\right) - \varepsilon_{yy}^{(1)}\left(\varepsilon_{zz}^{(1)} - N_y^2\right)\right]$, a parameter determination of which presents a fundamental interest.

This completes our analysis of the response in film–substrate system at polar magnetization in a medium with orthorhombic symmetry with one of the orthorhombic axes normal to the interfaces and the remaining two orthorhombic axes in the film and in the substrate confined to the interface plane. These situations occur in epitaxially grown magnetic films. The in-plane orthorhombic axes in the film may be rotated with respect to those in the substrate. The general solution automatically includes the case of a magnetic film on a nonmagnetic substrate as well as that of a nonmagnetic film on a magnetic substrate.

So far, our treatment has assumed the film and substrate media both characterized by the permittivity tensor of Eq. (4.177). This covers a much broader class of anisotropic media studied in the generalized ellipsometry than that with magnetic ordering of main concern here. In the following, we shall consider three simpler film–substrate cases, (1) the normal incidence on an axially symmetric magnetic system, (2) the oblique incidence on a nonmagnetic uniaxial system with a special crystallographic orientation, and (3) an approximate solution for the oblique incidence.

4.5.3 Normal Incidence

We assume $N_y = 0$, $\cos\varphi^{(0)} = 1$, $\sin\varphi^{(0)} = 0$, with $\varepsilon_{xy}^{(n)} = -\varepsilon_{yx}^{(n)}$ and $\varepsilon_{xx}^{(n)} = \varepsilon_{yy}^{(n)} = \varepsilon_{zz}^{(n)}$, for $n = 1, 2$. The proper polarization modes both in the film and in the substrate are circularly polarized. Therefore, there is no mode coupling upon reflection at the internal interfaces. This case has already been treated above in Section 4.2 under the assumption of circularly polarized proper polarization modes in the ambient isotropic half space $n = 0$. The solution will now be obtained for linearly polarized proper polarization modes in the ambient. We obtain from Eq. (4.235), Eq. (4.238), Eq. (4.239), Eq. (4.242), and Eq. (4.244)

$$\begin{aligned} \mathrm{M}_{11}\mathrm{M}_{33} - \mathrm{M}_{13}\mathrm{M}_{31} &= 1 + \mathrm{e}^{-2\mathrm{j}\beta_+} r_+^{(01)} r_+^{(12)} + \mathrm{e}^{-2\mathrm{j}\beta_-} r_-^{(01)} r_-^{(12)} \\ &\quad + \mathrm{e}^{-2\mathrm{j}\beta_+ + \beta_-} r_+^{(01)} r_-^{(01)} r_+^{(12)} r_-^{(12)} \\ &= \left(1 + r_+^{(01)} r_+^{(12)} \mathrm{e}^{-2\mathrm{j}\beta_+}\right)\left(1 + r_-^{(01)} r_-^{(12)} \mathrm{e}^{-2\mathrm{j}\beta_-}\right), \end{aligned} \tag{4.245a}$$

$$
\begin{aligned}
2(\mathrm{M}_{21}\mathrm{M}_{33} - \mathrm{M}_{23}\mathrm{M}_{31}) &= 2(\mathrm{M}_{11}\mathrm{M}_{43} - \mathrm{M}_{41}\mathrm{M}_{13}) \\
&= \left(r_+^{(01)} + r_-^{(01)}\right)\left[1 + r_+^{(12)} r_-^{(12)} \mathrm{e}^{-2\mathrm{j}(\beta_+ + \beta_-)}\right] \\
&\quad + \left(1 + r_+^{(01)} r_-^{(01)}\right)\left(r_+^{(12)} \mathrm{e}^{-2\mathrm{j}\beta_+} + r_-^{(12)} \mathrm{e}^{-2\mathrm{j}\beta_-}\right) \\
&= \left(r_+^{(01)} + r_+^{(12)} \mathrm{e}^{-2\mathrm{j}\beta_+}\right)\left(1 + r_-^{(01)} r_-^{(12)} \mathrm{e}^{-2\mathrm{j}\beta_-}\right) \\
&\quad + \left(1 + r_+^{(01)} r_+^{(12)} \mathrm{e}^{-2\mathrm{j}\beta_+}\right)\left(r_-^{(01)} + r_-^{(12)} \mathrm{e}^{-2\mathrm{j}\beta_-}\right),
\end{aligned} \tag{4.245b}
$$

$$
\begin{aligned}
2\mathrm{j}(\mathrm{M}_{11}\mathrm{M}_{23} - \mathrm{M}_{21}\mathrm{M}_{13}) &= -2\mathrm{j}(\mathrm{M}_{41}\mathrm{M}_{33} - \mathrm{M}_{43}\mathrm{M}_{31}) \\
&= -\left(r_+^{(01)} - r_-^{(01)}\right)\left[1 - r_+^{(12)} r_-^{(12)} \mathrm{e}^{-2\mathrm{j}(\beta_+ + \beta_-)}\right] \\
&\quad - \left(1 - r_+^{(01)} r_-^{(01)}\right)\left(r_+^{(12)} \mathrm{e}^{-2\mathrm{j}\beta_+} - r_-^{(12)} \mathrm{e}^{-2\mathrm{j}\beta_-}\right) \\
&= -\left(r_+^{(01)} + r_+^{(12)} \mathrm{e}^{-2\mathrm{j}\beta_+}\right)\left(1 + r_-^{(01)} r_-^{(12)} \mathrm{e}^{-2\mathrm{j}\beta_-}\right) \\
&\quad + \left(1 + r_+^{(01)} r_+^{(12)} \mathrm{e}^{-2\mathrm{j}\beta_+}\right)\left(r_-^{(01)} + r_-^{(12)} \mathrm{e}^{-2\mathrm{j}\beta_-}\right).
\end{aligned} \tag{4.245c}
$$

Here,

$$
r_\pm^{(01)} = \frac{N^{(0)} - N^{(1)\pm}}{N^{(0)} + N^{(1)\pm}}, \quad r_\pm^{(12)} = \frac{N^{(1)\pm} - N^{(2)\pm}}{N^{(1)\pm} + N^{(2)\pm}} \tag{4.246}
$$

$$
\left(N^{(n)\pm}\right)^2 = \varepsilon_0^{(n)} \pm \varepsilon_1^{(n)}, \quad n = 1, 2, \quad \beta^\pm = N^{(1)\pm} \frac{\omega}{c} d^{\,(1)}. \tag{4.247}
$$

Using Eqs. (3.72), we can establish the elements of the Jones reflection 2×2 matrix in Cartesian representation. Note that the elements of the reflection matrix following from Eq. (4.245) can be expressed as a sum of two components; each component characterizes a particular circular polarization. The results are consistent with those derived in Section 4.2.

4.5.4 Uniaxial Film on a Uniaxial Substrate

At zero off-diagonal permittivity tensor elements $\varepsilon_{xy}^{(n)} = -\varepsilon_{yx}^{(n)} = 0$, and in the absence of the in-plane anisotropy, $\varepsilon_{xx}^{(n)} = \varepsilon_{yy}^{(n)} = \varepsilon_0^{(n)}$ and $\varepsilon_{zz}^{(n)} = \varepsilon_3^{(n)}$, Eq. (4.235), Eq. (4.238), Eq. (4.239), Eq. (4.242), and Eq. (4.244) must reduce to the case of a uniaxial film on a uniaxial substrate in an isotropic ambient. The principal optical axes in both the film and substrate are normal to the interfaces. This situation is treated in the book by Azzam and Bashara [2]. We obtain from Eq. (4.182)

$$
N_{zs}^{(n)} = \left(\varepsilon_0^{(n)} - N_y^2\right)^{1/2}, \tag{4.248a}
$$

$$
N_{zp}^{(n)} = \left(\frac{\varepsilon_0^{(n)}}{\varepsilon_3^{(n)}}\right)^{1/2} \left(\varepsilon_3^{(n)} - N_y^2\right)^{1/2}. \tag{4.248b}
$$

The indices s and p distinguish the solution for two linear orthogonal polarizations normal and parallel to the plane of incidence. Up to a common factor, we obtain from Eq.(4.235), Eq. (4.238), and Eq. (4.242)

$$M_{11}M_{33} - M_{13}M_{31} = \left[1 + r_{ss}^{(01)} r_{ss}^{(12)} \exp\left(-2j\frac{\omega}{c} N_{zs}^{(1)} d^{(1)}\right)\right] \left[1 + r_{pp}^{(01)} r_{pp}^{(12)} \exp\left(-2j\frac{\omega}{c} N_{zp}^{(1)} d^{(1)}\right)\right], \tag{4.249a}$$

$$M_{11}M_{43} - M_{41}M_{13} = \left[r_{pp}^{(01)} + r_{pp}^{(12)} \exp\left(-2j\frac{\omega}{c} N_{zp}^{(1)} d^{(1)}\right)\right] \left[1 + r_{ss}^{(01)} r_{ss}^{(12)} \exp\left(-2j\frac{\omega}{c} N_{zs}^{(1)} d^{(1)}\right)\right], \tag{4.249b}$$

$$M_{21}M_{33} - M_{23}M_{31} = \left[r_{ss}^{(01)} + r_{ss}^{(12)} \exp\left(-2j\frac{\omega}{c} N_{zs}^{(1)} d^{(1)}\right)\right] \left[1 + r_{pp}^{(01)} r_{pp}^{(12)} \exp\left(-2j\frac{\omega}{c} N_{zp}^{(1)} d^{(1)}\right)\right]. \tag{4.249c}$$

The Cartesian Jones reflection matrix of this system is diagonal. The single-interface reflection coefficients are

$$r_{21}^{(01)} = r_{ss}^{(01)} = \frac{N^{(0)} \cos\varphi^{(0)} - \left(\varepsilon_0^{(1)} - N_y^2\right)^{1/2}}{N^{(0)} \cos\varphi^{(0)} + \left(\varepsilon_0^{(1)} - N_y^2\right)^{1/2}}, \tag{4.250a}$$

$$r_{21}^{(12)} = r_{ss}^{(12)} = \frac{\left(\varepsilon_0^{(1)} - N_y^2\right)^{1/2} - \left(\varepsilon_0^{(2)} - N_y^2\right)^{1/2}}{\left(\varepsilon_0^{(1)} - N_y^2\right)^{1/2} + \left(\varepsilon_0^{(2)} - N_y^2\right)^{1/2}}, \tag{4.250b}$$

$$r_{43}^{(12)} = r_{pp}^{(01)} = \frac{N^{(0)} \left(\varepsilon_3^{(1)} - N_y^2\right)^{1/2} - \left(\varepsilon_0^{(1)} \varepsilon_3^{(1)}\right)^{1/2} \cos\varphi^{(0)}}{N^{(0)} \left(\varepsilon_3^{(1)} - N_y^2\right)^{1/2} + \left(\varepsilon_0^{(1)} \varepsilon_3^{(1)}\right)^{1/2} \cos\varphi^{(0)}}, \tag{4.250c}$$

$$r_{43}^{(12)} = r_{pp}^{(12)} = \frac{\left[\varepsilon_0^{(1)} \varepsilon_3^{(1)} \left(\varepsilon_3^{(2)} - N_y^2\right)\right]^{1/2} - \left[\varepsilon_0^{(2)} \varepsilon_3^{(2)} \left(\varepsilon_3^{(1)} - N_y^2\right)\right]^{1/2}}{\left[\varepsilon_0^{(1)} \varepsilon_3^{(1)} \left(\varepsilon_3^{(2)} - N_y^2\right)\right]^{1/2} + \left[\varepsilon_0^{(2)} \varepsilon_3^{(2)} \left(\varepsilon_3^{(1)} - N_y^2\right)\right]^{1/2}}. \tag{4.250d}$$

For the proper polarization modes polarized in the plane of incidence, the sign of the single-interface reflection coefficients $r_{pp}^{(n)}$ of the system is a matter of convention. We systematically employ $r_{pp}^{(n)} = r_{43}^{(n)}$. Note that the convention with the opposite sign is established in the classical ellipsometry, giving $r_{pp}^{(n)} = -r_{43}^{(n)}$.

The special case of an orthorhombic nonmagnetic crystal with one of the orthorhombic axes oriented normal to the interface is considerably more involved. The solution starts from the tensor (4.177) with the elements expressed with Eq. (4.178) for $\varepsilon_1^{(n)} = 0$. Consequently, $\varepsilon_{xy}^{(n)} = \varepsilon_{yx}^{(n)}$ in Eq. (4.182). The procedure of obtaining the propagation and dynamical matrices for this case follows the lines described in Section 4.5.1.

4.5.5 Approximate Solution

We shall now give an approximate solution for the case of an arbitrary angle of incidence for a permittivity tensor with $\varepsilon_{xy}^{(n)} = -\varepsilon_{yx}^{(n)} = \mathrm{j}\,\varepsilon_1^{(n)}$ and $\varepsilon_{xx}^{(n)} = \varepsilon_{yy}^{(n)} = \varepsilon_{zz}^{(n)} = \varepsilon_0^{(n)}$ for $n = 1, 2$, corresponding to the polar magnetization and to the absence of both in-plane anisotropy and perpendicular anisotropies. This special case is easier to understand and provides a check of consistency. The approximation consists of retaining only the terms independent of magnetization and the terms linearly dependent on the magnetization both in the film and in the substrate. We therefore drop, among others, the terms containing the products of the terms linear in the film magnetization with those linear in the substrate magnetization. In other words, we keep only the terms of zero and first order in the off-diagonal tensor elements, $\varepsilon_1^{(1)}$ and $\varepsilon_1^{(2)}$. These quite realistic and widely used assumptions make the final expressions for the Jones reflection matrix much simpler. As we shall see below, to the first order in $\varepsilon_1^{(1)}$ and $\varepsilon_1^{(2)}$, $r_{21}^{(02)}$ and $r_{43}^{(02)}$, the diagonal elements of the Jones reflection matrix of the film substrate system do not contain $\varepsilon_1^{(1)}$ and $\varepsilon_1^{(2)}$. In this approximation, the off-diagonal elements of the Jones reflection matrix retain the terms containing the linear functions of either $\varepsilon_1^{(1)}$ or $\varepsilon^{(2)}$, only. From Eq. (4.182)

$$N_{z\pm}^{(n)2} = \left(\varepsilon_{zz}^{(n)} - N_y^2\right) \pm \frac{\sqrt{\varepsilon_{xy}^{(n)2}\left(\varepsilon_{zz}^{(n)} - N_y^2\right)}}{\varepsilon_{zz}^{(n)1/2}}. \tag{4.251}$$

Let us introduce the following convenient notation

$$N_{z0}^{(n)2} = \left(\varepsilon_0^{(n)} - N_y^2\right), \tag{4.252a}$$

$$N^{(n)} = \varepsilon_0^{(n)1/2}, \tag{4.252b}$$

$$\alpha_z^{(n)} = \left(\frac{\varepsilon_0^{(n)} - N_y^2}{\varepsilon_0^{(n)1/2}}\right)^{1/2} = \frac{N_{z0}^{(n)}}{N^{(n)}}. \tag{4.252c}$$

By convention, we assign the solutions of Eq. (4.251) as

$$\begin{aligned} N_{z\pm}^{(n)2} &= N_{z0}^{(n)2} \mp \mathrm{j}\,\varepsilon_{xy}^{(n)} \frac{N_{z0}^{(n)}}{\varepsilon_{zz}^{(n)1/2}} \\ &= \alpha_z^{(n)2} N^{(n)2} \pm \alpha_z^{(n)} \varepsilon_1^{(n)}. \end{aligned} \tag{4.253}$$

The restriction to the linearity in $\varepsilon_1^{(n)}$ simplifies Eq. (4.183) to

$$\begin{aligned} N_{z+}^{(n)2} + N_{z-}^{(n)2} &\approx 2N_{z0}^{(n)2} \\ &= 2N^{(n)2}\alpha_z^{(n)2}, \end{aligned} \tag{4.254a}$$

$$N_{z+}^{(n)2} N_{z-}^{(n)2} = \varepsilon_{zz}^{(n)-1}\left(\varepsilon_{zz}^{(n)} - N_y^2\right)\left[\varepsilon_{zz}^{(n)}\left(\varepsilon_{zz}^{(n)} - N_y^2\right) + \varepsilon_{xy}^{(n)2}\right]$$
$$= \alpha_z^{(n)2}\left(N^{(n)4}\alpha_z^{(n)2} - \varepsilon_1^{(n)2}\right)$$
$$N_{z+}^{(n)} N_{z-}^{(n)} \approx N^{(n)2}\alpha_z^{(n)2} - \frac{\varepsilon_1^{(n)2}}{2N^{(n)2}}, \tag{4.254b}$$

$$\varepsilon_1^{(n)} = \frac{N_{z+}^{(n)2} - N_{z-}^{(n)2}}{2\alpha_z^{(n)}}$$
$$\approx N^{(n)}\left(N_{z+}^{(n)} - N_{z-}^{(n)}\right). \tag{4.254c}$$

Eq. (4.224) yields

$$\mathcal{P}_+^{(n)} \approx 2N^{(n)4}\alpha_z^{(n)2}, \tag{4.255a}$$
$$\mathcal{P}_-^{(n)} \approx \frac{1}{2}\varepsilon_1^{(n)2} = \frac{1}{2}N^{(n)2}\left(N_{z+}^{(n)} - N_{z-}^{(n)}\right)^2, \tag{4.255b}$$
$$\mathcal{Q}_+^{(n)} \approx 2N^{(n)3}\alpha_z^{(n)3}, \tag{4.255c}$$
$$\mathcal{Q}_-^{(n)} \approx \left(N_{z+}^{(n)} - N_{z-}^{(n)}\right)N^{(n)2}\alpha_z^{(n)2} = \varepsilon_1^{(n)} N^{(n)}\alpha_z^{(n)2}, \tag{4.255d}$$
$$\mathcal{S}_+^{(n)} \approx 2N^{(n)5}\alpha_z^{(n)3}, \tag{4.255e}$$
$$\mathcal{S}_-^{(n)} \approx \left(N_{z+}^{(n)} - N_{z-}^{(n)}\right)N^{(n)4}\alpha_z^{(n)2} = \varepsilon_1^{(n)} N^{(n)3}\alpha_z^{(n)2}, \tag{4.255f}$$
$$\mathcal{T}_+^{(n)} \approx 2N^{(n)4}\alpha_z^{(n)4}, \tag{4.255g}$$
$$\mathcal{T}_-^{(n)} \approx -\frac{1}{2}\alpha_z^{(n)2}\varepsilon_1^{(n)2} = -\frac{1}{2}\alpha_z^{(n)2}N^{(n)2}\left(N_{z+}^{(n)} - N_{z-}^{(n)}\right)^2, \tag{4.255h}$$
$$\mathcal{U}_\pm^{(n)} \approx -\mathrm{j}\,\varepsilon_1^{(n)} N^{(n)2}\alpha_z^{(n)2} = -\mathrm{j}N^{(n)3}\alpha_z^{(n)2}\left(N_{z+}^{(n)} - N_{z-}^{(n)}\right), \tag{4.255i}$$
$$\mathcal{W}_\pm^{(n)} \approx \mathrm{j}\,\varepsilon_1^{(n)} N^{(n)2}\alpha_z^{(n)2} = \mathrm{j}N^{(n)3}\alpha_z^{(n)2}\left(N_{z+}^{(n)} - N_{z-}^{(n)}\right). \tag{4.255j}$$

We have included $\mathcal{P}_-^{(n)}$ and $\mathcal{T}_-^{(n)}$, which are of second order in $\varepsilon_1^{(n)}$. It turns out that in this approximation the off-diagonal elements of the global Jones reflection matrix differ in their signs only, *i.e.*, $M_{11}M_{23} - M_{13}M_{21} = -M_{41}M_{33} - M_{43}M_{31}$. The parameters $\mathcal{P}_-^{(n)}$ and $\mathcal{T}_-^{(n)}$ along with $\mathcal{W}_+^{(1)}\mathcal{U}_+^{(2)} + \mathcal{U}_+^{(1)}\mathcal{W}_+^{(2)}$ cannot be neglected in $M_{11}M_{23} - M_{13}M_{21} = -M_{41}M_{33} - M_{43}M_{31}$ as they give terms that are, after the removal of common factors, linear in $\varepsilon_1^{(n)}$. Therefore, they contribute to the off-diagonal reflection coefficients. The substitution according to Eqs. (4.252) to (4.255) into Eq. (4.235), Eq. (4.238), Eq. (4.239), Eq. (4.242), and Eq. (4.244) provides

$$M_{11}M_{33} - M_{13}M_{31}$$
$$= -4N^{(1)4}\alpha_z^{(1)4}N^{(2)3}\alpha_z^{(2)2}\left[2N^{(0)}N^{(1)}\left(\alpha_z^{(0)2} - \alpha_z^{(1)2}\right)\right]\varepsilon_1^{(1)2}$$
$$\times\left[2N^{(1)}N^{(2)}\left(\alpha_z^{(1)2} - \alpha_z^{(2)2}\right)\right] + 2e^{\mathrm{j}(\beta_+ + \beta_-)}\left(N_{z+}^{(1)} - N_{z-}^{(1)}\right)^2 N^{(1)3}\alpha_z^{(1)2}$$
$$\times\left(N^{(0)}\alpha_z^{(0)2}N^{(1)} + N^{(0)2}\alpha_z^{(0)}\alpha_z^{(1)} + N^{(1)2}\alpha_z^{(0)}\alpha_z^{(1)} + N^{(0)}\alpha_z^{(1)2}N^{(1)}\right)$$
$$\times 4N^{(1)3}\alpha_z^{(1)2}N^{(2)3}\alpha_z^{(2)2}$$

$$
\begin{aligned}
&\times\big(-N^{(1)}\alpha_z^{(1)2}N^{(2)} - N^{(1)2}\alpha_z^{(1)}\alpha_z^{(2)} - N^{(2)2}\alpha_z^{(1)}\alpha_z^{(2)} - N^{(1)}\alpha_z^{(2)2}N^{(2)}\big)\\
&+2\mathrm{e}^{-\mathrm{j}(\beta_+ + \beta_-)}\big(N_{z+}^{(1)} - N_{z-}^{(1)}\big)^2 N^{(1)3}\alpha_z^{(1)2}\\
&\times\big(-N^{(0)}\alpha_z^{(0)2}N^{(1)} + N^{(0)2}\alpha_z^{(0)}\alpha_z^{(1)} + N^{(1)2}\alpha_z^{(0)}\alpha_z^{(1)} - N^{(0)}\alpha_z^{(1)2}N^{(1)}\big)\\
&\times 4N^{(1)3}\alpha_z^{(1)2}N^{(2)3}\alpha_z^{(2)2}\\
&\times\big(N^{(1)}\alpha_z^{(1)2}N^{(2)} - N^{(1)2}\alpha_z^{(1)}\alpha_z^{(2)} - N^{(2)2}\alpha_z^{(1)}\alpha_z^{(2)} + N^{(1)}\alpha_z^{(2)2}N^{(2)}\big)\\
&+2N^{(1)2}\alpha_z^{(1)2}\big(N_{z+}^{(1)} - N_{z-}^{(1)}\big)^2\big(\mathrm{e}^{\mathrm{j}(\beta_+ - \beta_-)} + \mathrm{e}^{-\mathrm{j}(\beta_+ - \beta_-)}\big)\\
&\times N^{(1)2}\alpha_z^{(1)}\big[2\alpha_z^{(0)}\alpha_z^{(1)}\big(N^{(0)2} - N^{(1)2}\big)\big]\\
&\times N^{(1)2}\alpha_z^{(1)}N^{(2)3}\alpha_z^{(2)2}\big[-2\alpha_z^{(1)}\alpha_z^{(2)}\big(N^{(1)2} - N^{(2)2}\big)\big],
\end{aligned}
\tag{4.256a}
$$

$$
\begin{aligned}
&\mathrm{M}_{21}\mathrm{M}_{33} - \mathrm{M}_{23}\mathrm{M}_{31}\\
&\quad= -4N^{(1)4}\alpha_z^{(1)4}N^{(2)3}\alpha_z^{(2)2}\big[2N^{(0)}N^{(1)}\big(\alpha_z^{(0)2} + \alpha_z^{(1)2}\big)\big]\varepsilon_1^{(1)2}\big[2N^{(1)}N^{(2)}\\
&\quad\times\big(\alpha_z^{(1)2} - \alpha_z^{(2)2}\big)\big] + 2\mathrm{e}^{\mathrm{j}(\beta_+ + \beta_-)}\big(N_{z+}^{(1)} - N_{z-}^{(1)}\big)^2 N^{(1)3}\alpha_z^{(1)2}\\
&\quad\times\big(N^{(0)}\alpha_z^{(0)2}N^{(1)} + N^{(0)2}\alpha_z^{(0)}\alpha_z^{(1)} - N^{(1)2}\alpha_z^{(0)}\alpha_z^{(1)} - N^{(0)}\alpha_z^{(1)2}N^{(1)}\big)\\
&\quad\times 4N^{(1)3}\alpha_z^{(1)2}N^{(2)3}\alpha_z^{(2)2}\\
&\quad\times\big(-N^{(1)}\alpha_z^{(1)2}N^{(2)} - N^{(1)2}\alpha_z^{(1)}\alpha_z^{(2)} - N^{(2)2}\alpha_z^{(1)}\alpha_z^{(2)} - N^{(1)}\alpha_z^{(2)2}N^{(2)}\big)\\
&\quad+2\mathrm{e}^{-\mathrm{j}(\beta_+ + \beta_-)}\big(N_{z+}^{(1)} - N_{z-}^{(1)}\big)^2 N^{(1)3}\alpha_z^{(1)2}\\
&\quad\times\big(-N^{(0)}\alpha_z^{(0)2}N^{(1)} + N^{(0)2}\alpha_z^{(0)}\alpha_z^{(1)} - N^{(1)2}\alpha_z^{(0)}\alpha_z^{(1)} + N^{(0)}\alpha_z^{(1)2}N^{(1)}\big)\\
&\quad\times 4N^{(1)3}\alpha_z^{(1)2}N^{(2)3}\alpha_z^{(2)2}\\
&\quad\times\big(N^{(1)}\alpha_z^{(1)2}N^{(2)} - N^{(1)2}\alpha_z^{(1)}\alpha_z^{(2)} - N^{(2)2}\alpha_z^{(1)}\alpha_z^{(2)} + N^{(1)}\alpha_z^{(2)2}N^{(2)}\big)\\
&\quad+2N^{(1)2}\alpha_z^{(1)2}\big(N_{z+}^{(1)} - N_{z-}^{(1)}\big)^2\big(\mathrm{e}^{\mathrm{j}(\beta_+ - \beta_-)} + \mathrm{e}^{-\mathrm{j}(\beta_+ - \beta_-)}\big)\\
&\quad\times N^{(1)2}\alpha_z^{(1)}\big[2\alpha_z^{(0)}\alpha_z^{(1)}\big(N^{(0)2} + N^{(1)2}\big)\big]\\
&\quad\times N^{(1)2}\alpha_z^{(1)}N^{(2)3}\alpha_z^{(2)2}\big[-2\alpha_z^{(1)}\alpha_z^{(2)}\big(N^{(1)2} - N^{(2)2}\big)\big],
\end{aligned}
\tag{4.256b}
$$

$$
\begin{aligned}
&\mathrm{M}_{11}\mathrm{M}_{43} - \mathrm{M}_{13}\mathrm{M}_{41}\\
&\quad= 4N^{(1)4}\alpha_z^{(1)4}N^{(2)3}\alpha_z^{(2)2}\big[2N^{(0)}N^{(1)}\big(\alpha_z^{(0)2} + \alpha_z^{(1)2}\big)\big]\varepsilon_1^{(1)2}\big[2N^{(1)}N^{(2)}\\
&\quad\times\big(\alpha_z^{(1)2} - \alpha_z^{(2)2}\big)\big] + 2\mathrm{e}^{\mathrm{j}(\beta_+ + \beta_-)}\big(N_{z+}^{(1)} - N_{z-}^{(1)}\big)^2 N^{(1)3}\alpha_z^{(1)2}\\
&\quad\times\big(-N^{(0)}\alpha_z^{(0)2}N^{(1)} + N^{(0)2}\alpha_z^{(0)}\alpha_z^{(1)} - N^{(1)2}\alpha_z^{(0)}\alpha_z^{(1)} + N^{(0)}\alpha_z^{(1)2}N^{(1)}\big)\\
&\quad\times 4N^{(1)3}\alpha_z^{(1)2}N^{(2)3}\alpha_z^{(2)2}\\
&\quad\times\big(-N^{(1)}\alpha_z^{(1)2}N^{(2)} - N^{(1)2}\alpha_z^{(1)}\alpha_z^{(2)} - N^{(2)2}\alpha_z^{(1)}\alpha_z^{(2)} - N^{(1)}\alpha_z^{(2)2}N^{(2)}\big)\\
&\quad+2\mathrm{e}^{-\mathrm{j}(\beta_+ + \beta_-)}\big(N_{z+}^{(1)} - N_{z-}^{(1)}\big)^2 N^{(1)3}\alpha_z^{(1)2}\\
&\quad\times\big(N^{(0)}\alpha_z^{(0)2}N^{(1)} + N^{(0)2}\alpha_z^{(0)}\alpha_z^{(1)} - N^{(1)2}\alpha_z^{(0)}\alpha_z^{(1)} - N^{(0)}\alpha_z^{(1)2}N^{(1)}\big)\\
&\quad\times 4N^{(1)3}\alpha_z^{(1)2}N^{(2)3}\alpha_z^{(2)2}\\
&\quad\times\big(N^{(1)}\alpha_z^{(1)2}N^{(2)} - N^{(1)2}\alpha_z^{(1)}\alpha_z^{(2)} - N^{(2)2}\alpha_z^{(1)}\alpha_z^{(2)} + N^{(1)}\alpha_z^{(2)2}N^{(2)}\big)
\end{aligned}
$$

$$+2N^{(1)2}\alpha_z^{(1)2}\left(N_{z+}^{(1)}-N_{z-}^{(1)}\right)^2\left(e^{j(\beta_+-\beta_-)}+e^{-j(\beta_+-\beta_-)}\right)$$
$$\times N^{(1)2}\alpha_z^{(1)}\left[2\alpha_z^{(0)}\alpha_z^{(1)}\left(N^{(0)2}+N^{(1)2}\right)\right]$$
$$\times N^{(1)2}\alpha_z^{(1)}N^{(2)3}\alpha_z^{(2)2}\left[-2\alpha_z^{(1)}\alpha_z^{(2)}\left(N^{(1)2}-N^{(2)2}\right)\right], \tag{4.256c}$$

$$\begin{aligned}
\mathrm{M}_{41}\mathrm{M}_{33}-\mathrm{M}_{43}\mathrm{M}_{31} &= -\left(\mathrm{M}_{11}\mathrm{M}_{23}-\mathrm{M}_{13}\mathrm{M}_{21}\right)\\
&= 8j\,\varepsilon_1^{(1)}N^{(1)2}\alpha_z^{(1)2}N^{(0)}\alpha_z^{(0)}e^{j(\beta_++\beta_-)}\left(N_{z+}^{(1)}-N_{z-}^{(1)}\right)^2\\
&\times N^{(1)3}\alpha_z^{(1)2}N^{(2)3}\alpha_z^{(2)2}\\
&\times\left(-N^{(1)}\alpha_z^{(1)2}N^{(2)}-N^{(1)2}\alpha_z^{(1)}\alpha_z^{(2)}-N^{(2)2}\alpha_z^{(1)}\alpha_z^{(2)}-N^{(1)}\alpha_z^{(2)2}N^{(2)}\right)\\
&-8j\,\varepsilon_1^{(1)}N^{(1)2}\alpha_z^{(1)2}N^{(0)}\alpha_z^{(0)}e^{-j(\beta_++\beta_-)}\left(N_{z+}^{(1)}-N_{z-}^{(1)}\right)^2\\
&\times N^{(1)3}\alpha_z^{(1)2}N^{(2)3}\alpha_z^{(2)2}\\
&\times\left(N^{(1)}\alpha_z^{(1)2}N^{(2)}-N^{(1)2}\alpha_z^{(1)}\alpha_z^{(2)}-N^{(2)2}\alpha_z^{(1)}\alpha_z^{(2)}+N^{(1)}\alpha_z^{(2)2}N^{(2)}\right)\\
&+8j\,\varepsilon_1^{(1)}N^{(1)2}\alpha_z^{(1)2}N^{(0)}\alpha_z^{(0)}e^{j(\beta_+-\beta_-)}N^{(1)2}\alpha_z^{(1)2}\\
&\times\Big\{-2N^{(1)4}\alpha_z^{(1)2}\left(N_{z+}^{(1)}-N_{z-}^{(1)}\right)N^{(2)3}\alpha_z^{(2)3}+2N^{(1)2}\alpha_z^{(1)2}\left(N_{z+}^{(1)}-N_{z-}^{(1)}\right)\\
&\times N^{(2)5}\alpha_z^{(2)3}-2\varepsilon_1^{(1)}\varepsilon_1^{(2)}N^{(1)2}\alpha_z^{(1)2}N^{(2)2}\alpha_z^{(2)2}\\
&+\frac{1}{2}\left(N_{z+}^{(1)}-N_{z-}^{(1)}\right)^2\alpha_z^{(2)2}N^{(1)}N^{(2)3}\left[2N^{(1)}N^{(2)}\left(\alpha_z^{(1)2}+\alpha_z^{(2)2}\right)\right]\Big\}\\
&-8j\,\varepsilon_1^{(1)}N^{(1)2}\alpha_z^{(1)2}N^{(0)}\alpha_z^{(0)}e^{-j(\beta_+-\beta_-)}N^{(1)2}\alpha_z^{(1)2}\\
&\times\Big\{-2N^{(1)4}\alpha_z^{(1)2}\left(N_{z+}^{(1)}-N_{z-}^{(1)}\right)N^{(2)3}\alpha_z^{(2)3}+2N^{(1)2}\alpha_z^{(1)2}\left(N_{z+}^{(1)}-N_{z-}^{(1)}\right)\\
&\times N^{(2)5}\alpha_z^{(2)3}+2\varepsilon_1^{(1)}\varepsilon_1^{(2)}N^{(1)2}\alpha_z^{(1)2}N^{(2)2}\alpha_z^{(2)2}\\
&-\frac{1}{2}\left(N_{z+}^{(1)}-N_{z-}^{(1)}\right)^2\alpha_z^{(2)2}N^{(1)}N^{(2)3}\left[2N^{(1)}N^{(2)}\left(\alpha_z^{(1)2}+\alpha_z^{(2)2}\right)\right]\Big\}.
\end{aligned} \tag{4.256d}$$

The last two terms proportional to $e^{\pm j(\beta_+-\beta_-)}$ may be expressed as a sum of three sine and cosine components. The off-diagonal $\varepsilon_1^{(1)}$ may be replaced with Eq. (4.254c). Without a common factor

$$8N^{(1)6}\alpha_z^{(1)4}N^{(2)3}\alpha_z^{(2)2}\left(N_{z+}^{(1)}-N_{z-}^{(1)}\right)^2,$$

we have

$$\begin{aligned}
&+\left(j\,\varepsilon_1^{(1)}N^{(0)}\alpha_z^{(0)}N^{(1)-1}\right)\left[2N^{(1)}N^{(2)}\left(\alpha_z^{(1)2}+\alpha_z^{(2)2}\right)\right]\cos\left(\beta_+-\beta_-\right)\\
&+\frac{1}{2}\left(2N^{(0)}\alpha_z^{(0)}\right)\left(2N^{(1)}\alpha_z^{(1)}\right)\left[2\alpha_z^{(1)}\alpha_z^{(2)}\left(N^{(1)2}-N^{(2)2}\right)\right]\sin\left(\left(\beta_+-\beta_-\right)\right)\\
&-\left(j\,\varepsilon_1^{(2)}N^{(1)}\alpha_z^{(1)}N^{(2)-1}\right)\left(2N^{(0)}\alpha_z^{(0)}\right)\left(2N^{(1)}\alpha_z^{(1)}\right)\cos\left(\beta_+-\beta_-\right).
\end{aligned} \tag{4.257}$$

We remove this common factor from the remaining terms as well as from the other equations (4.254)

$$
\begin{aligned}
&M_{11}M_{33} - M_{13}M_{31} \\
&\quad = -\frac{1}{2}\left[2N^{(0)}N^{(1)}\left(\alpha_z^{(0)2} - \alpha_z^{(1)2}\right)\right]\left[2N^{(1)}N^{(2)}\left(\alpha_z^{(1)2} - \alpha_z^{(2)2}\right)\right] \\
&\quad - e^{j(\beta_+ + \beta_-)}\left[\left(N^{(0)}\alpha_z^{(0)} + N^{(1)}\alpha_z^{(1)}\right)\left(N^{(0)}\alpha_z^{(1)} + N^{(1)}\alpha_z^{(0)}\right)\right] \\
&\quad \times\left[\left(N^{(1)}\alpha_z^{(1)} + N^{(2)}\alpha_z^{(2)}\right)\left(N^{(1)}\alpha_z^{(2)} + N^{(2)}\alpha_z^{(1)}\right)\right] \\
&\quad - e^{-j(\beta_+ + \beta_-)}\left[\left(N^{(0)}\alpha_z^{(0)} - N^{(1)}\alpha_z^{(1)}\right)\left(N^{(0)}\alpha_z^{(1)} - N^{(1)}\alpha_z^{(0)}\right)\right] \\
&\quad \times\left[\left(N^{()}\alpha_z^{(1)} - N^{(2)}\alpha_z^{(2)}\right)\left(N^{(1)}\alpha_z^{(2)} - N^{(2)}\alpha_z^{(1)}\right)\right] \\
&\quad - \frac{1}{4}\left(e^{j(\beta_+ - \beta_-)} + e^{-j(\beta_+ - \beta_-)}\right)\left[2\alpha_z^{(0)}\alpha_z^{(1)}\left(N^{(0)2} - N^{(1)2}\right)\right] \\
&\quad \times\left[2\alpha_z^{(1)}\alpha_z^{(2)}\left(N^{(1)2} - N^{(2)2}\right)\right],
\end{aligned}
\tag{4.258a}
$$

$$
\begin{aligned}
&M_{21}M_{33} - M_{23}M_{31} \\
&\quad = -\frac{1}{2}\left[2N^{(0)}N^{(1)}\left(\alpha_z^{(0)2} + \alpha_z^{(1)2}\right)\right]\left[2N^{(1)}N^{(2)}\left(\alpha_z^{(1)2} - \alpha_z^{(2)2}\right)\right] \\
&\quad - e^{j(\beta_+ + \beta_-)}\left[\left(N^{(0)}\alpha_z^{(0)} - N^{(1)}\alpha_z^{(1)}\right)\left(N^{(0)}\alpha_z^{(1)} + N^{(1)}\alpha_z^{(0)}\right)\right] \\
&\quad \times\left[\left(N^{(1)}\alpha_z^{(1)} + N^{(2)}\alpha_z^{(2)}\right)\left(N^{(1)}\alpha_z^{(2)} + N^{(2)}\alpha_z^{(1)}\right)\right] \\
&\quad - e^{-j(\beta_+ + \beta_-)}\left[\left(N^{(0)}\alpha_z^{(0)} + N^{(1)}\alpha_z^{(1)}\right)\left(N^{(0)}\alpha_z^{(1)} - N^{(1)}\alpha_z^{(0)}\right)\right] \\
&\quad \times\left[\left(N^{(1)}\alpha_z^{(1)} - N^{(2)}\alpha_z^{(2)}\right)\left(N^{(1)}\alpha_z^{(2)} - N^{(2)}\alpha_z^{(1)}\right)\right] \\
&\quad - \frac{1}{4}\left(e^{j(\beta_+ - \beta_-)} + e^{-j(\beta_+ - \beta_-)}\right)\left[2\alpha_z^{(0)}\alpha_z^{(1)}\left(N^{(0)2} + N^{(1)2}\right)\right] \\
&\quad \times\left[2\alpha_z^{(1)}\alpha_z^{(2)}\left(N^{(1)2} - N^{(2)2}\right)\right],
\end{aligned}
\tag{4.258b}
$$

$$
\begin{aligned}
&M_{11}M_{43} - M_{13}M_{41} \\
&\quad = \frac{1}{2}\left[2N^{(0)}N^{(1)}\left(\alpha_z^{(0)2} + \alpha_z^{(1)2}\right)\right]\left[2N^{(1)}N^{(2)}\left(\alpha_z^{(1)2} - \alpha_z^{(2)2}\right)\right] \\
&\quad - e^{j(\beta_+ + \beta_-)}\left[\left(N^{(0)}\alpha_z^{(0)} + N^{(1)}\alpha_z^{(1)}\right)\left(N^{(0)}\alpha_z^{(1)} - N^{(1)}\alpha_z^{(0)}\right)\right] \\
&\quad \times\left[\left(N^{(1)}\alpha_z^{(1)} + N^{(2)}\alpha_z^{(2)}\right)\left(N^{(1)}\alpha_z^{(2)} + N^{(2)}\alpha_z^{(1)}\right)\right] \\
&\quad - e^{-j(\beta_+ + \beta_-)}\left[\left(N^{(0)}\alpha_z^{(0)} - N^{(1)}\alpha_z^{(1)}\right)\left(N^{(0)}\alpha_z^{(1)} + N^{(1)}\alpha_z^{(0)}\right)\right] \\
&\quad \times\left[\left(N^{(1)}\alpha_z^{(1)} - N^{(2)}\alpha_z^{(2)}\right)\left(N^{(1)}\alpha_z^{(2)} - N^{(2)}\alpha_z^{(1)}\right)\right] \\
&\quad - \frac{1}{4}\left(e^{j(\beta_+ - \beta_-)} + e^{-j(\beta_+ - \beta_-)}\right)\left[2\alpha_z^{(0)}\alpha_z^{(1)}\left(N^{(0)2} + N^{(1)2}\right)\right] \\
&\quad \times\left[2\alpha_z^{(1)}\alpha_z^{(2)}\left(N^{(1)2} - N^{(2)2}\right)\right],
\end{aligned}
\tag{4.258c}
$$

$$
\begin{aligned}
&\mathrm{M}_{41}\mathrm{M}_{33}-\mathrm{M}_{43}\mathrm{M}_{31}=-\left(\mathrm{M}_{11}\mathrm{M}_{23}-\mathrm{M}_{13}\mathrm{M}_{21}\right)\\
&=-\mathrm{j}\,\varepsilon_1^{(1)}N^{(1)-1}N^{(0)}\alpha_z^{(0)}\mathrm{e}^{\mathrm{j}(\beta_++\beta_-)}\left[\left(N^{(1)}\alpha_z^{(1)}+N^{(2)}\alpha_z^{(2)}\right)\left(N^{(1)}\alpha_z^{(2)}+N^{(2)}\alpha_z^{(1)}\right)\right]\\
&\quad+\mathrm{j}\,\varepsilon_1^{(1)}N^{(1)-1}N^{(0)}\alpha_z^{(0)}\mathrm{e}^{-\mathrm{j}(\beta_++\beta_-)}\left[\left(N^{(1)}\alpha_z^{(1)}-N^{(2)}\alpha_z^{(2)}\right)\left(N^{(1)}\alpha_z^{(2)}-N^{(2)}\alpha_z^{(1)}\right)\right]\\
&\quad+2N^{(0)}\alpha_z^{(0)}N^{(1)}\alpha_z^{(1)}\left[2\alpha_z^{(1)}\alpha_z^{(2)}\left(N^{(1)2}-N^{(2)2}\right)\right]\ \sin\left(\beta_+-\beta_-\right)\\
&\quad-4\mathrm{j}\,\varepsilon_1^{(2)}N^{(2)-1}N^{(0)}\alpha_z^{(0)}N^{(1)2}\alpha_z^{(1)2}\ \cos\left(\beta_+-\beta_-\right)\\
&\quad+\mathrm{j}\,\varepsilon_1^{(1)}N^{(1)-1}N^{(0)}\alpha_z^{(0)}\left[2N^{(1)}N^{(2)}\left(\alpha_z^{(1)2}+\alpha_z^{(2)2}\right)\right]\cos\left(\beta_+-\beta_-\right). \qquad (4.258d)
\end{aligned}
$$

With Eq. (4.252), we may express the reflection and transmission coefficients at the interface of the i-th and j-th media as

$$r_{21}^{(ij)}=r_{ss}^{(ij)}=\frac{N^{(i)}\alpha_z^{(i)}-N^{(j)}\alpha_z^{(j)}}{N^{(i)}\alpha_z^{(i)}+N^{(j)}\alpha_z^{(j)}}, \qquad (4.259a)$$

$$r_{43}^{(ij)}=r_{pp}^{(ij)}=\frac{N^{(i)}\alpha_z^{(j)}-N^{(j)}\alpha_z^{(i)}}{N^{(i)}\alpha_z^{(j)}+N^{(j)}\alpha_z^{(i)}}, \qquad (4.259b)$$

$$t_{11}^{(ij)}=t_{ss}^{(ij)}=\frac{2N^{(i)}\alpha_z^{(i)}}{N^{(i)}\alpha_z^{(i)}+N^{(j)}\alpha_z^{(j)}}, \qquad (4.259c)$$

$$t_{33}^{(ij)}=t_{pp}^{(ij)}=\frac{2N^{(i)}\alpha_z^{(i)}}{N^{(i)}\alpha_z^{(j)}+N^{(j)}\alpha_z^{(i)}}. \qquad (4.259d)$$

The obvious relations hold

$$
\begin{aligned}
r_{ss}^{(ij)}&=-r_{ss}^{(ji)}, &(4.260a)\\
r_{pp}^{(ij)}&=-r_{pp}^{(ji)}, &(4.260b)\\
t_{ss}^{(ij)}&=1+r_{ss}^{(ij)}, &(4.260c)\\
t_{ss}^{(ij)}&=1-r_{ss}^{(ij)}, &(4.260d)\\
t_{pp}^{(ij)}&=\frac{N^{(i)}}{N^{(j)}}\left(1-r_{pp}^{(ij)}\right), &(4.260e)\\
t_{pp}^{(ji)}&=\frac{N^{(j)}}{N^{(i)}}\left(1+r_{pp}^{(ij)}\right), &(4.260f)\\
t_{ss}^{(ij)}t_{pp}^{(ji)}&=t_{ss}^{(ji)}t_{pp}^{(ij)}. &(4.260g)
\end{aligned}
$$

The off-diagonal reflection coefficient for a single interface agrees with Eq. (1.114)

$$r_{ps}^{(ij)}=-r_{sp}^{(ij)}=\frac{\mathrm{j}\,\varepsilon_1^{(j)}\alpha_z^{(i)}N^{(i)}N^{(j)-1}}{\left(N^{(i)}\alpha_z^{(i)}+N^{(j)}\alpha_z^{(j)}\right)\left(N^{(i)}\alpha_z^{(j)}+N^{(j)}\alpha_z^{(i)}\right)}. \qquad (4.261)$$

We now divide by the factor

$$-e^{j(\beta_+ + \beta_-)}\left[\left(N^{(0)}\alpha_z^{(0)} + N^{(1)}\alpha_z^{(1)}\right)\left(N^{(0)}\alpha_z^{(1)} + N^{(1)}\alpha_z^{(0)}\right)\right] \\ \times\left[\left(N^{(1)}\alpha_z^{(1)} + N^{(2)}\alpha_z^{(2)}\right)\left(N^{(1)}\alpha_z^{(2)} + N^{(2)}\alpha_z^{(1)}\right)\right] \tag{4.262}$$

and employ the following relations valid for the classical Fresnel formulae

$$\begin{aligned} r_{ss}^{(ij)} + r_{pp}^{(ij)} &= \frac{N^{(i)}\alpha_z^{(i)} - N^{(j)}\alpha_z^{(j)}}{N^{(i)}\alpha_z^{(i)} + N^{(j)}\alpha_z^{(j)}} + \frac{N^{(i)}\alpha_z^{(j)} - N^{(j)}\alpha_z^{(i)}}{N^{(i)}\alpha_z^{(j)} + N^{(j)}\alpha_z^{(i)}} \\ &= \frac{2\alpha_z^{(i)}\alpha_z^{(j)}\left(N^{(i)2} - N^{(j)2}\right)}{\left(N^{(i)}\alpha_z^{(i)} + N^{(j)}\alpha_z^{(j)}\right)\left(N^{(i)}\alpha_z^{(j)} + N^{(j)}\alpha_z^{(i)}\right)}, \end{aligned} \tag{4.263a}$$

$$\begin{aligned} r_{ss}^{(ij)} - r_{pp}^{(ij)} &= \frac{N^{(i)}\alpha_z^{(i)} - N^{(j)}\alpha_z^{(j)}}{N^{(i)}\alpha_z^{(i)} + N^{(j)}\alpha_z^{(j)}} - \frac{N^{(i)}\alpha_z^{(j)} - N^{(j)}\alpha_z^{(i)}}{N^{(i)}\alpha_z^{(j)} + N^{(j)}\alpha_z^{(i)}} \\ &= \frac{2N^{(i)}N^{(j)}\left(\alpha_z^{(i)2} - \alpha_z^{(j)2}\right)}{\left(N^{(i)}\alpha_z^{(i)} + N^{(j)}\alpha_z^{(j)}\right)\left(N^{(i)}\alpha_z^{(j)} + N^{(j)}\alpha_z^{(i)}\right)}, \end{aligned} \tag{4.263b}$$

$$\begin{aligned} 1 + r_{ss}^{(ij)} r_{pp}^{(ij)} &= 1 + \frac{N^{(i)}\alpha_z^{(i)} - N^{(j)}\alpha_z^{(j)}}{N^{(i)}\alpha_z^{(i)} + N^{(j)}\alpha_z^{(j)}} \frac{N^{(i)}\alpha_z^{(j)} - N^{(j)}\alpha_z^{(i)}}{N^{(i)}\alpha_z^{(j)} + N^{(j)}\alpha_z^{(i)}} \\ &= \frac{2\alpha_z^{(i)}\alpha_z^{(j)}\left(N^{(i)2} + N^{(j)2}\right)}{\left(N^{(i)}\alpha_z^{(i)} + N^{(j)}\alpha_z^{(j)}\right)\left(N^{(i)}\alpha_z^{(j)} + N^{(j)}\alpha_z^{(i)}\right)}, \end{aligned} \tag{4.263c}$$

$$\begin{aligned} 1 - r_{ss}^{(ij)} r_{pp}^{(ij)} &= 1 - \frac{N^{(i)}\alpha_z^{(i)} - N^{(j)}\alpha_z^{(j)}}{N^{(i)}\alpha_z^{(i)} + N^{(j)}\alpha_z^{(j)}} \frac{N^{(i)}\alpha_z^{(j)} - N^{(j)}\alpha_z^{(i)}}{N^{(i)}\alpha_z^{(j)} + N^{(j)}\alpha_z^{(i)}} \\ &= \frac{2N^{(i)}N^{(j)}\left(\alpha_z^{(i)2} + \alpha_z^{(j)2}\right)}{\left(N^{(i)}\alpha_z^{(i)} + N^{(j)}\alpha_z^{(j)}\right)\left(N^{(i)}\alpha_z^{(j)} + N^{(j)}\alpha_z^{(i)}\right)}. \end{aligned} \tag{4.263d}$$

Making use of the above definitions of reflection and transmission coefficients, Eq. (4.258) can put in a concise form

$$\begin{aligned} &M_{11}M_{33} - M_{13}M_{31} \\ &\quad = 1 + e^{-2j(\beta_+ + \beta_-)} r_{ss}^{(01)} r_{pp}^{(01)} r_{ss}^{(12)} r_{pp}^{(12)} \\ &\quad + \frac{1}{2} e^{-j(\beta_+ + \beta_-)}\left[\left(r_{ss}^{(01)} - r_{pp}^{(01)}\right)\left(r_{ss}^{(12)} - r_{pp}^{(12)}\right)\right. \\ &\quad \left. + \cos\left(\beta_+ - \beta_-\right)\left(r_{ss}^{(01)} + r_{pp}^{(01)}\right)\left(r_{ss}^{(12)} + r_{pp}^{(12)}\right)\right], \end{aligned} \tag{4.264}$$

$$\begin{aligned} &M_{21}M_{33} - M_{23}M_{31} \\ &\quad = r_{ss}^{(01)} + e^{-2j(\beta_+ + \beta_-)} r_{pp}^{(01)} r_{ss}^{(12)} r_{pp}^{(12)} \\ &\quad + \frac{1}{2} e^{-j(\beta_+ + \beta_-)}\left[\left(1 - r_{ss}^{(01)} r_{pp}^{(01)}\right)\left(r_{ss}^{(12)} - r_{pp}^{(12)}\right)\right. \\ &\quad \left. + \cos\left(\beta_+ - \beta_-\right)\left(1 + r_{ss}^{(01)} r_{pp}^{(01)}\right)\left(r_{ss}^{(12)} + r_{pp}^{(12)}\right)\right], \end{aligned} \tag{4.265}$$

$$
\begin{aligned}
&\mathrm{M}_{11}\mathrm{M}_{43} - \mathrm{M}_{41}\mathrm{M}_{13} \\
&\quad = r_{pp}^{(01)} + \mathrm{e}^{-2\mathrm{j}(\beta_+ + \beta_-)} r_{ss}^{(01)} r_{ss}^{(12)} r_{pp}^{(12)} \\
&\quad + \frac{1}{2}\mathrm{e}^{-\mathrm{j}(\beta_+ + \beta_-)} \left[\left(1 - r_{ss}^{(01)} r_{pp}^{(01)}\right)\left(r_{pp}^{(12)} - r_{ss}^{(12)}\right) \right. \\
&\quad \left. + \cos\left(\beta_+ - \beta_-\right)\left(1 + r_{ss}^{(01)} r_{pp}^{(01)}\right)\left(r_{ss}^{(12)} + r_{pp}^{(12)}\right) \right],
\end{aligned} \tag{4.266}
$$

$$
\begin{aligned}
&\mathrm{M}_{41}\ \mathrm{M}_{33} - \mathrm{M}_{43}\mathrm{M}_{31} = -\left(\mathrm{M}_{11}\mathrm{M}_{23} - \mathrm{M}_{21}\mathrm{M}_{13}\right) \\
&\quad = r_{ps}^{(01)}\left(1 - r_{ss}^{(12)} r_{pp}^{(12)} \mathrm{e}^{-2\mathrm{j}(\beta_+ + \beta_-)}\right) \\
&\quad - r_{ps}^{(01)}\left(1 - r_{ss}^{(12)} r_{pp}^{(12)}\right)\cos\left(\beta_+ - \beta_-\right)\mathrm{e}^{-\mathrm{j}(\beta_+ + \beta_-)} \\
&\quad - t_{ss}^{(01)} t_{pp}^{(10)} \sin\left(\beta_+ - \beta_-\right)\left(r_{ss}^{(12)} + r_{pp}^{(12)}\right)\mathrm{e}^{-\mathrm{j}(\beta_+ + \beta_-)} \\
&\quad + t_{ss}^{(01)} t_{pp}^{(10)} r_{ps}^{(12)} \cos\left(\beta_+ - \beta_-\right)\mathrm{e}^{-\mathrm{j}(\beta_+ + \beta_-)}.
\end{aligned} \tag{4.267}
$$

The substitution into Eqs. (3.70a) to (3.70d) gives the desired Jones reflection matrix. The case treated in this section covers many practical situations in which $\varepsilon_0^{(n)} - \varepsilon_3^{(n)}$ and $\varepsilon_1^{(n)}$ are much smaller than $\varepsilon_0^{(n)}$ and at the same time in Eq. (4.182)

$$
\left[\varepsilon_{yy}^{(n)}\left(\varepsilon_{zz}^{(n)} - N_y^2\right) - \varepsilon_{zz}^{(n)}\left(\varepsilon_{xx}^{(n)} - N_y^2\right)\right]^2 \ll 4\varepsilon_{zz}^{(n)}\varepsilon_{xy}^{(n)}\varepsilon_{yx}^{(n)}\left(\varepsilon_{zz}^{(n)} - N_y^2\right). \tag{4.268}
$$

This condition is often met in practice.

A necessary condition for the consistency of Eqs. (4.264) to (4.267) is their ability to cover all simpler situations. For $\varepsilon_1^{(1)} = 0,\ \beta_+ = \beta_-$ we have to arrive at the case of a magnetic substrate covered by a nonmagnetic film, and similarly for $\varepsilon_1^{(2)} = 0$, we obtain the case of a magnetic film on a nonmagnetic substrate. The classical Airy formulae [2] result when $\varepsilon_1^{(1)} = \varepsilon_1^{(2)} = 0$.

References

1. Š. Višňovský, "Magneto-optical polar Kerr effect in a film substrate system," Czech. J. Phys. B **36**, 834–847, 1986.
2. R. M. A. Azzam and N. M. Bashara, *Ellipsometry and Polarized Light* (Elsevier, Amsterdam, 1987).
3. B. Donovan and T. Medcalf, "The inclusion of multiple reflections in the theory of the Faraday effect in semiconductors," Brit. J. Appl. Phys. **15**, 1139–1151, 1964.

4. H. Piller, "Effect of internal reflection on optical Faraday rotation," J.Appl. Phys. **37**, 763–767, 1966.
5. C. J. Sansalone, "Faraday rotation of a system of thin layers containing a thick layer," Appl. Opt. **10**, 2332–2335, 1971.
6. Z. Q. Qiu and S. D. Bader, "Surface magneto-optic effect (SMOKE)," J. Magn. Magn. Mat. **200**, 664–678, 1999.
7. D. G. Stearns, "The scattering of x-rays from nonideal multilayer structure," J. Appl. Phys. **65**, 491–506, 1989.
8. J. Lafait, T. Yamaguchi, J. M. Frigerio, A. Bichri, and K. Driss-Khodja, "Effective medium equivalent to a symmetric multilayer at oblique incidence," Applied Optics **29**, 2460–2465, 1990.
9. R. P. Hunt, "Magneto-optic scattering from thin solid films," J. Appl. Phys. **37**, 1652–1671, 1967.
10. Š. Višňovský, K. Postava, T. Yamaguchi, and R. Lopušník, "Magneto-optic ellipsometry in exchange-coupled films," Appl. Opt. **41**, 3950–3960, 2002.
11. P. B. Johnson and R. W. Christy, "Optical constants of the noble metals," Phys. Rev. B **6**, 4370–4379, 1972.
12. Š. Višňovský, M. Nývlt, V. Prosser, J. Ferré, G. Pénissard, D. Renard, and G. Sczigel, "Magnetooptical effects in Au/Co/Au ultrathin film sandwiches," J. Magn. Magn. Mater. **128**, 179–189 (1993).
13. Š. Višňovský, M. Nývlt, V. Prosser, R. Lopušník, R. Urban, J. Ferré, G. Pénissard, D. Renard, and R. Krishnan, "Polar magneto-optics in simple ultrathin-magnetic-film structures," Phys. Rev. B **52**, 1090–1106, 1995.
14. Š. Višňovský, R. Lopušník, M. Nývlt, V. Prosser, J. Ferré, C. Train, P. Beauvillain, D. Renard, R. Krishnan, and J. A. C. Bland, "Analytical expressions for polar magnetooptics in magnetic multilayers," Czech. J. Phys. B **50**, 857–882, 2000.
15. F. Abelès, "Recherches sur la propagation des ondes électromagnétiques sinusoïdales dans les milieux stratifiés. Application aux couches minces," Ann. Phys., Paris **5**, 596–640, 1950; M. Born and E. Wolf, *Principles of Optics* (Cambridge University Press, Cambridge, 1997) pp. 51–70.
16. Š. Višňovský, K. Postava, and T. Yamaguchi, "Magneto-optic polar Kerr and Faraday effects in periodic multilayers," Optics Express **9**, 158–171, 2001.
17. Š. Višňovský, K. Postava, and T. Yamaguchi, "Magneto-optic polar Kerr and Faraday effects in magnetic superlattices," Czech. J. Phys. B **51**, 917–949, 2001.

5

Longitudinal Magnetization

5.1 Introduction

This configuration is encountered in magnetic multilayers with zero or weak perpendicular magnetic anisotropy. The magnetization is restricted to the plane parallel to the interfaces and also to the plane of incidence (Section 1.6.2). We start from the permittivity tensor for an orthorhombic crystal that is free to rotate about the magnetization coincident with one of the orthorhombic axes [1]. The remaining two orthorhombic axes are restricted to the plane normal to the plane of incidence. In a special and interesting case, one of the two axes may be fixed perpendicular to the interfaces while the second is normal to the plane of incidence. Magnetically soft multilayer systems are easily magnetized in the longitudinal direction, which makes this configuration attractive for the *in situ* magnetooptic measurements. Small electromagnets, which can be put directly into a vacuum chamber, are often sufficient.

We provide the dynamical, propagation, and transfer matrices at an arbitrary angle of incidence required in the multiple angle ellipsometry. Another practical result is the condition for the guided wave propagation parallel to the magnetization axis. Effect of in-plane or perpendicular anisotropy on the guided mode degeneracy can be analyzed using this approach.

5.2 Transfer Matrix

5.2.1 Orthorhombic Anisotropy

In the longitudinal geometry, we choose the magnetization aligned along the y axis and parallel to an orthorhombic axis. The permittivity tensor in the n-th layer takes the form [1]

$$\varepsilon^{(n)} = \begin{pmatrix} \varepsilon_{xx}^{(n)} & 0 & \varepsilon_{xz}^{(n)} \\ 0 & \varepsilon_{yy}^{(n)} & 0 \\ \varepsilon_{zx}^{(n)} & 0 & \varepsilon_{zz}^{(n)} \end{pmatrix}, \tag{5.1}$$

where in a magnetic orthorhombic crystal rotated by an angle $\alpha^{(n)}$ about the y axis, the tensor elements may have the following meaning

$$\varepsilon_{xx}^{(n)} = \varepsilon_0^{(n)} + \varepsilon_2^{(n)} \cos 2\alpha^{(n)}, \tag{5.2a}$$

$$\varepsilon_{yy}^{(n)} = \varepsilon_3^{(n)}, \tag{5.2b}$$

$$\varepsilon_{zz}^{(n)} = \varepsilon_0^{(n)} - \varepsilon_2^{(n)} \cos 2\alpha^{(n)}, \tag{5.2c}$$

$$\varepsilon_{xz}^{(n)} = -\mathrm{j}\,\varepsilon_1^{(n)} - \varepsilon_2^{(n)} \sin 2\alpha^{(n)}, \tag{5.2d}$$

$$\varepsilon_{zx}^{(n)} = \mathrm{j}\,\varepsilon_1^{(n)} - \varepsilon_2^{(n)} \sin 2\alpha^{(n)}. \tag{5.2e}$$

The proper value equation (3.13) reduces to a biquadratic form

$$\begin{aligned} &\varepsilon_{zz}^{(n)} N_z^{(n)4} - \left[\varepsilon_{yy}^{(n)}\left(\varepsilon_{zz}^{(n)} - N_y^2\right) + \varepsilon_{zz}^{(n)}\left(\varepsilon_{xx}^{(n)} - N_y^2\right) - \varepsilon_{xz}^{(n)}\varepsilon_{zx}^{(n)}\right] N_z^{(n)2} \\ &+ \varepsilon_{yy}^{(n)}\left(\varepsilon_{xx}^{(n)} - N_y^2\right)\left(\varepsilon_{zz}^{(n)} - N_y^2\right) - \varepsilon_{yy}^{(n)}\varepsilon_{xz}^{(n)}\varepsilon_{zx}^{(n)} = 0, \end{aligned} \tag{5.3}$$

with the solutions

$$\begin{aligned} N_{z\pm}^{(n)2} &= \frac{\varepsilon_{yy}^{(n)}\left(\varepsilon_{zz}^{(n)} - N_y^2\right) + \varepsilon_{zz}^{(n)}\left(\varepsilon_{xx}^{(n)} - N_y^2\right) - \varepsilon_{xz}^{(n)}\varepsilon_{zx}^{(n)}}{2\varepsilon_{zz}^{(n)}} \\ &\pm \frac{\sqrt{\left[\varepsilon_{yy}^{(n)}\left(\varepsilon_{zz}^{(n)} - N_y^2\right) - \varepsilon_{zz}^{(n)}\left(\varepsilon_{xx}^{(n)} - N_y^2\right) + \varepsilon_{xz}^{(n)}\varepsilon_{zx}^{(n)}\right]^2 + 4N_y^2\varepsilon_{yy}^{(n)}\varepsilon_{xz}^{(n)}\varepsilon_{zx}^{(n)}}}{2\varepsilon_{zz}^{(n)}}. \end{aligned} \tag{5.4}$$

The proper polarizations follow from Eq. (3.16) and Eq. (3.19).

$$\boldsymbol{e}_j^{(n)} = C_j^{(n)} \begin{pmatrix} \varepsilon_{xz}^{(n)} N_y N_{zj}^{(n)} \\ \left(\varepsilon_{zz}^{(n)} - N_y^2\right)\left(\varepsilon_{xx}^{(n)} - N_y^2 - N_{zj}^{(n)2}\right) - \varepsilon_{zx}^{(n)}\varepsilon_{xz}^{(n)} \\ -N_y N_{zj}^{(n)}\left(\varepsilon_{xx}^{(n)} - N_y^2 - N_{zj}^{(n)2}\right) \end{pmatrix}, \tag{5.5a}$$

$$\boldsymbol{b}_j^{(n)} = C_j^{(n)} \begin{pmatrix} -N_{zj}^{(n)}\left[\varepsilon_{zz}^{(n)}\left(\varepsilon_{xx}^{(n)} - N_y^2 - N_{zj}^{(n)2}\right) - \varepsilon_{zx}^{(n)}\varepsilon_{xz}^{(n)}\right] \\ \varepsilon_{xz}^{(n)} N_y N_{zj}^{(n)2} \\ -\varepsilon_{xz}^{(n)} N_y^2 N_{zj}^{(n)} \end{pmatrix}. \tag{5.5b}$$

By the same procedure as in the polar case, we obtain from Eq. (3.30), taking into account Eq. (5.1), the following $\mathbf{D}^{(n)}$ matrix (up to a common factor)

$$\mathbf{D}^{(n)} = \begin{pmatrix} D_{11}^{(n)} & -D_{11}^{(n)} & D_{13}^{(n)} & -D_{13}^{(n)} \\ D_{21}^{(n)} & D_{21}^{(n)} & D_{23}^{(n)} & D_{23}^{(n)} \\ D_{31}^{(n)} & D_{31}^{(n)} & D_{33}^{(n)} & D_{33}^{(n)} \\ D_{41}^{(n)} & -D_{41}^{(n)} & D_{43}^{(n)} & -D_{43}^{(n)} \end{pmatrix}, \tag{5.6}$$

where

$$D_{11}^{(n)} = -D_{12}^{(n)} = \varepsilon_{xz}^{(n)} N_y N_{z+}^{(n)}, \tag{5.7a}$$

$$D_{13}^{(n)} = -D_{14}^{(n)} = \varepsilon_{xz}^{(n)} N_y N_{z-}^{(n)}, \tag{5.7b}$$

$$D_{21}^{(n)} = D_{22}^{(n)} = \varepsilon_{xz}^{(n)} N_y \left(N_{z+}^{(n)}\right)^2, \tag{5.7c}$$

$$D_{23}^{(n)} = D_{24}^{(n)} = \varepsilon_{xz}^{(n)} N_y \left(N_{z-}^{(n)}\right)^2, \tag{5.7d}$$

$$D_{31}^{(n)} = D_{32}^{(n)} = \left(\varepsilon_{zz}^{(n)} - N_y^2\right)\left[\varepsilon_{xx}^{(n)} - N_y^2 - \left(N_{z+}^{(n)}\right)^2\right] - \varepsilon_{xz}^{(n)}\varepsilon_{zx}^{(n)}, \tag{5.7e}$$

$$D_{33}^{(n)} = D_{34}^{(n)} = \left(\varepsilon_{zz}^{(n)} - N_y^2\right)\left[\varepsilon_{xx}^{(n)} - N_y^2 - \left(N_{z-}^{(n)}\right)^2\right] - \varepsilon_{xz}^{(n)}\varepsilon_{zx}^{(n)}, \tag{5.7f}$$

$$D_{41}^{(n)} = -D_{42}^{(n)} = -N_{z+}^{(n)}\left\{\varepsilon_{zz}^{(n)}\left[\varepsilon_{xx}^{(n)} - N_y^2 - \left(N_{z+}^{(n)}\right)^2\right] - \varepsilon_{xz}^{(n)}\varepsilon_{zx}^{(n)}\right\}, \tag{5.7g}$$

$$D_{43}^{(n)} = -D_{44}^{(n)} = -N_{z-}^{(n)}\left\{\varepsilon_{zz}^{(n)}\left[\varepsilon_{xx}^{(n)} - N_y^2 - \left(N_{z-}^{(n)}\right)^2\right] - \varepsilon_{xz}^{(n)}\varepsilon_{zx}^{(n)}\right\}. \tag{5.7h}$$

It is useful to determine the quantities $\mathcal{P}_\pm^{(n)}$, $\mathcal{Q}_\pm^{(n)}$, $\mathcal{S}_\pm^{(n)}$, $\mathcal{T}_\pm^{(n)}$, $\mathcal{U}_\pm^{(n)}$, and $\mathcal{W}_\pm^{(n)}$ specific for the symmetry characterized by the permittivity tensor in Eq. (5.1) and the chosen orientation of the interface plane ($z = \text{const}$) and the plane of incidence ($x = \text{const}$)

$$\begin{aligned} &D_{13}^{(n)} D_{41}^{(n)} - D_{11}^{(n)} D_{43}^{(n)} \\ &= \varepsilon_{xz}^{(n)}\varepsilon_{zz}^{(n)} N_y \left(N_{z+}^{(n)} N_{z-}^{(n)}\right)\left(N_{z+}^{(n)2} - N_{z-}^{(n)2}\right) = \varepsilon_{xz}^{(n)} N_y \left(N_{z+}^{(n)} \mp N_{z-}^{(n)}\right)\mathcal{P}_\pm^{(n)} \end{aligned} \tag{5.8a}$$

$$\begin{aligned} \pm D_{11}^{(n)} D_{33}^{(n)} - D_{13}^{(n)} D_{31}^{(n)} &= \varepsilon_{xz}^{(n)} N_y \left(N_{z+}^{(n)} \mp N_{z-}^{(n)}\right)\left\{\left(\varepsilon_{zz}^{(n)} - N_y^2\right)\left(N_{z+}^{(n)} N_{z-}^{(n)}\right)\right. \\ &\quad \left. \pm \left[\left(\varepsilon_{zz}^{(n)} - N_y^2\right)\left(\varepsilon_{xx}^{(n)} - N_y^2\right) - \varepsilon_{zx}^{(n)}\varepsilon_{xz}^{(n)}\right]\right\} \\ &= \varepsilon_{xz}^{(n)} N_y \left(N_{z+}^{(n)} \mp N_{z-}^{(n)}\right)\mathcal{Q}_\pm^{(n)} \end{aligned} \tag{5.8b}$$

$$
\begin{aligned}
&D_{41}^{(n)} D_{23}^{(n)} \mp D_{43}^{(n)} D_{21}^{(n)} \\
&\quad = \varepsilon_{xz}^{(n)} N_y \left(N_{z+}^{(n)} \mp N_{z-}^{(n)}\right) \left(N_{z+}^{(n)} N_{z-}^{(n)}\right) \left\{\varepsilon_{zz}^{(n)} \left(N_{z+}^{(n)} N_{z-}^{(n)}\right)\right. \\
&\quad \left. \pm \left[\varepsilon_{zz}^{(n)} \left(\varepsilon_{xx}^{(n)} - N_y^2\right) - \varepsilon_{zx}^{(n)} \varepsilon_{xz}^{(n)}\right]\right\} \\
&\quad = \varepsilon_{xz}^{(n)} N_y \left(N_{z+}^{(n)} \mp N_{z-}^{(n)}\right) \mathcal{S}_{\pm}^{(n)} \qquad (5.8c) \\
&D_{21}^{(n)} D_{33}^{(n)} - D_{23}^{(n)} D_{31}^{(n)} \\
&\quad = \varepsilon_{xz}^{(n)} N_y \left(N_{z+}^{(n)2} - N_{z-}^{(n)2}\right) \left[\left(\varepsilon_{zz}^{(n)} - N_y^2\right) \left(\varepsilon_{xx}^{(n)} - N_y^2\right) - \varepsilon_{zx}^{(n)} \varepsilon_{xz}^{(n)}\right] \\
&\quad = \varepsilon_{xz}^{(n)} N_y \left(N_{z+}^{(n)} \mp N_{z-}^{(n)}\right) \mathcal{T}_{\pm}^{(n)} \qquad (5.8d) \\
&D_{13}^{(n)} D_{21}^{(n)} \mp D_{11}^{(n)} D_{23}^{(n)} \\
&\quad = \varepsilon_{xz}^{(n)2} N_y^2 \left(N_{z+}^{(n)} \mp N_{z-}^{(n)}\right) \left(N_{z+}^{(n)} N_{z-}^{(n)}\right) = \varepsilon_{xz}^{(n)} N_y \left(N_{z+}^{(n)} \mp N_{z-}^{(n)}\right) \mathcal{U}_{\pm}^{(n)} \qquad (5.8e) \\
&\pm D_{33}^{(n)} D_{41}^{(n)} - D_{31}^{(n)} D_{43}^{(n)} \\
&\quad = -\varepsilon_{zx}^{(n)} \varepsilon_{xz}^{(n)} N_y^2 \left(N_{z+}^{(n)} \mp N_{z-}^{(n)}\right) \left(N_{z+}^{(n)} N_{z-}^{(n)}\right) \\
&\quad = \varepsilon_{xz}^{(n)} N_y \left(N_{z+}^{(n)} \mp N_{z-}^{(n)}\right) \mathcal{W}_{\pm}^{(n)}. \qquad (5.8f)
\end{aligned}
$$

In the last equation, the manipulation is a little more involved. Let us consider the first term of this expression.

$$
\begin{aligned}
&D_{41}^{(n)} D_{33}^{(n)} \\
&\quad = -N_{z+}^{(n)} \left\{\varepsilon_{zz}^{(n)} \left[\left(\varepsilon_{xx}^{(n)} - N_y^2\right) - N_{z+}^{(n)}\right] - \varepsilon_{zx}^{(n)} \varepsilon_{xz}^{(n)}\right\} \\
&\qquad \times \left\{\left(\varepsilon_{zz}^{(n)} - N_y^2\right) \left[\left(\varepsilon_{xx}^{(n)} - N_y^2\right) - N_{z-}^{(n)2}\right] - \varepsilon_{zx}^{(n)} \varepsilon_{xz}^{(n)}\right\} \\
&\quad = -N_{z+}^{(n)} \left[\varepsilon_{zz}^{(n)} \left(\varepsilon_{xx}^{(n)} - N_y^2\right) - \varepsilon_{zx}^{(n)} \varepsilon_{xz}^{(n)} - \varepsilon_{zz}^{(n)} N_{z+}^{(n)}\right] \\
&\qquad \times \left[\left(\varepsilon_{zz}^{(n)} - N_y^2\right) \left(\varepsilon_{xx}^{(n)} - N_y^2\right) - \varepsilon_{zx}^{(n)} \varepsilon_{xz}^{(n)} - \left(\varepsilon_{zz}^{(n)} - N_y^2\right) N_{z+}^{(n)}\right] \\
&\quad = -N_{z+}^{(n)} \left\{\left[\varepsilon_{zz}^{(n)} \left(\varepsilon_{xx}^{(n)} - N_y^2\right) - \varepsilon_{zx}^{(n)} \varepsilon_{xz}^{(n)}\right] \left[\left(\varepsilon_{zz}^{(n)} - N_y^2\right) \left(\varepsilon_{xx}^{(n)} - N_y^2\right) - \varepsilon_{zx}^{(n)} \varepsilon_{xz}^{(n)}\right]\right. \\
&\qquad - \varepsilon_{zz}^{(n)} N_{z+}^{(n)2} \left[\left(\varepsilon_{zz}^{(n)} - N_y^2\right) \left(\varepsilon_{xx}^{(n)} - N_y^2\right) - \varepsilon_{zx}^{(n)} \varepsilon_{xz}^{(n)}\right] \\
&\qquad - N_{z-}^{(n)2} \left[\varepsilon_{zz}^{(n)} \left(\varepsilon_{zz}^{(n)} - N_y^2\right) \left(\varepsilon_{xx}^{(n)} - N_y^2\right) - \left(\varepsilon_{zz}^{(n)} - N_y^2\right) \varepsilon_{zx}^{(n)} \varepsilon_{xz}^{(n)}\right] \\
&\qquad \left. + \left(\varepsilon_{zz}^{(n)} - N_y^2\right) \varepsilon_{zz}^{(n)} N_{z+}^{(n)2} N_{z-}^{(n)2}\right\} \\
&\quad = -N_{z+}^{(n)} \left\{\left[\varepsilon_{zz}^{(n)} \left(\varepsilon_{xx}^{(n)} - N_y^2\right) - \varepsilon_{zx}^{(n)} \varepsilon_{xz}^{(n)}\right] \left[\left(\varepsilon_{zz}^{(n)} - N_y^2\right) \left(\varepsilon_{xx}^{(n)} - N_y^2\right) - \varepsilon_{zx}^{(n)} \varepsilon_{xz}^{(n)}\right]\right. \\
&\qquad - \varepsilon_{zz}^{(n)} \left(N_{z+}^{(n)2} + N_{z-}^{(n)2}\right) \left[\left(\varepsilon_{zz}^{(n)} - N_y^2\right) \left(\varepsilon_{xx}^{(n)} - N_y^2\right) - \varepsilon_{zx}^{(n)} \varepsilon_{xz}^{(n)}\right] \\
&\qquad \left. - \varepsilon_{zx}^{(n)} \varepsilon_{xz}^{(n)} N_y^2 N_{z-}^{(n)2} + \left(\varepsilon_{zz}^{(n)} - N_y^2\right) \varepsilon_{zz}^{(n)} N_{z+}^{(n)2} N_{z-}^{(n)2}\right\}.
\end{aligned}
$$

Using the properties of the roots of the biquadratic characteristic equation (5.3), we obtain

$$
\begin{aligned}
&D_{41}^{(n)} D_{33}^{(n)} \\
&\quad = -N_{z+}^{(n)}\left\{\left[\varepsilon_{zz}^{(n)}\left(\varepsilon_{xx}^{(n)} - N_y^2\right) - \varepsilon_{zx}^{(n)}\varepsilon_{xz}^{(n)}\right]\left[\left(\varepsilon_{zz}^{(n)} - N_y^2\right)\left(\varepsilon_{xx}^{(n)} - N_y^2\right) - \varepsilon_{zx}^{(n)}\varepsilon_{xz}^{(n)}\right]\right. \\
&\qquad -\left[\varepsilon_{yy}^{(n)}\left(\varepsilon_{zz}^{(n)} - N_y^2\right) + \varepsilon_{zz}^{(n)}\left(\varepsilon_{xx}^{(n)} - N_y^2\right) - \varepsilon_{xz}^{(n)}\varepsilon_{zx}^{(n)}\right] \\
&\qquad \times\left[\left(\varepsilon_{zz}^{(n)} - N_y^2\right)\left(\varepsilon_{xx}^{(n)} - N_y^2\right) - \varepsilon_{zx}^{(n)}\varepsilon_{xz}^{(n)}\right] + \varepsilon_{yy}^{(n)}\left(\varepsilon_{zz}^{(n)} - N_y^2\right) \\
&\qquad \left.\times\left[\left(\varepsilon_{xx}^{(n)} - N_y^2\right)\left(\varepsilon_{zz}^{(n)} - N_y^2\right) - \varepsilon_{xz}^{(n)}\varepsilon_{zx}^{(n)}\right] - \varepsilon_{zx}^{(n)}\varepsilon_{xz}^{(n)} N_y^2 N_{z-}^{(n)2}\right\} \\
&\quad = \varepsilon_{zx}^{(n)}\varepsilon_{xz}^{(n)} N_y^2\left(N_{z+}^{(n)} N_{z-}^{(n)}\right) N_{z-}^{(n)}.
\end{aligned}
\tag{5.9a}
$$

In a similar way, we find

$$
D_{43}^{(n)} D_{31}^{(n)} = \varepsilon_{zx}^{(n)}\varepsilon_{xz}^{(n)} N_y^2\left(N_{z+}^{(n)} N_{z-}^{(n)}\right) N_{z+}^{(n)}. \tag{5.9b}
$$

We will further employ

$$
\begin{aligned}
&D_{31}^{(n)} D_{33}^{(n)} D_{41}^{(n)} D_{43}^{(n)} \\
&\quad = \varepsilon_{zx}^{(n)2}\varepsilon_{xz}^{(n)2}\varepsilon_{yy}^{(n)}\varepsilon_{zz}^{(n)-1} N_{z+}^{(n)} N_{z-}^{(n)} N_y^4\left[\left(\varepsilon_{zz}^{(n)} - N_y^2\right)\left(\varepsilon_{xx}^{(n)} - N_y^2\right) - \varepsilon_{zx}^{(n)}\varepsilon_{xz}^{(n)}\right]
\end{aligned}
\tag{5.10a}
$$

$$
\begin{aligned}
&D_{11}^{(n)} D_{13}^{(n)} D_{31}^{(n)} D_{33}^{(n)} \\
&\quad = -\varepsilon_{zx}^{(n)}\varepsilon_{xz}^{(n)3}\varepsilon_{zz}^{(n)-1} N_{z+}^{(n)} N_{z-}^{(n)} N_y^4\left[\left(\varepsilon_{zz}^{(n)} - N_y^2\right)\left(\varepsilon_{xx}^{(n)} - N_y^2\right) - \varepsilon_{zx}^{(n)}\varepsilon_{xz}^{(n)}\right]
\end{aligned}
\tag{5.10b}
$$

$$
\begin{aligned}
&D_{21}^{(n)} D_{23}^{(n)} D_{41}^{(n)} D_{43}^{(n)} \\
&\quad = -\varepsilon_{zx}^{(n)}\varepsilon_{xz}^{(n)3}\varepsilon_{yy}^{(n)2}\varepsilon_{zz}^{(n)-1} N_{z+}^{(n)} N_{z-}^{(n)} N_y^4\left[\left(\varepsilon_{zz}^{(n)} - N_y^2\right)\left(\varepsilon_{xx}^{(n)} - N_y^2\right) - \varepsilon_{zx}^{(n)}\varepsilon_{xz}^{(n)}\right]
\end{aligned}
\tag{5.10c}
$$

$$
\begin{aligned}
&D_{13}^{(n)} D_{23}^{(n)} D_{31}^{(n)} D_{41}^{(n)} + D_{11}^{(n)} D_{21}^{(n)} D_{33}^{(n)} D_{43}^{(n)} \\
&\quad = -\varepsilon_{xz}^{(n)2}\varepsilon_{zz}^{(n)-1} N_{z+}^{(n)} N_{z-}^{(n)} N_y^2\left[\left(\varepsilon_{zz}^{(n)} - N_y^2\right)\left(\varepsilon_{xx}^{(n)} - N_y^2\right) - \varepsilon_{zx}^{(n)}\varepsilon_{xz}^{(n)}\right] \\
&\qquad \times\left\{\left[\varepsilon_{zz}^{(n)}\left(\varepsilon_{xx}^{(n)} - N_y^2\right) - \varepsilon_{yy}^{(n)}\left(\varepsilon_{zz}^{(n)} - N_y^2\right) - \varepsilon_{xz}^{(n)}\varepsilon_{zx}^{(n)}\right]^2 + 2N_y^2\varepsilon_{yy}^{(n)}\varepsilon_{zx}^{(n)}\varepsilon_{xz}^{(n)}\right\}
\end{aligned}
\tag{5.10d}
$$

$$
\begin{aligned}
&D_{11}^{(n)} D_{23}^{(n)} D_{31}^{(n)} D_{43}^{(n)} + D_{13}^{(n)} D_{21}^{(n)} D_{33}^{(n)} D_{41}^{(n)} \\
&\quad = 2\varepsilon_{xz}^{(n)3}\varepsilon_{zx}^{(n)}\varepsilon_{yy}^{(n)}\varepsilon_{zz}^{(n)-1} N_{z+}^{(n)} N_{z-}^{(n)} N_y^4\left[\left(\varepsilon_{zz}^{(n)} - N_y^2\right)\left(\varepsilon_{xx}^{(n)} - N_y^2\right) - \varepsilon_{zx}^{(n)}\varepsilon_{xz}^{(n)}\right]
\end{aligned}
\tag{5.10e}
$$

$$D_{41}^{(n)} D_{43}^{(n)} \left(D_{23}^{(n)} D_{31}^{(n)} + D_{21}^{(n)} D_{33}^{(n)} \right)$$
$$= -\varepsilon_{xz}^{(n)2} \varepsilon_{zx}^{(n)} \varepsilon_{yy}^{(n)} \varepsilon_{zz}^{(n)-1} N_{z+}^{(n)} N_{z-}^{(n)} N_y^3 \left[\left(\varepsilon_{zz}^{(n)} - N_y^2 \right) \left(\varepsilon_{xx}^{(n)} - N_y^2 \right) - \varepsilon_{zx}^{(n)} \varepsilon_{xz}^{(n)} \right]$$
$$\times \left[\varepsilon_{zz}^{(n)} \left(\varepsilon_{xx}^{(n)} - N_y^2 \right) - \varepsilon_{yy}^{(n)} \left(\varepsilon_{zz}^{(n)} - N_y^2 \right) - \varepsilon_{xz}^{(n)} \varepsilon_{zx}^{(n)} \right] \tag{5.10f}$$

$$D_{11}^{(n)} D_{13}^{(n)} \left(D_{23}^{(n)} D_{31}^{(n)} + D_{21}^{(n)} D_{33}^{(n)} \right)$$
$$= \varepsilon_{xz}^{(n)3} \varepsilon_{zz}^{(n)-1} N_{z+}^{(n)} N_{z-}^{(n)} N_y^3 \left[\left(\varepsilon_{zz}^{(n)} - N_y^2 \right) \left(\varepsilon_{xx}^{(n)} - N_y^2 \right) - \varepsilon_{zx}^{(n)} \varepsilon_{xz}^{(n)} \right]$$
$$\times \left[\varepsilon_{zz}^{(n)} \left(\varepsilon_{xx}^{(n)} - N_y^2 \right) - \varepsilon_{yy}^{(n)} \left(\varepsilon_{zz}^{(n)} - N_y^2 \right) - \varepsilon_{xz}^{(n)} \varepsilon_{zx}^{(n)} \right] \tag{5.10g}$$

$$D_{31}^{(n)} D_{33}^{(n)} \left(D_{13}^{(n)} D_{41}^{(n)} + D_{11}^{(n)} D_{43}^{(n)} \right)$$
$$= \varepsilon_{xz}^{(n)2} \varepsilon_{zx}^{(n)} \varepsilon_{zz}^{(n)-1} N_{z+}^{(n)} N_{z-}^{(n)} N_y^3 \left[\left(\varepsilon_{zz}^{(n)} - N_y^2 \right) \left(\varepsilon_{xx}^{(n)} - N_y^2 \right) - \varepsilon_{zx}^{(n)} \varepsilon_{xz}^{(n)} \right]$$
$$\times \left[\varepsilon_{zz}^{(n)} \left(\varepsilon_{xx}^{(n)} - N_y^2 \right) - \varepsilon_{yy}^{(n)} \left(\varepsilon_{zz}^{(n)} - N_y^2 \right) - \varepsilon_{xz}^{(n)} \varepsilon_{zx}^{(n)} \right] \tag{5.10h}$$

$$D_{21}^{(n)} D_{23}^{(n)} \left(D_{13}^{(n)} D_{41}^{(n)} + D_{11}^{(n)} D_{43}^{(n)} \right)$$
$$= -\varepsilon_{xz}^{(n)3} \varepsilon_{yy}^{(n)} \varepsilon_{zz}^{(n)-1} N_{z+}^{(n)} N_{z-}^{(n)} N_y^3 \left[\left(\varepsilon_{zz}^{(n)} - N_y^2 \right) \left(\varepsilon_{xx}^{(n)} - N_y^2 \right) - \varepsilon_{zx}^{(n)} \varepsilon_{xz}^{(n)} \right]$$
$$\times \left[\varepsilon_{zz}^{(n)} \left(\varepsilon_{xx}^{(n)} - N_y^2 \right) - \varepsilon_{yy}^{(n)} \left(\varepsilon_{zz}^{(n)} - N_y^2 \right) - \varepsilon_{xz}^{(n)} \varepsilon_{zx}^{(n)} \right]. \tag{5.10i}$$

Let us summarize the quantities introduced in Eq. (5.8) pertinent to the longitudinal geometry[1]

$$\mathcal{P}_{\pm}^{(n)} = \varepsilon_{zz}^{(n)} \left(N_{z+}^{(n)} N_{z-}^{(n)} \right) \left(N_{z+}^{(n)} \pm N_{z-}^{(n)} \right) \tag{5.11a}$$
$$\mathcal{Q}_{\pm}^{(n)} = \left(\varepsilon_{zz}^{(n)} - N_y^2 \right) \left(N_{z+}^{(n)} N_{z-}^{(n)} \right) \pm \left[\left(\varepsilon_{zz}^{(n)} - N_y^2 \right) \left(\varepsilon_{xx}^{(n)} - N_y^2 \right) - \varepsilon_{zx}^{(n)} \varepsilon_{xz}^{(n)} \right] \tag{5.11b}$$
$$\mathcal{S}_{\pm}^{(n)} = \left(N_{z+}^{(n)} N_{z-}^{(n)} \right) \left\{ \varepsilon_{zz}^{(n)} \left(N_{z+}^{(n)} N_{z-}^{(n)} \right) \pm \left[\varepsilon_{zz}^{(n)} \left(\varepsilon_{xx}^{(n)} - N_y^2 \right) - \varepsilon_{zx}^{(n)} \varepsilon_{xz}^{(n)} \right] \right\} \tag{5.11c}$$
$$\mathcal{T}_{\pm}^{(n)} = \left(N_{z+}^{(n)} \pm N_{z-}^{(n)} \right) \left[\left(\varepsilon_{zz}^{(n)} - N_y^2 \right) \left(\varepsilon_{xx}^{(n)} - N_y^2 \right) - \varepsilon_{zx}^{(n)} \varepsilon_{xz}^{(n)} \right] \tag{5.11d}$$
$$\mathcal{U}_{\pm}^{(n)} = \varepsilon_{xz}^{(n)} N_y \left(N_{z+}^{(n)} N_{z-}^{(n)} \right) \tag{5.11e}$$
$$\mathcal{W}_{\pm}^{(n)} = -\varepsilon_{zx}^{(n)} N_y \left(N_{z+}^{(n)} N_{z-}^{(n)} \right). \tag{5.11f}$$

The inverse to the dynamical matrix is given by

$$\left(\mathbf{D}^{(n)} \right)^{-1} = \left[\det \left(\mathbf{D}^{(n)} \right) \right]^{-1} \begin{pmatrix} L_{11} & L_{12} & L_{13} & L_{14} \\ -L_{11} & L_{12} & L_{13} & -L_{14} \\ L_{31} & L_{32} & L_{33} & L_{34} \\ -L_{31} & L_{32} & L_{33} & -L_{34} \end{pmatrix}, \tag{5.12}$$

[1] We employ the same notation as in the polar geometry. In this chapter, however, these symbols have a different meaning.

where the determinant of $\mathbf{D}^{(n)}$,

$$\mathcal{D}_{\text{lon}}^{(n)} = \det\left(\mathbf{D}^{(n)}\right) = 4\left(D_{11}^{(n)} D_{43}^{(n)} - D_{13}^{(n)} D_{41}^{(n)}\right)\left(D_{23}^{(n)} D_{31}^{(n)} - D_{21}^{(n)} D_{33}^{(n)}\right). \quad (5.13)$$

The inverse dynamical matrix elements are, up to a common factor,

$$L_{11}^{(n)} = -2D_{43}^{(n)}\left(D_{21}^{(n)} D_{33}^{(n)} - D_{23}^{(n)} D_{31}^{(n)}\right), \quad (5.14a)$$

$$L_{31}^{(n)} = 2D_{41}^{(n)}\left(D_{21}^{(n)} D_{33}^{(n)} - D_{23}^{(n)} D_{31}^{(n)}\right), \quad (5.14b)$$

$$L_{12}^{(n)} = 2D_{33}^{(n)}\left(D_{13}^{(n)} D_{41}^{(n)} - D_{11}^{(n)} D_{43}^{(n)}\right), \quad (5.14c)$$

$$L_{32}^{(n)} = -2D_{31}^{(n)}\left(D_{13}^{(n)} D_{41}^{(n)} - D_{11}^{(n)} D_{43}^{(n)}\right), \quad (5.14d)$$

$$L_{13}^{(n)} = -2D_{23}^{(n)}\left(D_{13}^{(n)} D_{41}^{(n)} - D_{11}^{(n)} D_{43}^{(n)}\right), \quad (5.14e)$$

$$L_{33}^{(n)} = 2D_{21}^{(n)}\left(D_{13}^{(n)} D_{41}^{(n)} - D_{11}^{(n)} D_{43}^{(n)}\right), \quad (5.14f)$$

$$L_{14}^{(n)} = 2D_{13}^{(n)}\left(D_{21}^{(n)} D_{33}^{(n)} - D_{23}^{(n)} D_{31}^{(n)}\right), \quad (5.14g)$$

$$L_{34}^{(n)} = -2D_{11}^{(n)}\left(D_{21}^{(n)} D_{33}^{(n)} - D_{23}^{(n)} D_{31}^{(n)}\right). \quad (5.14h)$$

The next step is to obtain the product $\mathbf{F} = (\mathbf{D}^{(n-1)})^{-1}\mathbf{D}^{(n)}$, which represents the interface between two media characterized by the tensors of the form given in Eq. (5.1). Substituting for the matrices $(\mathbf{D}^{(n-1)})^{-1}$ and $\mathbf{D}^{(n)}$ according to Eq. (5.7) and Eq. (5.14), we arrive at the $\mathbf{F}$ matrix of the same structure as that for the polar case, given in Eq. (4.194). We form the matrix $\mathbf{F}$ as a product

$$\mathbf{F} = \begin{pmatrix} F_{11} & F_{12} & F_{13} & F_{14} \\ F_{12} & F_{11} & F_{14} & F_{13} \\ F_{31} & F_{32} & F_{33} & F_{34} \\ F_{32} & F_{31} & F_{34} & F_{33} \end{pmatrix}$$

$$= \begin{pmatrix} L_{11}^{(n-1)} & L_{12}^{(n-1)} & L_{13}^{(n-1)} & L_{14}^{(n-1)} \\ -L_{11}^{(n-1)} & L_{12}^{(n-1)} & L_{13}^{(n-1)} & -L_{14}^{(n-1)} \\ L_{31}^{(n-1)} & L_{32}^{(n-1)} & L_{13}^{(n-1)} & L_{34}^{(n-1)} \\ -L_{31}^{(n-1)} & L_{32}^{(n-1)} & -L_{13}^{(n-1)} & -L_{34}^{(n-1)} \end{pmatrix} \begin{pmatrix} D_{11}^{(n)} & -D_{11}^{(n)} & D_{13}^{(n)} & -D_{13}^{(n)} \\ D_{21}^{(n)} & D_{21}^{(n)} & D_{23}^{(n)} & D_{23}^{(n)} \\ D_{31}^{(n)} & D_{31}^{(n)} & D_{33}^{(n)} & D_{33}^{(n)} \\ D_{41}^{(n)} & -D_{41}^{(n)} & D_{43}^{(n)} & -D_{43}^{(n)} \end{pmatrix}. \quad (5.15)$$

The $\mathbf{F}$ matrix elements, up to a common factor, are

$$F_{\substack{11\\12}} = \pm L_{11}^{(n-1)} D_{11}^{(n)} + L_{12}^{(n-1)} D_{21}^{(n)} + L_{13}^{(n-1)} D_{31}^{(n)} \pm L_{14}^{(n-1)} D_{41}^{(n)}, \quad (5.16)$$

$$F_{\substack{13\\14}} = \pm L_{11}^{(n-1)} D_{13}^{(n)} + L_{12}^{(n-1)} D_{23}^{(n)} + L_{13}^{(n-1)} D_{33}^{(n)} \pm L_{14}^{(n-1)} D_{43}^{(n)}, \quad (5.17)$$

$$F_{\substack{31\\32}} = \pm L_{31}^{(n-1)} D_{11}^{(n)} + L_{32}^{(n-1)} D_{21}^{(n)} + L_{33}^{(n-1)} D_{31}^{(n)} \pm L_{34}^{(n-1)} D_{41}^{(n)}, \quad (5.18)$$

$$F_{\substack{33\\34}} = \pm L_{31}^{(n-1)} D_{13}^{(n)} + L_{32}^{(n-1)} D_{23}^{(n)} + L_{33}^{(n-1)} D_{33}^{(n)} \pm L_{34}^{(n-1)} D_{43}^{(n)}. \quad (5.19)$$

The matrix product $\mathbf{A} = (\mathbf{D}^{(0)})^{-1}\mathbf{D}^{(1)}$ given in Eq. (4.198) entering the transfer matrix $\mathbf{T}_{0,1}$ is obtained by the procedure similar to that employed for the polar case in obtaining Eq. (4.199) and Eq. (4.200). We obtain using the dielectric tensor pertinent to the longitudinal geometry (5.1)

$$\mathbf{A} = \begin{pmatrix} A_{11} & A_{12} & A_{13} & A_{14} \\ -A_{12} & -A_{11} & -A_{14} & -A_{13} \\ A_{31} & A_{32} & A_{33} & A_{34} \\ A_{32} & A_{31} & A_{34} & A_{33} \end{pmatrix}, \tag{5.20}$$

where

$$A_{\substack{11\\12}} = \left(2N^{(0)}\cos\varphi^{(0)}\right)^{-1}\left(\pm D_{11}^{(1)}N^{(0)}\cos\varphi^{(0)} + D_{21}^{(0)}\right), \tag{5.21}$$

$$A_{\substack{13\\14}} = \left(2N^{(0)}\cos\varphi^{(0)}\right)^{-1}\left(\pm D_{13}^{(1)}N^{(0)}\cos\varphi^{(0)} + D_{23}^{(0)}\right), \tag{5.22}$$

$$A_{\substack{31\\32}} = \left(2N^{(0)}\cos\varphi^{(0)}\right)^{-1}\left(D_{31}^{(1)}N^{(0)} \mp D_{41}^{(1)}\cos\varphi^{(0)}\right), \tag{5.23}$$

$$A_{\substack{33\\34}} = \left(2N^{(0)}\cos\varphi^{(0)}\right)^{-1}\left(D_{33}^{(1)}N^{(0)} \mp D_{43}^{(1)}\cos\varphi^{(0)}\right). \tag{5.24}$$

The product $\mathbf{Z} = \left(\mathbf{D}^{(\mathcal{N})}\right)^{-1}\mathbf{D}^{(\mathcal{N}+1)}$ has the following form (up to a multiple of the unit matrix)

$$\mathbf{Z} = \begin{pmatrix} Z_{11} & Z_{12} & Z_{13} & Z_{14} \\ -Z_{12} & -Z_{11} & Z_{14} & Z_{13} \\ Z_{31} & Z_{32} & Z_{33} & Z_{34} \\ -Z_{32} & -Z_{31} & Z_{34} & Z_{33} \end{pmatrix}, \tag{5.25}$$

with the elements given, up to a common factor, by

$$Z_{\substack{11\\12}} = L_{11}^{(\mathcal{N})} \pm L_{12}^{(\mathcal{N})}N^{(\mathcal{N}+1)}\cos\varphi^{(\mathcal{N}+1)}, \tag{5.26}$$

$$Z_{\substack{13\\14}} = L_{13}^{(\mathcal{N})}\cos\varphi^{(\mathcal{N}+1)} \mp L_{14}^{(\mathcal{N})}N^{(\mathcal{N}+1)}, \tag{5.27}$$

$$Z_{\substack{31\\32}} = L_{31}^{(\mathcal{N})} \pm L_{32}^{(\mathcal{N})}N^{(\mathcal{N}+1)}\cos\varphi^{(\mathcal{N}+1)}, \tag{5.28}$$

$$Z_{\substack{33\\34}} = L_{33}^{(\mathcal{N})}\cos\varphi^{(\mathcal{N}+1)} \mp L_{34}^{(\mathcal{N})}N^{(\mathcal{N}+1)}. \tag{5.29}$$

To obtain the proper polarizations of Eq. (5.5a) and Eq. (5.5b) we have started from the first and third equations of the wave equation system (3.13). We could use the second and third equations to get for electric fields

$$\mathbf{e}_j^{(n)} = C_j^{(n)} \begin{pmatrix} -\varepsilon_{zz}^{(n)}N_{zj}^{(n)2} + \varepsilon_{yy}^{(n)}\left(\varepsilon_{zz}^{(n)} - N_y^2\right) \\ \varepsilon_{zx}^{(n)}N_yN_{zj}^{(n)} \\ -\varepsilon_{zx}^{(n)}\left(\varepsilon_{yy}^{(n)} - N_{zj}^{(n)2}\right) \end{pmatrix} \tag{5.30}$$

and for magnetic fields

$$\mathbf{b}_j^{(n)} = C_j^{(n)} \begin{pmatrix} -\varepsilon_{zx}^{(n)}\varepsilon_{yy}^{(n)}N_y \\ -N_{zj}^{(n)}\left[\varepsilon_{zz}^{(n)}N_{zj}^{(n)2} - \varepsilon_{yy}^{(n)}\left(\varepsilon_{zz}^{(n)} - N_y^2\right)\right] \\ N_y\left[\varepsilon_{zz}^{(n)}N_{zj}^{(n)2} - \varepsilon_{yy}^{(n)}\left(\varepsilon_{zz}^{(n)} - N_y^2\right)\right] \end{pmatrix}. \tag{5.31}$$

The dynamical matrix would be

$$D^{(n)} = \begin{pmatrix} D_{11}^{(n)} & D_{11}^{(n)} & D_{13}^{(n)} & D_{13}^{(n)} \\ D_{21}^{(n)} & -D_{21}^{(n)} & D_{23}^{(n)} & -D_{23}^{(n)} \\ D_{31}^{(n)} & -D_{31}^{(n)} & D_{33}^{(n)} & -D_{33}^{(n)} \\ D_{41}^{(n)} & D_{41}^{(n)} & D_{41}^{(n)} & D_{41}^{(n)} \end{pmatrix}. \tag{5.32}$$

From the first and second equation of the system (3.13), we would get

$$\mathbf{e}_j^{(n)} = C_j^{(n)} \begin{pmatrix} -\varepsilon_{zx}^{(n)}\left(\varepsilon_{yy}^{(n)} - N_{zj}^{(n)2}\right) \\ -N_yN_{zj}^{(n)}\left(\varepsilon_{xx}^{(n)} - N_y^2 - N_{zj}^{(n)2}\right) \\ \left(\varepsilon_{yy}^{(n)} - N_{zj}^{(n)2}\right)\left(\varepsilon_{xx}^{(n)} - N_y^2 - N_{zj}^{(n)2}\right) \end{pmatrix} \tag{5.33}$$

and

$$\mathbf{b}_j^{(n)} = C_j^{(n)} \begin{pmatrix} \varepsilon_{yy}^{(n)}N_y\left(\varepsilon_{xx}^{(n)} - N_y^2 - N_{zj}^{(n)2}\right) \\ -\varepsilon_{xz}^{(n)}N_{zj}^{(n)}\left(\varepsilon_{yy}^{(n)} - N_{zj}^{(n)2}\right) \\ \varepsilon_{xz}^{(n)}N_y\left(\varepsilon_{yy}^{(n)} - N_{zj}^{(n)2}\right) \end{pmatrix}, \tag{5.34}$$

giving the dynamical matrix of the same form

$$\mathbf{D}^{(n)} = \begin{pmatrix} D_{11}^{(n)} & D_{11}^{(n)} & D_{13}^{(n)} & D_{13}^{(n)} \\ D_{21}^{(n)} & -D_{21}^{(n)} & D_{23}^{(n)} & -D_{23}^{(n)} \\ D_{31}^{(n)} & -D_{31}^{(n)} & D_{33}^{(n)} & -D_{33}^{(n)} \\ D_{41}^{(n)} & D_{41}^{(n)} & D_{43}^{(n)} & D_{43}^{(n)} \end{pmatrix}. \tag{5.35}$$

For example, the A matrix will assume the form

$$\mathbf{A} = \begin{pmatrix} A_{11} & A_{12} & A_{13} & A_{14} \\ A_{12} & A_{11} & A_{14} & A_{13} \\ A_{31} & A_{32} & A_{33} & A_{34} \\ -A_{32} & -A_{31} & -A_{34} & -A_{33} \end{pmatrix}, \tag{5.36}$$

where

$$\mathbf{A}^{(1)} = (\mathbf{D}^{(0)})^{-1}\mathbf{D}^{(1)} \tag{5.37}$$

and $\mathbf{D}^{(1)}$ is given either by Eq. (5.32) or Eq. (5.35). The final results, *i.e.*, the multilayer $\mathbf{M}$ matrix and the reflection and transmission coefficients, should not depend on the way we extract the proper polarizations from the wave equation system (3.13). However, an appropriate choice of starting equations may be helpful in the numerical analysis of particular cases.

The propagation in the film is described by the propagation matrix $\mathbf{P}^{(n)}$ given in Eq. (3.31). From Eq. (3.32), we determine a general transfer matrix for a multilayer at longitudinal magnetization

$$\mathbf{T}_{n-1,n} = \left(\mathbf{D}^{(n-1)}\right)^{-1}\mathbf{D}^{(n)}\mathbf{P}^{(n)} = \mathbf{F}^{(n)}\mathbf{P}^{(n)}. \tag{5.38}$$

In particular, the first transfer matrix in the product (3.33)

$$\mathbf{T}_{0,1} = (\mathbf{D}^{(0)})^{-1}\mathbf{D}^{(1)}\mathbf{P}^{(1)} = \mathbf{A}^{(1)}\mathbf{P}^{(1)}. \tag{5.39}$$

5.3 Magnetic Film-Magnetic Substrate System

5.3.1 M Matrix

The global matrix of the system is given by

$$\mathbf{M} = (\mathbf{D}^{(0)})^{-1}\mathbf{D}^{(1)}\mathbf{P}^{(1)}\left(\mathbf{D}^{(1)}\right)^{-1}\mathbf{D}^{(2)} = \mathbf{A}^{(1)}\mathbf{P}^{(1)}\mathbf{F}^{(2)}, \tag{5.40}$$

where the upper indices refer to the isotropic ambient (0), film (1), and substrate (2), respectively. The dynamical matrices $\mathbf{D}^{(1)}$ and $\mathbf{D}^{(2)}$ are given by Eq. (5.6). The matrix $(\mathbf{D}^{(0)})^{-1}$ for the proper polarization modes in the isotropic ambient linearly polarized perpendicular and parallel to the plane of incidence was obtained in Eq. (3.82b). We limit ourselves to the case of a sufficiently thick absorbing substrate and to the reflection characteristics of the system. To evaluate the $\mathbf{M}$ matrix, we make use of our previous results on $\mathbf{A}^{(1)} = \left(\mathbf{D}^{(0)}\right)^{-1}\mathbf{D}^{(1)}$ and $\mathbf{F}^{(1)} = \left(\mathbf{D}^{(1)}\right)^{-1}\mathbf{D}^{(2)}$

$$\begin{aligned}\mathbf{M} &= \mathbf{A}^{(1)}\mathbf{P}^{(1)}\mathbf{F}^{(2)}\\ &= \begin{pmatrix} A_{11} & A_{12} & A_{13} & A_{14}\\ -A_{12} & -A_{11} & -A_{14} & -A_{13}\\ A_{31} & A_{32} & A_{33} & A_{34}\\ A_{32} & A_{31} & A_{34} & A_{33}\end{pmatrix}\begin{pmatrix} e^{j\beta_+} & 0 & 0 & 0\\ 0 & e^{-j\beta_+} & 0 & 0\\ 0 & 0 & e^{j\beta_-} & 0\\ 0 & 0 & 0 & e^{-j\beta_-}\end{pmatrix}\\ &\quad\times\begin{pmatrix} F_{11} & F_{12} & F_{13} & F_{14}\\ F_{12} & F_{11} & F_{14} & F_{13}\\ F_{31} & F_{32} & F_{33} & F_{34}\\ F_{32} & F_{31} & F_{34} & F_{33}\end{pmatrix},\end{aligned} \tag{5.41}$$

where $\beta_\pm = \frac{\omega}{c} N_{z\pm}^{(n)} d^{(1)}$, $d^{(1)}$ being the thickness of the film.

Each matrix element consists of four Fourier components

$$\mathrm{M}_{\substack{11\\22}} = \pm A_{11} F_{11} \mathrm{e}^{\pm \mathrm{j}\beta_+} \pm A_{12} F_{12} \mathrm{e}^{\mp \mathrm{j}\beta_+} \pm A_{13} F_{31} \mathrm{e}^{\pm \mathrm{j}\beta_-} \pm A_{14} F_{32} \mathrm{e}^{\mp \mathrm{j}\beta_-}, \tag{5.42a}$$

$$\mathrm{M}_{\substack{12\\21}} = \pm A_{11} F_{12} \mathrm{e}^{\pm \mathrm{j}\beta_+} \pm A_{12} F_{11} \mathrm{e}^{\mp \mathrm{j}\beta_+} \pm A_{13} F_{32} \mathrm{e}^{\pm \mathrm{j}\beta_-} \pm A_{14} F_{31} \mathrm{e}^{\mp \mathrm{j}\beta_-}, \tag{5.42b}$$

$$\mathrm{M}_{\substack{13\\24}} = \pm A_{11} F_{13} \mathrm{e}^{\pm \mathrm{j}\beta_+} \pm A_{12} F_{14} \mathrm{e}^{\mp \mathrm{j}\beta_+} \pm A_{13} F_{33} \mathrm{e}^{\pm \mathrm{j}\beta_-} \pm A_{14} F_{34} \mathrm{e}^{\mp \mathrm{j}\beta_-}, \tag{5.42c}$$

$$\mathrm{M}_{\substack{14\\23}} = \pm A_{11} F_{14} \mathrm{e}^{\pm \mathrm{j}\beta_+} \pm A_{12} F_{13} \mathrm{e}^{\mp \mathrm{j}\beta_+} \pm A_{13} F_{34} \mathrm{e}^{\pm \mathrm{j}\beta_-} \pm A_{14} F_{33} \mathrm{e}^{\mp \mathrm{j}\beta_-}, \tag{5.42d}$$

$$\mathrm{M}_{\substack{31\\42}} = A_{31} F_{11} \mathrm{e}^{\pm \mathrm{j}\beta_+} + A_{32} F_{12} \mathrm{e}^{\mp \mathrm{j}\beta_+} + A_{33} F_{31} \mathrm{e}^{\pm \mathrm{j}\beta_-} + A_{34} F_{32} \mathrm{e}^{\mp \mathrm{j}\beta_-}, \tag{5.42e}$$

$$\mathrm{M}_{\substack{32\\41}} = A_{31} F_{12} \mathrm{e}^{\pm \mathrm{j}\beta_+} + A_{32} F_{11} \mathrm{e}^{\mp \mathrm{j}\beta_+} + A_{33} F_{32} \mathrm{e}^{\pm \mathrm{j}\beta_-} + A_{34} F_{31} \mathrm{e}^{\mp \mathrm{j}\beta_-}, \tag{5.42f}$$

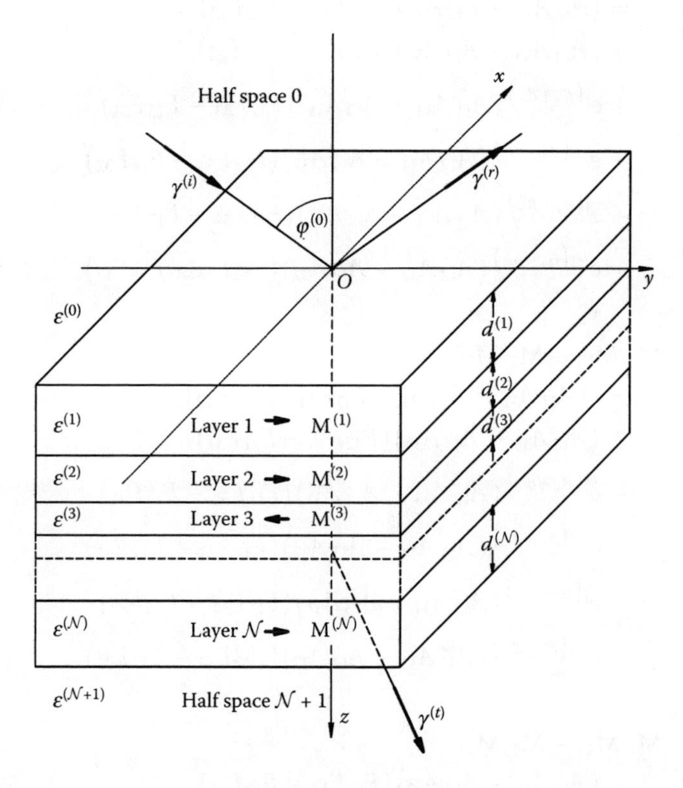

FIGURE 5.1
Oblique light incidence ($\varphi^{(0)} \neq 0$) at a multilayer consisting of $\mathcal{N}$ anisotropic layers with longitudinal magnetizations, $\boldsymbol{M}^{(n)}$ ($n = 1, \cdots, \mathcal{N}$). The layers are characterized by permittivity tensors $\varepsilon^{(n)}$. The incident, reflected, and transmitted propagation vectors in sandwiching isotropic nonabsorbing half spaces (0) and ($\mathcal{N}+1$) are denoted as $\gamma^{(i)}$, $\gamma^{(r)}$, and $\gamma^{(t)}$, respectively.

$$\mathrm{M}_{\substack{33\\44}} = A_{31}F_{13}\mathrm{e}^{\pm \mathrm{j}\beta_+} + A_{32}F_{14}\mathrm{e}^{\mp \mathrm{j}\beta_+} + A_{33}F_{33}\mathrm{e}^{\pm \mathrm{j}\beta_-} + A_{34}F_{34}\mathrm{e}^{\mp \mathrm{j}\beta_-}, \tag{5.42g}$$

$$\mathrm{M}_{\substack{34\\43}} = A_{31}F_{14}\mathrm{e}^{\pm \mathrm{j}\beta_+} + A_{32}F_{14}\mathrm{e}^{\mp \mathrm{j}\beta_+} + A_{33}F_{34}\mathrm{e}^{\pm \mathrm{j}\beta_-} + A_{34}F_{33}\mathrm{e}^{\mp \mathrm{j}\beta_-}. \tag{5.42h}$$

The **M** matrix given above is valid for the more general form of the permittivity tensor given in Eq. (5.1).

5.3.2 Reflection Matrix

We now evaluate the expressions that enter the reflection coefficients of the system. We first give them for the more general form of the permittivity tensor of Eq. (5.1) in terms of the $\mathbf{A}^{(1)}$ and $\mathbf{F}^{(2)}$ matrix elements

$$\begin{aligned}
&\mathrm{M}_{11}\mathrm{M}_{33} - \mathrm{M}_{13}\mathrm{M}_{31} \\
&\quad = (A_{11}A_{32} - A_{12}A_{31})(F_{11}F_{14} - F_{12}F_{13}) \\
&\quad + (A_{14}A_{33} - A_{13}A_{34})(F_{32}F_{33} - F_{31}F_{34}) \\
&\quad + \mathrm{e}^{\mathrm{j}(\beta_+ + \beta_-)}(A_{11}A_{33} - A_{13}A_{31})(F_{11}F_{33} - F_{13}F_{31}) \\
&\quad + \mathrm{e}^{-\mathrm{j}(\beta_+ + \beta_-)}(A_{12}A_{34} - A_{14}A_{32})(F_{12}F_{34} - F_{14}F_{32}) \\
&\quad + \mathrm{e}^{\mathrm{j}(\beta_+ - \beta_-)}(A_{11}A_{34} - A_{14}A_{31})(F_{11}F_{34} - F_{13}F_{32}) \\
&\quad + \mathrm{e}^{-\mathrm{j}(\beta_+ - \beta_-)}(A_{12}A_{33} - A_{13}A_{32})(F_{12}F_{33} - F_{31}F_{14}),
\end{aligned} \tag{5.43a}$$

$$\begin{aligned}
&\mathrm{M}_{21}\mathrm{M}_{33} - \mathrm{M}_{23}\mathrm{M}_{31} \\
&\quad = (A_{11}A_{31} - A_{12}A_{32})(F_{11}F_{14} - F_{13}F_{12}) \\
&\quad + (A_{14}A_{34} - A_{13}A_{33})(F_{32}F_{33} - F_{31}F_{34}) \\
&\quad - \mathrm{e}^{\mathrm{j}(\beta_+ + \beta_-)}(A_{12}A_{33} - A_{14}A_{31})(F_{11}F_{33} - F_{13}F_{31}) \\
&\quad - \mathrm{e}^{-\mathrm{j}(\beta_+ + \beta_-)}(A_{11}A_{34} - A_{13}A_{32})(F_{12}F_{34} - F_{14}F_{32}) \\
&\quad - \mathrm{e}^{\mathrm{j}(\beta_+ - \beta_-)}(A_{12}A_{34} - A_{13}A_{31})(F_{11}F_{34} - F_{13}F_{32}) \\
&\quad - \mathrm{e}^{-\mathrm{j}(\beta_+ - \beta_-)}(A_{11}A_{33} - A_{14}A_{32})(F_{12}F_{33} - F_{14}F_{31}),
\end{aligned} \tag{5.43b}$$

$$\begin{aligned}
&\mathrm{M}_{11}\mathrm{M}_{43} - \mathrm{M}_{13}\mathrm{M}_{41} \\
&\quad = (A_{11}A_{31} - A_{12}A_{32})(F_{11}F_{14} - F_{12}F_{13}) \\
&\quad + (A_{14}A_{34} - A_{13}A_{33})(F_{32}F_{33} - F_{31}F_{34}) \\
&\quad + \mathrm{e}^{\mathrm{j}(\beta_+ + \beta_-)}(A_{11}A_{34} - A_{13}A_{32})(F_{11}F_{33} - F_{13}F_{31})
\end{aligned}$$

$$
\begin{aligned}
&+ \mathrm{e}^{-\mathrm{j}\left(\beta_+ + \beta_-\right)} \left(A_{12}A_{33} - A_{14}A_{31}\right)\left(F_{12}F_{34} - F_{14}F_{32}\right) \\
&+ \mathrm{e}^{\mathrm{j}\left(\beta_+ - \beta_-\right)} \left(A_{11}A_{33} - A_{14}A_{32}\right)\left(F_{11}F_{34} - F_{13}F_{32}\right) \\
&+ \mathrm{e}^{-\mathrm{j}\left(\beta_+ - \beta_-\right)} \left(A_{12}A_{34} - A_{13}A_{31}\right)\left(F_{12}F_{33} - F_{14}F_{31}\right),
\end{aligned} \tag{5.43c}
$$

$$
\begin{aligned}
&\mathrm{M}_{41}\mathrm{M}_{33} - \mathrm{M}_{43}\mathrm{M}_{31} \\
&\quad = -\left(A_{31}^2 - A_{32}^2\right)\left(F_{11}F_{14} - F_{12}F_{13}\right) \\
&\quad + \left(A_{33}^2 - A_{34}^2\right)\left(F_{32}F_{33} - F_{31}F_{34}\right) \\
&\quad + \left(A_{32}A_{33} - A_{31}A_{34}\right)\left[\mathrm{e}^{\mathrm{j}\left(\beta_+ + \beta_-\right)}\left(F_{11}F_{33} - F_{13}F_{31}\right)\right. \\
&\quad \left. - \mathrm{e}^{-\mathrm{j}\left(\beta_+ + \beta_-\right)}\left(F_{12}F_{34} - F_{14}F_{32}\right)\right] + \left(A_{32}A_{34} - A_{31}A_{33}\right) \\
&\quad \times \left[\mathrm{e}^{\mathrm{j}\left(\beta_+ - \beta_-\right)}\left(F_{11}F_{34} - F_{13}F_{32}\right) - \mathrm{e}^{-\mathrm{j}\left(\beta_+ - \beta_-\right)}\left(F_{12}F_{33} - F_{14}F_{31}\right)\right],
\end{aligned} \tag{5.43d}
$$

$$
\begin{aligned}
&\mathrm{M}_{21}\mathrm{M}_{23} - \mathrm{M}_{13} \\
&\quad = -\left(A_{11}^2 - A_{12}^2\right)\left(F_{11}F_{14} - F_{12}F_{13}\right) \\
&\quad + \left(A_{13}^2 - A_{14}^2\right)\left(F_{32}F_{33} - F_{31}F_{34}\right) \\
&\quad - \left(A_{11}A_{14} - A_{12}A_{13}\right)\left[\mathrm{e}^{\mathrm{j}\left(\beta_+ + \beta_-\right)}\left(F_{11}F_{33} - F_{13}F_{31}\right)\right. \\
&\quad \left. - \mathrm{e}^{-\mathrm{j}\left(\beta_+ + \beta_-\right)}\left(F_{12}F_{34} - F_{14}F_{32}\right)\right] - \left(A_{11}A_{13} - A_{12}A_{14}\right) \\
&\quad \times \left[\mathrm{e}^{\mathrm{j}\left(\beta_+ - \beta_-\right)}\left(F_{11}F_{34} - F_{13}F_{32}\right) - \mathrm{e}^{-\mathrm{j}\left(\beta_+ - \beta_-\right)}\left(F_{12}F_{33} - F_{14}F_{31}\right)\right].
\end{aligned} \tag{5.43e}
$$

We denote $A_0 = A_{11}A_{33} - A_{13}A_{31}$. As in the polar case, we can identify the single-interface reflection coefficients. It turns out that they take the forms identical with those in Eq. (4.210) and Eq. (4.212) except

$$
r_{41}^{(01)} = \left(\frac{\mathrm{E}_{04}^{(0)}}{\mathrm{E}_{01}^{(0)}}\right)_{\mathrm{E}_{03}^{(0)}=0} = -\left(A_{32}A_{33} - A_{31}A_{34}\right)A_0^{-1}, \tag{5.44a}
$$

$$
r_{43}^{(01)} = \left(\frac{\mathrm{E}_{02}^{(0)}}{\mathrm{E}_{03}^{(0)}}\right)_{\mathrm{E}_{01}^{(0)}=0} = -\left(A_{11}A_{34} - A_{13}A_{32}\right)A_0^{-1}, \tag{5.44b}
$$

which appear with the opposite signs. With Eq. (4.200) and taking into account Eq. (3.99) or Eq. (4.189) we obtain, after skipping the common factor $(2N^{(0)}\cos\varphi^{(0)})^2$ on the left-hand side, the following useful relations

$$
\begin{aligned}
&\left(A_{11}A_{33} - A_{13}A_{31}\right) \\
&\quad = N^{(0)}\cos^2\varphi^{(0)}\left(D_{13}^{(1)}D_{41}^{(1)} - D_{11}^{(1)}D_{43}^{(1)}\right) + N^{(0)2}\cos\varphi^{(0)}\left(D_{11}^{(1)}D_{33}^{(1)} - D_{13}^{(1)}D_{31}^{(1)}\right) \\
&\quad + \cos\varphi^{(0)}\left(D_{41}^{(1)}D_{23}^{(1)} - D_{43}^{(1)}D_{21}^{(1)}\right) + N^{(0)}\left(D_{21}^{(1)}D_{33}^{(1)} - D_{23}^{(1)}D_{31}^{(1)}\right),
\end{aligned} \tag{5.45a}
$$

$$\begin{aligned}
&(A_{12}A_{33} - A_{14}A_{31}) \\
&= -N^{(0)}\cos^2\varphi^{(0)}\left(D_{13}^{(1)}D_{41}^{(1)} - D_{11}^{(1)}D_{43}^{(1)}\right) - N^{(0)2}\cos\varphi^{(0)}D_{11}^{(1)}\left(D_{11}^{(1)}D_{33}^{(1)} - D_{13}^{(1)}D_{31}^{(1)}\right) \\
&+\cos\varphi^{(0)}\left(D_{41}^{(1)}D_{23}^{(1)} - D_{43}^{(1)}D_{21}^{(1)}\right) + N^{(0)}\left(D_{21}^{(1)}D_{33}^{(1)} - D_{23}^{(1)}D_{31}^{(1)}\right),
\end{aligned} \tag{5.45b}$$

$$\begin{aligned}
&(A_{11}A_{34} - A_{13}A_{32}) \\
&= -N^{(0)}\cos^2\varphi^{(0)}\left(D_{13}^{(1)}D_{41}^{(1)} - D_{11}^{(1)}D_{43}^{(1)}\right) + N^{(0)2}\cos\varphi^{(0)}\left(D_{11}^{(1)}D_{33}^{(1)} - D_{13}^{(1)}D_{31}^{(1)}\right) \\
&-\cos\varphi^{(0)}\left(D_{41}^{(1)}D_{23}^{(1)} - D_{43}^{(1)}D_{21}^{(1)}\right) + N^{(0)}\left(D_{21}^{(1)}D_{33}^{(1)} - D_{23}^{(1)}D_{31}^{(1)}\right),
\end{aligned} \tag{5.45c}$$

$$\begin{aligned}
&(A_{12}A_{34} - A_{14}A_{32}) \\
&= N^{(0)}\cos^2\varphi^{(0)}\left(D_{13}^{(1)}D_{41}^{(1)} - D_{11}^{(1)}D_{43}^{(1)}\right) - N^{(0)2}\cos\varphi^{(0)}\left(D_{11}^{(1)}D_{33}^{(1)} - D_{13}^{(1)}D_{31}^{(1)}\right) \\
&-\cos\varphi^{(0)}\left(D_{41}^{(1)}D_{23}^{(1)} - D_{43}^{(1)}D_{21}^{(1)}\right) + N^{(0)}\left(D_{21}^{(1)}D_{33}^{(1)} - D_{23}^{(1)}D_{31}^{(1)}\right),
\end{aligned} \tag{5.45d}$$

$$\begin{aligned}
&(A_{11}A_{34} - A_{14}A_{31}) \\
&= -N^{(0)}\cos^2\varphi^{(0)}\left(D_{13}^{(1)}D_{41}^{(1)} - D_{11}^{(1)}D_{43}^{(1)}\right) + N^{(0)2}\cos\varphi^{(0)}\left(D_{11}^{(1)}D_{33}^{(1)} + D_{13}^{(1)}D_{31}^{(1)}\right) \\
&+\cos\varphi^{(0)}\left(D_{41}^{(1)}D_{23}^{(1)} + D_{43}^{(1)}D_{21}^{(1)}\right) + N^{(0)}\left(D_{21}^{(1)}D_{33}^{(1)} - D_{23}^{(1)}D_{31}^{(1)}\right),
\end{aligned} \tag{5.45e}$$

$$\begin{aligned}
&(A_{12}A_{33} - A_{13}A_{32}) \\
&= -N^{(0)}\cos^2\varphi^{(0)}\left(D_{13}^{(1)}D_{41}^{(1)} - D_{11}^{(1)}D_{43}^{(1)}\right) - N^{(0)2}\cos\varphi^{(0)}\left(D_{11}^{(1)}D_{33}^{(1)} + D_{13}^{(1)}D_{31}^{(1)}\right) \\
&-\cos\varphi^{(0)}\left(D_{41}^{(1)}D_{23}^{(1)} + D_{43}^{(1)}D_{21}^{(1)}\right) + N^{(0)}\left(D_{21}^{(1)}D_{33}^{(1)} - D_{23}^{(1)}D_{31}^{(1)}\right),
\end{aligned} \tag{5.45f}$$

$$\begin{aligned}
&(A_{12}A_{34} - A_{13}A_{31}) \\
&= N^{(0)}\cos^2\varphi^{(0)}\left(D_{13}^{(1)}D_{41}^{(1)} - D_{11}^{(1)}D_{43}^{(1)}\right) - N^{(0)2}\cos\varphi^{(0)}\left(D_{11}^{(1)}D_{33}^{(1)} + D_{13}^{(1)}D_{31}^{(1)}\right) \\
&+\cos\varphi^{(0)}\left(D_{41}^{(1)}D_{23}^{(1)} + D_{43}^{(1)}D_{21}^{(1)}\right) + N^{(0)}\left(D_{21}^{(1)}D_{33}^{(1)} - D_{23}^{(1)}D_{31}^{(1)}\right),
\end{aligned} \tag{5.45g}$$

$$\begin{aligned}
&(A_{11}A_{33} - A_{14}A_{32}) \\
&= N^{(0)}\cos^2\varphi^{(0)}\left(D_{13}^{(1)}D_{41}^{(1)} - D_{11}^{(1)}D_{43}^{(1)}\right) + N^{(0)2}\cos\varphi^{(0)}\left(D_{11}^{(1)}D_{33}^{(1)} + D_{13}^{(1)}D_{31}^{(1)}\right) \\
&-\cos\varphi^{(0)}\left(D_{41}^{(1)}D_{23}^{(1)} + D_{43}^{(1)}D_{21}^{(1)}\right) + N^{(0)}\left(D_{21}^{(1)}D_{33}^{(1)} - D_{23}^{(1)}D_{31}^{(1)}\right).
\end{aligned} \tag{5.45h}$$

We may further write

$$(A_{11}A_{14} - A_{12}A_{13}) = -2N^{(0)}\cos\varphi^{(0)}\left(D_{13}^{(1)}D_{21}^{(1)} - D_{11}^{(1)}D_{23}^{(1)}\right), \tag{5.45i}$$

$$(A_{11}A_{13} - A_{12}A_{14}) = 2N^{(0)}\cos\varphi^{(0)}\left(D_{13}^{(1)}D_{21}^{(1)} + D_{11}^{(1)}D_{23}^{(1)}\right), \tag{5.45j}$$

$$(A_{32}A_{33} - A_{31}A_{34}) = 2N^{(0)}\cos\varphi^{(0)}\left(D_{41}^{(1)}D_{33}^{(1)} - D_{43}^{(1)}D_{31}^{(1)}\right), \tag{5.45k}$$

$$(A_{32}A_{34} - A_{31}A_{33}) = 2N^{(0)}\cos\varphi^{(0)}\left(D_{41}^{(1)}D_{33}^{(1)} + D_{43}^{(1)}D_{31}^{(1)}\right), \tag{5.45l}$$

$$(A_{11}A_{31} - A_{12}A_{32}) = 2\cos\varphi^{(0)}\left(N^{(0)2}D_{11}^{(1)}D_{31}^{(1)} - D_{21}^{(1)}D_{41}^{(1)}\right), \tag{5.45m}$$

$$(A_{13}A_{33} - A_{14}A_{34}) = 2\cos\varphi^{(0)}\left(N^{(0)2}D_{13}^{(1)}D_{33}^{(1)} - D_{23}^{(1)}D_{43}^{(1)}\right), \tag{5.45n}$$

$$(A_{11}A_{32} - A_{12}A_{31}) = 2\cos\varphi^{(0)}\left(N^{(0)2}D_{11}^{(1)}D_{31}^{(1)} + D_{21}^{(1)}D_{41}^{(1)}\right), \tag{5.45o}$$

$$(A_{13}A_{34} - A_{14}A_{33}) = 2\cos\varphi^{(0)}\left(N^{(0)2}D_{13}^{(1)}D_{33}^{(1)} + D_{23}^{(1)}D_{43}^{(1)}\right). \tag{5.45p}$$

The differences of the squared **A** matrix elements can be expressed as

$$\left(A_{11}^2 - A_{12}^2\right) = 4N^{(0)} \cos\varphi^{(0)} D_{11}^{(1)} D_{21}^{(1)} \tag{5.45q}$$

$$\left(A_{13}^2 - A_{14}^2\right) = 4N^{(0)} \cos\varphi^{(0)} D_{13}^{(1)} D_{23}^{(1)} \tag{5.45r}$$

$$\left(A_{31}^2 - A_{32}^2\right) = -4N^{(0)} \cos\varphi^{(0)} D_{31}^{(1)} D_{41}^{(1)} \tag{5.45s}$$

$$\left(A_{33}^2 - A_{34}^2\right) = -4N^{(0)} \cos\varphi^{(0)} D_{33}^{(1)} D_{43}^{(1)}. \tag{5.45t}$$

Most of these expressions can be concisely expressed using Eq. (5.11) after skipping the common factor $\left(2N^{(0)} \cos\varphi^{(0)}\right)^{-2} \varepsilon_{xz}^{(1)} N_y$

$$\begin{aligned}&\left(A_{11} A_{33} - A_{13} A_{31}\right)\\ &\quad = \left(N_{z+}^{(1)} - N_{z-}^{(1)}\right)\left(\mathcal{T}^{(0)}\mathcal{P}_{+}^{(1)} + \mathcal{S}^{(0)}\mathcal{Q}_{+}^{(1)} + \mathcal{Q}^{(0)}\mathcal{S}_{+}^{(1)} + \mathcal{P}^{(0)}\mathcal{T}_{+}^{(1)}\right),\end{aligned} \tag{5.46a}$$

$$\begin{aligned}&\left(A_{12} A_{33} - A_{14} A_{31}\right)\\ &\quad = \left(N_{z+}^{(1)} - N_{z-}^{(1)}\right)\left(-\mathcal{T}^{(0)}\mathcal{P}_{+}^{(1)} - \mathcal{S}^{(0)}\mathcal{Q}_{+}^{(1)} + \mathcal{Q}^{(0)}\mathcal{S}_{+}^{(1)} + \mathcal{P}^{(0)}\mathcal{T}_{+}^{(1)}\right),\end{aligned} \tag{5.46b}$$

$$\begin{aligned}&\left(A_{11} A_{34} - A_{13} A_{32}\right)\\ &\quad = \left(N_{z+}^{(1)} - N_{z-}^{(1)}\right)\left(-\mathcal{T}^{(0)}\mathcal{P}_{+}^{(1)} + \mathcal{S}^{(0)}\mathcal{Q}_{+}^{(1)} - \mathcal{Q}^{(0)}\mathcal{S}_{+}^{(1)} + \mathcal{P}^{(0)}\mathcal{T}_{+}^{(1)}\right),\end{aligned} \tag{5.46c}$$

$$\begin{aligned}&\left(A_{12} A_{34} - A_{14} A_{32}\right)\\ &\quad = \left(N_{z+}^{(1)} - N_{z-}^{(1)}\right)\left(\mathcal{T}^{(0)}\mathcal{P}_{+}^{(1)} - \mathcal{S}^{(0)}\mathcal{Q}_{+}^{(1)} - \mathcal{Q}^{(0)}\mathcal{S}_{+}^{(1)} + \mathcal{P}^{(0)}\mathcal{T}_{+}^{(1)}\right),\end{aligned} \tag{5.46d}$$

$$\begin{aligned}&\left(A_{11} A_{34} - A_{14} A_{31}\right)\\ &\quad = \left(N_{z+}^{(1)} + N_{z-}^{(1)}\right)\left(-\mathcal{T}^{(0)}\mathcal{P}_{-}^{(1)} - \mathcal{S}^{(0)}\mathcal{Q}_{-}^{(1)} + \mathcal{Q}^{(0)}\mathcal{S}_{-}^{(1)} + \mathcal{P}^{(0)}\mathcal{T}_{-}^{(1)}\right),\end{aligned} \tag{5.46e}$$

$$\begin{aligned}&\left(A_{12} A_{33} - A_{13} A_{32}\right)\\ &\quad = \left(N_{z+}^{(1)} + N_{z-}^{(1)}\right)\left(-\mathcal{T}^{(0)}\mathcal{P}_{-}^{(1)} + \mathcal{S}^{(0)}\mathcal{Q}_{-}^{(1)} - \mathcal{Q}^{(0)}\mathcal{S}_{-}^{(1)} + \mathcal{P}^{(0)}\mathcal{T}_{-}^{(1)}\right),\end{aligned} \tag{5.46f}$$

$$\begin{aligned}&\left(A_{12} A_{34} - A_{13} A_{31}\right)\\ &\quad = \left(N_{z+}^{(1)} + N_{z-}^{(1)}\right)\left(\mathcal{T}^{(0)}\mathcal{P}_{-}^{(1)} + \mathcal{S}^{(0)}\mathcal{Q}_{-}^{(1)} + \mathcal{Q}^{(0)}\mathcal{S}_{-}^{(1)} + \mathcal{P}^{(0)}\mathcal{T}_{-}^{(1)}\right),\end{aligned} \tag{5.46g}$$

$$\begin{aligned}&\left(A_{11} A_{33} - A_{14} A_{32}\right)\\ &\quad = \left(N_{z+}^{(1)} + N_{z-}^{(1)}\right)\left(\mathcal{T}^{(0)}\mathcal{P}_{-}^{(1)} - \mathcal{S}^{(0)}\mathcal{Q}_{-}^{(1)} - \mathcal{Q}^{(0)}\mathcal{S}_{-}^{(1)} + \mathcal{P}^{(0)}\mathcal{T}_{-}^{(1)}\right),\end{aligned} \tag{5.46h}$$

$$\left(A_{11} A_{14} - A_{12} A_{13}\right) = -2N^{(0)} \cos\varphi^{(0)} \left(N_{z+}^{(1)} - N_{z-}^{(1)}\right)\mathcal{U}_{+}^{(1)}, \tag{5.46i}$$

$$\left(A_{11} A_{13} - A_{12} A_{14}\right) = 2N^{(0)} \cos\varphi^{(0)} \left(N_{z+}^{(1)} + N_{z-}^{(1)}\right)\mathcal{U}_{-}^{(1)}, \tag{5.46j}$$

$$\left(A_{32} A_{33} - A_{31} A_{34}\right) = 2N^{(0)} \cos\varphi^{(0)} \left(N_{z+}^{(1)} - N_{z-}^{(1)}\right)\mathcal{W}_{+}^{(1)}, \tag{5.46k}$$

$$\left(A_{32} A_{34} - A_{31} A_{33}\right) = -2N^{(0)} \cos\varphi^{(0)} \left(N_{z+}^{(1)} + N_{z-}^{(1)}\right)\mathcal{W}_{-}^{(1)}. \tag{5.46l}$$

As in Eq. (4.216) for the polar case, we express

$$
\begin{aligned}
&(F_{11}F_{14} - F_{12}F_{13}) \\
&\quad = -2L_{11}^{(n-1)}L_{12}^{(n-1)}\left(D_{13}^{(n)}D_{21}^{(n)} - D_{11}^{(n)}D_{23}^{(n)}\right) + 2L_{11}^{(n-1)}L_{13}^{(n-1)}\left(D_{11}^{(n)}D_{33}^{(n)} - D_{13}^{(n)}D_{31}^{(n)}\right) \\
&\quad + 2L_{12}^{(n-1)}L_{14}^{(n-1)}\left(D_{23}^{(n)}D_{41}^{(n)} - D_{21}^{(n)}D_{43}^{(n)}\right) + 2L_{13}^{(n-1)}L_{14}^{(n-1)}\left(D_{33}^{(n)}D_{41}^{(n)} - D_{31}^{(n)}D_{43}^{(n)}\right)
\end{aligned} \tag{5.47a}
$$

$$
\begin{aligned}
&(F_{32}F_{33} - F_{31}F_{34}) \\
&\quad = 2L_{31}^{(n-1)}L_{32}^{(n-1)}\left(D_{13}^{(n)}D_{21}^{(n)} - D_{11}^{(n)}D_{23}^{(n)}\right) - 2L_{31}^{(n-1)}L_{33}^{(n-1)}\left(D_{11}^{(n)}D_{33}^{(n)} - D_{13}^{(n)}D_{31}^{(n)}\right) \\
&\quad -2L_{32}^{(n-1)}L_{34}^{(n-1)}\left(D_{23}^{(n)}D_{41}^{(n)} - D_{21}^{(n)}D_{43}^{(n)}\right) - 2L_{33}^{(n-1)}L_{34}^{(n-1)}\left(D_{33}^{(n)}D_{41}^{(n)} - D_{31}^{(n)}D_{43}^{(n)}\right)
\end{aligned} \tag{5.47b}
$$

$$
\begin{aligned}
&(F_{11}F_{33} - F_{13}F_{31}) \\
&\quad = \left(L_{12}^{(n-1)}L_{31}^{(n-1)} - L_{11}^{(n-1)}L_{32}^{(n-1)}\right)\left(D_{13}^{(n)}D_{21}^{(n)} - D_{11}^{(n)}D_{23}^{(n)}\right) \\
&\quad +\left(L_{11}^{(n-1)}L_{33}^{(n-1)} - L_{13}^{(n-1)}L_{31}^{(n-1)}\right)\left(D_{11}^{(n)}D_{33}^{(n)} - D_{13}^{(n)}D_{31}^{(n)}\right) \\
&\quad +\left(L_{14}^{(n-1)}L_{31}^{(n-1)} - L_{11}^{(n-1)}L_{34}^{(n-1)}\right)\left(D_{13}^{(n)}D_{41}^{(n)} - D_{11}^{(n)}D_{43}^{(n)}\right) \\
&\quad +\left(L_{12}^{(n-1)}L_{33}^{(n-1)} - L_{13}^{(n-1)}L_{32}^{(n-1)}\right)\left(D_{21}^{(n)}D_{33}^{(n)} - D_{23}^{(n)}D_{31}^{(n)}\right) \\
&\quad +\left(L_{14}^{(n-1)}L_{32}^{(n-1)} - L_{12}^{(n-1)}L_{34}^{(n-1)}\right)\left(D_{23}^{(n)}D_{41}^{(n)} - D_{21}^{(n)}D_{43}^{(n)}\right) \\
&\quad +\left(L_{14}^{(n-1)}L_{33}^{(n-1)} - L_{13}^{(n-1)}L_{34}^{(n-1)}\right)\left(D_{33}^{(n)}D_{41}^{(n)} - D_{31}^{(n)}D_{43}^{(n)}\right)
\end{aligned} \tag{5.47c}
$$

$$
\begin{aligned}
&(F_{12}F_{34} - F_{14}F_{32}) \\
&\quad = -\left(L_{12}^{(n-1)}L_{31}^{(n-1)} - L_{11}^{(n-1)}L_{32}^{(n-1)}\right)\left(D_{13}^{(n)}D_{21}^{(n)} - D_{11}^{(n)}D_{23}^{(n)}\right) \\
&\quad -\left(L_{11}^{(n-1)}L_{33}^{(n-1)} - L_{13}^{(n-1)}L_{31}^{(n-1)}\right)\left(D_{11}^{(n)}D_{33}^{(n)} - D_{13}^{(n)}D_{31}^{(n)}\right) \\
&\quad +\left(L_{14}^{(n-1)}L_{31}^{(n-1)} - L_{11}^{(n-1)}L_{34}^{(n-1)}\right)\left(D_{13}^{(n)}D_{41}^{(n)} - D_{11}^{(n)}D_{43}^{(n)}\right) \\
&\quad +\left(L_{12}^{(n-1)}L_{33}^{(n-1)} - L_{13}^{(n-1)}L_{32}^{(n-1)}\right)\left(D_{21}^{(n)}D_{33}^{(n)} - D_{23}^{(n)}D_{31}^{(n)}\right) \\
&\quad -\left(L_{14}^{(n-1)}L_{32}^{(n-1)} - L_{12}^{(n-1)}L_{34}^{(n-1)}\right)\left(D_{23}^{(n)}D_{41}^{(n)} - D_{21}^{(n)}D_{43}^{(n)}\right) \\
&\quad -\left(L_{14}^{(n-1)}L_{33}^{(n-1)} - L_{13}^{(n-1)}L_{34}^{(n-1)}\right)\left(D_{33}^{(n)}D_{41}^{(n)} - D_{31}^{(n)}D_{43}^{(n)}\right)
\end{aligned} \tag{5.47d}
$$

$$
\begin{aligned}
&(F_{11}F_{34} - F_{13}F_{32}) \\
&\quad = -\left(L_{11}^{(n-1)}L_{32}^{(n-1)} + L_{12}^{(n-1)}L_{31}^{(n-1)}\right)\left(D_{13}^{(n)}D_{21}^{(n)} - D_{11}^{(n)}D_{23}^{(n)}\right) \\
&\quad +\left(L_{11}^{(n-1)}L_{33}^{(n-1)} + L_{13}^{(n-1)}L_{31}^{(n-1)}\right)\left(D_{11}^{(n)}D_{33}^{(n)} - D_{13}^{(n)}D_{31}^{(n)}\right) \\
&\quad -\left(L_{14}^{(n-1)}L_{31}^{(n-1)} - L_{11}^{(n-1)}L_{34}^{(n-1)}\right)\left(D_{13}^{(n)}D_{41}^{(n)} - D_{11}^{(n)}D_{43}^{(n)}\right)
\end{aligned}
$$

$$
\begin{aligned}
&+\left(L_{12}^{(n-1)} L_{33}^{(n-1)} - L_{13}^{(n-1)} L_{32}^{(n-1)}\right)\left(D_{21}^{(n)} D_{33}^{(n)} - D_{23}^{(n)} D_{31}^{(n)}\right) \\
&+\left(L_{12}^{(n-1)} L_{34}^{(n-1)} + L_{14}^{(n-1)} L_{32}^{(n-1)}\right)\left(D_{23}^{(n)} D_{41}^{(n)} - D_{21}^{(n)} D_{43}^{(n)}\right) \\
&+\left(L_{13}^{(n-1)} L_{34}^{(n-1)} + L_{14}^{(n-1)} L_{33}^{(n-1)}\right)\left(D_{33}^{(n)} D_{41}^{(n)} - D_{31}^{(n)} D_{43}^{(n)}\right)
\end{aligned} \tag{5.47e}
$$

$$
\begin{aligned}
&\left(F_{12} F_{33} - F_{14} F_{31}\right) \\
&= \left(L_{11}^{(n-1)} L_{32}^{(n-1)} + L_{12}^{(n-1)} L_{31}^{(n-1)}\right)\left(D_{13}^{(n)} D_{21}^{(n)} - D_{11}^{(n)} D_{23}^{(n)}\right) \\
&-\left(L_{11}^{(n-1)} L_{33}^{(n-1)} + L_{13}^{(n-1)} L_{31}^{(n-1)}\right)\left(D_{11}^{(n)} D_{33}^{(n)} - D_{13}^{(n)} D_{31}^{(n)}\right) \\
&-\left(L_{14}^{(n-1)} L_{31}^{(n-1)} - L_{11}^{(n-1)} L_{34}^{(n-1)}\right)\left(D_{13}^{(n)} D_{41}^{(n)} - D_{11}^{(n)} D_{43}^{(n)}\right) \\
&+\left(L_{12}^{(n-1)} L_{33}^{(n-1)} - L_{13}^{(n-1)} L_{32}^{(n-1)}\right)\left(D_{21}^{(n)} D_{33}^{(n)} - D_{23}^{(n)} D_{31}^{(n)}\right) \\
&-\left(L_{12}^{(n-1)} L_{34}^{(n-1)} + L_{14}^{(n-1)} L_{32}^{(n-1)}\right)\left(D_{23}^{(n)} D_{41}^{(n)} - D_{21}^{(n)} D_{43}^{(n)}\right) \\
&-\left(L_{13}^{(n-1)} L_{34}^{(n-1)} + L_{14}^{(n-1)} L_{33}^{(n-1)}\right)\left(D_{33}^{(n)} D_{41}^{(n)} - D_{31}^{(n)} D_{43}^{(n)}\right)
\end{aligned} \tag{5.47f}
$$

with $n = 2$. We can replace L_{ij} by the elements of the dynamical matrix using Eq. (5.14)

$$L_{11}^{(n-1)} L_{12}^{(n-1)} = -\mathcal{D}_{\text{lon}}^{(n-1)} D_{33}^{(n-1)} D_{43}^{(n-1)} \tag{5.48a}$$

$$L_{11}^{(n-1)} L_{13}^{(n-1)} = \mathcal{D}_{\text{lon}}^{(n-1)} D_{23}^{(n-1)} D_{43}^{(n-1)} \tag{5.48b}$$

$$L_{12}^{(n-1)} L_{14}^{(n-1)} = \mathcal{D}_{\text{lon}}^{(n-1)} D_{13}^{(n-1)} D_{33}^{(n-1)} \tag{5.48c}$$

$$L_{13}^{(n-1)} L_{14}^{(n-1)} = -\mathcal{D}_{\text{lon}}^{(n-1)} D_{13}^{(n-1)} D_{23}^{(n-1)} \tag{5.48d}$$

$$L_{31}^{(n-1)} L_{32}^{(n-1)} = -\mathcal{D}_{\text{lon}}^{(n-1)} D_{31}^{(n-1)} D_{41}^{(n-1)} \tag{5.48e}$$

$$L_{31}^{(n-1)} L_{33}^{(n-1)} = \mathcal{D}_{\text{lon}}^{(n-1)} D_{21}^{(n-1)} D_{41}^{(n-1)} \tag{5.48f}$$

$$L_{32}^{(n-1)} L_{34}^{(n-1)} = \mathcal{D}_{\text{lon}}^{(n-1)} D_{11}^{(n-1)} D_{31}^{(n-1)} \tag{5.48g}$$

$$L_{33}^{(n-1)} L_{34}^{(n-1)} = -\mathcal{D}_{\text{lon}}^{(n-1)} D_{11}^{(n-1)} D_{21}^{(n-1)} \tag{5.48h}$$

$$L_{12}^{(n-1)} L_{31}^{(n-1)} - L_{11}^{(n-1)} L_{32}^{(n-1)} = \mathcal{D}_{\text{lon}}^{(n-1)}\left(D_{33}^{(n-1)} D_{41}^{(n-1)} - D_{31}^{(n-1)} D_{43}^{(n-1)}\right) \tag{5.48i}$$

$$L_{11}^{(n-1)} L_{33}^{(n-1)} - L_{13}^{(n-1)} L_{31}^{(n-1)} = \mathcal{D}_{\text{lon}}^{(n-1)}\left(D_{23}^{(n-1)} D_{41}^{(n-1)} - D_{21}^{(n-1)} D_{43}^{(n-1)}\right) \tag{5.48j}$$

$$L_{14}^{(n-1)} L_{31}^{(n-1)} - L_{11}^{(n-1)} L_{34}^{(n-1)} = \mathcal{D}_{\text{lon}}^{(n-1)}\left(D_{21}^{(n-1)} D_{33}^{(n-1)} - D_{23}^{(n-1)} D_{31}^{(n-1)}\right) \tag{5.48k}$$

$$L_{12}^{(n-1)} L_{33}^{(n-1)} - L_{13}^{(n-1)} L_{32}^{(n-1)} = \mathcal{D}_{\text{lon}}^{(n-1)}\left(D_{13}^{(n-1)} D_{41}^{(n-1)} - D_{11}^{(n-1)} D_{43}^{(n-1)}\right) \tag{5.48l}$$

$$L_{14}^{(n-1)} L_{32}^{(n-1)} - L_{12}^{(n-1)} L_{34}^{(n-1)} = \mathcal{D}_{\text{lon}}^{(n-1)}\left(D_{11}^{(n-1)} D_{33}^{(n-1)} - D_{13}^{(n-1)} D_{31}^{(n-1)}\right) \tag{5.48m}$$

$$L_{14}^{(n-1)} L_{33}^{(n-1)} - L_{13}^{(n-1)} L_{34}^{(n-1)} = \mathcal{D}_{\text{lon}}^{(n-1)}\left(D_{13}^{(n-1)} D_{21}^{(n-1)} - D_{11}^{(n-1)} D_{23}^{(n-1)}\right) \tag{5.48n}$$

$$L_{11}^{(n-1)}L_{32}^{(n-1)} + L_{12}^{(n-1)}L_{31}^{(n-1)} = \mathcal{D}_{\text{lon}}^{(n-1)}\left(D_{41}^{(n-1)}D_{33}^{(n-1)} + D_{31}^{(n-1)}D_{43}^{(n-1)}\right) \quad (5.48\text{o})$$
$$L_{11}^{(n-1)}L_{33}^{(n-1)} + L_{13}^{(n-1)}L_{31}^{(n-1)} = -\mathcal{D}_{\text{lon}}^{(n-1)}\left(D_{21}^{(n-1)}D_{43}^{(n-1)} + D_{23}^{(n-1)}D_{41}^{(n-1)}\right) \quad (5.48\text{p})$$
$$L_{14}^{(n-1)}L_{32}^{(n-1)} + L_{12}^{(n-1)}L_{34}^{(n-1)} = -\mathcal{D}_{\text{lon}}^{(n-1)}\left(D_{11}^{(n-1)}D_{33}^{(n-1)} + D_{13}^{(n-1)}D_{31}^{(n-1)}\right) \quad (5.48\text{q})$$
$$L_{14}^{(n-1)}L_{33}^{(n-1)} + L_{13}^{(n-1)}L_{34}^{(n-1)} = \mathcal{D}_{\text{lon}}^{(n-1)}\left(D_{13}^{(n-1)}D_{21}^{(n-1)} + D_{11}^{(n-1)}D_{23}^{(n-1)}\right). \quad (5.48\text{r})$$

The substitution for the L_{ij} elements (with the common factor $\mathcal{D}_{\text{lon}}$ removed) provides

$$\begin{aligned}(F_{11}F_{14} - F_{12}F_{13})
&= 2D_{33}^{(n-1)}D_{43}^{(n-1)}\left(D_{13}^{(n)}D_{21}^{(n)} - D_{11}^{(n)}D_{23}^{(n)}\right) + 2D_{23}^{(n-1)}D_{43}^{(n-1)}\left(D_{11}^{(n)}D_{33}^{(n)} - D_{13}^{(n)}D_{31}^{(n)}\right)\\
&+ 2D_{13}^{(n-1)}D_{33}^{(n-1)}\left(D_{23}^{(n)}D_{41}^{(n)} - D_{21}^{(n)}D_{43}^{(n)}\right) - 2D_{13}^{(n-1)}D_{23}^{(n-1)}\left(D_{33}^{(n)}D_{41}^{(n)} - D_{31}^{(n)}D_{43}^{(n)}\right)\end{aligned} \quad (5.49\text{a})$$

$$\begin{aligned}(F_{32}F_{33} - F_{31}F_{34})
&= -2D_{31}^{(n-1)}D_{41}^{(n-1)}\left(D_{13}^{(n)}D_{21}^{(n)} - D_{11}^{(n)}D_{23}^{(n)}\right) - 2D_{21}^{(n-1)}D_{41}^{(n-1)}\left(D_{11}^{(n)}D_{33}^{(n)} - D_{13}^{(n)}D_{31}^{(n)}\right)\\
&-2D_{11}^{(n-1)}D_{31}^{(n-1)}\left(D_{23}^{(n)}D_{41}^{(n)} - D_{21}^{(n)}D_{43}^{(n)}\right) + 2D_{11}^{(n-1)}D_{21}^{(n-1)}\left(D_{33}^{(n)}D_{41}^{(n)} - D_{31}^{(n)}D_{43}^{(n)}\right)\end{aligned} \quad (5.49\text{b})$$

$$\begin{aligned}(F_{11}F_{33} - F_{13}F_{31})
&= \left(D_{33}^{(n-1)}D_{41}^{(n-1)} - D_{31}^{(n-1)}D_{43}^{(n-1)}\right)\left(D_{13}^{(n)}D_{21}^{(n)} - D_{11}^{(n)}D_{23}^{(n)}\right)\\
&+\left(D_{23}^{(n-1)}D_{41}^{(n-1)} - D_{21}^{(n-1)}D_{43}^{(n-1)}\right)\left(D_{11}^{(n)}D_{33}^{(n)} - D_{13}^{(n)}D_{31}^{(n)}\right)\\
&+\left(D_{21}^{(n-1)}D_{33}^{(n-1)} - D_{23}^{(n-1)}D_{31}^{(n-1)}\right)\left(D_{13}^{(n)}D_{41}^{(n)} - D_{11}^{(n)}D_{43}^{(n)}\right)\\
&+\left(D_{13}^{(n-1)}D_{41}^{(n-1)} - D_{11}^{(n-1)}D_{43}^{(n-1)}\right)\left(D_{21}^{(n)}D_{33}^{(n)} - D_{23}^{(n)}D_{31}^{(n)}\right)\\
&+\left(D_{11}^{(n-1)}D_{33}^{(n-1)} - D_{13}^{(n-1)}D_{31}^{(n-1)}\right)\left(D_{23}^{(n)}D_{41}^{(n)} - D_{21}^{(n)}D_{43}^{(n)}\right)\\
&+\left(D_{13}^{(n-1)}D_{21}^{(n-1)} - D_{11}^{(n-1)}D_{23}^{(n-1)}\right)\left(D_{33}^{(n)}D_{41}^{(n)} - D_{31}^{(n)}D_{43}^{(n)}\right)\end{aligned} \quad (5.49\text{c})$$

$$\begin{aligned}(F_{12}F_{34} - F_{14}F_{32})
&= -\left(D_{33}^{(n-1)}D_{41}^{(n-1)} - D_{31}^{(n-1)}D_{43}^{(n-1)}\right)\left(D_{13}^{(n)}D_{21}^{(n)} - D_{11}^{(n)}D_{23}^{(n)}\right)\\
&-\left(D_{23}^{(n-1)}D_{41}^{(n-1)} - D_{21}^{(n-1)}D_{43}^{(n-1)}\right)\left(D_{11}^{(n)}D_{33}^{(n)} - D_{13}^{(n)}D_{31}^{(n)}\right)\\
&+\left(D_{21}^{(n-1)}D_{33}^{(n-1)} - D_{23}^{(n-1)}D_{31}^{(n-1)}\right)\left(D_{13}^{(n)}D_{41}^{(n)} - D_{11}^{(n)}D_{43}^{(n)}\right)\\
&+\left(D_{13}^{(n-1)}D_{41}^{(n-1)} - D_{11}^{(n-1)}D_{43}^{(n-1)}\right)\left(D_{21}^{(n)}D_{33}^{(n)} - D_{23}^{(n)}D_{31}^{(n)}\right)\\
&-\left(D_{11}^{(n-1)}D_{33}^{(n-1)} - D_{13}^{(n-1)}D_{31}^{(n-1)}\right)\left(D_{23}^{(n)}D_{41}^{(n)} - D_{21}^{(n)}D_{43}^{(n)}\right)\\
&-\left(D_{13}^{(n-1)}D_{21}^{(n-1)} - D_{11}^{(n-1)}D_{23}^{(n-1)}\right)\left(D_{33}^{(n)}D_{41}^{(n)} - D_{31}^{(n)}D_{43}^{(n)}\right)\end{aligned} \quad (5.49\text{d})$$

$$
\begin{aligned}
&\left(F_{11}F_{34} - F_{13}F_{32}\right)\\
&\quad = -\left(D_{41}^{(n-1)}D_{33}^{(n-1)} + D_{31}^{(n-1)}D_{43}^{(n-1)}\right)\left(D_{13}^{(n)}D_{21}^{(n)} - D_{11}^{(n)}D_{23}^{(n)}\right)\\
&\quad -\left(D_{21}^{(n-1)}D_{43}^{(n-1)} + D_{23}^{(n-1)}D_{41}^{(n-1)}\right)\left(D_{11}^{(n)}D_{33}^{(n)} - D_{13}^{(n)}D_{31}^{(n)}\right)\\
&\quad -\left(D_{21}^{(n-1)}D_{33}^{(n-1)} - D_{23}^{(n-1)}D_{31}^{(n-1)}\right)\left(D_{13}^{(n)}D_{41}^{(n)} - D_{11}^{(n)}D_{43}^{(n)}\right)\\
&\quad +\left(D_{13}^{(n-1)}D_{41}^{(n-1)} - D_{11}^{(n-1)}D_{43}^{(n-1)}\right)\left(D_{21}^{(n)}D_{33}^{(n)} - D_{23}^{(n)}D_{31}^{(n)}\right)\\
&\quad -\left(D_{11}^{(n-1)}D_{33}^{(n-1)} + D_{13}^{(n-1)}D_{31}^{(n-1)}\right)\left(D_{23}^{(n)}D_{41}^{(n)} - D_{21}^{(n)}D_{43}^{(n)}\right)\\
&\quad -\left(D_{13}^{(n-1)}D_{21}^{(n-1)} + D_{11}^{(n-1)}D_{23}^{(n-1)}\right)\left(D_{33}^{(n)}D_{41}^{(n)} - D_{31}^{(n)}D_{43}^{(n)}\right)
\end{aligned}
\tag{5.49e}
$$

$$
\begin{aligned}
&\left(F_{12}F_{33} - F_{14}F_{31}\right)\\
&\quad = -\left(D_{41}^{(n-1)}D_{33}^{(n-1)} + D_{31}^{(n-1)}D_{43}^{(n-1)}\right)\left(D_{13}^{(n)}D_{21}^{(n)} - D_{11}^{(n)}D_{23}^{(n)}\right)\\
&\quad +\left(D_{21}^{(n-1)}D_{43}^{(n-1)} + D_{23}^{(n-1)}D_{41}^{(n-1)}\right)\left(D_{11}^{(n)}D_{33}^{(n)} - D_{13}^{(n)}D_{31}^{(n)}\right)\\
&\quad -\left(D_{21}^{(n-1)}D_{33}^{(n-1)} - D_{23}^{(n-1)}D_{41}^{(n-1)}\right)\left(D_{13}^{(n)}D_{41}^{(n)} - D_{11}^{(n)}D_{43}^{(n)}\right)\\
&\quad +\left(D_{13}^{(n-1)}D_{41}^{(n-1)} - D_{11}^{(n-1)}D_{43}^{(n-1)}\right)\left(D_{21}^{(n)}D_{33}^{(n)} - D_{23}^{(n)}D_{31}^{(n)}\right)\\
&\quad +\left(D_{11}^{(n-1)}D_{33}^{(n-1)} + D_{13}^{(n-1)}D_{31}^{(n-1)}\right)\left(D_{23}^{(n)}D_{41}^{(n)} - D_{21}^{(n)}D_{43}^{(n)}\right)\\
&\quad -\left(D_{13}^{(n-1)}D_{21}^{(n-1)} + D_{11}^{(n-1)}D_{23}^{(n-1)}\right)\left(D_{33}^{(n)}D_{41}^{(n)} - D_{31}^{(n)}D_{43}^{(n)}\right)
\end{aligned}
\tag{5.49f}
$$

with $n = 2$. This can now be expressed more concisely with help of the definitions collected in Eq. (5.11)

$$
\begin{aligned}
&\left(F_{11}F_{14} - F_{12}F_{13}\right)\\
&= 2\left[\mathcal{D}_{\text{lon}}^{(1)}\left(N_{z+}^{(2)} - N_{z-}^{(2)}\right)N_y\varepsilon_{xz}^{(2)}\right]\\
&\quad \times\left(D_{33}^{(1)}D_{43}^{(1)}\mathcal{U}_{+}^{(2)} + D_{23}^{(1)}D_{43}^{(1)}\mathcal{Q}_{+}^{(2)} + D_{13}^{(1)}D_{33}^{(1)}\mathcal{S}_{+}^{(2)} - D_{13}^{(1)}D_{23}^{(1)}\mathcal{W}_{+}^{(2)}\right)
\end{aligned}
\tag{5.50a}
$$

$$
\begin{aligned}
&\left(F_{32}F_{33} - F_{31}F_{34}\right)\\
&= 2\left[\mathcal{D}_{\text{lon}}^{(1)}\left(N_{z+}^{(2)} - N_{z-}^{(2)}\right)N_y\varepsilon_{xz}^{(2)}\right]\\
&\quad \times\left(-D_{31}^{(1)}D_{41}^{(1)}\mathcal{U}_{+}^{(2)} - D_{21}^{(1)}D_{41}^{(1)}\mathcal{Q}_{+}^{(2)} - D_{11}^{(1)}D_{31}^{(1)}\mathcal{S}_{+}^{(2)} + D_{11}^{(1)}D_{21}^{(1)}\mathcal{W}_{+}^{(2)}\right)
\end{aligned}
\tag{5.50b}
$$

$$
\begin{aligned}
&\left(F_{11}F_{33} - F_{13}F_{31}\right)\\
&= \left[\mathcal{D}_{\text{lon}}^{(1)}\left(N_{z+}^{(2)} - N_{z-}^{(2)}\right)N_y^2\varepsilon_{xz}^{(1)}\varepsilon_{xz}^{(2)}\right]\left(N_{z+}^{(1)} - N_{z-}^{(1)}\right)\\
&\quad \times\left(\mathcal{W}_{+}^{(1)}\mathcal{U}_{+}^{(2)} + \mathcal{U}_{+}^{(1)}\mathcal{W}_{+}^{(2)} + \mathcal{T}_{+}^{(1)}\mathcal{P}_{+}^{(2)} + \mathcal{S}_{+}^{(1)}\mathcal{Q}_{+}^{(2)} + \mathcal{Q}_{+}^{(1)}\mathcal{S}_{+}^{(2)} + \mathcal{P}_{+}^{(1)}\mathcal{T}_{+}^{(2)}\right)
\end{aligned}
\tag{5.50c}
$$

$$
\begin{aligned}
&\left(F_{12}F_{34} - F_{14}F_{32}\right) \\
&= \left[\mathcal{D}_{\mathrm{lon}}^{(1)}\left(N_{z+}^{(2)} - N_{z-}^{(2)}\right)N_y^2\varepsilon_{xz}^{(1)}\varepsilon_{xz}^{(2)}\right]\left(N_{z+}^{(1)} - N_{z-}^{(1)}\right) \\
&\quad\times\left(-\mathcal{W}_+^{(1)}\mathcal{U}_+^{(2)} - \mathcal{U}_+^{(1)}\mathcal{W}_+^{(2)} + \mathcal{T}_+^{(1)}\mathcal{P}_+^{(2)} - \mathcal{S}_+^{(1)}\mathcal{Q}_+^{(2)} - \mathcal{Q}_+^{(1)}\mathcal{S}_+^{(2)} + \mathcal{P}_+^{(1)}\mathcal{T}_+^{(2)}\right)
\end{aligned}
\tag{5.50d}
$$

$$
\begin{aligned}
&\left(F_{11}F_{34} - F_{13}F_{32}\right) \\
&= \left[\mathcal{D}_{\mathrm{lon}}^{(1)}\left(N_{z+}^{(2)} - N_{z-}^{(2)}\right)N_y^2\varepsilon_{xz}^{(1)}\varepsilon_{xz}^{(2)}\right]\left(N_{z+}^{(1)} + N_{z-}^{(1)}\right) \\
&\quad\times\left(\mathcal{W}_-^{(1)}\mathcal{U}_+^{(2)} + \mathcal{U}_-^{(1)}\mathcal{W}_+^{(2)} - \mathcal{T}_-^{(1)}\mathcal{P}_+^{(2)} - \mathcal{S}_-^{(1)}\mathcal{Q}_+^{(2)} + \mathcal{Q}_-^{(1)}\mathcal{S}_+^{(2)} + \mathcal{P}_-^{(1)}\mathcal{T}_+^{(2)}\right)
\end{aligned}
\tag{5.50e}
$$

$$
\begin{aligned}
&\left(F_{12}F_{33} - F_{14}F_{31}\right) \\
&= \left[\mathcal{D}_{\mathrm{lon}}^{(1)}\left(N_{z+}^{(2)} - N_{z-}^{(2)}\right)N_y^2\varepsilon_{xz}^{(1)}\varepsilon_{xz}^{(2)}\right]\left(N_{z+}^{(1)} + N_{z-}^{(1)}\right) \\
&\quad\times\left(-\mathcal{W}_-^{(1)}\mathcal{U}_+^{(2)} - \mathcal{U}_-^{(1)}\mathcal{W}_+^{(2)} - \mathcal{T}_-^{(1)}\mathcal{P}_+^{(2)} + \mathcal{S}_-^{(1)}\mathcal{Q}_+^{(2)} - \mathcal{Q}_-^{(1)}\mathcal{S}_+^{(2)} + \mathcal{P}_-^{(1)}\mathcal{T}_+^{(2)}\right).
\end{aligned}
\tag{5.50f}
$$

The common factor, which may be dropped in the reflection studies, was put into the brackets. We determine the first two terms in Eq. (5.43a) to Eq. (5.43e) describing a delicate mode coupling at the interfaces. Skipping the common factor

$$
\left(2N^{(0)}\cos\varphi^{(0)}\right)^{-2}\mathcal{D}_{\mathrm{lon}}^{(1)}\left(N_{z+}^{(2)} - N_{z-}^{(2)}\right)N_y\varepsilon_{xz}^{(2)}
$$

on the right-hand side of these equations, we have

$$
\begin{aligned}
&\left(A_{11}A_{32} - A_{12}A_{31}\right)\left(F_{11}F_{14} - F_{12}F_{13}\right) - \left(A_{13}A_{34} - A_{14}A_{33}\right)\left(F_{32}F_{33} - F_{31}F_{34}\right) \\
&= 4\cos\varphi^{(0)} \\
&\quad\times\Big\{\left[N^{(0)2}D_{31}^{(1)}D_{33}^{(1)}\left(D_{11}^{(1)}D_{43}^{(1)} + D_{13}^{(1)}D_{41}^{(1)}\right) + D_{41}^{(1)}D_{43}^{(1)}\left(D_{21}^{(1)}D_{33}^{(1)} + D_{23}^{(1)}D_{31}^{(1)}\right)\right]\mathcal{U}_+^{(2)} \\
&\quad-\left[N^{(0)2}D_{11}^{(1)}D_{13}^{(1)}\left(D_{21}^{(1)}D_{33}^{(1)} + D_{23}^{(1)}D_{31}^{(1)}\right) + D_{21}^{(1)}D_{23}^{(1)}\left(D_{13}^{(1)}D_{41}^{(1)} + D_{11}^{(1)}D_{43}^{(1)}\right)\right]\mathcal{W}_+^{(2)} \\
&\quad+\left[N^{(0)2}\left(D_{11}^{(1)}D_{23}^{(1)}D_{31}^{(1)}D_{43}^{(1)} + D_{13}^{(1)}D_{21}^{(1)}D_{33}^{(1)}D_{41}^{(1)}\right) + 2D_{21}^{(1)}D_{23}^{(1)}D_{41}^{(1)}D_{43}^{(1)}\right]\mathcal{Q}_+^{(2)} \\
&\quad+\left[2N^{(0)2}D_{11}^{(1)}D_{13}^{(1)}D_{31}^{(1)}D_{33}^{(1)} + \left(D_{11}^{(1)}D_{23}^{(1)}D_{31}^{(1)}D_{43}^{(1)} + D_{13}^{(1)}D_{21}^{(1)}D_{33}^{(1)}D_{41}^{(1)}\right)\right]\mathcal{S}_+^{(2)}\Big\} \\
&= 4\cos\varphi^{(0)}N_y\left[\varepsilon_{xz}^{(1)2}N_y^2\right] \\
&\quad\times N_{z+}^{(1)}N_{z-}^{(1)}\varepsilon_{zz}^{(1)-1}\left(N^{(0)2} - \varepsilon_{yy}^{(1)}\right)\left[\left(\varepsilon_{zz}^{(1)} - N_y^2\right)\left(\varepsilon_{xx}^{(1)} - N_y^2\right) - \varepsilon_{zx}^{(1)}\varepsilon_{xz}^{(1)}\right] \\
&\quad\times\Big\{\left[\varepsilon_{zz}^{(1)}\left(\varepsilon_{xx}^{(1)} - N_y^2\right) - \varepsilon_{yy}^{(1)}\left(\varepsilon_{zz}^{(1)} - N_y^2\right) - \varepsilon_{xz}^{(1)}\varepsilon_{zx}^{(1)}\right]\left(\varepsilon_{zx}^{(1)}\mathcal{U}_+^{(2)} - \varepsilon_{xz}^{(1)}\mathcal{W}_+^{(2)}\right) \\
&\quad+2\varepsilon_{zx}^{(1)}\varepsilon_{xz}^{(1)}N_y\left(\varepsilon_{yy}^{(1)}\mathcal{Q}_+^{(2)} - \mathcal{S}_+^{(2)}\right)\Big\}
\end{aligned}
\tag{5.51a}
$$

$$
\begin{aligned}
&\left(A_{11}A_{31}-A_{12}A_{32}\right)\left(F_{11}F_{14}-F_{12}F_{13}\right)-\left(A_{13}A_{33}-A_{14}A_{34}\right)\left(F_{32}F_{33}-F_{31}F_{34}\right)\\
&=4\cos\varphi^{(0)}\\
&\quad\times\left\{\left[N^{(0)2}D_{31}^{(1)}D_{33}^{(1)}\left(D_{11}^{(1)}D_{43}^{(1)}++D_{13}^{(1)}D_{41}^{(1)}\right)-D_{41}^{(1)}D_{43}^{(1)}\left(D_{21}^{(1)}D_{33}^{(1)}+D_{23}^{(1)}D_{31}^{(1)}\right)\right]\mathcal{U}_{+}^{(2)}\right.\\
&\quad-\left[N^{(0)2}D_{11}^{(1)}D_{13}^{(1)}\left(D_{21}^{(1)}D_{33}^{(1)}+D_{23}^{(1)}D_{31}^{(1)}\right)-D_{21}^{(1)}D_{23}^{(1)}\left(D_{13}^{(1)}D_{41}^{(1)}++D_{11}^{(1)}D_{43}^{(1)}\right)\right]\mathcal{W}_{+}^{(2)}\\
&\quad+\left[N^{(0)2}\left(D_{11}^{(1)}D_{23}^{(1)}D_{31}^{(1)}D_{43}^{(1)}+D_{13}^{(1)}D_{21}^{(1)}D_{33}^{(1)}D_{41}^{(1)}\right)-2D_{21}^{(1)}D_{23}^{(1)}D_{41}^{(1)}D_{43}^{(1)}\right]\mathcal{Q}_{+}^{(2)}\\
&\quad\left.+\left[2N^{(0)2}D_{11}^{(1)}D_{13}^{(1)}D_{31}^{(1)}D_{33}^{(1)}-\left(D_{11}^{(1)}D_{23}^{(1)}D_{31}^{(1)}D_{43}^{(1)}+D_{13}^{(1)}D_{21}^{(1)}D_{33}^{(1)}D_{41}^{(1)}\right)\right]\mathcal{S}_{+}^{(2)}\right\}\\
&=4\cos\varphi^{(0)}N_y\left[\varepsilon_{xz}^{(1)2}N_y^2\right]\\
&\quad\times N_{z+}^{(1)}N_{z-}^{(1)}\varepsilon_{zz}^{(1)-1}\left(N^{(0)2}+\varepsilon_{yy}^{(1)}\right)\left[\left(\varepsilon_{zz}^{(1)}-N_y^2\right)\left(\varepsilon_{xx}^{(1)}-N_y^2\right)-\varepsilon_{zx}^{(1)}\varepsilon_{xz}^{(1)}\right]\\
&\quad\times\left\{\left[\varepsilon_{zz}^{(1)}\left(\varepsilon_{xx}^{(1)}-N_y^2\right)-\varepsilon_{yy}^{(1)}\left(\varepsilon_{zz}^{(1)}-N_y^2\right)-\varepsilon_{xz}^{(1)}\varepsilon_{zx}^{(1)}\right]\left(\varepsilon_{zx}^{(1)}\mathcal{U}_{+}^{(2)}-\varepsilon_{xz}^{(1)}\mathcal{W}_{+}^{(2)}\right)\right.\\
&\quad\left.+2\varepsilon_{zx}^{(1)}\varepsilon_{xz}^{(1)}N_y\left(\varepsilon_{yy}^{(1)}\mathcal{Q}_{+}^{(2)}-\mathcal{S}_{+}^{(2)}\right)\right\}
\end{aligned}
\tag{5.51b}
$$

$$
\begin{aligned}
&\left(A_{31}^2-A_{32}^2\right)\left(F_{12}F_{13}-F_{11}F_{14}\right)+\left(A_{33}^2-A_{34}^2\right)\left(F_{32}F_{33}-F_{31}F_{34}\right)\\
&=8N^{(0)}\cos\varphi^{(0)}\\
&\quad\times\left[2D_{31}^{(1)}D_{33}^{(1)}D_{41}^{(1)}D_{43}^{(1)}\mathcal{U}_{+}^{(2)}-\left(D_{31}^{(1)}D_{41}^{(1)}D_{13}^{(1)}D_{23}^{(1)}+D_{33}^{(1)}D_{43}^{(1)}D_{11}^{(1)}D_{21}^{(1)}\right)\mathcal{W}_{+}^{(2)}\right.\\
&\quad\left.+D_{41}^{(1)}D_{43}^{(1)}\left(D_{21}^{(1)}D_{33}^{(1)}+D_{23}^{(1)}D_{31}^{(1)}\right)\mathcal{Q}_{+}^{(2)}+D_{31}^{(1)}D_{33}^{(1)}\left(D_{11}^{(1)}D_{43}^{(1)}+D_{13}^{(1)}D_{41}^{(1)}\right)\mathcal{S}_{+}^{(2)}\right]\\
&=8N^{(0)}\cos\varphi^{(0)}\left[\varepsilon_{xz}^{(1)2}N_y^2\right]\\
&\quad\times N_{z+}^{(1)}N_{z-}^{(1)}\varepsilon_{zz}^{(1)-1}\left[\left(\varepsilon_{zz}^{(1)}-N_y^2\right)\left(\varepsilon_{xx}^{(1)}-N_y^2\right)-\varepsilon_{zx}^{(1)}\varepsilon_{xz}^{(1)}\right]\\
&\quad\times\left\{2N_y^2\varepsilon_{yy}^{(1)}\varepsilon_{zx}^{(1)2}\mathcal{U}_{+}^{(2)}+2N_y^2\varepsilon_{yy}^{(1)}\varepsilon_{zx}^{(1)}\varepsilon_{xz}^{(1)}\mathcal{W}_{+}^{(2)}\right.\\
&\quad+\left[\varepsilon_{zz}^{(1)}\left(\varepsilon_{xx}^{(1)}-N_y^2\right)-\varepsilon_{yy}^{(1)}\left(\varepsilon_{zz}^{(1)}-N_y^2\right)-\varepsilon_{xz}^{(1)}\varepsilon_{zx}^{(1)}\right]^2\mathcal{W}_{+}^{(2)}\\
&\quad-\varepsilon_{zx}^{(1)}\varepsilon_{yy}^{(1)}N_y\left[\varepsilon_{zz}^{(1)}\left(\varepsilon_{xx}^{(1)}-N_y^2\right)-\varepsilon_{yy}^{(1)}\left(\varepsilon_{zz}^{(1)}-N_y^2\right)-\varepsilon_{xz}^{(1)}\varepsilon_{zx}^{(1)}\right]\mathcal{Q}_{+}^{(2)}\\
&\quad\left.+\varepsilon_{zx}^{(1)}N_y\left[\varepsilon_{zz}^{(1)}\left(\varepsilon_{xx}^{(1)}-N_y^2\right)-\varepsilon_{yy}^{(1)}\left(\varepsilon_{zz}^{(1)}-N_y^2\right)-\varepsilon_{xz}^{(1)}\varepsilon_{zx}^{(1)}\right]\mathcal{S}_{+}^{(2)}\right\}
\end{aligned}
\tag{5.51c}
$$

$$
\begin{aligned}
&-\left(A_{11}^2-A_{12}^2\right)\left(F_{11}F_{14}-F_{12}F_{13}\right)-\left(A_{13}^2-A_{14}^2\right)\left(F_{31}F_{34}-F_{32}F_{33}\right)\\
&=-4N^{(0)}\cos\varphi^{(0)}D_{11}^{(1)}D_{21}^{(1)}\\
&\quad\times\left(2D_{33}^{(1)}D_{43}^{(1)}\mathcal{U}_{+}^{(2)}+2D_{23}^{(1)}D_{43}^{(1)}\mathcal{Q}_{+}^{(2)}+2D_{13}^{(1)}D_{33}^{(1)}\mathcal{S}_{+}^{(2)}-2D_{13}^{(1)}D_{23}^{(1)}\mathcal{W}_{+}^{(2)}\right)\\
&\quad+4N^{(0)}\cos\varphi^{(0)}D_{13}^{(1)}D_{23}^{(1)}\varepsilon_{xz}^{(2)}N_y\left(N_{z+}^{(2)}-N_{z-}^{(2)}\right)\\
&\quad\times\left(-2D_{31}^{(1)}D_{41}^{(1)}\mathcal{U}_{+}^{(2)}-2D_{21}^{(1)}D_{41}^{(1)}\mathcal{Q}_{+}^{(2)}-2D_{11}^{(1)}D_{31}^{(1)}\mathcal{S}_{+}^{(2)}+2D_{11}^{(1)}D_{21}^{(1)}\mathcal{W}_{+}^{(2)}\right)\\
&=-8N^{(0)}\cos\varphi^{(0)}\left[\varepsilon_{xz}^{(1)2}N_y^2\right]\\
&\quad\times\left[\left(D_{11}^{(1)}D_{21}^{(1)}D_{33}^{(1)}D_{43}^{(1)}+D_{13}^{(1)}D_{23}^{(1)}D_{31}^{(1)}D_{41}^{(1)}\right)\mathcal{U}_{+}^{(2)}-2D_{11}^{(1)}D_{13}^{(1)}D_{21}^{(1)}D_{23}^{(1)}\mathcal{W}_{+}^{(2)}\right.\\
&\quad\left.+D_{21}^{(1)}D_{23}^{(1)}\left(D_{11}^{(1)}D_{43}^{(1)}+D_{13}^{(1)}D_{41}^{(1)}\right)\mathcal{Q}_{+}^{(2)}+D_{11}^{(1)}D_{13}^{(1)}\left(D_{21}^{(1)}D_{33}^{(1)}+D_{23}^{(1)}D_{31}^{(1)}\right)\mathcal{S}_{+}^{(2)}\right]\\
&=8N^{(0)}\cos\varphi^{(0)}N_{z+}^{(1)}N_{z-}^{(1)}\left[\varepsilon_{xz}^{(1)2}N_y^2\right]\varepsilon_{zz}^{(1)-1}\left[\left(\varepsilon_{zz}^{(1)}-N_y^2\right)\left(\varepsilon_{xx}^{(1)}-N_y^2\right)-\varepsilon_{zx}^{(1)}\varepsilon_{xz}^{(1)}\right]\\
&\quad\times\left\{2N_y^2\varepsilon_{yy}^{(1)}\varepsilon_{zx}^{(1)}\varepsilon_{xz}^{(1)}\mathcal{U}_{+}^{(2)}+2N_y^2\varepsilon_{yy}^{(1)}\varepsilon_{xz}^{(1)2}\mathcal{W}_{+}^{(2)}\right.
\end{aligned}
$$

$$+\left[\varepsilon_{zz}^{(1)}\left(\varepsilon_{xx}^{(1)}-N_y^2\right)-\varepsilon_{yy}^{(1)}\left(\varepsilon_{zz}^{(1)}-N_y^2\right)-\varepsilon_{xz}^{(1)}\varepsilon_{zx}^{(1)}\right]^2\mathcal{U}_+^{(2)}$$
$$+\varepsilon_{xz}^{(1)}\varepsilon_{yy}^{(1)}N_y\left[\varepsilon_{zz}^{(1)}\left(\varepsilon_{xx}^{(1)}-N_y^2\right)-\varepsilon_{yy}^{(1)}\left(\varepsilon_{zz}^{(1)}-N_y^2\right)-\varepsilon_{xz}^{(1)}\varepsilon_{zx}^{(1)}\right]\mathcal{Q}_+^{(2)}$$
$$-\varepsilon_{xz}^{(1)}N_y\left[\varepsilon_{zz}^{(1)}\left(\varepsilon_{xx}^{(1)}-N_y^2\right)-\varepsilon_{yy}^{(1)}\left(\varepsilon_{zz}^{(1)}-N_y^2\right)-\varepsilon_{xz}^{(1)}\varepsilon_{zx}^{(1)}\right]\mathcal{S}_+^{(2)}\}. \qquad (5.51d)$$

We finally establish the expressions defining the Cartesian Jones reflection matrix listed in Eq. (5.43). The leading terms have just been determined in Eq. (5.51). In the terms formed by the product of the $\mathbf{A}^{(1)}$ and $\mathbf{F}^{(2)}$ matrix elements in Eqs. (5.46a) to (5.46h) and Eqs. (5.50), we have already removed the irrelevant product of the common factors

$$\left(2N^{(0)}\cos\varphi^{(0)}\right)^{-2}N_y^3\varepsilon_{xz}^{(1)2}\mathcal{D}_{\text{lon}}^{(1)}\varepsilon_{xz}^{(2)}\left(N_{z+}^{(2)}-N_{z-}^{(2)}\right).$$

To adjust the same common factor in the leading terms of Eq. (5.51), it therefore remains to remove the common factor $\left[\varepsilon_{xz}^{(1)2}N_y^2\right]$ in these equations. We arrive at our final results

$$\begin{aligned}
&\mathrm{M}_{11}\mathrm{M}_{33}-\mathrm{M}_{13}\mathrm{M}_{31}\\
&=4\cos\varphi^{(0)}N_yN_{z+}^{(1)}N_{z-}^{(1)}\varepsilon_{zz}^{(1)-1}\left(N^{(0)2}-\varepsilon_{yy}^{(1)}\right)\left[\left(\varepsilon_{zz}^{(1)}-N_y^2\right)\left(\varepsilon_{xx}^{(1)}-N_y^2\right)-\varepsilon_{zx}^{(1)}\varepsilon_{xz}^{(1)}\right]\\
&\quad\times\left\{\left[\varepsilon_{zz}^{(1)}\left(\varepsilon_{xx}^{(1)}-N_y^2\right)-\varepsilon_{yy}^{(1)}\left(\varepsilon_{zz}^{(1)}-N_y^2\right)-\varepsilon_{xz}^{(1)}\varepsilon_{zx}^{(1)}\right]\left(\varepsilon_{zx}^{(1)}\mathcal{U}_+^{(2)}-\varepsilon_{xz}^{(1)}\mathcal{W}_+^{(2)}\right)\right.\\
&\quad\left.+2\varepsilon_{zx}^{(1)}\varepsilon_{xz}^{(1)}N_y\left(\varepsilon_{yy}^{(1)}\mathcal{Q}_+^{(2)}-\mathcal{S}_+^{(2)}\right)\right\}\\
&\quad+e^{j(\beta_++\beta_-)}\left(N_{z+}^{(1)}-N_{z-}^{(1)}\right)^2\left(\mathcal{T}^{(0)}\mathcal{P}_+^{(1)}+\mathcal{S}^{(0)}\mathcal{Q}_+^{(1)}+\mathcal{Q}^{(0)}\mathcal{S}_+^{(1)}+\mathcal{P}^{(0)}\mathcal{T}_+^{(1)}\right)\\
&\quad\times\left(\mathcal{W}_+^{(1)}\mathcal{U}_+^{(2)}+\mathcal{U}_+^{(1)}\mathcal{W}_+^{(2)}+\mathcal{T}_+^{(1)}\mathcal{P}_+^{(2)}+\mathcal{S}_+^{(1)}\mathcal{Q}_+^{(2)}+\mathcal{Q}_+^{(1)}\mathcal{S}_+^{(2)}+\mathcal{P}_+^{(1)}\mathcal{T}_+^{(2)}\right)\\
&\quad+e^{-j(\beta_++\beta_-)}\left(N_{z+}^{(1)}-N_{z-}^{(1)}\right)^2\left(\mathcal{T}^{(0)}\mathcal{P}_+^{(1)}-\mathcal{S}^{(0)}\mathcal{Q}_+^{(1)}-\mathcal{Q}^{(0)}\mathcal{S}_+^{(1)}+\mathcal{P}^{(0)}\mathcal{T}_+^{(1)}\right)\\
&\quad\times\left(-\mathcal{W}_+^{(1)}\mathcal{U}_+^{(2)}-\mathcal{U}_+^{(1)}\mathcal{W}_+^{(2)}+\mathcal{T}_+^{(1)}\mathcal{P}_+^{(2)}-\mathcal{S}_+^{(1)}\mathcal{Q}_+^{(2)}-\mathcal{Q}_+^{(1)}\mathcal{S}_+^{(2)}+\mathcal{P}_+^{(1)}\mathcal{T}_+^{(2)}\right)\\
&\quad+e^{j(\beta_+-\beta_-)}\left(N_{z+}^{(1)}+N_{z-}^{(1)}\right)^2\left(-\mathcal{T}^{(0)}\mathcal{P}_-^{(1)}-\mathcal{S}^{(0)}\mathcal{Q}_-^{(1)}+\mathcal{Q}^{(0)}\mathcal{S}_-^{(1)}+\mathcal{P}^{(0)}\mathcal{T}_-^{(1)}\right)\\
&\quad\times\left(\mathcal{W}_-^{(1)}\mathcal{U}_+^{(2)}+\mathcal{U}_-^{(1)}\mathcal{W}_+^{(2)}-\mathcal{T}_-^{(1)}\mathcal{P}_+^{(2)}-\mathcal{S}_-^{(1)}\mathcal{Q}_+^{(2)}+\mathcal{Q}_-^{(1)}\mathcal{S}_+^{(2)}+\mathcal{P}_-^{(1)}\mathcal{T}_+^{(2)}\right)\\
&\quad+e^{-j(\beta_+-\beta_-)}\left(N_{z+}^{(1)}+N_{z-}^{(1)}\right)^2\left(-\mathcal{T}^{(0)}\mathcal{P}_-^{(1)}+\mathcal{S}^{(0)}\mathcal{Q}_-^{(1)}-\mathcal{Q}^{(0)}\mathcal{S}_-^{(1)}+\mathcal{P}^{(0)}\mathcal{T}_-^{(1)}\right)\\
&\quad\times\left(-\mathcal{W}_-^{(1)}\mathcal{U}_+^{(2)}-\mathcal{U}_-^{(1)}\mathcal{W}_+^{(2)}-\mathcal{T}_-^{(1)}\mathcal{P}_+^{(2)}+\mathcal{S}_-^{(1)}\mathcal{Q}_+^{(2)}-\mathcal{Q}_-^{(1)}\mathcal{S}_+^{(2)}+\mathcal{P}_-^{(1)}\mathcal{T}_+^{(2)}\right)
\end{aligned} \qquad (5.52a)$$

$$\begin{aligned}
&\mathrm{M}_{21}\mathrm{M}_{33}-\mathrm{M}_{23}\mathrm{M}_{31}\\
&=4\cos\varphi^{(0)}N_yN_{z+}^{(1)}N_{z-}^{(1)}\varepsilon_{zz}^{(1)-1}\left(N^{(0)2}+\varepsilon_{yy}^{(1)}\right)\left[\left(\varepsilon_{zz}^{(1)}-N_y^2\right)\left(\varepsilon_{xx}^{(1)}-N_y^2\right)-\varepsilon_{zx}^{(1)}\varepsilon_{xz}^{(1)}\right]\\
&\quad\times\left\{\left[\varepsilon_{zz}^{(1)}\left(\varepsilon_{xx}^{(1)}-N_y^2\right)-\varepsilon_{yy}^{(1)}\left(\varepsilon_{zz}^{(1)}-N_y^2\right)-\varepsilon_{xz}^{(1)}\varepsilon_{zx}^{(1)}\right]\left(\varepsilon_{zx}^{(1)}\mathcal{U}_+^{(2)}-\varepsilon_{xz}^{(1)}\mathcal{W}_+^{(2)}\right)\right.\\
&\quad\left.+2\varepsilon_{zx}^{(1)}\varepsilon_{xz}^{(1)}N_y\left(\varepsilon_{yy}^{(1)}\mathcal{Q}_+^{(2)}-\mathcal{S}_+^{(2)}\right)\right\}\\
&\quad+e^{j(\beta_++\beta_-)}\left(N_{z+}^{(1)}-N_{z-}^{(1)}\right)^2\left(\mathcal{T}^{(0)}\mathcal{P}_+^{(1)}+\mathcal{S}^{(0)}\mathcal{Q}_+^{(1)}-\mathcal{Q}^{(0)}\mathcal{S}_+^{(1)}-\mathcal{P}^{(0)}\mathcal{T}_+^{(1)}\right)
\end{aligned}$$

$$\times\left(\mathcal{W}_+^{(1)}\mathcal{U}_+^{(2)}+\mathcal{U}_+^{(1)}\mathcal{W}_+^{(2)}+\mathcal{T}_+^{(1)}\mathcal{P}_+^{(2)}+\mathcal{S}_+^{(1)}\mathcal{Q}_+^{(2)}+\mathcal{Q}_+^{(1)}\mathcal{S}_+^{(2)}+\mathcal{P}_+^{(1)}\mathcal{T}_+^{(2)}\right)$$
$$+\mathrm{e}^{-\mathrm{j}(\beta_++\beta_-)}\left(N_{z+}^{(1)}-N_{z-}^{(1)}\right)^2\left(\mathcal{T}^{(0)}\mathcal{P}_+^{(1)}-\mathcal{S}^{(0)}\mathcal{Q}_+^{(1)}+\mathcal{Q}^{(0)}\mathcal{S}_+^{(1)}-\mathcal{P}^{(0)}\mathcal{T}_+^{(1)}\right)$$
$$\times\left(-\mathcal{W}_+^{(1)}\mathcal{U}_+^{(2)}-\mathcal{U}_+^{(1)}\mathcal{W}_+^{(2)}+\mathcal{T}_+^{(1)}\mathcal{P}_+^{(2)}-\mathcal{S}_+^{(1)}\mathcal{Q}_+^{(2)}-\mathcal{Q}_+^{(1)}\mathcal{S}_+^{(2)}+\mathcal{P}_+^{(1)}\mathcal{T}_+^{(2)}\right)$$
$$-\mathrm{e}^{\mathrm{j}(\beta_+-\beta_-)}\left(N_{z+}^{(1)}+N_{z-}^{(1)}\right)^2\left(\mathcal{T}^{(0)}\mathcal{P}_-^{(1)}+\mathcal{S}^{(0)}\mathcal{Q}_-^{(1)}+\mathcal{Q}^{(0)}\mathcal{S}_-^{(1)}+\mathcal{P}^{(0)}\mathcal{T}_-^{(1)}\right)$$
$$\times\left(\mathcal{W}_-^{(1)}\mathcal{U}_+^{(2)}+\mathcal{U}_-^{(1)}\mathcal{W}_+^{(2)}-\mathcal{T}_-^{(1)}\mathcal{P}_+^{(2)}-\mathcal{S}_-^{(1)}\mathcal{Q}_+^{(2)}+\mathcal{Q}_-^{(1)}\mathcal{S}_+^{(2)}+\mathcal{P}_-^{(1)}\mathcal{T}_+^{(2)}\right)$$
$$+\mathrm{e}^{-\mathrm{j}(\beta_+-\beta_-)}\left(N_{z+}^{(1)}+N_{z-}^{(1)}\right)^2\left(-\mathcal{T}^{(0)}\mathcal{P}_-^{(1)}+\mathcal{S}^{(0)}\mathcal{Q}_-^{(1)}+\mathcal{Q}^{(0)}\mathcal{S}_-^{(1)}-\mathcal{P}^{(0)}\mathcal{T}_-^{(1)}\right)$$
$$\times\left(-\mathcal{W}_-^{(1)}\mathcal{U}_+^{(2)}-\mathcal{U}_-^{(1)}\mathcal{W}_+^{(2)}-\mathcal{T}_-^{(1)}\mathcal{P}_+^{(2)}+\mathcal{S}_-^{(1)}\mathcal{Q}_+^{(2)}-\mathcal{Q}_-^{(1)}\mathcal{S}_+^{(2)}+\mathcal{P}_-^{(1)}\mathcal{T}_+^{(2)}\right)\tag{5.52b}$$

$$\mathrm{M}_{11}\mathrm{M}_{43}-\mathrm{M}_{13}\mathrm{M}_{41}$$
$$=4\cos\varphi^{(0)}N_yN_{z+}^{(1)}N_{z-}^{(1)}\varepsilon_{zz}^{(1)-1}\left(N^{(0)2}+\varepsilon_{yy}^{(1)}\right)\left[\left(\varepsilon_{zz}^{(1)}-N_y^2\right)\left(\varepsilon_{xx}^{(1)}-N_y^2\right)-\varepsilon_{zx}^{(1)}\varepsilon_{xz}^{(1)}\right]$$
$$\times\left\{\left[\varepsilon_{zz}^{(1)}\left(\varepsilon_{xx}^{(1)}-N_y^2\right)-\varepsilon_{yy}^{(1)}\left(\varepsilon_{zz}^{(1)}-N_y^2\right)-\varepsilon_{xz}^{(1)}\varepsilon_{zx}^{(1)}\right]\left(\varepsilon_{zx}^{(1)}\mathcal{U}_+^{(2)}-\varepsilon_{xz}^{(1)}\mathcal{W}_+^{(2)}\right)\right.$$
$$\left.+2\varepsilon_{zx}^{(1)}\varepsilon_{xz}^{(1)}N_y\left(\varepsilon_{yy}^{(1)}\mathcal{Q}_+^{(2)}-\mathcal{S}_+^{(2)}\right)\right\}$$
$$+\mathrm{e}^{\mathrm{j}(\beta_++\beta_-)}\left(N_{z+}^{(1)}-N_{z-}^{(1)}\right)^2\left(-\mathcal{T}^{(0)}\mathcal{P}_+^{(1)}+\mathcal{S}^{(0)}\mathcal{Q}_+^{(1)}-\mathcal{Q}^{(0)}\mathcal{S}_+^{(1)}+\mathcal{P}^{(0)}\mathcal{T}_+^{(1)}\right)$$
$$\times\left(\mathcal{W}_+^{(1)}\mathcal{U}_+^{(2)}+\mathcal{U}_+^{(1)}\mathcal{W}_+^{(2)}+\mathcal{T}_+^{(1)}\mathcal{P}_+^{(2)}+\mathcal{S}_+^{(1)}\mathcal{Q}_+^{(2)}+\mathcal{Q}_+^{(1)}\mathcal{S}_+^{(2)}+\mathcal{P}_+^{(1)}\mathcal{T}_+^{(2)}\right)$$
$$+\mathrm{e}^{-\mathrm{j}(\beta_++\beta_-)}\left(N_{z+}^{(1)}-N_{z-}^{(1)}\right)^2\left(-\mathcal{T}^{(0)}\mathcal{P}_+^{(1)}-\mathcal{S}^{(0)}\mathcal{Q}_+^{(1)}+\mathcal{Q}^{(0)}\mathcal{S}_+^{(1)}+\mathcal{P}^{(0)}\mathcal{T}_+^{(1)}\right)$$
$$\times\left(-\mathcal{W}_+^{(1)}\mathcal{U}_+^{(2)}-\mathcal{U}_+^{(1)}\mathcal{W}_+^{(2)}+\mathcal{T}_+^{(1)}\mathcal{P}_+^{(2)}-\mathcal{S}_+^{(1)}\mathcal{Q}_+^{(2)}-\mathcal{Q}_+^{(1)}\mathcal{S}_+^{(2)}+\mathcal{P}_+^{(1)}\mathcal{T}_+^{(2)}\right)$$
$$+\mathrm{e}^{\mathrm{j}(\beta_+-\beta_-)}\left(N_{z+}^{(1)}+N_{z-}^{(1)}\right)^2\left(\mathcal{T}^{(0)}\mathcal{P}_-^{(1)}-\mathcal{S}^{(0)}\mathcal{Q}_-^{(1)}-\mathcal{Q}^{(0)}\mathcal{S}_-^{(1)}+\mathcal{P}^{(0)}\mathcal{T}_-^{(1)}\right)$$
$$\times\left(\mathcal{W}_-^{(1)}\mathcal{U}_+^{(2)}+\mathcal{U}_-^{(1)}\mathcal{W}_+^{(2)}-\mathcal{T}_-^{(1)}\mathcal{P}_+^{(2)}-\mathcal{S}_-^{(1)}\mathcal{Q}_+^{(2)}+\mathcal{Q}_-^{(1)}\mathcal{S}_+^{(2)}+\mathcal{P}_-^{(1)}\mathcal{T}_+^{(2)}\right)$$
$$+\mathrm{e}^{-\mathrm{j}(\beta_+-\beta_-)}\left(N_{z+}^{(1)}+N_{z-}^{(1)}\right)^2\left(\mathcal{T}^{(0)}\mathcal{P}_-^{(1)}+\mathcal{S}^{(0)}\mathcal{Q}_-^{(1)}+\mathcal{Q}^{(0)}\mathcal{S}_-^{(1)}+\mathcal{P}^{(0)}\mathcal{T}_-^{(1)}\right)$$
$$\times\left(-\mathcal{W}_-^{(1)}\mathcal{U}_+^{(2)}-\mathcal{U}_-^{(1)}\mathcal{W}_+^{(2)}-\mathcal{T}_-^{(1)}\mathcal{P}_+^{(2)}+\mathcal{S}_-^{(1)}\mathcal{Q}_+^{(2)}-\mathcal{Q}_-^{(1)}\mathcal{S}_+^{(2)}+\mathcal{P}_-^{(1)}\mathcal{T}_+^{(2)}\right)\tag{5.52c}$$

$$\mathrm{M}_{41}\mathrm{M}_{33}-\mathrm{M}_{43}\mathrm{M}_{31}$$
$$=8N^{(0)}\cos\varphi^{(0)}N_{z+}^{(1)}N_{z-}^{(1)}\varepsilon_{zz}^{(1)-1}\left[\left(\varepsilon_{zz}^{(1)}-N_y^2\right)\left(\varepsilon_{xx}^{(1)}-N_y^2\right)-\varepsilon_{zx}^{(1)}\varepsilon_{xz}^{(1)}\right]$$
$$\times\left\{2N_y^2\varepsilon_{yy}^{(1)}\varepsilon_{zx}^{(1)}\left(\varepsilon_{zx}^{(1)}\mathcal{U}_+^{(2)}+\varepsilon_{xz}^{(1)}\mathcal{W}_+^{(2)}\right)\right.$$
$$+\left[\varepsilon_{zz}^{(1)}\left(\varepsilon_{xx}^{(1)}-N_y^2\right)-\varepsilon_{yy}^{(1)}\left(\varepsilon_{zz}^{(1)}-N_y^2\right)-\varepsilon_{xz}^{(1)}\varepsilon_{zx}^{(1)}\right]^2\mathcal{W}_+^{(2)}$$
$$\left.-\varepsilon_{zx}^{(1)}N_y\left[\varepsilon_{zz}^{(1)}\left(\varepsilon_{xx}^{(1)}-N_y^2\right)-\varepsilon_{yy}^{(1)}\left(\varepsilon_{zz}^{(1)}-N_y^2\right)-\varepsilon_{xz}^{(1)}\varepsilon_{zx}^{(1)}\right]\left(\varepsilon_{yy}^{(1)}\mathcal{Q}_+^{(2)}-\mathcal{S}_+^{(2)}\right)\right\}$$
$$+2N^{(0)}\cos\varphi^{(0)}\mathrm{e}^{\mathrm{j}(\beta_++\beta_-)}\left(N_{z+}^{(1)}-N_{z-}^{(1)}\right)^2\mathcal{W}_+^{(1)}$$
$$\times\left(\mathcal{W}_+^{(1)}\mathcal{U}_+^{(2)}+\mathcal{U}_+^{(1)}\mathcal{W}_+^{(2)}+\mathcal{T}_+^{(1)}\mathcal{P}_+^{(2)}+\mathcal{S}_+^{(1)}\mathcal{Q}_+^{(2)}+\mathcal{Q}_+^{(1)}\mathcal{S}_+^{(2)}+\mathcal{P}_+^{(1)}\mathcal{T}_+^{(2)}\right)$$
$$+2N^{(0)}\cos\varphi^{(0)}\mathrm{e}^{-\mathrm{j}(\beta_++\beta_-)}\left(N_{z+}^{(1)}-N_{z-}^{(1)}\right)^2\mathcal{W}_+^{(1)}$$
$$\times\left(-\mathcal{W}_+^{(1)}\mathcal{U}_+^{(2)}-\mathcal{U}_+^{(1)}\mathcal{W}_+^{(2)}+\mathcal{T}_+^{(1)}\mathcal{P}_+^{(2)}-\mathcal{S}_+^{(1)}\mathcal{Q}_+^{(2)}-\mathcal{Q}_+^{(1)}\mathcal{S}_+^{(2)}+\mathcal{P}_+^{(1)}\mathcal{T}_+^{(2)}\right)$$

$$
\begin{aligned}
&- 2N^{(0)} \cos\varphi^{(0)} \mathrm{e}^{\mathrm{j}(\beta_+ - \beta_-)} \left(N_{z+}^{(1)} + N_{z-}^{(1)}\right)^2 \mathcal{W}_-^{(1)} \\
&\times \left(\mathcal{W}_-^{(1)}\mathcal{U}_+^{(2)} + \mathcal{U}_-^{(1)}\mathcal{W}_+^{(2)} - \mathcal{T}_-^{(1)}\mathcal{P}_+^{(2)} - \mathcal{S}_-^{(1)}\mathcal{Q}_+^{(2)} + \mathcal{Q}_-^{(1)}\mathcal{S}_+^{(2)} + \mathcal{P}_-^{(1)}\mathcal{T}_+^{(2)}\right) \\
&+ 2N^{(0)} \cos\varphi^{(0)} \mathrm{e}^{-\mathrm{j}(\beta_+ - \beta_-)} \left(N_{z+}^{(1)} + N_{z-}^{(1)}\right)^2 \mathcal{W}_-^{(1)} \\
&\times \left(- \mathcal{W}_-^{(1)}\mathcal{U}_+^{(2)} - \mathcal{U}_-^{(1)}\mathcal{W}_+^{(2)} - \mathcal{T}_-^{(1)}\mathcal{P}_+^{(2)} + \mathcal{S}_-^{(1)}\mathcal{Q}_+^{(2)} - \mathcal{Q}_-^{(1)}\mathcal{S}_+^{(2)} + \mathcal{P}_-^{(1)}\mathcal{T}_+^{(2)}\right)
\end{aligned} \tag{5.52d}
$$

$$
\begin{aligned}
&\mathrm{M}_{11}\mathrm{M}_{23} - \mathrm{M}_{13}\mathrm{M}_{21} \\
&= 8N^{(0)} \cos\varphi^{(0)} N_{z+}^{(1)} N_{z-}^{(1)} \varepsilon_{zz}^{(1)-1} \left[\left(\varepsilon_{zz}^{(1)} - N_y^2\right)\left(\varepsilon_{xx}^{(1)} - N_y^2\right) - \varepsilon_{zx}^{(1)}\varepsilon_{xz}^{(1)}\right] \\
&\quad \times \left\{2N_y^2 \varepsilon_{yy}^{(1)} \varepsilon_{zx}^{(1)} \left(\varepsilon_{zx}^{(1)} \mathcal{U}_+^{(2)} + \varepsilon_{xz}^{(1)} \mathcal{W}_+^{(2)}\right)\right. \\
&\quad + \left[\varepsilon_{zz}^{(1)}\left(\varepsilon_{xx}^{(1)} - N_y^2\right) - \varepsilon_{yy}^{(1)}\left(\varepsilon_{zz}^{(1)} - N_y^2\right) - \varepsilon_{xz}^{(1)}\varepsilon_{zx}^{(1)}\right]^2 \mathcal{U}_+^{(2)} \\
&\quad \left. + \varepsilon_{xz}^{(1)} N_y \left[\varepsilon_{zz}^{(1)}\left(\varepsilon_{xx}^{(1)} - N_y^2\right) - \varepsilon_{yy}^{(1)}\left(\varepsilon_{zz}^{(1)} - N_y^2\right) - \varepsilon_{xz}^{(1)}\varepsilon_{zx}^{(1)}\right]\left(\varepsilon_{yy}^{(1)} \mathcal{Q}_+^{(2)} - \mathcal{S}_+^{(2)}\right)\right\} \\
&\quad + 2N^{(0)} \cos\varphi^{(0)} \mathrm{e}^{\mathrm{j}(\beta_+ + \beta_-)} \left(N_{z+}^{(1)} - N_{z-}^{(1)}\right)^2 \mathcal{U}_+^{(1)} \\
&\quad \times \left(\mathcal{W}_+^{(1)}\mathcal{U}_+^{(2)} + \mathcal{U}_+^{(1)}\mathcal{W}_+^{(2)} + \mathcal{T}_+^{(1)}\mathcal{P}_+^{(2)} + \mathcal{S}_+^{(1)}\mathcal{Q}_+^{(2)} + \mathcal{Q}_+^{(1)}\mathcal{S}_+^{(2)} + \mathcal{P}_+^{(1)}\mathcal{T}_+^{(2)}\right) \\
&\quad - 2N^{(0)} \cos\varphi^{(0)} \mathrm{e}^{-\mathrm{j}(\beta_+ + \beta_-)} \left(N_{z+}^{(1)} - N_{z-}^{(1)}\right)^2 \mathcal{U}_+^{(1)} \\
&\quad \times \left(- \mathcal{W}_+^{(1)}\mathcal{U}_+^{(2)} - \mathcal{U}_+^{(1)}\mathcal{W}_+^{(2)} + \mathcal{T}_+^{(1)}\mathcal{P}_+^{(2)} - \mathcal{S}_+^{(1)}\mathcal{Q}_+^{(2)} - \mathcal{Q}_+^{(1)}\mathcal{S}_+^{(2)} + \mathcal{P}_+^{(1)}\mathcal{T}_+^{(2)}\right) \\
&\quad - 2N^{(0)} \cos\varphi^{(0)} \mathrm{e}^{\mathrm{j}(\beta_+ - \beta_-)} \left(N_{z+}^{(1)} + N_{z-}^{(1)}\right)^2 \mathcal{U}_-^{(1)} \\
&\quad \times \left(\mathcal{W}_-^{(1)}\mathcal{U}_+^{(2)} + \mathcal{U}_-^{(1)}\mathcal{W}_+^{(2)} - \mathcal{T}_-^{(1)}\mathcal{P}_+^{(2)} - \mathcal{S}_-^{(1)}\mathcal{Q}_+^{(2)} + \mathcal{Q}_-^{(1)}\mathcal{S}_+^{(2)} + \mathcal{P}_-^{(1)}\mathcal{T}_+^{(2)}\right) \\
&\quad + 2N^{(0)} \cos\varphi^{(0)} \mathrm{e}^{-\mathrm{j}(\beta_+ - \beta_-)} \left(N_{z+}^{(1)} + N_{z-}^{(1)}\right)^2 \mathcal{U}_-^{(1)} \\
&\quad \times \left(- \mathcal{W}_-^{(1)}\mathcal{U}_+^{(2)} - \mathcal{U}_-^{(1)}\mathcal{W}_+^{(2)} - \mathcal{T}_-^{(1)}\mathcal{P}_+^{(2)} + \mathcal{S}_-^{(1)}\mathcal{Q}_+^{(2)} - \mathcal{Q}_-^{(1)}\mathcal{S}_+^{(2)} + \mathcal{P}_-^{(1)}\mathcal{T}_+^{(2)}\right).
\end{aligned} \tag{5.52e}
$$

This concludes our analysis of the reflection at an arbitrary angle of incidence in a film–substrate system with the media characterized by the permittivity tensor of Eq. (5.1). So far, no restrictions were made on the magnitude of the tensor element. This makes the results rather general. In the following, we shall consider several special cases. Figure 5.2 and Figure 5.3 display respectively the effect of the magnetic layer thickness on the longitudinal off-diagonal reflection matrix element and the longitudinal MO Kerr effect in a structure consisting of a Fe layer deposited on an optically thick Au buffer and capped by a 4-nm-thick Au layer. In the simulations, use is made of the literature values of the optical and magnetooptical constants for Fe and Au [2,3].

5.3.3 Normal Incidence

This is the simplest situation when the longitudinal Kerr effect vanishes. Let us evaluate the general expressions for the case of N_y going to zero.

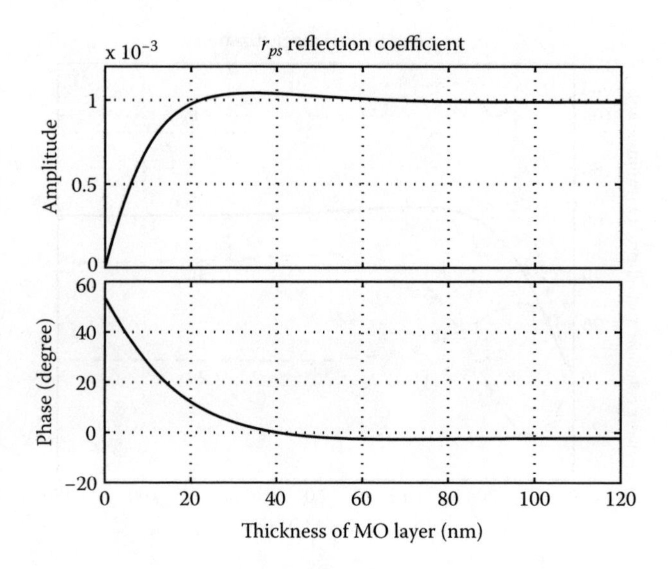

FIGURE 5.2
Effect of the ferromagnetic layer thickness, d, on the longitudinal off-diagonal reflection coefficient, r_{ps}, at an angle of incidence of 70 deg in a sandwich consisting of a Fe layer between optically thick Au buffer and the Au capping layer 4 nm thick. The protective Au capping layer separates the Fe(d) layer from the ambient medium of incidence (air). The simulation pertains to the wavelength of 632.8 nm and employs the diagonal and off-diagonal permittivity tensor elements for Fe given by $\varepsilon_{xx}^{(\mathrm{Fe})} = \varepsilon_0^{(\mathrm{Fe})} = -0.8845 - \mathrm{j}\,17.938$ and $\varepsilon_{xy}^{(\mathrm{Fe})} = \mathrm{j}\,\varepsilon_1^{(\mathrm{Fe})} = -0.6676 - \mathrm{j}\,0.08988$, respectively. The complex index of refraction for Au is given by $N^{(\mathrm{Au})} = 0.174 - \mathrm{j}\,3.65$. The differences in the diagonal permittivity tensor elements is set to zero.

For $\varepsilon_{zx}^{(n)} = -\varepsilon_{xz}^{(n)} = \mathrm{j}\,\varepsilon_1^{(n)}$, $\varepsilon_{zz}^{(n)} = \varepsilon_{xx}^{(n)} = \varepsilon_0^{(n)}$, and $\varepsilon_{yy}^{(n)} = \varepsilon_3^{(n)}$ $(n = 1, 2)$, the proper value equation (5.3) has the solutions

$$\left(N_{\parallel}^{(n)}\right)^2 = \varepsilon_3^{(n)} \tag{5.53}$$

$$\left(N_{\perp}^{(n)}\right)^2 = \varepsilon_0^{(n)-1}\left(\varepsilon_0^{(n)2} - \varepsilon_1^{(n)2}\right). \tag{5.54}$$

The proper polarization modes, obtained from Eq. (5.5), are orthogonal

$$\hat{\boldsymbol{e}}_{\parallel}^{(n)} = \begin{pmatrix} 0 \\ 1 \\ 0 \end{pmatrix} \tag{5.55}$$

and

$$\hat{\boldsymbol{e}}_{\perp}^{(n)} = \left(\left|\varepsilon_0^{(n)}\right|^2 + \left|\varepsilon_1^{(n)}\right|^2\right)^{-1/2} \begin{pmatrix} \varepsilon_0^{(n)} \\ 0 \\ \mathrm{j}\,\varepsilon_1^{(n)} \end{pmatrix}. \tag{5.56}$$

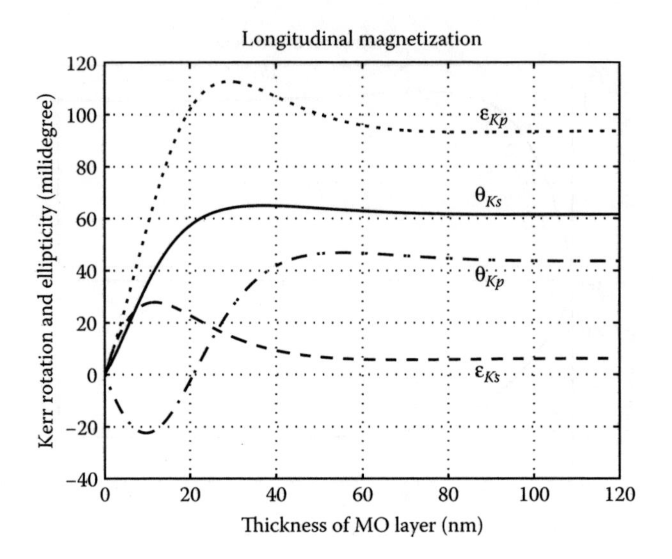

FIGURE 5.3
Effect of the ferromagnetic layer thickness, d, on the longitudinal MO Kerr rotation θ_{Ks} and θ_{Kp} and ellipticity ϵ_{Ks} and ϵ_{Kp} at an angle of incidence of 70 deg in a sandwich consisting of a Fe layer between optically thick Au buffer and the Au capping layer 4 nm thick. The protective Au capping layer separates the Fe(d) layer from the ambient medium of incidence (air). The subscripts s and p refer to the incident s and p polarizations, respectively. The situation corresponds to that of Figure 5.2.

Making use of Eq. (5.52), we obtain for the diagonal elements of the reflection matrix of the system defined by Eq. (3.70a) and Eq. (3.70c)

$$r_{\parallel}^{(0)} = \frac{r_{\parallel}^{(01)} + r_{\parallel}^{(12)} \mathrm{e}^{-2\mathrm{j}\beta_{\parallel}}}{1 + r_{\parallel}^{(01)} r_{\parallel}^{(12)} \mathrm{e}^{-2\mathrm{j}\beta_{\parallel}}} \tag{5.57}$$

for the waves polarized parallel and

$$r_{\perp}^{(0)} = \frac{r_{\perp}^{(01)} + r_{\perp}^{(12)} \mathrm{e}^{-2\mathrm{j}\beta_{\perp}}}{1 + r_{\perp}^{(01)} r_{\perp}^{(12)} \mathrm{e}^{-2\mathrm{j}\beta_{\perp}}} \tag{5.58}$$

for the waves polarized normal to the magnetization. The off-diagonal elements are zero. We have defined the single interface reflection coefficients at the ambient–film interface $r_{\parallel}^{(01)}$ and $r_{\perp}^{(01)}$ and at the film–substrate interface $r_{\parallel}^{(12)}$ and $r_{\perp}^{(12)}$ as follows:

$$r_{\parallel}^{(01)} = \frac{N^{(0)} - N_{\parallel}^{(1)}}{N^{(0)} + N_{\parallel}^{(1)}}, \tag{5.59a}$$

$$r_{\perp}^{(01)} = \frac{N^{(0)} - N^{(1)\perp}}{N^{(0)} + N_{\perp}^{(1)}}, \tag{5.59b}$$

$$r_{\parallel}^{(12)} = \frac{N_{\parallel}^{(1)} - N_{\parallel}^{(2)}}{N_{\parallel}^{(1)} + N_{\parallel}^{(2)}}, \tag{5.59c}$$

$$r_{\perp}^{(12)} = \frac{N_{\perp}^{(1)} - N_{\perp}^{(2)}}{N_{\perp}^{(1)} + N_{\perp}^{(2)}}. \tag{5.59d}$$

The symbols $\beta_{\parallel}$ and $\beta_{\perp}$ stand for $(\omega/c)N_{\parallel}^{(1)}d^{\,(1)}$ and $(\omega/c)N_{\perp}^{(n)}d^{\,(1)}$.

5.3.4 Uniaxial Film on a Uniaxial Substrate

Another example that is easy to solve occurs when $\varepsilon_{zx}^{(n)} = -\varepsilon_{xz}^{(n)} = \mathrm{j}\,\varepsilon_{1}^{(n)} = 0$, $\varepsilon_{zz}^{(n)} = \varepsilon_{xx}^{(n)} = \varepsilon_{0}^{(n)}$, and $\varepsilon_{yy}^{(n)} = \varepsilon_{3}^{(n)}$ $(n = 1, 2)$, in both the film and substrate. The permittivity tensor of Eq. (5.1) in both media reduces to a diagonal form. The proper value equation (5.3) has the solutions

$$\left(N_{z\parallel}^{(n)}\right)^{2} = \left(\varepsilon_{0}^{(n)}\right)^{-1}\varepsilon_{3}^{(n)}\left(\varepsilon_{0}^{(n)} - N_{y}^{2}\right) \tag{5.60}$$

and

$$\left(N_{z\perp}^{(n)}\right)^{2} = \left(\varepsilon_{0}^{(n)} - N_{y}^{2}\right). \tag{5.61}$$

The proper polarizations can be deduced from Eq. (5.5). The wave propagating with $N_{z\parallel}^{(n)}$ is polarized in the plane of incidence. We have, without a normalizing factor,

$$\mathbf{e}_{\parallel}^{(n)} = \begin{pmatrix} 0 \\ \left(\varepsilon_{0}^{(n)} - N_{y}^{2}\right)^{1/2} \\ N_{y}\left(\varepsilon_{3}^{(n)}/\varepsilon_{0}^{(n)}\right)^{1/2} \end{pmatrix}. \tag{5.62}$$

The wave propagating with $N_{z\perp}^{(n)}$ is polarized normal to the plane of incidence, *i.e.*,

$$\mathbf{e}_{\perp}^{(n)} = \begin{pmatrix} 1 \\ 0 \\ 0 \end{pmatrix}. \tag{5.63}$$

The proper polarization modes are orthogonal. As the optical axis is restricted to the plane of incidence, the reflection matrix is diagonal. Making use of Eq. (5.52), we obtain the expressions entering the reflection coefficients Eq. (3.70a) and Eq. (3.70c)

$$\begin{aligned} &\mathrm{M}_{11}\mathrm{M}_{33} - \mathrm{M}_{13}\mathrm{M}_{31} \\ &\quad = \mathrm{e}^{\mathrm{j}\left(\beta_{\parallel}+\beta_{\perp}\right)} + \mathrm{e}^{-\mathrm{j}\left(\beta_{\parallel}+\beta_{\perp}\right)} r_{ss}^{(01)} r_{pp}^{(01)} r_{ss}^{(12)} r_{pp}^{(12)} \\ &\quad + \mathrm{e}^{\mathrm{j}\left(\beta_{\parallel}-\beta_{\perp}\right)} r_{ss}^{(01)} r_{ss}^{(12)} + \mathrm{e}^{-\mathrm{j}\left(\beta_{\parallel}-\beta_{\perp}\right)} r_{pp}^{(01)} r_{pp}^{(12)}, \end{aligned} \tag{5.64}$$

$$
\begin{aligned}
&M_{21}M_{33} - M_{23}M_{31} \\
&\quad = e^{j\left(\beta_\parallel+\beta_\perp\right)} r_{ss}^{(01)} + e^{-j\left(\beta_\parallel+\beta_\perp\right)} r_{pp}^{(01)} r_{ss}^{(12)} r_{pp}^{(12)} \\
&\quad + e^{j\left(\beta_\parallel-\beta_\perp\right)} r_{ss}^{(12)} + e^{-j\left(\beta_\parallel-\beta_\perp\right)} r_{ss}^{(01)} r_{pp}^{(01)} r_{pp}^{(12)},
\end{aligned} \tag{5.65}
$$

$$
\begin{aligned}
&M_{11}M_{43} - M_{41}M_{13} \\
&\quad = e^{j\left(\beta_\parallel+\beta_\perp\right)} + e^{-j\left(\beta_\parallel+\beta_\perp\right)} r_{ss}^{(01)} r_{ss}^{(12)} r_{pp}^{(12)} \\
&\quad + e^{j\left(\beta_\parallel-\beta_\perp\right)} r_{ss}^{(01)} r_{pp}^{(01)} r_{ss}^{(12)} + e^{-j\left(\beta_\parallel-\beta_\perp\right)} r_{pp}^{(12)},
\end{aligned} \tag{5.66}
$$

where

$$
r_{21}^{(01)} = r_{ss}^{(01)} = \frac{N^{(0)}\cos\varphi^{(0)} - N_z^{(1)}}{N^{(0)}\cos\varphi^{(0)} + N_z^{(1)}}, \tag{5.67}
$$

$$
r_{43}^{(01)} = r_{pp}^{(01)} = \frac{N^{(0)}N_z^{(1)} - \cos\varphi^{(0)}\left(\varepsilon_0^{(1)}\varepsilon_3^{(1)}\right)^{1/2}}{N^{(0)}N_z^{(1)} + \cos\varphi^{(0)}\left(\varepsilon_0^{(1)}\varepsilon_3^{(1)}\right)^{1/2}}, \tag{5.68}
$$

$$
r_{21}^{(12)} = r_{ss}^{(12)} = \frac{N_z^{(1)} - N_z^{(2)}}{N_z^{(1)} + N_z^{(2)}}, \tag{5.69}
$$

$$
r_{43}^{(12)} = r_{pp}^{(12)} = \frac{N_z^{(2)}\left(\varepsilon_0^{(1)}\varepsilon_3^{(1)}\right)^{1/2} - N_z^{(1)}\left(\varepsilon_0^{(2)}\varepsilon_3^{(2)}\right)^{1/2}}{N_z^{(2)}\left(\varepsilon_0^{(1)}\varepsilon_3^{(1)}\right)^{1/2} + N_z^{(1)}\left(\varepsilon_0^{(2)}\varepsilon_3^{(2)}\right)^{1/2}}. \tag{5.70}
$$

Here, $N_z^{(n)} = \left(\varepsilon_0^{(n)} - N_y^2\right)^{1/2}$ in the film of the thickness $d^{(1)}$ $(n = 1)$ and in the substrate $(n = 2)$. We put $\beta_\parallel = (\omega/c)N_\parallel^{(1)}d_1$ and $\beta_\perp = (\omega/c)N_\perp^{(1)}d^{(1)}$.

5.4 Approximate Solution for the Oblique Incidence

In this section, we give an approximate solution for the case of an arbitrary angle of incidence for a permittivity tensor with $\varepsilon_{zx}^{(n)} = -\varepsilon_{xz}^{(n)} = j\,\varepsilon_1^{(n)}$ and $\varepsilon_{xx}^{(n)} = \varepsilon_{yy}^{(n)} = \varepsilon_{zz}^{(n)} = \varepsilon_0^{(n)}$ for $n = 1, 2$, corresponding to the longitudinal magnetization and to the absence of in-plane and perpendicular anisotropies. The approximation consists of retaining only the terms independent of magnetization and the terms linearly dependent on the magnetization either in the film or in the substrate. We therefore drop the terms containing the products of the terms linear in the film magnetization with those linear in the substrate magnetization. As in the polar case, all quadratic and higher order magnetooptic effects are neglected. In other words, we keep only the terms of zero and first order in the off-diagonal tensor elements, $\varepsilon_1^{(1)}$ and $\varepsilon_1^{(2)}$. These quite realistic and widely used assumptions make the final expressions for the Cartesian Jones reflection matrix much simpler. As we shall see below, to the first order in $\varepsilon_1^{(1)}$ and $\varepsilon_1^{(2)}$, $r_{21}^{(02)}$, and $r_{43}^{(02)}$, the diagonal elements of the Jones reflection matrix of the film–substrate system do not contain $\varepsilon_1^{(1)}$ and $\varepsilon_1^{(2)}$. In this approximation, the off–diagonal elements of the Jones reflection matrix retain the terms containing the linear functions of either $\varepsilon_1^{(1)}$ or $\varepsilon^{(2)}$, only. From Eq. (5.4)

$$N_{z\pm}^{(n)2} = \left(\varepsilon_0^{(n)} - N_y^2\right) - \frac{\varepsilon_1^{(n)2}}{2\varepsilon_0^{(n)}} \pm \frac{\varepsilon_1^{(n)}\sqrt{\varepsilon_1^{(n)2} + 4\varepsilon_0^{(n)} N_y^2}}{2\varepsilon_0^{(n)}}. \tag{5.71}$$

We observe

$$\left(N_{z+}^{(n)} - N_{z-}^{(n)}\right)^2 = 2\left(\varepsilon_0^{(n)} - N_y^2\right) - \frac{\varepsilon_1^{(n)2}}{\varepsilon_0^{(n)}} - 2\left(N_{z+}^{(n)} N_{z-}^{(n)}\right). \tag{5.72}$$

Assuming $\varepsilon_1^{(n)}$ much smaller than $\varepsilon_0^{(n)}$ and $\varepsilon_1^{(n)2} \ll 4\varepsilon_0^{(n)} N_y^2$, we have

$$N_{z+}^{(n)} - N_{z-}^{(n)} \approx \frac{\varepsilon_1^{(n)} N_y}{\varepsilon_0^{(n)1/2} N_z^{(n)}}. \tag{5.73}$$

We adopt the notation of Chapter 4 defined in Eq. (4.252). Eq. (5.71) further gives

$$N_{z+}^{(n)2} + N_{z-}^{(n)2} = 2N_{z0}^{(n)2} - \varepsilon_1^{(n)2}\varepsilon_0^{(n)-1} = 2N^{(n)2}\alpha_z^{(n)2} - \varepsilon_1^{(n)2}N^{(n)-2}, \tag{5.74a}$$

$$N_{z+}^{(n)2} N_{z-}^{(n)2} = N^{(n)4}\alpha_z^{(n)4} - \varepsilon_1^{(n)2}. \tag{5.74b}$$

Eq. (4.224) yields

$$\mathcal{P}_+^{(n)} \approx 2N^{(n)3}\alpha_z^{(n)}\left(N_{z+}^{(n)} N_{z-}^{(n)}\right), \tag{5.75a}$$

$$\mathcal{P}_-^{(n)} \approx N^{(n)2}\left(N_{z+}^{(n)} N_{z-}^{(n)}\right)\left(N_{z+}^{(n)} - N_{z-}^{(n)}\right), \tag{5.75b}$$

$$\mathcal{Q}_+^{(n)} \approx 2\left(N_{z+}^{(n)} N_{z-}^{(n)}\right) N^{(n)2}\alpha_z^{(n)2}, \tag{5.75c}$$

$$\begin{aligned}\mathcal{Q}_-^{(n)} &= \left(N_{z+}^{(n)} N_{z-}^{(n)}\right)\left(N^{(n)2}\alpha_z^{(n)2} - N_{z+}^{(n)} N_{z-}^{(n)}\right)\\ &\approx \frac{N^{(n)2}}{2N_y^2}\left(N_{z+}^{(n)} - N_{z-}^{(n)}\right)^2\left(N_{z+}^{(n)} N_{z-}^{(n)}\right),\end{aligned} \tag{5.75d}$$

$$\mathcal{S}_+^{(n)} \approx 2\left(N_{z+}^{(n)} N_{z-}^{(n)}\right) N^{(n)4}\alpha_z^{(n)2}, \tag{5.75e}$$

$$\begin{aligned}\mathcal{S}_-^{(n)} &= \left(N_{z+}^{(n)} N_{z-}^{(n)}\right)\left[N^{(n)2}\left(N_{z+}^{(n)} N_{z-}^{(n)}\right) - N^{(n)4}\alpha_z^{(n)2} + \varepsilon_1^{(n)2}\right]\\ &\approx \frac{N^{(n)4}}{2N_y^2}\left(N_{z+}^{(n)} - N_{z-}^{(n)}\right)^2\left(N_{z+}^{(n)} N_{z-}^{(n)}\right),\end{aligned} \tag{5.75f}$$

$$\mathcal{T}_{+}^{(n)} \approx 2\left(N_{z+}^{(n)} N_{z-}^{(n)}\right) N^{(n)3} \alpha_{z}^{(n)3}, \tag{5.75g}$$
$$\mathcal{T}_{-}^{(n)} \approx \left(N_{z+}^{(n)} N_{z-}^{(n)}\right)\left(N_{z+}^{(n)} - N_{z-}^{(n)}\right) N^{(n)2} \alpha_{z}^{(n)2}, \tag{5.75h}$$

$$\begin{aligned}\mathcal{U}_{\pm}^{(n)} &= \mathcal{W}_{\pm}^{(n)} \approx -\mathrm{j}\,\varepsilon_{1}^{(n)} N_{y}\left(N_{z+}^{(n)} N_{z-}^{(n)}\right) \\ &= -\mathrm{j} N^{(n)2} \alpha_{z}^{(n)}\left(N_{z+}^{(n)} - N_{z-}^{(n)}\right)\left(N_{z+}^{(n)} N_{z-}^{(n)}\right).\end{aligned} \tag{5.75i}$$

It turns out that in this approximation, the off-diagonal elements of the global Cartesian Jones reflection matrix are equal to each other, *i.e.*, $M_{11}M_{23} - M_{13}M_{21} = M_{41}M_{33} - M_{43}M_{31}$. We have included $\mathcal{Q}_{-}^{(n)}$ and $\mathcal{S}_{-}^{(n)}$, which are of second order in $\varepsilon_{1}^{(n)}$. The parameters $\mathcal{Q}_{-}^{(n)}$ and $\mathcal{S}_{-}^{(n)}$ along with $\mathcal{W}_{+}^{(1)}\mathcal{U}_{+}^{(2)} + \mathcal{U}_{+}^{(1)}\mathcal{W}_{+}^{(2)}$ cannot be neglected in $M_{11}M_{23} - M_{13}M_{21} = M_{41}M_{33} - M_{43}M_{31}$ as they give terms that are, after the removal of common factors, linear in $\varepsilon_{1}^{(n)}$. Eq. (5.75f) was obtained with $N_{z+}^{(n)} N_{z-}^{(n)}$ replaced according to Eq. (5.72). The use was made of Eq. (5.73) in Eq. (5.75d), Eq. (5.75f), and Eq. (5.75i).

From Eq. (5.52d)

$$\begin{aligned}
&M_{41}M_{33} - M_{43}M_{31} = M_{11}M_{23} - M_{13}M_{21} \\
&= 8\mathrm{j} N^{(0)} \cos\varphi^{(0)} N_{z+}^{(1)} N_{z-}^{(1)} \varepsilon_{0}^{(1)-1} N_{z0}^{(1)4} \varepsilon_{1}^{(1)3} N_{y}\left(N^{(1)2}\mathcal{Q}_{+}^{(2)} - \mathcal{S}_{+}^{(2)}\right) \\
&+ 2N^{(0)} \cos\varphi^{(0)} \left(N_{z+}^{(1)} - N_{z-}^{(1)}\right)^{2} \mathcal{W}_{+}^{(1)} \\
&\times\left[\left(\mathcal{T}_{+}^{(1)}\mathcal{P}_{+}^{(2)} + \mathcal{S}_{+}^{(1)}\mathcal{Q}_{+}^{(2)} + \mathcal{Q}_{+}^{(1)}\mathcal{S}_{+}^{(2)} + \mathcal{P}_{+}^{(1)}\mathcal{T}_{+}^{(2)}\right)\mathrm{e}^{\mathrm{j}(\beta_{+}+\beta_{-})}\right. \\
&\left.+\left(\mathcal{T}_{+}^{(1)}\mathcal{P}_{+}^{(2)} - \mathcal{S}_{+}^{(1)}\mathcal{Q}_{+}^{(2)} - \mathcal{Q}_{+}^{(1)}\mathcal{S}_{+}^{(2)} + \mathcal{P}_{+}^{(1)}\mathcal{T}_{+}^{(2)}\right)\mathrm{e}^{-\mathrm{j}(\beta_{+}+\beta_{-})}\right] \\
&- 8N^{(0)} \cos\varphi^{(0)} N_{z0}^{(1)2} \mathcal{W}_{-}^{(1)}\left(\mathcal{W}_{-}^{(1)}\mathcal{U}_{+}^{(2)} + \mathcal{U}_{-}^{(1)}\mathcal{W}_{+}^{(2)}\right)\left(\mathrm{e}^{\mathrm{j}(\beta_{+}-\beta_{-})} + \mathrm{e}^{-\mathrm{j}(\beta_{+}-\beta_{-})}\right) \\
&+ 8N^{(0)} \cos\varphi^{(0)} N_{z0}^{(1)2} \mathcal{W}_{-}^{(1)}\left(\mathcal{T}_{-}^{(1)}\mathcal{P}_{+}^{(2)} - \mathcal{P}_{-}^{(1)}\mathcal{T}_{+}^{(2)}\right)\left(\mathrm{e}^{\mathrm{j}(\beta_{+}-\beta_{-})} - \mathrm{e}^{-\mathrm{j}(\beta_{+}-\beta_{-})}\right) \\
&+ 8N^{(0)} \cos\varphi^{(0)} N_{z0}^{(1)2} \mathcal{W}_{-}^{(1)}\left(\mathcal{S}_{-}^{(1)}\mathcal{Q}_{+}^{(2)} - \mathcal{Q}_{-}^{(1)}\mathcal{S}_{+}^{(2)}\right)\left(\mathrm{e}^{\mathrm{j}(\beta_{+}-\beta_{-})} - \mathrm{e}^{-\mathrm{j}(\beta_{+}+\beta_{-})}\right).
\end{aligned} \tag{5.76}$$

We observe

$$\begin{aligned}
&\mathrm{j}\varepsilon_{0}^{(1)-1}\varepsilon_{1}^{(1)3} N_{y}\left(N^{(1)2}\mathcal{Q}_{+}^{(2)} - \mathcal{S}_{+}^{(2)}\right) \\
&= \mathrm{j}\,\varepsilon_{1}^{(1)} N_{y}\left(N_{z+}^{(1)} - N_{z-}^{(1)}\right)^{2} N^{(1)4}\alpha_{z}^{(1)} N^{(2)2}\alpha_{z}^{(2)}\left[2\alpha_{z}^{(1)}\alpha_{z}^{(2)}\left(N^{(1)2} - N^{(2)2}\right)\right] N_{y}^{-2}
\end{aligned} \tag{5.77a}$$

$$\begin{aligned}
&\mathcal{T}_{+}^{(1)}\mathcal{P}_{+}^{(2)} \pm \mathcal{S}_{+}^{(1)}\mathcal{Q}_{+}^{(2)} \pm \mathcal{Q}_{+}^{(1)}\mathcal{S}_{+}^{(2)} + \mathcal{P}_{+}^{(1)}\mathcal{T}_{+}^{(2)} \\
&= \pm 4N^{(1)2}N^{(2)2}\alpha_{z}^{(1)}\alpha_{z}^{(2)}\left(N^{(1)}\alpha_{z}^{(1)} \pm N^{(2)}\alpha_{z}^{(2)}\alpha_{z}^{(2)}\right)\left(N^{(1)}\alpha_{z}^{(2)} \pm N^{(2)}\alpha_{z}^{(2)}\alpha_{z}^{(1)}\right)
\end{aligned} \tag{5.77b}$$

$$\mathcal{W}_{-}^{(1)}\left(\mathcal{W}_{-}^{(1)}\mathcal{U}_{+}^{(2)} + \mathcal{U}_{-}^{(1)}\mathcal{W}_{+}^{(2)}\right) = 2\mathrm{j}N^{(1)4}\alpha_{z}^{(1)2}\left(N_{z+}^{(1)} - N_{z-}^{(1)}\right)^{2}\varepsilon_{1}^{(2)} N_{y} \tag{5.77c}$$

$$\mathcal{W}_{-}^{(1)}\left(\mathcal{T}_{-}^{(1)}\mathcal{P}_{+}^{(2)} - \mathcal{P}_{-}^{(1)}\mathcal{T}_{+}^{(2)}\right)$$
$$= -\mathrm{j}N^{(1)3}N^{(2)2}\alpha_z^{(1)}\alpha_z^{(2)}\left(N_{z+}^{(1)} - N_{z-}^{(1)}\right)^2\left[2N^{(1)}N^{(2)}\left(\alpha_z^{(1)2} - \alpha_z^{(2)2}\right)\right] \tag{5.77d}$$

$$N_{z0}^{(1)2}\mathcal{W}_{-}^{(1)}\left(\mathcal{S}_{-}^{(1)}\mathcal{Q}_{+}^{(2)} - \mathcal{Q}_{-}^{(1)}\mathcal{S}_{+}^{(2)}\right)$$
$$= -\mathrm{j}\,\varepsilon_1^{(1)}N_y\left(N_{z+}^{(1)} - N_{z-}^{(1)}\right)^2 N^{(1)4}\alpha_z^{(1)}N^{(2)2}\alpha_z^{(2)}\left[2\alpha_z^{(1)}\alpha_z^{(2)}\left(N^{(1)2} - N^{(2)2}\right)\right]$$
$$\times\left(2N_y\right)^{-2}. \tag{5.77e}$$

Let us substitute these results into Eq. (5.76). After the removal of the common factor

$$N_{z+}^{(1)2}N_{z-}^{(1)2}\left(N_{z+}^{(1)} - N_{z-}^{(1)}\right)^2 N_{z+}^{(2)}N_{z-}^{(2)},$$

and another common factor $\left[8N^{(1)4}N^{(1)2}\alpha_z^{(1)2}\alpha_z^{(2)}\right]$, we have

$$\mathrm{M}_{41}\mathrm{M}_{33} - \mathrm{M}_{43}\mathrm{M}_{31} = \mathrm{M}_{11}\mathrm{M}_{23} - \mathrm{M}_{13}\mathrm{M}_{21}$$
$$= -\left(-\mathrm{j}\,\varepsilon_1^{(1)}N_yN^{(0)}\alpha_z^{(0)}N^{(1)-2}\alpha_z^{(1)-1}\right)N^{(1)2}\left[2\alpha_z^{(1)}\alpha_z^{(2)}\left(N^{(1)2} - N^{(2)2}\right)\right]N_y^{-2}$$
$$+\left(-\mathrm{j}\,\varepsilon_1^{(1)}N_yN^{(0)}\alpha_z^{(0)}N^{(1)-2}\alpha_z^{(1)-1}\right)$$
$$\times\left[\mathrm{e}^{\mathrm{j}(\beta_+ + \beta_-)}\left(N^{(1)}\alpha_z^{(1)} + N^{(2)}\alpha_z^{(2)}\right)\left(N^{(1)}\alpha_z^{(2)} + N^{(2)}\alpha_z^{(1)}\right)\right.$$
$$\left.-\mathrm{e}^{-\mathrm{j}(\beta_+ + \beta_-)}\left(N^{(1)}\alpha_z^{(1)} - N^{(2)}\alpha_z^{(2)}\right)\left(N^{(1)}\alpha_z^{(2)} - N^{(2)}\alpha_z^{(1)}\right)\right]$$
$$+\left(-\mathrm{j}\,\varepsilon_1^{(2)}N_yN^{(1)}\alpha_z^{(1)}N^{(2)-2}\alpha_z^{(2)}\right)2N^{(0)}\alpha_z^{(0)}2N^{(1)}\alpha_z^{(1)}\cos\left(\beta_+ - \beta_-\right)$$
$$+\frac{1}{2}\left[2N^{(1)}N^{(2)}\left(\alpha_z^{(1)2} - \alpha_z^{(2)2}\right)\right]2N^{(0)}\alpha_z^{(0)}2N^{(1)}\alpha_z^{(1)}\sin\left(\beta_+ - \beta_-\right)$$
$$+\left(-\mathrm{j}\,\varepsilon_1^{(1)}N_yN^{(0)}\alpha_z^{(0)}N^{(1)-2}\alpha_z^{(1)-1}\right)\left[2\alpha_z^{(1)}\alpha_z^{(2)}\left(N^{(1)2} - N^{(2)2}\right)\right]$$
$$\times N^{(1)2}N_y^{-2}\cos\left(\beta_+ - \beta_-\right). \tag{5.78}$$

In the present approximation, Eq. (5.52a) simplifies to

$$\mathrm{M}_{11}\mathrm{M}_{33} - \mathrm{M}_{13}\mathrm{M}_{31}$$
$$= 4\cos\varphi^{(0)}N_yN_{z+}^{(1)}N_{z-}^{(1)}\varepsilon_0^{(1)-1}\left(N^{(0)2} - \varepsilon_0^{(1)}\right)\left(\varepsilon_0^{(1)} - N_y^2\right)^2$$
$$\times\left[2\varepsilon_1^{(1)2}N_y\left(\varepsilon_0^{(1)}\mathcal{Q}_{+}^{(2)} - \mathcal{S}_{+}^{(2)}\right)\right]$$
$$+\mathrm{e}^{\mathrm{j}(\beta_+ + \beta_-)}\left(N_{z+}^{(1)} - N_{z-}^{(1)}\right)^2$$
$$\times\left(\mathcal{T}^{(0)}\mathcal{P}_{+}^{(1)} + \mathcal{S}^{(0)}\mathcal{Q}_{+}^{(1)} + \mathcal{Q}^{(0)}\mathcal{S}_{+}^{(1)} + \mathcal{P}^{(0)}\mathcal{T}_{+}^{(1)}\right)$$
$$\times\left(\mathcal{T}_{+}^{(1)}\mathcal{P}_{+}^{(2)} + \mathcal{S}_{+}^{(1)}\mathcal{Q}_{+}^{(2)} + \mathcal{Q}_{+}^{(1)}\mathcal{S}_{+}^{(2)} + \mathcal{P}_{+}^{(1)}\mathcal{T}_{+}^{(2)}\right)$$
$$+\mathrm{e}^{-\mathrm{j}(\beta_+ + \beta_-)}\left(N_{z+}^{(1)} - N_{z-}^{(1)}\right)^2$$
$$\times\left(\mathcal{T}^{(0)}\mathcal{P}_{+}^{(1)} - \mathcal{S}^{(0)}\mathcal{Q}_{+}^{(1)} - \mathcal{Q}^{(0)}\mathcal{S}_{+}^{(1)} + \mathcal{P}^{(0)}\mathcal{T}_{+}^{(1)}\right)$$

$$\times\left(\mathcal{T}_+^{(1)}\mathcal{P}_+^{(2)} - \mathcal{S}_+^{(1)}\mathcal{Q}_+^{(2)} - \mathcal{Q}_+^{(1)}\mathcal{S}_+^{(2)} + \mathcal{P}_+^{(1)}\mathcal{T}_+^{(2)}\right)$$
$$+ e^{j(\beta_+-\beta_-)}\left(N_{z+}^{(1)} + N_{z-}^{(1)}\right)^2\left(-\mathcal{T}^{(0)}\mathcal{P}_-^{(1)} + \mathcal{P}^{(0)}\mathcal{T}_-^{(1)}\right)\left(-\mathcal{T}_-^{(1)}\mathcal{P}_+^{(2)} + \mathcal{P}_-^{(1)}\mathcal{T}_+^{(2)}\right)$$
$$+ e^{-j(\beta_+-\beta_-)}\left(N_{z+}^{(1)} + N_{z-}^{(1)}\right)^2\left(-\mathcal{T}^{(0)}\mathcal{P}_-^{(1)} + \mathcal{P}^{(0)}\mathcal{T}_-^{(1)}\right)\left(-\mathcal{T}_-^{(1)}\mathcal{P}_+^{(2)} + \mathcal{P}_-^{(1)}\mathcal{T}_+^{(2)}\right) \tag{5.79}$$

$$\begin{aligned}
&M_{21}M_{33} - M_{23}M_{31} \\
&= 4\cos\varphi^{(0)} N_y N_{z+}^{(1)} N_{z-}^{(1)} \varepsilon_0^{(1)-1}\left(N^{(0)2} + \varepsilon_0^{(1)}\right)\left(\varepsilon_0^{(1)} - N_y^2\right)^2 \\
&\quad \times\left[2\varepsilon_1^{(1)2} N_y\left(\varepsilon_0^{(1)}\mathcal{Q}_+^{(2)} - \mathcal{S}_+^{(2)}\right)\right] \\
&\quad + e^{j(\beta_++\beta_-)}\left(N_{z+}^{(1)} - N_{z-}^{(1)}\right)^2\left(\mathcal{T}^{(0)}\mathcal{P}_+^{(1)} + \mathcal{S}^{(0)}\mathcal{Q}_+^{(1)} - \mathcal{Q}^{(0)}\mathcal{S}_+^{(1)} - \mathcal{P}^{(0)}\mathcal{T}_+^{(1)}\right) \\
&\quad \times\left(\mathcal{T}_+^{(1)}\mathcal{P}_+^{(2)} + \mathcal{S}_+^{(1)}\mathcal{Q}_+^{(2)} + \mathcal{Q}_+^{(1)}\mathcal{S}_+^{(2)} + \mathcal{P}_+^{(1)}\mathcal{T}_+^{(2)}\right) \\
&\quad + e^{-j(\beta_++\beta_-)}\left(N_{z+}^{(1)} - N_{z-}^{(1)}\right)^2\left(\mathcal{T}^{(0)}\mathcal{P}_+^{(1)} - \mathcal{S}^{(0)}\mathcal{Q}_+^{(1)} + \mathcal{Q}^{(0)}\mathcal{S}_+^{(1)} - \mathcal{P}^{(0)}\mathcal{T}_+^{(1)}\right) \\
&\quad \times\left(\mathcal{T}_+^{(1)}\mathcal{P}_+^{(2)} - \mathcal{S}_+^{(1)}\mathcal{Q}_+^{(2)} - \mathcal{Q}_+^{(1)}\mathcal{S}_+^{(2)} + \mathcal{P}_+^{(1)}\mathcal{T}_+^{(2)}\right) \\
&\quad - e^{j(\beta_+-\beta_-)}\left(N_{z+}^{(1)} + N_{z-}^{(1)}\right)^2\left(\mathcal{T}^{(0)}\mathcal{P}_-^{(1)} + \mathcal{P}^{(0)}\mathcal{T}_-^{(1)}\right)\left(-\mathcal{T}_-^{(1)}\mathcal{P}_+^{(2)} + \mathcal{P}_-^{(1)}\mathcal{T}_+^{(2)}\right) \\
&\quad - e^{-j(\beta_+-\beta_-)}\left(N_{z+}^{(1)} + N_{z-}^{(1)}\right)^2\left(\mathcal{T}^{(0)}\mathcal{P}_-^{(1)} + \mathcal{P}^{(0)}\mathcal{T}_-^{(1)}\right) \\
&\quad \times\left(-\mathcal{T}_-^{(1)}\mathcal{P}_+^{(2)} + \mathcal{P}_-^{(1)}\mathcal{T}_+^{(2)}\right)
\end{aligned} \tag{5.80}$$

$$\begin{aligned}
&M_{11}M_{43} - M_{13}M_{41} \\
&= 4\cos\varphi^{(0)} N_y N_{z+}^{(1)} N_{z-}^{(1)} \varepsilon_0^{(1)-1}\left(N^{(0)2} + \varepsilon_0^{(1)}\right)\left[2\varepsilon_1^{(1)2} N_y\left(\varepsilon_0^{(1)}\mathcal{Q}_+^{(2)} - \mathcal{S}_+^{(2)}\right)\right] \\
&\quad + e^{j(\beta_++\beta_-)}\left(N_{z+}^{(1)} - N_{z-}^{(1)}\right)^2\left(-\mathcal{T}^{(0)}\mathcal{P}_+^{(1)} + \mathcal{S}^{(0)}\mathcal{Q}_+^{(1)} - \mathcal{Q}^{(0)}\mathcal{S}_+^{(1)} + \mathcal{P}^{(0)}\mathcal{T}_+^{(1)}\right) \\
&\quad \times\left(\mathcal{T}_+^{(1)}\mathcal{P}_+^{(2)} + \mathcal{S}_+^{(1)}\mathcal{Q}_+^{(2)} + \mathcal{Q}_+^{(1)}\mathcal{S}_+^{(2)} + \mathcal{P}_+^{(1)}\mathcal{T}_+^{(2)}\right) \\
&\quad + e^{-j(\beta_++\beta_-)}\left(N_{z+}^{(1)} - N_{z-}^{(1)}\right)^2\left(-\mathcal{T}^{(0)}\mathcal{P}_+^{(1)} - \mathcal{S}^{(0)}\mathcal{Q}_+^{(1)} + \mathcal{Q}^{(0)}\mathcal{S}_+^{(1)} + \mathcal{P}^{(0)}\mathcal{T}_+^{(1)}\right) \\
&\quad \times\left(\mathcal{T}_+^{(1)}\mathcal{P}_+^{(2)} - \mathcal{S}_+^{(1)}\mathcal{Q}_+^{(2)} - \mathcal{Q}_+^{(1)}\mathcal{S}_+^{(2)} + \mathcal{P}_+^{(1)}\mathcal{T}_+^{(2)}\right) \\
&\quad + e^{j(\beta_+-\beta_-)}\left(N_{z+}^{(1)} + N_{z-}^{(1)}\right)^2\left(\mathcal{T}^{(0)}\mathcal{P}_-^{(1)} + \mathcal{P}^{(0)}\mathcal{T}_-^{(1)}\right)\left(-\mathcal{T}_-^{(1)}\mathcal{P}_+^{(2)} + \mathcal{P}_-^{(1)}\mathcal{T}_+^{(2)}\right) \\
&\quad + e^{-j(\beta_+-\beta_-)}\left(N_{z+}^{(1)} + N_{z-}^{(1)}\right)^2\left(\mathcal{T}^{(0)}\mathcal{P}_-^{(1)} + \mathcal{P}^{(0)}\mathcal{T}_-^{(1)}\right) \\
&\quad \times\left(-\mathcal{T}_-^{(1)}\mathcal{P}_+^{(2)} + \mathcal{P}_-^{(1)}\mathcal{T}_+^{(2)}\right).
\end{aligned} \tag{5.81}$$

We obtain after skipping the common factor $8\left(N_{z+}^{(1)} - N_{z-}^{(1)}\right)^2 N^{(1)4} N^{(1)2} \alpha_z^{(1)2}\alpha_z^{(2)}$

$$\begin{aligned}
&M_{11}M_{33} - M_{13}M_{31} \\
&= \frac{1}{2}\left[2\alpha_z^{(0)}\alpha_z^{(1)}\left(N^{(0)2} - N^{(1)2}\right)\right]\left[2\alpha_z^{(1)}\alpha_z^{(2)}\left(N^{(1)2} - N^{(2)2}\right)\right] \\
&\quad + e^{j(\beta_++\beta_-)}\left(N^{(0)}\alpha_z^{(0)} + N^{(1)}\alpha_z^{(1)}\right)\left(N^{(0)}\alpha_z^{(1)} + N^{(1)}\alpha_z^{(0)}\right)
\end{aligned}$$

$$
\begin{aligned}
&\times\left(N^{(1)}\alpha_z^{(1)} + N^{(2)}\alpha_z^{(2)}\right)\left(N^{(1)}\alpha_z^{(2)} + N^{(2)}\alpha_z^{(1)}\right) \\
&+ e^{-j(\beta_+ + \beta_-)}\left(N^{(0)}\alpha_z^{(0)} - N^{(1)}\alpha_z^{(1)}\right)\left(N^{(0)}\alpha_z^{(1)} - N^{(1)}\alpha_z^{(0)}\right) \\
&\times\left(N^{(1)}\alpha_z^{(1)} - N^{(2)}\alpha_z^{(2)}\right)\left(N^{(1)}\alpha_z^{(2)} - N^{(2)}\alpha_z^{(1)}\right) \\
&+ \left(e^{j(\beta_+ - \beta_-)} + e^{-j(\beta_+ - \beta_-)}\right)\frac{1}{4}\left[2N^{(0)}N^{(1)}\left(\alpha_z^{(0)2} - \alpha_z^{(1)2}\right)\right] \\
&\quad\left[2N^{(1)}N^{(2)}\left(\alpha_z^{(1)2} - \alpha_z^{(2)2}\right)\right]
\end{aligned} \tag{5.82}
$$

$$
\begin{aligned}
&M_{21}M_{33} - M_{23}M_{31} \\
&= \frac{1}{2}\left[2\alpha_z^{(0)}\alpha_z^{(1)}\left(N^{(0)2} + N^{(1)2}\right)\right]\left[2\alpha_z^{(1)}\alpha_z^{(2)}\left(N^{(1)2} - N^{(2)2}\right)\right] \\
&\quad + e^{j(\beta_+ + \beta_-)} \\
&\quad \times\left(N^{(0)}\alpha_z^{(0)} - N^{(1)}\alpha_z^{(1)}\right)\left(N^{(0)}\alpha_z^{(1)} + N^{(1)}\alpha_z^{(0)}\right) \\
&\quad \times\left(N^{(1)}\alpha_z^{(1)} + N^{(2)}\alpha_z^{(2)}\right)\left(N^{(1)}\alpha_z^{(2)} + N^{(2)}\alpha_z^{(1)}\right) \\
&\quad + e^{-j(\beta_+ + \beta_-)} \\
&\quad \times\left(N^{(0)}\alpha_z^{(0)} + N^{(1)}\alpha_z^{(1)}\right)\left(N^{(0)}\alpha_z^{(1)} - N^{(1)}\alpha_z^{(0)}\right) \\
&\quad \times\left(N^{(1)}\alpha_z^{(1)} - N^{(2)}\alpha_z^{(2)}\right)\left(N^{(1)}\alpha_z^{(2)} - N^{(2)}\alpha_z^{(1)}\right) \\
&\quad + \left(e^{j(\beta_+ - \beta_-)} + e^{-j(\beta_+ - \beta_-)}\right) \\
&\quad \times\frac{1}{4}\left[2N^{(0)}N^{(1)}\left(\alpha_z^{(0)2} + \alpha_z^{(1)2}\right)\right]\left[2N^{(1)}N^{(2)}\left(\alpha_z^{(1)2} - \alpha_z^{(2)2}\right)\right]
\end{aligned} \tag{5.83}
$$

$$
\begin{aligned}
&M_{11}M_{43} - M_{13}M_{41} \\
&= \frac{1}{2}\left[2\alpha_z^{(0)}\alpha_z^{(1)}\left(N^{(0)2} + N^{(1)2}\right)\right]\left[2\alpha_z^{(1)}\alpha_z^{(2)}\left(N^{(1)2} - N^{(2)2}\right)\right] \\
&\quad + e^{j(\beta_+ + \beta_-)} \\
&\quad \times\left(N^{(0)}\alpha_z^{(0)} + N^{(1)}\alpha_z^{(1)}\right)\left(N^{(0)}\alpha_z^{(1)} - N^{(1)}\alpha_z^{(0)}\right) \\
&\quad \times\left(N^{(1)}\alpha_z^{(1)} + N^{(2)}\alpha_z^{(2)}\right)\left(N^{(1)}\alpha_z^{(2)} + N^{(2)}\alpha_z^{(1)}\right) \\
&\quad + e^{-j(\beta_+ + \beta_-)} \\
&\quad \times\left(N^{(0)}\alpha_z^{(0)} - N^{(1)}\alpha_z^{(1)}\right)\left(N^{(0)}\alpha_z^{(1)} + N^{(1)}\alpha_z^{(0)}\right) \\
&\quad \times\left(N^{(1)}\alpha_z^{(1)} - N^{(2)}\alpha_z^{(2)}\right)\left(N^{(1)}\alpha_z^{(2)} - N^{(2)}\alpha_z^{(1)}\right) \\
&\quad - \left(e^{j(\beta_+ - \beta_-)} + e^{-j(\beta_+ - \beta_-)}\right) \\
&\quad \times\frac{1}{4}\left[2N^{(0)}N^{(1)}\left(\alpha_z^{(0)2} + \alpha_z^{(1)2}\right)\right]\left[2N^{(1)}N^{(2)}\left(\alpha_z^{(1)2} - \alpha_z^{(2)2}\right)\right].
\end{aligned} \tag{5.84}
$$

With Eq. (4.259) and Eq. (4.263), and after removing the factor

$$
\begin{aligned}
&e^{j(\beta_+ + \beta_-)}\left[\left(N^{(0)}\alpha_z^{(0)} + N^{(1)}\alpha_z^{(1)}\right)\left(N^{(0)}\alpha_z^{(1)} + N^{(1)}\alpha_z^{(0)}\right)\right] \\
&\qquad \times\left[\left(N^{(1)}\alpha_z^{(1)} + N^{(2)}\alpha_z^{(2)}\right)\left(N^{(1)}\alpha_z^{(2)} + N^{(2)}\alpha_z^{(1)}\right)\right],
\end{aligned} \tag{5.85}
$$

we obtain

$$\begin{aligned}
&M_{11}M_{33} - M_{13}M_{31} \\
&\quad = 1 + r_{ss}^{(01)} r_{pp}^{(01)} r_{ss}^{(12)} r_{pp}^{(12)} e^{-2j\left(\beta_+ + \beta_-\right)} \\
&\quad + \frac{1}{2}\left(r_{ss}^{(01)} + r_{pp}^{(01)}\right)\left(r_{ss}^{(12)} + r_{pp}^{(12)}\right) e^{-j\left(\beta_+ + \beta_-\right)} \\
&\quad + \frac{1}{2}\left(r_{ss}^{(01)} - r_{pp}^{(01)}\right)\left(r_{ss}^{(12)} - r_{pp}^{(12)}\right) \cos\left(\beta_+ - \beta_-\right) e^{-j\left(\beta_+ + \beta_-\right)},
\end{aligned} \tag{5.86}$$

$$\begin{aligned}
&M_{21}M_{33} - M_{23}M_{31} \\
&\quad = r_{ss}^{(01)} + r_{pp}^{(01)} r_{ss}^{(12)} r_{pp}^{(12)} e^{-2j\left(\beta_+ + \beta_-\right)} \\
&\quad + \frac{1}{2}\left(1 + r_{ss}^{(01)} r_{pp}^{(01)}\right)\left(r_{ss}^{(12)} + r_{pp}^{(12)}\right) e^{-j\left(\beta_+ + \beta_-\right)} \\
&\quad + \frac{1}{2}\left(1 - r_{ss}^{(01)} r_{pp}^{(01)}\right)\left(r_{ss}^{(12)} - r_{pp}^{(12)}\right) \cos\left(\beta_+ - \beta_-\right) e^{-j\left(\beta_+ + \beta_-\right)},
\end{aligned} \tag{5.87}$$

$$\begin{aligned}
&M_{11}M_{43} - M_{41}M_{13} \\
&\quad = r_{pp}^{(01)} + r_{ss}^{(01)} r_{ss}^{(12)} r_{pp}^{(12)} e^{-2j\left(\beta_+ + \beta_-\right)} \\
&\quad + \frac{1}{2}\left(1 + r_{ss}^{(01)} r_{pp}^{(01)}\right)\left(r_{ss}^{(12)} + r_{pp}^{(12)}\right) e^{-j\left(\beta_+ + \beta_-\right)} \\
&\quad - \frac{1}{2}\left(1 - r_{ss}^{(01)} r_{pp}^{(01)}\right)\left(r_{ss}^{(12)} - r_{pp}^{(12)}\right) \cos\left(\beta_+ - \beta_-\right) e^{-j\left(\beta_+ + \beta_-\right)},
\end{aligned} \tag{5.88}$$

$$\begin{aligned}
&M_{11}M_{23} - M_{21}M_{13} = M_{41}M_{33} - M_{43}M_{31} \\
&\quad = r_{ps}^{(01)}\left(1 - r_{ss}^{(12)} r_{pp}^{(12)} e^{-2j\left(\beta_+ + \beta_-\right)}\right) \\
&\quad + \frac{1}{2} t_{pp}^{(01)} t_{ss}^{(01)} \left[\left(r_{ss}^{(12)} - r_{pp}^{(12)}\right) \sin\left(\beta_+ - \beta_-\right)\right. \\
&\quad \left. + r_{ps}^{(12)} \cos\left(\beta_+ - \beta_-\right)\right] e^{-j\left(\beta_+ + \beta_-\right)} \\
&\quad + r_{ps}^{(01)} N_y^{-2} N^{(1)2} \left[\cos\left(\beta_+ - \beta_-\right) - 1\right]\left(r_{ss}^{(12)} + r_{pp}^{(12)}\right) e^{-j\left(\beta_+ + \beta_-\right)},
\end{aligned} \tag{5.89}$$

where the single-interface isotropic Fresnel reflection and transmission coefficients are defined in Eq. (4.259). The longitudinal magnetooptic reflection coefficient, *i.e.*, the off-diagonal element of the reflection matrix, for an interface between the i-th layer (nonmagnetic) and the j-th layer (magnetic), is given in agreement with the definition (1.127) as

$$r_{23}^{(ij)} = r_{ps}^{(ij)} = r_{sp}^{(ij)} = \frac{-j\varepsilon_1^{(j)} N^{(i)} \cos\varphi^{(i)} \left(N^{(j)} \cos\varphi^{(j)}\right)^{-1} N_y N^{(j)-1}}{\left(N^{(i)} \cos\varphi^{(i)} + N^{(j)} \cos\varphi^{(j)}\right)\left(N^{(i)} \cos\varphi^{(j)} + N^{(j)} \cos\varphi^{(i)}\right)}. \tag{5.90}$$

At the normal incidence, the longitudinal Kerr effect vanishes. Consequently, the off-diagonal elements of the reflection matrix elements must be zero. This behavior is correctly reproduced even in the last term of Eq. (5.89), where the product $N_y^{-2}\left[\cos\left(\beta_+ - \beta_-\right) - 1\right]$ tends to zero with the angle of incidence $\varphi^{(0)}$.

References

1. Š. Višňovský, "Magneto-optical longitudinal Kerr effect in a film substrate system," Czech. J. Phys. B **36**, 1049–1057, 1986.
2. Š. Višňovský, K. Postava, T. Yamaguchi, and R. Lopušník, "Magneto-optic ellipsometry in exchange-coupled films," Appl. Opt. **41**, 3950–3960, 2002.
3. P. B. Johnson and R. W. Christy, "Optical constants of the noble metals," Phys. Rev. B **6**, 4370–4379, 1972.

6

Transverse Magnetization

6.1 Introduction

The transverse magnetization is confined to the plane parallel to the interfaces and perpendicular to the plane of incidence (Section 1.6.3). This configuration may be particularly advantageous in the studies of magnetically soft multilayers in the spectral regions where the efficient polarization optic elements are not available, *e.g.*, in the vacuum ultraviolet spectral region. The analysis is simpler than in the previous cases of the polar and longitudinal magnetizations. This is due to the fact that, to first order in the off-diagonal element of the permittivity tensor, only the TM wave (*i.e.*, the wave polarized parallel, p, to the plane of incidence) is affected by the transverse magnetization. The proper polarization modes are practically linearly polarized. They are, therefore, compatible with planar waveguide structures, which is not the case of the polar and longitudinal magnetization with the proper polarization modes elliptically polarized. Assuming a magnetic orthorhombic crystal with the crystallographic axes oriented normal and parallel to the interfaces, we provide the dynamical, propagation, and transfer matrices for an arbitrary angle of incidence leading to the quantities observed in the experiment, *i.e.*, the Cartesian reflection and transmission coefficients and corresponding MO ellipsometric parameters. Then, we focus on uniaxial magnetic media and analyze the reflection in a system consisting of a magnetic film on a magnetic substrate with the transverse magnetization. In the last section, we discuss planar waveguides with walls at transverse magnetization. Such waveguides exhibit nonreciprocal propagation and are exploited in isolators.

6.2 M Matrix

6.2.1 Orthorhombic Media

As before, we situate interface planes normal to the z axis and the plane of incidence perpendicular to the x axis, [1]. The magnetization is parallel to

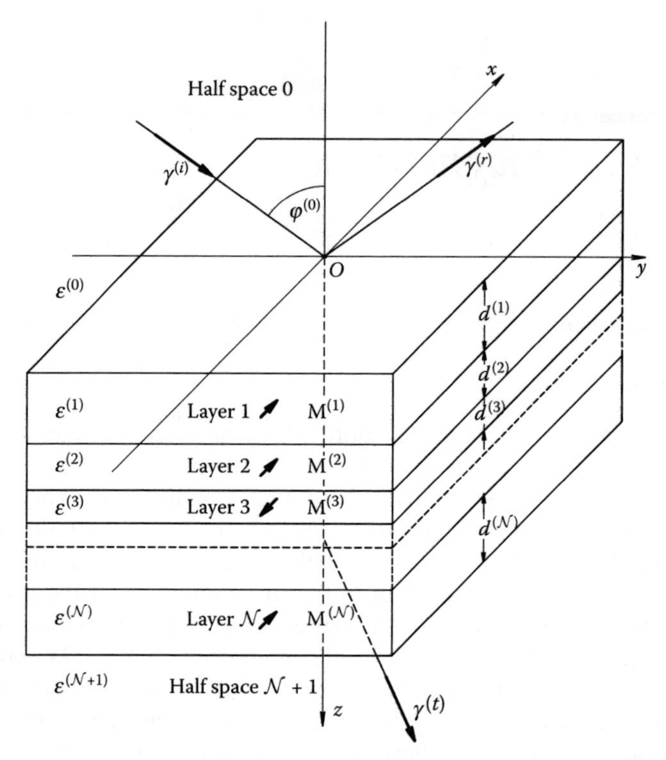

FIGURE 6.1
Oblique light incidence ($\varphi^{(0)} \neq 0$) at a multilayer consisting of $\mathcal{N}$ anisotropic layers with transverse magnetizations, $\boldsymbol{M}^{(n)}$ $(n = 1, \cdots, \mathcal{N})$. The layers are characterized by permittivity tensors $\varepsilon^{(n)}$. The incident, reflected, and transmitted propagation vectors in sandwiching isotropic nonabsorbing half spaces (0) and $(\mathcal{N}+1)$ are denoted as $\gamma^{(i)}$, $\gamma^{(r)}$, and $\gamma^{(t)}$, respectively.

the x axis, and the permittivity tensor assumes the form already employed in Eq. (1.129)

$$\varepsilon^{(n)} = \begin{pmatrix} \varepsilon_{xx}^{(n)} & 0 & 0 \\ 0 & \varepsilon_{yy}^{(n)} & \varepsilon_{yz}^{(n)} \\ 0 & \varepsilon_{zy}^{(n)} & \varepsilon_{zz}^{(n)} \end{pmatrix}. \tag{6.1}$$

It characterizes, *e.g.*, an orthorhombic crystal magnetized parallel to one of the three crystallographic axes that are free to rotate about the axis parallel to magnetization. The tensor elements may have the following meaning

$$\varepsilon_{xx}^{(n)} = \varepsilon_3^{(n)}, \tag{6.2a}$$

$$\varepsilon_{yy}^{(n)} = \varepsilon_0^{(n)} - \varepsilon_2^{(n)} \cos 2\alpha^{(n)}, \tag{6.2b}$$

$$\varepsilon_{zz}^{(n)} = \varepsilon_0^{(n)} + \varepsilon_2^{(n)} \cos 2\alpha^{(n)}, \tag{6.2c}$$

$$\varepsilon_{yz}^{(n)} = \mathrm{j}\,\varepsilon_1^{(n)} + \varepsilon_2^{(n)} \sin 2\alpha^{(n)}, \tag{6.2d}$$

$$\varepsilon_{zy}^{(n)} = -\mathrm{j}\,\varepsilon_1^{(n)} + \varepsilon_2^{(n)} \sin 2\alpha^{(n)}. \tag{6.2e}$$

From the inspection of Eqs. (4.178) and Eqs. (5.2), we can see that the parameters have the same meaning as in the polar and longitudinal geometry. Thus, $\varepsilon_1^{(n)}$ accounts for the effect of magnetization and nonzero $\varepsilon_3^{(n)} - \varepsilon_0^{(n)}$ and $\varepsilon_2^{(n)}$ at $\varepsilon_1^{(n)} = 0$ for the orthorhombic symmetry alone. At $\varepsilon_1^{(n)} \neq 0$, there is also a contribution in $\varepsilon_3^{(n)} - \varepsilon_0^{(n)}$ due to the uniaxial anisotropy introduced by the magnetic order. The angle by which the crystal is rotated about the x axis is denoted as $\alpha^{(n)}$. When $\alpha^{(n)} = 0$ or $\pi/2$, the orthorhombic and Cartesian system axes coincide.

The proper value equation (3.13) splits into two independent quadratic equations

$$\left[N_{zj}^{(n)2} - \left(\varepsilon_{xx}^{(n)} - N_y^2\right)\right] = 0, \tag{6.3a}$$

$$\left[\varepsilon_{zz}^{(n)} N_{zj}^{(n)2} + \left(\varepsilon_{yz}^{(n)} + \varepsilon_{zy}^{(n)}\right) N_y N_{zj}^{(n)} - \varepsilon_{yy}^{(n)}\left(\varepsilon_{zz}^{(n)} - N_y^2\right) + \varepsilon_{yz}^{(n)} \varepsilon_{zy}^{(n)}\right] = 0. \tag{6.3b}$$

The four solutions for the proper propagation vectors $N_{zj}^{(n)}$, with $j = 1, \ldots, 4$, expressed in the reduced form (*i.e.*, divided by the vacuum propagation constant, ω/c), become

$$N_{z1}^{(n)} = \pm\left(\varepsilon_{xx}^{(n)} - N_y^2\right)^{1/2}, \tag{6.4a}$$

$$N_{z3,4}^{(n)} = \left(2\varepsilon_{zz}^{(n)}\right)^{-1} \left\{-N_y\left(\varepsilon_{yz}^{(n)} + \varepsilon_{zy}^{(n)}\right) \pm \left[N_y^2\left(\varepsilon_{yz}^{(n)} + \varepsilon_{zy}^{(n)}\right)^2 + 4\varepsilon_{yy}^{(n)}\varepsilon_{zz}^{(n)}\left(\varepsilon_{zz}^{(n)} - N_y^2\right) - 4\varepsilon_{yz}^{(n)}\varepsilon_{zy}^{(n)}\varepsilon_{zz}^{(n)}\right]^{1/2}\right\}. \tag{6.4b}$$

For the proper values $N_{z1}^{(n)} = -N_{z2}^{(n)} = \left(\varepsilon_{xx}^{(n)} - N_y^2\right)$, the proper polarizations are

$$\mathbf{e}_{1,2}^{(n)} = \begin{pmatrix} 1 \\ 0 \\ 0 \end{pmatrix}. \tag{6.5}$$

For the other pair, *i.e.*, $N_{z3}^{(n)}$ and $N_{z4}^{(n)}$, we get

$$\boldsymbol{e}_{3,4}^{(n)} = C_{3,4}^{(n)} \begin{pmatrix} 0 \\ \varepsilon_{zz} - N_y^2 \\ -N_y N_{z3,4}^{(n)} - \varepsilon_{zy}^{(n)} \end{pmatrix}. \tag{6.6}$$

The corresponding magnetic field polarizations are obtained with Eq. (3.19)

$$\boldsymbol{b}_{1,2}^{(n)} = \begin{pmatrix} 0 \\ N_{z1,2}^{(n)} \\ -N_y \end{pmatrix} \tag{6.7}$$

and

$$\boldsymbol{b}_{3,4}^{(n)} = C_{3,4}^{(n)} \begin{pmatrix} -N_{z3,4}^{(n)}\varepsilon_{zz}^{(n)} - N_y\varepsilon_{zy}^{(n)} \\ 0 \\ 0 \end{pmatrix}. \tag{6.8}$$

From this information, we can construct the dynamical matrix for the transverse magnetization. It assumes a block diagonal form

$$\mathbf{D}^{(n)} = \begin{pmatrix} 1 & 1 & 0 & 0 \\ D_{21}^{(n)} & -D_{21}^{(n)} & 0 & 0 \\ 0 & 0 & D_{33}^{(n)} & D_{34}^{(n)} \\ 0 & 0 & D_{43}^{(n)} & D_{44}^{(n)} \end{pmatrix}, \tag{6.9}$$

where we have included the mode normalization. Then,

$$D_{21}^{(n)} = -D_{22}^{(n)} = N_{z1}^{(n)}, \tag{6.10a}$$
$$D_{33}^{(n)} = C_3^{(n)}\left(\varepsilon_{zz}^{(n)} - N_y^2\right), \tag{6.10b}$$
$$D_{34}^{(n)} = C_4^{(n)}\left(\varepsilon_{zz}^{(n)} - N_y^2\right), \tag{6.10c}$$
$$D_{43}^{(n)} = -C_3^{(n)}\left(N_y\varepsilon_{zy}^{(n)} + N_{z3}^{(n)}\varepsilon_{zz}^{(n)}\right), \tag{6.10d}$$
$$D_{44}^{(n)} = -C_4^{(n)}\left(N_y\varepsilon_{zy}^{(n)} + N_{z4}^{(n)}\varepsilon_{zz}^{(n)}\right). \tag{6.10e}$$

Here, $C_3^{(n)}$ and $C_4^{(n)}$ are the normalizing constants. We can write as in Eq. (4.188) for the normalized dynamical matrix in transverse magnetization

$$\mathbf{D}^{(n)} = \begin{pmatrix} 1 & 1 & 0 & 0 \\ N_{z1}^{(n)} & -N_{z1}^{(n)} & 0 & 0 \\ 0 & 0 & \left(\varepsilon_{zz}^{(n)} - N_y^2\right) & \left(\varepsilon_{zz}^{(n)} - N_y^2\right) \\ 0 & 0 & -\left(N_y\varepsilon_{zy}^{(n)} + N_{z3}^{(n)}\varepsilon_{zz}^{(n)}\right) & -\left(N_y\varepsilon_{zy}^{(n)} + N_{z4}^{(n)}\varepsilon_{zz}^{(n)}\right) \end{pmatrix}$$
$$\times \begin{pmatrix} 1 & 0 & 0 & 0 \\ 0 & 1 & 0 & 0 \\ 0 & 0 & C_3^{(n)} & 0 \\ 0 & 0 & 0 & C_4^{(n)} \end{pmatrix}. \tag{6.11}$$

We observe that the modes with $j = 1$ and 2 polarized perpendicular to the plane of incidence (s polarized or TE modes) have no coupling to the modes with $j = 3$ and 4 polarized parallel to the plane of incidence (p polarized or TM modes). The inverse matrix $(\mathbf{D}^{(n)})^{-1}$ is given by

$$(\mathbf{D}^{(n)})^{-1} = \begin{pmatrix} 2^{-1} & \left(2D_{21}^{(n)}\right)^{-1} & 0 & 0 \\ 2^{-1} & -\left(2D_{21}^{(n)}\right)^{-1} & 0 & 0 \\ 0 & 0 & L_{33}^{(n)} & L_{34}^{(n)} \\ 0 & 0 & L_{43}^{(n)} & L_{44}^{(n)} \end{pmatrix}, \tag{6.12}$$

where

$$L_{33}^{(n)} = D_{44}^{(n)}\left(D_{33}^{(n)}D_{44}^{(n)} - D_{34}^{(n)}D_{43}^{(n)}\right)^{-1}, \tag{6.13a}$$

$$L_{34}^{(n)} = -D_{34}^{(n)}\left(D_{33}^{(n)}D_{44}^{(n)} - D_{34}^{(n)}D_{43}^{(n)}\right)^{-1}, \tag{6.13b}$$

$$L_{43}^{(n)} = -D_{43}^{(n)}\left(D_{33}^{(n)}D_{44}^{(n)} - D_{34}^{(n)}D_{43}^{(n)}\right)^{-1}, \tag{6.13c}$$

$$L_{44}^{(n)} = D_{33}^{(n)}\left(D_{33}^{(n)}D_{44}^{(n)} - D_{34}^{(n)}D_{43}^{(n)}\right)^{-1}. \tag{6.13d}$$

Using the normal components of the proper propagation vectors given in Eq. (6.4), we can construct the propagation matrix, the general form of which was given in Eq. (3.31)

$$\mathbf{P}^{(n)} = \begin{pmatrix} \exp\left(j\frac{\omega}{c}N_{z1}^{(n)}d^{(n)}\right) & 0 & 0 & 0 \\ 0 & \exp\left(-j\frac{\omega}{c}N_{z1}^{(n)}d^{(n)}\right) & 0 & 0 \\ 0 & 0 & \exp\left(j\frac{\omega}{c}N_{z3}^{(n)}d^{(n)}\right) & 0 \\ 0 & 0 & 0 & \exp\left(j\frac{\omega}{c}N_{z4}^{(n)}d^{(n)}\right) \end{pmatrix}, \tag{6.14}$$

where $d^{(n)}$ is the thickness of the n-th layer. The $\mathbf{F}$ matrix characterizing an interface between two media at the transverse magnetization follows from the product $\mathbf{F} = \left(\mathbf{D}^{(n-1)}\right)^{-1}\mathbf{D}^{(n)}$

$$\mathbf{F} = \begin{pmatrix} F_{11} & F_{12} & 0 & 0 \\ F_{12} & F_{11} & 0 & 0 \\ 0 & 0 & F_{33} & F_{34} \\ 0 & 0 & F_{43} & F_{44} \end{pmatrix}, \tag{6.15}$$

where

$$F_{\substack{11\\12}} = \frac{D_{21}^{(n-1)} \pm D_{21}^{(n)}}{2D_{21}^{(n-1)}}, \tag{6.16a}$$

$$F_{33} = L_{33}^{(n-1)} D_{33}^{(n)} + L_{34}^{(n-1)} D_{43}^{(n)}, \tag{6.16b}$$

$$F_{34} = L_{33}^{(n-1)} D_{34}^{(n)} + L_{34}^{(n-1)} D_{44}^{(n)}, \tag{6.16c}$$

$$F_{43} = L_{43}^{(n-1)} D_{33}^{(n)} + L_{44}^{(n-1)} D_{43}^{(n)}, \tag{6.16d}$$

$$F_{44} = L_{43}^{(n-1)} D_{34}^{(n)} + L_{44}^{(n-1)} D_{44}^{(n)}. \tag{6.16e}$$

The product $\mathbf{A} = (\mathbf{D}^{(0)})^{-1}\mathbf{D}^{(1)}$ characterizes the interface between isotropic entrance medium and the first layer at the transverse magnetization. Multiplied from the right by the propagation matrix of the first layer, it becomes the first transfer matrix $\mathbf{T}_{01} = (\mathbf{D}^{(0)})^{-1}\mathbf{D}^{(1)}\mathbf{P}^{(1)}$ in the product (3.33) forming the $\mathbf{M}$ matrix. The product $\mathbf{Z} = (\mathbf{D}^{(\mathcal{N})})^{-1}\mathbf{D}^{(\mathcal{N}+1)}$ characterizes the last interface at the multilayer boundary with the isotropic half space and represents the last transfer matrix of the product (3.33). Both $\mathbf{A}$ and $\mathbf{Z}$ have the same block diagonal structure as the above $\mathbf{F}$ matrix, *i.e.*,

$$\mathbf{A} = \begin{pmatrix} A_{11} & A_{12} & 0 & 0 \\ A_{12} & A_{11} & 0 & 0 \\ 0 & 0 & A_{33} & A_{34} \\ 0 & 0 & A_{43} & A_{44} \end{pmatrix}, \tag{6.17}$$

$$\mathbf{Z} = \begin{pmatrix} Z_{11} & Z_{12} & 0 & 0 \\ Z_{12} & Z_{11} & 0 & 0 \\ 0 & 0 & Z_{33} & Z_{34} \\ 0 & 0 & Z_{43} & Z_{44} \end{pmatrix}. \tag{6.18}$$

The expression for $(\mathbf{D}^{(0)})^{-1}$ was already obtained in Eq. (3.82b)

$$(\mathbf{D}^{(0)})^{-1} = (2N^{(0)}\cos\varphi^{(0)})^{-1} \begin{pmatrix} N^{(0)}\cos\varphi^{(0)} & 1 & 0 & 0 \\ N^{(0)}\cos\varphi^{(0)} & -1 & 0 & 0 \\ 0 & 0 & N^{(0)} & -\cos\varphi^{(0)} \\ 0 & 0 & N^{(0)} & \cos\varphi^{(0)} \end{pmatrix}, \tag{6.19}$$

the matrix $\mathbf{D}^{(\mathcal{N}+1)}$ is given by Eq. (4.203), *i.e.*,

$$\mathbf{D}^{(\mathcal{N}+1)} = \begin{pmatrix} 1 & 1 & 0 & 0 \\ N^{(\mathcal{N}+1)}\cos\varphi^{(\mathcal{N}+1)} & -N^{(\mathcal{N}+1)}\cos\varphi^{(\mathcal{N}+1)} & 0 & 0 \\ 0 & 0 & \cos\varphi^{(\mathcal{N}+1)} & \cos\varphi^{(\mathcal{N}+1)} \\ 0 & 0 & -N^{(\mathcal{N}+1)} & N^{(\mathcal{N}+1)} \end{pmatrix}. \tag{6.20}$$

The nonzero elements of the **A** and **Z** matrices are as follows

$$A_{\substack{11\\12}} = N^{(0)}\cos\varphi^{(0)} \pm D_{21}^{(1)}, \tag{6.21a}$$

$$A_{\substack{33\\43}} = N^{(0)}D_{33}^{(1)} \mp \cos\varphi^{(0)}\,D_{43}^{(1)}, \tag{6.21b}$$

$$A_{\substack{34\\44}} = N^{(0)}D_{34}^{(1)} \mp \cos\varphi^{(0)}\,D_{44}^{(1)}, \tag{6.21c}$$

$$Z_{\substack{11\\12}} = L_{11}^{(\mathcal{N})} \mp L_{12}^{(\mathcal{N})}N^{(\mathcal{N}+1)}\cos\varphi^{(\mathcal{N}+1)}, \tag{6.21d}$$

$$Z_{\substack{33\\34}} = L_{33}^{(\mathcal{N})}\cos\varphi^{(\mathcal{N}+1)} \mp L_{34}^{(\mathcal{N})}N^{(\mathcal{N}+1)}, \tag{6.21e}$$

$$Z_{\substack{43\\44}} = L_{43}^{(\mathcal{N})}\cos\varphi^{(\mathcal{N}+1)} \mp L_{44}^{(\mathcal{N})}N^{(\mathcal{N}+1)}, \tag{6.21f}$$

the remaining elements being zero.

6.2.2 Uniaxial Media

We now focus on the special form of the permittivity tensor in Eq. (6.1) with the principal axis parallel to the magnetization [2], which takes place at $\varepsilon_2^{(n)} = 0$, *i.e.*,

$$\varepsilon^{(n)} = \begin{pmatrix} \varepsilon_3^{(n)} & 0 & 0 \\ 0 & \varepsilon_0 & \mathrm{j}\,\varepsilon_1^{(n)} \\ 0 & -\mathrm{j}\,\varepsilon_1^{(n)} & \varepsilon_0^{(n)} \end{pmatrix}. \tag{6.22}$$

Assuming the plane wave solution

$$\boldsymbol{E}^{(n)} = \boldsymbol{E}_0^{(n)}\exp\left[\mathrm{j}(\omega t - \boldsymbol{\gamma}^{(n)}\cdot\mathbf{r})\right] \tag{6.23}$$

and choosing the propagation vector $\boldsymbol{\gamma}$ oriented perpendicular to the x axis, *i.e.*,

$$\boldsymbol{\gamma}^{(n)} = \frac{\omega}{c}\left(N_y\hat{\boldsymbol{y}} + N_z^{(n)}\hat{\boldsymbol{z}}\right), \tag{6.24}$$

the wave equation (3.12) in the n-th layer medium becomes

$$\begin{pmatrix} \left(\varepsilon_3^{(n)} - N_y^2 - N_z^{(n)2}\right) & 0 & 0 \\ 0 & \left(\varepsilon_0^{(n)} - N_z^{(n)2}\right) & \left(N_y N_z^{(n)} + \mathrm{j}\,\varepsilon_1^{(n)}\right) \\ 0 & \left(N_y N_z^{(n)} - \mathrm{j}\,\varepsilon_1^{(n)}\right) & \left(\varepsilon_0^{(n)} - N_y^{(n)2}\right) \end{pmatrix} \begin{pmatrix} E_{0x}^{(n)} \\ E_{0y}^{(n)} \\ E_{0z}^{(n)} \end{pmatrix} = 0, \tag{6.25}$$

which provides the following proper value equation

$$\left(\varepsilon_3^{(n)} - N_y^2 - N_z^{(n)2}\right)\left[\varepsilon_0^{(n)2} - \left(N_z^{(n)2} + N_y^2\right)\varepsilon_0^{(n)} - \varepsilon_1^{(n)2}\right] = 0. \tag{6.26}$$

The four proper values are

$$N_{z1}^{(n)} = N_{z\perp}^{(n)} = \left(\varepsilon_3^{(n)} - N_y^2\right)^{1/2} \tag{6.27a}$$

$$N_{z2}^{(n)} = -N_{z\perp}^{(n)} = -\left(\varepsilon_3^{(n)} - N_y^2\right)^{1/2} \tag{6.27b}$$

$$N_{z3}^{(n)} = N_{z\parallel}^{(n)} = \left[\left(\varepsilon_0^{(n)} - N_y^2\right) - \varepsilon_1^{(n)2}\varepsilon_0^{(n)-1}\right]^{1/2} \tag{6.27c}$$

$$N_{z4}^{(n)} = -N_{z\parallel}^{(n)} = -\left[\left(\varepsilon_0^{(n)} - N_y^2\right) - \varepsilon_1^{(n)2}\varepsilon_0^{(n)-1}\right]^{1/2}. \tag{6.27d}$$

Their absolute values do not differ to first order in the off-diagonal tensor element. The associated proper polarization modes are either strictly linearly polarized normal to the plane of incidence (TE or s polarized), *i.e.*, parallel to the x axis ($\boldsymbol{e}_{1,2}^{(n)} = \hat{\boldsymbol{x}}$) and propagating with $N_{z\perp}^{(n)}$, or slightly elliptically polarized in the plane of incidence $x = 0$ and propagating with $N_{z\parallel}^{(n)}$

$$\boldsymbol{e}_{3,4}^{(n)} = C_{3,4}^{(n)} \begin{pmatrix} 0 \\ \varepsilon_0^{(n)} - N_y^2 \\ \mp N_y N_{z\parallel}^{(n)} + \mathrm{j}\,\varepsilon_1^{(n)} \end{pmatrix}. \tag{6.28}$$

The latter polarization modes are therefore TM (or p) polarized with the electric field approximately linearly polarized with a small electric field component proportional to the off-diagonal element $\varepsilon_1^{(n)}$ and parallel to the propagation vector $\gamma_{3,4}^{(n)} = \hat{\boldsymbol{y}} N_y \pm \hat{\boldsymbol{z}} N_{z\parallel}^{(n)}$. The corresponding magnetic fields are

$$\mathbf{b}_{1,2}^{(n)} = \begin{pmatrix} 0 \\ \pm N_{z\perp}^{(n)} \\ -N_y \end{pmatrix} \tag{6.29}$$

and

$$\mathbf{b}_{3,4}^{(n)} = C_{3,4}^{(n)}\left(\mp\,\varepsilon_{zz}^{(n)} N_{z\parallel}^{(n)} + \mathrm{j}\,\varepsilon_1^{(n)} N_y\right)\hat{\boldsymbol{x}}. \tag{6.30}$$

The dynamical and propagation matrices take the form

$$\mathbf{D}^{(n)} = \begin{pmatrix} 1 & 1 & 0 & 0 \\ N_{z\perp}^{(n)} & -N_{z\perp}^{(n)} & 0 & 0 \\ 0 & 0 & C_3^{(n)}\left(\varepsilon_0^{(n)} - N_y^2\right) & C_4^{(n)}\left(\varepsilon_0^{(n)} - N_y^2\right) \\ 0 & 0 & -C_3^{(n)}\left(\varepsilon_0^{(n)} N_{z\parallel}^{(n)} - \mathrm{j}\,\varepsilon_1^{(n)} N_y\right) & C_4^{(n)}\left(\varepsilon_0^{(n)} N_{z\parallel}^{(n)} + \mathrm{j}\,\varepsilon_1^{(n)} N_y\right) \end{pmatrix}, \tag{6.31}$$

$$\mathbf{P}^{(n)} = \begin{pmatrix} e^{j\beta_\perp^{(n)}} & 0 & 0 & 0 \\ 0 & e^{-j\beta_\perp^{(n)}} & 0 & 0 \\ 0 & 0 & e^{j\beta_\parallel^{(n)}} & 0 \\ 0 & 0 & 0 & e^{-j\beta_\parallel^{(n)}} \end{pmatrix}, \tag{6.32}$$

where $\beta_\perp^{(n)} = (\omega/c)N_{z\perp}^{(n)} d_n$ and $\beta_\parallel^{(n)} = (\omega/c)N_{z\parallel}^{(n)} d_n$, d_n being the thickness of the n-th layer. The inverse to the dynamical matrix of Eq. (6.31) is a special case of Eq. (6.12)

$$(\mathbf{D}^{(n)})^{-1} = \begin{pmatrix} 2^{-1} & \left(2N_{z\perp}^{(n)}\right)^{-1} & 0 & 0 \\ 2^{-1} & -\left(2N_{z\perp}^{(n)}\right)^{-1} & 0 & 0 \\ 0 & 0 & L_{33}^{(n)} & L_{34}^{(n)} \\ 0 & 0 & L_{43}^{(n)} & L_{44}^{(n)} \end{pmatrix}, \tag{6.33}$$

where

$$L_{33}^{(n)} = C_4^{(n)}\left(\varepsilon_0^{(n)} N_{z\parallel}^{(n)} + j\varepsilon_1^{(n)} N_y\right)\left[2C_3^{(n)}C_4^{(n)}\left(\varepsilon_0^{(n)} - N_y^2\right)\varepsilon_0^{(n)} N_{z\parallel}^{(n)}\right]^{-1}, \tag{6.34a}$$

$$L_{34}^{(n)} = -C_4^{(n)}\left(\varepsilon_0^{(n)} - N_y^2\right)\left[2C_3^{(n)}C_4^{(n)}\left(\varepsilon_0^{(n)} - N_y^2\right)\varepsilon_0^{(n)} N_{z\parallel}^{(n)}\right]^{-1}, \tag{6.34b}$$

$$L_{43}^{(n)} = C_3^{(n)}\left(\varepsilon_0^{(n)} N_{z\parallel}^{(n)} - j\varepsilon_1^{(n)} N_y\right)\left[2C_3^{(n)}C_4^{(n)}\left(\varepsilon_0^{(n)} - N_y^2\right)\varepsilon_0^{(n)} N_{z\parallel}^{(n)}\right]^{-1}, \tag{6.34c}$$

$$L_{44}^{(n)} = C_3^{(n)}\left(\varepsilon_0^{(n)} - N_y^2\right)\left[2C_3^{(n)}C_4^{(n)}\left(\varepsilon_0^{(n)} - N_y^2\right)\varepsilon_0^{(n)} N_{z\parallel}^{(n)}\right]^{-1}. \tag{6.34d}$$

The **F** matrix pertinent to the interface of two media at transverse magnetization, $\mathbf{F} = (\mathbf{D}^{(n-1)})^{-1}\mathbf{D}^{(n)}$, assumes the form given in Eq. (6.15). The matrices **A** and **Z** associated with the first and last multilayer interface follow from the procedure similar to that employed in Eq. (6.21).

6.3 Film-Substrate System at Transverse Magnetization

6.3.1 M Matrix of the System

The system consisting of isotropic half space, film, and substrate at transverse magnetization is represented by the matrix product

$$\mathbf{M} = (\mathbf{D}^{(0)})^{-1}\mathbf{D}^{(1)}\mathbf{P}^{(1)}(\mathbf{D}^{(1)})^{-1}\mathbf{D}^{(2)} = \mathbf{A}\mathbf{P}^{(1)}\mathbf{F}. \tag{6.35}$$

We assume the substrate sufficiently thick and absorbing so that there are no backward (retrograde) waves from the substrate that could affect reflection at the front multilayer boundary in the isotropic medium. We obtain for the matrix elements of $\mathbf{A} = (\mathbf{D}^{(0)})^{-1}\mathbf{D}^{(1)}$ defined in Eq. (6.17)

$$A_{11} = N^{(0)} \cos\varphi^{(0)} + N_{z\perp}^{(1)} \tag{6.36a}$$

$$A_{12} = N^{(0)} \cos\varphi^{(0)} - N_{z\perp}^{(1)} \tag{6.36b}$$

$$A_{33} = C_3^{(1)} \left[N^{(0)} N_z^{(1)2} + \left(\varepsilon_0^{(1)} N_{z\parallel}^{(1)} - \mathrm{j}\,\varepsilon_1^{(1)} N_y\right) \cos\varphi^{(0)}\right] \tag{6.36c}$$

$$A_{34} = C_4^{(1)} \left[N^{(0)} N_z^{(1)2} - \left(\varepsilon_0^{(1)} N_{z\parallel}^{(1)} + \mathrm{j}\,\varepsilon_1^{(1)} N_y\right) \cos\varphi^{(0)}\right] \tag{6.36d}$$

$$A_{43} = C_3^{(1)} \left[N^{(0)} N_z^{(1)2} - \left(\varepsilon_0^{(1)} N_{z\parallel}^{(1)} - \mathrm{j}\,\varepsilon_1^{(1)} N_y\right) \cos\varphi^{(0)}\right] \tag{6.36e}$$

$$A_{44} = C_4^{(1)} \left[N^{(0)} N_z^{(1)2} + \left(\varepsilon_0^{(1)} N_{z\parallel}^{(1)} + \mathrm{j}\,\varepsilon_1^{(1)} N_y\right) \cos\varphi^{(0)}\right], \tag{6.36f}$$

and for the elements of the matrix $\mathbf{F} = (\mathbf{D}^{(1)})^{-1}\mathbf{D}^{(2)}$, defined in Eq. (6.15) and pertaining to an interface between two media at the transverse magnetization,

$$F_{11} = \frac{1}{2N_{z\perp}^{(1)}} \left(N_{z\perp}^{(1)} + N_{z\perp}^{(2)}\right) \tag{6.37a}$$

$$F_{12} = \frac{1}{2N_{z\perp}^{(1)}} \left(N_{z\perp}^{(1)} - N_{z\perp}^{(2)}\right) \tag{6.37b}$$

$$F_{33} = C_3^{(2)} \left(2C_3^{(1)} N_{z0}^{(1)2} \varepsilon_0^{(1)} N_{z\parallel}^{(1)}\right)^{-1} \times \left[\left(\varepsilon_0^{(1)} N_{z\parallel}^{(1)} N_{z0}^{(2)2} + \varepsilon_0^{(2)} N_{z\parallel}^{(2)} N_{z0}^{(1)2}\right) + \mathrm{j}\left(\varepsilon_1^{(1)} N_{z0}^{(2)2} - \varepsilon_1^{(2)} N_{z0}^{(1)2}\right) N_y\right] \tag{6.37c}$$

$$F_{34} = C_4^{(2)} \left[2C_3^{(1)} \left(\varepsilon_0^{(1)} - N_y^2\right) \varepsilon_0^{(1)} N_{z\parallel}^{(1)}\right]^{-1} \times \left[\left(\varepsilon_0^{(1)} N_{z\parallel}^{(1)} + \mathrm{j}\,\varepsilon_1^{(1)} N_y\right) N_{z0}^{(2)2} - N_{z0}^{(1)2} \left(\varepsilon_0^{(2)} N_{z\parallel}^{(2)} + \mathrm{j}\,\varepsilon_1^{(2)} N_y\right)\right] \tag{6.37d}$$

$$F_{43} = C_3^{(2)} \left(2C_4^{(1)} N_{z0}^{(1)2} \varepsilon_0^{(1)} N_{z\parallel}^{(1)}\right)^{-1} \times \left[\left(\varepsilon_0^{(1)} N_{z\parallel}^{(1)} N_{z0}^{(2)2} - \varepsilon_0^{(2)} N_{z\parallel}^{(2)} N_{z0}^{(1)2}\right) - \mathrm{j}\left(\varepsilon_1^{(1)} N_{z0}^{(2)2} - \varepsilon_1^{(2)} N_{z0}^{(1)2}\right) N_y\right] \tag{6.37e}$$

$$F_{44} = C_4^{(2)} \left(2C_4^{(1)} N_{z0}^{(1)2} \varepsilon_0^{(1)} N_{z\parallel}^{(1)}\right)^{-1} \times \left[\left(\varepsilon_0^{(1)} N_{z\parallel}^{(1)} N_{z0}^{(2)2} + \varepsilon_0^{(2)} N_{z\parallel}^{(2)} N_{z0}^{(1)2}\right) - \mathrm{j}\left(\varepsilon_1^{(1)} N_{z0}^{(2)2} - \varepsilon_1^{(2)} N_{z0}^{(1)2}\right) N_y\right]. \tag{6.37f}$$

We have introduced the abbreviation $\varepsilon_0^{(n)} - N_y^2 = N_{z0}^{(n)2}$. The use was made of Eq. (6.16).

Another useful representation of the multilayer employs the characteristic matrix $\mathbf{S}^{(n)}$ defined as a product $\mathbf{D}^{(n)}\mathbf{P}^{(n)}(\mathbf{D}^{(n)})^{-1}$. We find from Eq. (6.9) to Eq. (6.14) using Eqs. (6.31) to (6.33)

$$\mathbf{S}^{(n)} = \begin{pmatrix} 1 & 1 & 0 & 0 \\ D_{21}^{(n)} & -D_{21}^{(n)} & 0 & 0 \\ 0 & 0 & D_{33}^{(n)} & D_{34}^{(n)} \\ 0 & 0 & D_{43}^{(n)} & D_{44}^{(n)} \end{pmatrix} \begin{pmatrix} e^{j\beta_{\perp}^{(n)}} & 0 & 0 & 0 \\ 0 & e^{-j\beta_{\perp}^{(n)}} & 0 & 0 \\ 0 & 0 & e^{j\beta_{\parallel}^{(n)}} & 0 \\ 0 & 0 & 0 & e^{-j\beta_{\parallel}^{(n)}} \end{pmatrix}$$

$$\times \begin{pmatrix} 2^{-1} & (2D_{21}^{(n)})^{-1} & 0 & 0 \\ 2^{-1} & -(2D_{21}^{(n)})^{-1} & 0 & 0 \\ 0 & 0 & L_{33}^{(n)} & L_{34}^{(n)} \\ 0 & 0 & L_{43}^{(n)} & L_{44}^{(n)} \end{pmatrix} = \begin{pmatrix} S_{11}^{(n)} & S_{12}^{(n)} & 0 & 0 \\ S_{21}^{(n)} & S_{22}^{(n)} & 0 & 0 \\ 0 & 0 & S_{33}^{(n)} & S_{34}^{(n)} \\ 0 & 0 & S_{43}^{(n)} & S_{44}^{(n)} \end{pmatrix}, \quad (6.38)$$

where

$$S_{11}^{(n)} = S_{22}^{(n)} = \cos\beta_{\perp}^{(n)} \quad (6.39a)$$

$$S_{12}^{(n)} = jD_{21}^{(1)-1}\sin\beta_{\perp}^{(n)} \quad (6.39b)$$

$$S_{21}^{(n)} = jD_{21}^{(1)}\sin\beta_{\perp}^{(n)} \quad (6.39c)$$

$$S_{33}^{(n)} = D_{33}^{(n)}L_{33}^{(n)}e^{j\beta_{\parallel}^{(n)}} + D_{34}^{(n)}L_{43}^{(n)}e^{-j\beta_{\parallel}^{(n)}} \quad (6.39d)$$

$$S_{34}^{(n)} = D_{33}^{(n)}L_{34}^{(n)}e^{j\beta_{\parallel}^{(n)}} + D_{34}^{(n)}L_{44}^{(n)}e^{-j\beta_{\parallel}^{(n)}} \quad (6.39e)$$

$$S_{43}^{(n)} = D_{43}^{(n)}L_{33}^{(n)}e^{j\beta_{\parallel}^{(n)}} + D_{44}^{(n)}L_{43}^{(n)}e^{-j\beta_{\parallel}^{(n)}} \quad (6.39f)$$

$$S_{44}^{(n)} = D_{43}^{(n)}L_{34}^{(n)}e^{j\beta_{\parallel}^{(n)}} + D_{44}^{(n)}L_{44}^{(n)}e^{-j\beta_{\parallel}^{(n)}}. \quad (6.39g)$$

Eq. (6.35) can then be written

$$\mathbf{M} = (\mathbf{D}^{(0)})^{-1}\mathbf{S}^{(1)}\mathbf{D}^{(2)} \quad (6.40)$$

The complete **M** matrix is block diagonal in the transversal geometry, and its matrix elements are given by

$$\mathrm{M}_{\substack{11\\22}} = A_{11}F_{11}e^{\pm j\beta_{\perp}^{(1)}} + A_{12}F_{12}e^{\mp j\beta_{\perp}^{(1)}}, \quad (6.41a)$$

$$\mathrm{M}_{\substack{12\\21}} = A_{11}F_{12}e^{\pm j\beta_{\perp}^{(1)}} + A_{12}F_{11}e^{\mp j\beta_{\perp}^{(1)}}, \quad (6.41b)$$

$$\mathrm{M}_{33} = A_{33}F_{33}e^{j\beta_{\parallel}^{(1)}} + A_{34}F_{43}e^{-j\beta_{\parallel}^{(1)}}, \quad (6.41c)$$

$$\mathrm{M}_{34} = A_{33}F_{34}e^{j\beta_{\parallel}^{(1)}} + A_{34}F_{44}e^{-j\beta_{\parallel}^{(1)}}, \quad (6.41d)$$

$$\mathrm{M}_{43} = A_{43}F_{33}e^{j\beta_{\parallel}^{(1)}} + A_{44}F_{43}e^{-j\beta_{\parallel}^{(1)}}, \quad (6.41e)$$

$$\mathrm{M}_{44} = A_{43}F_{34}e^{j\beta_{\parallel}^{(1)}} + A_{44}F_{44}e^{-j\beta_{\parallel}^{(1)}}. \quad (6.41f)$$

We have denoted $\beta_{\perp}^{(n)} = (\omega/c)N_{z\perp}^{(1)}d^{(1)}$ and $\beta_{\parallel}^{(n)} = (\omega/c)N_{z\parallel}^{(1)}d^{(1)}$, $d^{(1)}$ being the thickness of the film.

6.3.2 Reflection Characteristics

In the transverse geometry considered here, the Cartesian Jones reflection matrix is diagonal and the global reflection coefficients we find from Eq. (3.70a) and Eq. (3.70c)

$$r_{21}^{(0,\mathcal{N}+1)} = \left(\frac{\mathrm{E}_{02}^{(0)}}{\mathrm{E}_{01}^{(0)}}\right)_{\mathrm{E}_{03}^{(0)}=0} = \frac{\mathrm{M}_{21}\mathrm{M}_{33} - \mathrm{M}_{23}\mathrm{M}_{31}}{\mathrm{M}_{11}\mathrm{M}_{33} - \mathrm{M}_{13}\mathrm{M}_{31}}, \tag{6.42a}$$

$$r_{43}^{(0,\mathcal{N}+1)} = \left(\frac{\mathrm{E}_{04}^{(0)}}{\mathrm{E}_{03}^{(0)}}\right)_{\mathrm{E}_{01}^{(0)}=0} = \frac{\mathrm{M}_{11}\mathrm{M}_{43} - \mathrm{M}_{41}\mathrm{M}_{13}}{\mathrm{M}_{11}\mathrm{M}_{33} - \mathrm{M}_{13}\mathrm{M}_{31}}, \tag{6.42b}$$

for the transverse magnetization

$$r_{21}^{(02)} = r_{ss}^{(02)} = \frac{E_{0\perp}^{(r)}}{E_{0\perp}^{(i)}} = \frac{\mathrm{M}_{21}}{\mathrm{M}_{11}} \qquad \left(E_{0\parallel}^{(i)} = 0\right), \tag{6.43a}$$

$$r_{43}^{(02)} = r_{pp}^{(02)} = \frac{E_{0\parallel}^{(r)}}{E_{0\parallel}^{(i)}} = \frac{\mathrm{M}_{43}}{\mathrm{M}_{33}} \qquad \left(E_{0\perp}^{(i)} = 0\right). \tag{6.43b}$$

E_0 stands for the electric field amplitudes of the incident (i) and reflected (r) waves polarized either normal ($\perp$) or parallel ($\parallel$) to the plane of incidence. We can thus limit ourselves to the four elements M_{11}, M_{21}, M_{33}, and M_{43} in Eq. (6.41)

$$\mathrm{M}_{11} = A_{11}F_{11}\mathrm{e}^{\mathrm{j}\beta_{\perp}^{(1)}} + A_{12}F_{12}\mathrm{e}^{-\mathrm{j}\beta_{\perp}^{(1)}} \tag{6.44a}$$

$$\mathrm{M}_{21} = A_{11}F_{12}\mathrm{e}^{-\mathrm{j}\beta_{\perp}^{(1)}} + A_{12}F_{11}\mathrm{e}^{\mathrm{j}\beta_{\perp}^{(1)}} \tag{6.44b}$$

$$\mathrm{M}_{33} = A_{33}F_{33}\mathrm{e}^{\mathrm{j}\beta_{\parallel}^{(1)}} + A_{34}F_{43}\mathrm{e}^{-\mathrm{j}\beta_{\parallel}^{(1)}} \tag{6.44c}$$

$$\mathrm{M}_{43} = A_{43}F_{33}\mathrm{e}^{\mathrm{j}\beta_{\parallel}^{(1)}} + A_{44}F_{43}\mathrm{e}^{-\mathrm{j}\beta_{\parallel}^{(1)}} \tag{6.44d}$$

and obtain the global reflection coefficients $r_{21}^{(02)}$ and $r_{43}^{(02)}$

$$r_{21}^{(02)} = r_{ss}^{(02)} = \frac{r_{ss}^{(01)} + \mathrm{e}^{-2\mathrm{j}\beta_{\perp}^{(1)}} r_{ss}^{(12)}}{1 + \mathrm{e}^{-2\mathrm{j}\beta_{\perp}^{(1)}} r_{ss}^{(01)} r_{ss}^{(12)}}, \tag{6.45}$$

where

$$r_{ss}^{(01)} = \frac{N^{(0)}\cos\varphi^{(0)} - N_{z\perp}^{(1)}}{N^{(0)}\cos\varphi^{(0)} + N_{z\perp}^{(1)}} \tag{6.46a}$$

and

$$r_{ss}^{(12)} = \frac{N_{(z)}^{(1)\perp} - N_{z\perp}^{(2)}}{N_{(z)}^{(1)\perp} + N_{z\perp}^{(2)}} \tag{6.46b}$$

for the waves polarized normal to the plane of incidence. The reflection coefficient for the waves polarized in the plane of incidence has the same structure

$$r_{43}^{(02)} = r_{pp}^{(02)}\left(\varepsilon_1^{(1)}, \varepsilon_1^{(2)}\right) = \frac{\mathrm{M}_{43}}{\mathrm{M}_{33}} = \frac{\dfrac{A_{43}}{A_{33}} + \dfrac{A_{44}}{A_{33}}\dfrac{F_{43}}{F_{33}} \exp\left(-2\mathrm{j}\beta_{\parallel}^{(1)}\right)}{1 + \dfrac{A_{34}}{A_{33}}\dfrac{F_{43}}{F_{33}} \exp\left(-2\mathrm{j}\beta_{\parallel}^{(1)}\right)} \tag{6.47}$$

where the single-interface reflection coefficients perturbed by the transverse magnetization are

$$\frac{A_{43}}{A_{33}} = \frac{N^{(0)} N_{z0}^{(1)2} - \varepsilon_0^{(1)} N_{z\parallel}^{(1)} \left(1 - \dfrac{\mathrm{j}\,\varepsilon_1^{(1)} N_y}{\varepsilon_0^{(1)} N_{z\parallel}^{(1)}}\right) \cos\varphi^{(0)}}{N^{(0)} N_{z0}^{(1)2} + \varepsilon_0^{(1)} N_{z\parallel}^{(1)} \left(1 - \dfrac{\mathrm{j}\,\varepsilon_1^{(1)} N_y}{\varepsilon_0^{(1)} N_{z\parallel}^{(1)}}\right) \cos\varphi^{(0)}}, \tag{6.48a}$$

$$\frac{A_{44}}{A_{33}} = \frac{C_4^{(1)}}{C_3^{(1)}} \frac{N^{(0)} N_{z0}^{(1)2} + \varepsilon_0^{(1)} N_{z\parallel}^{(1)} \left(1 + \dfrac{\mathrm{j}\,\varepsilon_1^{(1)} N_y}{\varepsilon_0^{(1)} N_{z\parallel}^{(1)}}\right) \cos\varphi^{(0)}}{N^{(0)} N_{z0}^{(1)2} + \varepsilon_0^{(1)} N_{z\parallel}^{(1)} \left(1 - \dfrac{\mathrm{j}\,\varepsilon_1^{(1)} N_y}{\varepsilon_0^{(1)} N_{z\parallel}^{(1)}}\right) \cos\varphi^{(0)}}, \tag{6.48b}$$

$$\frac{A_{34}}{A_{33}} = \frac{C_4^{(1)}}{C_3^{(1)}} \frac{N^{(0)} N_{z0}^{(1)2} - \varepsilon_0^{(1)} N_{z\parallel}^{(1)} \left(1 + \dfrac{\mathrm{j}\,\varepsilon_1^{(1)} N_y}{\varepsilon_0^{(1)} N_{z\parallel}^{(1)}}\right) \cos\varphi^{(0)}}{N^{(0)} N_{z0}^{(1)2} + \varepsilon_0^{(1)} N_{z\parallel}^{(1)} \left(1 - \dfrac{\mathrm{j}\,\varepsilon_1^{(1)} N_y}{\varepsilon_0^{(1)} N_{z\parallel}^{(1)}}\right) \cos\varphi^{(0)}}, \tag{6.48c}$$

$$\frac{F_{43}}{F_{33}} = \frac{C_3^{(1)}}{C_4^{(1)}} \frac{\varepsilon_0^{(1)} N_{z\parallel}^{(1)} N_{z0}^{(2)2} \left(1 - \dfrac{\mathrm{j}\,\varepsilon_1^{(1)} N_y}{\varepsilon_0^{(1)} N_{z\parallel}^{(1)}}\right) - \varepsilon_0^{(2)} N_{z\parallel}^{(2)} N_{z0}^{(1)2} \left(1 - \dfrac{\mathrm{j}\,\varepsilon_1^{(2)} N_y}{\varepsilon_0^{(2)} N_{z\parallel}^{(2)}}\right)}{\varepsilon_0^{(1)} N_{z\parallel}^{(1)} N_{z0}^{(2)2} \left(1 + \dfrac{\mathrm{j}\,\varepsilon_1^{(1)} N_y}{\varepsilon_0^{(1)} N_{z\parallel}^{(1)}}\right) + \varepsilon_0^{(2)} N_{z\parallel}^{(2)} N_{z0}^{(1)2} \left(1 - \dfrac{\mathrm{j}\,\varepsilon_1^{(2)} N_y}{\varepsilon_0^{(2)} N_{z\parallel}^{(2)}}\right)}. \tag{6.48d}$$

The first equation represents the TM reflection coefficient for an interface between an isotropic ambient and a medium with transverse magnetization. In this way, the ratio $\frac{A_{43}}{A_{33}}$ contains complete information on the classic, *i.e.*, single-interface MO transverse (Kerr) effect. The last equation expresses a generalized Fresnel coefficient $\frac{F_{43}}{F_{33}}$ for an interface between two media with the transverse magnetization. We observe, *e.g.*, from Eqs. (6.37c) to (6.37f), that the MO contribution depends on the difference $\varepsilon_1^{(1)} N_{z0}^{(2)2} - \varepsilon_1^{(2)} N_{z0}^{(1)2}$. As expected, the MO contribution vanishes at an interface of two media with equal optical and MO parameters unless the media are antiferromagnetically coupled, *i.e.*, their transverse magnetizations are antiparallel to each other. Eq. (6.47) is independent of the normalizing coefficients $C_3^{(1)}$, $C_4^{(1)}$, $C_3^{(2)}$, and $C_4^{(2)}$, as expected from the general considerations on page 153.

At zero magnetizations in both the film and the substrate, $\varepsilon_1^{(1)} = 0$, $\varepsilon_1^{(2)} = 0$, and $\beta_\perp^{(1)} = \beta_\parallel^{(1)} = \beta^{(i)}$. Then Eq. (6.47) simplifies to the form corresponding to that for isotropic media

$$r_{pp}^{(02)}\left(\varepsilon_1^{(1)} = 0, \varepsilon_1^{(2)} = 0\right) = \frac{r_{pp}^{(01)} + r_{pp}^{(12)} e^{-2j\beta^{(1)}}}{1 + r_{pp}^{(01)} r_{pp}^{(12)} e^{-2j\beta^{(1)}}}, \tag{6.49}$$

where

$$r_{pp}^{(01)} = \frac{N^{(0)} N_{z0}^{(1)} - \varepsilon_0^{(1)} \cos\varphi^{(0)}}{N^{(0)} N_{z0}^{(1)} + \varepsilon_0^{(1)} \cos\varphi^{(0)}}, \tag{6.50a}$$

$$r_{pp}^{(12)} = \frac{\varepsilon_0^{(1)} N_{z0}^{(2)} - \varepsilon_0^{(2)} N_{z0}^{(1)}}{\varepsilon_0^{(1)} N_{z0}^{(2)} + \varepsilon_0^{(2)} N_{z0}^{(1)}}, \tag{6.50b}$$

are the classical Fresnel formulas for the reflection of TM waves at an interface of isotropic media. We extend our definition of Eq. (1.138) to

$$\Delta r_{pp}^{(\text{trans})} = r_{pp}^{(02)}\left(\varepsilon_1^{(1)}, \varepsilon_1^{(2)}\right) - r_{pp}^{(02)}\left(\varepsilon_1^{(1)} = 0, \varepsilon_1^{(2)} = 0\right). \tag{6.51}$$

The dispersion relations for guided wave propagation, *i.e.*, waveguiding condition in the system, follow from the conditions $M_{11} = 0$ and $M_{33} = 0$, which apply to TE and TM mode polarizations, respectively.

At vanishing off-diagonal tensor elements, $\varepsilon_1^{(1)} = \varepsilon_1^{(2)} = 0$, the permittivities are not necessarily scalar quantities. The diagonal permittivity tensor deduced from Eq. (6.22) corresponds to the case of a uniaxial film on a uniaxial substrate with the optical axes oriented normal to the plane of incidence [3]. Eq. (6.47) reduces to

$$r_{pp}^{(02)} = \frac{r_{pp}^{(01)} + r_{pp}^{(12)} e^{-2j\beta_\parallel^{(1)}}}{1 + r_{pp}^{(01)} r_{pp}^{(12)} e^{-2j\beta_\parallel^{(1)}}}, \tag{6.52}$$

where $\beta_{\parallel}^{(n)} = \frac{\omega}{c}\left(\varepsilon_0^{(1)} - N_y^2\right)^{1/2} d^{(1)}$, and

$$r_{pp}^{(01)} = \frac{N^{(0)} N_{z\parallel}^{(1)} - \varepsilon_0^{(1)} \cos\varphi^{(0)}}{N^{(0)} N_{z\parallel}^{(1)} + \varepsilon_0^{(1)} \cos\varphi^{(0)}}, \tag{6.53a}$$

$$r_{pp}^{(12)} = \frac{\varepsilon_0^{(1)} N_{z\parallel}^{(2)} - \varepsilon_0^{(2)} N_{z\parallel}^{(1)}}{\varepsilon_0^{(1)} N_{z\parallel}^{(2)} + \varepsilon_0^{(2)} N_{z\parallel}^{(1)}}, \tag{6.53b}$$

while $r_{ss}^{(02)}$ remains given by Eq. (6.45).

Figure 6.2 shows the simulated dependence of the MO transverse Kerr effect in an Au–Fe–Au sandwich computed with the literature values of the optical and magnetooptical constants for Fe and Au [4,5]. The effect is characterized in terms of Δr_{pp} defined in Eq. (6.51). The real and imaginary parts of the ratio $\Delta r_{pp}/r_{pp}$ are displayed in Figure 6.3. The amplitude and phase of the same quantity is displayed in Figure 6.4. At the Fe layer

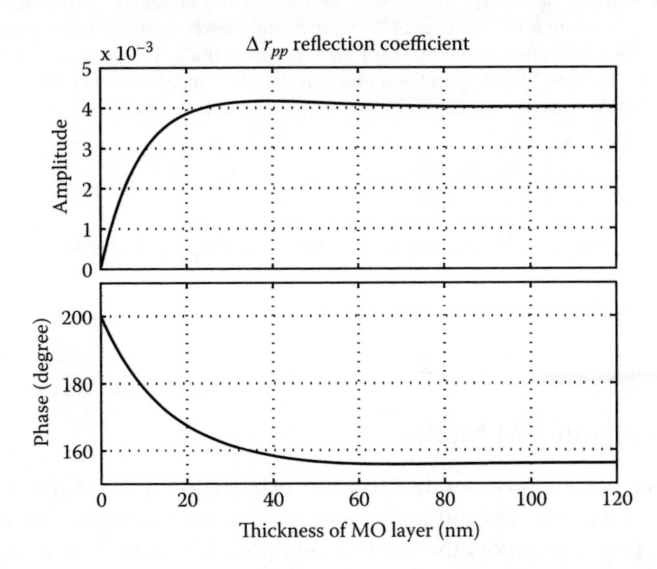

FIGURE 6.2
Effect of the ferromagnetic layer thickness, d, on Δr_{pp}, the difference between the reflection coefficient for the p polarized wave at transverse magnetization and that at zero magnetization. The angle of incidence is 70 deg in a sandwich consisting of a Fe layer between optically thick Au buffer and the Au capping layer 4 nm thick. The protective Au capping layer separates the Fe(d) layer from the ambient medium of incidence (air). The simulation pertains to the wavelength of 632.8 nm. The diagonal and off-diagonal permittivity tensor elements for Fe given by $\varepsilon_{xx}^{(Fe)} = \varepsilon_0^{(Fe)} = -0.8845 - \mathrm{j}\,17.938$ and $\varepsilon_{xy}^{(Fe)} = \mathrm{j}\,\varepsilon_1^{(Fe)} = -0.6676 - \mathrm{j}0.08988$, respectively. The differences in the diagonal permittivity tensor elements of Fe is set to zero. The complex index of refraction for Au is given by $N^{(Au)} = 0.174 - \mathrm{j}\,3.65$.

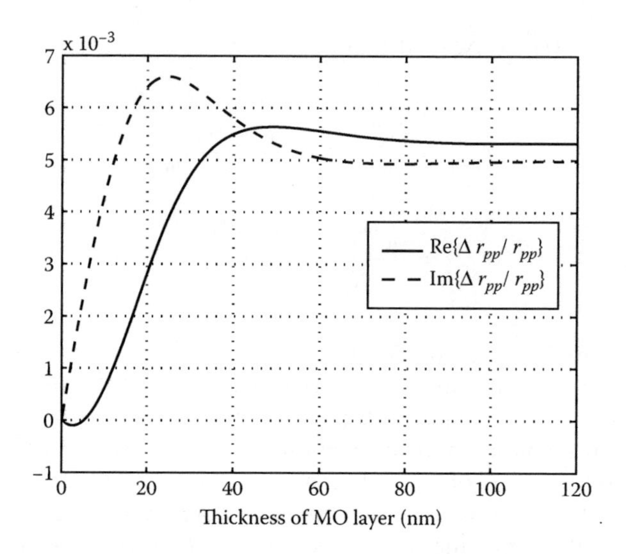

FIGURE 6.3
Effect of the ferromagnetic layer thickness, d, on the real and imaginary part of the complex ratio $\Delta r_{pp}/r_{pp}$ at an angle of incidence of 70 deg in a sandwich consisting of a Fe layer between an optically thick Au buffer and an Au capping layer 4 nm thick. The protective Au capping layer separates the Fe(d) layer from the ambient medium of incidence (air). The parameters employed in the simulation correspond to those of Figure 6.2.

thickness exceeding the radiation penetration depth, the transverse magnetooptic effect saturates.

6.4 Waveguide TM Modes

The proper polarization modes at transverse magnetization are compatible with the TE and TM guided modes in planar structures. The absence of mode coupling makes the analysis simpler. We devote this section to the demonstration of a nonreciprocal (unidirectional) propagation in planar waveguides with covers at transverse magnetization. The device is called an *isolator*. It is positioned between a laser and a fiber in optical communication systems to protect the laser source from spurious reflections in the transmission lines. A planar structure, utilizing the nonreciprocal phase shift originating from the transverse MO effect, consists of a central

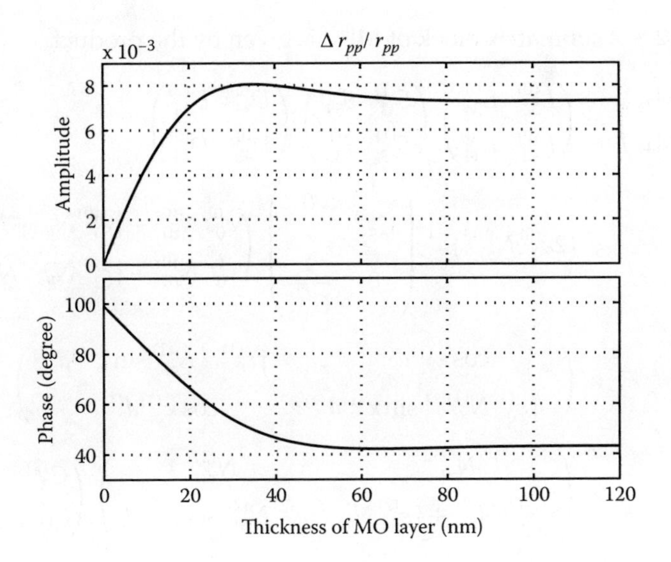

FIGURE 6.4
The effect of the ferromagnetic layer thickness, d, on the amplitude and phase complex ratio $\Delta r_{pp}/r_{pp}$ at an angle of incidence of 70 deg in a sandwich consisting of a Fe layer between optically thick Au buffer and the Au capping layer 4 nm thick. The protective Au capping layer separates the Fe(d) layer from the ambient medium of incidence (air). The parameters employed in the simulation correspond to those of Figure 6.2.

amplifying region sandwiched between buffer layers of lower real permittivity [6]. If the buffer layers were thick, the structure would correspond to a symmetrical waveguide designed for the operation on TM modes. However, in the MO nonreciprocal waveguide, the buffer layers are thin to provide an efficient evanescent wave coupling between the central region and ferromagnetic layers with transverse magnetization perpendicular to the plane of zigzag motion (*i.e.*, plane of incidence) of the guided waves.

To demonstrate the nonreciprocal transmission, we start with the analysis of TM modes in a simplified structure consisting of a dielectric film sandwiched between two half spaces at transverse magnetization. We restrict ourselves to terms of zero and first order in the off-diagonal permittivity tensor in magnetic media. The corresponding **M** matrix is given by the sequence

$$\begin{aligned}\mathbf{M}^{(02)} &= (\mathbf{D}^{(0)})^{-1}\mathbf{D}^{(1)}\mathbf{P}^{(1)}(\mathbf{D}^{(1)})^{-1}\mathbf{D}^{(2)}\\ &= (\mathbf{D}^{(0)})^{-1}\mathbf{S}^{(1)}\mathbf{D}^{(2)}.\end{aligned} \tag{6.54}$$

The TM 2×2 submatrix block of $\mathbf{M}^{(02)}$ is given by the product

$$\begin{pmatrix} M_{33} & M_{34} \\ M_{43} & M_{44} \end{pmatrix} = \begin{pmatrix} L_{33}^{(0)} & L_{34}^{(0)} \\ L_{43}^{(0)} & L_{44}^{(0)} \end{pmatrix} \begin{pmatrix} S_{33}^{(1)} & S_{34}^{(1)} \\ S_{43}^{(1)} & S_{33}^{(1)} \end{pmatrix} \begin{pmatrix} D_{33}^{(2)} & D_{34}^{(2)} \\ D_{43}^{(0)} & D_{44}^{(0)} \end{pmatrix}$$
$$= \left(2\varepsilon_0^{(0)} N_{z0}^{(0)3}\right)^{-1} \begin{pmatrix} \frac{1}{C_3^{(2)}} & 0 \\ 0 & \frac{1}{C_4^{(2)}} \end{pmatrix} \begin{pmatrix} \varepsilon_0^{(0)} N_{z0}^{(0)} + j\varepsilon_1^{(0)} N_y & -N_{z0}^{(0)2} \\ \varepsilon_0^{(0)} N_{z0}^{(0)} - j\varepsilon_1^{(0)} N_y & N_{z0}^{(0)2} \end{pmatrix}$$
$$\times \begin{pmatrix} \cos\kappa^{(1)} d^{(1)} & -j\varepsilon_0^{(1)-1} N_{z0}^{(1)} \sin\kappa^{(1)} d^{(1)} \\ -j\varepsilon_0^{(1)} N_{z0}^{(1)-1} \sin\kappa^{(1)} d^{(1)} & \cos\kappa^{(1)} d^{(1)} \end{pmatrix}$$
$$\times \begin{pmatrix} N_{z0}^{(2)2} & N_{z0}^{(2)2} \\ -\varepsilon_0^{(2)} N_{z0}^{(2)} + j\varepsilon_1^{(2)} N_y & \varepsilon_0^{(2)} N_{z0}^{(2)} + j\varepsilon_1^{(2)} N_y \end{pmatrix} \begin{pmatrix} C_3^{(2)} & 0 \\ 0 & C_4^{(2)} \end{pmatrix}, \tag{6.55}$$

where the use was made of Eq. (6.33) for the inverse of dynamical matrix in the magnetic half space (0), Eq. (6.31) for the dynamical matrix in the magnetic half space (2), and the characteristic matrix for isotropic medium (1)

$$\mathbf{S}^{(1)} = \begin{pmatrix} \cos\kappa^{(1)} d^{(1)} & jN_{z0}^{(1)-1} \sin\kappa^{(1)} d^{(1)} & 0 & 0 \\ jN_{z0}^{(1)} \sin\kappa^{(1)} d^{(1)} & \cos\kappa^{(1)} d^{(1)} & 0 & 0 \\ 0 & 0 & \cos\kappa^{(1)} d^{(1)} & -j\varepsilon_0^{(1)-1} N_{z0}^{(1)} \sin\kappa^{(1)} d^{(1)} \\ 0 & 0 & -j\varepsilon_0^{(1)} N_{z0}^{(1)-1} \sin\kappa^{(1)} d^{(1)} & \cos\kappa^{(1)} d^{(1)} \end{pmatrix}, \tag{6.56}$$

where $d^{(1)}$ is the thickness of the layer (1). The common factors are not relevant here, and we can write the characteristic equation expressing the waveguiding condition for TM modes as a special case of Eq. (3.76) as

$$M_{33} = \left(2\varepsilon_0^{(0)} N_{z0}^{(0)}\right)^{-1} \left[L_{33}\left(S_{33} D_{33} + S_{34} D_{43}\right) + L_{34}\left(S_{43} D_{33} + S_{33} D_{43}\right)\right] = 0,$$

i.e.,

$$\begin{aligned} &\left(\varepsilon_0^{(0)} N_{z0}^{(0)} + j\varepsilon_1^{(0)} N_y\right) \\ &\quad \times \left[\cos\kappa^{(1)} d^{(1)} N_{z0}^{(2)2} + j\varepsilon_0^{(1)-1} N_{z0}^{(1)} \sin\kappa^{(1)} d^{(1)} \left(\varepsilon_0^{(2)} N_{z0}^{(2)} - j\varepsilon_1^{(2)} N_y\right)\right] \\ &\quad + N_{z0}^{(0)2} \left[j\varepsilon_0^{(1)} N_{z0}^{(1)-1} \sin\kappa^{(1)} d^{(1)} N_{z0}^{(2)2} + \cos\kappa^{(1)} d^{(1)} \right. \\ &\quad \left. \times \left(\varepsilon_0^{(2)} N_{z0}^{(2)} - j\varepsilon_1^{(2)} N_y\right)\right] = 0. \end{aligned} \tag{6.57}$$

We replace the trigonometric functions by the exponentials and obtain

$$\begin{aligned}
M_{33} = {} & \left[\left(\varepsilon_0^{(0)} N_{z0}^{(0)} + \mathrm{j}\,\varepsilon_1^{(0)} N_y\right) N_{z0}^{(2)2} + \left(\varepsilon_0^{(0)} N_{z0}^{(0)} + \mathrm{j}\,\varepsilon_1^{(0)} N_y\right)\right. \\
& \times \varepsilon_0^{(1)-1} N_{z0}^{(1)} \left(\varepsilon_0^{(2)} N_{z0}^{(2)} - \mathrm{j}\,\varepsilon_1^{(2)} N_y\right) \\
& \left. + N_{z0}^{(0)2} \varepsilon_0^{(1)} N_{z0}^{(1)-1} N_{z0}^{(2)2} + N_{z0}^{(0)2} \left(\varepsilon_0^{(2)} N_{z0}^{(2)} - \mathrm{j}\,\varepsilon_1^{(2)} N_y\right)\right] \exp\left(\mathrm{j}\kappa^{(1)} d^{(1)}\right) \\
& + \left[\left(\varepsilon_0^{(0)} N_{z0}^{(0)} + \mathrm{j}\,\varepsilon_1^{(0)} N_y\right) N_{z0}^{(2)2} - \left(\varepsilon_0^{(0)} N_{z0}^{(0)} + \mathrm{j}\,\varepsilon_1^{(0)} N_y\right)\right. \\
& \times \varepsilon_0^{(1)-1} N_{z0}^{(1)} \left(\varepsilon_0^{(2)} N_{z0}^{(2)} - \mathrm{j}\,\varepsilon_1^{(2)} N_y\right) \\
& \left. - N_{z0}^{(0)2} \varepsilon_0^{(1)} N_{z0}^{(1)-1} N_{z0}^{(2)2} + N_{z0}^{(0)2} \left(\varepsilon_0^{(2)} N_{z0}^{(2)} - \mathrm{j}\,\varepsilon_1^{(2)} N_y\right)\right] \\
& \times \exp\left(-\mathrm{j}\kappa^{(1)} d^{(1)}\right) = 0. \qquad (6.58)
\end{aligned}$$

This can be transformed into a more transparent form

$$\begin{aligned}
1 + {} & \frac{\varepsilon_0^{(0)}\left(1 + \dfrac{\mathrm{j}\,\varepsilon_1^{(0)} N_y}{\varepsilon_0^{(0)} N_{z0}^{(0)}}\right) N_{z0}^{(1)} - \varepsilon_0^{(1)} N_{z0}^{(0)}}{\varepsilon_0^{(0)}\left(1 + \dfrac{\mathrm{j}\,\varepsilon_1^{(0)} N_y}{\varepsilon_0^{(0)} N_{z0}^{(0)}}\right) N_{z0}^{(1)} + \varepsilon_0^{(1)} N_{z0}^{(0)}} \; \frac{\varepsilon_0^{(1)} N_{z0}^{(2)} - N_{z0}^{(1)} \varepsilon_0^{(2)}\left(1 - \dfrac{\mathrm{j}\,\varepsilon_1^{(2)} N_y}{\varepsilon_0^{(2)} N_{z0}^{(2)}}\right)}{\varepsilon_0^{(1)} N_{z0}^{(2)} + N_{z0}^{(1)} \varepsilon_0^{(2)}\left(1 - \dfrac{\mathrm{j}\,\varepsilon_1^{(2)} N_y}{\varepsilon_0^{(2)} N_{z0}^{(2)}}\right)} \\
& \times \exp\left(-2\mathrm{j}\kappa^{(1)} d^{(1)}\right) = 0. \qquad (6.59)
\end{aligned}$$

We observe that the solution of this characteristic equation depends on the complex parameters characterizing the effect of transverse magnetization

$$q^{(0)} = \frac{\mathrm{j}\,\varepsilon_1^{(0)} N_y}{\varepsilon_0^{(0)} N_{z0}^{(0)}}, \qquad (6.60a)$$

$$q^{(2)} = \frac{\mathrm{j}\,\varepsilon_1^{(2)} N_y}{\varepsilon_0^{(2)} N_{z0}^{(2)}}. \qquad (6.60b)$$

Their form explains the nonreciprocal transmission. In fact, N_y changes sign when the propagation sense is changed to opposite one. Similarly, when the magnetizations are simultaneously switched to opposite ones, the sign of the corresponding off-diagonal permittivity tensor element, both $\varepsilon_1^{(0)}$ and $\varepsilon_1^{(2)}$ change to opposite one. Consequently, the waveguiding condition depends on the sense of propagation. If the waveguide parameters are adjusted for a maximum transmission for the forward modes, then the transmission of the backward modes would be reduced to a degree that depends on the magnitude of parameters $q^{(0)}$ and $q^{(2)}$ and on the coupling strength between the central region and the magnetic surface. As opposed to the TE modes, the TM modes are strongly attenuated upon the reflection at high zigzag angles. A strong coupling significantly perturbs the symmetric waveguide and requires high gain in the central region. On the other hand, a weak coupling may produce insufficient attenuation of the backward modes.

The characteristic equation, *i.e.*, waveguiding condition, can be expressed concisely as

$$1 + r_{pp}^{(01)} r_{pp}^{(12)} \exp\left(-2\mathrm{j}\kappa^{(1)} d^{(1)}\right) = 0, \tag{6.61}$$

where

$$r_{pp}^{(01)} = \frac{\varepsilon_0^{(0)} \left(1 + \dfrac{\mathrm{j}\,\varepsilon_1^{(0)} N_y}{\varepsilon_0^{(0)} N_{z0}^{(0)}}\right) N_{z0}^{(1)} - \varepsilon_0^{(1)} N_{z0}^{(0)}}{\varepsilon_0^{(0)} \left(1 + \dfrac{\mathrm{j}\,\varepsilon_1^{(0)} N_y}{\varepsilon_0^{(0)} N_{z0}^{(0)}}\right) N_{z0}^{(1)} + \varepsilon_0^{(1)} N_{z0}^{(0)}} \tag{6.62a}$$

$$r_{pp}^{(12)} = \frac{\varepsilon_0^{(1)} N_{z0}^{(2)} - N_{z0}^{(1)} \varepsilon_0^{(2)} \left(1 - \dfrac{\mathrm{j}\varepsilon_1^{(2)} N_y}{\varepsilon_0^{(2)} N_{z0}^{(2)}}\right)}{\varepsilon_0^{(1)} N_{z0}^{(2)} + N_{z0}^{(1)} \varepsilon_0^{(2)} \left(1 - \dfrac{\mathrm{j}\,\varepsilon_1^{(2)} N_y}{\varepsilon_0^{(2)} N_{z0}^{(2)}}\right)}. \tag{6.62b}$$

Note that in a symmetric waveguide, corresponding to $\varepsilon_0^{(0)} = \varepsilon_0^{(2)}$ and $N_{z0}^{(0)} = N_{z0}^{(2)}$, and for the parallel transverse magnetizations ($\varepsilon_1^{(0)}=\varepsilon_1^{(2)}$) in the sandwiching half spaces, the nonreciprocal phase shift disappears to first order in magnetization. On the other hand, for antiparallel transverse magnetizations, $\varepsilon_1^{(0)}=-\varepsilon_1^{(2)}$, the shift becomes doubled with respect to the case of a single magnetic interface. An independent control of the magnetization at the two interfaces makes various switching and modulation functions possible.

In a more realistic model, the central layer of the thickness $d^{(2)}$ will be separated from the magnetic cover and substrate by the buffer layers of the thickness $d^{(1)}$ and $d^{(3)}$, respectively. The corresponding **M** matrix assumes the form

$$\begin{aligned}\mathbf{M}^{(04)} &= (\mathbf{D}^{(0)})^{-1}\mathbf{D}^{(1)}\mathbf{P}^{(1)}(\mathbf{D}^{(1)})^{-1}\mathbf{D}^{(2)}\mathbf{P}^{(2)}(\mathbf{D}^{(2)})^{-1}\mathbf{D}^{(3)}\mathbf{P}^{(3)}(\mathbf{D}^{(3)})^{-1}\mathbf{D}^{(4)}\\ &= (\mathbf{D}^{(0)})^{-1}\mathbf{S}^{(1)}\mathbf{S}^{(2)}\mathbf{S}^{(3)}\mathbf{D}^{(4)}.\end{aligned} \tag{6.63}$$

The permittivity of the coupling buffer layers (1) and (3) must be lower with respect to that of the layer (2) in order to concentrate the fields in the amplifying layer (2). In addition, $d^{(1)}$ and $d^{(3)}$ should be properly adjusted to achieve an optimum nonreciprocal transmission. The waveguiding condition now becomes

$$1 + r_{pp}^{(01)} \left\{ \frac{r_{pp}^{(12)} + \left[\dfrac{r_{pp}^{(23)} + r_{pp}^{(34)} \exp\left(-2j\kappa^{(3)} d^{(2)}\right)}{1 + r_{pp}^{(23)} r_{pp}^{(34)} \exp\left(-2j\kappa^{(3)} d^{(3)}\right)} \right) \exp(-2j\kappa^{(2)} d^{(2)})}{1 + r_{pp}^{(12)} \left[\dfrac{r_{pp}^{(23)} + r_{pp}^{(34)} \exp\left(-2j\kappa^{(3)} d^{(3)}\right)}{1 + r_{pp}^{(23)} r_{pp}^{(34)} \exp\left(-2j\kappa^{(3)} d^{(2)}\right)} \right] \exp(-2j\kappa^{(2)} d^{(2)})} \right\}$$

$$\times \exp\left(-2j\kappa^{(1)} d^{(1)}\right) = 0, \tag{6.64}$$

where the reflection coefficients at the interfaces involving layers at transverse magnetization have the same meaning as in Eq. (6.62)

$$r_{pp}^{(01)} = \frac{\varepsilon_0^{(0)} \left(1 + \dfrac{j\,\varepsilon_1^{(0)} N_y}{\varepsilon_0^{(0)} N_{z0}^{(0)}}\right) N_{z0}^{(1)} - \varepsilon_0^{(1)} N_{z0}^{(0)}}{\varepsilon_0^{(0)} \left(1 + \dfrac{j\,\varepsilon_1^{(0)} N_y}{\varepsilon_0^{(0)} N_{z0}^{(0)}}\right) N_{z0}^{(1)} + \varepsilon_0^{(1)} N_{z0}^{(0)}} \tag{6.65a}$$

$$r_{pp}^{(34)} = \frac{\varepsilon_0^{(3)} N_{z0}^{(4)} - N_{z0}^{(3)} \varepsilon_0^{(4)} \left(1 - \dfrac{j\,\varepsilon_1^{(4)} N_y}{\varepsilon_0^{(4)} N_{z0}^{(4)}}\right)}{\varepsilon_0^{(3)} N_{z0}^{(4)} + N_{z0}^{(3)} \varepsilon_0^{(4)} \left(1 - \dfrac{j\,\varepsilon_1^{(4)} N_y}{\varepsilon_0^{(4)} N_{z0}^{(4)}}\right)}. \tag{6.65b}$$

In a technologically feasible structure, one of the sandwiching magnetic interfaces is removed, and the thickness of the buffer layer is increased above the penetration depth of the TM guided modes. The **M** matrix of this system, consisting of a semiconductor cover (0), central amplifying region (1), buffer (2), and an optically thick magnetic film (3), becomes

$$\begin{aligned} \mathbf{M}^{(03)} &= [\mathbf{D}^{(0)}]^{-1} \mathbf{D}^{(1)} \mathbf{P}^{(1)} [\mathbf{D}^{(1)}]^{-1} \mathbf{D}^{(2)} \mathbf{P}^{(2)} [\mathbf{D}^{(2)}]^{-1} \mathbf{D}^{(3)} \\ &= [\mathbf{D}^{(0)}]^{-1} \mathbf{S}^{(1)} \mathbf{S}^{(2)} \mathbf{D}^{(3)}. \end{aligned} \tag{6.66}$$

The TM (2×2) block of $\mathbf{M}^{(03)}$ is given by the product

$$\begin{bmatrix} M_{33} & M_{34} \\ M_{43} & M_{44} \end{bmatrix} = \left(2N_{z0}^{(0)}\right)^{-1} \begin{bmatrix} N^{(0)} & -\cos\varphi^{(0)} \\ N^{(0)} & \cos\varphi^{(0)} \end{bmatrix}$$

$$\times \begin{bmatrix} \cos\kappa^{(1)} d^{(1)} & -j\,\varepsilon_0^{(1)-1} N_{z0}^{(1)} \sin\kappa^{(1)} d^{(1)} \\ -j\,\varepsilon_0^{(1)} N_{z0}^{(1)-1} \sin\kappa^{(1)} d^{(1)} & \cos\kappa^{(1)} d^{(1)} \end{bmatrix}$$

$$\times \begin{bmatrix} \cos\kappa^{(2)} d^{(2)} & -j\,\varepsilon_0^{(2)-1} N_{z0}^{(2)} \sin\kappa^{(2)} d^{(2)} \\ -j\,\varepsilon_0^{(2)} N_{z0}^{(2)-1} \sin\kappa^{(2)} d^{(2)} & \cos\kappa^{(2)} d^{(2)} \end{bmatrix}$$

$$\times \begin{bmatrix} N_{z0}^{(3)2} & N_{z0}^{(3)2} \\ j\,\varepsilon_1^{(3)} N_y - \varepsilon_0^{(3)} N_{z0}^{(3)} & j\,\varepsilon_1^{(3)} N_y + \varepsilon_0^{(3)} N_{z0}^{(3)} \end{bmatrix}.$$

The characteristic equation (waveguiding condition) takes the form

$$1+\left(\frac{\varepsilon_0^{(0)}N_{z0}^{(1)}-\varepsilon_0^{(1)}N_{z0}^{(0)}}{\varepsilon_0^{(0)}N_{z0}^{(1)}+\varepsilon_0^{(1)}N_{z0}^{(0)}}\right)\left(\frac{\varepsilon_0^{(1)}N_{z0}^{(2)}-\varepsilon_0^{(2)}N_{z0}^{(1)}}{\varepsilon_0^{(1)}N_{z0}^{(2)}+\varepsilon_0^{(2)}N_{z0}^{(1)}}\right)\exp(-2\mathrm{j}\kappa^{(1)}d^{(1)})$$
$$+\left(\frac{\varepsilon_0^{(1)}N_{z0}^{(2)}-\varepsilon_0^{(2)}N_{z0}^{(1)}}{\varepsilon_0^{(1)}N_{z0}^{(2)}+\varepsilon_0^{(2)}N_{z0}^{(1)}}\right)\left[\frac{\varepsilon_0^{(2)}N_{z0}^{(3)}-N_{z0}^{(2)}\varepsilon_0^{(3)}\left(1-\frac{\mathrm{j}\,\varepsilon_1^{(3)}N_y}{\varepsilon_0^{(3)}N_{z0}^{(3)}}\right)}{\varepsilon_0^{(2)}N_{z0}^{(3)}+N_{z0}^{(2)}\varepsilon_0^{(3)}\left(1-\frac{\mathrm{j}\,\varepsilon_1^{(3)}N_y}{\varepsilon_0^{(3)}N_{z0}^{(3)}}\right)}\right]$$
$$\times\exp\left(-2\mathrm{j}\kappa^{(2)}d^{(2)}\right)$$
$$+\left(\frac{\varepsilon_0^{(0)}N_{z0}^{(1)}-\varepsilon_0^{(1)}N_{z0}^{(0)}}{\varepsilon_0^{(0)}N_{z0}^{(1)}+\varepsilon_0^{(1)}N_{z0}^{(0)}}\right)\left[\frac{\varepsilon_0^{(2)}N_{z0}^{(3)}-N_{z0}^{(2)}\varepsilon_0^{(3)}\left(1-\frac{\mathrm{j}\,\varepsilon_1^{(3)}N_y}{\varepsilon_0^{(3)}N_{z0}^{(3)}}\right)}{\varepsilon_0^{(2)}N_{z0}^{(3)}+N_{z0}^{(2)}\varepsilon_0^{(3)}\left(1-\frac{\mathrm{j}\,\varepsilon_1^{(3)}N_y}{\varepsilon_0^{(3)}N_{z0}^{(3)}}\right)}\right]$$
$$\times\exp\left[-2\mathrm{j}\left(\kappa^{(1)}d^{(1)}+\kappa^{(2)}d^{(2)}\right)\right]=0. \tag{6.67}$$

The equation is to be solved with respect to N_y. Here, $\kappa^{(1)}=\frac{\omega}{c}(\varepsilon_0^{(1)}-N_y^2)^{1/2}$, $\kappa^{(2)}=\frac{\omega}{c}(\varepsilon_0^{(2)}-N_y^2)^{1/2}$, and $N_{z0}^{(i)}=(\varepsilon_0^{(i)}-N_y^2)^{1/2}$ for $i=0,\ldots,3$. The characteristic equation can be concisely expressed in the following form

$$\begin{aligned}1&+r_{pp}^{(01)}r_{pp}^{(12)}\exp(-2\mathrm{j}\kappa^{(1)}d^{(1)})\\&+r_{pp}^{(12)}r_{pp}^{(23)}\left(\mp\mathrm{j}\,\varepsilon_1^{(3)}\right)\exp(-2\mathrm{j}\kappa^{(2)}d^{(2)})\\&+r_{pp}^{(01)}r_{pp}^{(23)}\left(\mp\mathrm{j}\,\varepsilon_1^{(3)}\right)\exp\left[-2\mathrm{j}\left(\kappa^{(1)}d^{(1)}+\kappa^{(2)}d^{(2)}\right)\right]=0,\end{aligned} \tag{6.68}$$

where it is emphasized that the reflection coefficient $r_{pp}^{(23)}(\mp\mathrm{j}\,\varepsilon_1^{(3)})$ is an explicit function of the nondiagonal permittivity tensor element $\mathrm{j}\,\varepsilon_1^{(3)}$ induced by the magnetization. In the single-mode regime, there is only one pair of solutions $N_y^{\mp}$ with the TM polarization, one for the forward and the other for the backward (retrograde) propagation

$$1+r_{pp}^{(01)}\left[\frac{r_{pp}^{(12)}+r_{pp}^{(23)}\exp(-2\mathrm{j}\kappa^{(2)}d^{(2)})}{1+r_{pp}^{(12)}r_{pp}^{(23)}\exp(-2\mathrm{j}\kappa^{(2)}d^{(2)})}\right]\exp\left(-2\mathrm{j}\kappa^{(1)}d^{(1)}\right)=0. \tag{6.69}$$

We now characterize the exponential decaying with the distance evanescent waves in both the cover (0) and buffer (2) by the transverse attenuation constants

$$j\kappa^{(0)} = \gamma^{(0)} = \left[\beta^2 - \left(\frac{2\pi}{\lambda_{\text{vac}}} N^{(0)}\right)^2\right]^{1/2} = \frac{\omega}{c}\left(N_y^2 - \varepsilon_0^{(0)}\right)^{1/2} = j\frac{\omega}{c} N_{z0}^{(0)}, \quad (6.70a)$$

$$j\kappa^{(2)} = \gamma^{(2)} = \left[\beta^2 - \left(\frac{2\pi}{\lambda_{\text{vac}}} N^{(2)}\right)^2\right]^{1/2} = \frac{\omega}{c}\left(N_y^2 - \varepsilon_0^{(2)}\right)^{1/2} = j\frac{\omega}{c} N_{z0}^{(2)}, \quad (6.70b)$$

and express the characteristic equation in terms of the unknown parameter N_y.

$$\begin{aligned}
1 &- \left[\frac{\varepsilon_0^{(0)}\left(\varepsilon_0^{(1)} - N_y^2\right)^{1/2} + j\,\varepsilon_0^{(1)}\left(N_y^2 - \varepsilon_0^{(0)}\right)^{1/2}}{\varepsilon_0^{(0)}\left(\varepsilon_0^{(1)} - N_y^2\right)^{1/2} - j\,\varepsilon_0^{(1)}\left(N_y^2 - \varepsilon_0^{(0)}\right)^{1/2}}\right] \\
&\times \left[\frac{\varepsilon_0^{(2)}\left(\varepsilon_0^{(1)} - N_y^2\right)^{1/2} + j\,\varepsilon_0^{(1)}\left(N_y^2 - \varepsilon_0^{(2)}\right)^{1/2}}{\varepsilon_0^{(2)}\left(\varepsilon_0^{(1)} - N_y^2\right)^{1/2} - j\,\varepsilon_0^{(1)}\left(N_y^2 - \varepsilon_0^{(2)}\right)^{1/2}}\right] \\
&\times \exp\left[-2j\frac{\omega}{c}\left(\varepsilon_0^{(1)} - N_y^2\right)^{1/2} d^{(1)}\right] \\
&- \left[\frac{\varepsilon_0^{(2)}\left(\varepsilon_0^{(1)} - N_y^2\right)^{1/2} + j\,\varepsilon_0^{(1)}\left(N_y^2 - \varepsilon_0^{(2)}\right)^{1/2}}{\varepsilon_0^{(2)}\left(\varepsilon_0^{(1)} - N_y^2\right)^{1/2} - j\,\varepsilon_0^{(1)}\left(N_y^2 - \varepsilon_0^{(2)}\right)^{1/2}}\right] \\
&\times \left[\frac{\varepsilon_0^{(2)} N_{z0}^{(3)} + j\left(N_y^2 - \varepsilon_0^{(2)}\right)^{1/2}\left(\varepsilon_0^{(3)} - \dfrac{j\,\varepsilon_1^{(3)} N_y}{N_{z0}^{(3)}}\right)}{\varepsilon_0^{(2)} N_{z0}^{(3)} - j\left(N_y^2 - \varepsilon_0^{(2)}\right)^{1/2}\left(\varepsilon_0^{(3)} - \dfrac{j\,\varepsilon_1^{(3)} N_y}{N_{z0}^{(3)}}\right)}\right] \\
&\times \exp\left[-2\frac{\omega}{c}\left(N_y^2 - \varepsilon_0^{(2)}\right)^{1/2} d^{(2)}\right] \\
&+ \left[\frac{\varepsilon_0^{(0)}\left(\varepsilon_0^{(1)} - N_y^2\right)^{1/2} + j\varepsilon_0^{(1)}\left(N_y^2 - \varepsilon_0^{(0)}\right)^{1/2}}{\varepsilon_0^{(0)}\left(\varepsilon_0^{(1)} - N_y^2\right)^{1/2} - j\,\varepsilon_0^{(1)}\left(N_y^2 - \varepsilon_0^{(0)}\right)^{1/2}}\right] \\
&\times \left[\frac{\varepsilon_0^{(2)} N_{z0}^{(3)} + j\left(N_y^2 - \varepsilon_0^{(2)}\right)^{1/2}\left(\varepsilon_0^{(3)} - \dfrac{j\,\varepsilon_1^{(3)} N_y}{N_{z0}^{(3)}}\right)}{\varepsilon_0^{(2)} N_{z0}^{(3)} - j\left(N_y^2 - \varepsilon_0^{(2)}\right)^{1/2}\left(\varepsilon_0^{(3)} - \dfrac{j\,\varepsilon_1^{(3)} N_y}{N_{z0}^{(3)}}\right)}\right] \\
&\times \exp\left[-2j\frac{\omega}{c}\left(\varepsilon_0^{(1)} - N_y^2\right)^{1/2} d^{(1)}\right] \\
&\times \exp\left[-2\frac{\omega}{c}\left(N_y^2 - \varepsilon_0^{(2)}\right)^{1/2} d^{(2)}\right] = 0. \quad (6.71)
\end{aligned}$$

We look for the solution corresponding to $N_y > 0$ and $N_y < 0$ and the value of the gain in the central layer (1), characterized by the imaginary part of $\varepsilon_0^{(1)}$ providing real $N_y > 0$ in the forward direction. From the solution for $N_y < 0$, we then evaluate the rejection of the backward mode. Due to much higher reflectivity of TE modes at the metallic interface at high zigzag angles, the TE modes are only moderately attenuated in the metal clad waveguides. In addition, the ratios of the permittivities from the central region and buffers, $\varepsilon^{(1)}/\varepsilon^{(0)}$ and $\varepsilon^{(1)}/\varepsilon^{(2)}$, are only slightly higher than one, which brings the solutions of the TE and TM characteristic equations close to the mode degeneracy. All these aspects must be accounted for in order to achieve an optimum performance of the device. Various designs of nonreciprocal devices for integrated optoelectronics were proposed [11] to [17].

References

1. Š. Višňovský, "Magneto-optical transverse Kerr effect in a film substrate system," Czech. J. Phys. B **36**, 1203–1208, 1986.
2. K. Postava, J. Pištora, and Š. Višňovský, "Magneto-optical effects in ultrathin structures at transversal magnetization," Czech. J. Phys. B 49, 1185–1204, 1999.
3. R. M. A. Azzam and N. M. Bashara, *Ellipsometry and Polarized Light* (Elsevier, Amsterdam, 1987), Chapter 4.
4. Š. Višňovský, K. Postava, T. Yamaguchi, and R. Lopušník, "Magneto-optic ellipsometry in exchange-coupled films," Appl. Opt. **41**, 3950–3960, 2002.
5. P. B. Johnson and R. W. Christy, "Optical constants of the noble metals," Phys. Rev. B **6**, 4370–4379, 1972.
6. W. Zaets and Koji Ando, "Optical waveguide isolator based on nonreciprocal loss/gain of amplifier covered by ferromagnetic layer," IEEE Photonics Technology Letters, **11**, 1012–1014, 1999.
7. K. Postava, S. Visnovsky, M. Veis, V. Kolinsky, J. Pistora, D. Ciprian, P. Gogol, and P. Beauvillain, "Optimization of a magneto-optical integrated isolator," J. Magn. Magn. Mater. 2319-232, 2004.
8. K. Postava, M. Vangwolleghem, D. Van Thourhout, R. Baets, Š. Višňovský, P. Beauvillain, and J. Pištora, "Modeling a novel InP-based monolithically integrated magneto-optical waveguide isolator," J. Opt. Soc. Am. B **22**, 2005.
9. Š. Višňovský, T. Yamaguchi, J. Pištora, K. Postava, P. Beauvillain, and P. Gogol, "Waveguide isolator based on magneto-optic transverse effect," to be submitted to Czech. J. Phys.
10. Š. Višňovský, T. Yamaguchi, J. Pištora, K. Postava, P. Beauvillain, and M. Veis "Physics of magneto-optic waveguide isolator," 10th International Symposium on Microwave and Optical Technology (ISMOT2005, Fukuoka, Japan).
11. N. Bahlmann, M. Lohmeyer, O. Zhuromskyy, H. Dötsch, and P. Hertel, "Nonreciprocal coupled waveguides for integrated optical isolators and circulators for TM-modes," Optics Comm **161**, 330–337, 1999.

12. M. Inoue, K. Arai, T. Fujii, and M. Abe, "One-dimensional magnetophotonic crystals," J. Appl. Phys. **85**, 5768–5770.
13. J. Fujita, M. Levy, R. M. Osgood, Jr., L. Wilkens, and H. Dötsch, "Waveguide optical isolator based on Mach-Zender interferometer," Appl. Phys. Lett. **76**, 2158–2170, 2000.
14. H. Kato, M. Inoue, "Reflection mode operation of one-dimensional magnetophotonic crystals for use in film-based magneto-optical isolator devices," J. Appl. Phys. **91**, 7017–7019, 2002.
15. H. Shimizu and M. Tanaka, "Design of semiconductor-waveguide-type optical isolators using the nonreciprocal loss/gain in the magneto-optical waveguides having MnAs nanoclusters," Appl. Phys. Lett. **81**, 5246–5248, 2002.
16. H. Kato, T. Matsushita, A. Takayama, M. Egawa, K. Nishimura, and M. Inoue, "Theoretical analysis of optical and magneto-optical properties of one-dimensional magnetophotonic crystals," J. Appl. Phys. **93**, 3906–3911, 2002.
17. D. C. Hutchings, "Prospects for the implementation of magneto-optic elements in optoelectronic integrated circuits: A personal perspective," J. Phys. D: Appl. Phys. **36**, 2222–2229, 2003.

7

Normal Incidence

7.1 Introduction

At the normal light incidence, the distinction between the perpendicular and parallel polarizations with respect to the plane of incidence disappears. In isotropic media, the ellipsometric methods can provide useful information at oblique incidence only, and the normal incidence is employed in the polarization independent reflectance and transmittance spectroscopy. On the other hand, in anisotropic media the response even at normal light incidence may be sensitive to the incident wave polarization. In Chapter 4, we investigated the anisotropy induced by the polar magnetization, which resulted in a different multilayer response to the circular polarizations with different handedness. In fact, the configuration with the magnetization and incident, transmitted, or reflected propagation vectors all oriented normal to the interfaces is the most important one for its experimental convenience and simple relations between the response and material parameters.

The transmission MO azimuth rotation and ellipticity experiments are mostly performed at the normal incidence of a sensing beam. Similarly, in the MO reflection (mainly polar Kerr) experiments, we prefer the normal or nearly normal incidence. Additional advantages of this arrangement include a strong MO signal, a reduced effect of surface roughness on the noise level, and the possibility to focus the beam on a small area. Disadvantages include a more complicated optical access to the probed sample through the drilled pole piece of an electromagnet and higher requirements on the applied field to achieve magnetic saturation in the samples with high magnetization or in those with an easy in-plane magnetization axis. However, for practical reasons, magnetic materials with perpendicular magnetic anisotropy, *i.e.*, with an easy magnetization axis normal to the sample interfaces, deserve attention. There, the magnetization is often of moderate values thanks to a partial compensation of sublattice magnetic moments in oxide or in metal ferrimagnets.

An easy saturation with the polar magnetization during the writing process and the stability of the particular magnetization polarity (representing the logical zero or one) is required for the applications in the MO recording technology [1]. This technology, exploiting perpendicular magnetic uniaxial anisotropy, is consistent with a concept of perpendicular magnetic recording, proposed by Bell Laboratories at the end of the 1960s. With the perpendicular magnetization, an optimum can in principle be achieved from the point of view of the recording density and the efficiency of writing and reading processes.

In the absence of any in-plane optical anisotropy, the response includes the Faraday effect and normal incidence MO polar Kerr effect as special cases. In the limits, the Faraday effect becomes a pure propagation effect as opposed to the MO polar Kerr effect, which becomes a pure interface effect. In magnetic multilayers with individual layer thickness in nanometer range, the analysis becomes more involved, as discussed in Chapter 4. Both interface and propagation contributions to the response were properly accounted for using the Yeh's 4×4 matrix formalism.

In uniaxial media with the principal (or optical) axis inclined with respect to magnetization, the proper polarization modes are no more circularly polarized waves. As a result, Faraday effect becomes quenched unless the propagation vector coincides with the optical axis. In optically biaxial media, no special orientation of the propagation vector allowing for CP proper modes can be found. As an example of optically biaxial magnetic media, we mention ferric oxides with perovskite structure (orthoferrites) that display orthorhombic symmetry [2].

The quenching of Faraday effect also takes place in other situations where uniaxial anisotropy is operative. We list two of them. In planar optical waveguides, uniaxial anisotropy originates from the presence of boundaries. Its axis is parallel to boundary planes and perpendicular to the propagation vector. The uniaxial anisotropy results in different propagation conditions for TE and TM proper polarized waveguide modes. This presents an obstacle for an efficient magnetooptic mode conversion. A ray optics zigzag model [3] provides an insight into the phenomenon. Strain-induced optical anisotropy in magnetooptic multilayer disks may impair the reading process with a focused beam. The difficulties are related to propagation effects. Note that in reflection at a single interface, the polar MO Kerr effect and the in-plane optical anisotropy can be separated and measured almost independently [4].

In another limiting case (always at normal light incidence), the magnetization is restricted to the interface plane. Then, the magnetic linear birefringence and magnetic linear dichroism in both transmission (*e.g.*, Voigt effect) and reflection become observable.

This chapter is devoted to the response in anisotropic multilayers at normal incidence. The absence of an in-plane component of the propagation vector (corresponding to $N_y = 0$) considerably simplifies the analysis and extends the class of problems that can be reasonably treated analytically.

We first give the solution for the general form of the permittivity tensor and then discuss several practically important special cases. The first one is represented by a single plate characterized by a general permittivity tensor that reduces to a well-known special case of isotropic plate when the anisotropy contributions are removed. Next is the case of the magnetization oriented perpendicularly at zero in-plane anisotropy with the circular proper polarizations. As shown in Chapter 4, the transfer matrices are block diagonal, and the analysis of the optical response becomes simple. Then, we analyze the case of an orthorhombic crystal with one of three orthorhombic axes normal to the interface and parallel to the magnetization vector. The situation models orthoferrite crystals with the crystallographic orthorhombic c axis and magnetization normal to its natural face.

In the last section, we discuss the arrangement for the detection of the quadratic MO effects in reflection, *i.e.* magnetic linear birefringence and magnetic linear dichroism. As the MO experiments require a precise determination of small ellipsometric angles, we have to minimize several systematic errors. One of the remedies reduces to a minimum the number of optical elements positioned between the polarizer (or analyzer) and the sample. On the other hand, the MO reflection experiments at strictly zero angle of incidence would require a beam-splitter. Its incorporation cannot be avoided in the MO Kerr microscopy. There, however, the monitoring is often qualitative looking for optimum magnetic domain contrast. In addition, because of a significant volume occupied in the optical sequence by prism polarizers, focusing elements between the polarizer and the sample must be incorporated in the MO microscopy. Note that the situation may soon improve because of the progress in wire grid thin film polarizers for microellipsometry with performance extended to the near ultraviolet region.

The inconvenient beam-splitting is removed in MO spectroscopic and MO magnetometric setups using the configurations with small angles of incidence. This allows a separation of the incident and reflected beams. Small deviations from the zero angle of incidence do not significantly perturb the MO polar Kerr measurements. Experimental data may in principle be treated with formulae pertinent to nonzero angle of incidence. More often, the formulae for normal incidence are employed, as this does not introduce an important error.

The error becomes more critical where the magnetization is parallel to interfaces. At strictly normal incidence of the collimated beam, MO transverse and longitudinal Kerr effects vanish and the response exclusively contains the information on the reflection magnetic linear birefringence and dichroism (MLB and MLD in reflection). However, even at very small angles of incidence, the MO Kerr effects (odd in magnetization) contribute to the response. The MO effects linear in magnetization must be eliminated in order to determine MLB and MLD (even in magnetization).

7.2 Wave Equation

Let us assume that an electromagnetic plane wave propagates in an anisotropic n-th layer medium with an arbitrary orientation of magnetization [5] characterized by the permittivity tensor

$$\varepsilon^{(n)} = \begin{pmatrix} \varepsilon_{xx}^{(n)} & \varepsilon_{xy}^{(n)} & \varepsilon_{xz}^{(n)} \\ \varepsilon_{yx}^{(n)} & \varepsilon_{yy}^{(n)} & \varepsilon_{yz}^{(n)} \\ \varepsilon_{zx}^{(n)} & \varepsilon_{zy}^{(n)} & \varepsilon_{zz}^{(n)} \end{pmatrix} \tag{7.1}$$

perpendicular to the planar interfaces $z = z^{(n-1)}$ and $z = z^{(n)}$. The electric field of the plane wave becomes

$$\mathbf{E}^{(n)} = \mathbf{E}_0^{(n)} \left[\mathrm{j}\left(\omega t - \gamma_z^{(n)} z\right)\right], \tag{7.2}$$

with the propagation vector parallel to the z axis. The substitution of $N_y = 0$ in the wave equation (3.12) provides

$$\begin{pmatrix} \varepsilon_{xx}^{(n)} - N_z^{(n)2} & \varepsilon_{xy}^{(n)} & \varepsilon_{xz}^{(n)} \\ \varepsilon_{yx}^{(n)} & \varepsilon_{yy}^{(n)} - N_z^{(n)2} & \varepsilon_{yz}^{(n)} \\ \varepsilon_{zx}^{(n)} & \varepsilon_{zy}^{(n)} & \varepsilon_{zz}^{(n)} \end{pmatrix} \begin{pmatrix} E_{0x}^{(n)} \\ E_{0y}^{(n)} \\ E_{0z}^{(n)} \end{pmatrix} = 0, \tag{7.3}$$

where the only nonzero propagation vector component is $(\omega/c)\, N_z^{(n)}$. Its proper values follow from the condition for zero determinant in the equation system (7.3), which is a special case of Eq. (3.13), *i.e.*,

$$\begin{aligned} &\varepsilon_{zz}^{(n)} N_z^{(n)4} - \left(\varepsilon_{xx}^{(n)}\varepsilon_{zz}^{(n)} + \varepsilon_{yy}^{(n)}\varepsilon_{zz}^{(n)} - \varepsilon_{zx}^{(n)}\varepsilon_{xz}^{(n)} - \varepsilon_{yz}^{(n)}\varepsilon_{zy}^{(n)}\right) N_z^{(n)} \\ &\times \varepsilon_{xx}^{(n)}\varepsilon_{yy}^{(n)}\varepsilon_{zz}^{(n)} - \varepsilon_{xx}^{(n)}\varepsilon_{yz}^{(n)}\varepsilon_{zy}^{(n)} - \varepsilon_{yy}^{(n)}\varepsilon_{zx}^{(n)}\varepsilon_{xz}^{(n)} - \varepsilon_{zz}^{(n)}\varepsilon_{xy}^{(n)}\varepsilon_{yx}^{(n)} \\ &+ \varepsilon_{xy}^{(n)}\varepsilon_{yz}^{(n)}\varepsilon_{zx}^{(n)} + \varepsilon_{yx}^{(n)}\varepsilon_{zy}^{(n)}\varepsilon_{xz}^{(n)} = 0. \end{aligned} \tag{7.4}$$

This is a biquadratic equation with the solutions $N_{z1}^{(n)} = N_+^{(n)}$, $N_{z2}^{(n)} = -N_+^{(n)}$, $N_{z3}^{(n)} = N_-^{(n)}$, and $N_{z4}^{(n)} = -N_-^{(n)}$, *i.e.*,

$$\begin{aligned} \left(N_\pm^{(n)}\right)^2 &= \left(2\,\varepsilon_{zz}^{(n)}\right)^{-1} \left[\varepsilon_{zz}^{(n)}\left(\varepsilon_{xx}^{(n)} + \varepsilon_{yy}^{(n)}\right) - \left(\varepsilon_{zx}^{(n)}\varepsilon_{xz}^{(n)} + \varepsilon_{yz}^{(n)}\varepsilon_{zy}^{(n)}\right)\right] \pm \left(2\,\varepsilon_{zz}^{(n)}\right)^{-1} \\ &\times \left[\varepsilon_{zz}^{(n)2}\left(\varepsilon_{xx}^{(n)} - \varepsilon_{yy}^{(n)}\right)^2 - 2\left(\varepsilon_{xx}^{(n)} - \varepsilon_{yy}^{(n)}\right)\left(\varepsilon_{zx}^{(n)}\varepsilon_{xz}^{(n)} - \varepsilon_{yz}^{(n)}\varepsilon_{zy}^{(n)}\right) + \left(\varepsilon_{zx}^{(n)}\varepsilon_{xz}^{(n)}\right.\right. \\ &\left.\left. + \varepsilon_{yz}^{(n)}\varepsilon_{zy}^{(n)}\right)^2 + 4\varepsilon_{zz}^{(n)}\left(\varepsilon_{zz}^{(n)}\varepsilon_{xy}^{(n)}\varepsilon_{yx}^{(n)} - \varepsilon_{xy}^{(n)}\varepsilon_{yz}^{(n)}\varepsilon_{zx}^{(n)} - \varepsilon_{yx}^{(n)}\varepsilon_{zy}^{(n)}\varepsilon_{xz}^{(n)}\right)\right]. \end{aligned} \tag{7.5}$$

The proper polarizations are given by the unit electric field Cartesian vectors following from Eq. (3.16)

$$\hat{\boldsymbol{e}}_{\pm}^{(n)} = C_{\pm}^{(n)} \begin{pmatrix} -\varepsilon_{xy}^{(n)}\varepsilon_{zz}^{(n)} + \varepsilon_{xz}^{(n)}\varepsilon_{zy}^{(n)} \\ \varepsilon_{zz}^{(n)}\left[\varepsilon_{xx}^{(n)} - \left(N_{\pm}^{(n)}\right)^2\right] - \varepsilon_{zx}^{(n)}\varepsilon_{xz}^{(n)} \\ -\varepsilon_{zy}^{(n)}\left[\varepsilon_{xx}^{(n)} - \left(N_{\pm}^{(n)}\right)^2\right] + \varepsilon_{xy}^{(n)}\varepsilon_{zx}^{(n)} \end{pmatrix}, \tag{7.6}$$

where $C_{\pm}^{(n)}$ are the normalizing coefficients. Note that, in general, there is a nonzero longitudinal electric field component parallel to the z axis. The corresponding magnetic field proper polarizations follow from Eq. (3.19)

$$\boldsymbol{b}_j^{(n)} = N_{zj}^{(n)}\hat{\boldsymbol{z}} \times \hat{\boldsymbol{e}}_j^{(n)} = -N_{zj}^{(n)}e_{yj}^{(n)}\hat{\boldsymbol{x}} + N_{zj}^{(n)}e_{xj}^{(n)}\hat{\boldsymbol{y}}. \tag{7.7}$$

We express the proper polarizations of the electric fields from Eq. (7.6) as $\hat{\boldsymbol{e}}_j^{(n)} = \hat{\boldsymbol{x}}e_{xj} + \hat{\boldsymbol{y}}e_{yj} + \hat{\boldsymbol{z}}e_{zj}$, $j = 1, 2, 3$, and 4 with $\hat{\boldsymbol{e}}_1^{(n)} = \hat{\boldsymbol{e}}_2^{(n)} = \hat{\boldsymbol{e}}_+^{(n)}$ and $\mathbf{e}_3^{(n)} = \mathbf{e}_4^{(n)} = \mathbf{e}_-^{(n)}$

$$\boldsymbol{b}_{1,2}^{(n)} = \begin{pmatrix} \mp N_+^{(n)}\left\{\varepsilon_{zz}^{(n)}\left[\varepsilon_{xx}^{(n)} - \left(N_+^{(n)}\right)^2\right] - \varepsilon_{zx}^{(n)}\varepsilon_{xz}^{(n)}\right\} \\ \mp N_+^{(n)}\left(\varepsilon_{xy}^{(n)}\varepsilon_{zz}^{(n)} - \varepsilon_{xz}^{(n)}\varepsilon_{zy}^{(n)}\right) \\ 0 \end{pmatrix} \tag{7.8a}$$

$$\boldsymbol{b}_{3,4}^{(n)} = \begin{pmatrix} \mp N_-^{(n)}\left\{\varepsilon_{zz}^{(n)}\left[\varepsilon_{xx}^{(n)} - \left(N_-^{(n)}\right)^2\right] - \varepsilon_{zx}^{(n)}\varepsilon_{xz}^{(n)}\right\} \\ \mp N_-^{(n)}\left(\varepsilon_{xy}^{(n)}\varepsilon_{zz}^{(n)} - \varepsilon_{xz}^{(n)}\varepsilon_{zy}^{(n)}\right) \\ 0 \end{pmatrix}. \tag{7.8b}$$

The magnetic permeability assumes its scalar vacuum value, and consequently there is no magnetic field component parallel to the propagation vector.

7.3 General M Matrix

We can construct the dynamical matrix, a special case of Eq. (3.29)

$$\mathbf{D}^{(n)} = \begin{pmatrix} D_{11}^{(n)} & D_{11}^{(n)} & D_{11}^{(n)} & D_{11}^{(n)} \\ D_{21}^{(n)} & -D_{21}^{(n)} & D_{23}^{(n)} & -D_{23}^{(n)} \\ D_{31}^{(n)} & D_{31}^{(n)} & D_{33}^{(n)} & D_{33}^{(n)} \\ D_{41}^{(n)} & -D_{41}^{(n)} & D_{43}^{(n)} & -D_{43}^{(n)} \end{pmatrix}, \tag{7.9}$$

where according to Eqs. (3.30)

$$D_{11}^{(n)} = \left(-\varepsilon_{xy}^{(n)}\varepsilon_{zz}^{(n)} + \varepsilon_{xz}^{(n)}\varepsilon_{zy}^{(n)}\right) \tag{7.10a}$$

$$D_{21}^{(n)} = -N_{+}^{(n)}\left(\varepsilon_{xy}^{(n)}\varepsilon_{zz}^{(n)} - \varepsilon_{xz}^{(n)}\varepsilon_{zy}^{(n)}\right) \tag{7.10b}$$

$$D_{23}^{(n)} = -N_{-}^{(n)}\left(\varepsilon_{xy}^{(n)}\varepsilon_{zz}^{(n)} - \varepsilon_{xz}^{(n)}\varepsilon_{zy}^{(n)}\right) \tag{7.10c}$$

$$D_{31}^{(n)} = \varepsilon_{zz}^{(n)}\left[\varepsilon_{xx}^{(n)} - \left(N_{+}^{(n)}\right)^2\right] - \varepsilon_{zx}^{(n)}\varepsilon_{xz}^{(n)} \tag{7.10d}$$

$$D_{33}^{(n)} = \varepsilon_{zz}^{(n)}\left[\varepsilon_{xx}^{(n)} - \left(N_{-}^{(n)}\right)^2\right] - \varepsilon_{zx}^{(n)}\varepsilon_{xz}^{(n)} \tag{7.10e}$$

$$D_{41}^{(n)} = -N_{+}^{(n)}\left\{\varepsilon_{zz}^{(n)}\left[\varepsilon_{xx}^{(n)} - \left(N_{+}^{(n)}\right)^2\right] - \varepsilon_{zx}^{(n)}\varepsilon_{xz}^{(n)}\right\} \tag{7.10f}$$

$$D_{43}^{(n)} = -N_{-}^{(n)}\left\{\varepsilon_{zz}^{(n)}\left[\varepsilon_{xx}^{(n)} - \left(N_{-}^{(n)}\right)^2\right] - \varepsilon_{zx}^{(n)}\varepsilon_{xz}^{(n)}\right\}. \tag{7.10g}$$

In this simple form, $\mathbf{D}^{(n)}$ employs the proper polarization modes that are not normalized. The mode normalization can be included, if needed, by the multiplication from right of the dynamical matrix, $\mathbf{D}^{(n)}$, by the normalization matrix of proper polarization as in Eq. (3.39)

$$\mathbf{C}^{(n)} = \begin{pmatrix} C_{+}^{(n)} & 0 & 0 & 0 \\ 0 & C_{+}^{(n)} & 0 & 0 \\ 0 & 0 & C_{-}^{(n)} & 0 \\ 0 & 0 & 0 & C_{-}^{(n)} \end{pmatrix}, \tag{7.11}$$

with

$$C_{\pm}^{(n)} = \left\{\left|-\varepsilon_{xy}^{(n)}\varepsilon_{zz}^{(n)} + \varepsilon_{xz}^{(n)}\varepsilon_{zy}^{(n)}\right|^2 + \left|\varepsilon_{zz}^{(n)}\left[\varepsilon_{xx}^{(n)} - \left(N_{\pm}^{(n)}\right)^2\right] - \varepsilon_{zx}^{(n)}\varepsilon_{xz}^{(n)}\right|^2 \right.$$
$$\left. + \left|-\varepsilon_{zy}^{(n)}\left[\varepsilon_{xx}^{(n)} - \left(N_{\pm}^{(n)}\right)^2\right] + \varepsilon_{xy}^{(n)}\varepsilon_{zx}^{(n)}\right|^2\right\}^{-1/2}, \tag{7.12}$$

following Eq. (7.6). We wish to write down the dynamical matrix, $\mathbf{D}^{(n)}$, in a more concise form with the abbreviations

$$a = D_{11}^{(n)}, \tag{7.13a}$$

$$b_{+} = D_{31}^{(n)}, \tag{7.13b}$$

$$b_{-} = D_{33}^{(n)}. \tag{7.13c}$$

The normalized dynamical matrix would be obtained from the product

$$\mathbf{D}_{\text{norm}}^{(n)} = \begin{pmatrix} a & a & a & a \\ N_{+}^{(n)}a & -N_{+}^{(n)}a & N_{-}^{(n)}a & -N_{-}^{(n)}a \\ b_{+} & b_{+} & b_{-} & b_{-} \\ -N_{+}^{(n)}b_{+} & N_{+}^{(n)}b_{+} & -N_{-}^{(n)}b_{-} & N_{-}^{(n)}b_{-} \end{pmatrix}\mathbf{C}^{(n)}. \tag{7.14}$$

The inverse of the dynamical matrix becomes

$$\left(\mathbf{D}^{(n)}_{\text{norm}}\right)^{-1} = (\mathbf{C}^{(n)})^{-1}\left[2a\,(b_- - b_+)N^{(n)}_+ N^{(n)}_-\right]^{-1}$$
$$\times \begin{pmatrix} N^{(n)}_+ N^{(n)}_- b_- & N^{(n)}_- b_- & -N^{(n)}_+ N^{(n)}_- a & N^{(n)}_- a \\ N^{(n)}_+ N^{(n)}_- b_- & -N^{(n)}_- b_- & -N^{(n)}_+ N^{(n)}_- a & -N^{(n)}_- a \\ -N^{(n)}_+ N^{(n)}_- b_+ & -N^{(n)}_+ b_+ & N^{(n)}_+ N^{(n)}_- a & -N^{(n)}_+ a \\ -N^{(n)}_+ N^{(n)}_- b_+ & N^{(n)}_+ b_+ & N^{(n)}_+ N^{(n)}_- a & N^{(n)}_+ a \end{pmatrix}. \quad (7.15)$$

The propagation matrix takes the form

$$\mathbf{P}^{(n)} = \begin{pmatrix} e^{j\beta^{(n)}_+} & 0 & 0 & 0 \\ 0 & e^{-j\beta^{(n)}_+} & 0 & 0 \\ 0 & 0 & e^{j\beta^{(n)}_-} & 0 \\ 0 & 0 & 0 & e^{-j\beta^{(n)}_-} \end{pmatrix}, \quad (7.16)$$

where

$$\beta^{(n)}_\pm = \frac{\omega}{c} N^{(n)}_\pm d^{\,(n)} \quad (7.17)$$

and $d^{\,(n)} = z^{(n-1)} - z^{(n)}$ is the thickness of the n-th layer.

To completely specify the product (3.33) that defines the multilayer **M** matrix, it remains to list the normalized dynamical matrices in the isotropic media on both sides of the multilayer system. We thus need, assuming linearly polarized orthogonal and normalized proper polarizations in media $n = 0$ and $\mathcal{N}+1$, the inverse of the dynamical matrix in the entrance isotropic medium and the dynamic matrix in the exit isotropic medium. According to Eqs. (3.82), we can write

$$(\mathbf{D}^{(0)})^{-1} = \frac{1}{2N^{(0)}} \begin{pmatrix} N^{(0)} & 1 & 0 & 0 \\ N^{(0)} & -1 & 0 & 0 \\ 0 & 0 & N^{(0)} & -1 \\ 0 & 0 & N^{(0)} & 1 \end{pmatrix} \quad (7.18)$$

and

$$\mathbf{D}^{(\mathcal{N}+1)} = \begin{pmatrix} 1 & 1 & 0 & 0 \\ N^{(\mathcal{N}+1)} & -N^{(\mathcal{N}+1)} & 0 & 0 \\ 0 & 0 & 1 & 1 \\ 0 & 0 & -N^{(\mathcal{N}+1)} & N^{(\mathcal{N}+1)} \end{pmatrix}. \quad (7.19)$$

An important special case is represented by the system consisting of a single crystal magnetic thin film epitaxially grown on single crystal thick nonmagnetic substrate represented by the product

$$\begin{aligned} \mathbf{M} &= (\mathbf{D}^{(0)})^{-1}\mathbf{D}^{(1)}\mathbf{P}^{(1)}(\mathbf{D}^{(1)})^{-1}\mathbf{D}^{(2)} \\ &= (\mathbf{D}^{(0)})^{-1}\mathbf{S}^{(1)}\mathbf{D}^{(2)}, \end{aligned} \quad (7.20)$$

where $\mathbf{S}^{(1)} = \mathbf{D}^{(1)}\mathbf{P}^{(1)}(\mathbf{D}^{(1)})^{-1}$ stands for the characteristic matrix of the film,

$$\mathbf{S}^{(1)} = \begin{pmatrix} S_{11}^{(1)} & S_{12}^{(1)} & S_{13}^{(1)} & S_{14}^{(1)} \\ S_{21}^{(1)} & S_{11}^{(1)} & S_{23}^{(1)} & -S_{13}^{(1)} \\ S_{31}^{(1)} & S_{32}^{(1)} & S_{33}^{(1)} & S_{34}^{(1)} \\ S_{41}^{(1)} & -S_{31}^{(1)} & S_{43}^{(1)} & S_{33}^{(1)} \end{pmatrix}, \tag{7.21}$$

with (up to a common factor)

$$S_{11}^{(1)} = a N_+^{(1)} N_-^{(1)} \left[b_- \left(e^{j\beta_+^{(1)}} + e^{-j\beta_+^{(1)}} \right) - b_+ \left(e^{j\beta_-^{(1)}} + e^{-j\beta_-^{(1)}} \right) \right], \tag{7.22a}$$

$$S_{12}^{(1)} = a \left[N_-^{(1)} b_- \left(e^{j\beta_+^{(1)}} - e^{-j\beta_+^{(1)}} \right) - N_+^{(1)} b_+ \left(e^{j\beta_-^{(1)}} - e^{-j\beta_-^{(1)}} \right) \right], \tag{7.22b}$$

$$S_{13}^{(1)} = a^2 N_+^{(1)} N_-^{(1)} \left[- \left(e^{j\beta_+^{(1)}} + e^{-j\beta_+^{(1)}} \right) + \left(e^{j\beta_-^{(1)}} + e^{-j\beta_-^{(1)}} \right) \right], \tag{7.22c}$$

$$S_{14}^{(1)} = a^2 \left[N_-^{(1)} \left(e^{j\beta_+^{(1)}} - e^{-j\beta_+^{(1)}} \right) - N_+^{(1)} \left(e^{j\beta_-^{(1)}} - e^{-j\beta_-^{(1)}} \right) \right], \tag{7.22d}$$

$$S_{21}^{(1)} = a N_+^{(1)} N_-^{(1)} \left[N_+^{(1)} b_- \left(e^{j\beta_+^{(1)}} - e^{-j\beta_+^{(1)}} \right) - N_-^{(1)} b_+ \left(e^{j\beta_-^{(1)}} - e^{-j\beta_-^{(1)}} \right) \right], \tag{7.22e}$$

$$S_{23}^{(1)} = a^2 N_+^{(1)} N_-^{(1)} \left[-N_+^{(1)} \left(e^{j\beta_+^{(1)}} - e^{-j\beta_+^{(1)}} \right) + N_-^{(1)} \left(e^{j\beta_-^{(1)}} - e^{-j\beta_-^{(1)}} \right) \right], \tag{7.22f}$$

$$S_{31}^{(1)} = b_+ b_- N_+^{(1)} N_-^{(1)} \left[\left(e^{j\beta_+^{(1)}} + e^{-j\beta_+^{(1)}} \right) - \left(e^{j\beta_-^{(1)}} + e^{-j\beta_-^{(1)}} \right) \right], \tag{7.22g}$$

$$S_{32}^{(1)} = b_+ b_- \left[N_-^{(1)} \left(e^{j\beta_+^{(1)}} - e^{-j\beta_+^{(1)}} \right) - N_+^{(1)} \left(e^{j\beta_-^{(1)}} - e^{-j\beta_-^{(1)}} \right) \right], \tag{7.22h}$$

$$S_{33}^{(1)} = a N_+^{(1)} N_-^{(1)} \left[-b_+ \left(e^{j\beta_+^{(1)}} + e^{-j\beta_+^{(1)}} \right) + b_- \left(e^{j\beta_-^{(1)}} + e^{-j\beta_-^{(1)}} \right) \right], \tag{7.22i}$$

$$S_{34}^{(1)} = a \left[N_-^{(1)} b_+ \left(e^{j\beta_+^{(1)}} - e^{-j\beta_+^{(1)}} \right) - N_+^{(1)} b_- \left(e^{j\beta_-^{(1)}} - e^{-j\beta_-^{(1)}} \right) \right], \tag{7.22j}$$

$$S_{41}^{(1)} = b_+ b_- N_+^{(1)} N_-^{(1)} \left[-N_+^{(1)} \left(e^{j\beta_+^{(1)}} - e^{-j\beta_+^{(1)}} \right) + N_-^{(1)} \left(e^{j\beta_-^{(1)}} - e^{-j\beta_-^{(1)}} \right) \right], \tag{7.22k}$$

$$S_{43}^{(1)} = a N_+^{(1)} N_-^{(1)} \left[N_+^{(1)} b_+ \left(e^{j\beta_+^{(1)}} - e^{-j\beta_+^{(1)}} \right) - N_-^{(1)} b_- \left(e^{j\beta_-^{(1)}} - e^{-j\beta_-^{(1)}} \right) \right]. \tag{7.22l}$$

7.4 Examples

7.4.1 General Anisotropy

To illustrate the application of the 4 × 4 matrix formalism, we first treat the response of a polarized electromagnetic plane wave in a plate (1), specified by the permittivity tensor (7.1), surrounded by an isotropic ambient, specified by the refractive index $N^{(0)} = N^{(2)}$. The proper polarization amplitudes of electric fields at the front (0) interface and the corresponding amplitudes at the back (2) interface at $z = d^{(1)}$ are defined in the relation

$$\begin{pmatrix} E_{0x+}^{(0)} \\ E_{0x-}^{(0)} \\ E_{0y+}^{(0)} \\ E_{0y-}^{(0)} \end{pmatrix} = \mathbf{M} \begin{pmatrix} E_{0x+}^{(2)} \\ E_{0x-}^{(2)} \\ E_{0y+}^{(2)} \\ E_{0y-}^{(2)} \end{pmatrix}, \tag{7.23}$$

where $d^{(1)}$ is the thickness of the plate and the $+$ and $-$ signs distinguish the wave's propagation in the positive and negative senses of the z axis. The **M** matrix takes the form

$$\mathbf{M} = (\mathbf{D}^{(0)})^{-1}\mathbf{D}^{(1)}\mathbf{P}^{(1)}(\mathbf{D}^{(1)})^{-1}\mathbf{D}^{(0)}. \tag{7.24}$$

In the isotropic ambient, any pair of the orthonormal polarizations can serve as proper polarizations. It is often advantageous, however, to choose the proper polarizations that agree with those of the system considered. Choosing the proper polarizations linearly polarized along the Cartesian x and y axes, we obtain for the **M** matrix elements that enter the reflection and transmission coefficients

$$M_{11} = q^{-1}a(b_- A_+ - b_+ A_-), \tag{7.25a}$$
$$M_{21} = q^{-1}a(b_- B_+ - b_+ B_-), \tag{7.25b}$$
$$M_{31} = q^{-1}b_+ b_-(A_+ - A_-), \tag{7.25c}$$
$$M_{41} = q^{-1}b_+ b_-(B_+ - B_-), \tag{7.25d}$$
$$M_{13} = -q^{-1}a^2(A_+ - A_-), \tag{7.25e}$$
$$M_{23} = -q^{-1}a^2(B_+ - B_-), \tag{7.25f}$$
$$M_{33} = -q^{-1}a(b_+ A_+ - b_- A_-), \tag{7.25g}$$
$$M_{43} = -q^{-1}a(b_+ B_+ - b_- B_-). \tag{7.25h}$$

In this case, the relevant expressions entering the definitions of the global reflection and transmission coefficients (3.70) become

$$M_{11}M_{33} - M_{13}M_{31} = q^{-2}a^2(b_+ - b_-)^2 A_+ A_-, \tag{7.26a}$$
$$M_{21}M_{33} - M_{23}M_{31} = q^{-2}a^2(b_+ - b_-)(b_+ A_+ B_- - b_- A_- B_+), \tag{7.26b}$$
$$M_{41}M_{33} - M_{43}M_{31} = q^{-2}ab_+ b_-(b_+ - b_-)(A_+ B_- - A_- B_+), \tag{7.26c}$$
$$M_{11}M_{23} - M_{21}M_{13} = q^{-2}a^3(b_+ - b_-)(A_- B_+ - A_+ B_-), \tag{7.26d}$$
$$M_{11}M_{43} - M_{41}M_{13} = q^{-2}a^2(b_+ - b_-)(b_+ A_- B_+ - b_- A_+ B_-). \tag{7.26e}$$

We have denoted

$$q = -4a(b_+ - b_-)N^{(0)}N_+^{(1)}N_-^{(1)} \tag{7.27}$$

and

$$A_\pm = N_\mp^{(1)}\left[\left(N^{(0)} + N_\pm^{(1)}\right)^2 e^{j\beta_\pm^{(1)}} - \left(N^{(0)} - N_\pm^{(1)}\right)^2 e^{-j\beta_\pm^{(1)}}\right], \tag{7.28a}$$

$$B_\pm = N_\mp^{(1)}\left[(N^{(0)})^2 - \left(N_\pm^{(1)}\right)^2\right]\left(e^{j\beta_\pm^{(1)}} - e^{-j\beta_\pm^{(1)}}\right). \tag{7.28b}$$

The Fresnel reflection coefficients at the ambient–film interface can be defined as

$$r_\pm^{(01)} = -r_\pm^{(10)} = \frac{N^{(0)} - N_\pm^{(1)}}{N^{(0)} + N_\pm^{(1)}}. \tag{7.29}$$

Note that from Eq. (7.13b) and Eq. (7.13c) it follows

$$b_+ b_- = -\varepsilon_{zz}^{(1)2}\varepsilon_{xy}^{(1)}\varepsilon_{yx}^{(1)} + \varepsilon_{zz}^{(1)}(\varepsilon_{xy}^{(1)}\varepsilon_{yz}^{(1)}\varepsilon_{zx}^{(1)} + \varepsilon_{xz}^{(1)}\varepsilon_{yx}^{(1)}\varepsilon_{zy}^{(1)}) - \varepsilon_{zx}^{(1)}\varepsilon_{xz}^{(1)}\varepsilon_{yz}^{(1)}\varepsilon_{zy}^{(1)}. \tag{7.30}$$

We can now compute the elements of the reflection and transmission matrices. Using Eqs. (7.26) and assuming that there are no waves travelling from the isotropic ambient at $z > d^{(1)}$ toward the back interface $z = d^{(1)}$, *i.e.*, $E_{0x-}^{(2)} = E_{0y-}^{(2)} = 0$, we have

$$r_{xx}^{(02)} = \frac{E_{0x-}^{(0)}}{E_{0x+}^{(0)}} = \frac{M_{21}M_{33} - M_{23}M_{31}}{M_{11}M_{33} - M_{13}M_{31}} = \frac{b_+A_+B_- - b_-A_-B_+}{(b_+ - b_-)A_+A_-}, \quad (E_{0y+}^{(0)} = 0), \tag{7.31a}$$

$$r_{yx}^{(02)} = \frac{E_{0y-}^{(0)}}{E_{0x+}^{(0)}} = \frac{M_{41}M_{33} - M_{43}M_{31}}{M_{11}M_{33} - M_{13}M_{31}} = \frac{b_+b_-(A_+B_- - A_-B_+)}{a(b_+ - b_-)A_+A_-}, \quad (E_{0y+}^{(0)} = 0), \tag{7.31b}$$

$$r_{xy}^{(02)} = \frac{E_{0x-}^{(0)}}{E_{0y+}^{(0)}} = \frac{M_{11}M_{23} - M_{21}M_{13}}{M_{11}M_{33} - M_{13}M_{31}} = \frac{a\,(A_-B_+ - A_+B_-)}{(b_+ - b_-)\,A_+A_-}, \quad (E_{0x+}^{(0)} = 0), \tag{7.31c}$$

$$r_{yy}^{(02)} = \frac{E_{0y-}^{(0)}}{E_{0y+}^{(0)}} = \frac{M_{11}M_{43} - M_{41}M_{13}}{M_{11}M_{33} - M_{13}M_{31}} = \frac{b_+A_-B_+ - b_-A_+B_-}{(b_+ - b_-)A_+A_-}, \quad (E_{0x+}^{(0)} = 0), \tag{7.31d}$$

$$t_{xx}^{(02)} = \frac{E_{0x+}^{(2)}}{E_{0x+}^{(0)}} = \frac{M_{33}}{M_{11}M_{33} - M_{13}M_{31}} = q\frac{-b_+A_+ + b_-A_-}{a(b_+ - b_-)^2A_+A_-}, \quad (E_{0y+}^{(0)} = 0), \tag{7.31e}$$

$$t_{yx}^{(02)} = \frac{E_{0y-}^{(2)}}{E_{0x+}^{(0)}} = \frac{-M_{31}}{M_{11}M_{33} - M_{13}M_{31}} = q\frac{-b_+b_-(A_+ - A_-)}{a^2(b_+ - b_-)^2A_+A_-}, \quad (E_{0y+}^{(0)} = 0), \tag{7.31f}$$

$$t_{xy}^{(02)} = \frac{E_{0x+}^{(2)}}{E_{0y+}^{(0)}} = \frac{-M_{31}}{M_{11}M_{33} - M_{13}M_{31}} = q\frac{A_+ - A_-}{(b_+ - b_-)^2 A_+ A_-}, \quad (E_{0x+}^{(0)} = 0), \tag{7.31g}$$

$$t_{yy}^{(02)} = \frac{E_{0y+}^{(2)}}{E_{0y+}^{(0)}} = \frac{M_{11}}{M_{11}M_{33} - M_{13}M_{31}} = q\frac{b_- A_+ - b_+ A_-}{a(b_+ - b_-)^2 A_+ A_-}, \quad (E_{0x+}^{(0)} = 0). \tag{7.31h}$$

These equations represent the general solution to the problem. In the following sections, we illustrate the applications of these results on a few simple cases.

7.4.2 Isotropic Nonmagnetic Plate

This is the simplest case. We have $N_+^{(1)} = N_-^{(1)} = N^{(1)}$, $A_+ = A_-$, $B_+ = B_-$, and $\beta_+^{(1)} = \beta_-^{(1)} = \beta^{(1)}$. The Cartesian Jones reflection and transmission matrices are diagonal [6], and we get for their elements

$$r_{xx}^{(02)} = r_{yy}^{(02)} = \frac{r^{(01)} + r^{(10)} e^{-2j\beta^{(1)}}}{1 + r^{(01)} r^{(10)} e^{-2j\beta^{(1)}}}, \tag{7.32a}$$

$$r_{xy}^{(02)} = r_{yx}^{(02)} = 0, \tag{7.32b}$$

$$t_{xx}^{(02)} = t_{yy}^{(02)} = \frac{t^{(01)} t^{(10)} e^{-2j\beta^{(1)}}}{1 + r^{(01)} r^{(10)} e^{-2j\beta^{(1)}}}, \tag{7.32c}$$

$$t_{xy}^{(02)} = t_{yx}^{(02)} = 0. \tag{7.32d}$$

The single-interface Fresnel transmission coefficients are defined by

$$t^{(01)} = \frac{2N^{(0)}}{N^{(0)} + N^{(1)}}, \tag{7.33a}$$

and

$$t^{(10)} = \frac{2N^{(1)}}{N^{(0)} + N^{(1)}}, \tag{7.33b}$$

$r^{(01)}$ and $r^{(10)}$ are given by Eq. (7.29) with $N_+^{(1)} = N_-^{(1)} = N^{(1)}$.

7.4.3 Isotropic Plate Magnetized Normal to the Interfaces

The problem was considered by Donovan and Medcalf [7], Piller [8], and Sansalone [9]. The nonzero elements of the permittivity tensor are given by

$$\varepsilon_{xx}^{(1)} = \varepsilon_{yy}^{(1)} = \varepsilon_0^{(1)}, \tag{7.34a}$$

$$\varepsilon_{zz}^{(1)} = \varepsilon_3^{(1)}, \tag{7.34b}$$

$$\varepsilon_{xy}^{(1)} = -\varepsilon_{yx}^{(1)} = \mathrm{j}\,\varepsilon_1^{(1)}. \tag{7.34c}$$

We assign the solutions of the proper value equation (7.4) given in Eq. (7.5) as follows

$$\left(N_{\pm}^{(1)}\right)^2 = \varepsilon_0^{(1)} \mp \varepsilon_1^{(1)}, \tag{7.35}$$

and from Eq. (7.13) we obtain

$$\mathrm{j}a = -b_+ = b_- = \varepsilon_1^{(1)}\varepsilon_3^{(1)}. \tag{7.36}$$

The Jones transmission matrix elements according to Eqs. (7.31e) to (7.31h) take the form

$$t_{xx} = t_{yy} = 2N^{(0)}N_+^{(1)}N_-^{(1)}\frac{A_+ + A_-}{A_+A_-} = \frac{1}{2}\frac{t_+^{(01)}t_+^{(10)}\mathrm{e}^{-\mathrm{j}\beta_{\pm}^{(1)}}}{1 + r_+^{(01)}r_+^{(10)}\mathrm{e}^{-2\mathrm{j}\beta^{(1)}}} + \frac{1}{2}\frac{t_-^{(01)}t_-^{(10)}\mathrm{e}^{-\mathrm{j}\beta_-^{(1)}}}{1 + r_-^{(01)}r_-^{(10)}\mathrm{e}^{-2\mathrm{j}\beta_-^{(1)}}}, \tag{7.37a}$$

$$t_{yx} = -t_{xy} = -2\mathrm{j}N^{(0)}N_+^{(1)}N_-^{(1)}\frac{A_+ - A_-}{A_+A_-} = \frac{1}{2}\mathrm{j}\frac{t_+^{(01)}t_+^{(10)}\mathrm{e}^{-\mathrm{j}\beta_+^{(1)}}}{1 + r_+^{(01)}r_+^{(10)}\mathrm{e}^{-2\mathrm{j}\beta_+^{(1)}}} - \frac{1}{2}\mathrm{j}\frac{t_-^{(01)}t_-^{(10)}\mathrm{e}^{-\mathrm{j}\beta_-^{(1)}}}{1 + r_-^{(01)}r_-^{(10)}\mathrm{e}^{-2\mathrm{j}\beta_-^{(1)}}}, \tag{7.37b}$$

where $r_{\pm}^{(01)}$ and $r_{\pm}^{(10)}$ are defined by Eq. (7.29) and

$$t_{\pm}^{(01)} = \frac{2N^{(0)}}{N^{(0)} + N_{\pm}^{(1)}}, \tag{7.38a}$$

$$t_{\pm}^{(10)} = \frac{2N_{\pm}^{(1)}}{N^{(0)} + N_{\pm}^{(1)}}. \tag{7.38b}$$

The reflection coefficients of the system follow from Eqs. (7.31)

$$r_{xx} = r_{yy} = \frac{1}{2}\left(\frac{B_+}{A_+} + \frac{B_-}{A_-}\right), \tag{7.39a}$$

$$r_{yx} = -r_{xy} = \frac{1}{2}\mathrm{j}\left(\frac{B_-}{A_-} - \frac{B_+}{A_+}\right), \tag{7.39b}$$

where A_+ and B_+ are given by Eqs. (7.28). This gives the expected result for the elements of Cartesian Jones reflection matrix

$$r_{xx} = r_{yy} = \frac{1}{2}\left(\frac{r_+^{(01)} + r_+^{(10)}\mathrm{e}^{-2\mathrm{j}\beta_+^{(1)}}}{1 + r_+^{(01)}r_+^{(10)}\mathrm{e}^{-2\mathrm{j}\beta_+^{(1)}}} + \frac{r_-^{(01)} + r_-^{(10)}\mathrm{e}^{-2\mathrm{j}\beta_-^{(1)}}}{1 + r_-^{(01)}r_-^{(10)}\mathrm{e}^{-2\mathrm{j}\beta_-^{(1)}}}\right), \tag{7.40a}$$

$$r_{yx} = -r_{xy} = \frac{1}{2}\mathrm{j}\left(\frac{r_-^{(01)} + r_-^{(10)}\mathrm{e}^{-2\mathrm{j}\beta_-^{(1)}}}{1 + r_-^{(01)}r_-^{(10)}\mathrm{e}^{-2\mathrm{j}\beta_-^{(1)}}} - \frac{r_+^{(01)} + r_+^{(10)}\mathrm{e}^{-2\mathrm{j}\beta_+^{(1)}}}{1 + r_+^{(01)}r_+^{(10)}\mathrm{e}^{-2\mathrm{j}\beta_+^{(1)}}}\right). \tag{7.40b}$$

Let us define the complex ratios characterizing the Faraday and the normal incidence magnetooptic polar Kerr effect, respectively

$$\chi_F = \frac{t_{yx}}{t_{xx}} = -\frac{t_{xy}}{t_{yy}} \tag{7.41a}$$

and

$$\chi_K = \frac{r_{yx}}{r_{xx}} = -\frac{r_{xy}}{r_{yy}}. \tag{7.41b}$$

The Faraday and MO Kerr azimuth rotations θ and the corresponding ellipticities ϵ can be obtained from the relations [6]

$$\tan(2\theta) = \frac{2\mathrm{Re}(\chi)}{1 - |\chi|^2}, \tag{7.42a}$$

$$\sin(2\varepsilon) = \frac{2\mathrm{Im}(\chi)}{1 + |\chi|^2}. \tag{7.42b}$$

In plates with the thickness in the range of a few micrometers, the contribution to the observed magnetooptic azimuth rotation and ellipticity coming from the propagation effect strongly dominates over that originating from interfaces. In such situations, we can put[1]

$$t_+^{(01)} = t_-^{(01)}, \tag{7.43a}$$

$$t_+^{(10)} = t_-^{(10)}, \tag{7.43b}$$

and

$$r_+^{(01)} = r_-^{(01)} = -r_+^{(10)} = -r_-^{(10)}. \tag{7.44}$$

The ratio χ_F can be written as

$$\chi_F = \frac{1 + r^{(01)2}\mathrm{e}^{-\mathrm{j}(\beta_+^{(1)} + \beta_-^{(1)})}}{1 - r^{(01)2}\mathrm{e}^{-\mathrm{j}(\beta_+^{(1)} + \beta_-^{(1)})}} \tan\left[\frac{1}{2}\left(\beta_+^{(1)} - \beta_-^{(1)}\right)\right]. \tag{7.45}$$

[1] This approximation would not be justified in the nanometer range where both propagation and interface effect may be of the same order of magnitude.

In the regions of low absorption, the fraction in this equation is greater than one corresponding to a value of the Faraday rotation higher than that corresponding to a single pass of the wave across the plate. If the plate absorbs strongly, *i.e.*,

$$r^{(01)2} e^{-j(\beta_+^{(1)}+\beta_-^{(1)})} \ll 1, \tag{7.46}$$

the effect of multiple reflections disappears and the Faraday effect is given simply by a complex angle $(\beta_+^{(1)} - \beta_-^{(1)})/2$, for which we have

$$\chi_F = \tan\left[(\beta_+^{(1)} - \beta_-^{(1)})/2\right]. \tag{7.47}$$

For reflection, when the condition (7.46) is met, we obtain from Eq. (7.39a), Eq. (7.39b), and Eq. (7.41b)

$$\chi_K = -j\frac{r_-^{(01)} - r_+^{(01)}}{r_-^{(01)} + r_+^{(01)}} = -j\frac{N^{(0)}\left(N_+^{(1)} - N_-^{(1)}\right)}{(N^{(0)})^2 - N_+^{(1)}N_-^{(1)}} \approx \frac{j\,N^{(0)}\varepsilon_1^{(1)}}{\varepsilon_0^{(n)1/2}\left[(N^{(0)2}) - \varepsilon_0^{(1)}\right]}. \tag{7.48}$$

Here, the use was made of the approximation of Eq. (7.35)

$$N_\pm^{(1)} \approx \varepsilon_0^{(n)1/2} \mp \frac{1}{2}\frac{\varepsilon_1^{(1)}}{\varepsilon_0^{(n)1/2}}. \tag{7.49}$$

7.4.4 Orthorhombic Crystal

The orthorhombic a and b axes are assumed parallel to the Cartesian x and y axes. The c axis and magnetization coincide with the interface normal and the z axis. The nonzero tensor elements are

$$\varepsilon_{xx}^{(1)} = \varepsilon_a^{(1)}, \tag{7.50a}$$
$$\varepsilon_{yy}^{(1)} = \varepsilon_b^{(1)}, \tag{7.50b}$$
$$\varepsilon_{zz}^{(1)} = \varepsilon_c^{(1)}, \tag{7.50c}$$
$$\varepsilon_{xy}^{(1)} = -\varepsilon_{yx}^{(1)} = j\,\varepsilon_1^{(1)}, \tag{7.50d}$$

and the permittivity tensor assumes, at this specific crystallographic orientation, the form

$$\varepsilon^{(n)} = \begin{pmatrix} \varepsilon_a^{(1)} & j\,\varepsilon_1^{(1)} & 0 \\ -j\,\varepsilon_1^{(1)} & \varepsilon_b^{(1)} & 0 \\ 0 & 0 & \varepsilon_c \end{pmatrix}. \tag{7.51}$$

From Eq. (7.4) and Eq. (7.13) we obtain

$$N_{\pm}^{(1)2} = \frac{1}{2}\left\{\left(\varepsilon_a^{(1)} + \varepsilon_b^{(1)}\right) \pm \left[\left(\varepsilon_a^{(1)} - \varepsilon_b^{(n)2}\right) + 4\varepsilon_1^{(n)2}\right]^{1/2}\right\}, \tag{7.52a}$$

$$a = -\mathrm{j}\,\varepsilon_1^{(1)}\varepsilon_c^{(1)}, \tag{7.52b}$$

$$b = \frac{1}{2}\varepsilon_c^{(1)}\left\{\left(\varepsilon_a^{(1)} - \varepsilon_b^{(1)}\right) - \left[\left(\varepsilon_a^{(1)} - \varepsilon_b^{(n)2}\right) + 4\varepsilon_1^{(n)2}\right]^{1/2}\right\}, \tag{7.52c}$$

$$c = \frac{1}{2}\varepsilon_c^{(1)}\left\{\left(\varepsilon_a^{(1)} - \varepsilon_b^{(1)}\right) + \left[\left(\varepsilon_a^{(1)} - \varepsilon_b^{(n)2}\right) + 4\varepsilon_1^{(n)2}\right]^{1/2}\right\}. \tag{7.52d}$$

This yields elliptical proper polarizations

$$\hat{\mathbf{e}}_{\pm}^{(1)} = C_{\pm}^{(1)}\begin{pmatrix} \mathrm{j}\,\varepsilon_1^{(1)} \\ \left(\varepsilon_b^{(1)} - \varepsilon_a^{(1)}\right) \pm \left[\left(\varepsilon_b^{(1)} - \varepsilon_a^{(1)}\right)^2 + \varepsilon_1^{(1)}\right]^{1/2} \\ 0 \end{pmatrix}, \tag{7.53}$$

which reduce to circular ones at zero in-plane anisotropy, *i.e.*, for $\varepsilon_b^{(1)} - \varepsilon_a^{(1)} = 0$. The elements of Jones transmission matrix follow from Eq. (7.31e) to Eq. (7.31h)

$$\begin{aligned} t_{xx,yy} = \pm & \frac{\varepsilon_a^{(1)} - \varepsilon_b^{(1)}}{\left[\left(\varepsilon_a^{(1)} - \varepsilon_b^{(1)}\right)^2 + 4\varepsilon_1^{(n)2}\right]^{1/2}} \\ & \times \left(\frac{t_-^{(01)}t_-^{(10)}\mathrm{e}^{-\mathrm{j}\beta_-^{(1)}}}{1 + r_-^{(01)}r_-^{(10)}\mathrm{e}^{-2\mathrm{j}\beta_-^{(1)}}} - \frac{t_+^{(01)}t_+^{(10)}\mathrm{e}^{-\mathrm{j}\beta_+^{(1)}}}{1 + r_+^{(01)}r_+^{(10)}\mathrm{e}^{-2\mathrm{j}\beta_+^{(1)}}}\right) \\ & + \frac{1}{2}\left(\frac{t_-^{(01)}t_-^{(10)}\mathrm{e}^{-\mathrm{j}\beta_-^{(1)}}}{1 + r_-^{(01)}r_-^{(10)}\mathrm{e}^{-2\mathrm{j}\beta_-^{(1)}}} + \frac{t_+^{(01)}t_+^{(10)}\mathrm{e}^{-\mathrm{j}\beta_+^{(1)}}}{1 + r_+^{(01)}r_+^{(10)}\mathrm{e}^{-2\mathrm{j}\beta_+^{(1)}}}\right), \end{aligned} \tag{7.54a}$$

$$\begin{aligned} t_{yx} = & \frac{-\mathrm{j}\,\varepsilon_1^{(1)}}{\left[\left(\varepsilon_a^{(1)} - \varepsilon_b^{(1)}\right)^2 + 4\varepsilon_1^{(n)2}\right]^{1/2}} \\ & \times \left(\frac{t_-^{(01)}t_-^{(10)}\mathrm{e}^{-\mathrm{j}\beta_-^{(1)}}}{1 + r_-^{(01)}r_-^{(10)}\mathrm{e}^{-2\mathrm{j}\beta_-^{(1)}}} - \frac{t_+^{(01)}t_+^{(10)}\mathrm{e}^{-\mathrm{j}\beta_+^{(1)}}}{1 + r_+^{(01)}r_+^{(10)}\mathrm{e}^{-2\mathrm{j}\beta_+^{(1)}}}\right). \end{aligned} \tag{7.54b}$$

In a similar way, we obtain from Eqs. (7.31) the Cartesian Jones reflection matrix. We have, *e.g.*,

$$r_{xx} = \frac{-\left(\varepsilon_a^{(1)} - \varepsilon_b^{(1)}\right)(A_+B_- - A_-B_+) + \left[\left(\varepsilon_a^{(1)} - \varepsilon_b^{(1)}\right)^2 + 4\varepsilon_1^{(n)2}\right]^{1/2}(A_+B_- + A_-B_+)}{2\left[\left(\varepsilon_a^{(1)} - \varepsilon_b^{(1)}\right)^2 + 4\varepsilon_1^{(n)2}\right]^{1/2} A_+A_-}, \tag{7.55a}$$

$$r_{yx} = \frac{\mathrm{j}\,\varepsilon_1^{(1)}(A_+B_- - A_-B_+)}{\left[\left(\varepsilon_a^{(1)} - \varepsilon_b^{(1)}\right)^2 + 4\varepsilon_1^{(n)2}\right]^{1/2} A_+A_-}, \tag{7.55b}$$

where $A_\pm$ and $B_\pm$ are defined in Eq. (7.28). Under the simplifying assumptions of Eq. (7.43) and Eq. (7.46), the formulas (7.54) and (7.55) lead to the previously established expressions for the Faraday and MO Kerr polar effects in magnetic orthorhombic magnetic crystals [2,4]. In transmission, we have, *e.g.*, from Eq. (7.54)

$$\frac{t_{yx}}{t_{xx}} = \frac{2\mathrm{j}\,\varepsilon_1^{(1)} \tan\left[\frac{1}{2}\left(\beta_+^{(1)} - \beta_-^{(1)}\right)\right]}{\left(\varepsilon_a^{(1)} - \varepsilon_b^{(1)}\right)\tan\left[\frac{1}{2}\left(\beta_+^{(1)} - \beta_-^{(1)}\right)\right] - \mathrm{j}\left[\left(\varepsilon_a^{(1)} - \varepsilon_b^{(n)2}\right) + 4\varepsilon_1^{(n)2}\right]^{1/2}}. \tag{7.56}$$

The substitution of this ratio into Eq. (7.42) provides information on the observed azimuth rotation and ellipticity. At the zero in-plane anisotropy, *i.e.*, for $\varepsilon_b^{(1)} - \varepsilon_a^{(1)} = 0$, we get the previous result of Eq. (7.47).

7.4.5 Voigt Effect in Cubic Crystals

Let us consider the case where the fourfold axes of a cubic crystal are parallel to the Cartesian axes, and the magnetization is parallel to the y axis and to the interface, *i.e.*, oriented normal to the propagation vector of the incident wave. The permittivity tensor (7.1) is reduced to

$$\varepsilon_{xx} = \varepsilon_{zz} = \varepsilon_0, \tag{7.57a}$$

$$\varepsilon_{yy} = \varepsilon_3, \tag{7.57b}$$

$$\varepsilon_{zx} = -\varepsilon_{xz} = \mathrm{j}\,\varepsilon_1. \tag{7.57c}$$

The proper value equation (7.4) provides in agreement with Eq. (7.5)

$$\left(N_+^{(1)}\right)^2 = \varepsilon_0^{-1}\left(\varepsilon_0^2 - \varepsilon_1^2\right), \tag{7.58a}$$

$$\left(N_-^{(1)}\right)^2 = \varepsilon_3. \tag{7.58b}$$

The corresponding proper polarizations can be expressed (neglecting the mode normalization) as

$$e_+ = \begin{pmatrix} \varepsilon_0 \\ 0 \\ \mathrm{j}\,\varepsilon_1 \end{pmatrix}, \tag{7.59a}$$

$$e_- = \begin{pmatrix} 0 \\ \varepsilon_3 \\ 0 \end{pmatrix}. \tag{7.59b}$$

From Eq. (7.13), we obtain $a = b_\pm = 0$ and $c = \varepsilon_0(\varepsilon_0 - \varepsilon_3) - \varepsilon_1^2$. The substitution into Eqs. (7.31e) to (7.31h) yields the Jones transmission matrix. We have, *e.g.*, for the ratio t_{yy}/t_{xx}

$$\frac{t_{yy}}{t_{xx}} = \frac{t_-^{(01)} t_-^{(10)}}{t_+^{(01)} t_+^{(10)}} \, \mathrm{e}^{\mathrm{j}\left(\beta_+^{(1)} - \beta^{(1)}{}_-\right)} \, \frac{1 + r_+^{(01)} r_+^{(10)} \mathrm{e}^{-2\mathrm{j}\beta_+^{(1)}}}{1 + r_-^{(01)} r_-^{(10)} \mathrm{e}^{-2\mathrm{j}\beta_-^{(1)}}}, \tag{7.60}$$

where $t_\pm^{(01)}$, $t_\pm^{(10)}$, $r_\pm^{(01)}$, and $r_\pm^{(10)}$ are defined in Eq. (7.29) and Eq. (7.38). The substitution of Eq. (7.60) into Eq. (7.42a) and Eq. (7.42b) provides the information on the azimuth rotation and ellipticity of the wave issued from the plate. In Eq. (7.60), the incident wave is assumed linearly polarized at an angle of $\frac{1}{4}\pi$ with respect to the magnetization. In most cases, the conditions (7.43) and (7.44) are met and Eq. (7.60) is reduced to the classical expression for the Voigt effect [10,11] discussed in Section 1.4.

7.5 Nearly Normal Incidence

In this section, we are concerned with the determination of the even magnetooptic effects in reflection from experiments at small but nonzero angles of incidence, $\varphi^{(0)}$. In this arrangement, the experiments are easier to accomplish than at strictly zero angle of incidence. As before, the interface is normal to the z axis. The plane of incidence is chosen parallel to the yz plane. For the general permittivity tensor of Eq. (7.1), the Cartesian Jones reflection matrix was obtained in Eq. (3.96).

We assume that the crystal is thick and absorbing, so that the reflections from the back interface play no role. We further limit ourselves to the case of a cubic crystal, the fourfold axes of which coincide with those of our Cartesian system. We further restrict the magnetization, $\boldsymbol{M}$, to the interface

xy plane. The permittivity tensor that includes the effect of magnetization up to the second order takes the form

$$\varepsilon_{ij} = \varepsilon_{ij}(\boldsymbol{M}^0) + \varepsilon_{ij}(\boldsymbol{M}^1) + \varepsilon_{ij}(\boldsymbol{M}^2), \tag{7.61}$$

where i, $j = x$, y, and z. In cubic crystals $\varepsilon_{ij}(0)$ reduces to the scalar quantity, which we denote ε_0. The contribution from the linear magnetooptic effects is

$$\begin{pmatrix} \varepsilon_{yz}(\boldsymbol{M}^1) \\ \varepsilon_{zx}(\boldsymbol{M}^1) \\ \varepsilon_{xy}(\boldsymbol{M}^1) \end{pmatrix} = \begin{pmatrix} K & 0 & 0 \\ 0 & K & 0 \\ 0 & 0 & K \end{pmatrix} M \begin{pmatrix} \cos\psi_M \\ \sin\psi_M \\ 0 \end{pmatrix} = -\mathrm{j}\,\varepsilon_1 \begin{pmatrix} \cos\psi_M \\ \sin\psi_M \\ 0 \end{pmatrix}, \tag{7.62}$$

where $\varepsilon_{ij}(\boldsymbol{M}^1) = -\varepsilon_{ji}(\boldsymbol{M}^1)$. Here, ψ_M is the angle between the magnetization confined to the xy plane and the x axis. The contribution quadratic in magnetization follows from Eq. (2.65)

$$\begin{pmatrix} \varepsilon_{xx}(\boldsymbol{M}^2) \\ \varepsilon_{yy}(\boldsymbol{M}^2) \\ \varepsilon_{zz}(\boldsymbol{M}^2) \\ \varepsilon_{yz}(\boldsymbol{M}^2) \\ \varepsilon_{zx}(\boldsymbol{M}^2) \\ \varepsilon_{xy}(\boldsymbol{M}^2) \end{pmatrix} = \begin{pmatrix} G_{xxxx} & G_{xxyy} & G_{xxyy} & 0 & 0 & 0 \\ G_{xxyy} & G_{xxxx} & G_{xxyy} & 0 & 0 & 0 \\ G_{xxyy} & G_{xxyy} & G_{xxxx} & 0 & 0 & 0 \\ 0 & 0 & 0 & 2G_{yzyz} & 0 & 0 \\ 0 & 0 & 0 & 0 & 2G_{yzyz} & 0 \\ 0 & 0 & 0 & 0 & 0 & 2G_{yzyz} \end{pmatrix}$$

$$\times\, M^2 \begin{pmatrix} \cos^2\psi_M \\ \sin^2\psi_M \\ 0 \\ 0 \\ 0 \\ \cos\psi_M \sin\psi_M \end{pmatrix} = \begin{pmatrix} \varepsilon_2\cos^2\psi_M + \varepsilon_3\sin^2\psi_M \\ \varepsilon_3\cos^2\psi_M + \varepsilon_2\sin^2\psi_M \\ \varepsilon_3 \\ 0 \\ 0 \\ \varepsilon_4\cos\psi_M\sin\psi_M \end{pmatrix}, \tag{7.63}$$

where $\varepsilon_{ij}(\boldsymbol{M}^2) = \varepsilon_{ji}(\boldsymbol{M}^2)$. The permittivity tensor elements are therefore given by

$$\varepsilon_{xx} = \varepsilon_0 + \varepsilon_2\cos^2\psi_M + \varepsilon_3\sin^2\psi_M, \tag{7.64a}$$

$$\varepsilon_{yy} = \varepsilon_0 + \varepsilon_2\sin^2\psi_M + \varepsilon_3\cos^2\psi_M, \tag{7.64b}$$

$$\varepsilon_{zz} = \varepsilon_0 + \varepsilon_3, \tag{7.64c}$$

$$\varepsilon_{yz} = -\varepsilon_{zy} = \mathrm{j}\,\varepsilon_1\cos\psi_M, \tag{7.64d}$$

$$\varepsilon_{zx} = -\varepsilon_{xz} = \mathrm{j}\,\varepsilon_1\sin\psi_M, \tag{7.64e}$$

$$\varepsilon_{xy} = \varepsilon_{yx} = \varepsilon_4\cos\psi_M\sin\psi_M. \tag{7.64f}$$

We note that ε_0 is of zeroth order; ε_1 is of first order; and ε_2, ε_3, and ε_4 are of second order in magnetization. This information allows us to obtain approximate expressions for the Cartesian Jones reflection matrix. With the restriction to the terms up to the second order and at small angles of incidence, we have from Eq. (3.100)

$$\mathcal{P}^{(1)} \approx 2N_z^2 \varepsilon_{zz} (\mathcal{V} + \mathcal{W}), \tag{7.65a}$$

$$\mathcal{Q}^{(1)} \approx 2N_z^3 (\mathcal{V} + \mathcal{W}), \tag{7.65b}$$

$$\mathcal{S}^{(1)} \approx 2N_z^3 \varepsilon_{zz} (\mathcal{V} + \mathcal{W}), \tag{7.65c}$$

$$\mathcal{T}^{(1)} \approx 2N_z^4 (\mathcal{V} + \mathcal{W}), \tag{7.65d}$$

where $N_z = \varepsilon_{zz} - N_y^2 \approx \varepsilon_0 - N_y^2$, $\mathcal{V} = \varepsilon_{xz} N_y N_z$, and $\mathcal{W} = -\varepsilon_{xy} N_z^2 + \varepsilon_{xz}\varepsilon_{zy}$. The substitution into Eq. (3.96) provides the required Cartesian Jones reflection matrix elements

$$r_{xx} \approx \frac{N^{(0)} \cos\varphi^{(0)} - N_z}{N^{(0)} \cos\varphi^{(0)} + N_z}, \tag{7.66a}$$

$$r_{xy} = -r_{yx} \approx \frac{-2N^{(0)} \cos\varphi^{(0)} (\mathcal{V} + \mathcal{W})}{(N^{(0)} \cos\varphi^{(0)} + N_z)(N^{(0)} N_z + \varepsilon_0 \cos\varphi^{(0)})}, \tag{7.66b}$$

$$r_{yy} \approx \frac{N^{(0)} N_z - \varepsilon_0 \cos\varphi^{(0)}}{N^{(0)} N_z + \varepsilon_0 \cos\varphi^{(0)}}. \tag{7.66c}$$

The information on the even magnetooptic effects can conveniently be obtained from the experiments at $\psi_M = \frac{1}{4}\pi,\ \frac{4}{4}\pi,\ \frac{5}{4}\pi$, and $\frac{7}{4}\pi$. The factor $(\mathcal{V} + \mathcal{W})$ in the off-diagonal elements takes the values

$$(\mathcal{V} + \mathcal{W})_{\frac{1}{4}\pi} = -\mathrm{j}\frac{\sqrt{2}}{2}\varepsilon_1 N_y N_z - \frac{1}{2}(\varepsilon_4 N_z^2 + \varepsilon_1^2), \tag{7.67a}$$

$$(\mathcal{V} + \mathcal{W})_{\frac{3}{4}\pi} = -\mathrm{j}\frac{\sqrt{2}}{2}\varepsilon_1 N_y N_z + \frac{1}{2}(\varepsilon_4 N_z^2 + \varepsilon_1^2), \tag{7.67b}$$

$$(\mathcal{V} + \mathcal{W})_{\frac{5}{4}\pi} = \mathrm{j}\frac{\sqrt{2}}{2}\varepsilon_1 N_y N_z - \frac{1}{2}(\varepsilon_4 N_z^2 + \varepsilon_1^2), \tag{7.67c}$$

$$(\mathcal{V} + \mathcal{W})_{\frac{7}{4}\pi} = \mathrm{j}\frac{\sqrt{2}}{2}\varepsilon_1 N_y N_z + \frac{1}{2}(\varepsilon_4 N_z^2 + \varepsilon_1^2). \tag{7.67d}$$

Note that $\mathcal{V}$ is of the first order, whereas $\mathcal{W}$ is of the second order in magnetization. We have to eliminate the $\mathcal{V}$ term, which represents the longitudinal Kerr effect. The transverse Kerr effect does not contribute to the off-diagonal reflection matrix elements. The required information is therefore obtained from the difference in the ratios of the off-diagonal to diagonal elements taken at either $\psi_M = \frac{1}{4}\pi$ and $\frac{3}{4}\pi$ or $\psi_M = \frac{5}{4}\pi$ and $\frac{7}{4}\pi$. Assuming ε_1 and ε_0 known, the element G_{yzyz} in the quadratic magnetooptic tensor is obtained from $\mathcal{W}$. If the crystal is rotated by an angle ζ about the z axis, the element transforms according to

$$G'_{yzyz} = G_{yzyz} + 2\cos^2\zeta[(G_{xxxx} - G_{xxyy}) - 2G_{yzyz}]. \tag{7.68}$$

To determine the difference $(G_{xxxx} - G_{xxyy})$, it is thus sufficient to rotate the crystal to the position $\zeta = \frac{1}{4}\pi$. Here, we have treated the case of a single interface in a magnetic cubic crystal in a special orientation. The approach can be extended, with the general form of reflection matrix, to other crystallographic orientations, to lower symmetries, as well as to multilayer systems.

References

1. M. Mansuripur, *The Physical Principles of Magneto-optical Recording* (Cambridge University Press, London, 1996).
2. W. J. Tabor and F. S. Chen, "Electromagnetic propagation through materials possessing both Faraday rotation and birefingence: Experiments with ytterbium orthoferrite," J. Appl. Phys. **40**, 2760–2765, 1969.
3. P. K. Tien, "Integrated optics and new wave phenomena in optical waveguides," Rev. Mod. Phys. **49**, 361–420, 1977.
4. F. J. Kahn, P. S. Pershan, and J. P. Remeika, "Ultraviolet magneto-optical properties of single-crystal orthoferrites, garnets, and other ferric oxide compounds," Phys. Rev. **186**, 891–918, 1969.
5. Š. Višňovský, "Magnetooptical effects in crystals at the normal incidence," Czech. J. Phys. B **37** 218–231, 1987.
6. R. M. A. Azzam, and N. M. Bashara, *Ellipsometry and Polarized Light* (Elsevier, Amsterdam, 1987).
7. B. Donovan and T. Medcalf, "The inclusion of multiple reflections in the theory of the Faraday effect in semiconductors," Brit. J. Appl. Phys. **15**, 1139–1151, 1964.
8. H. Piller, "Effect of internal reflection on optical faraday rotation," J. Appl. Phys. **37**, 763–767, 1966.
9. C. J. Sansalone, "Faraday rotation of a system of thin layers containing a thick layer," Appl. Opt. **10**, 2332–2335, 1971.
10. J. F. Dillon, Jr., in *Physics of Magnetic Garnets*, Proc. Int. School of Physics Enrico Fermi, A. Paoletti, ed. (North-Holland, Amsterdam, 1978).
11. G. A. Smolenskii, R. V. Pisarev, and I. G. Sinii, Usp. Fiz. Nauk **116**, 231, 1975.

8

Arbitrary Magnetization

8.1 Introduction

This chapter represents the generalization of the results obtained for magnetic multilayers polar, longitudinal, and transverse magnetization. The analytical representations will be provided in terms of Jones reflection and transmission matrices for a single-interface, magnetic film on a nonmagnetic substrate, and a magnetic film on a magnetic substrate separated by a nonmagnetic spacer with arbitrary and independent orientations of magnetization. Finally, we write the analytical formulae for the transmission through a film with arbitrary magnetization sandwiched between two isotropic media. We assume that a uniform magnetization, ***M***, was established in originally isotropic media and confine ourselves to the terms linear in the off-diagonal permittivity tensor elements. The restriction to the linearity is essential here, as it allows a significant simplification of the analysis. A general approach of Chapter 3 will be necessary in more involved situations where, for example, the MO effects quadratic in magnetization are to be determined [1–4].

The analysis given below should be useful for the explanation of experiments in spectroscopic MO ellipsometry and MO vector magnetometry of magnetic multilayers [5–11]. In ultrathin films with high interface quality, magnetooptics can provide information on effects related to exchange coupling and to reduced dimensionality. The fundamental results have been obtained on the structures deposited on iron whiskers, which present the best flat metallic surfaces available [6]. Another widely studied structure is formed by two ultrathin magnetic films exchange coupled across a nonmagnetic spacer. Typical examples are represented by ultrathin film sandwiches Fe–Cu–Fe, Fe–Cr–Fe, and Fe–Au–Fe [11–14].

The main goal of the magnetooptic vector magnetometry is to determine the development of magnetization in three dimensions. The three principal components in magnetooptic response, *i.e.*, polar, longitudinal, and transverse ones, contribute with different weights. For a single interface between a nonmagnetic medium and a medium with arbitrary (but uniform) orientation of ***M***, it is in principle possible to deduce the direction cosines of $\boldsymbol{M}$ from MO studies. In the case of a finite thickness of magnetic layer(s) in a multilayer system, the situation becomes more involved because several layers and interfaces contribute. The explanation of the MO response requires a model of the multilayer system based on the structural information and using computer simulations.

The symmetry properties and relations between various multilayer parameters and MO response can be best explained with simple systems that can be represented by analytical formulae. These provide an insight into the influence of various parameters, *i.e.*, layer thickness, orientation of $\boldsymbol{M}$, photon energy, and angle of incidence. Thanks to the linearity, the global MO response in multilayers can be expressed as a sum of contributions originating from individual magnetic layers. Each layer contribution can be further considered as a weighted sum of the polar, longitudinal, and transverse magnetization components. The polar and longitudinal components can be separated into interface and propagation parts.

8.2 Matrix Representation

The permittivity tensor in an originally isotropic medium of the n-th layer subjected to a uniform magnetization $\boldsymbol{M}$ assumes the form [15,16] that to first order in the magnetization is given by the sum of contributions coming from the permittivity in nonmagnetic isotropic medium and that from the linear magnetooptic tensor, as discussed in Section 2.2.2 and Section 2.2.3

$$\varepsilon^{(n)} = \begin{pmatrix} \varepsilon_0^{(n)} & \mathrm{j}\varepsilon_1^{(n)} \cos\theta_M^{(n)} & -\mathrm{j}\varepsilon_1^{(n)} \sin\theta_M^{(n)} \sin\varphi_M^{(n)} \\ -\mathrm{j}\varepsilon_1^{(n)} \cos\theta_M^{(n)} & \varepsilon_0^{(n)} & \mathrm{j}\varepsilon_1^{(n)} \sin\theta_M^{(n)} \cos\varphi_M^{(n)} \\ \mathrm{j}\varepsilon_1^{(n)} \sin\theta_M^{(n)} \sin\varphi_M^{(n)} & -\mathrm{j}\,\varepsilon_1^{(n)} \sin\theta_M^{(n)} \cos\varphi_M^{(n)} & \varepsilon_0^{(n)} \end{pmatrix}. \tag{8.1}$$

The orientation of $\boldsymbol{M}$ is specified by spherical coordinate angles $\theta_M^{(n)}$ and $\phi_M^{(n)}$. Three magnetization components are distinguished: the transversal one, $M_x = M \sin\theta_M^{(n)} \cos\varphi_M^{(n)}$, longitudinal one, $M_y = M \sin\theta_M^{(n)} \sin\varphi_M^{(n)}$, and polar one, $M_z = M \cos\theta_M^{(n)}$. For a plane wave solution (3.2), the wave equation in a medium characterized by $\varepsilon^{(n)}$ assumes the form given by Eq. (3.3). The normal component of the propagation vector, $N_z^{(n)}$, becomes a solution of the proper value equation obtained from the condition for nontrivial solutions of Eq. (3.3). For $|\varepsilon_1^{(n)}| \ll |\varepsilon_0^{(n)}|$, the solutions $N_{zj}^{(n)}$, $(j = 1, 2, 3, 4)$, to second order in $\varepsilon_1^{(n)}$, of the proper value equation become

$$\begin{aligned} N_{z1,3}^{(n)} = {} & N_{z0}^{(n)}\left(1 - \frac{\varepsilon_1^{(n)2}}{4\varepsilon_0^{(n)} N_{z0}^{(n)2}}\right) \\ & \pm \frac{\varepsilon_1^{(n)}}{2\varepsilon_0^{(n)1/2} N_{z0}^{(n)}}\left(N_{z0}^{(n)} \cos\theta_M^{(n)} + N_y \sin\theta_M^{(n)} \sin\varphi_M^{(n)}\right) \\ & + \frac{\varepsilon_1^{(n)2}}{8\varepsilon_0^{(n)} N_{z0}^{(n)3}}\left(N_{z0}^{(n)2} \cos^2\theta_M^{(n)} - N_y^2 \sin\theta_M^{(n)2} \sin\varphi_M^{(n)2}\right), \end{aligned} \tag{8.2a}$$

for the two modes propagating in the positive z direction, and

$$\begin{aligned} N_{z2,4}^{(n)} = {} & -N_{z0}^{(n)}\left(1 - \frac{\varepsilon_1^{(n)2}}{4\varepsilon_0^{(n)} N_{z0}^{(n)2}}\right) \\ & \mp \frac{\varepsilon_1^{(n)}}{2\varepsilon_0^{(n)1/2} N_{z0}^{(n)}}\left(N_{z0}^{(n)} \cos\theta_M^{(n)} - N_y \sin\theta_M^{(n)} \sin\varphi_M^{(n)}\right) \\ & - \frac{\varepsilon_1^{(n)2}}{8\varepsilon_0^{(n)} N_{z0}^{(n)3}}\left(N_{z0}^{(n)2} \cos^2\theta_M^{(n)} - N_y^2 \sin\theta_M^{(n)2} \sin\varphi_M^{(n)2}\right), \end{aligned} \tag{8.2b}$$

for the two modes propagating in the negative z direction. We have denoted $\varepsilon_0^{(n)} - N_y^2 = N_{z0}^{(n)2}$. The solutions are valid to the second order in $\varepsilon_1^{(n)}$, provided either $\cos\theta_M^{(n)}$ or $\sin\phi_M^{(n)}$ is not equal to zero. Therefore, the second order perturbation to the complex index of refraction in the transverse configuration of magnetization (where $\cos\theta_M^{(n)} = 0$ and $\sin\phi_M^{(n)} = 0$, simultaneously) is not included. In this case, the proper values of $N_z^{(n)}$ can be obtained directly from the proper value equation, as shown below. At a general orientation of $\boldsymbol{M}$ with respect to the propagation vector, the proper polarization modes are elliptically polarized and become linearly polarized for $\boldsymbol{M}$ normal to $\gamma^{(n)}$. The proper value equation is invariant with respect to the reversal of $\boldsymbol{M}$ represented either by $\varepsilon_1^{(n)}(\boldsymbol{M}) \to \varepsilon_1^{(n)}(-\boldsymbol{M}) = -\varepsilon_1^{(n)}(\boldsymbol{M})$ or by $\theta_M^{(n)} \to \pi - \theta_M^{(n)}$ and $\varphi_M^{(n)} \to \varphi_M^{(n)} + \pi$. Note that $\varepsilon_1^{(n)}(M)$ is an odd function of $\boldsymbol{M}$. In the special cases of the polar ($\theta_M = 0$), longitudinal ($\theta_M = \pi/2$, $\phi = \pi/2$ or $3\pi/2$), or transverse ($\theta_M = \pi/2$, $\phi = 0$ or π) magnetization, the proper value equation reduces to a biquadratic one. It also becomes a biquadratic one in two more general cases, where $\boldsymbol{M}$ is restricted either to the interface plane ($\theta_M = \pi/2$)

$$N_{z\pm}^{(n)2} = N_{z0}^{(n)2} - \frac{\varepsilon_1^{(n)2}}{\left(2\varepsilon_0^{(n)}\right)} \pm \left[\varepsilon_1^{(n)2}\varepsilon_0^{(n)-1} N_y^2 \sin^2\phi_M^{(n)} + \left(\frac{\varepsilon_1^{(n)2}}{2\varepsilon_0}\right)^2\right]^{1/2}, \tag{8.3}$$

or to the plane normal to the plane of incidence ($\phi_M^{(n)} = 0$ or π)

$$N_{z\pm}^{(n)2} = N_{z0}^{(n)2} - \frac{\varepsilon_1^{(n)2}\sin^2\theta_M^{(n)}}{(2\varepsilon_0^{(n)})} \pm \left[\varepsilon_1^{(n)2}\varepsilon_0^{(n)-1}N_{z0}^{(n)2}\cos^2\theta_M^{(n)} + \left(\frac{\varepsilon_1^{(n)2}\sin^2\theta_M^{(n)}}{2\varepsilon_0^{(n)}}\right)^2\right]^{1/2}. \quad (8.4)$$

Thus, for the transverse magnetization, corresponding to $\theta_M^{(n)} = \pi/2$ and $\phi_M^{(n)} = 0$ or π, it follows for these equations

$$N_{z\pm}^{(n)2} = N_{z0}^{(n)2} - \frac{\varepsilon_1^{(n)2}}{2\varepsilon_0^{(n)}} \pm \frac{\varepsilon_1^{(n)2}}{2\varepsilon_0^{(n)}}. \quad (8.5)$$

For the other configurations, the proper value equation for $N_z^{(n)}$ becomes quartic. Its classical analytical solution to the general case was applied by Schubert *et al.* [17]. We have included the terms of the second order in $\varepsilon_1^{(n)}$ for illustration only, and we drop them from now on. The relation between the four proper polarization amplitudes of the electric field in the half spaces sandwiching the multilayer is given by Eq. (3.33)

$$\mathbf{E}_0^{(0)} = \begin{pmatrix} E_{01}^{(0)} \\ E_{02}^{(0)} \\ E_{03}^{(0)} \\ E_{04}^{(0)} \end{pmatrix} = \mathbf{M}\,\mathbf{E}_0^{(\mathcal{N}+1)} = \mathbf{M}\begin{pmatrix} E_{01}^{(\mathcal{N}+1)} \\ E_{02}^{(\mathcal{N}+1)} \\ E_{03}^{(\mathcal{N}+1)} \\ E_{04}^{(\mathcal{N}+1)} \end{pmatrix}, \quad (8.6)$$

where the **M** matrix is given by the matrix product [18]

$$\mathbf{M} = [\mathbf{D}^{(0)}]^{-1}\mathbf{D}^{(1)}\mathbf{P}^{(1)}[\mathbf{D}^{(1)}]^{-1}\cdots[\mathbf{D}^{(n-1)}]^{-1}\mathbf{D}^{(n)}\mathbf{P}^{(n)}[\mathbf{D}^{(n)}]^{-1}\mathbf{D}^{(n+1)} \cdots\mathbf{D}^{(\mathcal{N})}\mathbf{P}^{(\mathcal{N})}[\mathbf{D}^{(\mathcal{N})}]^{-1}[\mathbf{D}^{(\mathcal{N}+1)}]. \quad (8.7)$$

Here, $\mathbf{D}^{(n)}$ and $\mathbf{P}^{(n)}$ (for $n = 1, \cdots, \mathcal{N}$) are the dynamical and propagation matrices, respectively. The Jones reflection matrix (3.74a), defined for $E_{02}^{(\mathcal{N}+1)} = E_{04}^{(\mathcal{N}+1)} = 0$, relates the complex electric field amplitudes of the incident (i) and reflected (r) waves linearly polarized perpendicular (s) and parallel (p) with respect to the plane of wave incidence [19]

$$\begin{pmatrix} E_{0s}^{(r)} \\ E_{0p}^{(r)} \end{pmatrix} = \begin{pmatrix} r_{ss} & r_{sp} \\ r_{ps} & r_{pp} \end{pmatrix}\begin{pmatrix} E_{0s}^{(i)} \\ E_{0p}^{(i)} \end{pmatrix}. \quad (8.8)$$

In terms of the **M** matrix elements, the elements of the Jones reflection matrix are expressed by the relations obtained from Eq. (3.70)

$$r_{ss}^{(0,\mathcal{N}+1)} = \left(\frac{E_{0s}^{(r)}}{E_{0s}^{(i)}}\right)_{E_{0p}^{(i)}=0} = \frac{M_{21}M_{33} - M_{23}M_{31}}{M_{11}M_{33} - M_{13}M_{31}}, \quad (8.9a)$$

$$r_{ps}^{(0,\mathcal{N}+1)} = \left(\frac{E_{0p}^{(r)}}{E_{0s}^{(i)}}\right)_{E_{0p}^{(i)}=0} = \frac{M_{41}M_{33} - M_{43}M_{31}}{M_{11}M_{33} - M_{13}M_{31}}, \tag{8.9b}$$

$$r_{sp}^{(0,\mathcal{N}+1)} = \left(\frac{E_{0s}^{(r)}}{E_{0p}^{(i)}}\right)_{E_{0s}^{(i)}=0} = \frac{M_{11}M_{23} - M_{13}M_{21}}{M_{11}M_{33} - M_{13}M_{31}}, \tag{8.9c}$$

$$r_{pp}^{(0,\mathcal{N}+1)} = \left(\frac{E_{0p}^{(r)}}{E_{0p}^{(i)}}\right)_{E_{0s}^{(i)}=0} = \frac{M_{11}M_{43} - M_{13}M_{41}}{M_{11}M_{33} - M_{13}M_{31}}. \tag{8.9d}$$

We have identified the complex amplitudes of linearly polarized proper polarization modes in the sandwiching half spaces related by **M** in Eq. (8.6) as follows

$$\mathbf{E}_0^{(0)} = \begin{pmatrix} E_{01}^{(0)} \\ E_{02}^{(0)} \\ E_{03}^{(0)} \\ E_{04}^{(0)} \end{pmatrix} = \begin{pmatrix} E_{0s}^{(i)} \\ E_{0s}^{(r)} \\ E_{0p}^{(i)} \\ E_{0p}^{(r)} \end{pmatrix} \tag{8.10a}$$

$$\mathbf{E}_0^{(\mathcal{N}+1)} = \begin{pmatrix} E_{01}^{(\mathcal{N}+1)} \\ E_{02}^{(\mathcal{N}+1)} \\ E_{03}^{(\mathcal{N}+1)} \\ E_{04}^{(\mathcal{N}+1)} \end{pmatrix} = \begin{pmatrix} E_{0s}^{(t)} \\ 0 \\ E_{0p}^{(t)} \\ 0 \end{pmatrix}. \tag{8.10b}$$

Up to a common factor, the elements of the dynamical matrix (3.29) for a medium with arbitrary $\boldsymbol{M}$ are given by [20,21]

$$\begin{aligned} D_{1j}^{(n)} &= -\mathrm{j}\,\varepsilon_1^{(n)} N_{z0}^{(n)2} \cos\theta_M^{(n)} - \mathrm{j}\,\varepsilon_1^{(n)} N_y N_{zj}^{(n)} \sin\theta_M^{(n)} \sin\varphi_M^{(n)} \\ &\quad - \varepsilon_1^{(n)2} \sin^2\theta_M^{(n)} \cos\varphi_M^{(n)} \sin\varphi_M^{(n)}, \end{aligned} \tag{8.11a}$$

$$\begin{aligned} D_{2j}^{(n)} &= N_{zj}^{(n)} \big(-\mathrm{j}\,\varepsilon_1^{(n)} N_{z0}^{(n)2} \cos\theta_M^{(n)} - \mathrm{j}\,\varepsilon_1^{(n)} N_y N_{zj}^{(n)} \sin\theta_M^{(n)} \sin\varphi_M^{(n)} \\ &\quad - \varepsilon_1^{(n)2} \sin^2\theta_M^{(n)} \cos\varphi_M^{(n)} \sin\varphi_M^{(n)}\big) \end{aligned} \tag{8.11b}$$

$$D_{3j}^{(n)} = N_{z0}^{(n)2}\big(N_{z0}^{(n)2} - N_{zj}^{(n)2}\big) - \varepsilon_1^{(n)2} \sin^2\theta_M^{(n)} \sin^2\varphi_M^{(n)}, \tag{8.11c}$$

$$\begin{aligned} D_{4j}^{(n)} &= -\big(\varepsilon_0^{(n)} N_{zj}^{(n)} - \mathrm{j}\,\varepsilon_1^{(n)} N_y \sin\theta_M^{(n)} \cos\varphi_M^{(n)}\big)\big(N_{z0}^{(n)2} - N_{zj}^{(n)2}\big) \\ &\quad + \varepsilon_1^{(n)2} \sin\theta_M^{(n)} \sin\varphi_M^{(n)} \big(N_{zj}^{(n)} \sin\theta_M^{(n)} \sin\varphi_M^{(n)} - N_y \cos\theta_M^{(n)}\big). \end{aligned} \tag{8.11d}$$

The diagonal propagation matrix $\mathbf{P}^{(n)}$ takes the form of Eq. (3.31)

$$P_{jj}^{(n)} = \exp\left(\mathrm{j}\frac{\omega}{c} N_{zj}^{(n)} d^{(n)}\right), \qquad j = 1, \ldots, 4, \tag{8.12}$$

where $d^{(n)}$ is the thickness of the n–th layer. In the isotropic layers, the dynamical matrix is given by

$$\mathbf{D}^{(n)} = \begin{pmatrix} 1 & 1 & 0 & 0 \\ N_{z0}^{(n)} & -N_{z0}^{(n)} & 0 & 0 \\ 0 & 0 & N_{z0}^{(n)}\left(\varepsilon_0^{(n)}\right)^{-1/2} & N_{z0}^{(n)}\left(\varepsilon_0^{(n)}\right)^{-1/2} \\ 0 & 0 & -\left(\varepsilon_0^{(n)}\right)^{-1/2} & \left(\varepsilon_0^{(n)}\right)^{-1/2} \end{pmatrix}. \tag{8.13}$$

The dynamical matrix and its inverse were given in Eq. (3.82). As before, we have adopted the sign convention for the electric fields of the incident and reflected waves according to which both incident and reflected electric field amplitudes are assumed oriented in the same direction at the limit of normal incidence. The matrices given above provide sufficient information for the numerical evaluation up to first order in $\varepsilon_1^{(n)}$ of the electromagnetic response in a multilayer consisting of layers with arbitrary $\boldsymbol{M}$. The layers are characterized by the simplest form of the permittivity tensor in a uniformly magnetized medium obtained from the consideration that led to Eq. (1.3) and which were reproduced in Eq. (8.1). Note that the response in multilayers with magnetization continuously varying parallel to the axis normal to the interfaces can be studied by splitting the medium to an appropriate number of homogeneous layers (multilayer or staircase approximation).

8.3 Single Interface

The knowledge of the dynamical matrix in the magnetic medium and that in the isotropic medium, as given by Eq. (8.13) and obtained previously in Eq. (3.82a), is sufficient for the solution of the reflection problem at a single interface. Note that the transmission across the interface, which is not considered here, would require the normalization of the proper modes in the magnetic medium. The **M** matrix for the interface between an isotropic ambient and a medium with an arbitrary orientation of magnetization is given by

$$\mathbf{M} = (\mathbf{D}^{(0)})^{-1}\mathbf{D}^{(1)}, \tag{8.14}$$

where $\mathbf{D}^{(0)}$ is given by Eq. (8.13). Its inverse becomes after Eq. (3.82b)

$$(\mathbf{D}^{(0)})^{-1} = \left(2N^{(0)}\cos\varphi^{(0)}\right)^{-1} \begin{pmatrix} N^{(0)}\cos\varphi^{(0)} & 1 & 0 & 0 \\ N^{(0)}\cos\varphi^{(0)} & -1 & 0 & 0 \\ 0 & 0 & N^{(0)} & -\cos\varphi^{(0)} \\ 0 & 0 & N^{(0)} & \cos\varphi^{(0)} \end{pmatrix}, \tag{8.15}$$

where $N^{(0)} = (\varepsilon_0^{(0)})^{1/2}$ and $\varphi^{(0)}$ are the refractive index and the angle of wave incidence in the entrance medium, respectively. The dynamical matrix of the magnetic medium is given by Eq. (8.11). After the substitution according to Eq. (8.11) and Eq. (8.2) into Eq. (3.99), we obtain the elements of Jones reflection matrix at arbitrary $\boldsymbol{M}$ from Eq. (3.96) for the incident wave with s polarization

$$r_{ss}^{(01)} = \frac{N^{(0)}\cos\varphi^{(0)} - N_{z0}^{(1)}}{N^{(0)}\cos\varphi^{(0)} + N_{z0}^{(1)}}, \tag{8.16a}$$

$$r_{ps}^{(01)}\left(\varepsilon_1^{(1)}\right) = \frac{\mathrm{j}\varepsilon_1^{(1)} N^{(0)}\cos\varphi^{(0)}\left(N_{z0}^{(1)}\cos\theta_M^{(1)} - N_y\sin\theta_M^{(1)}\sin\phi_M^{(1)}\right)}{N_{z0}^{(1)}\left(N^{(0)}N_{z0}^{(1)} + \varepsilon_0^{(1)}\cos\varphi^{(0)}\right)\left(N^{(0)}\cos\varphi^{(0)} + N_{z0}^{(1)}\right)}, \tag{8.16b}$$

and for the incident wave with p polarization

$$r_{sp}^{(01)}\left(\varepsilon_1^{(1)}\right) = \frac{-\mathrm{j}\varepsilon_1^{(1)} N^{(0)}\cos\varphi^{(0)}\left(N_{z0}^{(1)}\cos\theta_M^{(1)} + N_y\sin\theta_M^{(1)}\sin\phi_M^{(1)}\right)}{N_{z0}^{(1)}\left(N^{(0)}N_{z0}^{(1)} + \varepsilon_0^{(1)}\cos\varphi^{(0)}\right)\left(N^{(0)}\cos\varphi^{(0)} + N_{z0}^{(1)}\right)}, \tag{8.16c}$$

$$r_{pp}^{(01)}\left(\varepsilon_1^{(1)}\right) = \frac{N^{(0)}N_{z0}^{(1)} - \varepsilon_0^{(1)}\cos\varphi^{(0)}}{N^{(0)}N_{z0}^{(1)} + \varepsilon_0^{(1)}\cos\varphi^{(0)}} + \frac{2\mathrm{j}\varepsilon_1^{(1)} N^{(0)}\cos\varphi^{(0)} N_y\sin\theta_M^{(1)}\cos\phi_M^{(1)}}{\left(N^{(0)}N_{z0}^{(1)} + \varepsilon_0^{(1)}\cos\varphi^{(0)}\right)^2}. \tag{8.16d}$$

Except for $r_{ss}^{(01)}$, the Jones reflection matrix elements depend on the off-diagonal permittivity tensor element $\varepsilon_1^{(1)}$. These results may be obtained directly as special cases from a more general solution for an interface between an isotropic medium and a medium characterized by a general permittivity tensor [22] discussed in Section 1.6 and Section 3.6.

8.4 Characteristic Matrix

The characteristic matrix is defined by the product $\mathbf{S}^{(n)} = \mathbf{D}^{(n)}\mathbf{P}^{(n)}(\mathbf{D}^{(n)})^{-1}$ [23–25]. For the present purpose it is, however, sufficient to consider only the terms linear in $\varepsilon_1^{(n)}$. In a layer with arbitrary magnetization, the characteristic matrix assumes the form

$$\mathbf{S}^{(n)} = \begin{pmatrix} S_{11}^{(n)} & S_{12}^{(n)} & S_{13}^{(n)} & S_{14}^{(n)} \\ S_{21}^{(n)} & S_{11}^{(n)} & S_{23}^{(n)} & S_{24}^{(n)} \\ S_{24}^{(n)} & S_{14}^{(n)} & S_{33}^{(n)} & S_{34}^{(n)} \\ S_{23}^{(n)} & S_{13}^{(n)} & S_{43}^{(n)} & S_{44}^{(n)} \end{pmatrix}, \tag{8.17}$$

where

$$S_{11}^{(n)} = d_n, \tag{8.18a}$$

$$S_{12}^{(n)} = N_{z0}^{(n)-1} c_n, \tag{8.18b}$$

$$S_{21}^{(n)} = N_{z0}^{(n)} c_n, \tag{8.18c}$$

$$S_{34}^{(n)} = -N_{z0}^{(n)} \varepsilon_0^{(n)-1} c_n, \tag{8.18d}$$

$$S_{43}^{(n)} = -N_{z0}^{(n)-1} \varepsilon_0^{(n)} c_n, \tag{8.18e}$$

$$S_{33}^{(n)} = d_n + jq^{(n)} c_n, \tag{8.18f}$$

$$S_{44}^{(n)} = d_n - jq^{(n)} c_n, \tag{8.18g}$$

$$S_{13}^{(n)} = jN_{z0}^{(n)-1} \varepsilon_0^{(n)1/2} \left(a_n - l^{(n)} c_n\right), \tag{8.18h}$$

$$S_{14}^{(n)} = -j\varepsilon_0^{(n)-1/2} \left(b_n - p^{(n)} c_n\right), \tag{8.18i}$$

$$S_{23}^{(n)} = j\varepsilon_0^{(n)1/2} \left(b_n + p^{(n)} c_n\right), \tag{8.18j}$$

$$S_{24}^{(n)} = -jN_{z0}^{(n)} \varepsilon_0^{(n)-1/2} \left(a_n + l^{(n)} c_n\right). \tag{8.18k}$$

Here,

$$a_n = \frac{1}{4} \left\{ \exp(j\beta^{(n)}) \left[\exp(j\Delta^{(n)+}) - \exp(-j\Delta^{(n)+})\right] - \exp(-j\beta^{(n)}) \left[\exp(j\Delta^{(n)-}) - \exp(-j\Delta^{(n)-})\right] \right\}, \tag{8.19a}$$

$$b_n = \frac{1}{4} \left\{ \exp(j\beta^{(n)}) \left[\exp(j\Delta^{(n)+}) - \exp(-j\Delta^{(n)+})\right] + \exp(-j\beta^{(n)}) \left[\exp(j\Delta^{(n)-}) - \exp(-j\Delta^{(n)-})\right] \right\}, \tag{8.19b}$$

$$c_n = \frac{1}{4} \left\{ \exp(j\beta^{(n)}) \left[\exp(j\Delta^{(n)+}) + \exp(-j\Delta^{(n)+})\right] - \exp(-j\beta^{(n)}) \left[\exp(j\Delta^{(n)-}) + \exp(-j\Delta^{(n)-})\right] \right\}, \tag{8.19c}$$

$$d_n = \frac{1}{4} \left\{ \exp(j\beta^{(n)}) \left[\exp(j\Delta^{(n)+}) + \exp(-j\Delta^{(n)+})\right] + \exp(-j\beta^{(n)}) \left[\exp(j\Delta^{(n)-}) + \exp(-j\Delta^{(n)-})\right] \right\}, \tag{8.19d}$$

with

$$\beta^{(n)} = \frac{\omega}{c} d^{(n)} N_{z0}^{(n)}, \tag{8.20a}$$

$$\Delta^{(n)\pm} = \frac{\omega}{2c} d^{(n)} \varepsilon_1^{(n)} \varepsilon_0^{(n)-1/2} N_{z0}^{(n)-1} \left(N_{z0}^{(n)} \cos\theta_M^{(n)} \pm N_y \sin\theta_M^{(n)} \sin\varphi_M^{(n)}\right). \tag{8.20b}$$

The parameters $p^{(n)}, l^{(n)}$, and $q^{(n)}$ are proportional to $\varepsilon_1^{(n)}$

$$p^{(n)} = \frac{\varepsilon_1^{(n)}\left(N_{z0}^{(n)} \cos\theta_M^{(n)}\right)}{2\varepsilon_0^{(n)1/2} N_{z0}^{(n)2}}, \tag{8.21a}$$

$$l^{(n)} = \frac{\varepsilon_1^{(n)}\left(N_y \sin\theta_M^{(n)} \sin\varphi_M^{(n)}\right)}{2\varepsilon_0^{(n)1/2} N_{z0}^{(n)2}}, \tag{8.21b}$$

$$q^{(n)} = \frac{\varepsilon_1^{(n)}\left(N_y \sin\theta_M^{(n)} \cos\varphi_M^{(n)}\right)}{\varepsilon_0^{(n)} N_{z0}^{(n)}}. \tag{8.21c}$$

Then,

$$\Delta^{(n)\pm} = \frac{\omega}{c} d^{(n)} N_{z0}^{(n)} (p^{(n)} \pm l^{(n)}) = \beta^{(n)} (p^{(n)} \pm l^{(n)}). \tag{8.22}$$

In ultrathin films, $\frac{\omega}{c} d^{(n)} \ll 1$. There, in most cases, the quantities $\Delta^{(n)\pm}$ are small, and we adopt the approximations

$$\exp[\mathrm{j}\Delta^{(n)\pm}] - \exp[-\mathrm{j}\Delta^{(n)\pm}] \approx 2\mathrm{j}\Delta^{(n)\pm}, \tag{8.23a}$$

$$\exp[\mathrm{j}\Delta^{(n)\pm}] + \exp[-\mathrm{j}\Delta^{(n)\pm}] \approx 2. \tag{8.23b}$$

8.5 Magnetic Film

Before approaching the configuration of exchange coupled magnetic film, let us consider the system consisting of a single magnetic film with arbitrary magnetization on a nonmagnetic substrate. The electromagnetic plane wave response is represented by

$$\mathbf{M} = \left(\mathbf{D}^{(0)}\right)^{-1} \mathbf{S}^{(1)} \mathbf{D}^{(2)}, \tag{8.24}$$

where $\mathbf{D}^{(0)}$ and $\mathbf{D}^{(2)}$ are given by Eq. (8.13) and $\mathbf{S}^{(1)}$ follows from Eq. (8.17). Using Eq. (8.9a) to Eq. (8.9d), we obtain for small MO azimuth rotations and ellipticities in reflection

$$r_{ss}^{(02)} = \frac{r_{ss}^{(01)} + r_{ss}^{(12)} e^{-2\mathrm{j}\beta^{(1)}}}{1 + r_{ss}^{(01)} r_{ss}^{(12)} e^{-2\mathrm{j}\beta^{(1)}}}, \tag{8.25a}$$

$$\begin{aligned} r_{ps,sp}^{(02)} = t_{ss}^{(01)} t_{pp}^{(10)} &\left\{ \beta^{(1)} e^{-2\mathrm{j}\beta^{(1)}} \left[\mp p^{(1)} \left(r_{ss}^{(12)} + r_{pp}^{(12)} \right) + l^{(1)} \left(r_{ss}^{(12)} - r_{pp}^{(12)} \right) \right] \right. \\ &+ \frac{\mathrm{j}}{2} \left(1 - e^{-2\mathrm{j}\beta^{(1)}} \right) \left[\pm p^{(1)} \left(1 + r_{ss}^{(12)} r_{pp}^{(12)} e^{-2\mathrm{j}\beta^{(1)}} \right) \right. \\ &\left. \left. - l^{(1)} \left(1 - r_{ss}^{(12)} r_{pp}^{(12)} e^{-2\mathrm{j}\beta^{(1)}} \right) \right] \right\} \\ &\times \left[\left(1 + r_{ss}^{(01)} r_{ss}^{(12)} e^{-2\mathrm{j}\beta^{(1)}} \right) \left(1 + r_{pp}^{(01)} r_{pp}^{(12)} e^{-2\mathrm{j}\beta^{(1)}} \right) \right]^{-1}, \end{aligned} \tag{8.25b}$$

$$r_{pp}^{(02)} = \frac{r_{pp}^{(01)} + r_{pp}^{(12)} e^{-2j\beta^{(1)}}}{1 + r_{pp}^{(01)} r_{pp}^{(12)} e^{-2j\beta^{(1)}}} + \frac{j}{2} q^{(1)} t_{pp}^{(01)} t_{pp}^{(10)} \left(1 - e^{-2j\beta^{(1)}}\right) \frac{1 - r_{pp}^{(12)2} e^{-2j\beta^{(1)}}}{\left(1 + r_{pp}^{(01)} r_{pp}^{(12)} e^{-2j\beta^{(1)}}\right)^2}. \tag{8.25c}$$

In Eq. (8.25b) for the off-diagonal elements, the upper sign pertains to $r_{ps}^{(02)}$ (the incident wave with s polarization) and lower one to $r_{sp}^{(02)}$ (the incident wave with p polarization). The element $r_{pp}^{(02)}$ of Eq. (8.25c) consists of an isotropic part $r_{pp}^{(02)}(\varepsilon_1^{(1)} = 0)$ and a perturbation $\Delta r_{pp}^{(02)}$ proportional to $\varepsilon_1^{(1)}$ induced by transverse magnetization

$$\Delta r_{pp}^{(02)} = \frac{j}{2} q^{(1)} t_{pp}^{(01)} t_{pp}^{(10)} \left(1 - e^{-2j\beta^{(1)}}\right) \frac{1 - r_{pp}^{(12)2} e^{-2j\beta^{(1)}}}{\left(1 + r_{pp}^{(01)} r_{pp}^{(12)} e^{-2j\beta^{(1)}}\right)^2}. \tag{8.26}$$

The single-interface reflection and transmission coefficients are defined by the relations

$$r_{ps}^{(01,\text{pol})} = \frac{j\varepsilon_1^{(1)} N^{(0)} \cos\varphi^{(0)} N_{z0}^{(1)}}{N_{z0}^{(1)} \left(N^{(0)} \cos\varphi^{(0)} + N_{z0}^{(1)}\right) \left(N^{(0)} N_{z0}^{(1)} + \varepsilon_0^{(1)} \cos\varphi^{(0)}\right)} = \frac{j}{2} p^{(1)} t_{ss}^{(01)} t_{pp}^{(10)} \tag{8.27a}$$

$$r_{ps}^{(01,\text{lon})} = \frac{-j\varepsilon_1^{(1)} N^{(0)} \cos\varphi^{(0)} N_y}{N_{z0}^{(1)} \left(N^{(0)} \cos\varphi^{(0)} + N_{z0}^{(1)}\right) \left(N^{(0)} N_{z0}^{(1)} + \varepsilon_0^{(1)} \cos\varphi^{(0)}\right)} = -\frac{j}{2} l^{(1)} t_{ss}^{(01)} t_{pp}^{(10)} \tag{8.27b}$$

$$r_{ss}^{(ij)} = \frac{N_{z0}^{(i)} - N_{z0}^{(j)}}{N_{z0}^{(i)} + N_{z0}^{(j)}} \tag{8.27c}$$

$$r_{pp}^{(ij)} = \frac{\varepsilon_0^{(i)} N_{z0}^{(j)} - \varepsilon_0^{(j)} N_{z0}^{(i)}}{\varepsilon_0^{(i)} N_{z0}^{(j)} + \varepsilon_0^{(j)} N_{z0}^{(i)}} \tag{8.27d}$$

$$t_{ss}^{(ij)} = 1 + r_{ss}^{(ij)} \tag{8.27e}$$

$$t_{pp}^{(ij)} = \left(\varepsilon_0^{(i)} / \varepsilon_0^{(j)}\right)^{1/2} \left(1 - r_{pp}^{(ij)}\right), \tag{8.27f}$$

where $i, j = 0, 1, 2$, and $N_{z0}^{(0)} = N^{(0)} \cos\varphi^{(0)}$, $\varphi^{(0)}$ is the angle of incidence, $N_{z0}^{(1)} = (\varepsilon_0^{(1)} - N_y^2)^{1/2}$ and $N_{z0}^{(2)} = (\varepsilon_0^{(2)} - N_y^2)^{1/2}$. The single-interface off-diagonal elements of the Jones reflection matrix in the polar and longitudinal configurations are denoted as $r_{ps}^{(01,\text{pol})}$ and $r_{ps}^{(01,\text{lon})}$, respectively. From the Jones reflection matrix elements in Eq. (8.25), expressed in terms of the single-interface Fresnel coefficients, the anomalies in the region of the principal angle of incidence are easily understood. The diagonal and off-diagonal elements are of zero and first order in $\varepsilon_1^{(1)}$, respectively.

Note that the first term in the curly brackets of $r_{ps,sp}^{(02)}$, proportional to $\beta^{(1)}$, represents the propagation across the film, whereas the second one, proportional to $(1 - e^{-2j\beta^{(1)}})$, accounts for the interface effects. There is an interface term but no propagation term included in $\Delta r_{pp}^{(02)}$, the MO transverse perturbation of $r_{pp}^{(02)}$. From the formulae, the symmetry of the linear MO Kerr effects can be appreciated. They confirm vanishing of transverse and longitudinal component at normal incidence ($\varphi^{(0)} = 0$). The off-diagonal elements $r_{ps}^{(02)}$ and $r_{sp}^{(02)}$ change the sign when the magnetization is reversed. The transverse MO component $\Delta r_{pp}^{(02)}$ also changes its sign when the magnetization is reversed. When the thickness of the absorbing magnetic layer is much higher than the penetration depth of the radiation, $|\exp(-2j\beta^{(1)})| \ll 1$, Eqs. (8.25) reduce to the case of a single interface between a nonmagnetic half space and a half space at arbitrary magnetization summarized in Eqs. (8.16).

8.6 Film–Spacer System

We consider a system consisting of a magnetic film with arbitrary $\boldsymbol{M}^{(1)}$ separated from a magnetic substrate with another arbitrary $\boldsymbol{M}^{(3)}$ by a nonmagnetic spacer. The electromagnetic plane wave response is represented by the matrix product

$$\mathbf{M} = (\mathbf{D}^{(0)})^{-1}\mathbf{S}^{(1)}\mathbf{S}^{(2)}\mathbf{D}^{(3)} \tag{8.28}$$

obtained from Eq. (8.7), Eq. (8.11), Eq. (8.13), and Eq. (8.17). We focus on the reflection case. The diagonal element of the Jones reflection matrix, $r_{pp}^{(03)}$, corresponding to an isotropic system, may be written as [19]

$$r_{pp}^{(03)}\left(\varepsilon_1^{(1)} = \varepsilon_1^{(3)} = 0\right) = \frac{r_{pp}^{(01)} + r_{pp}^{(13)} e^{-2j\beta^{(1)}}}{1 + r_{pp}^{(01)} r_{pp}^{(13)} e^{-2j\beta^{(1)}}}, \tag{8.29}$$

where

$$r_{pp}^{(13)} = \frac{r_{pp}^{(12)} + r_{pp}^{(23)} e^{-2j\beta^{(2)}}}{1 + r_{pp}^{(12)} r_{pp}^{(23)} e^{-2j\beta^{(2)}}} \tag{8.30}$$

represents an isotropic reflection coefficient restricted to the spacer (2)–substrate (3) system, the incident medium being characterized by the optical constants of the magnetic film (1). The expression for $r_{ss}^{(03)}$ is formally obtained by replacing the subscripts pp by ss in Eq. (8.29) and Eq. (8.30). Section C.2 of Appendix C provides a detailed derivation of $r_{ss}^{(03)}$ and $r_{pp}^{(03)}$. The single-interface isotropic Fresnel reflection coefficients are defined by Eq. (8.27).

As before, $r_{ss}^{(03)}$ is not perturbed to first order in the off-diagonal elements of the permittivity tensor when $\varepsilon_1^{(1)} \neq 0$ and $\varepsilon_1^{(3)} \neq 0$, while $r_{pp}^{(03)}$ acquires small additional terms linear in $\varepsilon_1^{(1)}$ and $\varepsilon_1^{(3)}$, respectively

$$r_{pp}^{(03)} = r_{pp}\left(\varepsilon_1^{(1)} = \varepsilon_1^{(3)} = 0\right) + \frac{\mathrm{j}}{2} t_{pp}^{(01)} t_{pp}^{(10)} \left[q^{(1)} \left(1 - \mathrm{e}^{-2\mathrm{j}\beta^{(1)}}\right) \left(1 - r_{pp}^{(13)2} \mathrm{e}^{-2\mathrm{j}\beta^{(1)}}\right) \right.$$
$$\left. + t_{pp}^{(12)} t_{pp}^{(21)} \mathrm{e}^{-2\mathrm{j}(\beta^{(1)}+\beta^{(2)})} q^{(3)} t_{pp}^{(23)} t_{pp}^{(32)} \left(1 + r_{pp}^{(12)} r_{pp}^{(23)} \mathrm{e}^{-2\mathrm{j}\beta^{(2)}}\right)^{-2} \right]$$
$$\times \left(1 + r_{pp}^{(01)} r_{pp}^{(13)} \mathrm{e}^{-2\mathrm{j}\beta^{(1)}}\right)^{-2}. \tag{8.31}$$

The off-diagonal Jones reflection matrix elements are given by

$$r_{ps,sp}^{(03)} = t_{ss}^{(01)} t_{pp}^{(10)} \left\{ \beta^{(1)} \mathrm{e}^{-2\mathrm{j}\beta^{(1)}} \left[\mp p^{(1)} \left(r_{ss}^{(13)} + r_{pp}^{(13)}\right) + l^{(1)} \left(r_{ss}^{(13)} - r_{pp}^{(13)}\right) \right] \right.$$
$$+ \frac{\mathrm{j}}{2} \left(1 - \mathrm{e}^{-2\mathrm{j}\beta^{(1)}}\right) \left[\pm p^{(1)} \left(1 + r_{ss}^{(13)} r_{pp}^{(13)} \mathrm{e}^{-2\mathrm{j}\beta^{(1)}}\right) \right.$$
$$\left. - l^{(1)} \left(1 - r_{ss}^{(13)} r_{pp}^{(13)} \mathrm{e}^{-2\mathrm{j}\beta^{(1)}}\right) \right] + \frac{\mathrm{j}}{2} t_{ss}^{(12)} t_{pp}^{(21)} \mathrm{e}^{-2\mathrm{j}(\beta^{(1)}+\beta^{(2)})} \left(\pm p^{(3)} - l^{(3)}\right)$$
$$\left. \times t_{ss}^{(23)} t_{pp}^{(32)} \left(1 + r_{ss}^{(12)} r_{ss}^{(23)} \mathrm{e}^{-2\mathrm{j}\beta^{(2)}}\right)^{-1} \left(1 + r_{pp}^{(12)} r_{pp}^{(23)} \mathrm{e}^{-2\mathrm{j}\beta^{(2)}}\right)^{-1} \right\}$$
$$\times \left[\left(1 + r_{ss}^{(01)} r_{ss}^{(13)} \mathrm{e}^{-2\mathrm{j}\beta^{(1)}}\right) \left(1 + r_{pp}^{(01)} r_{pp}^{(13)} \mathrm{e}^{-2\mathrm{j}\beta^{(1)}}\right) \right]^{-1}. \tag{8.32}$$

The upper and lower signs in Eq. (8.32) pertain to $r_{ps}^{(03)}$ and $r_{sp}^{(03)}$, respectively. These Jones reflection matrix off-diagonal elements differ only in sign at the polar components $p^{(1)}$ and $p^{(3)}$. We see again that $r_{ps}^{(03)}$ and $r_{sp}^{(03)}$ consist of two contributions, of which the first one represents the effect of magnetic film and the second one that of magnetic substrate. Each contribution is formed by the polar and longitudinal terms proportional to $p^{(n)}$ and $l^{(n)}$ ($n = 1$ and 3), respectively. Further, each term in the magnetic film contribution can be split into two parts, one proportional to $\beta^{(1)} \exp(-2\mathrm{j}\beta^{(1)})$ and another one to $\left[1 - \exp(-2\mathrm{j}\beta^{(1)})\right]$. In the limits, these account approximately for radiation propagation across the magnetic film and for the front (incident medium/magnetic film) interface effect, respectively.

Indeed, in the limit of high $d^{(1)}$ and low absorption, $\left|\exp(-2j\beta^{(1)})\right| \approx 1$, the term proportional to $\beta^{(1)}\exp(-2j\beta^{(1)})$, characterizing the propagation, dominates. The propagation part depends not only on the absorption in the magnetic film but also on the reflection at the back (magnetic film/spacer) interface.

On the other hand, in the limit of high $d^{(1)}$ and high absorption, $|\exp(-2j\beta^{(1)})| \ll 1$. The penetration depth in the magnetic film is not high enough for the radiation to reach the second (back) interface of the film, the term $r_{ss}^{(13)} r_{pp}^{(13)} e^{-2j\beta^{(1)}} \approx 0$ and the interface part is practically exclusively due to the front interface of the film. The propagation part disappears, too ($|\beta^{(1)}\exp(-2j\beta^{(1)})| \approx 0$), $[1-\exp(-2j\beta^{(1)})]$ reduces to unity, and other factors $\exp(-2j\beta^{(1)})$ eliminate the dependence on $r_{ss}^{(13)}$ and $r_{pp}^{(13)}$. In this way, both the effect of the second interface and that of propagation are eliminated at $|\exp(-2j\beta^{(1)})| \ll 1$. This limiting case pertains to a single interface between an isotropic incident medium and a half space at arbitrary magnetization. The transverse magnetization contribution in r_{pp} contains only the interface parts. Note that single-interface MO effects with arbitrary magnetization can be concisely characterized with the parameters introduced in Eq. (8.21a), Eq. (8.21b), and Eq. (8.21c)

$$r_{ps,sp}^{(01)} = \frac{j}{2}(\pm p^{(1)} - l^{(1)}) t_{ss}^{(01)} t_{pp}^{(10)}, \tag{8.33a}$$

$$\Delta r_{pp}^{(01)} = \frac{j}{2} q^{(1)} t_{pp}^{(01)} t_{pp}^{(10)} = \frac{j}{2} q^{(1)} \left(1 - r_{pp}^{(01)\,2}\right). \tag{8.33b}$$

The magnetic film–nonmagnetic spacer–magnetic substrate system includes several special cases, *e.g.*, one or two nonmagnetic films on a magnetic substrate, a magnetic film on a magnetic substrate $\left(\beta^{(2)} = 0\right)$, *etc*.

In Eq. (8.32), the relation between the polar and longitudinal components in the substrate, proportional respectively to $p^{(3)}$ and $l^{(3)}$, is invariant with respect to the optical constants of the upper layers, as $p^{(3)}$ and $l^{(3)}$ enter Eq. (8.32) with the same weight. This simplifies the modelling of their contributions in the MO response. The situation in the magnetic film is more involved. Here, the corresponding components (proportional to $p^{(1)}$ and $l^{(1)}$) show different dependence on $r_{ss}^{(13)}$, $r_{pp}^{(13)}$, and $\beta^{(1)}$. An appropriate model is also needed in order to explain the spacer-dependent relations between the components originating from the magnetic film and those from the magnetic substrate.

Eq. (8.31) and Eq. (8.32) can be simplified using the ultrathin film approximation applied to the magnetic film (1) justified in the limit $d^{(1)} \to 0$.

Only terms of the lowest order in $\beta^{(1)}$ are retained

$$r^{(03)}_{ps,sp} = \left\{ \frac{4N^{(0)}\cos\varphi^{(0)}\beta^{(1)}}{\left(N^{(0)}\cos\varphi^{(0)} + N^{(2)}_{z0}\right)\left(N^{(0)}N^{(2)}_{z0}\varepsilon_0^{(2)-1/2} + \varepsilon_0^{(2)1/2}\cos\varphi^{(0)}\right)} \right.$$
$$\times \left(1 + r^{(23)}_{ss}\mathrm{e}^{-2\mathrm{j}\beta^{(2)}}\right)\left[\mp p^{(1)}\varepsilon_0^{(1)1/2}N^{(2)}_{z0}\varepsilon_0^{(2)-1/2}\left(1 + r^{(23)}_{pp}\mathrm{e}^{-2\mathrm{j}\beta^{(2)}}\right)\right.$$
$$\left. + l^{(1)}N^{(1)}_{z0}\varepsilon_0^{(1)-1/2}\varepsilon_0^{(2)1/2}\left(1 - r^{(23)}_{pp}\mathrm{e}^{-2\mathrm{j}\beta^{(2)}}\right)\right]$$
$$\left. + \frac{\mathrm{j}}{2}(\pm p^{(3)} - l^{(3)})t^{(23)}_{ss}t^{(32)}_{pp}t^{(02)}_{ss}t^{(20)}_{pp}\mathrm{e}^{-2\mathrm{j}\beta^{(2)}} \right\}$$
$$\times \left[\left(1 + r^{(02)}_{ss}r^{(23)}_{ss}\mathrm{e}^{-2\mathrm{j}\beta^{(2)}}\right)\left(1 + r^{(02)}_{pp}r^{(23)}_{pp}\mathrm{e}^{-2\mathrm{j}\beta^{(2)}}\right)\right]^{-1} \tag{8.34a}$$

$$r^{(03)}_{pp} = \frac{r^{(02)}_{pp} + r^{(23)}_{pp}\mathrm{e}^{-2\mathrm{j}\beta^{(2)}}}{1 + r^{(02)}_{pp}r^{(23)}_{pp}\mathrm{e}^{-2\mathrm{j}\beta^{(2)}}} + \frac{t^{(02)}_{pp}t^{(20)}_{pp}}{\left(1 + r^{(02)}_{pp}r^{(23)}_{pp}\mathrm{e}^{-2\mathrm{j}\beta^{(2)}}\right)^2}$$
$$\times \left[-\beta^{(1)}q^{(1)}\left(1 - r^{(23)2}_{pp}\mathrm{e}^{-4\mathrm{j}\beta^{(2)}}\right) + \frac{1}{2}\mathrm{j}q^{(3)}t^{(23)}_{pp}t^{(32)}_{pp}\mathrm{e}^{-2\mathrm{j}\beta^{(2)}}\right] \tag{8.34b}$$

$$r^{(03)}_{ss} = \frac{r^{(02)}_{ss} + r^{(23)}_{ss}\mathrm{e}^{-2\mathrm{j}\beta^{(2)}}}{1 + r^{(02)}_{ss}r^{(23)}_{ss}\mathrm{e}^{-2\mathrm{j}\beta^{(2)}}}, \tag{8.34c}$$

where by setting $\beta^{(1)} = 0$, the coefficients $t^{(02)}_{ss}$, $t^{(20)}_{pp}$, $t^{(02)}_{ss}$, $t^{(20)}_{pp}$, $r^{(02)}_{ss}$, and $r^{(20)}_{pp}$ are reduced to the single interface Fresnel transmission and reflection coefficients of Eq. (8.27c) to Eq. (8.27f) for a virtual interface between the ambient (0) and the spacer (2), *e.g.*,

$$r^{(02)}_{pp} = \frac{\varepsilon_0^{(0)}N^{(2)}_{z0} - \varepsilon_0^{(2)}N^{(0)}_{z0}}{\varepsilon_0^{(0)}N^{(2)}_{z0} + \varepsilon_0^{(2)}N^{(0)}_{z0}}$$
$$= \frac{N^{(0)}_0 N^{(2)}_{z0} - \varepsilon_0^{(2)}\cos\varphi^{(0)}}{N^{(0)}_0 N^{(2)}_{z0} + \varepsilon_0^{(2)}\cos\varphi^{(0)}}. \tag{8.35}$$

The upper and lower signs in Eq. (8.34a) pertain again to r_{ps} and r_{sp}, respectively. A small magnetization-independent term proportional to $\beta^{(1)} = (\omega/c)d^{(1)}N^{(1)}_{z0}$ in $r^{(03)}_{ss}$ and $r^{(03)}_{pp}$ was neglected. The propagation and interface parts in the magnetic film component of $r^{(03)}_{ps}$ and $r^{(03)}_{sp}$ are united into a single term and cannot therefore be distinguished.

8.7 Transmission in a Film Substrate System

Here, we consider a magnetic film sandwiched between isotropic nonmagnetic incident and exit media. The Jones transmission matrix (3.74b) relates the complex electric field amplitudes of the incident (i) and transmitted (t) waves

$$\begin{pmatrix} E_{0s}^{(t)} \\ E_{0p}^{(t)} \end{pmatrix} = \begin{pmatrix} t_{ss} & t_{sp} \\ t_{ps} & t_{pp} \end{pmatrix} \begin{pmatrix} E_{0s}^{(\mathrm{j})} \\ E_{0p}^{(i)} \end{pmatrix}. \tag{8.36}$$

The elements of the Jones transmission matrix follow from Eq. (3.70e) to Eq. (3.70h)

$$t_{ss}^{(0,\mathcal{N}+1)} = \left(\frac{E_{0s}^{(t)}}{E_{0s}^{(i)}}\right)_{E_{0p}^{(i)}=0} = \frac{M_{33}}{M_{11}M_{33} - M_{13}M_{31}}, \tag{8.37a}$$

$$t_{ps}^{(0,\mathcal{N}+1)} = \left(\frac{E_{0p}^{(t)}}{E_{0s}^{(\mathrm{j})}}\right)_{E_{0p}^{(\mathrm{j})}=0} = \frac{-M_{31}}{M_{11}M_{33} - M_{13}M_{31}}, \tag{8.37b}$$

$$t_{sp}^{(0,\mathcal{N}+1)} = \left(\frac{E_{0s}^{(t)}}{E_{0p}^{(i)}}\right)_{E_{0s}^{(\mathrm{j})}=0} = \frac{-M_{13}}{M_{11}M_{33} - M_{13}M_{31}}, \tag{8.37c}$$

$$t_{pp}^{(0,\mathcal{N}+1)} = \left(\frac{E_{0p}^{(t)}}{E_{0p}^{(i)}}\right)_{E_{0s}^{(i)}=0} = \frac{M_{11}}{M_{11}M_{33} - M_{13}M_{31}}. \tag{8.37d}$$

In the small MO ellipsometric angle approximation and to first order in $\varepsilon_1^{(1)}$, they are given by the relations

$$t_{ss}^{(02)} = \frac{t_{ss}^{(01)}t_{ss}^{(12)}\mathrm{e}^{-\mathrm{j}\beta^{(1)}}}{1 + r_{ss}^{(01)}r_{ss}^{(12)}\mathrm{e}^{-2\mathrm{j}\beta^{(1)}}}, \tag{8.38a}$$

$$\begin{aligned} t_{ps}^{(02)} = &- \mathrm{e}^{-\mathrm{j}\beta^{(1)}} t_{ss}^{(01)} t_{pp}^{(12)} \left\{ \beta^{(1)} \left[p^{(1)} \left(1 - r_{pp}^{(01)} r_{ss}^{(12)} \mathrm{e}^{-2\mathrm{j}\beta^{(1)}}\right) \right.\right. \\ &\left. + l^{(1)} \left(1 + r_{pp}^{(01)} r_{ss}^{(12)} \mathrm{e}^{-2\mathrm{j}\beta^{(1)}}\right)\right] \\ &\left. + \frac{\mathrm{j}}{2}\left(1 - \mathrm{e}^{-2\mathrm{j}\beta^{(1)}}\right)\left[p^{(1)}\left(r_{pp}^{(01)} - r_{ss}^{(12)}\right) - l^{(1)}\left(r_{pp}^{(01)} + r_{ss}^{(12)}\right)\right]\right\} \\ &\times \left[\left(1 + r_{ss}^{(01)} r_{ss}^{(12)} \mathrm{e}^{-2\mathrm{j}\beta^{(1)}}\right)\left(1 + r_{pp}^{(01)} r_{pp}^{(12)} \mathrm{e}^{-2\mathrm{j}\beta^{(1)}}\right)\right]^{-1}, \end{aligned} \tag{8.38b}$$

$$\begin{aligned} t_{sp}^{(02)} = &\, \mathrm{e}^{-\mathrm{j}\beta^{(1)}} t_{pp}^{(01)} t_{ss}^{(12)} \left\{ \beta^{(1)} \left[p^{(1)} \left(1 - r_{ss}^{(01)} r_{pp}^{(12)} \mathrm{e}^{-2\mathrm{j}\beta^{(1)}}\right) \right.\right. \\ &\left. + l^{(1)} \left(1 + r_{ss}^{(01)} r_{pp}^{(12)} \mathrm{e}^{-2\mathrm{j}\beta^{(1)}}\right)\right] \end{aligned}$$

$$+\frac{\mathrm{j}}{2}\left(1-\mathrm{e}^{-2\mathrm{j}\beta^{(1)}}\right)\left[p^{(1)}\left(r_{ss}^{(01)}-r_{pp}^{(12)}\right)+l^{(1)}\left(r_{ss}^{(01)}+r_{pp}^{(12)}\right)\right]\Big\}$$

$$\times\left[\left(1+r_{ss}^{(01)}r_{ss}^{(12)}\mathrm{e}^{-2\mathrm{j}\beta^{(1)}}\right)\left(1+r_{pp}^{(01)}r_{pp}^{(12)}\mathrm{e}^{-2\mathrm{j}\beta^{(1)}}\right)\right]^{-1}, \tag{8.38c}$$

$$t_{pp}^{(02)}=\frac{t_{pp}^{(01)}t_{pp}^{(12)}\mathrm{e}^{-\mathrm{j}\beta^{(1)}}}{1+r_{pp}^{(01)}r_{pp}^{(12)}\mathrm{e}^{-2\mathrm{j}\beta^{(1)}}}-\frac{\mathrm{j}q^{(1)}N_{z0}^{(1)}\varepsilon_0^{(1)}\left(N^{(0)}N_{z0}^{(2)}-\varepsilon_0^{(2)}\cos\varphi^{(0)}\right)}{\left(N^{(0)}N_{z0}^{(1)}+\varepsilon_0^{(1)}\cos\varphi^{(0)}\right)\left(\varepsilon_0^{(1)}N_{z0}^{(2)}+\varepsilon_0^{(2)}N_{z0}^{(1)}\right)}$$

$$\times\frac{t_{pp}^{(01)}t_{pp}^{(12)}\mathrm{e}^{-\mathrm{j}\beta^{(1)}}\left(1-\mathrm{e}^{-2\mathrm{j}\beta^{(1)}}\right)}{\left(1+r_{pp}^{(01)}r_{pp}^{(12)}\mathrm{e}^{-2\mathrm{j}\beta^{(1)}}\right)^2}, \tag{8.38d}$$

where the single-interface isotropic Fresnel coefficients are given by Eq. (8.27). In Eq. (8.38b) and Eq. (8.38c) for the off-diagonal Cartesian transmission matrix elements, the first term in the braces represents the propagation across the film, whereas the second one accounts for the interface effects. For the same incident and exit media, $t_{ps}=-t_{sp}$. In addition, the contribution from transverse $\boldsymbol{M}$ to t_{pp} proportional to $q^{(1)}(N^{(0)}N_{z0}^{(2)}-\varepsilon_0^{(2)}\cos\varphi^{(0)})$ vanishes. The formulas are useful for the evaluation of the transmission of refracted waves across a plate with an arbitrary orientation of the magnetization vector with respect to the interface normal.

References

1. K. Postava, H. Jaffres, A. Schuhl, F. Nguyen Van Dau, M. Goiran, and A. R. Fert, "Linear and quadratic magneto-optical measurements of the spin reorientation in epitaxial Fe films on MgO," J. Magn. Magn. Mater. **172**, 199–208, 1997.
2. K. Postava, J. Pištora, D. Ciprian, D. Hrabovský, M. Lesňák, and A. R. Fert, "Linear and quadratic magneto-optical effects in reflection from a medium with an arbitrary direction of magnetization," Proc. of SPIE **3820**, 412–422, 1999.
3. R. M. Osgood, III, S. D. Bader, B. M. Clemens, R. L. White, and H. Matsuyama, "Second order magneto-optic effects in anisotropic thin films," J. Magn. Magn. Mater. **182**, 297–323, 1998.
4. S.-S. Yan, R. Schreiber, P. Grünberg, and R. Schäfer, "Magnetization reversal in (001)Fe thin film studied by combining domain images and MOKE hysteresis loops," J. Magn. Magn. Mater. **210**, 309–315, 2000.
5. J. Zak, E. R. Moog, C. Liu, and S. D. Bader, "Magneto-optics of multilayers with arbitrary magnetization directions," Phys. Rev. B **43**, 6423–6429, 1991.
6. B. Heinrich and J. A. C. Bland, Eds. *Ultrathin Magnetic Structures* (Springer Verlag, Berlin, 1994).

7. Z. Q. Qiu and S. D. Bader, "Surface magneto-optic effect (SMOKE)," J. Magn. Magn. Mat. **200**, 664–678, 1999.
8. M. R. Pufall, C. Platt, and A. Berger, "Layer resolved magnetometry of magnetic bilayer using the magneto-optical Kerr effect with varying angle of incidence," J. Appl. Phys. **85**, 4818–4820, 1999.
9. A. Berger and M. R. Pufall, "Quantitative vector magnetometry using generalized magneto-optical ellipsometry," J. Appl. Phys. **85**, 4583–4585, 1999.
10. H. F. Ding, S. Pütter, H. P. Oepen, and J. Kirschner, "Experimental method for separating longitudinal and polar Kerr signals," J. Magn. Magn. Mat. **212**, L5–L11, 2000.
11. B. Heinrich, "Magnetic nanostructures. From physical principles to spintronics," Can. J. Phys. **78**, 161–199, 2000.
12. B. Heinrich, J. F. Cochran, T. Monchesky, and R. Urban, "Exchange coupling through spin-density waves in Cr(001) structures: Fe-whisker/Cr/Fe(001) studies," Phys. Rev. B **59**, 14520–14532, 1999.
13. D. T. Pierce, J. Unguris, R. J. Celotta, and M. D. Stiles, "Effect of roughness, frustration, and antiferromagnetic order on magnetic coupling of Fe/Cr multilayers," J. Magn. Magn. Mat. **200**, 290–321, 1999.
14. J. Gřondilová, M. Rickart, J. Mistrík, K. Postava, S. Višňovský, T. Yamaguchi, R. Lopušník, S. O. Demokritov, and B. Hillebrands, "Anisotropy of magneto-optical spectra in ultrathin Fe/Au/Fe bilayers," J. Appl. Phys. **91**, 8246–8248, 2002.
15. W. Wettling, "Magneto-optics of ferrites," J. Magn. Magn. Mat. **3**, 147–160, 1976.
16. Š. Višňovský, "Magneto-optical permittivity tensor in crystals," Czech. J. Phys. B **36**, 1424–1433, 1986.
17. M. Schubert, T. E. Tiwald, and J. A. Woollam, "Explicit solutions for the optical properties of arbitrary magneto-optic materials in generalized ellipsometry," Appl. Opt. **38**, 177–187, 1999.
18. P. Yeh, "Optics of anisotropic layered media: A new 4×4 matrix algebra," Surf. Sci. **96**, 41–53, 1980.
19. R. M. A. Azzam and N. M. Bashara, *Ellipsometry and Polarized Light* (Elsevier, Amsterdam, 1987).
20. S. Višňovský, R. Lopušník, M. Bauer, J. Bok, J. Fassbender, and B. Hillebrands, "Magneto-optic ellipsometry in multilayers at arbitrary magnetization." Optics Express **9**, 121–135, 2001.
21. Š. Višňovský, K. Postava, T. Yamaguchi, and R. Lopušník, "Magneto-optic ellipsometry in exchange-coupled films," Appl. Opt. **41**, 3950–3960, 2002.
22. Š. Višňovský, "Magneto-optical longitudinal and transverse Kerr and birefringence effects in orthorhombic crystals," Czech. J. Phys. B **34**,969–980,1984.
23. F. Abelès, "Recherches sur la propagation des ondes électromagnétiques sinusoïdales dans les milieux stratifiés. Application aux couches minces," Ann. Phys. Paris **5**, 596–640, 1950.
24. M. Born and E. Wolf, *Principles of Optics* (Cambridge University Press, Cambridge, 1997) pp. 51–70.
25. J. Lafait, T. Yamaguchi, J. M. Frigerio, A. Bichri, and K. Driss-Khodja, "Effective medium equivalent to a symmetric multilayer at oblique incidence," Applied Optics **29**, 2460–2465, 1990.

9

Anisotropic Multilayer Gratings

9.1 Introduction

So far, we were concerned with the response of multilayer systems consisting of plane parallel layers each characterized by a general permittivity tensor *homogeneous* in the region of the layer. The analysis can be generalized to laterally structured periodic (both isotropic and anisotropic) multilayer systems. The approach finds applications in the design of various frequency selective elements including photonic crystals, in the spectroscopic ellipsometry for modelling surface roughness, and in-depth profiles. It forms the basis for the solution of inverse problem in the scatterometry of periodic nanostructures including semiconductor chips; liquid crystal structures; optic, magnetic, and magnetooptic recording disks; *etc*.

The development of the subject can be traced from 1950 when the *first* step in the treatment of the response from a stack consisting of an arbitrary number of layers to monochromatic polarized electromagnetic plane waves was made by Abelès [1]. The individual layers were assumed to consist of *homogeneous* media. To characterize the response in isotropic multilayers with plane parallel interfaces, Abelès applied the Fourier modal method. He introduced the 4×4 matrix formalism and the notion of the characteristic matrix. Because of the absence of the TE↔TM mode coupling, Abelès characteristic matrix became quasidiagonal, consisting of the 2×2 submatrices. Consequently, the TE polarization modes propagating in forward and backward (retrograde) directions could be treated separately from the corresponding TM polarization modes. Abelès extended the analysis to multilayers periodic on the multilayer axis (in both infinite or restricted ranges) which, in the present context, corresponds to periodic structures with a grating vector normal to the interfaces.

The Abelès approach is not restricted to the TE and TM proper polarization modes. Any anisotropic multilayer can be treated, provided the symmetry of individual layers allows the propagation of two independent pairs of forward- and backward-going, in general elliptically polarized, proper polarization modes common to all layers of the multilayer system. Examples are (periodic or nonperiodic) multilayers consisting of orthorhombic layers at the same orientation of orthorhombic axes with one orthorhombic axis normal to the interfaces and the second one in the plane of incidence (linearly polarized TE and TM proper modes), magnetic multilayers with polar magnetization and normal incidence (circularly polarized proper modes) [2,3], and magnetic multilayers at transverse, *i.e.*, normal to the plane of incidence magnetization (linearly polarized TE and TM proper modes) [4]. The Abelès formalism can be extended to quantum problems of one-dimensional potential wells of arbitrary profile and to infinite or finite periodic arrays of such quantum wells [5].

The *second* step was made by Berreman, who extended the Abelès analysis to multilayers consisting of layers with arbitrary anisotropies [6,7]. The off-diagonal 2×2 blocks in 4×4 characteristic matrices of individual layers in general do not vanish. Inspired by Berreman, Yeh developed another 4×4 matrix formalism that represents the multilayer in terms of dynamical and propagation matrices combined into transfer matrices [8]. This was employed in the previous chapters.

The generalization to multilayer media with periodic variations of the permittivity tensor in each of the layers by Rokushima and Yamakita [9] represents the *third* step in this sequence. Their analytical Fourier modal formalism treats the Fraunhofer diffraction in planar *multilayer anisotropic gratings*. It represents the extension of the Berreman 4×4 matrix formalism combined with the extension of the Kogelnik coupled wave theory [10] to anisotropic media.

In the Rokushima formalism, all regions of the system including the sandwiching half spaces may be characterized by general permittivity tensors. The grating vectors, which define the tensor periodicity in a given region, may be arbitrarily oriented in a plane normal to the interfaces (one-dimensional multilayer anisotropic gratings). There are no restrictions as for the period-to-wavelength ratio. So far, the formalism was applied to both specific cases and to the development of more general theories treating electromagnetic plane wave effects in periodic media from millimeter and microwave to x-ray wavelengths [11,12]. In particular, an improved phenomenological description of the material response at optical frequencies, *e.g.*, for chiral media, was obtained [19,20]. The formalism was extended to two or three dimensions [13–15] and to matter wave diffraction [16]. The subject forms the basis for the solution of *inverse problems* employed, *e.g.*, in *scatterometry* [17], also called optical critical dimension (OCD) measurement method [18]. The latter represents a nondestructive and noncontact analytical method that provides the information on the objects from diffracted plane waves. The technique finds a growing number of applications in the diagnostics of periodic nanostructures in the data recording and semiconductor industries, *e.g.*, in the metrology of optical and magnetooptic disks, semiconductor chips, *etc*. It seriously competes with a more expensive and destructive technique of scanning electron microscopy (SEM).

Several aspects of electromagnetic optics are covered as special cases of the Rokushima formalism [9] from a more general point of view, *e.g.*, plane wave propagation in isotropic media; reflection and refraction at interfaces; optics of thin films and multilayers [1,36]; slab dielectric optical waveguides [49,50,52]; crystal optics [37]; ellipsometry [38]; acousto-, electro-, and magnetooptics [39,40]; diffraction gratings; *etc*.

The formalism by Rokushima and Yamakita can be used as a systematic introduction to the analysis of the optical response in periodic nanostructures, and we devote the present chapter to its detailed explanation. We follow the patterns of the original work [9] and illustrate the approach on simple situations. To preserve a consistency with the previous chapters, we transform the problem geometry to a coordinate system with the z axis normal to the interfaces. The formalism starts from the solution of the Maxwell equations for monochromatic plane waves assuming periodic space dependence of the permittivity tensor in the periodic region. Each layer is characterized by ε, a *Hermitian permittivity tensor* (corresponding to a nonabsorbing medium), which periodically varies inside the layer. The axis of periodic variation of ε, which defines the direction of the grating vector, can be either perpendicular (corresponding to periodic multilayers) or parallel to the layer normal (corresponding, *e.g.*, to diffraction gratings used, *e.g.*, in spectroscopy) or of arbitrary orientation corresponding to *slanted* diffraction gratings. The latter are of interest, for example, in anisotropic vicinal surface studies.

The problem leads to an equation for proper values of the normal components of the propagation vectors of all considered diffraction orders. Any field in a particular region can then be expressed as a linear superposition of the proper polarizations consistent with the continuity of the electric and magnetic fields at interfaces. The Hermitian permittivity tensor-based exposition, appropriate for the electric dipole approximation for parameters of lossless media, can be extended to multilayer systems with absorbing regions characterized by the permittivity tensor consisting of both Hermitian and anti-Hermitian parts. The Rokushima formalism avoids the use of secondary parameters introduced in the early classical optics of nonabsorbing crystals, *e.g.*, optical axes, ordinary and extraordinary rays, *etc*. However, these can always be deduced from the present analysis.

9.2 Fields in the Grating Region

We first consider a single anisotropic periodic region with plane parallel boundaries sandwiched between isotropic half spaces. Polarized monochromatic electromagnetic plane wave impinges, at an oblique angle of incidence, $\varphi_i^{(0)}$, on a structure consisting of a lossless anisotropic grating

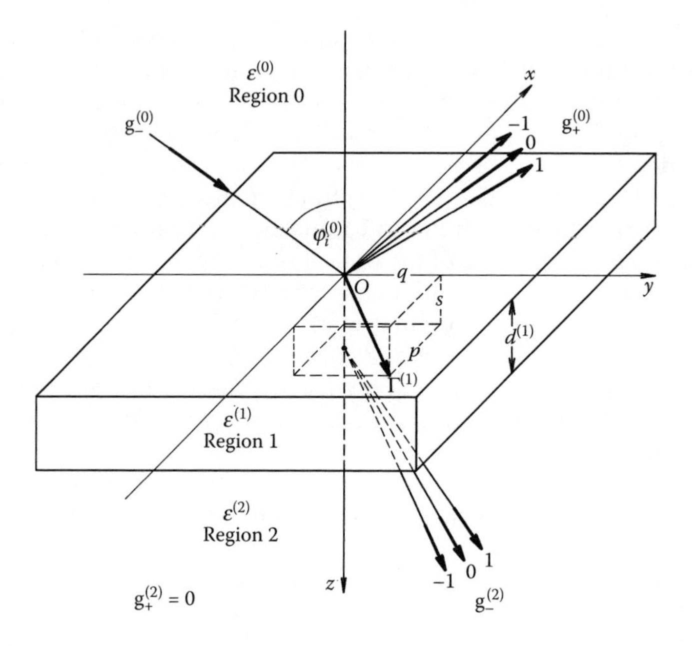

FIGURE 9.1
Anisotropic grating region (1) of the thickness $d^{(1)}$ sandwiched between two homogeneous isotropic semi-infinite regions (0) and (2). The figure shows the grating vector $\mathbf{\Gamma}$, the propagation vectors of the incident wave, and the propagation vectors of reflected and transmitted waves of zero and ± 1 diffraction orders.

sandwiched between two, in general different, homogeneous isotropic lossless media (Figure 9.1). The incident wave generates diffracted both transmitted and reflected plane waves. The relation between the incident and diffracted waves in the structure is given by Maxwell equations. We use the SI units and choose a Cartesian coordinate system with the z axis normal to interfaces. The time harmonic dependence is described with $\exp\left(j\omega t\right)$. The entrance and exit regions, denoted as region (0) and (2), respectively, are characterized by scalar relative permittivities $\varepsilon^{(0)}$ and $\varepsilon^{(2)}$, respectively. Below, we show a generalization, already included in the original formalism [9], to the case where the anisotropic periodic region is in contact with an anisotropic homogeneous one. We situate the plane of incidence of an electromagnetic plane wave normal to the x axis, *i.e.*, in the yz plane, without any loss in generality. In fact, the isotropic regions are invariant with respect to the rotation about the axis (z axis) normal to the interfaces. Further, we have stated that the anisotropic periodic region is characterized by a Hermitian relative permittivity tensor $\boldsymbol{\varepsilon}^{(1)}$ of the most general form. Then, the rotation about the axis cannot change the number of independent tensor components. The incident plane vector wave has a general elliptic polarization. Its electric and magnetic fields change with the position $\boldsymbol{r}$ and time t as

$$\exp\left[\mathrm{j}\omega t - \mathrm{j}\,\varepsilon^{(0)1/2}\frac{\omega}{c}\left(y\sin\varphi_i^{(0)} + z\cos\varphi_i^{(0)}\right)\right],$$

where $\omega/c = (\varepsilon_{\text{vac}}\mu_{\text{vac}})^{1/2}\,\omega = 2\pi/\lambda_{\text{vac}}$ and λ_{vac} are the propagation constant and radiation wavelength in vacuum, respectively. The propagation vector of the incident wave is given by

$$\gamma^{(0)} = \varepsilon^{(0)1/2}\frac{\omega}{c}\left(\hat{\boldsymbol{y}}\sin\varphi_i^{(0)} + \hat{\boldsymbol{z}}\cos\varphi_i^{(0)}\right).$$

The grating region, *i.e.*, the region (1), is filled with a periodic anisotropic dielectric. Its periodicity is characterized by a *grating vector* with a general orientation

$$\boldsymbol{\Gamma}^{(1)} = \hat{\boldsymbol{x}}\Gamma_x^{(1)} + \hat{\boldsymbol{y}}\Gamma_y^{(1)} + \hat{\boldsymbol{z}}\Gamma_z^{(1)}, \tag{9.1}$$

where

$$|\boldsymbol{\Gamma}^{(1)}| = \Gamma^{(1)} = 2\pi/\Lambda^{(1)}, \tag{9.2}$$

$$\Gamma_x^{(1)} = 2\pi/\Lambda_x^{(1)}, \qquad \Gamma_y^{(1)} = 2\pi/\Lambda_y^{(1)}, \qquad \Gamma_z^{(1)} = 2\pi/\Lambda_z^{(1)}. \tag{9.3}$$

We have

$$\frac{1}{\Lambda^{(1)2}} = \frac{1}{\Lambda_x^{(1)2}} + \frac{1}{\Lambda_y^{(1)2}} + \frac{1}{\Lambda_z^{(1)2}},$$

and

$$\Lambda^{(1)2} = \frac{\Lambda_x^{(1)2}\Lambda_y^{(1)2}\Lambda_z^{(1)2}}{\Lambda_y^{(1)2}\Lambda_z^{(1)2} + \Lambda_z^{(1)2}\Lambda_x^{(1)2} + \Lambda_x^{(1)2}\Lambda_y^{(1)2}}.$$

Here, $\hat{\boldsymbol{x}}, \hat{\boldsymbol{y}}$, and $\hat{\boldsymbol{z}}$ are the unit vectors parallel to the Cartesian axes x, y, and z, respectively. The symbol $\Lambda^{(1)}$ denotes the grating period and $\Lambda_i^{(1)}$ denotes the grating period parallel to the i-th axis, $i = x, y,$ or z. Subscripts (i, j) will denote coordinates, while the subscript $(l, m,$ and $n)$ are reserved for space harmonics. The indices (I, J) distinguish the regions. In the cases where no confusion can occur, the region subscripts will be dropped.

The Hermitian relative permittivity tensor, a characteristic of the anisotropic periodic region in frame of the electric dipole approximation, takes the form corresponding to the most general anisotropy

$$\boldsymbol{\varepsilon} = (\varepsilon_{ij}) = \begin{pmatrix} \varepsilon_{xx} & \varepsilon_{xy} & \varepsilon_{xz} \\ \varepsilon_{yx} & \varepsilon_{yy} & \varepsilon_{yz} \\ \varepsilon_{zx} & \varepsilon_{zy} & \varepsilon_{zz} \end{pmatrix}, \tag{9.4}$$

where the $\boldsymbol{\varepsilon}$ tensor elements are periodic functions of position $\boldsymbol{r}$

$$\varepsilon_{ij} = \varepsilon_{ij}(\boldsymbol{r}) = \sum_l \exp(\mathrm{j}\,l\boldsymbol{\Gamma}\cdot\boldsymbol{r})\varepsilon_{ij,l}. \tag{9.5}$$

Here, $\varepsilon_{ij,l}$ are the Fourier expansion coefficients, which define a given grating. The index l specifies terms in the Fourier expansion. The relative magnetic permeability could be assumed a constant scalar within a given region. In frame of the electric dipole approximation used here, it is set equal to unity in all regions.

We transform the space coordinates with the help of the vacuum propagation constant, $\frac{\omega}{c}$, to a dimensionless one by setting $\overline{r} = \frac{\omega}{c}\boldsymbol{r}$, or in components, $\overline{x} = \frac{\omega}{c}x$, $\overline{y} = \frac{\omega}{c}y$, and $\overline{z} = \frac{\omega}{c}z$, where the bar over the symbol denotes the normalized coordinates.

We start from the Maxwell equations for fields harmonic in time with the angular frequency ω

$$\nabla \times \boldsymbol{E} = -\mathrm{j}\omega \boldsymbol{B} = -\mathrm{j}\omega\mu_{\text{vac}}\boldsymbol{H} \tag{9.6}$$

$$\nabla \times \boldsymbol{B} = \mathrm{j}\omega\varepsilon_{\text{vac}}\mu_{\text{vac}}\varepsilon\, \boldsymbol{E}, \tag{9.7}$$

where $\boldsymbol{B} = \mu_{\text{vac}}\boldsymbol{H}$. We shall express them using the relations

$$\varepsilon_{\text{vac}}\mu_{\text{vac}}c^2 = 1, \qquad \frac{\mu_{\text{vac}}}{\varepsilon_{\text{vac}}} = Z_{\text{vac}}^2 = \frac{1}{Y_{\text{vac}}^2}, \tag{9.8}$$

where Z_{vac}, resp. Y_{vac} are the vacuum wave impedance and the vacuum wave admittance, respectively. We obtain

$$\begin{aligned}\nabla \times \boldsymbol{E} &= -\mathrm{j}\frac{\mu_{\text{vac}}\omega}{(\mu_{\text{vac}}\varepsilon_{\text{vac}})^{1/2}c}\boldsymbol{H} = -\mathrm{j}\frac{\omega}{c}\left(\frac{\mu_{\text{vac}}}{\varepsilon_{\text{vac}}}\right)^{1/2}\boldsymbol{H} \\ &= -\mathrm{j}\frac{\omega}{c}\left(\frac{\mu_{\text{vac}}}{\varepsilon_{\text{vac}}}\right)^{1/4}\left(\frac{\mu_{\text{vac}}}{\varepsilon_{\text{vac}}}\right)^{1/4}\boldsymbol{H} = -\mathrm{j}\frac{\omega}{c}\left(\frac{Z_{\text{vac}}}{Y_{\text{vac}}}\right)^{1/2}\boldsymbol{H},\end{aligned} \tag{9.9}$$

$$\begin{aligned}\nabla \times \boldsymbol{H} &= \mathrm{j}\frac{\varepsilon_{\text{vac}}\omega}{(\mu_{\text{vac}}\varepsilon_{\text{vac}})^{1/2}c}\varepsilon\, \boldsymbol{E} = \mathrm{j}\frac{\omega}{c}\left(\frac{\varepsilon_{\text{vac}}}{\mu_{\text{vac}}}\right)^{1/2}\varepsilon\, \boldsymbol{E} \\ &= \mathrm{j}\frac{\omega}{c}\left(\frac{Y_{\text{vac}}}{Z_{\text{vac}}}\right)^{1/2}\varepsilon\, \boldsymbol{E}.\end{aligned} \tag{9.10}$$

We denote

$$\begin{aligned}\overline{\nabla} = \left(\frac{\omega}{c}\right)^{-1}\nabla &= \hat{\boldsymbol{x}}\frac{\partial}{\partial\left(\frac{\omega}{c}x\right)} + \hat{\boldsymbol{y}}\frac{\partial}{\partial\left(\frac{\omega}{c}y\right)} + \hat{\boldsymbol{z}}\frac{\partial}{\partial\left(\frac{\omega}{c}z\right)} \\ &= \hat{\boldsymbol{x}}\frac{\partial}{\partial\overline{x}} + \hat{\boldsymbol{y}}\frac{\partial}{\partial\overline{y}} + \hat{\boldsymbol{z}}\frac{\partial}{\partial\overline{z}}.\end{aligned} \tag{9.11}$$

This allows us to express the Maxwell equations with the normalized coordinates in the following form

$$\overline{\nabla} \times \sqrt{Y_{\text{vac}}}\boldsymbol{E} = -\mathrm{j}\sqrt{Z_{\text{vac}}}\boldsymbol{H}, \tag{9.12}$$

$$\overline{\nabla} \times \sqrt{Z_{\text{vac}}}\boldsymbol{H} = \mathrm{j}\sqrt{Y_{\text{vac}}}\varepsilon\, \boldsymbol{E}. \tag{9.13}$$

We next express the Fourier expansions for the components of the permittivity tensor with normalized coordinates

$$\varepsilon_{ij} = \varepsilon_{ij}(\bar{\boldsymbol{r}}) = \sum_{l} \varepsilon_{ij,l} \exp(\mathrm{j}\, l \boldsymbol{n}_{\Gamma} \cdot \bar{\boldsymbol{r}}). \tag{9.14}$$

Here, we have denoted $\boldsymbol{\Gamma} = \dfrac{\omega}{c}\boldsymbol{n}_{\Gamma}$, where $\boldsymbol{n}_{\Gamma}$

$$\begin{aligned} \boldsymbol{n}_{\Gamma} &= \left(\frac{\omega}{c}\right)^{-1} (\hat{\boldsymbol{x}}\Gamma_x + \hat{\boldsymbol{y}}\Gamma_y + \hat{\boldsymbol{z}}\Gamma_z) \\ &= \hat{\boldsymbol{x}}\frac{\lambda_0}{\Lambda_x} + \hat{\boldsymbol{y}}\frac{\lambda_0}{\Lambda_y} + \hat{\boldsymbol{z}}\frac{\lambda_0}{\Lambda_z} \\ &= \hat{\boldsymbol{x}}p + \hat{\boldsymbol{y}}q + \hat{\boldsymbol{z}}s \end{aligned} \tag{9.15}$$

is the *normalized grating vector* (Figure (9.2)). Its magnitude and its components are given by

$$|\boldsymbol{n}_{\Gamma}| = n_{\Gamma} = \frac{\lambda_0}{\Lambda}, \tag{9.16}$$

$$p = \frac{\lambda_0}{\Lambda_x}, \qquad q = \frac{\lambda_0}{\Lambda_y}, \qquad s = \frac{\lambda_0}{\Lambda_z}, \tag{9.17}$$

where we have used Eq. (9.2). This magnitude $|\boldsymbol{n}_{\Gamma}|$ will be called the *effective grating index*.

According to the Floquet theorem, the electric and magnetic fields of waves can be expressed as

$$\sqrt{Y_{\mathrm{vac}}}\boldsymbol{E} = \sum_{m} \boldsymbol{e}_m(\bar{z}) \exp(-\mathrm{j}\boldsymbol{n}_m \cdot \bar{\boldsymbol{r}}), \tag{9.18}$$

$$\sqrt{Z_{\mathrm{vac}}}\boldsymbol{H} = \sum_{m} \boldsymbol{h}_m(\bar{z}) \exp(-\mathrm{j}\boldsymbol{n}_m \cdot \bar{\boldsymbol{r}}), \tag{9.19}$$

where

$$\boldsymbol{e}_m(\bar{z}) = \hat{\boldsymbol{x}}\boldsymbol{e}_{xm}(\bar{z}) + \hat{\boldsymbol{y}}\boldsymbol{e}_{ym}(\bar{z}) + \hat{\boldsymbol{z}}\boldsymbol{e}_{zm}(\bar{z}), \tag{9.20}$$

$$\boldsymbol{h}_m(\bar{z}) = \hat{\boldsymbol{x}}\boldsymbol{h}_{xm}(\bar{z}) + \hat{\boldsymbol{y}}\boldsymbol{h}_{ym}(\bar{z}) + \hat{\boldsymbol{z}}\boldsymbol{h}_{zm}(\bar{z}), \tag{9.21}$$

and the normalized propagation vector of the m-th Fourier components

$$\boldsymbol{n}_m = \hat{\boldsymbol{x}}p_m + \hat{\boldsymbol{y}}q_m + \hat{\boldsymbol{z}}s_m. \tag{9.22}$$

Here,

$$p_m = mp, \qquad q_m = mq + q_0, \qquad s_m = ms, \tag{9.23}$$

with

$$q_0 = \sqrt{\varepsilon^{(0)}} \sin\varphi_i^{(0)}. \tag{9.24}$$

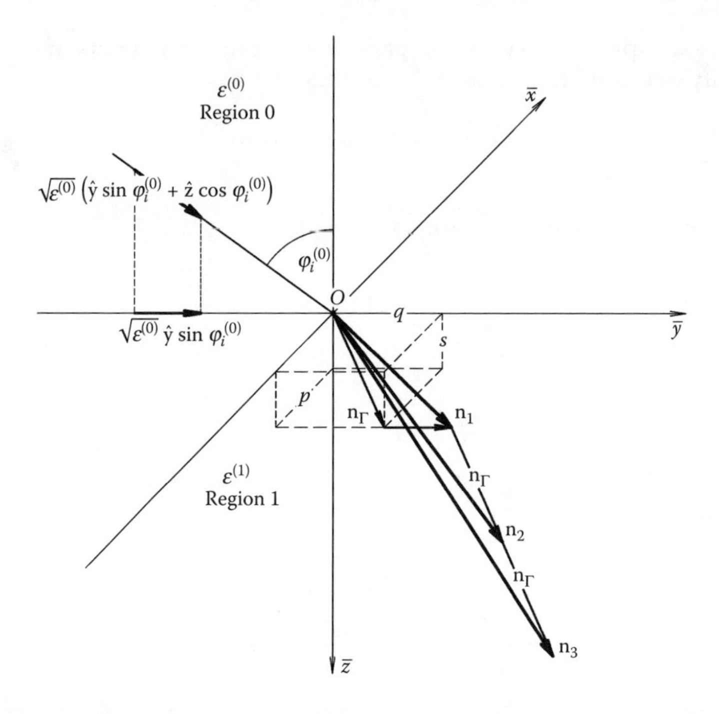

FIGURE 9.2
The normalized grating vector $\boldsymbol{n}_\Gamma$ and the normalized propagation vector of the incident wave $\sqrt{\varepsilon^{(0)}}(\hat{\boldsymbol{y}} \sin\varphi_i^{(0)} + \hat{\boldsymbol{z}} \cos\varphi_i^{(0)})$. The component parallel to the interface plane $z = 0$ of the normalized propagation vector of the incident wave is represented by $\hat{\boldsymbol{y}} q_0 = \hat{\boldsymbol{y}} \sqrt{\varepsilon^{(0)}} \sin\varphi_i^{(0)}$. The vectors $\boldsymbol{n}_m = \hat{\boldsymbol{x}} p_m + \hat{\boldsymbol{y}} q_m + \hat{\boldsymbol{z}} s_m$ for $m = 1, 2,$ and 3 employed in the Floquet development are also shown.

The summing index m runs over all integer m

$$m = 0, \pm 1, \pm 2, \ldots.$$

The two-vector Maxwell vector equations expressed in their components become

$$\overline{\nabla} \times \sqrt{Y_{\text{vac}}} \boldsymbol{E} = -\mathrm{j}\sqrt{Z_{\text{vac}}} \boldsymbol{H},$$

i.e.,

$$\frac{\partial}{\partial \bar{y}}(\sqrt{Y_{\text{vac}}} E_z) - \frac{\partial}{\partial \bar{z}}(\sqrt{Y_{\text{vac}}} E_y) = -\mathrm{j}(\sqrt{Z_{\text{vac}}} H_x), \tag{9.25a}$$

$$\frac{\partial}{\partial \bar{z}}(\sqrt{Y_{\text{vac}}} E_x) - \frac{\partial}{\partial \bar{x}}(\sqrt{Y_{\text{vac}}} E_z) = -\mathrm{j}(\sqrt{Z_{\text{vac}}} H_y), \tag{9.25b}$$

$$\frac{\partial}{\partial \bar{x}}(\sqrt{Y_{\text{vac}}} E_y) - \frac{\partial}{\partial \bar{y}}(\sqrt{Y_{\text{vac}}} E_x) = -\mathrm{j}(\sqrt{Z_{\text{vac}}} H_x), \tag{9.25c}$$

and

$$\bar{\nabla} \times \sqrt{Z_{\text{vac}}}\boldsymbol{H} = \mathrm{j}\varepsilon(\bar{\boldsymbol{r}})\sqrt{Y_{\text{vac}}}\boldsymbol{E},$$

i.e.,

$$\frac{\partial}{\partial \bar{y}}(\sqrt{Z_{\text{vac}}}H_z) - \frac{\partial}{\partial \bar{z}}(\sqrt{Z_{\text{vac}}}H_y)$$
$$= \mathrm{j}[\varepsilon_{xx}(\sqrt{Y_{\text{vac}}}E_x) + \varepsilon_{xy}(\sqrt{Y_{\text{vac}}}E_y)\varepsilon_{xz}(\sqrt{Y_{\text{vac}}}E_z)], \tag{9.26a}$$

$$\frac{\partial}{\partial \bar{z}}(\sqrt{Z_{\text{vac}}}H_x) - \frac{\partial}{\partial \bar{x}}(\sqrt{Z_{\text{vac}}}H_z)$$
$$= \mathrm{j}[\varepsilon_{yx}(\sqrt{Y_{\text{vac}}}E_x) + \varepsilon_{yy}(\sqrt{Y_{\text{vac}}}E_y)\varepsilon_{yz}(\sqrt{Y_{\text{vac}}}E_z)], \tag{9.26b}$$

$$\frac{\partial}{\partial \bar{x}}(\sqrt{Z_{\text{vac}}}H_y) - \frac{\partial}{\partial \bar{y}}(\sqrt{Z_{\text{vac}}}H_x)$$
$$= \mathrm{j}[\varepsilon_{zx}(\sqrt{Y_{\text{vac}}}E_x) + \varepsilon_{zy}(\sqrt{Y_{\text{vac}}}E_y)\varepsilon_{zz}(\sqrt{Y_{\text{vac}}}E_z)]. \tag{9.26c}$$

Next, we substitute the Floquet series for the fields according to Eq. (9.18) and Eq. (9.19) and the Fourier expansions for the permittivity tensor elements according to Eq. (9.5). We start with the substitution into the left-hand side of these equations. For the moment, it is convenient to use a more explicit expression for the exponentials, *i.e.*,

$$\exp[-\mathrm{j}(\boldsymbol{n}_m \cdot \bar{\boldsymbol{r}})] = \exp[-\mathrm{j}m(p\bar{x} + q\bar{y} + s\bar{z}) - \mathrm{j}q_0\bar{y}]. \tag{9.27}$$

From Eq. (9.25a) for $-\mathrm{j}(\sqrt{Z_{\text{vac}}}H_x)$

$$\frac{\partial}{\partial \bar{y}}(\sqrt{Y_{\text{vac}}}E_z) - \frac{\partial}{\partial \bar{z}}(\sqrt{Y_{\text{vac}}}E_y)$$
$$= \sum_m e_{zm}(\bar{z})\exp[-\mathrm{j}m(p\bar{x} + s\bar{z})]\frac{\partial}{\partial \bar{y}}\{\exp[-\mathrm{j}(mq + q_0)\bar{y}]\}$$
$$-\sum_m \exp\left[-\mathrm{j}m(p\bar{x} + q\bar{y}) - \mathrm{j}q_0\bar{y}\right]\frac{\partial}{\partial \bar{z}}[e_{ym}(\bar{z})\exp(-\mathrm{j}ms\bar{z})]$$
$$= \sum_m [-\mathrm{j}(mq + q_0)]e_{zm}(\bar{z})\exp[-\mathrm{j}m(p\bar{x} + q\bar{y} + s\bar{z}) - \mathrm{j}q_0\bar{y}]$$
$$-\sum_m \left\{\left[\frac{\mathrm{d}}{\mathrm{d}\bar{z}}e_{ym}(\bar{z})\right] - \mathrm{j}ms\,e_{ym}(\bar{z})\right\}\exp[-\mathrm{j}m(p\bar{x} + q\bar{y} + s\bar{z}) - \mathrm{j}q_0\bar{y}]. \tag{9.28}$$

From Eq. (9.25b) for $-\mathrm{j}(\sqrt{Z_{\text{vac}}}H_y)$

$$\frac{\partial}{\partial \bar{z}}(\sqrt{Y_{\text{vac}}}E_x) - \frac{\partial}{\partial \bar{x}}(\sqrt{Y_{\text{vac}}}E_z)$$
$$= \exp(-\mathrm{j}q_0\bar{y})\sum_m \exp\left[-\mathrm{j}m(p\bar{x} + q\bar{y})\right]\frac{\partial}{\partial \bar{z}}\left[e_{xm}(\bar{z})\exp(-\mathrm{j}ms\bar{z})\right]$$

$$- \exp(-jq_0\bar{y}) \sum_m e_{zm}(\bar{z}) \exp\left[-jm(q\,\bar{y} + s\bar{z})\right] \frac{\partial}{\partial \bar{x}} \left[\exp(-jmp\bar{x})\right]$$
$$= \sum_m \left\{ \left[\frac{\mathrm{d}}{\mathrm{d}\bar{z}} e_{xm}(\bar{z})\right] - \mathrm{j}ms\, e_{xm}(\bar{z}) \right\} \exp[-\mathrm{j}m(p\bar{x} + q\,\bar{y} + s\bar{z}) - \mathrm{j}q_0\bar{y}]$$
$$- \sum_m (-\mathrm{j}mp) e_{zm}(\bar{z}) \exp[-\mathrm{j}m(p\bar{x} + q\,\bar{y} + s\bar{z}) - \mathrm{j}q_0\bar{y}]. \tag{9.29}$$

From Eq. (9.25c) for $-\mathrm{j}(\sqrt{Z_{\mathrm{vac}}} H_z)$

$$\frac{\partial}{\partial \bar{x}}(\sqrt{Y_{\mathrm{vac}}} E_y) - \frac{\partial}{\partial \bar{y}}(\sqrt{Y_{\mathrm{vac}}} E_x)$$
$$= \sum_m (-\mathrm{j}mp) e_{ym}(\bar{z}) \exp[-\mathrm{j}m(p\bar{x} + q\,\bar{y} + s\bar{z}) - \mathrm{j}q_0\bar{y}]$$
$$- \sum_m [-\mathrm{j}(mq + q_0)] e_{xm}(\bar{z}) \exp[-\mathrm{j}m(p\bar{x} + q\,\bar{y} + s\bar{z}) - \mathrm{j}q_0\bar{y}]. \tag{9.30}$$

From Eq. (9.26a) for $\mathrm{j}(\sqrt{Y_{\mathrm{vac}}} \varepsilon \boldsymbol{E})_x$

$$\frac{\partial}{\partial \bar{y}}(\sqrt{Z_{\mathrm{vac}}} H_z) - \frac{\partial}{\partial \bar{z}}(\sqrt{Z_{\mathrm{vac}}} H_y)$$
$$= \sum_m [-\mathrm{j}(mq + q_0)] h_{zm}(\bar{z}) \exp[-\mathrm{j}m(p\bar{x} + q\,\bar{y} + s\bar{z}) - \mathrm{j}q_0\bar{y}]$$
$$- \sum_m \exp[-\mathrm{j}m(p\bar{x} + q\,\bar{y}) - \mathrm{j}q_0\bar{y}] \frac{\partial}{\partial \bar{z}} [h_{ym}(\bar{z}) \exp(-\mathrm{j}ms\bar{z})]$$
$$= \sum_m [-\mathrm{j}(mq + q_0)] h_{zm}(\bar{z}) \exp[-\mathrm{j}m(p\bar{x} + q\,\bar{y} + s\bar{z}) - \mathrm{j}q_0\bar{y}]$$
$$- \sum_m \left\{ \left[\frac{\mathrm{d}}{\mathrm{d}\bar{z}} h_{ym}(\bar{z})\right] - \mathrm{j}ms\, h_{ym}(\bar{z}) \right\} \exp[-\mathrm{j}m(p\bar{x} + q\,\bar{y} + s\bar{z}) - \mathrm{j}q_0\bar{y}]. \tag{9.31}$$

From Eq. (9.26b) for $\mathrm{j}(\sqrt{Y_{\mathrm{vac}}} \varepsilon \boldsymbol{E})_y$

$$\frac{\partial}{\partial \bar{z}}(\sqrt{Z_{\mathrm{vac}}} H_x) - \frac{\partial}{\partial \bar{x}}(\sqrt{Z_{\mathrm{vac}}} H_z)$$
$$= \sum_m \exp[-\mathrm{j}m(p\bar{x} + q\,\bar{y}) - \mathrm{j}q_0\bar{y}] \frac{\partial}{\partial \bar{z}} [h_{xm}(\bar{z}) \exp(-\mathrm{j}ms\bar{z})]$$
$$- \sum_m (-\mathrm{j}mp) h_{zm}(\bar{z}) \exp[-\mathrm{j}m(p\bar{x} + q\,\bar{y} + s\bar{z}) - \mathrm{j}q_0\bar{y}]$$
$$= \sum_m \left\{ \left[\frac{\mathrm{d}}{\mathrm{d}\bar{z}} h_{xm}(\bar{z})\right] - \mathrm{j}ms\, h_{xm}(\bar{z}) \right\} \exp[-\mathrm{j}m(p\bar{x} + q\,\bar{y} + s\bar{z}) - \mathrm{j}q_0\bar{y}]$$
$$- \sum_m (-\mathrm{j}mp) h_{zm}(\bar{z}) \exp[-\mathrm{j}m(p\bar{x} + q\,\bar{y} + s\bar{z}) - \mathrm{j}q_0\bar{y}]. \tag{9.32}$$

From Eq. (9.26c) for $\mathrm{j}(\sqrt{Y_{\text{vac}}}\varepsilon\,\boldsymbol{E})_z$

$$\frac{\partial}{\partial\bar{x}}(\sqrt{Z_{\text{vac}}}H_y) - \frac{\partial}{\partial\bar{y}}(\sqrt{Z_{\text{vac}}}H_x)$$
$$= \sum_m (-\mathrm{j}mp)h_{ym}(\bar{z})\exp[-\mathrm{j}m(p\bar{x}+q\bar{y}+s\bar{z}) - \mathrm{j}q_0\bar{y}]$$
$$- \sum_m [-\mathrm{j}(mq+q_0)]h_{xm}(\bar{z})\exp[-\mathrm{j}m(p\bar{x}+q\bar{y}+s\bar{z}) - \mathrm{j}q_0\bar{y}]. \quad (9.33)$$

We start the substitution into the right-hand side of Eq. (9.25) and Eq. (9.26) with the simpler equations (9.25) expressing the Faraday law (9.12),

$$\bar{\nabla}\times\sqrt{Y_{\text{vac}}}\boldsymbol{E} = -\mathrm{j}\sqrt{Z_{\text{vac}}}\boldsymbol{H}.$$

Eq. (9.25c) for $-\mathrm{j}(\sqrt{Z_{\text{vac}}}H_z)$ with Eq. (9.30) becomes

$$\frac{\partial}{\partial\bar{x}}(\sqrt{Y_{\text{vac}}}E_y) - \frac{\partial}{\partial\bar{y}}(\sqrt{Y_{\text{vac}}}E_x) = -\mathrm{j}(\sqrt{Z_{\text{vac}}}H_z),$$
$$\times\exp(-\mathrm{j}q_0\bar{y})\sum_m[(mq+q_0)e_{xm}(\bar{z}) - mp\,e_{ym}(\bar{z})]\exp[-\mathrm{j}m(p\,\bar{z}+q\bar{x}+s\bar{y})]$$
$$= -\exp(-\mathrm{j}q_0\bar{y})\sum_m h_{zm}(\bar{z})\exp[-\mathrm{j}m(p\bar{z}+q\bar{x}+s\bar{y})]. \quad (9.34)$$

We cancel $\exp(-\mathrm{j}q_0\bar{y})$ and multiply both sides of the equation with $\exp[\mathrm{j}(\boldsymbol{n}_{m'}\cdot\bar{\boldsymbol{r}})]$, exchange the indices m and m', and integrate over the normalized periods $(\Lambda_x/\lambda_0) = 2\pi/p$ and $(\Lambda_y/\lambda_0) = 2\pi/q$

$$\sum_{m'}[(m'q+q_0)e_{xm'}(\bar{z}) - m'p\,e_{ym'}(\bar{z}) + h_{zm'}(\bar{z})]$$
$$\times\int_{\bar{x}_0}^{\bar{x}_0+2\pi/p}\mathrm{d}\bar{x}\int_{\bar{y}_0}^{\bar{y}_0+2\pi/q}\mathrm{d}\bar{y}\exp[-\mathrm{j}(m-m')p - \mathrm{j}(m-m')q\bar{y} - \mathrm{j}(m-m')s\bar{z}] = 0.$$
$$(9.35)$$

From this, we obtain

$$\sum_{m'}\delta_{mm'}[(m'q+q_0)e_{xm'}(\bar{z}) - m'p\,e_{ym'}(\bar{z}) + h_{zm'}(\bar{z})] = 0, \quad (9.36)$$

where $\delta_{mm'}$ is the Kronecker delta, $\delta_{mm'} = 1$, if $m = m'$ and $\delta_{mm'} = 0$ otherwise. The equation holds, provided

$$mp\,e_{ym}(\bar{z}) - (mq+q_0)e_{xm}(\bar{z}) = h_{zm}(\bar{z}). \quad (9.37)$$

A similar procedure applied to Eq. (9.25a) for $-\mathrm{j}(\sqrt{Z_{\mathrm{vac}}}H_x)$ using Eq. (9.28) gives

$$\frac{\partial}{\partial \bar{y}}(\sqrt{Y_{\mathrm{vac}}}E_z) - \frac{\partial}{\partial \bar{z}}(\sqrt{Y_{\mathrm{vac}}}E_y) = -\mathrm{j}(\sqrt{Z_{\mathrm{vac}}}H_x),$$

$$\sum_m (mq + q_0)e_{zm}(\bar{z}) \exp[-\mathrm{j}m(p\bar{x} + q\bar{y} + s\bar{z}) - \mathrm{j}q_0\bar{y}]$$
$$+\sum_m \left\{\left[-\mathrm{j}\frac{\mathrm{d}}{\mathrm{d}\bar{z}}\, e_{ym}(\bar{z})\right] - mse_{ym}(\bar{z})\right\} \exp[-\mathrm{j}m(p\bar{x} + q\bar{y} + s\bar{z}) - \mathrm{j}q_0\bar{y}]$$
$$= \sum_m h_{xm}(\bar{z}) \exp[-\mathrm{j}m(p\bar{x} + q\bar{y} + s\bar{z}) - \mathrm{j}q_0\bar{y}] \tag{9.38}$$

and after rearranging, we have

$$\frac{\mathrm{d}}{\mathrm{d}\bar{z}}\, e_{ym}(\bar{z}) - \mathrm{j}mse_{zm}(\bar{z}) + \mathrm{j}(mq + q_0)e_{zxm}(\bar{z}) = \mathrm{j}h_{xm}(\bar{z}). \tag{9.39}$$

From Eq. (9.25b) for $-\mathrm{j}(\sqrt{Z_{\mathrm{vac}}}H_y)$ using Eq. (9.29)

$$\frac{\partial}{\partial \bar{z}}(\sqrt{Y_{\mathrm{vac}}}E_x) - \frac{\partial}{\partial \bar{x}}(\sqrt{Y_{\mathrm{vac}}}E_z) = -\mathrm{j}(\sqrt{Z_{\mathrm{vac}}}H_y),$$

we obtain in a similar way

$$\frac{\mathrm{d}}{\mathrm{d}\bar{z}}\, e_{xm}(\bar{z}) - \mathrm{j}ms\, e_{xm}(\bar{z}) + \mathrm{j}mp\, e_{zm}(\bar{z}) = -\mathrm{j}h_{ym}(\bar{z}). \tag{9.40}$$

The substitution into the second Maxwell equation (9.13),

$$\overline{\nabla} \times \sqrt{Z_{\mathrm{vac}}}\boldsymbol{H} = \mathrm{j}\sqrt{Y_{\mathrm{vac}}}\, \boldsymbol{\varepsilon}\, \boldsymbol{E},$$

is more involved in view of the tensor nature of the permittivity. It is, therefore, useful to consider a simple case first before considering the general procedure.

We consider the case where the Fourier expansion of permittivity in Eq. (9.14)

$$\varepsilon_{ij}(\bar{\boldsymbol{r}}) = \sum_l \varepsilon_{ij,l} \exp(\mathrm{j}\, l\boldsymbol{n}_\Gamma \cdot \bar{\boldsymbol{r}}) = \sum_l \varepsilon_{ij,l} \exp[\mathrm{j}l(p\bar{x} + q\bar{y} + s\bar{z})]$$

includes only the terms $l = 0, \pm 1$, *i.e.*,

$$\varepsilon_{ij}(\boldsymbol{r}) = \varepsilon_{ij,0} + \varepsilon_{ij,1} \exp(\mathrm{j}\,\boldsymbol{n}_\Gamma \cdot \bar{\boldsymbol{r}}) + \varepsilon_{ij,0} + \varepsilon_{ij,-1} \exp(-\mathrm{j}\,\boldsymbol{n}_\Gamma \cdot \bar{\boldsymbol{r}}). \tag{9.41}$$

This truncated expansion will be substituted into Eq. (9.13). The $\bar{z}$ component of Eq. (9.13) is expressed in Eq. (9.26a)

$$-\mathrm{j}\left[\frac{\partial}{\partial \bar{x}}(\sqrt{Z_{\mathrm{vac}}}H_y) - \frac{\partial}{\partial \bar{y}}(\sqrt{Z_{\mathrm{vac}}}H_x)\right]$$
$$= \varepsilon_{zx}(\sqrt{Y_{\mathrm{vac}}}E_x) + \varepsilon_{zy}(\sqrt{Y_{\mathrm{vac}}}E_y)\varepsilon_{zz}(\sqrt{Y_{\mathrm{vac}}}E_z).$$

The substitution for the components of the permittivity tensor $\varepsilon_{ij}(\boldsymbol{r})$ according to Eq. (9.41) gives

$$\begin{aligned}
&\sum_m [(mq + q_0)h_{xm}(\bar{z}) - mp\, h_{ym}(\bar{z})] \exp[-\mathrm{j}(\boldsymbol{n}_m \cdot \bar{\boldsymbol{r}})] \\
&\quad = \varepsilon_{zx,0} \sum_m e_{xm}(\bar{z}) \exp[-\mathrm{j}(\boldsymbol{n}_m \cdot \bar{\boldsymbol{r}})] + \varepsilon_{zy,0} \sum_m e_{ym}(\bar{z}) \exp[-\mathrm{j}(\boldsymbol{n}_m \cdot \bar{\boldsymbol{r}})] \\
&\qquad + \varepsilon_{zz,0} \sum_m e_{zm}(\bar{z}) \exp[-\mathrm{j}(\boldsymbol{n}_m \cdot \bar{\boldsymbol{r}})] \\
&\qquad + \varepsilon_{zx,1} \exp[\mathrm{j}(p\bar{x} + q\bar{y} + s\bar{z})] \sum_m e_{xm}(\bar{z}) \exp[-\mathrm{j}(\boldsymbol{n}_m \cdot \bar{\boldsymbol{r}})] \\
&\qquad + \varepsilon_{zy,1} \exp[\mathrm{j}(p\bar{x} + q\bar{y} + s\bar{z})] \sum_m e_{ym}(\bar{z}) \exp[-\mathrm{j}(\boldsymbol{n}_m \cdot \bar{\boldsymbol{r}})] \\
&\qquad + \varepsilon_{zz,1} \exp[\mathrm{j}(p\bar{x} + q\bar{y} + s\bar{z})] \sum_m e_{zm}(\bar{z}) \exp[-\mathrm{j}(\boldsymbol{n}_m \cdot \bar{\boldsymbol{r}})] \\
&\qquad + \varepsilon_{zx,-1} \exp[-\mathrm{j}(p\bar{x} + q\bar{y} + s\bar{z})] \sum_m e_{xm}(\bar{z}) \exp[-\mathrm{j}(\boldsymbol{n}_m \cdot \bar{\boldsymbol{r}})] \\
&\qquad + \varepsilon_{zy,-1} \exp[-\mathrm{j}(p\bar{x} + q\bar{y} + s\bar{z})] \sum_m e_{ym}(\bar{z}) \exp[-\mathrm{j}(\boldsymbol{n}_m \cdot \bar{\boldsymbol{r}})] \\
&\qquad + \varepsilon_{zz,-1} \exp[-\mathrm{j}(p\bar{x} + q\bar{y} + s\bar{z})] \sum_m e_{zm}(\bar{z}) \exp[-\mathrm{j}(\boldsymbol{n}_m \cdot \bar{\boldsymbol{r}})], \qquad (9.42)
\end{aligned}$$

where $\exp[-\mathrm{j}(\boldsymbol{n}_m \cdot \bar{\boldsymbol{r}})]$ is given by Eq. (9.27). We multiply both sides of this equation with $\exp\left[\mathrm{j}\,(\boldsymbol{n}_{m'} \cdot \bar{\boldsymbol{r}})\right] = \exp\left[\mathrm{j}m'(p\bar{x} + q\bar{y} + s\bar{z})\right]$ and obtain

$$\begin{aligned}
&\sum_m [(mq + q_0)h_{xm}(\bar{z}) - mp\, h_{ym}(\bar{z})] \exp[-\mathrm{j}(m - m')(p\bar{x} + q\bar{y} + s\bar{z})] \\
&\quad = \sum_m [\varepsilon_{zx,0} e_{xm}(\bar{z}) + \varepsilon_{zy,0} e_{ym}(\bar{z}) + \varepsilon_{zz,0} e_{zm}(\bar{z})] \\
&\qquad \times \exp[-\mathrm{j}m(m - m')(p\bar{x} + q\bar{y} + s\bar{z})] \\
&\qquad + \sum_m [\varepsilon_{zx,1} e_{xm}(\bar{z}) + \varepsilon_{zy,1} e_{ym}(\bar{z}) + \varepsilon_{zz,1} e_{zm}(\bar{z})] \\
&\qquad \times \exp[-\mathrm{j}(m - m' - 1)(p\bar{x} + q\bar{y} + s\bar{z})] \\
&\qquad + \sum_m [\varepsilon_{zx,-1} e_{xm}(\bar{z}) + \varepsilon_{zy,-1} e_{ym}(\bar{z}) + \varepsilon_{zz,-1} e_{zm}(\bar{z})] \\
&\qquad \times \exp[-\mathrm{j}(m - m' + 1)(p\bar{x} + q\bar{y} + s\bar{z})]. \qquad (9.43)
\end{aligned}$$

The integration from $\bar{x}_0$ to $\bar{x}_0 + 2\pi/p$ and from $\bar{y}_0$ to $\bar{y}_0 + 2\pi/q$ after an exchange of m and m' gives

$$\begin{aligned}
&(\varepsilon_{zx,0} e_{xm} + \varepsilon_{zy,0} e_{ym} + \varepsilon_{zz,0} e_{zm}) + (\varepsilon_{zx,1} e_{x,m+1} + \varepsilon_{zy,1} e_{y,m+1} + \varepsilon_{zz,1} e_{z,m+1}) \\
&\quad + (\varepsilon_{zx,-1} e_{x,m-1} + \varepsilon_{zy,-1} e_{y,m-1} + \varepsilon_{zz,-1} e_{z,m-1}) = (mq + q_0)h_{xm} - mp\, h_{ym}. \qquad (9.44)
\end{aligned}$$

We obtain for the case of $m = 1$

$$(\varepsilon_{zx,0}e_{x1} + \varepsilon_{zy,0}e_{y1} + \varepsilon_{zz,0}e_{z1}) + (\varepsilon_{zx,1}e_{x,2} + \varepsilon_{zy,1}e_{y,2} + \varepsilon_{zz,1}e_{z,2}) + (\varepsilon_{zx,-1}e_{x,0} + \varepsilon_{zy,-1}e_{y,0} + \varepsilon_{zz,-1}e_{z,0}) = (q + q_0)h_{x,1} - ph_{y,1}. \quad (9.45)$$

For the case of $m = 0$

$$(\varepsilon_{zx,0}e_{x0} + \varepsilon_{zy,0}e_{y0} + \varepsilon_{zz,0}e_{z0}) + (\varepsilon_{zx,1}e_{x,1} + \varepsilon_{zy,1}e_{y,1} + \varepsilon_{zz,1}e_{z,1}) + (\varepsilon_{zx,-1}e_{x,-1} + \varepsilon_{zy,-1}e_{y,-1} + \varepsilon_{zz,-1}e_{z,-1}) = q_0 h_{y0}. \quad (9.46)$$

For the case of $m = -1$

$$(\varepsilon_{zx,0}e_{x,-1} + \varepsilon_{zy,0}e_{y,-1} + \varepsilon_{zz,0}e_{z,-1}) + (\varepsilon_{zx,1}e_{z,0} + \varepsilon_{zy,1}e_{y,0} + \varepsilon_{zz,1}e_{z,0}) + (\varepsilon_{zx,-1}e_{x,-2} + \varepsilon_{zy,-1}e_{y,m-1} + \varepsilon_{zz,-1}e_{z,m-1}) = (-q + q_0)h_{x,-1} + ph_{y,-1}. \quad (9.47)$$

In the description of the permittivity tensor, we have confined ourselves to the first three terms in the development, *i.e.*, to the terms $\varepsilon_{ij,0}$ and $\varepsilon_{ij,\pm 1}$. Correspondingly, we also restrict the development of the wave electric fields to the first three terms, *i.e.*, $e_{j,0}$ and $e_{j,\pm 1}$ and, in the same way, the development of the wave magnetic field to the first three terms, *i.e.*, $h_{j,0}$ and $h_{j,\pm 1}$. The accuracy of the permittivity development can be improved by including higher order terms. This would require a corresponding extension in the number of the terms involved in the developments for the fields.

9.3 Product of Series

Before the treatment of the general case, we first recall the rules for the product of two absolutely convergent infinite series. The product of the series

$$\sum_{n=1}^{\infty} a_n = a_1 + a_2 + a_3 + \cdots + a_n + \cdots \quad (9.48)$$

$$\sum_{n=1}^{\infty} b_n = b_1 + b_2 + b_3 + \cdots + b_n + \cdots \quad (9.49)$$

is obtained by multiplication of each term from the first series with each term of the second series

$$\left(\sum_{n=1}^{\infty} a_n\right)\left(\sum_{n=1}^{\infty} b_n\right) =$$

$$\begin{array}{ccccccccc}
a_1b_1+ & a_1b_2+ & a_1b_3+ & a_1b_4+ & \cdots & \cdots & \cdots + & a_1b_n+ & \cdots \\
a_2b_1+ & a_2b_2+ & a_2b_3+ & a_2b_4+ & \cdots & \cdots + & a_2b_{n-1}+ & a_2b_n+ & \cdots \\
a_3b_1+ & a_3b_2+ & a_3b_3+ & a_3b_4+ & \cdots + & a_3b_{n-2}+ & a_3b_{n-1}+ & a_3b_n+ & \cdots \\
a_4b_1+ & a_4b_2+ & a_4b_3+ & a_4b_4+ & \cdots & \cdots & \cdots & \cdots & \cdots \\
\cdots & \cdots & \cdots & \cdots & \cdots & \cdots & \cdots & \cdots & \cdots \\
a_{n-2}b_1+ & a_{n-2}b_2+ & a_{n-2}b_3+ & a_{n-2}b_4+ & \cdots & \cdots & \cdots + & a_{n-2}b_n+ & \cdots \\
a_{n-1}b_1+ & a_{n-1}b_2+ & a_{n-1}b_3+ & a_{n-1}b_4+ & \cdots & \cdots & \cdots + & a_{n-1}b_n+ & \cdots \\
a_nb_1+ & a_nb_2+ & a_nb_3+ & a_nb_4+ & \cdots & \cdots & \cdots + & a_nb_n+ & \cdots \\
\cdots & \cdots & \cdots & \cdots & \cdots & \cdots & \cdots & \cdots & \cdots
\end{array}$$

This pattern represents an infinite number of infinite series. We can build a single series by grouping the terms with the same sum of indices. In the above pattern, these terms are positioned on diagonals

$$\begin{aligned}
&a_1b_1 + (a_1b_2 + a_2b_1) + (a_1b_3 + a_2b_2 + a_3b_1) + (a_1b_4 + a_2b_3 + a_3b_2 + a_4b_1) \\
&\quad + \cdots + (a_1b_n + a_2b_{n-1} + a_3b_{n-2} + a_4b_{n-3} + a_5b_{n-4} + \cdots \\
&\quad + a_{n-2}b_3 + a_{n-1}b_2 + a_nb_1) + \cdots = \sum_{n=1}^{\infty}\left(\sum_{\lambda_0=1}^{n} a_{\lambda_0} b_{n+1-\lambda_0}\right).
\end{aligned} \tag{9.50}$$

The terms in the series were numbered with positive integers. More details on the expression for the product of two absolutely convergent series can be found in standard mathematics textbooks.

For the present purpose, this result is more conveniently expressed using the term numbering with zero and positive and negative integers

$$\sum_{l} \varepsilon_{ij,l} \exp[\mathrm{j}l(p\bar{x} + q\bar{y} + s\bar{z})] \sum_{m} e_{jm} \exp[-\mathrm{j}m(p\bar{x} + q\bar{y} + s\bar{z}) - \mathrm{j}q_0\bar{y}]. \tag{9.51}$$

We denote

$$a_l = \varepsilon_{ij,l} \exp[\mathrm{j}l(p\bar{x} + q\bar{y} + s\bar{z})] = \varepsilon_{ij,l} \exp(\mathrm{j}l\boldsymbol{n}_\Gamma \cdot \bar{\boldsymbol{r}}), \tag{9.52a}$$

$$b_m = e_{jm} \exp[-\mathrm{j}m(p\bar{x} + q\bar{y} + s\bar{z}) - \mathrm{j}q_0\bar{y}] = e_{jm} \exp(-\mathrm{j}m\boldsymbol{n}_\Gamma \cdot \bar{\boldsymbol{r}}). \tag{9.52b}$$

The product of the series $\sum_l a_l \sum_m b_m$, where $l, m = 0, \pm 1, \pm 2, \pm 3, \ldots$, can be expressed in a square matrix form, which for a few first terms becomes

$$\begin{pmatrix} a_3 b_{-3} & a_3 b_{-2} & a_3 b_{-1} & a_3 b_0 & a_3 b_1 & a_3 b_2 & a_3 b_3 \\ a_2 b_{-3} & a_2 b_{-2} & a_2 b_{-1} & a_2 b_0 & a_2 b_1 & a_2 b_2 & a_2 b_3 \\ a_1 b_{-3} & a_1 b_{-2} & a_1 b_{-1} & a_1 b_0 & a_1 b_1 & a_1 b_2 & a_1 b_3 \\ a_0 b_{-3} & a_0 b_{-2} & a_0 b_{-1} & a_0 b_0 & a_0 b_1 & a_0 b_2 & a_0 b_3 \\ a_{-1} b_{-3} & a_{-1} b_{-2} & a_{-1} b_{-1} & a_{-1} b_0 & a_{-1} b_1 & a_{-1} b_2 & a_{-1} b_3 \\ a_{-2} b_{-3} & a_{-2} b_{-2} & a_{-2} b_{-1} & a_{-2} b_0 & a_{-2} b_1 & a_{-2} b_2 & a_{-2} b_3 \\ a_{-3} b_{-3} & a_{-3} b_{-2} & a_{-3} b_{-1} & a_{-3} b_0 & a_{-3} b_1 & a_{-3} b_2 & a_{-3} b_3 \end{pmatrix}. \tag{9.53}$$

A generic matrix element represents the product given by Eq. (9.52)

$$a_l b_m = \varepsilon_{ij,l} e_{jm} \exp[\mathrm{j}\,(l - m)\,(p\bar{x} + q\bar{y} + s\bar{z})]. \tag{9.54}$$

In the numerical simulation, the product of two infinite series is replaced by a finite sum of products. The dimension of the (truncated) matrix representing this finite sum is characterized by a positive integer M. We write for a general block forming the matrix of the dimension $2M + 1$

$$\sum_{q=-M}^{M} \sum_{p=-q}^{q} a_{q-p} b_p, \tag{9.55}$$

and skip the terms that do not fulfill the condition

$$-q \leq q - p \leq q. \tag{9.56}$$

For the case of $M = 0$

$$(a_0 b_0). \tag{9.57}$$

For the case $M = 1$

$$\begin{pmatrix} 0 & a_1 b_0 & a_1 b_1 \\ a_0 b_{-1} & a_0 b_0 & a_0 b_1 \\ a_{-1} b_{-1} & a_{-1} b_0 & 0 \end{pmatrix}. \tag{9.58}$$

For the case of $M = 2$

$$\begin{pmatrix} 0 & 0 & a_2 b_0 & a_2 b_1 & a_2 b_2 \\ 0 & a_1 b_{-1} & a_1 b_0 & a_1 b_1 & a_1 b_2 \\ a_0 b_{-2} & a_0 b_{-1} & a_0 b_0 & a_0 b_1 & a_0 b_2 \\ a_{-1} b_{-2} & a_{-1} b_{-1} & a_{-1} b_0 & a_{-1} b_1 & 0 \\ a_{-2} b_{-2} & a_{-2} b_{-1} & a_{-2} b_0 & 0 & 0 \end{pmatrix}. \tag{9.59}$$

9.4 Matrix Representation

We now return to the general case of the development of the permittivity tensor components (9.14) and substitute into Eq. (9.13). We start with z component given by Eq. (9.26c)

$$\frac{\partial}{\partial \bar{x}}(\sqrt{Z_{\text{vac}}}H_y) - \frac{\partial}{\partial \bar{y}}(\sqrt{Z_{\text{vac}}}H_x) = \text{j}[\varepsilon_{zx}(\sqrt{Y_{\text{vac}}}E_x) + \varepsilon_{zy}(\sqrt{Y_{\text{vac}}}E_y)\varepsilon_{zz}(\sqrt{Y_{\text{vac}}}E_z)]$$

and obtain with Eq. (9.33)

$$\begin{aligned}
&\sum_m [(mq + q_0)h_{xm}(\bar{z}) - mph_{ym}(\bar{z})]\exp[-\text{j}m(p\bar{x} + q\bar{y} + s\bar{z}) - \text{j}q_0\bar{y}] \\
&= \sum_l \varepsilon_{zx,l}\exp[\text{j}l(p\bar{x} + q\bar{y} + s\bar{z})]\sum_m e_{xm}(\bar{z})\exp[-\text{j}m(p\bar{x} + q\bar{y} + s\bar{z}) - \text{j}q_0\bar{y}] \\
&+ \sum_l \varepsilon_{zy,l}\exp[\text{j}l(p\bar{x} + q\bar{y} + s\bar{z})]\sum_m e_{ym}(\bar{z})\exp[-\text{j}m(p\bar{x} + q\bar{y} + s\bar{z}) - \text{j}q_0\bar{y}] \\
&+ \sum_l \varepsilon_{zz,l}\exp[\text{j}l(p\bar{x} + q\bar{y} + s\bar{z})]\sum_m e_{zm}(\bar{z})\exp[-\text{j}m(p\bar{x} + q\bar{y} + s\bar{z}) - \text{j}q_0\bar{y}],
\end{aligned} \tag{9.60}$$

or more concisely,

$$\sum_m [(mq + q_0)h_{xm}(\bar{z}) - mph_{ym}(\bar{z})]\exp[-\text{j}m(p\bar{x} + q\bar{y} + s\bar{z})] = \sum_{i,l,m} \varepsilon_{zi,l}e_{im}(\bar{z})\exp[\text{j}\,(l - m)\,(p\bar{x} + q\bar{y} + s\bar{z})], \tag{9.61}$$

where $i = x, y, z$. We multiply both sides of the equation with $\exp[\text{j}(\boldsymbol{n}_{m'} \cdot \bar{\boldsymbol{r}})] = \exp[\text{j}m'(p\bar{x} + q\bar{y} + s\bar{z})]$

$$\sum_m [(mq + q_0)h_{xm} - mp\,h_{ym}]\exp[-\text{j}(m - m')(p\bar{x} + q\bar{y} + s\bar{z})] = \sum_{i,l,m} \varepsilon_{zi,l}e_{im}\exp[\text{j}(l - m + m')(p\bar{x} + q\bar{y} + s\bar{z})], \tag{9.62}$$

and integrate over the periods $\langle \bar{x}_0, \bar{x}_0 + 2\pi/p \rangle$ and $\langle \bar{y}_0, \bar{y}_0 + 2\pi/q \rangle$. We obtain

$$\sum_i \sum_{l,m} \varepsilon_{zi,l}e_{im}\delta_{l,m-m'} = \sum_m [(mq + q_0)h_{xm} - mp\,h_{ym}]\delta_{m,m'}, \tag{9.63}$$

after performing the sums

$$\sum_i \sum_m \varepsilon_{zi,m-m'} e_{im} = (m'q + q_0) h_{xm'} - m' p\, h_{ym'}. \tag{9.64}$$

We exchange the summing indices m for l and m' for m

$$(mq + q_0) h_{xm} - mp\, h_{ym} = \sum_l \sum_i \varepsilon_{zi,l-m} e_{il}. \tag{9.65}$$

In a similar way, we find for the x component from Eq. (9.26a) with Eq. (9.31)

$$\frac{dh_{ym}}{d\bar{z}} - jms\, h_{ym} + j(mq + q_0) h_{zm} = -j \sum_l \sum_i \varepsilon_{xi,l-m} e_{il} \tag{9.66}$$

and for the y component from Eq. (9.26b) with Eq. (9.32)

$$\frac{dh_{xm}}{d\bar{z}} - jms\, h_{xm} + jmp\, h_{zm} = j \sum_l \sum_i \varepsilon_{yi,l-m} e_{il}. \tag{9.67}$$

Eq. (9.37), Eq. (9.39), Eq. (9.40), and Eqs. (9.65) to (9.67) were obtained from the Maxwell equations, and we now summarize them (in a rearranged order of terms) employing the notation of Eq. (9.23), *i.e.*, $mp = p_m$, $(mq + q_0) = q_m$, $ms = s_m$

$$\frac{de_{ym}}{d\bar{z}} = js_m e_{ym} - jq_m e_{zm} + jh_{xm}, \tag{9.68a}$$

$$\frac{de_{xm}}{d\bar{z}} = js_m e_{xm} - jp_m e_{zm} - jh_{ym}, \tag{9.68b}$$

$$0 = p_m e_{ym} - q_m e_{xm} - h_{zm}, \tag{9.68c}$$

$$\frac{dh_{ym}}{d\bar{z}} = js_m\, h_{ym} - jq_m h_{zm} - j\sum_l \varepsilon_{xx,l-m} e_{xl} - j\sum_l \varepsilon_{xy,l-m} e_{yl} - j\sum_l \varepsilon_{xz,l-m} e_{zl}, \tag{9.68d}$$

$$\frac{dh_{xm}}{d\bar{z}} = js_m h_{xm} - jp_m h_{zm} + j\sum_l \varepsilon_{yx,l-m} e_{xl} + j\sum_l \varepsilon_{yy,l-m} e_{yl} + j\sum_l \varepsilon_{yz,l-m} e_{zl}, \tag{9.68e}$$

$$0 = \sum_l \varepsilon_{zx,l-m} e_{xl} + \sum_l \varepsilon_{zy,l-m} e_{yl} + \sum_l \varepsilon_{zz,l-m} e_{zl} + p_m h_{ym} - q_m h_{xm}. \tag{9.68f}$$

These results should now be transformed into a matrix form. To this purpose, we express the last equation, *i.e.*, Eq. (9.68f), obtained from Eq. (9.65), in the form

$$(mq + q_0) h_{xm} - mp\, h_{ym} = \sum_l \varepsilon_{zx,l-m} e_{xl} + \sum_l \varepsilon_{zy,l-m} e_{yl} + \sum_l \varepsilon_{zz,l-m} e_{zl}. \tag{9.69}$$

We denote

$$\varepsilon_{ij,l-n} = \varepsilon_{ij,nl} \tag{9.70}$$

and restrict the number of first terms in the development of the permittivity tensor components (9.14) in practical computations by choosing the range for the integer l as

$$l \in \langle -M, M \rangle, \qquad i.e. \qquad l = -M, -M+1, \ldots, 0, \ldots, M-1, M. \tag{9.71}$$

The symbol $\varepsilon_{ij,nl}$ denotes a general element of a matrix of dimension $(2M+1) \times (2M+1)$. The restriction to a finite number of terms in the development of permittivity tensor elements in Eq. (9.69) can be explicitly written as

$$(ms + q_0)\, h_{ym} - mq\, h_{zm} = \sum_{l=-M}^{M} \varepsilon_{xx,l-m} e_{xl} + \sum_{l=-M}^{M} \varepsilon_{xy,l-m} e_{yl} + \sum_{l=-M}^{M} \varepsilon_{xz,l-m} e_{zl}. \tag{9.72}$$

In order to illustrate the procedure of transforming the results into a matrix form, let us start with the case $M = 0$, which corresponds to the tensor permittivity homogeneous in space

$$q_0 h_{x0} - \varepsilon_{zx;0,0} e_{x0} - \varepsilon_{zy;0,0} e_{y0} = \varepsilon_{zz;0,0} e_{z0}. \tag{9.73}$$

A single equation is sufficient for the determination of the field e_{x0}. We go on with the case of $M = 1$, which corresponds to the harmonic space dependence of the permittivity tensor components. Accordingly, in the developments for $M = 1$ of the fields given by Eq. (9.18) and Eq. (9.19), we are restricted to the terms $m = 0, \pm 1$. We obtain
for $m = 1$

$$\begin{aligned}
&(q + q_0)h_{x1} - ph_{y1} - (\varepsilon_{zy;1,1} e_{yl} + \varepsilon_{zy;1,0} e_{y0} + \varepsilon_{zy;1,-1} e_{y,-l}) \\
&\quad - (\varepsilon_{zz;1,1} e_{zl} + \varepsilon_{zz;1,0} e_{z0} + \varepsilon_{zz;1,-1} e_{z,-l}) \\
&\quad = \varepsilon_{zx;1,1} e_{x1} + \varepsilon_{zx;1,0} e_{x0} + \varepsilon_{zx;1,-1} e_{x,-1},
\end{aligned} \tag{9.74}$$

for $m = 0$

$$\begin{aligned}
&q_0 h_{x0} - (\varepsilon_{zy;0,1} e_{yl} + \varepsilon_{zy;0,0} e_{y0} + \varepsilon_{zy;0,-1} e_{y,-l}) \\
&\quad - (\varepsilon_{zz;0,1} e_{zl} + \varepsilon_{zz;0,0} e_{z0} + \varepsilon_{zz;0,-1} e_{z,-l}) \\
&\quad = \varepsilon_{zx;0,1} e_{x1} + \varepsilon_{zx;0,0} e_{x0} + \varepsilon_{zx;0,-1} e_{x,-1},
\end{aligned} \tag{9.75}$$

and for $m = -1$

$$\begin{aligned}
&(-q + q_0)h_{x,-1} + ph_{z,-1} - (\varepsilon_{zy;-1,1} e_{yl} + \varepsilon_{zy;-1,0} e_{y0} + \varepsilon_{zy;-1,-1} e_{y,-l}) \\
&\quad - (\varepsilon_{zz;-1,1} e_{zl} + \varepsilon_{zz;-1,0} e_{z0} + \varepsilon_{zz;-1,-1} e_{z,-l}) \\
&\quad = \varepsilon_{zx;-1,1} e_{x1} + \varepsilon_{zx;-1,0} e_{x0} + \varepsilon_{zx;-1,-1} e_{x,-1}.
\end{aligned} \tag{9.76}$$

The equations (9.74) to (9.76) for $M = 1$ can be expressed in a matrix form as follows

$$\begin{pmatrix} q_0+q & 0 & 0 \\ 0 & q_0 & 0 \\ 0 & 0 & q_0-q \end{pmatrix} \begin{pmatrix} h_{x,1} \\ h_{x,0} \\ h_{x,-1} \end{pmatrix} - \begin{pmatrix} p & 0 & 0 \\ 0 & 0 & 0 \\ 0 & 0 & -p \end{pmatrix} \begin{pmatrix} h_{y,1} \\ h_{y,0} \\ h_{y,-1} \end{pmatrix}$$
$$- \begin{pmatrix} \varepsilon_{zx;\,0} & \varepsilon_{zx;-1} & \varepsilon_{zx;-2} \\ \varepsilon_{zx;1} & \varepsilon_{zx;\,0} & \varepsilon_{zx;-1} \\ \varepsilon_{zx;\,2} & \varepsilon_{zx;1} & \varepsilon_{zx;\,0} \end{pmatrix} \begin{pmatrix} e_{x,1} \\ e_{x,0} \\ e_{x,-1} \end{pmatrix} - \begin{pmatrix} \varepsilon_{zy;\,0} & \varepsilon_{zy;-1} & \varepsilon_{zy;-2} \\ \varepsilon_{zy;1} & \varepsilon_{zy;\,0} & \varepsilon_{zy;-1} \\ \varepsilon_{zy;\,2} & \varepsilon_{zy;1} & \varepsilon_{zy;\,0} \end{pmatrix} \begin{pmatrix} e_{y,1} \\ e_{y,0} \\ e_{y,-1} \end{pmatrix}$$
$$= \begin{pmatrix} \varepsilon_{zz;\,0} & \varepsilon_{zz;-1} & \varepsilon_{zz;-2} \\ \varepsilon_{zz;1} & \varepsilon_{zz;\,0} & \varepsilon_{zz;-1} \\ \varepsilon_{zz;\,2} & \varepsilon_{zz;1} & \varepsilon_{zz;\,0} \end{pmatrix} \begin{pmatrix} e_{z,1} \\ e_{z,0} \\ e_{z,-1} \end{pmatrix}. \tag{9.77}$$

This equation generalized to an arbitrary integer M can be concisely expressed as

$$(q)(h_x) - (p)(h_y) - (\varepsilon_{zx})(e_x) - (\varepsilon_{zy})(e_y) = (\varepsilon_{zz})(e_z). \tag{9.78}$$

The symbols with parentheses represent matrices and column vectors in the space of dimension $(2M+1)$. This is the desired matrix representation of Eq. (9.68f). In a similar way, we find

$$(h_z) = (p)(e_y) - (q)(e_x). \tag{9.79}$$

We separate (e_z) and (h_z) from Eq. (9.78) and Eq. (9.79)

$$\begin{aligned} (e_z) &= -(\varepsilon_{zz})^{-1}(\varepsilon_{zx})(e_x) - (\varepsilon_{zz})^{-1}(p)(h_y) - (\varepsilon_{zz})^{-1}(\varepsilon_{zy})(e_y) + (\varepsilon_{zz})^{-1}(q)(h_x) \\ (h_z) &= \quad -(q)(e_x) \quad + (p)(e_y). \end{aligned}$$

This equation system (9.80) forms a generalized matrix equation, in which the matrix elements themselves represent matrices or column vectors

$$\begin{pmatrix} \mathbf{e}_z \\ \mathbf{h}_z \end{pmatrix} = \begin{pmatrix} -\varepsilon_{zz}^{-1}\varepsilon_{zx} & -\varepsilon_{zz}^{-1}\mathbf{p} & -\varepsilon_{zz}^{-1}\varepsilon_{zy} & \varepsilon_{zz}^{-1}\mathbf{q} \\ -\mathbf{q} & \mathbf{0} & \mathbf{p} & \mathbf{0} \end{pmatrix} \begin{pmatrix} \mathbf{e}_x \\ \mathbf{h}_y \\ \mathbf{e}_y \\ \mathbf{h}_x \end{pmatrix}. \tag{9.80}$$

The elements are now denoted with Roman bold letter symbols in place of the symbols with parentheses employed in Eqs. (9.78) to (9.80).

Eq. (9.68b) can be transformed into a matrix form

$$\frac{\mathrm{d}}{\mathrm{d}\bar{z}}(e_x) = \mathrm{j}[(s)(e_x) - (h_y) - (p)(e_z)]. \tag{9.81}$$

We eliminate (e_z) using the first equation of the system (9.80)

$$\frac{\mathrm{d}}{\mathrm{d}\bar{z}}(e_x) = \mathrm{j}(s)(e_x) - \mathrm{j}(h_y) - \mathrm{j}(p)\big[-(\varepsilon_{zz})^{-1}(\varepsilon_{zx})(e_x) - (\varepsilon_{zz})^{-1}(p)(h_y) \\ -(\varepsilon_{zz})^{-1}(\varepsilon_{zy})(e_y) + (\varepsilon_{zz})^{-1}(q)(h_x)\big] \tag{9.82}$$

and after changing the order of the terms

$$\frac{\mathrm{d}}{\mathrm{d}\bar{z}}(e_x) = \mathrm{j}\left\{\left[(p)(\varepsilon_{zz})^{-1}(\varepsilon_{zx}) + (s)\right](e_x) + \left[(p)(\varepsilon_{zz})^{-1}(p) - (1)\right](h_y) \\ + (p)(\varepsilon_{zz})^{-1}(\varepsilon_{zy})(e_y) - (p)(\varepsilon_{zz})^{-1}(q)(h_x)\right\}. \tag{9.83}$$

This equation forms the first row of another generalized matrix equation that we are constructing. The second row follows from the equation (9.68d). The equation assumes the following matrix form

$$\frac{\mathrm{d}}{\mathrm{d}\bar{z}}(h_y) = \mathrm{j}[(s)(h_y) - (q)(h_z) - (xz)(e_z) - (\varepsilon_{xx})(e_x) - (\varepsilon_{xy})(e_y)]. \tag{9.84}$$

We now eliminate (e_x) and (h_x), using the equation system (9.80)

$$\frac{\mathrm{d}}{\mathrm{d}\bar{z}}(h_y) = \mathrm{j}\left\{(s)(h_y) - (q)[-(q)(e_x) + (p)(e_y)] - (xz)(\varepsilon_{zz})^{-1} \\ \times[-(\varepsilon_{zx})(e_x) - (p)(h_y) - (\varepsilon_{zy})(e_y) + (q)(h_x)] \\ - (\varepsilon_{xx})(e_x) - (\varepsilon_{xy})(e_y)\right\}, \tag{9.85}$$

and after changing the order of the terms into the standard form of Eq. (9.80), we obtain

$$\frac{\mathrm{d}}{\mathrm{d}\bar{z}}(h_y) = \mathrm{j}\left\{\left[(\varepsilon_{xz})(\varepsilon_{zz})^{-1}(\varepsilon_{zx}) - (\varepsilon_{xx}) + (q)^2\right](e_x) \\ + \left[(\varepsilon_{xz})(\varepsilon_{zz})^{-1}(p) + (s)\right](h_y) + \left[(\varepsilon_{xz})(\varepsilon_{zz})^{-1}(\varepsilon_{zy}) \\ - (\varepsilon_{xy}) - (q)(p)\right](e_y) - (\varepsilon_{xz})(\varepsilon_{zz})^{-1}(q)(h_x)\right\}. \tag{9.86}$$

The third row is constructed with Eq. (9.68a) transformed into the matrix form

$$\frac{\mathrm{d}}{\mathrm{d}\bar{z}}(e_y) = \mathrm{j}[(s)(e_y) + (h_x) - (q)(e_z)]. \tag{9.87}$$

To eliminate (e_z), we again make use of the first equation of the system (9.80)

$$\frac{\mathrm{d}}{\mathrm{d}\bar{z}}(e_y) = \mathrm{j}\left\{(s)(e_y) + (h_x) - (q)(\varepsilon_{zz})^{-1}\left[-(\varepsilon_{zx})(e_x) - (p)(h_y) \\ - (\varepsilon_{zy})(e_y) + (q)(h_x)\right\}\right], \tag{9.88}$$

and after rearranging

$$\frac{d}{d\bar{z}}(e_y) = j\{(q)(\varepsilon_{zz})^{-1}(\varepsilon_{zx})(e_x) + (q)(\varepsilon_{zz})^{-1}(p)(h_y) + [(q)(\varepsilon_{zz})^{-1}(\varepsilon_{zy}) + (s)](e_y) + [-(q)(\varepsilon_{zz})^{-1}(q) + (1)](h_x)\}. \quad (9.89)$$

Finally, the fourth row is obtained from Eq. (9.68e) transformed into the matrix form

$$\frac{d}{d\bar{z}}(h_x) = j[(s)(h_x) - (p)(h_y) + (\varepsilon_{yz})(e_z) + (\varepsilon_{yx})(e_x) - (\varepsilon_{yy})(e_y)]. \quad (9.90)$$

Using the system (9.80), we eliminate (e_z) and (h_z)

$$\frac{d}{d\bar{z}}(h_x) = j\{(s)(h_x) - (p)[-(q)(e_x) + (p)(e_y)] + (\varepsilon_{yz})(\varepsilon_{zz})^{-1}[-(\varepsilon_{zx})(e_x) - (p)(h_y) - (\varepsilon_{zy})(e_y) + (q)(h_x)] + (\varepsilon_{yx})(e_x) + (\varepsilon_{yy})(e_y)\}, \quad (9.91)$$

and after rearranging

$$\frac{d}{d\bar{z}}(h_x) = j\{[(\varepsilon_{yx}) - (\varepsilon_{yz})(\varepsilon_{zz})^{-1}(\varepsilon_{zx}) + (p)(q)](e_x) - (\varepsilon_{yz})(\varepsilon_{zz})^{-1}(p)(h_y) + [(\varepsilon_{yy}) - (\varepsilon_{yz})(\varepsilon_{zz})^{-1}(\varepsilon_{zy}) - (p)^2](e_y) + (\varepsilon_{yz})(\varepsilon_{zz})^{-1}(q)(h_x)\}. \quad (9.92)$$

The matrix equations (9.83), (9.84), (9.89), and (9.92) can be arranged into generalized matrix equations that relate the tangential and normal (with respect to the xy interface plane) field components denoted respectively as

$$\mathbf{f}_{\mathfrak{t}} = \begin{pmatrix} \mathbf{e}_x \\ \mathbf{h}_y \\ \mathbf{e}_y \\ \mathbf{h}_x \end{pmatrix}, \qquad \mathbf{f}_{\mathfrak{n}} = \begin{pmatrix} \mathbf{e}_z \\ \mathbf{h}_z \end{pmatrix}, \quad (9.93)$$

where, as before, $\mathbf{e}_i = \mathbf{e}_i(\bar{z})$ and $\mathbf{h}_i = \mathbf{h}_i(\bar{z})$, $i = x, y, z$, are the column vectors with elements $e_{im}(\bar{z})$ and $h_{im}(\bar{z})$, $m = 0, \pm1, \pm2, \ldots \pm M$. By substituting Eq. (9.14), Eq. (9.18), and Eq. (9.19) into the Maxwell equations (9.12) and Eq. (9.13), we have arrived at an equation system with an infinite number of equations (assuming $M \to \infty$) for *coupled waves*. They can be concisely expressed in generalized matrix forms

$$\frac{d}{d\bar{z}}\mathbf{f}_{\mathfrak{t}} = j\mathbf{C}\,\mathbf{f}_{\mathfrak{t}}, \quad (9.94a)$$

$$\mathbf{f}_{\mathfrak{n}} = \mathbf{D}\,\mathbf{f}_{\mathfrak{t}}, \quad (9.94b)$$

where

$$\mathbf{C} = \begin{pmatrix} \mathbf{p}\varepsilon_{zz}^{-1}\varepsilon_{zx} + \mathbf{s} & \mathbf{p}\varepsilon_{zz}^{-1}\mathbf{p} - \mathbf{1} & \mathbf{p}\varepsilon_{zz}^{-1}\varepsilon_{zy} & -\mathbf{p}\varepsilon_{zz}^{-1}\mathbf{q} \\ \varepsilon_{xz}\varepsilon_{zz}^{-1}\varepsilon_{zx} - \varepsilon_{xx} + \mathbf{q}^2 & \varepsilon_{xz}\varepsilon_{zz}^{-1}\mathbf{p} + \mathbf{s} & \varepsilon_{xz}\varepsilon_{zz}^{-1}\varepsilon_{zy} - \varepsilon_{xy} - \mathbf{qp} & -\varepsilon_{xz}\varepsilon_{zz}^{-1}\mathbf{q} \\ \mathbf{q}\varepsilon_{zz}^{-1}\varepsilon_{zx} & \mathbf{q}\varepsilon_{zz}^{-1}\mathbf{p} & \mathbf{q}\varepsilon_{zz}^{-1}\varepsilon_{zy} + \mathbf{s} & -\mathbf{q}\varepsilon_{zz}^{-1}\mathbf{q} + \mathbf{1} \\ \varepsilon_{yx} - \varepsilon_{yz}\varepsilon_{zz}^{-1}\varepsilon_{zx} + \mathbf{pq} & -\varepsilon_{yz}\varepsilon_{zz}^{-1}\mathbf{p} & \varepsilon_{yy} - \varepsilon_{yz}\varepsilon_{zz}^{-1}\varepsilon_{zy} - \mathbf{p}^2 & \varepsilon_{yz}\varepsilon_{zz}^{-1}\mathbf{q} + \mathbf{s} \end{pmatrix}, \tag{9.95}$$

$$\mathbf{D} = \begin{pmatrix} -\varepsilon_{zz}^{-1}\varepsilon_{zx} & -\varepsilon_{zz}^{-1}\mathbf{p} & -\varepsilon_{zz}^{-1}\varepsilon_{zy} & \varepsilon_{zz}^{-1}\mathbf{q} \\ -\mathbf{q} & \mathbf{0} & \mathbf{p} & \mathbf{0} \end{pmatrix}. \tag{9.96}$$

In summary, the matrix **C** was obtained from Eq. (9.83), Eq. (9.89), Eq. (9.91), and Eq. (9.92). This is one of the most important results of the formalism [9]. It has found many applications, *e.g.*, Reference 28. The Rokushima matrix **C** given in Eq. (9.95) represents a generalization of the Berreman matrix for planar layered anisotropic homogeneous media [6] to periodic, both isotropic and anisotropic, layered media and a generalization of the coupled mode theory for isotropic media by Kogelnik [10] to coupled mode theory incorporating anisotropic media.

We have already obtained the matrix **D** in connection with Eq. (9.80). Eq. (9.95) was obtained from the Maxwell equations for a general periodic anisotropic assuming monochromatic plane wave solutions in a specific Cartesian coordinate system. The elements of the generalized matrices **C** and **D** are themselves matrix expressions. Here, $\varepsilon_{ij} = \left(\varepsilon_{ij,nl}\right)$ are the submatrices of dimension $(2M+1) \times (2M+1)$, with elements given by Eq. (9.70)

$$\varepsilon_{ij,l-n} = \varepsilon_{ij,nl}.$$

As seen from Eq. (9.95), the matrix **C** contains ε_{zz}^{-1}, an inverse matrix of the diagonal element of the permittivity tensor. For $M = 1$, this assumes the form

$$\varepsilon_{zz}^{-1} = \left(\varepsilon_{zz}^{-1}\right) = \begin{pmatrix} \varepsilon_{zz,0} & \varepsilon_{zz,-1} & \varepsilon_{zz,-2} \\ \varepsilon_{zz,1} & \varepsilon_{zz,0} & \varepsilon_{zz,-1} \\ \varepsilon_{zz,2} & \varepsilon_{zz,1} & \varepsilon_{zz,0} \end{pmatrix}^{-1}$$

$$= \left(\varepsilon_{zz,0}^3 + \varepsilon_{zz,-1}^2\varepsilon_{zz,2} + \varepsilon_{zz,1}^2\varepsilon_{zz,-2} - \varepsilon_{zz,0}\varepsilon_{zz,-2}\varepsilon_{zz,2} - 2\varepsilon_{zz,0}\varepsilon_{zz,-1}\varepsilon_{zz,1}\right)^{-1}$$

$$\times \begin{pmatrix} \varepsilon_{zz,0}^2 - \varepsilon_{zz,-1}\varepsilon_{zz,1} & -\varepsilon_{zz,-1}\varepsilon_{zz,0} + \varepsilon_{zz,1}\varepsilon_{zz,-2} & \varepsilon_{zz,-1}^2 - \varepsilon_{zz,0}\varepsilon_{zz,-2} \\ -\varepsilon_{zz,1}\varepsilon_{zz,0} + \varepsilon_{zz,2}\varepsilon_{zz,-1} & \varepsilon_{zz,0}^2 - \varepsilon_{zz,-2}\varepsilon_{zz,2} & -\varepsilon_{zz,0}\varepsilon_{zz,-1} + \varepsilon_{zz,1}\varepsilon_{zz,-2} \\ \varepsilon_{zz,1}^2 - \varepsilon_{zz,0}\varepsilon_{zz,2} & -\varepsilon_{zz,0}\varepsilon_{zz,1} + \varepsilon_{zz,-1}\varepsilon_{zz,2} & \varepsilon_{zz,0}^2 - \varepsilon_{zz,-1}\varepsilon_{zz,1} \end{pmatrix}$$

$$\approx \left[\left(\varepsilon_{zz,0}^2 - 2\varepsilon_{zz,1}\varepsilon_{zz,-1}\right)\varepsilon_{zz,0}\right]^{-1}$$
$$\times \begin{pmatrix} \varepsilon_{zz,0}^2 - \varepsilon_{zz,-1}\varepsilon_{zz,1} & -\varepsilon_{zz,-1}\varepsilon_{zz,0} & \varepsilon_{zz,-1}^2 \\ -\varepsilon_{zz,1}\varepsilon_{zz,0} & \varepsilon_{zz,0}^2 & -\varepsilon_{zz,0}\varepsilon_{zz,-1} \\ \varepsilon_{zz,1}^2 & -\varepsilon_{zz,0}\varepsilon_{zz,1} & \varepsilon_{zz,0}^2 - \varepsilon_{zz,-1}\varepsilon_{zz,1}, \end{pmatrix}, \quad (9.97)$$

where, in the last step, we have dropped the terms containing the elements $\varepsilon_{zz,\pm 2}$ of the matrix (ε_{xx}) the most distant from the diagonal of (ε_{xx}).

The other matrices are diagonal

$$\mathbf{p} = (p) = (\delta_{nl}\, p_l), \qquad \mathbf{q} = (q) = (\delta_{nl} q_l),$$
$$\mathbf{s} = (s) = (\delta_{nl} s_l), \qquad \mathbf{1} = (1)(\delta_{nl}), \quad (9.98)$$

where δ_{nl} is the Kronecker delta. The matrices $\mathbf{C}$ and $\mathbf{D}$ characterize the coupling of all space harmonics, in a general way, including the mode coupling from TE to TM waves and *vice versa* in the generally anisotropic grating region. Subsequent studies have shown that an improved convergence in the numerical treatment of the problems can be achieved if the matrices representing the elements of the permittivity tensor in anisotropic periodic region are expressed in a modified form [43–47]. The procedure was summarized by Pagani *et al.* [28].

In a special case of isotropic gratings, the tensor permittivity reduces to a scalar one, $\varepsilon_{ij} = \delta_{ij}\varepsilon$. Then,

$$\varepsilon = \varepsilon(\bar{\mathbf{r}}) = \sum_l \varepsilon_l \exp\left(\mathrm{j} l \mathbf{n}_\Gamma \cdot \bar{\mathbf{r}}\right). \quad (9.99)$$

The $\mathbf{C}$ and $\mathbf{D}$ matrices simplify to

$$\mathbf{C} = \begin{pmatrix} \mathbf{s} & \mathbf{p}\boldsymbol{\varepsilon}^{-1}\mathbf{p} - \mathbf{1} & \mathbf{0} & -\mathbf{p}\boldsymbol{\varepsilon}^{-1}\mathbf{q} \\ \mathbf{s}^2 - \boldsymbol{\varepsilon} & \mathbf{s} & -\mathbf{q}\mathbf{p} & \mathbf{0} \\ \mathbf{0} & \mathbf{q}\boldsymbol{\varepsilon}^{-1}\mathbf{p} & \mathbf{s} & \mathbf{1} - \mathbf{q}\boldsymbol{\varepsilon}^{-1}\mathbf{q} \\ \mathbf{p}\mathbf{q} & \mathbf{0} & \boldsymbol{\varepsilon} - \mathbf{p}^2 & \mathbf{s} \end{pmatrix}, \quad (9.100)$$

$$\mathbf{D} = \begin{pmatrix} \mathbf{0} & -\boldsymbol{\varepsilon}^{-1}\mathbf{p} & \mathbf{0} & \boldsymbol{\varepsilon}^{-1}\mathbf{q} \\ -\mathbf{q} & \mathbf{0} & \mathbf{p} & \mathbf{0} \end{pmatrix}. \quad (9.101)$$

With restriction to the case $M = 1$ and to an isotropic periodic medium, the matrix $\boldsymbol{\varepsilon} = (\varepsilon)$ can easily be written down with the definition (9.70)

$$\boldsymbol{\varepsilon} = \begin{pmatrix} \varepsilon_{1,1} & \varepsilon_{1,0} & \varepsilon_{1,-1} \\ \varepsilon_{0,1} & \varepsilon_{0,0} & \varepsilon_{0,-1} \\ \varepsilon_{-1,1} & \varepsilon_{-1,0} & \varepsilon_{-1,-1} \end{pmatrix} = \begin{pmatrix} \varepsilon_0 & \varepsilon_{-1} & \varepsilon_{-2} \\ \varepsilon_1 & \varepsilon_0 & \varepsilon_{-1} \\ \varepsilon_2 & \varepsilon_1 & \varepsilon_0 \end{pmatrix}. \quad (9.102)$$

In addition, if the grating vector $\mathbf{n}_\Gamma$ is restricted to the plane of incidence yz, we have $\mathbf{p} = 0$, and the matrix $\mathbf{C}$ further simplifies

$$\mathbf{C} = \begin{pmatrix} \mathbf{C}_{\mathrm{TE}} & \mathbf{0} \\ \mathbf{0} & \mathbf{C}_{\mathrm{TM}} \end{pmatrix}, \quad (9.103)$$

where

$$\mathbf{C}_{\mathrm{TE}} = \begin{pmatrix} \mathbf{s} & -\mathbf{1} \\ \mathbf{q}^2 - \boldsymbol{\varepsilon} & \mathbf{s} \end{pmatrix}, \qquad \mathbf{C}_{\mathrm{TM}} = \begin{pmatrix} \mathbf{s} & \mathbf{1} - \mathbf{q}\boldsymbol{\varepsilon}^{-1}\mathbf{q} \\ \boldsymbol{\varepsilon} & \mathbf{s} \end{pmatrix}. \tag{9.104}$$

In this case, Eq. (9.94a) splits into two independent equations

$$\frac{\mathrm{d}}{\mathrm{d}\bar{x}}\mathbf{f}_{\hat{t}_{\mathrm{TE}}} = \mathrm{j}\mathbf{C}_{\mathrm{TE}}\mathbf{f}_{\hat{t}_{\mathrm{TE}}}, \qquad \mathbf{f}_{\hat{t}_{\mathrm{TE}}} = \begin{pmatrix} \mathbf{e}_x \\ \mathbf{h}_y \end{pmatrix}, \qquad \mathbf{f}_{\hat{n}_{\mathrm{TE}}} = (\,\mathbf{h}_z\,), \tag{9.105a}$$

$$\frac{\mathrm{d}}{\mathrm{d}\bar{z}}\mathbf{f}_{\hat{t}_{\mathrm{TM}}} = \mathrm{j}\mathbf{C}_{\mathrm{TM}}\mathbf{f}_{\hat{t}_{\mathrm{TM}}}, \qquad \mathbf{f}_{\hat{t}_{\mathrm{TM}}} = \begin{pmatrix} \mathbf{e}_y \\ \mathbf{h}_x \end{pmatrix}, \qquad \mathbf{f}_{\hat{n}_{\mathrm{TM}}} = (\,\mathbf{e}_z\,). \tag{9.105b}$$

From this, it follows that there is no TE $\leftrightarrow$ TM wave mode coupling for the grating vector parallel to the plane of incidence ($\mathbf{p} = 0$). For $M = 1$, the product $\mathbf{C}_{\mathrm{TE}}\mathbf{f}_{\hat{t}_{\mathrm{TE}}}$, pertinent to the TE waves, becomes

$$\mathbf{C}_{\mathrm{TE}}\mathbf{f}_{\hat{t}_{\mathrm{TE}}} = \begin{pmatrix} s & 0 & 0 & -1 & 0 & 0 \\ 0 & 0 & 0 & 0 & -1 & 0 \\ 0 & 0 & -s & 0 & 0 & -1 \\ (q+q_0)^2 - \varepsilon_0 & -\varepsilon_{-1} & -\varepsilon_{-2} & s & 0 & 0 \\ -\varepsilon_1 & p_0^2 - \varepsilon_0 & -\varepsilon_{-1} & 0 & 0 & 0 \\ -\varepsilon_1 & -\varepsilon_1 & (q-q_0)^2 - \varepsilon_0 & 0 & 0 & -s \end{pmatrix} \begin{pmatrix} e_{x,1} \\ e_{x,0} \\ e_{x,-1} \\ h_{y,1} \\ h_{y,0} \\ h_{y,-1} \end{pmatrix}. \tag{9.106}$$

The matrix $\mathbf{C}_{\mathrm{TM}}$ contains the inverse matrix of scalar permittivity. For $M = 1$, this can be found in a way similar to that employed in Eq. (9.94a)

$$\begin{aligned} \varepsilon^{-1} = (\varepsilon)^{-1} &= \begin{pmatrix} \varepsilon_0 & \varepsilon_{-1} & \varepsilon_{-2} \\ \varepsilon_1 & \varepsilon_0 & \varepsilon_{-1} \\ \varepsilon_2 & \varepsilon_1 & \varepsilon_0 \end{pmatrix}^{-1} \\ &= \left(\varepsilon_0^3 + \varepsilon_{-1}^2\varepsilon_2 + \varepsilon_1^2\varepsilon_{-2} - \varepsilon_0\varepsilon_{-2}\varepsilon_2 - 2\varepsilon_0\varepsilon_{-1}\varepsilon_1\right)^{-1} \\ &\quad \times \begin{pmatrix} \varepsilon_0^2 - \varepsilon_{-1}\varepsilon_1 & -\varepsilon_{-1}\varepsilon_0 + \varepsilon_1\varepsilon_{-2} & \varepsilon_{-1}^2 - \varepsilon_0\varepsilon_{-2} \\ -\varepsilon_1\varepsilon_0 + \varepsilon_2\varepsilon_{-1} & \varepsilon_0^2 - \varepsilon_{-2}\varepsilon_2 & -\varepsilon_0\varepsilon_{-1} + \varepsilon_1\varepsilon_{-2} \\ \varepsilon_1^2 - \varepsilon_0\varepsilon_2 & -\varepsilon_0\varepsilon_1 + \varepsilon_{-1}\varepsilon_2 & \varepsilon_0^2 - \varepsilon_{-1}\varepsilon_1 \end{pmatrix} \\ &\approx \left[\left(\varepsilon_0^2 - 2\varepsilon_1\varepsilon_{-1}\right)\varepsilon_0\right]^{-1} \begin{pmatrix} \varepsilon_0^2 - \varepsilon_{-1}\varepsilon_1 & -\varepsilon_{-1}\varepsilon_0 & \varepsilon_{-1}^2 \\ -\varepsilon_1\varepsilon_0 & \varepsilon_0^2 & -\varepsilon_0\varepsilon_{-1} \\ \varepsilon_1^2 & -\varepsilon_0\varepsilon_1 & \varepsilon_0^2 - \varepsilon_{-1}\varepsilon_1, \end{pmatrix}, \end{aligned} \tag{9.107}$$

where, in the last step, we have dropped the terms containing the elements $\varepsilon_{\pm 2}$ of the matrix (ε) the most distant from the diagonal of (ε). In particular, the submatrix $\mathbf{1} - \mathbf{q}\varepsilon^{-1}\mathbf{q}$ of $\mathbf{C}$ and $\mathbf{C}_{\mathrm{TM}}$ in Eq. (9.100) and Eq. (9.103), respectively, becomes

$$\mathbf{1} - \mathbf{q}\varepsilon^{-1}\mathbf{q} \approx \begin{pmatrix} 1 & 0 & 0 \\ 0 & 1 & 0 \\ 0 & 0 & 1 \end{pmatrix} - \left[\left(\varepsilon_0^2 - 2\varepsilon_1\varepsilon_{-1}\right)\varepsilon_0\right]^{-1}$$
$$\times \begin{pmatrix} \left(\varepsilon_0^2 - \varepsilon_{-1}\varepsilon_1\right)(q_0+q)^2 & -\varepsilon_{-1}\varepsilon_0\,(q_0+q)\,q_0 & \varepsilon_{-1}^2\left(q_0^2 - q^2\right) \\ -\varepsilon_1\varepsilon_0\,(q_0+q)\,q_0 & \varepsilon_0^2 q_0^2 & -\varepsilon_0\varepsilon_1\,(q_0-q)\,q_0 \\ \varepsilon_1^2\left(q_0^2 - q^2\right) & -\varepsilon_0\varepsilon_1\,(q_0-q)\,q_0 & \left(\varepsilon_0^2 - \varepsilon_{-1}\varepsilon_1\right)(q_0-q)^2 \end{pmatrix}. \tag{9.108}$$

The remaining submatrices in $\mathbf{C}_{\mathrm{TM}}$ can be deduced from those of $\mathbf{C}_{\mathrm{TE}}$ in Eq. (9.104)

$$\mathbf{s} = \begin{pmatrix} s & 0 & 0 \\ 0 & 0 & 0 \\ 0 & 0 & -s \end{pmatrix}, \qquad \varepsilon = \begin{pmatrix} \varepsilon_0 & \varepsilon_{-1} & \varepsilon_{-2} \\ \varepsilon_1 & \varepsilon_0 & \varepsilon_{-1} \\ \varepsilon_2 & \varepsilon_1 & \varepsilon_0 \end{pmatrix}. \tag{9.109}$$

The matrix $\mathbf{D}$ for an isotropic grating with the grating vector parallel to the plane of incidence ($\mathbf{p} = 0$) simplifies, and the relation $\mathbf{f}_{\hat{n}} = \mathbf{D}\mathbf{f}_{\hat{t}}$, given by the general equation (9.80), reduces to

$$\begin{pmatrix} \mathbf{e}_z \\ \mathbf{h}_z \end{pmatrix} = \begin{pmatrix} \mathbf{0} & \mathbf{D}_{\mathrm{TM}} \\ \mathbf{D}_{\mathrm{TE}} & \mathbf{0} \end{pmatrix} \begin{pmatrix} \mathbf{f}_{\hat{t}\,TE} \\ \mathbf{f}_{\hat{t}\,TM} \end{pmatrix} = \begin{pmatrix} \mathbf{0} & \mathbf{0} & \mathbf{0} & \varepsilon^{-1}\mathbf{q} \\ -\mathbf{q} & \mathbf{0} & \mathbf{0} & \mathbf{0} \end{pmatrix} \begin{pmatrix} \mathbf{e}_x \\ \mathbf{h}_y \\ \mathbf{e}_y \\ \mathbf{h}_x \end{pmatrix}. \tag{9.110}$$

In the homogeneous isotropic regions (0) and (2), the permittivity is a constant scalar quantity. The permittivity matrix $\varepsilon = (\varepsilon)$ entering Eq. (9.99), which defines $\mathbf{C}$ for isotropic regions, reduces to a multiple of a unit matrix $\varepsilon = \varepsilon\mathbf{1}$. The homogeneous regions are in contact with the grating region with periodic permittivity. From the boundary conditions at the interfaces between homogeneous and periodic (grating) regions, it follows that the fields in the homogeneous regions should contain higher space harmonics generated in the periodic (grating) region. Equations (9.94) for $M = 1$, *i.e.*, $2M + 1 = 3$, can be expressed in a closed form

$$
-\mathrm{j}\frac{\mathrm{d}}{\mathrm{d}\bar{x}}\begin{pmatrix} e_{x,1} \\ e_{x,0} \\ e_{x,-1} \\ h_{y,1} \\ h_{y,0} \\ h_{y,-1} \\ e_{y,1} \\ e_{y,0} \\ e_{y,-1} \\ h_{x,1} \\ h_{x,0} \\ h_{x,-1} \end{pmatrix}
= \begin{pmatrix}
s & 0 & 0 & \frac{p^2}{\varepsilon}-1 & 0 & 0 & 0 & 0 & 0 & -\frac{pq_+}{\varepsilon} & 0 & 0 \\
0 & 0 & 0 & 0 & -1 & 0 & 0 & 0 & 0 & 0 & 0 & 0 \\
0 & 0 & -s & 0 & 0 & \frac{p^2}{\varepsilon}-1 & 0 & 0 & 0 & 0 & 0 & -\frac{pq_-}{\varepsilon} \\
q_+^2-\varepsilon & 0 & 0 & p & 0 & 0 & -pq_+ & 0 & 0 & 0 & 0 & 0 \\
0 & q_0^2-\varepsilon & 0 & 0 & 0 & 0 & 0 & 0 & 0 & 0 & 0 & 0 \\
0 & 0 & q_-^2-\varepsilon & 0 & 0 & -s & 0 & 0 & pq_- & 0 & 0 & 0 \\
0 & 0 & 0 & \frac{pq_+}{\varepsilon} & 0 & 0 & s & 0 & 0 & 1-\frac{q_+^2}{\varepsilon} & 0 & 0 \\
0 & 0 & 0 & 0 & 0 & 0 & 0 & 0 & 0 & 0 & 1-\frac{q_0^2}{\varepsilon} & 0 \\
0 & 0 & 0 & 0 & 0 & -\frac{pq_-}{\varepsilon} & 0 & 0 & -s & 0 & 0 & 1-\frac{q_-^2}{\varepsilon} \\
pq_+ & 0 & 0 & 0 & 0 & 0 & \varepsilon-p^2 & 0 & 0 & s & 0 & 0 \\
0 & 0 & 0 & 0 & 0 & 0 & 0 & \varepsilon & 0 & 0 & 0 & 0 \\
0 & 0 & -pq_- & 0 & 0 & 0 & 0 & 0 & \varepsilon-p^2 & 0 & 0 & -s
\end{pmatrix}
\times \begin{pmatrix} e_{x,1} \\ e_{x,0} \\ e_{x,-1} \\ h_{y,1} \\ h_{y,0} \\ h_{y,-1} \\ e_{y,1} \\ e_{y,0} \\ e_{y,-1} \\ h_{y,1} \\ h_{y,0} \\ h_{y,-1} \end{pmatrix}, \quad (9.111)
$$

$$
\begin{pmatrix} e_{z,1} \\ e_{z,0} \\ e_{z,-1} \\ h_{z,1} \\ h_{z,0} \\ h_{z,-1} \end{pmatrix}
= \begin{pmatrix}
0 & 0 & 0 & -\frac{p}{\varepsilon} & 0 & 0 & 0 & 0 & 0 & \frac{q_+}{\varepsilon} & 0 & 0 \\
0 & 0 & 0 & 0 & 0 & 0 & 0 & 0 & 0 & 0 & \frac{q_0}{\varepsilon} & 0 \\
0 & 0 & 0 & 0 & 0 & +\frac{p}{\varepsilon} & 0 & 0 & 0 & 0 & 0 & \frac{q_-}{\varepsilon} \\
-q_+ & 0 & 0 & 0 & 0 & 0 & +p & 0 & 0 & 0 & 0 & 0 \\
0 & -q_0 & 0 & 0 & 0 & 0 & 0 & 0 & 0 & 0 & 0 & 0 \\
0 & 0 & -q_- & 0 & 0 & 0 & 0 & 0 & -p & 0 & 0 & 0
\end{pmatrix}
\times \begin{pmatrix} e_{x,1} \\ e_{x,0} \\ e_{x,-1} \\ h_{y,1} \\ h_{y,0} \\ h_{y,-1} \\ e_{y,1} \\ e_{y,0} \\ e_{y,-1} \\ h_{x,1} \\ h_{x,0} \\ h_{x,-1} \end{pmatrix}. \tag{9.112}
$$

The matrices are built of diagonal 3×3 submatrices, *i.e.*, the off-diagonal elements of the submatrices are zero. Here, $q_\pm = q_0 \pm q$. In the homogeneous isotropic regions, there is no interaction between space harmonics. Consequently, each of the space harmonics separately should represent a solution to the Maxwell equations. Eqs. (9.94a and 9.94b) split into $2M + 1$ independent equations

$$
\frac{\mathrm{d}}{\mathrm{d}\bar{x}} \mathbf{f}^{(u)}_{\hat{t}m} = \mathrm{j}\mathbf{C}^{(u)}_{m} \mathbf{f}^{(u)}_{\hat{t}m}, \tag{9.113a}
$$

$$
\mathbf{f}^{(u)}_{\hat{n}m} = \mathrm{j}\mathbf{D}^{(u)}_{m} \mathbf{f}^{(u)}_{\hat{t}m}, \tag{9.113b}
$$

where

$$\mathbf{f}_{\bar{t}m}^{(u)} = \begin{pmatrix} e_{xm} \\ h_{ym} \\ e_{ym} \\ h_{xm} \end{pmatrix}, \qquad \mathbf{f}_{\bar{n}m}^{(u)} = \begin{pmatrix} \mathbf{e}_{zm} \\ \mathbf{h}_{zm} \end{pmatrix}. \tag{9.114}$$

Eq. (9.113a) can be expressed in more detail as

$$\frac{\mathrm{d}}{\mathrm{d}\bar{x}} \begin{pmatrix} e_{xm} \\ h_{ym} \\ e_{ym} \\ h_{xm} \end{pmatrix} = \mathrm{j} \begin{pmatrix} s_m & p_m^2/\varepsilon - 1 & 0 & -p_m q_m/\varepsilon \\ q_m^2 - \varepsilon & s_m & -p_m q_m & 0 \\ 0 & p_m q_m/\varepsilon & s_m & 1 - q_m^2/\varepsilon \\ p_m q_m & 0 & \varepsilon - p_m^2 & s_m \end{pmatrix} \begin{pmatrix} e_{xm} \\ h_{ym} \\ e_{ym} \\ h_{xm} \end{pmatrix}. \tag{9.115}$$

Matrix $\mathbf{C}_m^{(u)}$ of dimension 4×4 is defined as

$$\mathbf{C}_m^{(u)} = \begin{pmatrix} s_m & p_m^2/\varepsilon - 1 & 0 & -p_m q_m/\varepsilon \\ q_m^2 - \varepsilon & s_m & -p_m q_m & 0 \\ 0 & p_m q_m/\varepsilon & s_m & 1 - q_m^2/\varepsilon \\ p_m q_m & 0 & \varepsilon - p_m^2 & s_m \end{pmatrix}. \tag{9.116}$$

Eq. (9.113b) becomes

$$\begin{pmatrix} e_{zm} \\ h_{zm} \end{pmatrix} = \begin{pmatrix} 0 & -p_m/\varepsilon & 0 & q_m/\varepsilon \\ -q_m & 0 & p_m & 0 \end{pmatrix} \begin{pmatrix} e_{xm} \\ h_{ym} \\ e_{ym} \\ h_{xm} \end{pmatrix}. \tag{9.117}$$

Matrix $\mathbf{D}_m^{(u)}$ of dimension 4×2 is then defined as

$$\mathbf{D}_m^{(u)} = \begin{pmatrix} 0 & -p_m/\varepsilon & 0 & q_m/\varepsilon \\ -q_m & 0 & p_m & 0 \end{pmatrix}. \tag{9.118}$$

Eq. (9.111) and Eq. (9.112) pertaining to the case $M = 1$ may be rearranged by changing the order of rows and columns in the following way

$$-\mathrm{j}\frac{\mathrm{d}}{\mathrm{d}\bar{z}}\begin{pmatrix} e_{x,1} \\ h_{y,1} \\ e_{y,1} \\ h_{x,1} \\ e_{x,0} \\ h_{y,0} \\ e_{y,0} \\ h_{x,0} \\ e_{x,-1} \\ h_{y,-1} \\ e_{y,-1} \\ h_{x,-1} \end{pmatrix} = \left(\begin{array}{cccccccccccc} s & \frac{p^2}{\varepsilon}-1 & 0 & -\frac{pq_+}{\varepsilon} & & & & & & & & \\ q_+^2-\varepsilon & s & -pq_+ & 0 & & & & & & & & \\ 0 & \frac{pq_+}{\varepsilon} & s & 1-\frac{q_+^2}{\varepsilon} & & & & & & & & \\ pq_+ & 0 & \varepsilon-p^2 & s & & & & & & & & \\ & & & & 0 & -1 & 0 & 0 & & & & \\ & & & & p_0^2-\varepsilon & 0 & 0 & 0 & & & & \\ & & & & 0 & 0 & 0 & 1-\frac{p_0^2}{\varepsilon} & & & & \\ & & & & 0 & 0 & \varepsilon & 0 & & & & \\ & & & & & & & & -s & \frac{p^2}{\varepsilon}-1 & 0 & -\frac{pq_-}{\varepsilon} \\ & & & & & & & & q_-^2-\varepsilon & -s & pq_- & 0 \\ & & & & & & & & 0 & -\frac{pq_-}{\varepsilon} & -s & 1-\frac{q_-^2}{\varepsilon} \\ & & & & & & & & -pq_- & 0 & \varepsilon-p^2 & -s \end{array}\right) \times \begin{pmatrix} e_{x,1} \\ h_{y,1} \\ e_{y,1} \\ h_{x,1} \\ e_{x,0} \\ h_{y,0} \\ e_{y,0} \\ h_{x,0} \\ e_{x,-1} \\ h_{y,-1} \\ e_{y,-1} \\ h_{x,-1} \end{pmatrix}. \quad (9.119)$$

Matrix $\mathbf{C}^{(u)}$ is here quasidiagonal with 4×4 submatrices $\mathbf{C}_0^{(u)}$ and $\mathbf{C}_{\pm 1}^{(u)}$. The off-diagonal 4×4 submatrices are zero. The quasidiagonal form of this matrix is due to the absence of the interaction between different space harmonics. Eq. (9.112) for $M = 1$ can be expressed in a quasidiagonal form, in a similar way

$$\begin{pmatrix} e_{z,1} \\ h_{z,1} \\ e_{z,0} \\ h_{z,0} \\ e_{z,-1} \\ h_{z,-1} \end{pmatrix} = \begin{pmatrix} 0 & -\frac{p}{\varepsilon} & 0 & \frac{q_+}{\varepsilon} & & & & & & & & \\ -q_+ & 0 & p & 0 & & & & & & & & \\ & & & & 0 & 0 & 0 & \frac{q_0}{\varepsilon} & & & & \\ & & & & -q_0 & 0 & 0 & 0 & & & & \\ & & & & & & & & 0 & \frac{p}{\varepsilon} & 0 & \frac{q_-}{\varepsilon} \\ & & & & & & & & -q_- & 0 & -p & 0 \end{pmatrix} \begin{pmatrix} e_{x,1} \\ h_{y,1} \\ e_{y,1} \\ h_{x,1} \\ e_{x,0} \\ h_{y,0} \\ e_{y,0} \\ h_{x,0} \\ e_{x,-1} \\ h_{y,-1} \\ e_{y,-1} \\ h_{x,-1} \end{pmatrix}. \tag{9.120}$$

Here, the matrix $\mathbf{D}^{(u)}$ consists of submatrices $\mathbf{D}_0^{(u)}$ and $\mathbf{D}_{\pm 1}^{(u)}$ of the dimension 4×2. The off-diagonal 4×2 submatrices are zero.

9.5 Matrix Formulation of the Solution

The solution to the equation system (9.105a) for coupled waves reduces to the problem of finding proper values of the matrix $\mathbf{C}$ of dimension $n \times n$, where $n = 4\,(2M + 1)$. Let κ_r denote a proper value determined from the characteristic equation

$$\det\left(\mathbf{C} - \kappa \mathbf{1}\right) = 0. \tag{9.121}$$

The corresponding (column) proper vector will be denoted as $\boldsymbol{\nu}_r$ (with elements ν_{rl}). The matrix $\mathbf{T}$ transforming $\mathbf{C}$ to a diagonal one is built of these proper vectors, *i.e.*, $\mathbf{T} = (\boldsymbol{\nu}_1, \boldsymbol{\nu}_2, \ldots, \boldsymbol{\nu}_n)$, where the subscript numbers columns of $\mathbf{T}$. We define a vector $\mathbf{g}$, which multiplied by $\mathbf{T}$ gives $\mathbf{f}_t$

$$\mathbf{f}_t = \mathbf{T}\mathbf{g}. \tag{9.122}$$

The multiplication from the left of Eq. (9.94a) by matrix $\mathbf{T}^{-1}$, independent of $\bar{z}$, provides

$$\frac{d}{d\bar{z}}\mathbf{T}^{-1}\mathbf{f}_{\dagger} = j\mathbf{T}^{-1}\mathbf{C}\mathbf{T}\mathbf{T}^{-1}\mathbf{f}_{\dagger}$$
$$\frac{d}{d\bar{z}}\mathbf{g} = j\boldsymbol{\kappa}\mathbf{g}, \tag{9.123}$$

where $\boldsymbol{\kappa} = (\delta_{rl}\kappa_l)$ is a diagonal matrix and $\mathbf{g}$ is a column vector with elements g_r. Eq. (9.122) has the solutions of the form

$$g_r(\bar{z}) = g_r(\bar{z}^{(0)})\exp[j\kappa_r(\bar{z} - \bar{z}^{(0)})], \tag{9.124}$$

where the factor $g_r(0)$ is to be determined from the boundary conditions. The corresponding column vector $\mathbf{f}_{\dagger r}$ is obtained from

$$\mathbf{f}_{\dagger r} = \begin{pmatrix} \nu_{11} & \nu_{12} & \cdot & \cdot & \nu_{1r} & \cdot & \cdot & \cdot & \nu_{1n} \\ \nu_{21} & \nu_{22} & \cdot & \cdot & \nu_{2r} & \cdot & \cdot & \cdot & \nu_{2n} \\ \cdot & \cdot & \cdot & \cdot & \cdot & \cdot & \cdot & \cdot & \cdot \\ \cdot & \cdot & \cdot & \cdot & \cdot & \cdot & \cdot & \cdot & \cdot \\ \nu_{r1} & \nu_{r2} & \cdot & \cdot & \nu_{rr} & \cdot & \cdot & \cdot & \nu_{rn} \\ \cdot & \cdot & \cdot & \cdot & \cdot & \cdot & \cdot & \cdot & \cdot \\ \cdot & \cdot & \cdot & \cdot & \cdot & \cdot & \cdot & \cdot & \cdot \\ \cdot & \cdot & \cdot & \cdot & \cdot & \cdot & \cdot & \cdot & \cdot \\ \nu_{n1} & \nu & \cdot & \cdot & \nu_{nr} & \cdot & \cdot & \cdot & \nu_{nn} \end{pmatrix} \begin{pmatrix} 0 \\ 0 \\ \cdot \\ 0 \\ g_r(\bar{z}^{(0)})e^{j\kappa_r(\bar{z}-\bar{z}^{(0)})} \\ 0 \\ \cdot \\ \cdot \\ 0 \end{pmatrix}$$
$$= \begin{pmatrix} \nu_{1r} \\ \nu_{1r} \\ \cdot \\ \cdot \\ \nu_{1r} \\ \cdot \\ \cdot \\ \cdot \\ \nu_{1r} \end{pmatrix} g_r(\bar{z}^{(0)})e^{j\kappa_r(\bar{z}-\bar{z}^{(0)})}$$
$$= \boldsymbol{\nu}_r g_r(\bar{z}^{(0)})e^{j\kappa_r(\bar{z}-\bar{z}^{(0)})}. \tag{9.125}$$

Writing down this equation for the general case, we obtain

$$\mathbf{f}_{\bar{t}} = \begin{pmatrix} \nu_{11} & \nu_{12} & \cdot & \cdot & \nu_{1r} & \cdot & \cdot & \cdot & \nu_{1n} \\ \nu_{21} & \nu_{22} & \cdot & \cdot & \nu_{2r} & \cdot & \cdot & \cdot & \nu_{2n} \\ \cdot & \cdot & \cdot & \cdot & \cdot & \cdot & \cdot & \cdot & \cdot \\ \nu_{r-1,1} & \nu_{r-1,2} & \cdot & \nu_{r-1,r} & \cdot & \cdot & \cdot & \cdot & \nu_{r-1,n} \\ \nu_{r1} & \nu_{r2} & \cdot & \cdot & \nu_{rr} & \cdot & \cdot & \cdot & \nu_{rn} \\ \nu_{r+1,1} & \nu_{r+1,2} & \cdot & \cdot & \nu_{r+1,r} & \cdot & \cdot & \cdot & \nu_{r+1,n} \\ & \cdot & \cdot & \cdot & \cdot & \cdot & \cdot & \cdot & \cdot \\ & \cdot & \cdot & \cdot & \cdot & \cdot & \cdot & \cdot & \cdot \\ \nu_{n1} & \nu_{n2} & \cdot & \cdot & \nu_{nr} & \cdot & \cdot & \cdot & \nu_{nn} \end{pmatrix}$$

$$\times \begin{pmatrix} g_1(\bar{z}^{(0)}) \exp[\mathrm{j}\kappa_1(\bar{z} - \bar{z}^{(0)})] \\ g_2(\bar{z}^{(0)}) \exp[\mathrm{j}\kappa_2(\bar{z} - \bar{z}^{(0)})] \\ \cdot \\ g_{r-1}(\bar{z}^{(0)}) \exp[\mathrm{j}\kappa_{r-1}(\bar{z} - \bar{z}^{(0)})] \\ g_r(\bar{z}^{(0)}) \exp[\mathrm{j}\kappa_r(\bar{z} - \bar{z}^{(0)})] \\ g_{r+1}(\bar{z}^{(0)}) \exp[\mathrm{j}\kappa_{r+1}(\bar{z} - \bar{z}^{(0)})] \\ \cdot \\ \cdot \\ g_n(\bar{z}^{(0)}) \exp[\mathrm{j}\kappa_n(\bar{z} - \bar{z}^{(0)})] \end{pmatrix}. \tag{9.126}$$

In isotropic homogeneous region, Eq. (9.123) becomes

$$\frac{\mathrm{d}}{\mathrm{d}\bar{z}}\mathbf{g}_m^{(u)} = \mathrm{j}\boldsymbol{\kappa}_m^{(u)}\mathbf{g}_m^{(u)}. \tag{9.127}$$

Here, the proper values κ_r, the proper vectors $\mathbf{f}_{\bar{t}r}$, and the matrix $\mathbf{T}^{(u)}$ transforming $\mathbf{C}$ to a diagonal one can be expressed explicitly. From Eq. (9.116) and Eq. (9.121)

$$\det\left(\mathbf{C}_m^{(u)} - \kappa_m^{(u)}\mathbf{1}\right) = \begin{vmatrix} s_m - \kappa_m^{(u)} & p_m^2/\varepsilon - 1 & 0 & -p_m q_m/\varepsilon \\ q_m^2 - \varepsilon & s_m - \kappa_m^{(u)} & -p_m q_m & 0 \\ 0 & p_m q_m/\varepsilon & s_m - \kappa_m^{(u)} & 1 - q_m^2/\varepsilon \\ p_m q_m & 0 & \varepsilon - p_m^2 & s_m - \kappa_m^{(u)} \end{vmatrix}. \tag{9.128}$$

In order to simplify the computation of the determinant, we introduce the temporary abbreviations

$$\det\left(\mathbf{C}_m^{(u)} - \kappa_m^{(u)}\mathbf{1}\right) = \begin{vmatrix} a_{11} & a_{12} & 0 & a_{14} \\ a_{21} & a_{11} & -a_{41} & 0 \\ 0 & -a_{14} & a_{11} & a_{34} \\ a_{41} & 0 & a_{43} & a_{11} \end{vmatrix}, \tag{9.129}$$

where

$$\begin{aligned}
a_{11} &= s_m - \kappa_m^{(u)}, \\
a_{12} &= \frac{p_m^2}{\varepsilon} - 1, \\
a_{21} &= q_m^2 - \varepsilon, \\
a_{41} &= p_m q_m, \\
a_{14} &= -\frac{p_m q_m}{\varepsilon}, \\
a_{34} &= 1 - \frac{q_m^2}{\varepsilon}, \\
a_{43} &= \varepsilon - p_m^2.
\end{aligned}$$

Developing the determinant gives

$$\det\left(\mathbf{C}_m^{(u)} - \kappa_m^{(u)}\mathbf{1}\right) = a_{11}\begin{vmatrix} a_{11} & -a_{41} & 0 \\ -a_{14} & a_{11} & a_{34} \\ 0 & a_{43} & a_{11} \end{vmatrix} - a_{21}\begin{vmatrix} a_{12} & 0 & a_{14} \\ -a_{14} & a_{11} & a_{34} \\ 0 & a_{43} & a_{11} \end{vmatrix}$$
$$- a_{41}\begin{vmatrix} a_{12} & 0 & a_{14} \\ a_{11} & -a_{41} & 0 \\ -a_{14} & a_{11} & a_{34} \end{vmatrix}.$$

After the evaluation of the subdeterminants

$$\begin{aligned}
\det\left(\mathbf{C}_m^{(u)} - \kappa_m^{(u)}\mathbf{1}\right) &= a_{11}\left(a_{11}^3 - a_{11}a_{14}a_{41} - a_{11}a_{34}a_{43}\right) \\
&\quad - a_{21}\left(a_{11}^2 a_{12} - a_{14}^2 a_{43} - a_{12}a_{34}a_{43}\right) \\
&\quad - a_{41}\left(-a_{12}a_{41}a_{34} + a_{11}^2 a_{14} - a_{14}^2 a_{41}\right) \\
&= a_{11}^4 - a_{11}^2\left(a_{34}a_{43} + a_{12}a_{21} + 2a_{14}a_{41}\right) \\
&\quad + a_{14}^2 a_{21}a_{43} + a_{21}a_{12}a_{34}a_{43} + a_{41}^2 a_{12}a_{34} + a_{14}^2 a_{41}^2.
\end{aligned}$$

The condition for vanishing of the determinant provides the proper values $\kappa_m^{(u)}$

$$\left(s_m - \kappa_m^{(u)}\right)^4 - 2\left(s_m - \kappa_m^{(u)}\right)^2 \varepsilon^{-1}\left[\left(\varepsilon - p_m^2\right)\left(\varepsilon - q_m^2\right) - p_m^2 q_m^2\right]$$
$$+\left[p_m^2 q_m^2 - \left(\varepsilon - p_m^2\right)\left(\varepsilon - q_m^2\right)\right]\left[p_m^2 q_m^2 \varepsilon^{-2} - \left(\varepsilon - p_m^2\right)\left(\varepsilon - q_m^2\right)\varepsilon^{-2}\right] = 0$$
$$\left(s_m - \kappa_m^{(u)}\right)^4 - 2\left(s_m - \kappa_m^{(u)}\right)^2\left[\varepsilon - \left(p_m^2 + q_m^2\right)\right] + \left[\varepsilon - \left(p_m^2 + q_m^2\right)\right]^2 = 0,$$

which finally simplifies to

$$\left(s_m - \kappa_m^{(u)}\right)^2 = \varepsilon - \left(p_m^2 + q_m^2\right). \tag{9.130}$$

The $\bar{z}$ component of the vector $\boldsymbol{n}_\Gamma = \hat{x}p + \hat{y}q + \hat{z}s$ defined in Eq. (9.15) characterizes the periodicity of the permittivity in the direction normal to the interfaces and becomes zero in homogeneous regions. We therefore have $s_m = 0$. The proper values are then given by

$$\kappa_m^{(u)} = \pm\left(\varepsilon - p_m^2 - q_m^2\right)^{1/2}. \tag{9.131}$$

In homogeneous isotropic media, monochromatic waves propagate with equal indices of refraction. The corresponding proper polarizations may be chosen orthogonal. We denote the waves propagating with increasing and decreasing $\bar{z}$ as

$$\kappa_{m1}^{(u)} = \kappa_{m2}^{(u)} = -\left(\varepsilon - p_m^2 - q_m^2\right)^{1/2}, \tag{9.132a}$$
$$\kappa_{m3}^{(u)} = \kappa_{m4}^{(u)} = +\left(\varepsilon - p_m^2 - q_m^2\right)^{1/2}, \tag{9.132b}$$

respectively. We next look for the four proper vectors of the matrix $\mathbf{C}_m^{(u)}$ associated to its two proper values. Another criterion is needed to distinguish the two proper vectors belonging to the same proper value. These will be chosen as the TE and TM waves.

To determine the proper vectors, we first substitute into equation $\mathbf{C} = \kappa\mathbf{1}$ the proper value $\kappa_{m1}^{(u)} = \kappa_{m2}^{(u)} = s_m - \left(\varepsilon - q_m^2 - s_m^2\right)^{1/2}$ given by Eq. (9.132a). We assume $s_m = 0$ and obtain a homogeneous system of four equations represented in the following extended matrix form

$$\left(\begin{array}{cccc|c} \left(\varepsilon - p_m^2 - q_m^2\right)^{1/2} & p_m^2/\varepsilon - 1 & 0 & -p_m q_m/\varepsilon & 0 \\ q_m^2 - \varepsilon & \left(\varepsilon - p_m^2 - q_m^2\right)^{1/2} & -p_m q_m & 0 & 0 \\ 0 & p_m q_m/\varepsilon & \left(\varepsilon - p_m^2 - q_m^2\right)^{1/2} & 1 - q_m^2/\varepsilon & 0 \\ p_m q_m & 0 & \varepsilon - p_m^2 & \left(\varepsilon - p_m^2 - q_m^2\right)^{1/2} & 0 \end{array}\right).$$

We multiply the second row by $(\varepsilon - p_m^2 - q_m^2)^{1/2}$ and subtract from the result the first row multiplied by $(q_m^2 - \varepsilon)$

$$\left(\begin{array}{cccc|c} \left(\varepsilon - p_m^2 - q_m^2\right)^{1/2} & p_m^2/\varepsilon - 1 & 0 & -p_m q_m/\varepsilon & 0 \\ 0 & -p_m^2 q_m^2/\varepsilon & -p_m q_m \left(\varepsilon - p_m^2 - q_m^2\right)^{1/2} & \left(q_m^2 - \varepsilon\right) p_m q_m/\varepsilon & 0 \\ 0 & p_m q_m/\varepsilon & \left(\varepsilon - p_m^2 - q_m^2\right)^{1/2} & 1 - q_m^2/\varepsilon & 0 \\ p_m q_m & 0 & \varepsilon - p_m^2 & \left(\varepsilon - p_m^2 - q_m^2\right)^{1/2} & 0 \end{array}\right).$$

We next multiply the first and third row by $p_m q_m$ and the fourth row by $(\varepsilon - p_m^2 - q_m^2)^{1/2}$

$$\left(\begin{array}{cccc|c} p_m q_m \left(\varepsilon - p_m^2 - q_m^2\right)^{1/2} & p_m q_m \left(p_m^2/\varepsilon - 1\right) & 0 & -p_m^2 q_m^2/\varepsilon & 0 \\ 0 & -p_m^2 q_m^2/\varepsilon & -p_m q_m \left(\varepsilon - p_m^2 - q_m^2\right)^{1/2} & \left(q_m^2 - \varepsilon\right) p_m q_m/\varepsilon & 0 \\ 0 & p_m^2 q_m^2/\varepsilon & p_m q_m \left(\varepsilon - p_m^2 - q_m^2\right)^{1/2} & p_m q_m \left(1 - q_m^2/\varepsilon\right) & 0 \\ p_m q_m \left(\varepsilon - p_m^2 - q_m^2\right)^{1/2} & 0 & \left(\varepsilon - p_m^2\right)\left(\varepsilon - p_m^2 - q_m^2\right)^{1/2} & \left(\varepsilon - p_m^2 - q_m^2\right) & 0 \end{array}\right).$$

We add the third row to the second one.

$$\left(\begin{array}{cccc|c} \left(\varepsilon - p_m^2 - q_m^2\right)^{1/2} & \left(p_m^2/\varepsilon - 1\right) & 0 & -p_m q_m/\varepsilon & 0 \\ 0 & 0 & 0 & 0 & 0 \\ 0 & p_m^2 q_m^2/\varepsilon & p_m q_m \left(\varepsilon - p_m^2 - q_m^2\right)^{1/2} & p_m q_m \left(1 - q_m^2/\varepsilon\right) & 0 \\ p_m q_m \left(\varepsilon - p_m^2 - q_m^2\right)^{1/2} & 0 & \left(\varepsilon - p_m^2\right)\left(\varepsilon - p_m^2 - q_m^2\right)^{1/2} & \left(\varepsilon - p_m^2 - q_m^2\right) & 0 \end{array}\right).$$

We subtract the first row multiplied by $p_m q_m$ from the fourth one and divide the third row by $p_m q_m$. We observe that the fourth row differs from the third one by a common factor of $(\varepsilon - p_m^2)$, only. The matrix reduces to the form

$$\left(\begin{array}{cccc|c} \left(\varepsilon - p_m^2 - q_m^2\right)^{1/2} & \left(p_m^2/\varepsilon - 1\right) & 0 & -p_m q_m/\varepsilon & 0 \\ 0 & 0 & 0 & 0 & 0 \\ 0 & p_m q_m/\varepsilon & \left(\varepsilon - p_m^2 - q_m^2\right)^{1/2} & \left(1 - q_m^2/\varepsilon\right) & 0 \\ 0 & 0 & 0 & 0 & 0 \end{array}\right).$$

In order to determine the four unknown quantities x_1, x_2, x_3, and x_4, which give the proper vector belonging to the proper value $\kappa_{m1}^{(u)}$ of the matrix $\mathbf{C}_m^{(u)}$, we have two equations, only

$$\begin{pmatrix} \left(\varepsilon - p_m^2 - q_m^2\right)^{1/2} & \left(p_m^2/\varepsilon - 1\right) & 0 & -p_m q_m/\varepsilon \\ 0 & 0 & 0 & 0 \\ 0 & p_m q_m/\varepsilon & \left(\varepsilon - p_m^2 - q_m^2\right)^{1/2} & \left(1 - q_m^2/\varepsilon\right) \\ 0 & 0 & 0 & 0 \end{pmatrix} \begin{pmatrix} x_1 \\ x_2 \\ x_3 \\ x_4 \end{pmatrix} = \begin{pmatrix} 0 \\ 0 \\ 0 \\ 0 \end{pmatrix}.$$

Two of the four unknowns can be chosen arbitrarily

$$x_2 = \alpha, \qquad x_4 = \beta. \tag{9.133}$$

We obtain

$$\left(\varepsilon - p_m^2 - q_m^2\right)^{1/2} x_1 = \left(1 - p_m^2/\varepsilon\right)\alpha + (p_m q_m/\varepsilon)\beta \tag{9.134a}$$

$$\left(\varepsilon - p_m^2 - q_m^2\right)^{1/2} x_3 = -(p_m q_m/\varepsilon)\alpha - \left(1 - q_m^2/\varepsilon\right)\beta. \tag{9.134b}$$

This gives

$$x_1 = \frac{\left(\varepsilon - p_m^2\right)\alpha + (p_m q_m)\beta}{\varepsilon\left(\varepsilon - p_m^2 - q_m^2\right)^{1/2}}, \tag{9.135a}$$

$$x_3 = \frac{-(p_m q_m)\alpha - \left(\varepsilon - q_m^2\right)\beta}{\varepsilon\left(\varepsilon - p_m^2 - q_m^2\right)^{1/2}}. \tag{9.135b}$$

The proper vectors belonging to the proper values $\kappa_{m1}^{(u)} = \kappa_{m2}^{(u)}$ will be, up to a common factor,

$$\boldsymbol{\nu}_m^{(u)} = \begin{pmatrix} \left(\varepsilon - p_m^2\right)\alpha + (p_m q_m)\beta \\ \varepsilon\left(\varepsilon - p_m^2 - q_m^2\right)^{1/2}\alpha \\ -(p_m q_m)\alpha - \left(\varepsilon - q_m^2\right)\beta \\ \varepsilon\left(\varepsilon - p_m^2 - q_m^2\right)^{1/2}\beta \end{pmatrix}. \tag{9.136}$$

We substitute this result into Eq. (9.117)

$$\begin{pmatrix} e_{xm} \\ h_{xm} \end{pmatrix} = \begin{pmatrix} 0 & -p_m/\varepsilon & 0 & q_m/\varepsilon \\ -q_m & 0 & p_m & 0 \end{pmatrix} \begin{pmatrix} \left(\varepsilon - p_m^2\right)\alpha + (p_m q_m)\beta \\ \varepsilon\left(\varepsilon - p_m^2 - q_m^2\right)^{1/2}\alpha \\ -(p_m q_m)\alpha - \left(\varepsilon - q_m^2\right)\beta \\ \varepsilon\left(\varepsilon - p_m^2 - q_m^2\right)^{1/2}\beta \end{pmatrix}$$

$$= \begin{pmatrix} -(p_m/\varepsilon)\,\varepsilon\left(\varepsilon - p_m^2 - q_m^2\right)^{1/2}\alpha + (q_m/\varepsilon)\,\varepsilon\left(\varepsilon - p_m^2 - q_m^2\right)^{1/2}\beta \\ -q_m\left[\left(\varepsilon - p_m^2\right)\alpha + (p_m q_m)\beta\right] - p_m\left[(p_m q_m)\alpha + \left(\varepsilon - q_m^2\right)\beta\right] \end{pmatrix}$$

$$= \begin{pmatrix} \left(\varepsilon - p_m^2 - q_m^2\right)^{1/2}(-p_m\alpha + q_m\beta) \\ -\varepsilon\,(q_m\alpha + p_m\beta) \end{pmatrix}. \tag{9.137}$$

We now choose the arbitrary parameters α and β in a way to achieve either $e_{zm} = 0$, which corresponds to TE waves, or $h_{zm} = 0$, which corresponds to TM waves. The first case ($e_{zm} = 0$) is obtained with the choice $\alpha = (\frac{q_m}{p_m}\beta)$ in the expression for the proper vector (9.136)

$$\begin{pmatrix} \left(\varepsilon - p_m^2\right)\left(\dfrac{q_m}{p_m}\beta\right) + (p_m q_m)\beta \\ \varepsilon\left(\varepsilon - p_m^2 - q_m^2\right)^{1/2}\left(\dfrac{q_m}{p_m}\beta\right) \\ -(p_m q_m)\left(\dfrac{q_m}{p_m}\beta\right) - \left(\varepsilon - q_m^2\right)\beta \\ \varepsilon\left(\varepsilon - p_m^2 - q_m^2\right)^{1/2}\beta \end{pmatrix} = \beta \begin{pmatrix} \varepsilon\left(\dfrac{q_m}{p_m}\right) \\ \varepsilon\left(\varepsilon - p_m^2 - q_m^2\right)^{1/2}\left(\dfrac{q_m}{p_m}\right) \\ -\varepsilon \\ \varepsilon\left(\varepsilon - p_m^2 - q_m^2\right)^{1/2} \end{pmatrix}$$

$$= \frac{\beta\varepsilon}{p_m}\begin{pmatrix} q_m \\ q_m\xi_m \\ -p_m \\ p_m\xi_m \end{pmatrix}, \tag{9.138}$$

where $\xi_m = (\varepsilon - p_m^2 - q_m^2)^{1/2}$.

The second case ($h_{zm} = 0$) is obtained with the choice $\alpha = (-\frac{p_m}{q_m}\beta)$ in the equation for the proper vector (9.136)

$$\begin{pmatrix} (\varepsilon - p_m^2)\left(-\dfrac{p_m}{q_m}\beta\right) + (p_m q_m)\,\beta \\ \varepsilon\left(\varepsilon - p_m^2 - q_m^2\right)^{1/2}\left(-\dfrac{p_m}{q_m}\beta\right) \\ -(p_m q_m)\left(-\dfrac{p_m}{q_m}\beta\right) - (\varepsilon - q_m^2)\,\beta \\ \varepsilon\left(\varepsilon - p_m^2 - q_m^2\right)^{1/2}\beta \end{pmatrix}$$

$$= \beta \begin{pmatrix} p_m\left[-(\varepsilon - p_m^2 - q_m^2)/q_m\right] \\ \varepsilon p_m\left(\varepsilon - p_m^2 - q_m^2\right)^{-1/2}\left[-(\varepsilon - p_m^2 - q_m^2)/q_m\right] \\ q_m\left[-(\varepsilon - p_m^2 - q_m^2)/q_m\right] \\ -\varepsilon q_m\left(\varepsilon - p_m^2 - q_m^2\right)^{-1/2}\left[-(\varepsilon - p_m^2 - q_m^2)/q_m\right] \end{pmatrix}$$

$$= \beta\left[-(\varepsilon - p_m^2 - q_m^2)/q_m\right] \begin{pmatrix} p_m \\ \varepsilon p_m \xi_m^{-1} \\ q_m \\ -\varepsilon q_m \xi_m^{-1} \end{pmatrix}. \tag{9.139}$$

These are the proper vectors of the TM waves propagating with the normalized propagation vector $\boldsymbol{n}_m^{(u)} = \xi_m \hat{\boldsymbol{x}} + p_m \hat{\boldsymbol{y}} + q_m \hat{\boldsymbol{z}}$. As mentioned above, the proper vectors are determined up to an arbitrary common factor. We shall choose the proper vectors in the following way. In the first case of TE waves ($e_{zm} = 0$)

$$\boldsymbol{\nu}_{m1}^{(u)} = \left(p_m^2 + q_m^2\right)^{-1/2} \begin{pmatrix} q_m \\ q_m \xi_m \\ -p_m \\ p_m \xi_m \end{pmatrix}, \tag{9.140}$$

in the second case of TM waves ($h_{zm} = 0$)

$$\boldsymbol{\nu}_{m2}^{(u)} = \left(p_m^2 + q_m^2\right)^{-1/2} \begin{pmatrix} p_m \\ \varepsilon p_m \xi_m^{-1} \\ q_m \\ -\varepsilon p_m \xi_m^{-1} \end{pmatrix}. \tag{9.141}$$

Next, we find the proper vectors belonging to the proper value $\kappa_{m3}^{(u)} = \kappa_{m4}^{(u)} = +(\varepsilon - p_m^2 - q_m^2)^{1/2} = +\xi_m$ obtained in Eq. (9.132b). We arrive at a homogeneous equation system represented in the extended matrix form as

$$\left(\begin{array}{cccc|c} -(\varepsilon - p_m^2 - q_m^2)^{1/2} & p_m^2/\varepsilon - 1 & 0 & -p_m q_m/\varepsilon & 0 \\ q_m^2 - \varepsilon & -\left(\varepsilon - p_m^2 - q_m^2\right)^{1/2} & -p_m q_m & 0 & 0 \\ 0 & p_m q_m/\varepsilon & -\left(\varepsilon - p_m^2 - q_m^2\right)^{1/2} & 1 - q_m^2/\varepsilon & 0 \\ p_m q_m & 0 & \varepsilon - p_m^2 & -\left(\varepsilon - p_m^2 - q_m^2\right)^{1/2} & 0 \end{array}\right).$$

We multiply the first row by $(q_m^2 - \varepsilon)$ and add this to the second row multiplied by $\xi_m = (\varepsilon - p_m^2 - q_m^2)^{1/2}$

$$\left(\begin{array}{cccc|c} -\xi_m & (p_m^2/\varepsilon - 1) & 0 & -(p_m q_m/\varepsilon) & 0 \\ 0 & (p_m^2 q_m^2/\varepsilon) & -p_m q_m \xi_m & (p_m q_m/\varepsilon)\,(q_m^2 - \varepsilon) & 0 \\ 0 & (p_m q_m/\varepsilon) & -\xi_m & (1 - q_m^2/\varepsilon) & 0 \\ p_m q_m & 0 & (\varepsilon - p_m^2) & -\xi_m & 0 \end{array}\right).$$

We add the first row multiplied by $p_m q_m$ to the fourth row multiplied by ξ_m. Next, we add the third row multiplied by $-p_m q_m$ to the second one and obtain

$$\left(\begin{array}{cccc|c} -\xi_m & (p_m^2/\varepsilon - 1) & 0 & -(p_m q_m/\varepsilon) & 0 \\ 0 & 0 & 0 & 0 & 0 \\ 0 & (p_m q_m/\varepsilon) & -\xi_m & (1 - q_m^2/\varepsilon) & 0 \\ 0 & p_m q_m\,(p_m^2/\varepsilon - 1) & \xi_m\,(\varepsilon - p_m^2) & -\varepsilon + p_m^2 + q_m^2 - p_m^2 q_m^2/\varepsilon & 0 \end{array}\right).$$

Note that $-\varepsilon + p_m^2 + q_m^2 - p_m^2 q_m^2/\varepsilon = -(1 - q_m^2/\varepsilon)(\varepsilon - p_m^2)$. Finally, we multiply the third row by $(\varepsilon - p_m^2)$ and add this to the fourth one

$$\left(\begin{array}{cccc|c} -\xi_m & (p_m^2/\varepsilon - 1) & 0 & -(p_m q_m/\varepsilon) & 0 \\ 0 & 0 & 0 & 0 & 0 \\ 0 & (p_m q_m/\varepsilon) & -\xi_m & (1 - q_m^2/\varepsilon) & 0 \\ 0 & 0 & 0 & 0 & 0 \end{array}\right).$$

We have to compute the four unknowns x_1, x_2, x_3, and x_4, which define the proper vector belonging to the proper value $\kappa_{m3}^{(u)} = \kappa_{m4}^{(u)} = p_m + \xi_m$ of the matrix $\mathbf{C}_m^{(u)}$ from the two equations

$$\begin{pmatrix} -\xi_m & (p_m^2/\varepsilon - 1) & 0 & -(p_m q_m/\varepsilon) \\ 0 & 0 & 0 & 0 \\ 0 & (p_m q_m/\varepsilon) & -\xi_m & (1 - q_m^2/\varepsilon) \\ 0 & 0 & 0 & 0 \end{pmatrix} \begin{pmatrix} x_1 \\ x_2 \\ x_3 \\ x_4 \end{pmatrix} = \begin{pmatrix} 0 \\ 0 \\ 0 \\ 0 \end{pmatrix}.$$

Two unknowns can be chosen arbitrarily

$$x_2 = \gamma, \qquad x_4 = \delta. \tag{9.142}$$

We obtain

$$-\xi_m x_1 = -\left(p_m^2/\varepsilon - 1\right)\gamma + \left(p_m q_m/\varepsilon\right)\delta \tag{9.143a}$$

$$-\xi_m x_3 = -\left(p_m q_m/\varepsilon\right)\gamma - \left(1 - q_m^2/\varepsilon\right)\delta. \tag{9.143b}$$

The proper vectors belonging to the proper value $\kappa_{m3}^{(u)} = \kappa_{m4}^{(u)} = p_m + \xi_m$ should be constructed from

$$\boldsymbol{\nu}_m^{(u)} = \begin{pmatrix} -\left(\varepsilon - p_m^2\right)\gamma - \left(p_m q_m\right)\delta \\ \varepsilon\left(\varepsilon - p_m^2 - q_m^2\right)^{1/2}\gamma \\ \left(p_m q_m\right)\gamma + \left(\varepsilon - q_m^2\right)\delta \\ \varepsilon\left(\varepsilon - p_m^2 - q_m^2\right)^{1/2}\delta \end{pmatrix}. \tag{9.144}$$

The choice of γ and δ is again given by the requirement that proper vectors fulfill either $e_{zm} = 0$ (TE waves) or $h_{zm} = 0$ (TM waves). To this aim, we substitute the vector (9.144) into Eq. (9.117)

$$\begin{aligned} \begin{pmatrix} e_{zm} \\ h_{zm} \end{pmatrix} &= \begin{pmatrix} 0 & -p_m/\varepsilon & 0 & q_m/\varepsilon \\ -q_m & 0 & p_m & 0 \end{pmatrix} \begin{pmatrix} -\left(\varepsilon - p_m^2\right)\gamma - \left(p_m q_m\right)\delta \\ \varepsilon\left(\varepsilon - p_m^2 - q_m^2\right)^{1/2}\gamma \\ \left(p_m q_m\right)\gamma + \left(\varepsilon - q_m^2\right)\delta \\ \varepsilon\left(\varepsilon - p_m^2 - q_m^2\right)^{1/2}\delta \end{pmatrix} \\ &= \begin{pmatrix} -\left(p_m/\varepsilon\right)\varepsilon\xi_m\gamma + \left(q_m/\varepsilon\right)\varepsilon\xi_m\delta \\ q_m\left(\varepsilon - q_m^2\right)\gamma + p_m q_m^2\delta + p_m^2 q_m\gamma + p_m\left(\varepsilon - q_m^2\right)\delta \end{pmatrix} \\ &= \begin{pmatrix} \left(-p_m\gamma + q_m\delta\right)\xi_m \\ \varepsilon\left(q_m\gamma + p_m\delta\right) \end{pmatrix}. \end{aligned} \tag{9.145}$$

The condition $e_{xm} = 0$ (TE waves) is fulfilled, provided $\gamma = -q_m/\varepsilon$ and $\delta = -p_m/\varepsilon$

$$\boldsymbol{\nu}_{m3}^{(u)} = \left(p_m^2 + q_m^2\right)^{-1/2} \begin{pmatrix} q_m \\ -q_m\xi_m \\ -p_m \\ -p_m\xi_m \end{pmatrix}. \tag{9.146}$$

The condition $h_{xm} = 0$ (TM waves) requires $\gamma = -p_m \xi_m^{-2}$ and $\delta = q_m \xi_m^{-2}$ $(h_{xm} = 0)$

$$\boldsymbol{\nu}_{m4}^{(u)} = \left(p_m^2 + q_m^2\right)^{-1/2} \begin{pmatrix} p_m \\ -\varepsilon p_m \xi_m^{-1} \\ q_m \\ \varepsilon q_m \xi_m^{-1} \end{pmatrix}. \tag{9.147}$$

This completes our analysis of the proper vectors in homogeneous isotropic regions. We can now construct the matrix $\mathbf{T}^{(u)}$ transforming $\mathbf{C}^{(u)}$ to the diagonal form

$$\mathbf{T}_m^{(u)} = \left(\boldsymbol{\nu}_{m1}^{(u)}, \boldsymbol{\nu}_{m2}^{(u)}, \boldsymbol{\nu}_{m3}^{(u)}, \boldsymbol{\nu}_{m4}^{(u)}\right), \tag{9.148}$$

or in more detail

$$\mathbf{T}_m^{(u)} = \begin{pmatrix} \dot{q}_m & \dot{q}_m & \dot{q}_m & \dot{q}_m \\ \dot{q}_m \xi_m & \varepsilon \dot{q}_m \xi_m^{-1} & -\dot{q}_m \xi_m & -\varepsilon \dot{q}_m \xi_m^{-1} \\ -\dot{q}_m & \dot{q}_m & -\dot{q}_m & \dot{q}_m \\ \dot{q}_m \xi_m & -\varepsilon \dot{q}_m \xi_m^{-1} & -\dot{q}_m \xi_m & \varepsilon \dot{q}_m \xi_m^{-1} \end{pmatrix}, \tag{9.149}$$

where

$$\xi_m = \left(\varepsilon - n_{\dagger m}^2\right)^{1/2}, \tag{9.150a}$$

$$n_{\dagger m} = \left(p_m^2 + q_m^2\right)^{1/2}, \tag{9.150b}$$

$$\dot{p}_m = p_m / n_{\dagger m}, \tag{9.150c}$$

$$\dot{q}_m = q_m / n_{\dagger m}. \tag{9.150d}$$

Note that the determinant of the matrix $\mathbf{T}_m^{(u)}$ and the matrix inverse to $\mathbf{T}_m^{(u)}$ become, respectively,

$$\det \mathbf{T}_m^{(u)} = 4\varepsilon \tag{9.151}$$

$$\left[\mathbf{T}_m^{(u)}\right]^{-1} = \frac{1}{2\varepsilon} \begin{pmatrix} \dot{q}_m \varepsilon & \dot{q}_m \varepsilon \xi_m^{-1} & -\dot{p}_m \varepsilon & \dot{p}_m \varepsilon \xi_m^{-1} \\ \dot{p}_m \varepsilon & \dot{p}_m \xi_m & \dot{q}_m \varepsilon & -\dot{q}_m \xi_m \\ \dot{q}_m \varepsilon & -\dot{q}_m \varepsilon \xi_m^{-1} & -\dot{p}_m \varepsilon & -\dot{p}_m \varepsilon \xi_m^{-1} \\ \dot{p}_m \varepsilon & -\dot{p}_m \xi_m & \dot{q}_m \varepsilon & \dot{q}_m \xi_m \end{pmatrix}. \tag{9.152}$$

We consider the solution of Eq. (9.127) corresponding to TE waves propagating with the proper values $\kappa_{m1} = -\xi_m$

$$g_m(\bar{z}) = \exp[\mathrm{j}(\kappa_{m1}\bar{z})] = \exp[-\mathrm{j}(\xi_m \bar{z})]. \tag{9.153}$$

Note that, according to Eq. (9.132a), the proper values for the TM waves are the same, *i.e.*, $\kappa_{m2} = -\xi_m$. The column vector $\mathbf{f}^{(u)}_{\bar{t}m1}$ defined in Eq. (9.122) is obtained from the proper vector $\boldsymbol{\nu}^{(u)}_{m1}$. The proper vector $\boldsymbol{\nu}^{(u)}_{m1}$ enters the matrix $\mathbf{T}^{(u)}_m$ given by Eq. (9.149). We obtain

$$\mathbf{f}^{(u)}_{\bar{t}m1} = \begin{pmatrix} e_{xm1}(\bar{z}) \\ h_{ym1}(\bar{z}) \\ e_{ym1}(\bar{z}) \\ h_{xm1}(\bar{z}) \end{pmatrix} = \mathbf{T}^{(u)}_m \mathbf{g}^{(u)}_{m1}$$
$$= \begin{pmatrix} \dot{q}_m & \dot{p}_m & \dot{q}_m & \dot{p}_m \\ \dot{q}_m\xi_m & \varepsilon\dot{p}_m\xi_m^{-1} & -\dot{q}_m\xi_m & -\varepsilon\dot{p}_m\xi_m^{-1} \\ -\dot{p}_m & \dot{q}_m & -\dot{p}_m & \dot{q}_m \\ \dot{p}_m\xi_m & -\varepsilon\dot{q}_m\xi_m^{-1} & -\dot{p}_m\xi_m & \varepsilon\dot{q}_m\xi_m^{-1} \end{pmatrix} \begin{pmatrix} 1 \\ 0 \\ 0 \\ 0 \end{pmatrix} \exp(-\mathrm{j}\xi_m\bar{z})$$
$$= \begin{pmatrix} \dot{q}_m \\ \dot{q}_m\xi_m \\ -\dot{p}_m \\ \dot{p}_m\xi_m \end{pmatrix} \exp(-\mathrm{j}\xi_m\bar{z}). \tag{9.154}$$

We shall show that this solution is consistent with Eq. (9.113a). On the left-hand side, we have

$$\frac{\mathrm{d}}{\mathrm{d}\bar{z}}\mathbf{f}^{(u)}_{\bar{t}m1} = -\mathrm{j}\xi_m\mathbf{f}^{(u)}_{\bar{t}m1}. \tag{9.155}$$

The right-hand side of Eq. (9.115) with $s_m = 0$ gives the same expression

$$\mathrm{j}\mathbf{C}^{(u)}_m\mathbf{f}^{(u)}_{\bar{t}m1}$$
$$= \mathrm{j}\begin{pmatrix} 0 & p_m^2/\varepsilon - 1 & 0 & -p_mq_m/\varepsilon \\ q_m^2 - \varepsilon & 0 & -p_mq_m & 0 \\ 0 & p_mq_m/\varepsilon & 0 & 1 - q_m^2/\varepsilon \\ q_ms_m & 0 & \varepsilon - q_m^2 & 0 \end{pmatrix} \begin{pmatrix} \dot{q}_m \\ \dot{q}_m\xi_m \\ -\dot{p}_m \\ \dot{p}_m\xi_m \end{pmatrix} \exp(-\mathrm{j}\xi_m\bar{z})$$
$$= \mathrm{j}\begin{pmatrix} (p_m^2/\varepsilon - 1)\,\dot{q}_m\xi_m - (p_mq_m/\varepsilon)\,\dot{p}_m\xi_m \\ (q_m^2 - \varepsilon)\,\dot{q}_m + p_mq_m\dot{p}_m \\ (p_mq_m/\varepsilon)\,\dot{q}_m\xi_m + (1 - q_m^2/\varepsilon)\,\dot{p}_m\xi_m \\ p_mq_m\dot{q}_m - (\varepsilon - p_m^2)\,\dot{p}_m \end{pmatrix} \exp(-\mathrm{j}\xi_m\bar{z})$$
$$= \mathrm{j}\begin{pmatrix} -\dot{q}_m\xi_m \\ -\dot{q}_m\xi_m^2 \\ \dot{p}_m\xi_m \\ -\dot{p}_m\xi_m^2 \end{pmatrix} \exp(-\mathrm{j}\xi_m\bar{z}) = -\mathrm{j}\xi_m\mathbf{f}^{(u)}_{\bar{t}m1} = -\mathrm{j}\xi_m \begin{pmatrix} e_{xm1}(\bar{z}) \\ h_{ym1}(\bar{z}) \\ e_{ym1}(\bar{z}) \\ h_{xm1}(\bar{z}) \end{pmatrix}. \tag{9.156}$$

The components normal with respect to the interfaces, *i.e.*, the $\bar{z}$ components in the present choice of the Cartesian coordinate system, follow from Eq. (9.113b). As expected, using Eq. (9.117), we obtain $e_{zm1}(\bar{z}) = 0$

$$f_{\hat{n}m1}^{(u)} = \begin{pmatrix} e_{zm1}(\bar{z}) \\ h_{zm1}(\bar{z}) \end{pmatrix} = \begin{pmatrix} 0 & -p_m/\varepsilon & 0 & q_m/\varepsilon \\ -q_m & 0 & p_m & 0 \end{pmatrix} \begin{pmatrix} \dot{q}_m \\ \dot{q}_m\xi_m \\ -\dot{p}_m \\ \dot{p}_m\xi_m \end{pmatrix} \exp(-\mathrm{j}\xi_m\bar{z})$$

$$= \begin{pmatrix} -(p_m/\varepsilon)\,\dot{q}_m\xi_m + (q_m/\varepsilon)\,\dot{p}_m\xi_m \\ -q_m\dot{q}_m - p_m\dot{p}_m \end{pmatrix} \exp(-\mathrm{j}\xi_m\bar{z})$$

$$= \begin{pmatrix} 0 \\ -(q_m\dot{q}_m + p_m\dot{p}_m) \end{pmatrix} \exp(-\mathrm{j}\xi_m\bar{z})$$

$$= \begin{pmatrix} 0 \\ -n_{\hat{t}m} \end{pmatrix} \exp(-\mathrm{j}\xi_m\bar{z}). \tag{9.157}$$

The characteristic field, with the factor $\exp(-\mathrm{j}\xi_m\bar{z})$ skipped, is given by

$$\mathbf{e}_{m1} = \mathbf{e}_{\hat{t}m1} = \hat{x}\dot{q}_m - \hat{y}\dot{p}_m, \tag{9.158a}$$

$$\boldsymbol{h}_{m1} = \mathbf{h}_{\hat{t}m1} + \hat{z}h_{zm1} = \xi_m(\hat{x}\dot{p}_m + \hat{y}\dot{q}_m) - \hat{z}n_{\hat{t}m}. \tag{9.158b}$$

The Poynting vector is proportional to

$$\begin{aligned} \boldsymbol{e}_{m1} \times \boldsymbol{h}_{m1} &= (\hat{x}\dot{q}_m - \hat{y}\dot{p}_m) \times [\xi_m(\hat{x}\dot{p}_m + \hat{y}\dot{q}_m) - \hat{z}n_{\hat{t}m}] \\ &= \hat{z}\left(\dot{q}_m^2 + \dot{p}_m^2\right)\xi_m + n_{\hat{t}m}(\hat{x}\dot{p}_m + \hat{y}\dot{q}_m) \\ &= \hat{z}\left(\dot{q}_m^2 + \dot{p}_m^2\right)\xi_m + n_{\hat{t}m}(\hat{x}\dot{p}_m + \hat{y}\dot{q}_m) \\ &= \hat{z}\xi_m + \hat{x}p_m + \hat{y}q_m. \end{aligned} \tag{9.159}$$

The waves propagate parallel to $\boldsymbol{n}_m^{(u)+} = \hat{x}p_m + \hat{y}q_m + \hat{z}\xi_m$. These are the TE waves,[1] and their electric field is polarized normal with respect to the plane of incidence specified by the vector normal to the interfaces and the vector $\mathbf{n}_m^{(u)+}$.

In homogeneous isotropic media, these waves are transverse with respect to the propagation direction determined by $\mathbf{n}_m^{(u)+}$. The scalar product

[1] Note that the definition of TE and TM wave becomes here more general than that employed in the case of the grating vector in the plane of incidence, where $q = 0$, as well as that employed in the homogeneous isotropic waveguides.

of the electric field and $\mathbf{n}_m^{(u)+}$ is then zero

$$\begin{aligned}\mathbf{e}_{m1} \cdot \boldsymbol{n}_m^{(u)+} &= (\hat{x}\dot{q}_m - \hat{y}\dot{p}_m) \cdot (\hat{x}p_m + \hat{y}q_m + \hat{z}\xi_m) \\ &= \dot{p}_m q_m - \dot{p}_m q_m = 0,\end{aligned} \tag{9.160}$$

and the same condition is valid for the magnetic fields

$$\begin{aligned}\mathbf{h}_{m1} \cdot \boldsymbol{n}_m^{(u)+} &= [-\hat{z}(q_m\dot{q}_m + p_m\dot{p}_m) + \xi_m(\hat{x}\dot{p}_m + \hat{y}\dot{q}_m)] \cdot (\hat{z}\xi_m + \hat{x}p_m + \hat{y}q_m) \\ &= -n_{\hat{t}m}\xi_m + (p_m\dot{p}_m + q_m\dot{q}_m)\xi_m \\ &= \left[-\left(\dot{p}_m^2 + \dot{q}_m^2\right)^{1/2}\xi_m + \frac{(p_m^2 + q_m^2)}{\left(p_m^2 + q_m^2\right)^{1/2}}\xi_m\right] = 0.\end{aligned} \tag{9.161}$$

In an isotropic and homogeneous region, the vectors of the electric and magnetic wave fields are perpendicular to each other and satisfy $\mathbf{e}_{m1} \cdot \mathbf{h}_{m1} = 0$, as shown below

$$\begin{aligned}&(\hat{x}s\dot{q}_m - \hat{y}\dot{p}_m) \cdot [-\hat{z}(q_m\dot{q}_m + p_m\dot{p}_m) + \xi_m(\hat{x}\dot{p}_m + \hat{y}\dot{q}_m)] \\ &\quad = (\dot{p}_m\dot{q}_m - \dot{p}_m\dot{q}_m)\xi_m = 0.\end{aligned} \tag{9.162}$$

The electric field of TE waves in the isotropic and homogeneous regions is normalized to the unit amplitude thanks to the choice of the common factor in Eq. (9.140)

$$|\mathbf{e}_{m1}| = |\hat{x}\dot{q}_m - \hat{y}\dot{p}_m| = \left(\dot{q}_m^2 + \dot{p}_m^2\right)^{1/2} = \left(\frac{q_m^2 + p_m^2}{q_m^2 + p_m^2}\right)^{1/2} = 1. \tag{9.163}$$

The magnetic field of TE waves in the isotropic and homogeneous regions is normalized to the square root of the relative permittivity

$$\begin{aligned}|\mathbf{h}_{m1}| &= |-\hat{z}n_{\hat{t}m} + \xi_m(\hat{x}\dot{p}_m + \hat{y}\dot{q}_m)| \\ &= \left[\left(q_m^2 + p_m^2\right) + \left(\varepsilon - p_m^2 - q_m^2\right) \cdot 1\right]^{1/2} = \varepsilon^{1/2}.\end{aligned} \tag{9.164}$$

The solution of Eq. (9.127) corresponding to the TM waves is given by

$$g_{m2}(\bar{z}) = \exp[\mathrm{j}(\kappa_{m2}\bar{z})] = \exp(-\mathrm{j}(\xi_m\bar{z})]. \tag{9.165}$$

The column vector $\mathbf{f}_{\hat{t}m2}^{(u)}$ defined in Eq. (9.122) is obtained from the proper vector $\boldsymbol{\nu}_{m2}^{(u)}$ for TM waves associated to the proper value $\kappa_{m2} = -\xi_m$ entering the matrix $\mathbf{T}_m^{(u)}$ given in Eq. (9.149).

We obtain for $\mathbf{f}_{\hat{t}m2}^{(u)}$

$$\mathbf{f}_{\hat{t}m2}^{(u)} = \begin{pmatrix} e_{xm2}(\bar{z}) \\ h_{ym2}(\bar{z}) \\ e_{ym2}(\bar{z}) \\ h_{xm2}(\bar{z}) \end{pmatrix} = \mathbf{T}_m^{(u)} \mathbf{g}_{m2}^{(u)}$$

$$= \begin{pmatrix} \dot{q}_m & \dot{p}_m & \dot{q}_m & \dot{p}_m \\ \dot{q}_m \xi_m & \varepsilon \dot{p}_m \xi_m^{-1} & -\dot{q}_m \xi_m & -\varepsilon \dot{p}_m \xi_m^{-1} \\ -\dot{p}_m & \dot{q}_m & -\dot{p}_m & \dot{q}_m \\ \dot{p}_m \xi_m & -\varepsilon \dot{q}_m \xi_m^{-1} & -\dot{p}_m \xi_m & \varepsilon \dot{q}_m \xi_m^{-1} \end{pmatrix} \begin{pmatrix} 0 \\ 1 \\ 0 \\ 0 \end{pmatrix} \exp(-\mathrm{j}\xi_m \bar{z})$$

$$= \begin{pmatrix} \dot{p}_m \\ \varepsilon \dot{p}_m / \xi_m \\ \dot{q}_m \\ -\varepsilon \dot{q}_m / \xi_m \end{pmatrix} \exp(-\mathrm{j}\xi_m \bar{z}). \tag{9.166}$$

We wish to show that this solution is consistent with Eq. (9.113a). On the left-hand side, we have

$$\frac{\mathrm{d}}{\mathrm{d}\bar{z}} \mathbf{f}_{\hat{t}m2}^{(u)} = -\mathrm{j}\xi_m \mathbf{f}_{\hat{t}m2}^{(u)}. \tag{9.167}$$

The right-hand side gives, according to Eq. (9.115) with $s_m = 0$,

$$\mathrm{j}\mathbf{C}_m^{(u)} \mathbf{f}_{\hat{t}m2}^{(u)}$$

$$= \mathrm{j} \begin{pmatrix} 0 & p_m^2/\varepsilon - 1 & 0 & -p_m q_m/\varepsilon \\ q_m^2 - \varepsilon & 0 & -p_m q_m & 0 \\ 0 & p_m q_m/\varepsilon & 0 & 1 - q_m^2/\varepsilon \\ p_m q_m & 0 & \varepsilon - p_m^2 & 0 \end{pmatrix} \begin{pmatrix} \dot{p}_m \\ \varepsilon \dot{p}_m / \xi_m \\ \dot{q}_m \\ -\varepsilon \dot{q}_m / \xi_m \end{pmatrix} \exp(-\mathrm{j}\xi_m \bar{z})$$

$$= \mathrm{j} \begin{pmatrix} \left(p_m^2/\varepsilon - 1\right) \varepsilon \dot{p}_m \xi_m + (p_m q_m/\varepsilon)\, \varepsilon \dot{q}_m / \xi_m \\ \left(q_m^2 - \varepsilon\right) \dot{p}_m - p_m q_m \dot{q}_m \\ (p_m q_m/\varepsilon)\, \varepsilon \dot{p}_m / \xi_m - \left(1 - q_m^2/\varepsilon\right) \varepsilon \dot{q}_m / \xi_m \\ p_m q_m \dot{p}_m + \left(\varepsilon - p_m^2\right) \dot{q}_m \end{pmatrix} \exp(-\mathrm{j}\xi_m \bar{z})$$

$$= \mathrm{j} \begin{pmatrix} -\dot{p}_m \xi_m \\ -\varepsilon \dot{p}_m \xi_m \\ -\dot{q}_m \xi_m \\ \varepsilon \dot{q}_m \xi_m \end{pmatrix} \exp(-\mathrm{j}\xi_m \bar{z}) = -\mathrm{j}\xi_m \mathbf{f}_{\hat{t}m2}^{(u)} = -\mathrm{j}\xi_m \begin{pmatrix} e_{xm2}(\bar{z}) \\ h_{ym2}(\bar{z}) \\ e_{ym2}(\bar{z}) \\ h_{xm2}(\bar{z}) \end{pmatrix}. \tag{9.168}$$

In this way, we have shown that the vector $\mathbf{f}_{\hat{t}m2}^{(u)}$ represents the solution of Eq. (9.113a) corresponding to the proper value $\kappa_{m2} = -\xi_m$. For the normal field components, we obtain from Eq. (9.113b) with Eq. (9.117) for the $\bar{z}$ components the expected result $h_{zm2}(\bar{z}) = 0$

$$f_{\mathfrak{h}m2}^{(u)} = \begin{pmatrix} e_{zm2}(\bar{z}) \\ h_{zm2}(\bar{z}) \end{pmatrix}$$

$$= \begin{pmatrix} 0 & -p_m/\varepsilon & 0 & q_m/\varepsilon \\ -q_m & 0 & p_m & 0 \end{pmatrix} \begin{pmatrix} \dot{p}_m \\ \varepsilon \dot{p}_m/\xi_m \\ \dot{q}_m \\ -\varepsilon \dot{q}_m/\xi_m \end{pmatrix} \exp(-\mathrm{j}\xi_m \bar{z})$$

$$= \begin{pmatrix} -p_m \dot{p}_m/\xi_m - q_m \dot{q}_m/\xi_m \\ -q_m \dot{p}_m + p_m \dot{q}_m \end{pmatrix} \exp(-\mathrm{j}\xi_m \bar{z})$$

$$= \begin{pmatrix} -n_{\mathfrak{t}m}/\xi_m \\ 0 \end{pmatrix} \exp(-\mathrm{j}\xi_m \bar{z}). \tag{9.169}$$

The characteristic electric and magnetic fields, with the factor $\exp(-\mathrm{j}\xi_m \bar{z})$ skipped, become

$$\mathbf{e}_{m2} = = -\hat{z} n_{\mathfrak{t}m}/\xi_m + (\hat{x}\dot{p}_m + \hat{y}\dot{q}_m), \tag{9.170a}$$

$$\mathbf{h}_{m2} = (-\hat{x}\dot{q}_m + \hat{y}\dot{p}_m)\varepsilon/\xi_m, \tag{9.170b}$$

where the z component $h_{zm2} = 0$, as we are concerned with TM waves. The Poynting vector is proportional to

$$\begin{aligned} \mathbf{e}_{m2} \times \mathbf{h}_{m2} &= [-\hat{z} n_{\mathfrak{t}m}/\xi_m + (\hat{x}\dot{p}_m + \hat{y}\dot{q}_m)] \times (-\hat{x}\dot{q}_m + \hat{y}\dot{p}_m)\varepsilon/\xi_m \\ &= \hat{z}\left(\dot{q}_m^2 + \dot{p}_m^2\right)\varepsilon/\xi_m + n_{\mathfrak{t}m}(\hat{x}\dot{p}_m + \hat{y}\dot{q}_m)\varepsilon/\xi_m^2 \\ &= \frac{\varepsilon}{\xi_m}\hat{z} + n_{\mathfrak{t}m}\frac{\varepsilon}{\xi_m^2}(\hat{x}\dot{q}_m + \hat{y}\dot{q}_m) \\ &= \frac{\varepsilon}{\xi_m^2}(\hat{x}p_m + \hat{y}q_m + \hat{z}\xi_m). \end{aligned} \tag{9.171}$$

The wave electric and magnetic fields are normal to $\mathbf{n}_m^{(u)+} = \hat{z}\xi_m + \hat{x}p_m + \hat{y}q_m$. The magnetic field of TM waves is polarized normal to the plane determined by the normal to the interfaces and the propagation vector $\mathbf{n}_m^{(u)+}$. The wave fields in isotropic homogeneous media are transverse with respect to the direction of the propagation vector $\mathbf{n}_m^{(u)+}$ as seen from the scalar product for electric fields

$$\begin{aligned} \mathbf{e}_{m2} \cdot \mathbf{n}_m^{(u)+} &= [-\hat{z} n_{\mathfrak{t}m}/\xi_m + (\hat{x}\dot{p}_m + \hat{y}\dot{q}_m)] \cdot (\hat{z}\xi_m + \hat{x}p_m + \hat{y}q_m) \\ &= (-n_{\mathfrak{t}m}/\xi_m)\xi_m + p_m\dot{p}_m + q_m\dot{q}_m = 0 \end{aligned} \tag{9.172}$$

and in a similar way for magnetic fields

$$\begin{aligned} \mathbf{h}_{m2} \cdot \mathbf{n}_m^{(u)+} &= [(-\hat{x}\dot{q}_m + \hat{y}\dot{p}_m)\varepsilon/\xi_m] \cdot (\hat{z}\xi_m + \hat{x}p_m + \hat{y}q_m) \\ &= -p_m\dot{q}_m + \dot{p}_m q_m = 0. \end{aligned} \tag{9.173}$$

The amplitude of the electric field for TM waves in homogeneous and isotropic regions is normalized according to Eq. (9.141) to the value

$$|\mathbf{e}_{m2}| = |-\hat{z}n_{\bar{t}m}/\xi_m + (\hat{x}\dot{p}_m + \hat{y}\dot{q}_m)|$$
$$= \left(\frac{p_m^2 + q_m^2}{\varepsilon - p_m^2 - q_m^2} + 1\right)^{1/2} = \left(\frac{p_m^2 + q_m^2 + \varepsilon - p_m^2 - q_m^2}{\varepsilon - p_m^2 - q_m^2}\right)^{1/2} = \frac{\varepsilon^{1/2}}{|\xi_m|}. \tag{9.174}$$

The amplitude of magnetic field in TM waves in homogeneous and isotropic regions is normalized to

$$|\mathbf{h}_{m2}| = |(-\hat{x}\dot{q}_m + \hat{y}\dot{p}_m)\varepsilon/\xi_m| = \frac{\varepsilon}{|\xi_m|}. \tag{9.175}$$

The TE and TM waves propagate as $\exp(-j\xi_m\bar{z})$ in the direction of $\mathbf{n}_m^{(u)+}$ polarized normal to each other

$$\mathbf{e}_{m1} \cdot \mathbf{e}_{m2} = (\hat{x}\dot{q}_m - \hat{y}\dot{p}_m) \cdot [-\hat{z}n_{\bar{t}m}/\xi_m + (\hat{x}\dot{p}_m + \hat{y}\dot{q}_m)] = 0, \tag{9.176a}$$

$$\mathbf{h}_{m1} \cdot \mathbf{h}_{m2} = [-\hat{z}n_{\bar{t}m} + (\hat{x}\dot{p}_m + \hat{y}\dot{q}_m)\xi_m] \cdot (\hat{x}\dot{q}_m - \hat{y}\dot{p}_m)\varepsilon/\xi_m = 0, \tag{9.176b}$$

$$\mathbf{e}_{m1} \times \mathbf{h}_{m2} = 0, \qquad \mathbf{e}_{m2} \times \mathbf{h}_{m1} = 0. \tag{9.176c}$$

The solution of Eq. (9.127) for TE waves corresponding to the proper value $\kappa_{m3} = \xi_m$ is given by

$$g_{m3}(\bar{z}) = \exp[j(\kappa_{m3}\bar{z})] = \exp(j(\xi_m\bar{z})]. \tag{9.177}$$

The column vector $\mathbf{f}_{\bar{t}m3}^{(u)}$ is obtained using the proper vector ν_{m3} from Eq. (9.146) associated to the proper value $\kappa_{m3} = \xi_m$

$$\mathbf{f}_{\bar{t}m3}^{(u)} = \begin{pmatrix} e_{xm3}(\bar{z}) \\ h_{ym3}(\bar{z}) \\ e_{ym3}(\bar{z}) \\ h_{xm3}(\bar{z}) \end{pmatrix} = \mathbf{T}_m^{(u)}\mathbf{g}_{m3}^{(u)}$$
$$= \begin{pmatrix} \dot{q}_m & \dot{p}_m & \dot{q}_m & \dot{p}_m \\ \dot{q}_m\xi_m & \varepsilon\dot{p}_m\xi_m^{-1} & -\dot{q}_m\xi_m & -\varepsilon\dot{p}_m\xi_m^{-1} \\ -\dot{p}_m & \dot{q}_m & -\dot{p}_m & \dot{q}_m \\ \dot{p}_m\xi_m & -\varepsilon\dot{q}_m\xi_m^{-1} & -\dot{p}_m\xi_m & \varepsilon\dot{q}_m\xi_m^{-1} \end{pmatrix} \begin{pmatrix} 0 \\ 0 \\ 1 \\ 0 \end{pmatrix} \exp(j\xi_m\bar{z})$$
$$= \begin{pmatrix} \dot{q}_m \\ -\dot{q}_m\xi_m \\ -\dot{p}_m \\ -\dot{p}_m\xi_m \end{pmatrix} \exp(j\xi_m\bar{z}). \tag{9.178}$$

This solution is consistent with Eq. (9.113a). On the left-hand side, we have

$$\frac{\mathrm{d}}{\mathrm{d}\bar{z}}\mathbf{f}^{(u)}_{\hat{t}m3} = \mathrm{j}\xi_m \mathbf{f}^{(u)}_{\hat{t}m3} \tag{9.179}$$

and this is to be compared with the right-hand side of Eq. (9.115), always with $p_m = 0$

$$\begin{aligned}
\mathrm{j}\mathbf{C}^{(u)}_m \mathbf{f}^{(u)}_{\hat{t}m3} &= \begin{pmatrix} 0 & p_m^2/\varepsilon - 1 & 0 & -p_m s_m/\varepsilon \\ q_m^2 - \varepsilon & 0 & -p_m q_m & 0 \\ 0 & p_m q_m/\varepsilon & 0 & 1 - q_m^2/\varepsilon \\ p_m q_m & 0 & \varepsilon - p_m^2 & 0 \end{pmatrix} \begin{pmatrix} \dot{q}_m \\ -\dot{q}_m \xi_m \\ -\dot{p}_m \\ -\dot{p}_m \xi_m \end{pmatrix} \exp(\mathrm{j}\xi_m \bar{z}) \\
&= \mathrm{j} \begin{pmatrix} \left(p_m^2/\varepsilon - 1\right) \dot{q}_m \xi_m + \left(p_m q_m/\varepsilon\right) \dot{p}_m \xi_m \\ \left(q_m^2 - \varepsilon\right) \dot{q}_m + p_m q_m \dot{p}_m \\ -\left(p_m q_m/\varepsilon\right) \dot{q}_m \xi_m - \left(1 - q_m^2/\varepsilon\right) \dot{p}_m \xi_m \\ p_m q_m \dot{q}_m + \left(\varepsilon - p_m^2\right) \dot{p}_m \end{pmatrix} \exp(\mathrm{j}\xi_m \bar{z}) \\
&= \mathrm{j} \begin{pmatrix} \dot{q}_m \xi_m \\ -\dot{q}_m \xi_m^2 \\ -\dot{p}_m \xi_m \\ -\dot{p}_m \xi_m^2 \end{pmatrix} \exp(\mathrm{j}\xi_m \bar{z}) = \mathrm{j}\xi_m \mathbf{f}^{(u)}_{\hat{t}m3} = \mathrm{j}\xi_m \begin{pmatrix} e_{xm3}(\bar{z}) \\ h_{ym3}(\bar{z}) \\ e_{ym3}(\bar{z}) \\ h_{xm3}(\bar{z}) \end{pmatrix}.
\end{aligned} \tag{9.180}$$

In this way, we have shown that Eq. (9.127) is satisfied for the proper value $\kappa_{m3} = \xi_m$ and the proper vector $\mathbf{f}^{(u)}_{\hat{t}m3}$. The components normal to the interfaces, *i.e.*, the x components of the fields, follow from Eq. (9.113b). As expected, we have with Eq. (9.117) the result $h_{xm3}(\bar{z}) = 0$

$$\begin{aligned}
f^{(u)}_{\hat{n}m3} = \begin{pmatrix} e_{xm3}(\bar{z}) \\ h_{xm3}(\bar{z}) \end{pmatrix} &= \begin{pmatrix} 0 & -p_m/\varepsilon & 0 & q_m/\varepsilon \\ -q_m & 0 & p_m & 0 \end{pmatrix} \begin{pmatrix} \dot{q}_m \\ -\dot{q}_m \xi_m \\ -\dot{p}_m \\ -\dot{p}_m \xi_m \end{pmatrix} \exp(\mathrm{j}\xi_m \bar{z}) \\
&= \begin{pmatrix} -p_m \dot{p}_m/\xi_m - q_m \dot{q}_m/\xi_m \\ -q_m \dot{p}_m + p_m \dot{q}_m \end{pmatrix} \exp(\mathrm{j}\xi_m \bar{z}) \\
&= \begin{pmatrix} -n_{\hat{t}m}/\xi_m \\ 0 \end{pmatrix} \exp(\mathrm{j}\xi_m \bar{z}).
\end{aligned} \tag{9.181}$$

The characteristic fields propagating with $\exp(\mathrm{j}\xi_m \bar{z})$ are given, with the factor $\exp(\mathrm{j}\xi_m \bar{z})$ skipped, by

$$\mathbf{e}_{m3} = \mathbf{e}_{\hat{t}m3} = \hat{\boldsymbol{x}}\dot{q}_m - \hat{\boldsymbol{y}}\dot{p}_m, \tag{9.182a}$$

$$\mathbf{h}_{m3} = \hat{\boldsymbol{z}} h_{xm3} + \mathbf{h}_{\hat{t}m3} = -\hat{\boldsymbol{z}} n_{\hat{t}m} - \xi_m(\hat{\boldsymbol{x}}\dot{p}_m + \hat{\boldsymbol{y}}\dot{q}_m). \tag{9.182b}$$

The Poynting vector is proportional to

$$\begin{aligned}\mathbf{e}_{m3} \times \mathbf{h}_{m3} &= (\hat{x}\dot{q}_m - \hat{y}\dot{p}_m) \times [\hat{z}n_{\hat{t}m} - \xi_m(\hat{x}\dot{p}_m + \hat{y}\dot{q}_m)] \\ &= -\hat{z}\left(\dot{q}_m^2 + \dot{p}_m^2\right)\xi_m + n_{\hat{t}m}(\hat{x}\dot{q}_m + \hat{y}\dot{q}_m) \\ &= -\hat{z}\left(\dot{q}_m^2 + \dot{p}_m^2\right)\xi_m + n_{\hat{t}m}(\hat{x}\dot{p}_m + \hat{z}\dot{q}_m) \\ &= -\hat{z}\xi_m + \hat{x}p_m + \hat{y}q_m. \end{aligned} \tag{9.183}$$

The waves propagate parallel to $\mathbf{n}_m^{(u)-} = \hat{x}q_m + \hat{y}s_m - \hat{z}\xi_m$. These are the TE waves, and their electric field is polarized normal with respect to the plane of incidence specified by the vector normal to the interfaces and the vector $\mathbf{n}_m^{(u)-}$. In homogeneous isotropic media, these waves are transverse with respect to the propagation direction determined by $\mathbf{n}_m^{(u)-}$

$$\begin{aligned}\mathbf{e}_{m3} \cdot \mathbf{n}_m^{(u)-} &= (\hat{x}\dot{q}_m - \hat{y}\dot{p}_m) \cdot (-\hat{z}\xi_m + \hat{x}q_m + \hat{y}q_m) \\ &= \dot{p}_m q_m - \dot{p}_m q_m = 0, \end{aligned} \tag{9.184}$$

$$\begin{aligned}\mathbf{h}_{m3} \cdot \mathbf{n}_m^{(u)-} &= [-\hat{z}(q_m\dot{q}_m + p_m\dot{p}_m) - \xi_m(\hat{x}\dot{p}_m + \hat{y}\dot{q}_m)] \cdot (-\hat{z}\xi_m + \hat{x}p_m + \hat{y}q_m) \\ &= n_{\hat{t}m}\xi_m - (p_m\dot{p}_m + q_m\dot{q}_m)\xi_m \\ &= \left[\left(\dot{p}_m^2 + \dot{q}_m^2\right)^{1/2}\xi_m - \frac{\left(p_m^2 + q_m^2\right)}{\left(p_m^2 + q_m^2\right)^{1/2}}\xi_m\right] = 0. \end{aligned} \tag{9.185}$$

The solution of Eq. (9.127) corresponding to TM waves propagating in the direction of $\mathbf{n}_m^{(u)-} = \hat{x}p_m + \hat{y}q_m - \hat{z}\xi_m$ is given by

$$g_{m4}(\bar{z}) = \exp[\mathrm{j}(\kappa_{m4}\bar{z})] = \exp(\mathrm{j}(\xi_m\bar{z})]. \tag{9.186}$$

The column vector $\mathbf{f}_{\hat{t}m4}^{(u)}$ is obtained using the proper vector $v_{m4}^{(u)}$ from Eq. (9.147) associated to the proper value $\kappa_{m4} = \xi_m$. Eq. (9.122) gives for $\mathbf{f}_{\hat{t}m4}^{(u)}$

$$\begin{aligned}\mathbf{f}_{\hat{t}m4}^{(u)} &= \begin{pmatrix} e_{xm4}(\bar{z}) \\ h_{ym4}(\bar{z}) \\ e_{ym4}(\bar{z}) \\ h_{xm4}(\bar{z}) \end{pmatrix} = \mathbf{T}_m^{(u)}\mathbf{g}_{m4}^{(u)} \\ &= \begin{pmatrix} \dot{q}_m & \dot{p}_m & \dot{q}_m & \dot{p}_m \\ \dot{q}_m\xi_m & \varepsilon\dot{p}_m\xi_m^{-1} & -\dot{q}_m\xi_m & -\varepsilon\dot{p}_m\xi_m^{-1} \\ -\dot{p}_m & \dot{q}_m & -\dot{p}_m & \dot{q}_m \\ \dot{p}_m\xi_m & -\varepsilon\dot{q}_m\xi_m^{-1} & -\dot{p}_m\xi_m & \varepsilon\dot{q}_m\xi_m^{-1} \end{pmatrix} \begin{pmatrix} 0 \\ 0 \\ 0 \\ 1 \end{pmatrix} \exp(\mathrm{j}\xi_m\bar{z}) \\ &= \begin{pmatrix} \dot{p}_m \\ -\varepsilon\dot{p}_m/\xi_m \\ \dot{q}_m \\ \varepsilon\dot{q}_m/\xi_m \end{pmatrix} \exp(\mathrm{j}\xi_m\bar{z}). \end{aligned} \tag{9.187}$$

This solution satisfies Eq. (9.127). We have on the left-hand side

$$\frac{\mathrm{d}}{\mathrm{d}\bar{z}}\mathbf{f}^{(u)}_{\hat{t}m4} = \mathrm{j}\xi_m \mathbf{f}^{(u)}_{\hat{t}m4}. \tag{9.188}$$

The column vector $\mathbf{f}^{(u)}_{\hat{t}m4}$ is obtained using the proper vector ν_{m4} from Eq. (9.147). We find from Eq. (9.115), with $s_m = 0$,

$$\begin{aligned}
\mathrm{j}\mathbf{C}^{(u)}_m \mathbf{f}^{(u)}_{\hat{t}m4} &= \begin{pmatrix} 0 & p_m^2/\varepsilon - 1 & 0 & -p_m q_m/\varepsilon \\ q_m^2 - \varepsilon & 0 & -p_m q_m & 0 \\ 0 & p_m q_m/\varepsilon & 0 & 1 - q_m^2/\varepsilon \\ p_m q_m & 0 & \varepsilon - p_m^2 & 0 \end{pmatrix} \begin{pmatrix} \dot{p}_m \\ -\varepsilon \dot{p}_m/\xi_m \\ \dot{q}_m \\ \varepsilon \dot{q}_m/\xi_m \end{pmatrix} \exp(\mathrm{j}\xi_m \bar{z}) \\
&= \mathrm{j}\begin{pmatrix} (p_m^2/\varepsilon - 1)\,\varepsilon \dot{p}_m \xi_m + (p_m q_m/\varepsilon)\,\varepsilon \dot{q}_m/\xi_m \\ (q_m^2 - \varepsilon)\,\dot{p}_m - p_m q_m \dot{q}_m \\ (p_m q_m/\varepsilon)\,\varepsilon \dot{p}_m/\xi_m - (1 - q_m^2/\varepsilon)\,\varepsilon \dot{q}_m/\xi_m \\ p_m q_m \dot{q}_m + (\varepsilon - p_m^2)\,\dot{q}_m \end{pmatrix} \exp(\mathrm{j}\xi_m \bar{z}) \\
&= \mathrm{j}\begin{pmatrix} \dot{q}_m \xi_m \\ -\varepsilon \dot{q}_m \xi_m \\ \dot{q}_m \xi_m \\ \varepsilon \dot{q}_m \xi_m \end{pmatrix} \exp(\mathrm{j}\xi_m \bar{z}) = \mathrm{j}\xi_m \mathbf{f}^{(u)}_{\hat{t}m4} = \mathrm{j}\xi_m \begin{pmatrix} e_{ym4}(\bar{z}) \\ h_{zm4}(\bar{z}) \\ e_{zm4}(\bar{z}) \\ h_{ym4}(\bar{z}) \end{pmatrix}.
\end{aligned} \tag{9.189}$$

We have shown that Eq. (9.127) is satisfied with $\kappa_{m4} = \xi_m$ and vector $\mathbf{f}^{(u)}_{\hat{t}m4}$. The normal components follow from Eq. (9.113b). We get using Eq. (9.117) the TM waves

$$\begin{aligned}
f^{(u)}_{\hat{n}m4} = \begin{pmatrix} e_{xm4}(\bar{z}) \\ h_{xm4}(\bar{z}) \end{pmatrix} &= \begin{pmatrix} 0 & -p_m/\varepsilon & 0 & q_m/\varepsilon \\ -q_m & 0 & p_m & 0 \end{pmatrix} \begin{pmatrix} \dot{q}_m \\ -\varepsilon \dot{q}_m/\xi_m \\ \dot{q}_m \\ \varepsilon \dot{q}_m/\xi_m \end{pmatrix} \exp(\mathrm{j}\xi_m \bar{z}) \\
&= \begin{pmatrix} p_m \dot{q}_m/\xi_m + q_m \dot{q}_m/\xi_m \\ -q_m \dot{q}_m + p_m \dot{q}_m \end{pmatrix} \exp(\mathrm{j}\xi_m \bar{z}) \\
&= \begin{pmatrix} n_{\hat{t}m}/\xi_m \\ 0 \end{pmatrix} \exp(\mathrm{j}\xi_m \bar{z}).
\end{aligned} \tag{9.190}$$

The characteristic electric and magnetic fields, with the factor $\exp(\mathrm{j}\xi_m \bar{z})$ skipped, become

$$\mathbf{e}_{m4} = \hat{x} e_{xm4} + \mathbf{e}_{\hat{t}m4} = \hat{x} n_{\hat{t}m}/\xi_m + (\hat{y}\dot{q}_m + \hat{z}\dot{q}_m), \tag{9.191a}$$

$$\mathbf{h}_{m4} = \mathbf{h}_{\hat{t}m4} = (\hat{y}\dot{q}_m - \hat{z}\dot{q}_m)\varepsilon/\xi_m, \tag{9.191b}$$

where $h_{xm4} = 0$, which characterizes TM waves. In homogeneous isotropic regions, the wave electric and magnetic fields are normal to the direction of the propagation vector $\mathbf{n}_m^{(u)-}$. The Poynting vector is proportional to

$$\begin{aligned}
\mathbf{e}_{m4} \times \mathbf{h}_{m4} &= [\hat{x} n_{\dagger m}/\xi_m + (\hat{y}\dot{q}_m + \hat{z}\dot{q}_m)] \times (\hat{y}\dot{q}_m - \hat{z}\dot{q}_m)\varepsilon/\xi_m \\
&= -\hat{x}\left(\dot{q}_m^2 + \dot{q}_m^2\right)\varepsilon/\xi_m + n_{\dagger m}(\hat{y}\dot{q}_m + \hat{z}\dot{q}_m)\varepsilon/\xi_m^2 \\
&= -\frac{\varepsilon}{\xi_m}\hat{x} + n_{\dagger m}\frac{\varepsilon}{\xi_m^2}(\hat{y}\dot{q}_m + \hat{z}\dot{q}_m) \\
&= \frac{\varepsilon}{\xi_m^2}(-\hat{x}\xi_m + \hat{y}p_m + \hat{z}q_m).
\end{aligned} \tag{9.192}$$

The wave electric and magnetic fields are normal to the direction of the propagation vector determined by $\mathbf{n}_m^{(u)-} = -\hat{x}\xi_m + \hat{y}p_m + \hat{z}q_m$. The magnetic field of TM waves is polarized normal to the plane determined by the normal to the interfaces and the propagation vector $\mathbf{n}_m^{(u)-}$. The TM waves propagating in homogeneous isotropic regions are transverse with respect to $\mathbf{n}_m^{(u)-}$. As in the previous three cases, we have, for example,

$$\mathbf{e}_{m4} \cdot \mathbf{h}_{m4} = \mathbf{e}_{m4} \cdot \mathbf{n}_m^{(u)-} = \mathbf{h}_{m4} \cdot \mathbf{n}_m^{(u)-} = 0, \tag{9.193a}$$

$$\mathbf{e}_{m3} \cdot \mathbf{e}_{m4} = \mathbf{h}_{m3} \cdot \mathbf{h}_{m4} = 0. \tag{9.193b}$$

It is instructive to write down Eq. (9.149) and Eq. (9.152) for the special case of homogeneous region and $m = 0$. We have

$$\mathbf{T}_0^{(u)} = \begin{pmatrix} 1 & 0 & 1 & 0 \\ \left(\varepsilon - q_0^2\right)^{1/2} & 0 & -\left(\varepsilon - q_0^2\right)^{1/2} & 0 \\ 0 & 1 & 0 & 1 \\ 0 & -\varepsilon\left(\varepsilon - q_0^2\right)^{-1/2} & 0 & \varepsilon\left(\varepsilon - q_0^2\right)^{-1/2} \end{pmatrix}, \tag{9.194}$$

$$\left[\mathbf{T}_0^{(u)}\right]^{-1} = \frac{1}{2}\left(\varepsilon - q_0^2\right)^{-1/2} \begin{pmatrix} \left(\varepsilon - q_0^2\right)^{1/2} & 1 & 0 & 0 \\ 0 & 0 & \left(\varepsilon - q_0^2\right)^{1/2} & -\left(\varepsilon - q_0^2\right)/\varepsilon \\ \left(\varepsilon - q_0^2\right)^{1/2} & -1 & 0 & 0 \\ 0 & 0 & \left(\varepsilon - q_0^2\right)^{1/2} & \left(\varepsilon - q_0^2\right)/\varepsilon \end{pmatrix}. \tag{9.195}$$

The matrix $\mathbf{T}_0^{(u)}$ recalls the Yeh dynamical matrix $\mathcal{D}^{(u)}$ for a homogeneous isotropic region [8]. The essential difference consists in the normalization of the electric field amplitudes of the TM proper vectors $\boldsymbol{\nu}_{m2}^{(u)}$ and $\boldsymbol{\nu}_{m4}^{(u)}$. For $m = 0$, these are normalized to $[\varepsilon/(\varepsilon - q_0^2)]^{1/2}$ according to Eq. (9.174). In addition, in the Yeh dynamical matrix, the second and third columns are exchanged due to the different convention in the ordering of the proper vectors. The Cartesian Yeh formalism uses the following sequence of the proper vectors: the forward TE, the backward TE, the forward TM, and the backward TM modes. The Yeh dynamical matrix can then be obtained from

$$\mathcal{D}^{(u)} = \left(\boldsymbol{\nu}^{(u)}_{m1}, \boldsymbol{\nu}^{(u)}_{m3}, \frac{\xi_m}{\varepsilon^{1/2}}\boldsymbol{\nu}^{(u)}_{m2}, \frac{\xi_m}{\varepsilon^{1/2}}\boldsymbol{\nu}^{(u)}_{m4}\right)$$
$$= \begin{pmatrix} 1 & 1 & 0 & 0 \\ \left(\varepsilon - q_0^2\right)^{1/2} & -\left(\varepsilon - q_0^2\right)^{1/2} & 0 & 0 \\ 0 & 0 & \left(\dfrac{\varepsilon - q_0^2}{\varepsilon}\right)^{1/2} & \left(\dfrac{\varepsilon - q_0^2}{\varepsilon}\right)^{1/2} \\ 0 & 0 & -\varepsilon^{1/2} & \varepsilon^{1/2} \end{pmatrix}, \tag{9.196}$$

and its inverse

$$[\mathcal{D}^{(u)}]^{-1} = \frac{1}{2}\left(\varepsilon - q_0^2\right)^{-1/2} \begin{pmatrix} \left(\varepsilon - q_0^2\right)^{1/2} & 1 & 0 & 0 \\ \left(\varepsilon - q_0^2\right)^{1/2} & -1 & 0 & 0 \\ 0 & 0 & \varepsilon^{1/2} & -\left(\dfrac{\varepsilon - q_0^2}{\varepsilon}\right)^{1/2} \\ 0 & 0 & \varepsilon^{1/2} & \left(\dfrac{\varepsilon - q_0^2}{\varepsilon}\right)^{1/2} \end{pmatrix}. \tag{9.197}$$

In the situation where the TE↔TM mode coupling is relevant, it might be convenient to introduce the diagonalizing matrix with the electric fields normalized to the unity for both TE and TM proper vectors as follows

$$\overline{\mathbf{T}}^{(u)}_m = \begin{pmatrix} \dot{q}_m & \dot{p}_m\xi_m\varepsilon^{-1/2} & \dot{q}_m & \dot{p}_m\xi_m\varepsilon^{-1/2} \\ \dot{q}_m\xi_m & \dot{p}_m\varepsilon^{1/2} & -\dot{q}_m\xi_m & -\dot{p}_m\varepsilon^{1/2} \\ -\dot{p}_m & \dot{q}_m\xi_m\varepsilon^{-1/2} & -\dot{p}_m & \dot{q}_m\xi_m\varepsilon^{-1/2} \\ \dot{p}_m\xi_m & -\dot{q}_m\varepsilon^{1/2} & -\dot{p}_m\xi_m & \dot{q}_m\varepsilon^{1/2} \end{pmatrix}. \tag{9.198}$$

Note that the determinant of the matrix $\overline{\mathbf{T}}^{(u)}_m$ and the matrix inverse to $\overline{\mathbf{T}}^{(u)}_m$ become

$$\det \overline{\mathbf{T}}^{(u)}_m = 4\xi_m^2 \tag{9.199}$$

$$\left[\overline{\mathbf{T}}^{(u)}_m\right]^{-1} = \frac{1}{2\xi_m} \begin{pmatrix} \dot{q}_m\xi_m & \dot{q}_m & -\dot{p}_m\xi_m & \dot{p}_m \\ \dot{p}_m\varepsilon^{1/2} & \dot{p}_m\xi_m\varepsilon^{-1/2} & \dot{q}_m\varepsilon^{1/2} & -\dot{q}_m\xi_m\varepsilon^{-1/2} \\ \dot{q}_m\xi_m & -\dot{q}_m & -\dot{p}_m\xi_m & -\dot{p}_m \\ \dot{p}_m\varepsilon^{1/2} & -\dot{p}_m\xi_m\varepsilon^{-1/2} & \dot{q}_m\varepsilon^{1/2} & \dot{q}_m\xi_m\varepsilon^{-1/2} \end{pmatrix}. \tag{9.200}$$

These matrices transform $C_m^{(u)}$ to the diagonal form

$$\boldsymbol{\kappa}_m^{(u)} = \begin{pmatrix} \kappa_{m1} & 0 & 0 & 0 \\ 0 & \kappa_{m2} & 0 & 0 \\ 0 & 0 & \kappa_{m3} & 0 \\ 0 & 0 & 0 & \kappa_{m4} \end{pmatrix} = \overline{\mathbf{T}}_m^{(u)-1} \mathbf{C}_m^{(u)} \overline{\mathbf{T}}_u^{(u)}$$

$$= \frac{1}{2\xi_m} \begin{pmatrix} \dot{q}_m\xi_m & \dot{q}_m & -\dot{p}_m\xi_m & \dot{p}_m \\ \dot{p}_m\varepsilon^{1/2} & \dot{p}_m\xi_m\varepsilon^{-1/2} & \dot{q}_m\varepsilon^{1/2} & -\dot{q}_m\xi_m\varepsilon^{-1/2} \\ \dot{q}_m\xi_m & -\dot{q}_m & -\dot{p}_m\xi_m & -\dot{p}_m \\ \dot{p}_m\varepsilon^{1/2} & -\dot{p}_m\xi_m\varepsilon^{-1/2} & \dot{q}_m\varepsilon^{1/2} & \dot{q}_m\xi_m\varepsilon^{-1/2} \end{pmatrix}$$

$$\times \begin{pmatrix} 0 & p_m^2/\varepsilon - 1 & 0 & -p_m q_m/\varepsilon \\ q_m^2 - \varepsilon & 0 & -p_m q_m & 0 \\ 0 & p_m q_m/\varepsilon & 0 & 1 - q_m^2/\varepsilon \\ p_m q_m & 0 & \varepsilon - p_m^2 & 0 \end{pmatrix} \overline{\mathbf{T}}_u^{(u)}$$

$$= \frac{1}{2\xi_m} \begin{pmatrix} -\dot{q}_m\xi_m^2 & -\dot{q}_m\xi_m & \dot{p}_m\xi_m^2 & -\dot{p}_m\xi_m \\ -\dot{p}_m\xi_m\varepsilon^{1/2} & -\dot{p}_m\xi_m^2\varepsilon^{-1/2} & -\dot{q}_m\xi_m\varepsilon^{1/2} & \dot{q}_m\xi_m^2\varepsilon^{-1/2} \\ \dot{q}_m\xi_m^2 & -\dot{q}_m\xi_m & -\dot{p}_m\xi_m^2 & -\dot{p}_m\xi_m \\ \dot{p}_m\xi_m\varepsilon^{1/2} & -\dot{p}_m\xi_m^2\varepsilon^{-1/2} & \dot{q}_m\xi_m\varepsilon^{1/2} & \dot{q}_m\xi_m^2\varepsilon^{-1/2} \end{pmatrix}$$

$$\times \begin{pmatrix} \dot{q}_m & \dot{p}_m\xi_m\varepsilon^{-1/2} & \dot{q}_m & \dot{p}_m\xi_m\varepsilon^{-1/2} \\ \dot{q}_m\xi_m & \dot{p}_m\varepsilon^{1/2} & -\dot{q}_m\xi_m & -\dot{p}_m\varepsilon^{1/2} \\ -\dot{p}_m & \dot{q}_m\xi_m\varepsilon^{-1/2} & -\dot{p}_m & \dot{q}_m\xi_m\varepsilon^{-1/2} \\ \dot{p}_m\xi_m & -\dot{q}_m\varepsilon^{1/2} & -\dot{p}_m\xi_m & \dot{q}_m\varepsilon^{1/2} \end{pmatrix}$$

$$= \frac{1}{2\xi_m} \begin{pmatrix} -2\xi_m^2 & 0 & 0 & 0 \\ 0 & -2\xi_m^2 & 0 & 0 \\ 0 & 0 & 2\xi_m^2 & 0 \\ 0 & 0 & 0 & 2\xi_m^2 \end{pmatrix} = \begin{pmatrix} -\xi_m & 0 & 0 & 0 \\ 0 & -\xi_m & 0 & 0 \\ 0 & 0 & \xi_m & 0 \\ 0 & 0 & 0 & \xi_m \end{pmatrix}, \tag{9.201}$$

as expected.

The proper vectors $\overline{\mathbf{f}}_{\bar{t}m2}^{(u)}$ and $\overline{\mathbf{f}}_{\bar{t}m4}^{(u)}$ should also be consistent with Eq. (9.113a). On the left-hand side, we have, using Eq. (9.198),

$$\frac{\mathrm{d}}{\mathrm{d}\bar{z}} \overline{\mathbf{f}}_{\bar{t}m2}^{(u)} = -\mathrm{j}\xi_m \overline{\mathbf{f}}_{\bar{t}m2}^{(u)}. \tag{9.202}$$

The right-hand side gives, according to Eq. (9.115) with $s_m = 0$,

$$\mathbf{j}\mathbf{C}_m^{(u)}\overline{\mathbf{f}}_{\hat{t}m2}^{(u)} = \begin{pmatrix} 0 & p_m^2/\varepsilon - 1 & 0 & -p_m q_m/\varepsilon \\ q_m^2 - \varepsilon & 0 & -p_m q_m & 0 \\ 0 & p_m q_m/\varepsilon & 0 & 1 - q_m^2/\varepsilon \\ p_m q_m & 0 & \varepsilon - p_m^2 & 0 \end{pmatrix} \begin{pmatrix} \varepsilon^{-1/2}\dot{p}_m\xi_m \\ \varepsilon^{1/2}\dot{p}_m \\ \varepsilon^{-1/2}\dot{q}_m\xi_m \\ -\varepsilon^{1/2}\dot{q}_m \end{pmatrix} \exp(-\mathrm{j}\xi_m\bar{z})$$

$$= -\mathrm{j}\xi_m \begin{pmatrix} \varepsilon^{-1/2}\dot{p}_m\xi_m \\ \varepsilon^{1/2}\dot{p}_m \\ \varepsilon^{-1/2}\dot{q}_m\xi_m \\ -\varepsilon^{1/2}\dot{q}_m \end{pmatrix} \exp(-\mathrm{j}\xi_m\bar{z}) = -\mathrm{j}\xi_m\overline{\mathbf{f}}_{\hat{t}m2}^{(u)} = -\mathrm{j}\xi_m \begin{pmatrix} \bar{e}_{xm2}(\bar{z}) \\ \bar{h}_{ym2}(\bar{z}) \\ \bar{e}_{ym2}(\bar{z}) \\ \bar{h}_{xm2}(\bar{z}) \end{pmatrix}. \tag{9.203}$$

The vector $\overline{\mathbf{f}}_{\hat{t}m2}^{(u)}$ also represents the solution of Eq. (9.113a) corresponding to the proper value $\kappa_{m2} = \xi_m$. For the normal or the z field components, we obtain from Eq. (9.113b) with Eq. (9.117) the expected result $h_{zm2}(\bar{z}) = 0$

$$\overline{\mathbf{f}}_{\hat{n}m2}^{(u)} = \begin{pmatrix} \bar{e}_{zm2}(\bar{z}) \\ \bar{h}_{zm2}(\bar{z}) \end{pmatrix}$$

$$= \begin{pmatrix} 0 & -p_m/\varepsilon & 0 & q_m/\varepsilon \\ -q_m & 0 & p_m & 0 \end{pmatrix} \begin{pmatrix} \varepsilon^{-1/2}\dot{p}_m\xi_m \\ \varepsilon^{1/2}\dot{p}_m \\ \varepsilon^{-1/2}\dot{q}_m\xi_m \\ -\varepsilon^{1/2}\dot{q}_m \end{pmatrix} \exp(-\mathrm{j}\xi_m\bar{z})$$

$$= \begin{pmatrix} -\varepsilon^{-1/2}n_{\hat{t}m} \\ 0 \end{pmatrix} \exp(-\mathrm{j}\xi_m\bar{z}). \tag{9.204}$$

The characteristic electric and magnetic fields, with the factor $\exp(-\mathrm{j}\xi_m\bar{z})$ skipped, become

$$\overline{\mathbf{e}}_{m2} = \varepsilon^{-1/2}\xi_m(\hat{\boldsymbol{x}}\dot{p}_m + \hat{\boldsymbol{y}}\dot{q}_m) - \hat{\boldsymbol{z}}\varepsilon^{-1/2}n_{\hat{t}m}, \tag{9.205a}$$

$$\overline{\mathbf{h}}_{m2} = (-\hat{\boldsymbol{x}}\dot{q}_m + \hat{\boldsymbol{y}}\dot{p}_m)\varepsilon^{1/2}, \tag{9.205b}$$

where the z component $\bar{h}_{zm2} = 0$, as we are concerned with TM waves. The Poynting vector is proportional to

$$\begin{aligned} \overline{\mathbf{e}}_{m2} \times \overline{\mathbf{h}}_{m2} &= \left[\varepsilon^{-1/2}\xi_m(\hat{\boldsymbol{x}}\dot{p}_m + \hat{\boldsymbol{y}}\dot{q}_m) - \hat{\boldsymbol{z}}\varepsilon^{-1/2}n_{\hat{t}m}\right] \times (-\hat{\boldsymbol{x}}\dot{q}_m + \hat{\boldsymbol{y}}\dot{p}_m)\varepsilon^{1/2} \\ &= \hat{\boldsymbol{x}}p_m + \hat{\boldsymbol{y}}q_m + \hat{\boldsymbol{z}}\xi_m. \end{aligned} \tag{9.206}$$

The amplitude of the electric field for the TM waves in homogeneous and isotropic regions can be normalized to the unity using Eq. (9.141)

$$\begin{aligned} |\overline{\mathbf{e}}_{m2}| &= \left|\varepsilon^{-1/2}\xi_m(\hat{\boldsymbol{x}}\dot{p}_m + \hat{\boldsymbol{y}}\dot{q}_m) - \hat{\boldsymbol{z}}\varepsilon^{-1/2}n_{\hat{t}m}\right| \\ &= \varepsilon^{-1/2}\left[n_{\hat{t}m}^2 + \xi^2\left(\dot{p}_m^2 + \dot{q}_m^2\right)\right]^{1/2} = 1. \end{aligned} \tag{9.207}$$

The amplitude of magnetic field in TM waves in homogeneous and isotropic regions is normalized to

$$|\overline{\mathbf{h}}_{m2}| = \left|(-\hat{x}\dot{q}_m + \hat{y}\dot{p}_m)\varepsilon^{1/2}\right| = \varepsilon^{1/2}. \tag{9.208}$$

The same procedure applied to $\overline{\mathbf{f}}^{(u)}_{\hat{t}m4}$ provides, using Eq. (9.198),

$$\frac{\mathrm{d}}{\mathrm{d}\bar{z}}\overline{\mathbf{f}}^{(u)}_{\hat{t}m4} = \mathrm{j}\xi_m \overline{\mathbf{f}}^{(u)}_{\hat{t}m4}. \tag{9.209}$$

The right-hand side gives, according to Eq. (9.115) with $s_m = 0$,

$$\begin{aligned}
&\mathrm{j}\mathbf{C}^{(u)}_m \overline{\mathbf{f}}^{(u)}_{\hat{t}m4} \\
&= \begin{pmatrix} 0 & p_m^2/\varepsilon - 1 & 0 & -p_m q_m/\varepsilon \\ q_m^2 - \varepsilon & 0 & -p_m q_m & 0 \\ 0 & p_m q_m/\varepsilon & 0 & 1 - q_m^2/\varepsilon \\ p_m q_m & 0 & \varepsilon - p_m^2 & 0 \end{pmatrix} \begin{pmatrix} \varepsilon^{-1/2}\dot{p}_m\xi_m \\ -\varepsilon^{1/2}\dot{p}_m \\ \varepsilon^{-1/2}\dot{q}_m\xi_m \\ \varepsilon^{1/2}\dot{q}_m \end{pmatrix} \exp(\mathrm{j}\xi_m\bar{z}) \\
&= \mathrm{j}\xi_m \begin{pmatrix} \varepsilon^{-1/2}\dot{p}_m\xi_m \\ -\varepsilon^{1/2}\dot{p}_m \\ \varepsilon^{-1/2}\dot{q}_m\xi_m \\ \varepsilon^{1/2}\dot{q}_m \end{pmatrix} \exp(\mathrm{j}\xi_m\bar{z}) \\
&= \mathrm{j}\xi_m \overline{\mathbf{f}}^{(u)}_{\hat{t}m4} = \mathrm{j}\xi_m \begin{pmatrix} \overline{e}_{xm4}(\bar{z}) \\ \overline{h}_{ym4}(\bar{z}) \\ \overline{e}_{ym4}(\bar{z}) \\ \overline{h}_{xm4}(\bar{z}) \end{pmatrix}.
\end{aligned} \tag{9.210}$$

The vector $\overline{\mathbf{f}}^{(u)}_{\hat{t}m4}$ represents the solution of Eq. (9.113a) corresponding to the proper value $\kappa_{m4} = \xi_m$. For the normal or z field components, we obtain from Eq. (9.113b) with Eq. (9.117) the expected result $h_{zm4}(\bar{z}) = 0$

$$\begin{aligned}
\overline{\mathbf{f}}^{(u)}_{\hat{n}m4} &= \begin{pmatrix} \overline{e}_{zm4}(\bar{z}) \\ \overline{h}_{zm4}(\bar{z}) \end{pmatrix} = \begin{pmatrix} 0 & -p_m/\varepsilon & 0 & q_m/\varepsilon \\ -q_m & 0 & p_m & 0 \end{pmatrix} \begin{pmatrix} \varepsilon^{-1/2}\dot{p}_m\xi_m \\ -\varepsilon^{1/2}\dot{p}_m \\ \varepsilon^{-1/2}\dot{q}_m\xi_m \\ \varepsilon^{1/2}\dot{q}_m \end{pmatrix} \exp(\mathrm{j}\xi_m\bar{z}) \\
&= \begin{pmatrix} \varepsilon^{-1/2} n_{\hat{t}m} \\ 0 \end{pmatrix} \exp(-\mathrm{j}\xi_m\bar{z}).
\end{aligned} \tag{9.211}$$

The characteristic electric and magnetic fields, with the factor $\exp(-\mathrm{j}\xi_m\bar{z})$ skipped, become

$$\overline{\mathbf{e}}_{m4} = \varepsilon^{-1/2}\xi_m(\hat{x}\dot{p}_m + \hat{y}\dot{q}_m) + \hat{z}\varepsilon^{-1/2} n_{\hat{t}m}, \tag{9.212a}$$

$$\overline{\mathbf{h}}_{m4} = (\hat{x}\dot{q}_m - \hat{y}\dot{p}_m)\varepsilon^{1/2}, \tag{9.212b}$$

where the z component $\bar{h}_{zm4} = 0$, as we are concerned with TM waves. The Poynting vector is proportional to

$$\begin{aligned}\bar{\mathbf{e}}_{m4} \times \bar{\mathbf{h}}_{m4} &= \left[\varepsilon^{-1/2}\xi_m(\hat{x}\dot{p}_m + \hat{y}\dot{q}_m) - \hat{z}\varepsilon^{-1/2}n_{\hat{t}m}\right] \times (\hat{x}\dot{q}_m - \hat{y}\dot{p}_m)\varepsilon^{1/2} \\ &= \hat{x}p_m + \hat{y}q_m - \hat{z}\xi_m. \end{aligned} \tag{9.213}$$

The amplitude of the electric field for the TM waves in homogeneous and isotropic regions now becomes normalized, according to Eq. (9.141), to the unity

$$\begin{aligned}|\bar{\mathbf{e}}_{m4}| &= \left|\varepsilon^{-1/2}\xi_m(\hat{x}\dot{p}_m + \hat{y}\dot{q}_m) + \hat{z}\varepsilon^{-1/2}n_{\hat{t}m}\right| \\ &= \varepsilon^{-1/2}\left[n_{\hat{t}m}^2 + \xi_m^2(\dot{p}^2 + \dot{q}^2)\right]^{1/2} = 1. \end{aligned} \tag{9.214}$$

The amplitude of magnetic field in TM waves in homogeneous and isotropic regions is normalized to

$$|\bar{\mathbf{h}}_{m4}| = \left|(\hat{x}\dot{q}_m - \hat{y}\dot{p}_m)\varepsilon^{1/2}\right| = \varepsilon^{1/2}. \tag{9.215}$$

For $m = 0$, we obtain from Eq. (9.198) and Eq. (9.200)

$$\bar{\mathbf{T}}_0^{(u)} = \begin{pmatrix} 1 & 0 & 1 & 0 \\ \left(\varepsilon - q_0^2\right)^{1/2} & 0 & -\left(\varepsilon - q_0^2\right)^{1/2} & 0 \\ 0 & \left(\varepsilon - q_0^2\right)^{1/2}\varepsilon^{-1/2} & 0 & \left(\varepsilon - q_0^2\right)^{1/2}\varepsilon^{-1/2} \\ 0 & -\varepsilon^{1/2} & 0 & \varepsilon^{1/2} \end{pmatrix}, \tag{9.216}$$

$$\left[\bar{\mathbf{T}}_0^{(u)}\right]^{-1} = \frac{1}{2}\left(\varepsilon - q_0^2\right)^{-1/2} \begin{pmatrix} \left(\varepsilon - q_0^2\right)^{1/2} & 1 & 0 & 0 \\ 0 & 0 & \varepsilon^{1/2} & -\left(\varepsilon - q_0^2\right)^{1/2}\varepsilon^{-1/2} \\ \left(\varepsilon - q_0^2\right)^{1/2} & -1 & 0 & 0 \\ 0 & 0 & \varepsilon^{1/2} & \left(\varepsilon - q_0^2\right)^{1/2}\varepsilon^{-1/2} \end{pmatrix}. \tag{9.217}$$

We now return to the ordering of rows and columns used in Eq. (9.93), Eq. (9.95), and Eq. (9.96) compatible with that for anisotropic grating regions. In isotropic homogeneous regions, Eq. (9.123) reduces to Eq. (9.127). The matrix $\mathbf{T}^{(u)}$ transforming the matrix $\mathbf{C}^{(u)}$ to a diagonal one in the homogeneous and isotropic regions for $-M \leq m \leq M$ takes the form

$$\mathbf{T}^{(u)} = \begin{pmatrix} \dot{\mathbf{q}} & \dot{\mathbf{p}} & \dot{\mathbf{q}} & \dot{\mathbf{p}} \\ \dot{\mathbf{q}}\boldsymbol{\xi} & \varepsilon\dot{\mathbf{p}}\boldsymbol{\xi}^{-1} & -\dot{\mathbf{q}}\boldsymbol{\xi} & -\varepsilon\dot{\mathbf{p}}\boldsymbol{\xi}^{-1} \\ -\dot{\mathbf{p}} & \dot{\mathbf{q}} & -\dot{\mathbf{p}} & \dot{\mathbf{q}} \\ \dot{\mathbf{p}}\boldsymbol{\xi} & -\varepsilon\dot{\mathbf{q}}\boldsymbol{\xi}^{-1} & -\dot{\mathbf{p}}\boldsymbol{\xi} & \varepsilon\dot{\mathbf{q}}\boldsymbol{\xi}^{-1} \end{pmatrix}, \tag{9.218}$$

where the diagonal submatrices are given by $\dot{\mathbf{p}} = (\delta_{nl}\ \dot{p}_l)$, $\dot{\mathbf{q}} = (\delta_{nl}\ \dot{q}_l)$, and $\boldsymbol{\xi}_m = (\delta_{nl}\xi_l)$. The corresponding column vector $\mathbf{g}$ becomes

$$\mathbf{g} = \begin{pmatrix} \mathbf{g}^+ \\ \mathbf{g}^- \end{pmatrix} = \begin{pmatrix} {}^{TE}\mathbf{g}^+ \\ {}^{TM}\mathbf{g}^+ \\ {}^{TE}\mathbf{g}^- \\ {}^{TM}\mathbf{g}^- \end{pmatrix}, \tag{9.219}$$

where the elements of the column vectors on the right-hand side are column vectors themselves

$${}^{TE}\mathbf{g}^{\pm} = \begin{pmatrix} {}^{TE}g_M^{\pm} \\ \cdot \\ \cdot \\ {}^{TE}g_0^{\pm} \\ \cdot \\ \cdot \\ {}^{TE}g_{-M}^{\pm} \end{pmatrix}, \tag{9.220a}$$

$${}^{TM}\mathbf{g}^{\pm} = \begin{pmatrix} {}^{TM}g_M^{\pm} \\ \cdot \\ \cdot \\ {}^{TM}g_0^{\pm} \\ \cdot \\ \cdot \\ {}^{TM}g_{-M}^{\pm} \end{pmatrix}. \tag{9.220b}$$

Here, the signs + and − distinguish the waves propagating with the positive and negative $\bar{z}$ component of the Poynting vector as $\exp(-\mathrm{j}\xi_m\bar{z})$ and $\exp(\mathrm{j}\xi_m\bar{z})$, respectively.

The matrix equation (9.122) in homogeneous isotropic regions

$$\mathbf{f}_t^{(u)} = \mathbf{T}^{(u)}\mathbf{g}^{(u)} \tag{9.221}$$

can be written down for $M = 1$

$$
\begin{pmatrix} e_{x,1} \\ e_{x,0} \\ e_{x,-1} \\ h_{y,1} \\ h_{y,0} \\ h_{y,-1} \\ e_{y,1} \\ e_{y,0} \\ e_{y,-1} \\ h_{x,1} \\ h_{x,0} \\ h_{x,-1} \end{pmatrix}
= \begin{pmatrix}
\dot{q} & & & \dot{p} & & & \dot{q} & & & \dot{p} & & \\
 & \dot{q}_0 & & & 0 & & & \dot{q}_0 & & & 0 & \\
 & & -\dot{q} & & & -\dot{p} & & & \dot{q} & & & -\dot{p} \\
\dot{q}\xi_1 & & & \varepsilon\dot{p}\xi_1^{-1} & & & -\dot{q}\xi_1 & & & -\varepsilon\dot{p}\xi_1^{-1} & & \\
 & \dot{q}_0\xi_0 & & & 0 & & & -\dot{q}_0\xi_0 & & & 0 & \\
 & & -\dot{q}\xi_{-1} & & & -\varepsilon\dot{p}\xi_{-1}^{-1} & & & qs_- & & & \varepsilon\dot{p}\xi_{-1}^{-1} \\
-\dot{p} & & & \dot{q} & & & -\dot{p} & & & \dot{q} & & \\
 & 0 & & & \dot{q}_0 & & & 0 & & & \dot{q}_0 & \\
 & & \dot{p} & & & -\dot{q} & & & -\dot{p} & & & -\dot{q} \\
q\xi_1 & & & -\varepsilon\dot{q}\xi_1^{-1} & & & -\dot{p}\xi_1 & & & \varepsilon\dot{q}\xi_1^{-1} & & \\
 & 0 & & & -\varepsilon\dot{q}_0\xi_0^{-1} & & & 0 & & & \varepsilon\dot{q}_0\xi_0^{-1} & \\
 & & -q\xi_{-1} & & & \varepsilon\dot{q}\xi_{-1}^{-1} & & & \dot{p}\xi_{-1} & & & -\varepsilon\dot{q}\xi_{-1}^{-1}
\end{pmatrix}
\times \begin{pmatrix} {}^{TE}g^{(1)+} \\ {}^{TE}g_0^+ \\ {}^{TE}g_{-1}^+ \\ {}^{TM}g^{(1)+} \\ {}^{TM}g_0^+ \\ {}^{TM}g_{-1}^+ \\ {}^{TE}g^{(0)-} \\ {}^{TE}g_0^- \\ {}^{TE}g_{-1}^- \\ {}^{TM}g^{(0)-} \\ {}^{TM}g_0^- \\ {}^{TM}g_{-1}^- \end{pmatrix}. \tag{9.222}
$$

All the 3×3 submatrices are diagonal, and the empty spaces are to be filled in with zeros. Note that $|\dot{q}_0| = 1$.

9.6 Homogeneous Anisotropic Region

In homogeneous either isotropic or anisotropic regions, the grating vector is zero, $\mathbf{n}_\Gamma = 0$. We now focus on the case of homogeneous anisotropic region, first with restriction to the integer $M = 0$. The Floquet developments of fields (9.18) and (9.19) reduce to a single term

$$\sqrt{Y_0}\mathbf{E} = \mathbf{e}_0(\bar{z})\exp(-\mathrm{j}q_0\bar{y}), \tag{9.223a}$$

$$\sqrt{Z_0}\mathbf{H} = \mathbf{h}_0(\bar{z})\exp(-\mathrm{j}q_0\bar{y}). \tag{9.223b}$$

The subscript $m = 0$ will therefore be skipped. We substitute this into the components of the Maxwell equations (9.25) and (9.26) and take into account that, thanks to homogeneity, we can put $(\frac{\partial}{\partial\bar{x}} = 0)$

$$\frac{\partial}{\partial\bar{x}}(\sqrt{Y_0}E_y) - \frac{\partial}{\partial\bar{y}}(\sqrt{Y_0}E_x) = -\mathrm{j}(\sqrt{Z_0}H_z),$$
$$\mathrm{j}q_0e_x = -\mathrm{j}h_z, \tag{9.224a}$$

$$\frac{\partial}{\partial\bar{y}}(\sqrt{Y_0}E_z) - \frac{\partial}{\partial\bar{z}}(\sqrt{Y_0}E_y) = -\mathrm{j}(\sqrt{Z_0}H_x),$$
$$-\mathrm{j}q_0e_z - \frac{\mathrm{d}}{\mathrm{d}\bar{z}}e_y = -\mathrm{j}h_x, \tag{9.224b}$$

$$\frac{\partial}{\partial\bar{z}}(\sqrt{Y_0}E_x) - \frac{\partial}{\partial\bar{x}}(\sqrt{Y_0}E_z) = -\mathrm{j}(\sqrt{Z_0}H_y),$$
$$\frac{\mathrm{d}}{\mathrm{d}\bar{z}}e_x = -\mathrm{j}h_y, \tag{9.224c}$$

$$\frac{\partial}{\partial\bar{x}}(\sqrt{Z_0}H_y) - \frac{\partial}{\partial\bar{y}}(\sqrt{Z_0}H_x) = \mathrm{j}[\varepsilon_{zx}(\sqrt{Y_0}E_x) + \varepsilon_{zy}(\sqrt{Y_0}E_y)\varepsilon_{zz}(\sqrt{Y_0}E_z)],$$
$$\mathrm{j}q_0h_z = \mathrm{j}(\varepsilon_{zx}e_x + \varepsilon_{zy}e_y + \varepsilon_{zz}e_z), \tag{9.224d}$$

$$\frac{\partial}{\partial\bar{y}}(\sqrt{Z_0}H_z) - \frac{\partial}{\partial\bar{z}}(\sqrt{Z_0}H_y) = \mathrm{j}[\varepsilon_{xx}(\sqrt{Y_0}E_x) + \varepsilon_{xy}(\sqrt{Y_0}E_y)\varepsilon_{xz}(\sqrt{Y_0}E_z)],$$
$$-\mathrm{j}p_0h_z - \frac{\mathrm{d}}{\mathrm{d}\bar{z}}h_y = \mathrm{j}(\varepsilon_{xx}e_x + \varepsilon_{xy}e_y + \varepsilon_{xz}e_z), \tag{9.224e}$$

$$\frac{\partial}{\partial\bar{z}}(\sqrt{Z_0}H_x) - \frac{\partial}{\partial\bar{x}}(\sqrt{Z_0}H_z) = \mathrm{j}[\varepsilon_{yx}(\sqrt{Y_0}E_x) + \varepsilon_{yy}(\sqrt{Y_0}E_y)\varepsilon_{yz}(\sqrt{Y_0}E_z)],$$
$$\frac{\mathrm{d}}{\mathrm{d}\bar{z}}h_x = \mathrm{j}(\varepsilon_{yx}e_x + \varepsilon_{yy}e_y + \varepsilon_{yz}e_z). \tag{9.224f}$$

We eliminate the fields e_z and h_z and arrive at a special case of Eq. (9.80)

$$\begin{pmatrix} e_z \\ h_z \end{pmatrix} = \begin{pmatrix} -\varepsilon_{zz}^{-1}\,\varepsilon_{zx} & 0 & -\varepsilon_{zz}^{-1}\,\varepsilon_{zy} & \varepsilon_{zz}^{-1}q_0 \\ -q_0 & 0 & 0 & 0 \end{pmatrix} \begin{pmatrix} e_x \\ h_y \\ e_y \\ h_x \end{pmatrix}. \tag{9.225}$$

By substituting into the equations containing $\dfrac{\mathrm{d}}{\mathrm{d}\bar{z}}$ we obtain

$$\frac{\mathrm{d}}{\mathrm{d}\bar{z}} e_x = -\mathrm{j} h_y, \tag{9.226}$$

[a special case of Eq. (9.83)]

$$\frac{\mathrm{d}}{\mathrm{d}\bar{x}} h_z = \mathrm{j} s_0^2\, e_y - \mathrm{j}\,\varepsilon_{yx}\varepsilon_{xx}^{-1}(s_0\, h_y - \varepsilon_{xy}\, e_y - \varepsilon_{xz}\, e_z) - \mathrm{j}\,\varepsilon_{yy}\, e_y - \mathrm{j}\,\varepsilon_{yz}\, e_z, \tag{9.227}$$

[a special case of Eq. (9.86)]

$$\frac{\mathrm{d}}{\mathrm{d}\bar{x}} e_z = \mathrm{j} h_y - \mathrm{j} s_0 \varepsilon_{xx}^{-1}(s_0 h_y - \varepsilon_{xy} e_y - \varepsilon_{xz} e_z), \tag{9.228}$$

[a special case of Eq. (9.89)]

$$\frac{\mathrm{d}}{\mathrm{d}\bar{z}} h_x = \mathrm{j}\varepsilon_{yz}\varepsilon_{zz}^{-1}(q_0 h_x - \varepsilon_{zx}\, e_x - \varepsilon_{zy}\, e_y) + \mathrm{j}\,\varepsilon_{yx}\, e_x + \mathrm{j}\,\varepsilon_{yy}\, e_y, \tag{9.229}$$

[a special case of Eq. (9.92)]. These four equations can be united into a single matrix equation

$$\frac{\mathrm{d}}{\mathrm{d}\bar{x}} \begin{pmatrix} e_x(\bar{z}) \\ h_y(\bar{z}) \\ e_y(\bar{z}) \\ h_x(\bar{z}) \end{pmatrix} = \mathrm{j} \begin{pmatrix} 0 & -1 & 0 & 0 \\ \varepsilon_{xz}\,\varepsilon_{zz}^{-1}\,\varepsilon_{zx} - \varepsilon_{xx} + q_0^2 & 0 & \varepsilon_{xz}\,\varepsilon_{zz}^{-1}\,\varepsilon_{zy} - \varepsilon_{xy} & -\varepsilon_{xz}\,\varepsilon_{zz}^{-1}q_0 \\ q_0\,\varepsilon_{zz}^{-1}\,\varepsilon_{zx} & 0 & s_0\,\varepsilon_{zz}^{-1}\,\varepsilon_{zy} & -q_0^2\,\varepsilon_{zz}^{-1} + 1 \\ \varepsilon_{yx} - \varepsilon_{yz}\,\varepsilon_{zz}^{-1}\,\varepsilon_{zx} & 0 & \varepsilon_{yy} - \varepsilon_{yz}\,\varepsilon_{zz}^{-1}\,\varepsilon_{zy} & \varepsilon_{yz}\,\varepsilon_{zz}^{-1}q_0 \end{pmatrix} \begin{pmatrix} e_x(\bar{z}) \\ h_y(\bar{z}) \\ e_y(\bar{z}) \\ h_x(\bar{z}) \end{pmatrix}, \tag{9.230}$$

where the 4×4 matrix represents a special case of the matrix $\mathbf{C}$ from Eq. (9.95). Correspondingly, Eq. (9.230) represents a special case of Eq. (9.94a)

$$\frac{\mathrm{d}}{\mathrm{d}\bar{x}} \mathbf{f}_{\dagger}^{(ua)} = \mathrm{j}\mathbf{C}_0^{(ua)} \mathbf{f}_{\dagger}^{(ua)}. \tag{9.231}$$

In a similar way, Eq. (9.225) represents a special case of Eq. (9.94b)

$$\mathbf{f}_{\hat{n}0}^{(ua)} = \mathrm{j}\mathbf{D}_0^{(ua)} \mathbf{f}_{\hat{t}0}^{(ua)}. \tag{9.232}$$

We denote the elements of the matrix $\mathbf{C}_0^{(ua)}$ in Eq. (9.230) as

$$a_{21} = \varepsilon_{xz}\, \varepsilon_{zz}^{-1}\, \varepsilon_{zx} - \varepsilon_{xx} + q_0^2, \tag{9.233a}$$
$$a_{23} = \varepsilon_{zy}\, \varepsilon_{zz}^{-1}\, \varepsilon_{zy} - \varepsilon_{xy}, \tag{9.233b}$$
$$a_{24} = -\varepsilon_{zy}\, \varepsilon_{zz}^{-1} q_0, \tag{9.233c}$$
$$a_{31} = \varepsilon_{zz}^{-1}\, \varepsilon_{zx} q_0, \tag{9.233d}$$
$$a_{33} = \varepsilon_{zz}^{-1}\, \varepsilon_{zy} q_0, \tag{9.233e}$$
$$a_{34} = -\varepsilon_{zz}^{-1} q_0^2 + 1, \tag{9.233f}$$
$$a_{41} = \varepsilon_{yx} - \varepsilon_{yz}\, \varepsilon_{zz}^{-1}\, \varepsilon_{zx}, \tag{9.233g}$$
$$a_{43} = \varepsilon_{zz} - \varepsilon_{yz}\, \varepsilon_{zz}^{-1}\, \varepsilon_{zy}, \tag{9.233h}$$
$$a_{24} = \varepsilon_{yz}\, \varepsilon_{zz}^{-1} q_0. \tag{9.233i}$$

The proper values κ_0 of the matrix $\mathbf{C}_0^{(ua)}$ follow from the condition for the determinant $\det(\mathbf{C}_0^{(ua)} - \kappa_0 \mathbf{1})$ to be zero

$$\det\left(\mathbf{C}_0^{(ua)} - \kappa_0 \mathbf{1}\right) = \begin{vmatrix} -\kappa_0 & -1 & 0 & 0 \\ a_{21} & -\kappa_0 & a_{23} & a_{24} \\ a_{31} & 0 & a_{33} - \kappa_0 & a_{34} \\ a_{41} & 0 & a_{43} & a_{44} - \kappa_0 \end{vmatrix}. \tag{9.234}$$

The development of the determinant gives

$$\det\left(\mathbf{C}_0^{(ua)} - \kappa_0 \mathbf{1}\right) = \kappa_0 \begin{vmatrix} \kappa_0 & a_{23} & a_{24} \\ 0 & a_{33} - \kappa_0 & a_{34} \\ 0 & a_{43} & a_{44} - \kappa_0 \end{vmatrix} + a_{21} \begin{vmatrix} 1 & 0 & 0 \\ 0 & a_{33} - \kappa_0 & a_{34} \\ 0 & a_{43} & a_{44} - \kappa_0 \end{vmatrix}$$
$$- a_{31} \begin{vmatrix} 1 & 0 & 0 \\ \kappa_0 & a_{23} & a_{24} \\ 0 & a_{43} & a_{44} - \kappa_0 \end{vmatrix} + a_{41} \begin{vmatrix} 1 & 0 & 0 \\ \kappa_0 & a_{23} & a_{24} \\ 0 & a_{33} - \kappa_0 & a_{34} \end{vmatrix} = 0 \tag{9.235}$$

which results in an explicit equation for the proper values

$$\kappa_0^2[(\kappa_0 - a_{33})(\kappa_0 - a_{44}) - a_{34}a_{43}] + a_{21}[(\kappa_0 - a_{33})(\kappa_0 - a_{44}) - a_{34}a_{43}]$$
$$+ a_{31}[a_{23}(\kappa_0 - a_{44}) + a_{24}a_{43}] + a_{41}[a_{23}a_{34} - a_{24}(\kappa_0 - a_{33})] = 0. \tag{9.236}$$

Further manipulations on the right-hand side of the proper value equation give

$$\begin{aligned}
&\kappa_0^4 - (a_{33} + a_{44})\kappa_0^3 + [(a_{33}a_{44} - a_{34}a_{43}) + a_{21}]\,\kappa_0^2 - [a_{21}(a_{33} + a_{44}) \\
&\quad - a_{23}a_{31} - a_{24}a_{41}]\kappa_0 + a_{21}(a_{33}a_{44} - a_{34}a_{43})a_{34} + a_{31}(a_{24}a_{43} - a_{23}a_{44}) \\
&\quad + a_{41}(a_{23}a_{34} - a_{24}a_{33}) = 0.
\end{aligned} \tag{9.237}$$

The substitutions according to Eq. (9.233) provide

$$\begin{aligned}
&\kappa_0^4 - \varepsilon_{zz}^{-1}(\varepsilon_{yz} + \varepsilon_{zy})q_0\kappa_0^3 - \varepsilon_{zz}^{-1}\left[\varepsilon_{yy}\left(\varepsilon_{zz} - q_0^2\right) + \varepsilon_{zz}\left(\varepsilon_{xx} - q_0^2\right) - \varepsilon_{yz}\varepsilon_{zy}\right. \\
&\quad \left. - \varepsilon_{zx}\varepsilon_{xz}\right]\kappa_0^2 + \varepsilon_{zz}^{-1}\left[\left(\varepsilon_{xx} - q_0^2\right)(\varepsilon_{yz} + \varepsilon_{zy}) - \varepsilon_{zx}\varepsilon_{xy} - \varepsilon_{xz}\varepsilon_{yx}\right]q_0\kappa_0 \\
&\quad + \varepsilon_{yy}\varepsilon_{zz}^{-1}\left[\left(\varepsilon_{xx} - q_0^2\right)\varepsilon_{zz} - \varepsilon_{zx}\varepsilon_{xz}\right] - \varepsilon_{zz}^{-1}\left(\varepsilon_{xx} - p_0^2\right)\left(\varepsilon_{yy}q_0^2 + \varepsilon_{yz}\varepsilon_{zy}\right) \\
&\quad + \varepsilon_{zz}^{-1}\left(\varepsilon_{zx}\varepsilon_{yz}\varepsilon_{xy} + \varepsilon_{xz}\varepsilon_{zy}\varepsilon_{yx} - \varepsilon_{zz}\varepsilon_{xy}\varepsilon_{yx} + \varepsilon_{yx}\varepsilon_{xy}q_0^2\right) = 0.
\end{aligned} \tag{9.238}$$

After the multiplication by ε_{zz}

$$\begin{aligned}
&\varepsilon_{zz}\kappa_0^4 - (\varepsilon_{yz} + \varepsilon_{zy})q_0\kappa_0^3 - \left[\varepsilon_{yy}\left(\varepsilon_{zz} - q_0^2\right) + \varepsilon_{zz}\left(\varepsilon_{xx} - q_0^2\right) - \varepsilon_{yz}\varepsilon_{zy} - \varepsilon_{zx}\varepsilon_{xz}\right]\kappa_0^2 \\
&\quad + \left[\left(\varepsilon_{xx} - q_0^2\right)(\varepsilon_{yz} + \varepsilon_{zy}) - \varepsilon_{zx}\varepsilon_{xy} - \varepsilon_{xz}\varepsilon_{yx}\right]q_0\kappa_0 \\
&\quad + \varepsilon_{yy}\left(\varepsilon_{zz} - q_0^2\right)\left(\varepsilon_{xx} - q_0^2\right) - \varepsilon_{yy}\varepsilon_{zx}\varepsilon_{xz} \\
&\quad - \left(\varepsilon_{xx} - q_0^2\right)\varepsilon_{yz}\varepsilon_{zy} - \left(\varepsilon_{zz} - q_0^2\right)\varepsilon_{xy}\varepsilon_{yx} + \varepsilon_{zx}\varepsilon_{yz}\varepsilon_{xy} + \varepsilon_{xz}\varepsilon_{zy}\varepsilon_{yx} = 0.
\end{aligned} \tag{9.239}$$

This quartic equation for $m = 0$ provides four, in general different, proper values $\kappa_{1,0}, \kappa_{2,0}, \kappa_{3,0}$, and $\kappa_{4,0}$. The four proper vectors are in general different and elliptically polarized. For a real and symmetric permittivity tensor, they are linearly polarized but they cannot in general be classified as TE and TM.

This result should be generalized to the case of the homogeneous anisotropic region in contact with a region periodic in space. Then, in general all the terms of the Floquet developments of Eq. (9.18) and Eq. (9.19) should be accounted for. They are determined by the form of the corresponding (normalized) grating vector $\mathbf{n}_\Gamma$ from Eq. (9.15), which specifies the region with periodic either scalar or tensor permittivity. For an m-th term of the Floquet development in the homogeneous anisotropic region, we obtain from the Maxwell equations (9.68b) and (9.68f)

$$\frac{de_{xm}}{d\bar{z}} = js_m e_{xm} + jp_m e_{zm} - jh_{ym}, \tag{9.240a}$$

$$\frac{de_{ym}}{d\bar{z}} = js_m e_{ym} - jq_m e_{zm} + jh_{xm}, \tag{9.240b}$$

$$\frac{dh_{xm}}{d\bar{z}} = js_m h_{xm} - jp_m h_{zm} + j(\varepsilon_{yz}e_{zm} + \varepsilon_{yx}e_{xm} + \varepsilon_{yy}e_{ym}), \tag{9.240c}$$

$$\frac{dh_{ym}}{d\bar{z}} = js_m h_{ym} - jq_m h_{zm} - j(\varepsilon_{xz}e_{zm} + \varepsilon_{xx}e_{xm} + \varepsilon_{xy}e_{ym}), \tag{9.240d}$$

$$h_{zm} = p_m e_{ym} - q_m e_{xm}, \tag{9.240e}$$

$$(\varepsilon_{zx}e_{xm} + \varepsilon_{zy}e_{ym} + \varepsilon_{zz}e_{zm}) = -p_m h_{ym} + q_m h_{xm}. \tag{9.240f}$$

The last two equations allow us to eliminate e_{zm} and h_{zm}. To this purpose, we transform Eq. (9.240f) into the form

$$e_{zm} = -\varepsilon_{zz}^{-1}(p_m h_{ym} - q_m h_{xm} + \varepsilon_{zx} e_{xm} + \varepsilon_{zy} e_{ym}). \tag{9.241}$$

The substitution according to Eq. (9.240e) and Eq. (9.241) into Eq. (9.240a) to Eq. (9.240d) containing $\frac{\mathrm{d}}{\mathrm{d}\bar{z}}$

$$\frac{\mathrm{d}e_{xm}}{\mathrm{d}\bar{z}} = \mathrm{j}s_m e_{xm} + \mathrm{j}p_m\left(-\varepsilon_{zz}^{-1}(p_m h_{ym} - q_m h_{xm} + \varepsilon_{zx} e_{xm} + \varepsilon_{zy} e_{ym})\right) - \mathrm{j}h_{ym}, \tag{9.242a}$$

$$\frac{\mathrm{d}e_{ym}}{\mathrm{d}\bar{z}} = \mathrm{j}s_m e_{ym} - \mathrm{j}q_m\left(-\varepsilon_{xx}^{-1}(p_m h_{ym} - q_m h_{xm} + \varepsilon_{zx} e_{xm} + \varepsilon_{zy} e_{ym})\right)e_{zm} + \mathrm{j}h_{xm}, \tag{9.242b}$$

$$\frac{\mathrm{d}h_{xm}}{\mathrm{d}\bar{z}} = \mathrm{j}s_m h_{xm} - \mathrm{j}p_m(p_m e_{ym} - q_m e_{xm}) + \mathrm{j}\left[-\varepsilon_{yz}\varepsilon_{zz}^{-1}(p_m h_{ym} - q_m h_{xm} + \varepsilon_{zx} e_{xm} + \varepsilon_{zy} e_{ym}) + \varepsilon_{yx} e_{xm} + \varepsilon_{yy} e_{ym}\right], \tag{9.242c}$$

$$\frac{\mathrm{d}h_{ym}}{\mathrm{d}\bar{z}} = \mathrm{j}s_m h_{ym} - \mathrm{j}q_m(p_m e_{ym} - q_m e_{xm}) - \mathrm{j}\left\{\varepsilon_{xz}\left[-\varepsilon_{zz}^{-1}(p_m h_{ym} - q_m h_{xm} + \varepsilon_{zx} e_{xm} + \varepsilon_{zy} e_{ym})\right] + \varepsilon_{xx} e_{xm} + \varepsilon_{xy} e_{ym}\right\}. \tag{9.242d}$$

After rearranging

$$\frac{\mathrm{d}e_{xm}}{\mathrm{d}\bar{z}} = \mathrm{j}(s_m + p_m\varepsilon_{zz}^{-1})e_{xm} + \mathrm{j}(p_m^2\varepsilon_{zz}^{-1} - 1)h_{ym} + \mathrm{j}\varepsilon_{zz}^{-1}\varepsilon_{zy} e_{ym} - \mathrm{j}p_m\varepsilon_{zz}^{-1}q_m h_{xm}, \tag{9.243a}$$

$$\frac{\mathrm{d}h_{ym}}{\mathrm{d}\bar{z}} = \mathrm{j}(q_m^2 - \varepsilon_{xx} + \varepsilon_{xz}\varepsilon_{zz}^{-1}\varepsilon_{zx})e_{xm} + \mathrm{j}(s_m + \varepsilon_{xz}\varepsilon_{zz}^{-1}p_m)\,h_{ym} + \mathrm{j}(\varepsilon_{xz}\varepsilon_{zz}^{-1}\varepsilon_{zy} - \varepsilon_{xy} - p_m q_m)e_{ym} - \mathrm{j}\,\varepsilon_{xz}\varepsilon_{zz}^{-1}q_m h_{xm}, \tag{9.243b}$$

$$\frac{\mathrm{d}e_{ym}}{\mathrm{d}\bar{z}} = \mathrm{j}q_m\varepsilon_{zz}^{-1}\varepsilon_{zx} e_{xm} + \mathrm{j}q_m\varepsilon_{zz}^{-1}p_m h_{ym} + \mathrm{j}(s_m + q_m\varepsilon_{zz}^{-1}\varepsilon_{zy})e_{ym} + \mathrm{j}(1 - q_m^2\varepsilon_{zz}^{-1})h_{xm}, \tag{9.243c}$$

$$\frac{\mathrm{d}h_{xm}}{\mathrm{d}\bar{z}} = \mathrm{j}(p_m q_m + \varepsilon_{yx} - \varepsilon_{yz}\varepsilon_{zz}^{-1}\varepsilon_{zx})e_{xm} - \mathrm{j}\varepsilon_{yz}\varepsilon_{zz}^{-1}p_m h_{ym} + \mathrm{j}(-p_m^2 + \varepsilon_{yy} - \varepsilon_{yz}\varepsilon_{zz}^{-1}\varepsilon_{zy})e_{ym} + \mathrm{j}(s_m + \varepsilon_{yz}\varepsilon_{zz}^{-1}q_m)h_{xm}. \tag{9.243d}$$

Eq. (9.240e), Eq. (9.241), and Eq. (9.243) can be transformed into a matrix form. In view of the space uniformity of the permittivity tensor, the Fourier developments for the permittivity tensor, components reduce to single terms. Consequently, Eq. (9.94a) and Eq. (9.94b) split to m independent matrix equations with scalar matrix elements

$$\frac{\mathrm{d}}{\mathrm{d}\bar{x}}\mathbf{f}_{\hat{t}m}^{(ua)} = \mathrm{j}\mathbf{C}_m^{(ua)}\mathbf{f}_{\hat{t}m}^{(ua)}, \tag{9.244a}$$

$$\mathbf{f}_{\hat{n}m}^{(ua)} = \mathrm{j}\mathbf{D}_m^{(ua)}\mathbf{f}_{\hat{t}m}^{(ua)}, \tag{9.244b}$$

where

$$\mathbf{C}_m^{(ua)} = \begin{pmatrix} p_m\varepsilon_{zz}^{-1}\varepsilon_{zx} + s_m & p_m^2\varepsilon_{zz}^{-1} - 1 & p_m\varepsilon_{zz}^{-1}\varepsilon_{zy} & -p_m\varepsilon_{zz}^{-1}q_m \\ \varepsilon_{xz}\varepsilon_{zz}^{-1}\varepsilon_{zx} - \varepsilon_{xx} + q_m^2 & \varepsilon_{xz}\varepsilon_{zz}^{-1}p_m + s_m & \varepsilon_{xz}\varepsilon_{zz}^{-1}\varepsilon_{zy} - \varepsilon_{xy} - q_m p_m & -\varepsilon_{xz}\varepsilon_{zz}^{-1}q_m \\ q_m\varepsilon_{zz}^{-1}\varepsilon_{zx} & q_m\varepsilon_{zz}^{-1}p_m & q_m\varepsilon_{zz}^{-1}\varepsilon_{zy} + s_m & 1 - q_m^2\varepsilon_{zz}^{-1} \\ \varepsilon_{yx} - \varepsilon_{yz}\varepsilon_{zz}^{-1}\varepsilon_{zx} + p_m q_m & -\varepsilon_{yz}\varepsilon_{zz}^{-1}p_m & \varepsilon_{yy} - \varepsilon_{yz}\varepsilon_{zz}^{-1}\varepsilon_{zy} - p_m^2 & \varepsilon_{yz}\varepsilon_{zz}^{-1}q_m + s_m \end{pmatrix}, \tag{9.245}$$

$$\mathbf{D}_m^{(ua)} = \begin{pmatrix} -\varepsilon_{zz}^{-1}\varepsilon_{zx} & -\varepsilon_{zz}^{-1}p_m & -\varepsilon_{zz}^{-1}\varepsilon_{zy} & \varepsilon_{zz}^{-1}q_m \\ -q_m & 0 & p_m & 0 \end{pmatrix}. \tag{9.246}$$

The proper values κ_m of the matrix $\mathbf{C}_m^{(ua)}$ follow from the condition for the determinant $\det\left(\mathbf{C}_m^{(ua)} - \kappa_m\mathbf{1}\right)$ to be zero. The calculation is performed in Appendix F. After the evaluation of the determinants, we arrive at

$$\begin{aligned}
&(s_m - \kappa_m)^4 + (s_m - \kappa_m)^3\varepsilon_{zz}^{-1}[p_m(\varepsilon_{zx} + \varepsilon_{xz}) + q_m(\varepsilon_{yz} + \varepsilon_{zy})] \\
&\quad + (s_m - \kappa_m)^2\big[-\left(\varepsilon_{xx} - p_m^2\right) - \left(\varepsilon_{yy} - q_m^2\right) + \varepsilon_{zz}^{-1}\left(p_m^2\varepsilon_{xx} + q_m^2\varepsilon_{yy}\right) \\
&\quad + p_m q_m\varepsilon_{zz}^{-1}(\varepsilon_{xy} + \varepsilon_{yx}) + \varepsilon_{zz}^{-1}(\varepsilon_{zx}\varepsilon_{xz} + \varepsilon_{yz}\varepsilon_{zy})\big] \\
&\quad + (s_m - \kappa_m)\big[-\varepsilon_{zz}^{-1}p_m\left(\varepsilon_{yy} - p_m^2\right)(\varepsilon_{zx} + \varepsilon_{xz}) - \varepsilon_{zz}^{-1}q_m\left(\varepsilon_{xx} - q_m^2\right)(\varepsilon_{yz} + \varepsilon_{zy}) \\
&\quad + \varepsilon_{zz}^{-1}(\varepsilon_{xy} + p_m q_m)(q_m\varepsilon_{zx} + p_m\varepsilon_{yz}) + \varepsilon_{zz}^{-1}(\varepsilon_{yx} + p_m q_m)(q_m\varepsilon_{xz} + p_m\varepsilon_{zy})\big] \\
&\quad + \left[1 - \left(p_m^2 + q_m^2\right)\varepsilon_{zz}^{-1}\right]\left[\left(\varepsilon_{xx} - q_m^2\right)\left(\varepsilon_{yy} - p_m^2\right) - (\varepsilon_{xy} + p_m q_m)(\varepsilon_{yx} + p_m q_m)\right] \\
&\quad - \varepsilon_{zz}^{-1}\left[\varepsilon_{yz}\varepsilon_{zy}\left(\varepsilon_{xx} - q_m^2\right) + \varepsilon_{zx}\varepsilon_{xz}\left(\varepsilon_{yy} - p_m^2\right)\right] + p_m q_m\varepsilon_{zz}^{-1}(\varepsilon_{zx}\varepsilon_{yz} + \varepsilon_{xz}\varepsilon_{zy}) \\
&\quad + (\varepsilon_{zx}\varepsilon_{xy}\varepsilon_{yz} + \varepsilon_{xz}\varepsilon_{yx}\varepsilon_{zy}) = 0.
\end{aligned} \tag{9.247}$$

This is a generalization of Eq. (9.239) to homogeneous anisotropic media in contact with periodic regions.

9.7 Transmission and Interface Matrices

So far, we have considered unbound media. We now account for interfaces assuming that the field propagation in bounded regions remains characterized by the proper values and proper polarization modes of plane waves found for unbound media. This assumption is consistent with the Fraunhofer approximation.

Within the region I (I = 0, 1, 2), the solution to the differential equation (9.123)

$$\frac{\mathrm{d}}{\mathrm{d}\bar{z}}\mathbf{g}^{(\mathrm{I})} = \mathrm{j}\boldsymbol{\kappa}^{(\mathrm{I})}\mathbf{g}^{(\mathrm{I})}$$

is given by

$$\mathbf{g}^{(\mathrm{I})}(\bar{z}) = \mathbf{U}^{(\mathrm{I})}(\bar{z} - \bar{z}^{(0)})\mathbf{g}^{(\mathrm{I})}(\bar{z}^{0}), \tag{9.248}$$

where $\mathbf{U}^{(\mathrm{I})}(\bar{z} - \bar{z}^{(0)})$ denotes a diagonal matrix of dimension $(2M + 1)$

$$\mathbf{U}^{(\mathrm{I})}(\bar{z} - \bar{z}^{(0)}) = \exp[\mathrm{j}\boldsymbol{\kappa}^{(\mathrm{I})}(\bar{z} - \bar{z}^{(0)})], \tag{9.249}$$

where

$$\exp[\mathrm{j}\boldsymbol{\kappa}^{(\mathrm{I})}(\bar{z} - \bar{z}^{(0)})] = \left\{\delta_{nl}\exp\left[\mathrm{j}\kappa_l^{(\mathrm{I})}(\bar{z} - \bar{z}^{(0)})\right]\right\}. \tag{9.250}$$

The $\mathbf{U}^{(\mathrm{I})}(\bar{z} - \bar{z}^{(0)})$ matrix contains the information on the distance $\bar{z} - \bar{z}^{(0)}$ on the z axis but no reference to the position $\bar{z}$. The importance of this feature becomes more apparent below, where we consider multilayer anisotropic gratings.

The constant column vector $\mathbf{g}^{(\mathrm{I})}(\bar{z}^{0})$ pertains to the normalized coordinate $\bar{z} = \bar{z}^{(0)}$. The diagonal matrix $\mathbf{U}^{(\mathrm{I})}(\bar{z} - \bar{z}^{(0)})$ is termed *transmission matrix* and represents the field progression normal to the interfaces, *i.e.*, parallel to $\pm\hat{z}$, of the field $\mathbf{g}^{(\mathrm{I})}(\bar{z})$ in the region I. The corresponding matrix in the Yeh formalism [8] is called the propagation matrix $\mathcal{P}$. We remind that the product $\mathcal{S} = \mathcal{D}\mathcal{P}\mathcal{D}^{-1}$ defines the Abelès characteristic matrix $\mathcal{S}$ in isotropic media [1,37].

On traversing the interface planes ($\bar{z} = 0$ and $\bar{z} = \bar{d}_1$), which separate the grating region from the homogeneous regions, the tangential field components must remain continuous. From Eq. (9.18) and Eq. (9.19) for the field vectors in the three-dimensional space,

$$\sqrt{Y_0}\mathbf{E}\cdot\hat{x} = \sum_m \mathbf{e}_m(\bar{z})\cdot\hat{x}\exp(-\mathrm{j}s_m\bar{z})\exp[-\mathrm{j}(p_m\bar{x} + q_m\bar{y})], \tag{9.251a}$$

$$\sqrt{Y_0}\mathbf{E}\cdot\hat{y} = \sum_m \mathbf{e}_m(\bar{z})\cdot\hat{y}\exp(-\mathrm{j}s_m\bar{z})\exp[-\mathrm{j}(p_m\bar{x} + q_m\bar{y})], \tag{9.251b}$$

$$\sqrt{Z_0}\mathbf{H}\cdot\hat{x} = \sum_m \mathbf{h}_m(\bar{z})\cdot\hat{x}\exp(-\mathrm{j}s_m\bar{z})\exp[-\mathrm{j}(p_m\bar{x} + q_m\bar{y})], \tag{9.251c}$$

$$\sqrt{Z_0}\mathbf{H}\cdot\hat{y} = \sum_m \mathbf{h}_m(\bar{z})\cdot\hat{y}\exp(-\mathrm{j}s_m\bar{z})\exp[-\mathrm{j}(p_m\bar{x} + q_m\bar{y})], \tag{9.251d}$$

and from the definitions (9.93),

$$\mathbf{f}_{\hat{t}}(\bar{z}) = \begin{pmatrix} \mathbf{e}_x(\bar{z}) \\ \mathbf{h}_y(\bar{z}) \\ \mathbf{e}_y(\bar{z}) \\ \mathbf{h}_x(\bar{z}) \end{pmatrix}, \qquad \mathbf{f}_{\hat{n}}(\bar{z}) = \begin{pmatrix} \mathbf{e}_z(\bar{z}) \\ \mathbf{h}_z(\bar{z}) \end{pmatrix},$$

it follows that the column matrix $\mathrm{EXP}(-\mathrm{j}\,\mathbf{s}^{(\mathrm{I})}\bar{z})\mathbf{f}_{\hat{t}}^{\mathrm{I}}(\bar{z})$ must be continuous at the interfaces. As a result, each single term in the Floquet development should be continuous at the interfaces. Here, the diagonal matrix $\mathrm{EXP}(-\mathrm{j}\,\mathbf{s}^{(\mathrm{I})}\bar{z})$ of $4 \times (2M+1)$ dimension contains four diagonal matrices $\exp(-\mathrm{j}\,\mathbf{s}^{(\mathrm{I})}\bar{z}) = [\delta_{nl}\exp(-\mathrm{j}\,s_{nl}^{(\mathrm{I})}\bar{z})]$. These are associated with the four column vectors $\mathbf{e}_y$, $\mathbf{h}_z$, $\mathbf{e}_z$, $\mathbf{h}_y$, which represent the tangential field components, respectively

$$\mathrm{EXP}\left(-\mathrm{j}\,\mathbf{s}^{(\mathrm{I})}\bar{z}\right) = \begin{pmatrix} \exp(-\mathrm{j}\,\mathbf{s}^{(\mathrm{I})}\bar{z}) & \mathbf{0} & \mathbf{0} & \mathbf{0} \\ \mathbf{0} & \exp(-\mathrm{j}\,\mathbf{s}^{(\mathrm{I})}\bar{z}) & \mathbf{0} & \mathbf{0} \\ \mathbf{0} & \mathbf{0} & \exp(-\mathrm{j}\,\mathbf{s}^{(\mathrm{I})}\bar{z}) & \mathbf{0} \\ \mathbf{0} & \mathbf{0} & \mathbf{0} & \exp(-\mathrm{j}\,\mathbf{s}^{(\mathrm{I})}\bar{z}) \end{pmatrix}. \tag{9.252}$$

The conditions for the continuity of the tangential field components at the interface $\bar{x}_{\mathrm{I}-1}$ between the region $\mathrm{I}-1$ and the region I, therefore, require

$$\mathrm{EXP}(-\mathrm{j}\,\mathbf{s}^{(\mathrm{I}-1)}\bar{z}^{(\mathrm{I}-1)})\mathbf{f}_{\hat{t}}^{(\mathrm{I}-1)}(\bar{z}^{(\mathrm{I}-1)}) = \mathrm{EXP}(-\mathrm{j}\,\mathbf{s}^{(\mathrm{I})}\bar{z}^{(\mathrm{I}-1)})\mathbf{f}_{\hat{t}}^{\mathrm{I}}(\bar{z}^{(\mathrm{I}-1)}), \tag{9.253}$$

i.e.,

$$\mathrm{EXP}(-\mathrm{j}\,\mathbf{s}^{(\mathrm{I}-1)}\bar{z}^{(\mathrm{I}-1)})\mathbf{T}_{(\mathrm{I}-1)}\mathbf{g}^{(\mathrm{I}-1)}(\bar{z}^{(\mathrm{I}-1)}) = \mathrm{EXP}(-\mathrm{j}\,\mathbf{s}^{(\mathrm{I})}\bar{z}^{(\mathrm{I}-1)})\mathbf{T}^{(\mathrm{I})}\mathbf{g}^{(\mathrm{I})}(\bar{z}^{(\mathrm{I}-1)}). \tag{9.254}$$

In homogeneous regions $I = 0, 2$, we put $s_m^{(\mathrm{I})} = 0$. The conditions for the continuity of the tangential field components at the interface $\bar{z} = 0$ between the entrance homogeneous isotropic region (0) and the periodic region (1) can then be expressed as

$$\mathbf{f}_{\hat{t}}^{(0u)}(\bar{z} = 0) = \mathbf{f}_{\hat{t}}^{(1)}(\bar{z} = 0), \tag{9.255}$$

i.e.,

$$\mathbf{T}^{(0u)}\mathbf{g}^{(0u)}(0) = \mathbf{T}^{(1)}\mathbf{g}^{(1)}(0), \tag{9.256}$$

where we have used $\mathbf{s}^{(0)} = 0$ in the isotropic homogeneous region (0) and $\mathrm{EXP}(-\mathrm{j}\,\mathbf{s}^{(1)}\bar{z}^{(0)}) = 1$ for $\bar{z}^{(0)} = 0$. In a similar way for the interface $\bar{z} = \bar{d}$

(for the thickness of the region (1) we have $\bar{d} > 0$) between the periodic region (1) and the homogeneous isotropic exit region (2)

$$\mathrm{EXP}(\mathrm{j}\,\mathbf{s}^{(1)}\bar{d}_1)\mathbf{f}_t^{(1)}(\bar{z} = \bar{d}_1) = \mathbf{f}_t^{(2u)}(\bar{z} = \bar{d}_1),$$

i.e.,

$$\mathrm{EXP}(\mathrm{j}\,\mathbf{s}^{(1)}\bar{d}_1)\mathbf{T}^{(1)}\mathbf{g}^{(1)}(\bar{d}_1) = \mathbf{T}^{(2u)}\mathbf{g}^{(2u)}(\bar{d}_1), \tag{9.257}$$

because $\mathbf{s}^{(2)} = 0$ in the homogeneous region (2). We have used Eq. (9.122).

The boundary conditions can be expressed in terms of the so-called *interface matrices* $\mathbf{B}^{(\mathrm{I},\mathrm{I}+1)}$. These relate vectors $\mathbf{g}^{(\mathrm{I})}$ and $\mathbf{g}^{(\mathrm{I}+1)}$ at the interfaces. Specifically,

$$\mathbf{g}^{(0)}(0) = \mathbf{B}^{(01)}\,\mathbf{g}^{(1)}(0), \qquad \mathbf{g}^{(1)}(\bar{d}_1) = \mathbf{B}^{(12)}\,\mathbf{g}^{(2)}(\bar{d}_1). \tag{9.258}$$

According to Eq. (9.248) and Eq. (9.249)

$$\mathbf{g}^{(1)}(\bar{d}_1) = \mathbf{U}^{(1)}(\bar{d}_1)\mathbf{g}^{(1)}(0), \tag{9.259}$$

where

$$\mathbf{U}_2(\bar{d}_1) = \exp(\mathrm{j}\,\boldsymbol{\kappa}^{(1)}\bar{d}_1). \tag{9.260}$$

The interface matrices are defined by the relations

$$\mathbf{B}^{(01)} = [\mathbf{T}^{(0u)}]^{-1}\mathbf{T}^{(1)}, \tag{9.261a}$$

$$\mathbf{B}^{(12)} = [\mathbf{T}^{(1)}]^{-1}\mathrm{EXP}(-\mathrm{j}\,\mathbf{s}^{(1)}\bar{d}_1)\mathbf{T}^{(2u)}. \tag{9.261b}$$

The matrix relation that couples the field $\mathbf{g}^{(1)}(0)$ at the interface $\bar{z} = 0$ and the field $\mathbf{g}^{(1)}(\bar{d}_1)$ at the interface $\bar{z} = \bar{d}_1$ can be expressed with the transmission matrix and the interface matrix

$$\mathbf{g}^{(0)} = \begin{pmatrix} \mathbf{g}^{(0)+} \\ \mathbf{g}^{(0)-} \end{pmatrix} = \begin{pmatrix} \mathbf{W}_1 & \mathbf{W}_2 \\ \mathbf{W}_3 & \mathbf{W}_4 \end{pmatrix} \begin{pmatrix} \mathbf{g}^{(2)+} \\ \mathbf{g}^{(2)-} \end{pmatrix} = \mathbf{W}\mathbf{g}^{(2)}, \tag{9.262}$$

where $\mathbf{W} = \mathbf{B}^{(01)}\mathbf{U}^{(1)}(\bar{d}_1)\mathbf{B}^{(12)}$. Here, $\mathbf{W}$ represents the *transmission matrix* of the grating and $\mathbf{W}_k$ $k = 1, 2, 3, 4$ denote its submatrices. The use was made of Eq. (9.255) to Eq. (9.261)

$$\begin{aligned} \mathbf{g}^{(0)}(0) &= [\mathbf{T}^{(0u)}]^{-1}\mathbf{T}^{(1)}\mathbf{g}^{(1)}(0) = [\mathbf{T}^{(0u)}]^{-1}\mathbf{T}^{(1)}\left[\mathbf{U}^{(1)}(\bar{d}_1)\right]^{-1}\mathbf{g}^{(1)}(\bar{d}_1) \\ &= [\mathbf{T}^{(0u)}]^{-1}\mathbf{T}^{(1)}\mathbf{U}^{(1)}(-\bar{d}_1)[\mathbf{T}^{(1)}]^{-1}\mathrm{EXP}(-\mathrm{j}\,\mathbf{s}^{(1)}\bar{d}_1)\mathbf{T}^{(2u)}\mathbf{g}^{(2)}(\bar{d}_1) \\ &= \mathbf{B}^{(01)}\mathbf{U}^{(1)}(-\bar{d}_1)\mathbf{B}^{(12)}\mathbf{g}^{(2)}(\bar{d}_1). \end{aligned} \tag{9.263}$$

In Eq. (9.262), the column vector $\mathbf{g}^{(0)-}$ represents the wave incident from the homogeneous region (0) at the interface $\bar{z} = 0$. The column vector

$\mathbf{g}^{(0)+}$ represents the coherent sum of the wave reflected at this interface $\bar{z} = 0$ and the waves traversing the interface $\bar{z} = 0$ and arriving from the regions (1) and (2). The wave exiting from the grating region (1) into the region (2) is represented by $\mathbf{g}^{(2)-}$. In the case of a wave proceeding from the region (2) towards the interface $\bar{z} = \bar{d}_1$, this would be represented by $\mathbf{g}^{(2)+}$. Consequently, the wave $\mathbf{g}^{(2)-}$ could contain reflected waves generated by $\mathbf{g}^{(2)+} \neq 0$.

The present formalism can be extended to absorbing media. Then, in practice, for higher order space harmonics, large values of exponential functions in thick layers may cause overflow problems in numerical calculations. These can be avoided by using suitable normalizations proposed by Rokushima and Yamakita [48].

9.8 Wave Diffraction on the Grating

The wave of unit amplitude incident from the region (0) onto the grating (1) with the TE wave polarization is defined by

$$^{TE}\mathbf{g}^{(0)-} = \begin{pmatrix} 0 \\ \cdot \\ \cdot \\ 0 \\ 1 \\ 0 \\ \cdot \\ \cdot \\ 0 \end{pmatrix}, \qquad ^{TM}\mathbf{g}^{(0)-} = \mathbf{0}. \tag{9.264}$$

For the TM polarization

$$^{TM}\mathbf{g}^{(0)-} = \begin{pmatrix} 0 \\ \cdot \\ \cdot \\ 0 \\ 1 \\ 0 \\ \cdot \\ \cdot \\ 0 \end{pmatrix}, \qquad ^{TE}\mathbf{g}^{(0)-} = \mathbf{0}. \tag{9.265}$$

We shall assume that there is no wave propagating from the region (2) towards the grating region (1)

$$\mathbf{g}^{(2)+} = \mathbf{0}. \tag{9.266}$$

The incident wave generates reflected ($\mathbf{g}^{(1)+}$) and transmitted ($\mathbf{g}^{(2)-}$) waves. We get from Eq. (9.262)

$$\mathbf{g}^{(0)-} = \mathbf{W}_4\mathbf{g}^{(2)-}, \qquad \text{or} \qquad \mathbf{g}^{(2)-} = (\mathbf{W}_4)^{-1}\,\mathbf{g}^{(0)-}, \tag{9.267a}$$

$$\mathbf{g}^{(0)+} = \mathbf{W}_2\mathbf{g}^{(2)-} = \mathbf{W}_2\,(\mathbf{W}_4)^{-1}\,\mathbf{g}^{(0)-}. \tag{9.267b}$$

The power carried by a wave of the m-th diffraction order propagating with the normal, *i.e.*, the $\hat{z}$ component of the normalized propagation vector oriented parallel or antiparallel with respect to $\hat{z}$ ($\pm\hat{z}$) in the region I (I = 0, 2) with TE and TM polarizations, respectively, is given by

$${}^{TE}\mathrm{p}_m^{(\mathrm{I})\pm} = \left|\hat{z}\cdot{}^{TE}\boldsymbol{\pi}_m^{(\mathrm{I})\pm}\right| = \Re\left(\xi_m^{(\mathrm{I})}\right)\left|{}^{TE}g_m^{(\mathrm{I})\pm}\right|^2, \tag{9.268a}$$

$${}^{TM}\mathrm{p}_m^{(\mathrm{I})\pm} = \left|\hat{z}\cdot{}^{TM}\boldsymbol{\pi}_m^{(\mathrm{I})\pm}\right| = \Re\left(\frac{\varepsilon}{\xi_m^{(\mathrm{I})}}\right)\left|{}^{TM}g_m^{(\mathrm{I})\pm}\right|^2, \tag{9.268b}$$

where $\xi_m^{(\mathrm{I})} = (\varepsilon^{(\mathrm{I})} - p_m^{(\mathrm{I})2} - q_m^{(\mathrm{I})2})^{1/2}$.

This result follows from the $\hat{z}$ component of the Poynting vector. The contributions to this come from the fields e_{xm}, h_{ym}, e_{ym}, and h_{xm}. The Poynting vector of TE waves propagating with the normal component of the propagation vector in the direction of $\hat{z}$ follows from Eq. (9.158a) and Eq. (9.158b)

$$\begin{aligned}
&{}^{TE}\boldsymbol{\pi}_m^{(\mathrm{I})+}\\
&\quad= \Re\left({}^{TE}\mathbf{e}_m^{(\mathrm{I})+}\times{}^{TE}\,\mathbf{h}_{Im}^{(\mathrm{I})+*}\right)\\
&\quad= \Re\left\{(\hat{x}\dot{q}_m - \hat{y}\dot{p}_m)\times\left[-\hat{z}n_{\mathfrak{t}m} + \xi_m^{(\mathrm{I})*}(\hat{x}\dot{p}_m + \hat{y}\dot{q}_m)\right]\right\}\left|{}^{TE}g_m^{(\mathrm{I})+}\right|^2\\
&\quad= \Re\left[\hat{z}\xi_m^{(\mathrm{I})*}\left(\dot{p}_m^2 + \dot{q}_m^2\right) + n_{\mathfrak{t}m}\left(\hat{x}\dot{p}_m + \hat{y}\dot{q}_m\right)\right]\left|{}^{TE}g_m^{(\mathrm{I})+}\right|^2\\
&\quad= \Re\left[\hat{z}\xi_m^{(\mathrm{I})*} + n_{\mathfrak{t}m}(\hat{x}\dot{p}_m + \hat{y}\dot{q}_m)\right]\left|{}^{TE}g_m^{(\mathrm{I})+}\right|^2.
\end{aligned} \tag{9.269}$$

The asterisk, *, indicates complex conjugate. We note that $(\dot{p}_m^2 + \dot{q}_m^2) = 1$. Then, the $\hat{z}$ component corresponds to that of Eq. (9.268a) with superscript (+).

The Poynting vector of TM waves propagating with the normal component of the propagation vector in the direction of $\hat{z}$ follows from Eq. (9.170a) and Eq. (9.170b)

$$\begin{aligned}
&{}^{TM}\boldsymbol{\pi}_m^{(\mathrm{I})+}\\
&= \Re\left({}^{TM}\mathbf{e}_m^{(\mathrm{I})+}\times{}^{TM}\,\mathbf{h}_m^{(\mathrm{I})+*}\right)\\
&= \Re\left\{\left[-\hat{z}n_{\mathfrak{t}m}/\xi_m^{(\mathrm{I})} + (\hat{x}\dot{p}_m + \hat{y}\dot{q}_m)\right]\times(-\hat{x}\dot{q}_m + \hat{y}\dot{p}_m)\left(\varepsilon^{(\mathrm{I})*}/\xi_m^{(\mathrm{I})*}\right)\right\}\left|{}^{TM}g_m^{(\mathrm{I})+}\right|^2
\end{aligned}$$

$$= \Re\left[\hat{z}\left(\varepsilon^{(\mathrm{I})*}/\xi_m^{(\mathrm{I})*}\right)\left(\dot{p}_m^2+\dot{q}_m^2\right)+n_{\bar{t}m}\left(\varepsilon^{(\mathrm{I})*}/\left|\xi_m^{(\mathrm{I})}\right|^2\right)(\hat{x}\dot{p}_m+\hat{y}\dot{q}_m)\right]\left|{}^{TM}g_m^{(\mathrm{I})+}\right|^2$$
$$= \Re\left[\hat{z}\left(\varepsilon^{(\mathrm{I})*}/\xi_m^{(\mathrm{I})*}\right)+n_{\bar{t}m}\left(\varepsilon^{(\mathrm{I})*}/\left|\xi_m^{(\mathrm{I})}\right|^2\right)(\hat{x}\dot{p}_m+\hat{y}\dot{q}_m)\right]\left|{}^{TM}g_m^{(\mathrm{I})+}\right|^2. \tag{9.270}$$

The $\bar{z}$ component corresponds to that of Eq. (9.268b) with superscript (+).

The Poynting vector of TE waves propagating with the normal component of the propagation vector in the direction of $-\hat{z}$ follows from Eq. (9.182a) and Eq. (9.182b)

$$\begin{aligned}
&{}^{TE}\boldsymbol{\pi}_m^{(\mathrm{I})-}\\
&= \Re\left({}^{TE}\mathbf{e}_m^{(\mathrm{I})-}\times{}^{TE}\,\mathbf{h}_m^{(\mathrm{I})-*}\right)\\
&= \Re\left\{(\hat{x}\dot{q}_m-\hat{y}\dot{p}_m)\times\left[-\hat{z}n_{\bar{t}m}-\xi_m^{\mathrm{I}*}(\hat{x}\dot{p}_m+\hat{y}\dot{q}_m)\right]\right\}\left|{}^{TE}g_m^{(\mathrm{I})-}\right|^2\\
&= \Re\left[-\hat{x}\xi_m^{(\mathrm{I})*}\left(\dot{q}_m^2+\dot{s}_m^2\right)+n_{\bar{t}m}(\hat{y}\dot{q}_m+\hat{z}\dot{s}_m)\right]\left|{}^{TE}g_m^{(I)-}\right|^2\\
&= \Re\left[-\hat{x}\xi_m^{(\mathrm{I})*}+n_{\bar{t}m}(\hat{y}\dot{q}_m+\hat{z}\dot{s}_m)\right]\left|{}^{TE}g_m^{(I)-}\right|^2.
\end{aligned} \tag{9.271}$$

The x component corresponds to that of Eq. (9.268a) with superscript (−).

The Poynting vector of TM waves propagating with the normal component of the propagation vector in the direction of $-\hat{z}$ follows from Eq. (9.191a) and Eq. (9.191b)

$$\begin{aligned}
&{}^{TM}\boldsymbol{\pi}_m^{(\mathrm{I})-}\\
&= \Re\left({}^{TM}\mathbf{e}_m^{(\mathrm{I})-}\times{}^{TM}\,\mathbf{h}^{(\mathrm{I})-*}\right)\\
&= \Re\left\{\left[\hat{z}n_{\bar{t}m}/\xi_m^{(\mathrm{I})}+(\hat{x}\dot{p}_m+\hat{y}\dot{q}_m)\right]\times(\hat{x}\dot{q}_m-\hat{y}\dot{p}_m)\left(\varepsilon^{(\mathrm{I})*}/\xi_m^{(\mathrm{I})*}\right)\right\}\left|{}^{TM}g_m^{(\mathrm{I})-}\right|^2\\
&= \Re\left[-\hat{z}\left(\varepsilon^{(\mathrm{I})*}/\xi_m^{(\mathrm{I})*}\right)\left(\dot{p}_m^2+\dot{q}_m^2\right)+n_{\bar{t}m}\left(\varepsilon^{(\mathrm{I})*}/\left|\xi_m^{(\mathrm{I})}\right|^2\right)(\hat{x}\dot{p}_m+\hat{y}\dot{q}_m)\right]\left|{}^{TM}g_m^{(\mathrm{I})-}\right|^2\\
&= \Re\left[-\hat{z}\left(\varepsilon^{(\mathrm{I})*}/\xi_m^{(\mathrm{I})*}\right)+n_{\bar{t}m}\left(\varepsilon^{(\mathrm{I})*}/\left|\xi_m^{(\mathrm{I})}\right|^2\right)(\hat{x}\dot{p}_m+\hat{y}\dot{q}_m)\right]\left|{}^{TM}g_m^{(\mathrm{I})-}\right|^2.
\end{aligned} \tag{9.272}$$

The z component corresponds to that of Eq. (9.268b) with superscript (−).

The expressions for the fields in homogeneous regions employed in obtaining their Poynting vectors, Eq. (9.158a) and Eq. (9.158b), Eq. (9.170a), and Eq. (9.170b), Eq. (9.182a) and Eq. (9.182b), Eq. (9.191a) and Eq. (9.191b), may be concisely summarized as

$$\begin{pmatrix} e_{xm}(\bar{z})\\ h_{ym}(\bar{z})\\ e_{ym}(\bar{z})\\ h_{xm}(\bar{z})\end{pmatrix} = \begin{pmatrix} \dot{q}_m & \dot{p}_m & \dot{q}_m & \dot{p}_m\\ \dot{q}_m\xi_m & \varepsilon\dot{p}_m\xi_m^{-1} & -\dot{q}_m\xi_m & -\varepsilon\dot{p}_m\xi_m^{-1}\\ -\dot{p}_m & \dot{q}_m & -\dot{p}_m & \dot{q}_m\\ \dot{p}_m\xi_m & -\varepsilon\dot{q}_m\xi_m^{-1} & -\dot{p}_m\xi_m & \varepsilon\dot{q}_m\xi_m^{-1}\end{pmatrix}\begin{pmatrix} {}^{TE}g_m^{(\mathrm{I})+}(\bar{z})\\ {}^{TM}g_m^{(\mathrm{I})+}(\bar{z})\\ {}^{TE}g_m^{(\mathrm{I})-}(\bar{z})\\ {}^{TM}g_m^{(\mathrm{I})-}(\bar{z})\end{pmatrix}, \tag{9.273}$$

where according to Eq. (9.248), $\mathbf{g}_m^{(\mathrm{I})}(\bar{z}) = \mathbf{U}_m^{(\mathrm{I}u)}(\bar{z} - \bar{z}^{(0)})\mathbf{g}_m^{(\mathrm{I})}(\bar{z}^0)$ in homogeneous isotropic regions

$$\begin{pmatrix} {}^{TE}g_m^{(\mathrm{I})+}(\bar{z}) \\ {}^{TM}g_m^{(\mathrm{I})+}(\bar{z}) \\ {}^{TE}g_m^{(\mathrm{I})-}(\bar{z}) \\ {}^{TM}g_m^{(\mathrm{I})-}(\bar{z}) \end{pmatrix}$$

$$= \begin{pmatrix} \exp\left[\mathrm{j}\kappa_{m1}^{(\mathrm{I}u)}(\bar{z} - \bar{z}^{(0)})\right] & 0 & 0 & 0 \\ 0 & \exp\left[\mathrm{j}\kappa_{m2}^{(\mathrm{I}u)}(\bar{z} - \bar{z}^{(0)})\right] & 0 & 0 \\ 0 & 0 & \exp\left[\mathrm{j}\kappa_{m3}^{(\mathrm{I}u)}(\bar{z} - \bar{z}^{(0)})\right] & 0 \\ 0 & 0 & 0 & \exp\left[\mathrm{j}\kappa_{m4}^{(\mathrm{I}u)}(\bar{z} - \bar{z}^{(0)})\right] \end{pmatrix}$$

$$\times \begin{pmatrix} {}^{TE}g_m^{(\mathrm{I})+}(\bar{z}^0) \\ {}^{TM}g_m^{(\mathrm{I})+}(\bar{z}^0) \\ {}^{TE}g_m^{(\mathrm{I})-}(\bar{z}^0) \\ {}^{TM}g_m^{(\mathrm{I})-}(\bar{z}^0) \end{pmatrix} \tag{9.274}$$

and for $\bar{z}^{(0)} = 0$

$$\begin{aligned}
{}^{TE}g_m^{(\mathrm{I})+}(\bar{z}) &= {}^{TE}g_m^{(\mathrm{I})+}(0)\exp\left(\mathrm{j}\kappa_{m1}^{(\mathrm{I}u)}\bar{z}\right) = {}^{TE}g_m^{(\mathrm{I})+}(0)\exp\left(-\mathrm{j}\xi_m^{(\mathrm{I})}\bar{z}\right), \\
{}^{TM}g_{\mathrm{I}m}^{+}(\bar{z}) &= {}^{TM}g_m^{(\mathrm{I})+}(0)\exp\left(\mathrm{j}\kappa_{m2}^{(\mathrm{I}u)}\bar{z}\right) = {}^{TM}g_m^{(\mathrm{I})+}(0)\exp\left(-\mathrm{j}\xi_m^{(\mathrm{I})}\bar{z}\right), \\
{}^{TE}g_{\mathrm{I}m}^{-}(\bar{z}) &= {}^{TE}g_m^{(\mathrm{I})-}(0)\exp\left(\mathrm{j}\kappa_{m3}^{(\mathrm{I}u)}\bar{z}\right) = {}^{TE}g_m^{(\mathrm{I})-}(0)\exp\left(\mathrm{j}\xi_m^{(\mathrm{I})}\bar{z}\right), \\
{}^{TM}g_{\mathrm{I}m}^{-}(\bar{z}) &= {}^{TE}g_m^{(\mathrm{I})-}(0)\exp\left(\mathrm{j}\kappa_{m4}^{(\mathrm{I}u)}\bar{z}\right) = {}^{TE}g_m^{(\mathrm{I})-}(0)\exp\left(\mathrm{j}\xi_m^{(\mathrm{I})}\bar{z}\right).
\end{aligned}$$

The diffraction efficiencies for reflected (r) and transmitted (t) waves of the m-th diffraction order are given by the ratio of the $\bar{z}$ component of the Poynting vector of the m-th diffraction order to the same $\bar{z}$ component of the Poynting vector of the incident wave with $m = 0$. We obtain for the incident TE wave

$${}^{EE}\eta_m^{(r)} = \left|\frac{\hat{\mathbf{z}}\cdot\Re\left({}^{TE}\mathbf{e}_m^{(0)+}\times{}^{TE}\mathbf{h}_m^{(0)+*}\right)}{\hat{\mathbf{z}}\cdot\Re\left({}^{TE}\mathbf{e}_0^{(0)-}\times{}^{TE}\mathbf{h}_0^{(0)-*}\right)}\right| = \frac{{}^{TE}\mathrm{p}_m^{(0)+}}{\left|\xi_0^{(0)}\right|}, \tag{9.275a}$$

$${}^{EE}\eta_m^{(t)} = \left|\frac{\hat{\mathbf{z}}\cdot\Re\left({}^{TE}\mathbf{e}_m^{(2)-}\times{}^{TE}\mathbf{h}_m^{(2)-*}\right)}{\hat{\mathbf{z}}\cdot\Re\left({}^{TE}\mathbf{e}_0^{(0)-}\times{}^{TE}\mathbf{h}_0^{(0)-*}\right)}\right| = \frac{{}^{TE}\mathrm{p}_m^{(2)-}}{\left|\xi_0^{(0)}\right|}, \tag{9.275b}$$

$$ {}^{EM}\eta_m^{(r)} = \left| \frac{\hat{\mathbf{z}} \cdot \Re\left({}^{TM}\mathbf{e}_m^{(0)+} \times {}^{TM}\mathbf{h}_m^{(0)+*}\right)}{\hat{\mathbf{z}} \cdot \Re\left({}^{TE}\mathbf{e}_0^{(0)-} \times {}^{TE}\mathbf{h}_0^{(0)-*}\right)} \right| = \frac{{}^{TM}\mathrm{p}_m^{(0)+}}{|\xi_0^{(0)}|}, \tag{9.275c} $$

$$ {}^{EM}\eta_m^{(t)} = \left| \frac{\hat{\mathbf{z}} \cdot \Re\left({}^{TM}\mathbf{e}_m^{(2)-} \times {}^{TM}\mathbf{h}_m^{(2)-*}\right)}{\hat{\mathbf{z}} \cdot \Re\left({}^{TE}\mathbf{e}_0^{(0)-} \times {}^{TE}\mathbf{h}_0^{(0)-*}\right)} \right| = \frac{{}^{TM}\mathrm{p}_m^{(2)-}}{|\xi_0^{(0)}|}. \tag{9.275d} $$

Here, the z component of the Poynting vector for the incident TE wave, according to Eq. (9.271) with $I = 0$ and $m = 0$, is given by

$$ {}^{TE}\mathrm{p}_0^{(0)-} = \left|\hat{\mathbf{z}} \cdot \Re\left({}^{TE}\mathbf{e}_0^{(0)-} \times {}^{TE}\mathbf{h}_0^{(0)-*}\right)\right| = |-\xi^{(0)*}| = |\xi^{(0)}|. \tag{9.276} $$

For the incident TM wave, according to Eq. (9.272) with $I = 1$ and $m = 0$, we have

$$ {}^{TM}\mathrm{p}_0^{(0)-} = \left|\hat{\mathbf{z}} \cdot \Re\left({}^{TM}\mathbf{e}_0^{(0)-} \times {}^{TM}\mathbf{h}_0^{(0)-*}\right)\right| = \left|-\frac{\varepsilon_1^*}{\xi_{10}^*}\right| = \left|\frac{\varepsilon_1}{\xi_{10}}\right|, \tag{9.277} $$

and the diffraction efficiencies are given by

$$ {}^{ME}\eta_m^{(r)} = \left| \frac{\hat{\mathbf{z}} \cdot \Re\left({}^{TE}\mathbf{e}_m^{(0)+} \times {}^{TE}\mathbf{h}_m^{(0)+*}\right)}{\hat{\mathbf{z}} \cdot \Re\left({}^{TM}\mathbf{e}_0^{(0)-} \times {}^{TM}\mathbf{h}_0^{(0)-*}\right)} \right| = {}^{TE}\mathrm{p}_m^{(0)+} \left|\frac{\xi_0^{(0)}}{\varepsilon^{(0)}}\right|, \tag{9.278a} $$

$$ {}^{ME}\eta_m^{(t)} = \left| \frac{\hat{\mathbf{z}} \cdot \Re\left({}^{TE}\mathbf{e}_m^{(2)-} \times {}^{TE}\mathbf{h}_m^{(2)-*}\right)}{\hat{\mathbf{z}} \cdot \Re\left({}^{TM}\mathbf{e}_0^{(0)-} \times {}^{TM}\mathbf{h}_0^{(0)-*}\right)} \right| = {}^{TE}\mathrm{p}_m^{(2)-} \left|\frac{\xi_0^{(0)}}{\varepsilon^{(0)}}\right|, \tag{9.278b} $$

$$ {}^{MM}\eta_m^{(r)} = \left| \frac{\hat{\mathbf{z}} \cdot \Re\left({}^{TM}\mathbf{e}_m^{(0)+} \times {}^{TM}\mathbf{h}_m^{(0)+*}\right)}{\hat{\mathbf{z}} \cdot \Re\left({}^{TM}\mathbf{e}_0^{(0)-} \times {}^{TM}\mathbf{h}_0^{(0)-*}\right)} \right| = {}^{TM}\mathrm{p}_m^{(0)+} \left|\frac{\xi_0^{(0)}}{\varepsilon^{(0)}}\right|, \tag{9.278c} $$

$$ {}^{MM}\eta_m^{(t)} = \left| \frac{\hat{\mathbf{z}} \cdot \Re\left({}^{TM}\mathbf{e}_m^{(2)-} \times {}^{TM}\mathbf{h}_m^{(2)-*}\right)}{\hat{\mathbf{z}} \cdot \Re\left({}^{TM}\mathbf{e}_0^{(0)-} \times {}^{TM}\mathbf{h}_0^{(0)-*}\right)} \right| = {}^{TM}\mathrm{p}_m^{(2)-} \left|\frac{\xi_0^{(0)}}{\varepsilon^{(0)}}\right|, \tag{9.278d} $$

where the superscript EM and ME denotes the coupling TE $\rightarrow$ TM and TM $\rightarrow$ TE generated in the diffraction process, respectively. The energy conservation in lossless dielectric gratings requires

$$ \sum_m \sum_{TE,TM} \left({}^{\alpha\beta}\eta_m^{(r)} + {}^{\alpha\beta}\eta_m^{(t)}\right) = 1, \tag{9.279} $$

where $\alpha, \beta = TE, TM$.

The above analysis for TE and TM waves may be extended to the waves of general elliptic polarization. The incident elliptically polarized wave is characterized by

$$^{TE}\mathbf{g}^{(0)-} = \begin{pmatrix} 0 \\ \cdot \\ a \\ \cdot \\ 0 \end{pmatrix}, \qquad ^{TM}\mathbf{g}^{(0)-} = \begin{pmatrix} 0 \\ \cdot \\ b \\ \cdot \\ 0 \end{pmatrix}, \tag{9.280}$$

where $|a|^2 + |b|^2 = 1$. The complex ratio a/b specifies a general elliptic polarization. The diffraction of transmitted $\mathbf{g}^{(2)-}$ and reflected $\mathbf{g}^{(1)+}$ waves induced by the wave with general elliptic polarization can be obtained from Eq. (9.267a) and Eq. (9.267b). For example, the diffraction efficiencies for reflected TE and TM waves of the m-th order are given by

$$^{TE}\eta_m^{(r)} = \frac{^{TE}p_m^{(0)+}}{p_0^{(0)-}}, \qquad ^{TM}\eta_m^{(r)} = \frac{^{TM}p_m^{(0)+}}{p_0^{(0)-}}, \tag{9.281}$$

where

$$p_0^{(0)-} = |a|^2 \xi_0^{(0)} + |b|^2 \frac{\varepsilon^{(0)}}{\xi_0^{(0)}}. \tag{9.282}$$

9.9 Multilayer Periodic Structures

Up to present, we were limited to single-layer anisotropic gratings. The results can be extended to multilayer anisotropic gratings with the same periodicity in the planes parallel to interfaces (Figure 9.3). This is, however, only a practical limitation regarding the number of space harmonics, which must be included. In principle, there are no restrictions on the periodicity of individual layers in the multilayer stack, provided their grating vectors are all in a common plane normal to the interfaces, *i.e.*, parallel to a plane containing the z axis of the Cartesian coordinate system chosen here. Moreover, there are no restrictions on the permittivity tensor in the sandwiching entrance and exit half spaces. One or both half spaces may be isotropic or anisotropic, homogeneous or periodic. For simplicity, we discuss the case with the identical components of the grating vectors parallel to the interfaces; *i.e.*, the parameters $p^{(\mathrm{I})}$ and $q^{(\mathrm{I})}$ are assumed to be the same for all $\mathrm{I} = 0, \ldots, \mathcal{N}+1$.

Figure 9.3 shows the structure consisting of $\mathcal{N}$ periodic anisotropic layers. Here, the grating vectors of individual layers are all restricted to a single plane normal to the vector $-\hat{x}q + \hat{y}p$. The structure is sandwiched between isotropic homogeneous semi-infinite media (half spaces) indexed

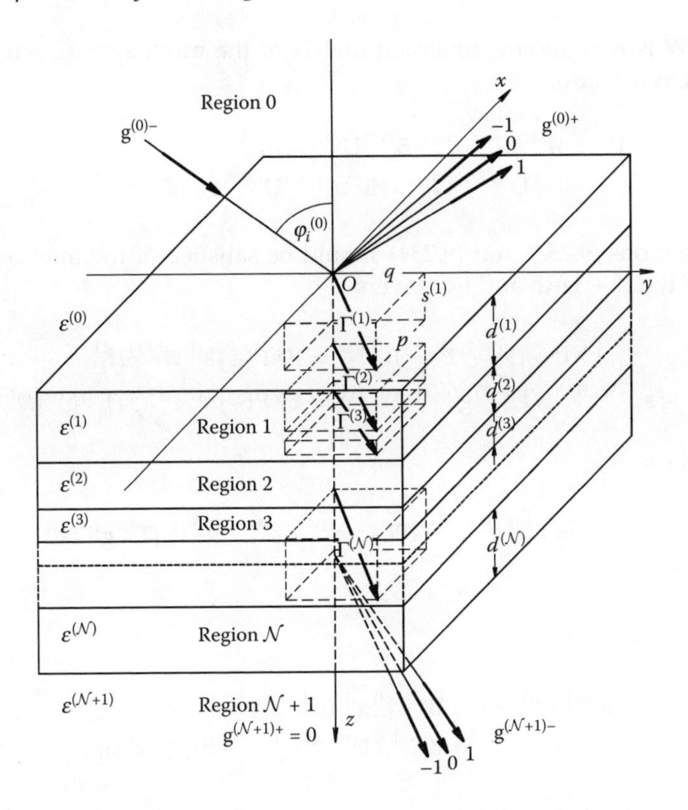

FIGURE 9.3
The cross section of an anisotropic periodic multilayer structure consisting of $\mathcal{N}$ periodic anisotropic layers. The grating vectors of individual layers are all restricted to a single plane normal to the vector $-\hat{x}q + \hat{y}p$.

(0) and $(\mathcal{N}+1)$. The I-th periodic anisotropic layer is located between the planes $z = z^{(\mathrm{I}-1)}$ and $z = z^{(\mathrm{I})}$. Its thickness $d^{(\mathrm{I})}$ is given by the difference $d^{(\mathrm{I})} = \left|z^{(\mathrm{I})} - z^{(\mathrm{I}-1)}\right|$. The normal components of the grating vector in layers are in general different, $\Gamma^{(\mathrm{I})} = \hat{x}\,p + \hat{y}\,q + \hat{z}\,s^{(\mathrm{I})}$. The extension to multilayer gratings allows, among others, to analyze the gratings with complicated profiles by splitting the grating region to a convenient number of layers that can be assumed homogeneous parallel to the normal to the interfaces. This is so-called *multilayer approximation*.

The relation between the incident $\mathbf{g}^{(0)}$ and exit $\mathbf{g}^{(\mathcal{N}+1)}$ vectors can be expressed in terms of the transmission and interface matrices as an extension of Eq. (9.262)

$$\mathbf{g}^{(0)} = \begin{pmatrix} \mathbf{g}^{(0+)} \\ \mathbf{g}^{(0)-} \end{pmatrix} = \begin{pmatrix} \mathbf{W}_1 & \mathbf{W}_2 \\ \mathbf{W}_3 & \mathbf{W}_4 \end{pmatrix} \begin{pmatrix} \mathbf{g}^{(\mathcal{N}+1)+} \\ \mathbf{g}^{(\mathcal{N}+1)-} \end{pmatrix} = \mathbf{W}\mathbf{g}^{(\mathcal{N}+1)}, \tag{9.283}$$

where $\mathbf{W}$ is now the transmission matrix of the multilayer structure expressed as a product

$$\mathbf{W} = \mathbf{B}^{(01)}\mathbf{U}^{(1)}(\bar{d}^{(1)})\mathbf{B}^{(12)}\mathbf{U}^{(2)}(\bar{d}_3)\mathbf{B}^{(34)} \\ \cdots \mathbf{U}^{(\mathcal{N}-1)}(\bar{d}_{\mathcal{N}-1})\mathbf{B}^{(\mathcal{N}-1,\mathcal{N})}\mathbf{U}^{(\mathcal{N})}(\bar{d}_{\mathcal{N}})\mathbf{B}^{(\mathcal{N},\mathcal{N}+1)}. \tag{9.284}$$

The equations (9.253) and (9.254) should be satisfied at the interface $\bar{z} = \bar{z}^{(\mathrm{I}-1)}$ of the $(\mathrm{I}-1)$-th and I-th layers

$$\mathrm{EXP}(-\mathrm{j}\,\mathbf{s}^{(\mathrm{I}-1)}\bar{z}^{(\mathrm{I}-1)})\mathbf{f}_{\hat{t}}^{(\mathrm{I}-1)} = \mathrm{EXP}(-\mathrm{j}\,\mathbf{s}^{(\mathrm{I})}\bar{z}^{(\mathrm{I}-1)})\mathbf{f}_{\hat{t}}^{(\mathrm{I})}$$
$$\mathrm{EXP}(-\mathrm{j}\,\mathbf{s}^{(\mathrm{I}-1)}\bar{z}^{(\mathrm{I}-1)})\mathbf{T}^{(\mathrm{I}-1)}\mathbf{g}^{(\mathrm{I}-1)}(\bar{z}^{(\mathrm{I}-1)}) = \mathrm{EXP}(-\mathrm{j}\,\mathbf{s}^{(\mathrm{I})}\bar{z}^{(\mathrm{I}-1)})\mathbf{T}^{(\mathrm{I})}\mathbf{g}^{(\mathrm{I})}(\bar{z}^{(\mathrm{I}-1)}).$$

From this

$$\mathbf{g}^{(\mathrm{I}-1)}(\bar{z}^{(\mathrm{I}-1)}) = [\mathbf{T}^{(\mathrm{I}-1)}]^{-1}\mathrm{EXP}[-\mathrm{j}\,(\mathbf{s}^{(\mathrm{I})} - \mathbf{s}^{(\mathrm{I}-1)})\bar{z}^{(\mathrm{I}-1)}]\mathbf{T}^{(\mathrm{I})}\mathbf{g}^{(\mathrm{I})}(\bar{z}^{(\mathrm{I}-1)}), \tag{9.285}$$

or with Eq. (9.248)

$$\mathbf{g}^{(\mathrm{I}-1)}(\bar{z}^{(\mathrm{I}-1)}) = \mathbf{B}^{(\mathrm{I}-1,\mathrm{I})}\mathbf{g}^{(\mathrm{I})}(\bar{z}^{(\mathrm{I}-1)}) \\ = \mathbf{B}^{(\mathrm{I}-1,\mathrm{I})}\mathbf{U}^{(\mathrm{I})}(\bar{z}^{(\mathrm{I}-1)} - \bar{z}^{(\mathrm{I})})\mathbf{g}^{(\mathrm{I})}(\bar{z}^{(\mathrm{I})}), \tag{9.286}$$

where the interface matrix

$$\mathbf{B}^{(\mathrm{I}-1,\mathrm{I})} = [\mathbf{T}^{(\mathrm{I}-1)}]^{-1}\mathrm{EXP}[-\mathrm{j}\,(\mathbf{s}^{(\mathrm{I})} - \mathbf{s}^{(\mathrm{I}-1)})\bar{z}^{(\mathrm{I}-1)}]\mathbf{T}^{(\mathrm{I})}. \tag{9.287}$$

This represents the extension to an arbitrary number of layers.

9.10 Isotropic Layers at Normal Incidence

We consider a homogeneous isotropic layer at the normal light incidence $s_0 = 0$ and limit ourselves to the case $m = 0$. Then, $p_m = q_m = s_m = 0$. The matrix $\mathbf{C}_m^{(u)} = \mathbf{C}_0^{(u)}$ and the corresponding column vector $\mathbf{f}_{\hat{t}m}^{(u)} = \mathbf{f}_{\hat{t}0}^{(u)}$ become

$$\mathbf{C}_0^{(u)} = \begin{pmatrix} 0 & -1 & 0 & 0 \\ -\varepsilon & 0 & 0 & 0 \\ 0 & 0 & 0 & 1 \\ 0 & 0 & \varepsilon & 0 \end{pmatrix}, \qquad \mathbf{f}_{\hat{t}0}^{(u)} = \begin{pmatrix} e_{y0} \\ h_{z0} \\ e_{z0} \\ h_{y0} \end{pmatrix}. \tag{9.288}$$

The proper values κ of the matrix $\mathbf{C}_0^{(u)}$ follow from the condition for vanishing of the following determinant

$$\det\left(\mathbf{C}_0^{(u)} - \kappa\mathbf{1}\right) = \begin{vmatrix} -\kappa & -1 & 0 & 0 \\ -\varepsilon & -\kappa & 0 & 0 \\ 0 & 0 & -\kappa & 1 \\ 0 & 0 & \varepsilon & -\kappa \end{vmatrix} = (\kappa^2 - \varepsilon)^2. \tag{9.289}$$

The matrix $\mathbf{C}_0^{(u)}$ can be transformed to a diagonal one using Eq. (9.194) and Eq. (9.195) for $\mathbf{T}_0^{(u)}$ and $[\mathbf{T}_0^{(u)}]^{-1}$ with $s_0 = 0$

$$\begin{aligned}
&\left[\mathbf{T}_0^{(u)}\right]^{-1}\mathbf{C}_0^{(u)}\mathbf{T}_0^{(u)} \\
&= \frac{1}{2\sqrt{\varepsilon}}\begin{pmatrix} \sqrt{\varepsilon} & 1 & 0 & 0 \\ 0 & 0 & \sqrt{\varepsilon} & -1 \\ \sqrt{\varepsilon} & -1 & 0 & 0 \\ 0 & 0 & \sqrt{\varepsilon} & 1 \end{pmatrix}\begin{pmatrix} 0 & -1 & 0 & 0 \\ -\varepsilon & 0 & 0 & 0 \\ 0 & 0 & 0 & 1 \\ 0 & 0 & \varepsilon & 0 \end{pmatrix}\begin{pmatrix} 1 & 0 & 1 & 0 \\ \sqrt{\varepsilon} & 0 & -\sqrt{\varepsilon} & 0 \\ 0 & 1 & 0 & 1 \\ 0 & -\sqrt{\varepsilon} & 0 & \sqrt{\varepsilon} \end{pmatrix} \\
&= \frac{1}{2\sqrt{\varepsilon}}\begin{pmatrix} -\varepsilon & -\sqrt{\varepsilon} & 0 & 0 \\ 0 & 0 & -\varepsilon & \sqrt{\varepsilon} \\ \varepsilon & -\sqrt{\varepsilon} & 0 & 0 \\ 0 & 0 & \varepsilon & \sqrt{\varepsilon} \end{pmatrix}\begin{pmatrix} 1 & 0 & 1 & 0 \\ \sqrt{\varepsilon} & 0 & -\sqrt{\varepsilon} & 0 \\ 0 & 1 & 0 & 1 \\ 0 & -\sqrt{\varepsilon} & 0 & \sqrt{\varepsilon} \end{pmatrix}
\end{aligned} \tag{9.290}$$

with the result

$$\boldsymbol{\kappa}_0^{(u)} = \left[\mathbf{T}_0^{(u)}\right]^{-1}\mathbf{C}_0^{(u)}\mathbf{T}_0^{(u)} = \frac{1}{2\sqrt{\varepsilon}}\begin{pmatrix} -\sqrt{\varepsilon} & 0 & 0 & 0 \\ 0 & -\sqrt{\varepsilon} & 0 & 0 \\ 0 & 0 & \sqrt{\varepsilon} & 0 \\ 0 & 0 & 0 & \sqrt{\varepsilon} \end{pmatrix}. \tag{9.291}$$

From Eq. (9.221), we have

$$\begin{aligned}
&\mathbf{f}_{t0}^{(u)} = \mathbf{T}_0^{(u)}\mathbf{g}_0^{(u)}, \\
&\begin{pmatrix} e_{x0} \\ h_{y0} \\ e_{y0} \\ h_{x0} \end{pmatrix} = \begin{pmatrix} 1 & 0 & 1 & 0 \\ \sqrt{\varepsilon} & 0 & -\sqrt{\varepsilon} & 0 \\ 0 & 1 & 0 & 1 \\ 0 & -\sqrt{\varepsilon} & 0 & \sqrt{\varepsilon} \end{pmatrix}\begin{pmatrix} {}^{TE}g_0^+ \\ {}^{TM}g_0^+ \\ {}^{TE}g_0^- \\ {}^{TM}g_0^- \end{pmatrix}.
\end{aligned} \tag{9.292}$$

At the normal incidence, the plane of incidence cannot be defined, and the TE and TM classification is no more meaningful. The indices $E(M)$ merely denote the orientation of the electric and magnetic fields with respect to the unit vectors $\hat{x}$ and $\hat{y}$ of the chosen Cartesian coordinate system, respectively

$$^{TE}g_0^+(\bar{z}) = {}^{TE}g_0^+(\bar{z}^0)\exp[-j\sqrt{\varepsilon}(\bar{z}-\bar{z}^{(0)})], \tag{9.293a}$$
$$^{TM}g_0^+(\bar{z}) = {}^{TM}g_0^+(\bar{z}^0)\exp[-j\sqrt{\varepsilon}(\bar{z}-\bar{z}^{(0)})], \tag{9.293b}$$
$$^{TE}g_0^-(\bar{z}) = {}^{TE}g_0^-(\bar{z}^0)\exp[j\sqrt{\varepsilon}(\bar{z}-\bar{z}^{(0)})], \tag{9.293c}$$
$$^{TM}g_0^-(\bar{z}) = {}^{TM}g_0^+(\bar{z}^0)\exp[j\sqrt{\varepsilon}(\bar{z}-\bar{z}^{(0)})], \tag{9.293d}$$

in agreement with Eq. (9.127). The equation for the transmission matrix (9.248) for the special case of $m = 0$ can be written as

$$\begin{pmatrix} ^{TE}g_0^{(I)+}(\bar{z}) \\ ^{TM}g_0^{(I)+}(\bar{z}) \\ ^{TE}g_0^{(I)-}(\bar{z}) \\ ^{TM}g_0^{(I)-}(\bar{z}) \end{pmatrix} = \begin{pmatrix} \exp[-j\sqrt{\varepsilon^{(I)}}(\bar{z}-\bar{z}^{(0)})] & 0 & 0 & 0 \\ 0 & \exp[-j\sqrt{\varepsilon^{(I)}}(\bar{z}-\bar{z}^{(0)})] & 0 & 0 \\ 0 & 0 & \exp[j\sqrt{\varepsilon^{(I)}}(\bar{z}-\bar{z}^{(0)})] & 0 \\ 0 & 0 & 0 & \exp[j\sqrt{\varepsilon^{(I)}}(\bar{z}-\bar{z}^{(0)})] \end{pmatrix} \times \begin{pmatrix} ^{TE}g_0^{(I)+}(\bar{z}^0) \\ ^{TM}g_0^{(I)+}(\bar{z}^0) \\ ^{TE}g_0^{(I)-}(\bar{z}^0) \\ ^{TM}g_0^{(I)-}(\bar{z}^0) \end{pmatrix}. \tag{9.294}$$

We have included the index (I) distinguishing the regions (or the layers). At interfaces, the continuity for the tangential field components should be observed, and we have from Eq. (9.253) and Eq. (9.254) for the special case of an interface between homogeneous regions

$$\mathbf{f}_{\hat{t}0}^{(I-1,u)}(\bar{z}^{(I-1)}) = \mathbf{f}_{\hat{t}0}^{(I,u)}(\bar{z}^{(I-1)}),$$
$$\mathbf{T}_0^{(I-1,u)}\mathbf{g}_0^{(I-1,u)}(\bar{z}^{(I-1)}) = \mathbf{T}_0^{(I,u)}\mathbf{g}_0^{(I,u)}(\bar{z}^{(I-1)}),$$

$$\begin{pmatrix} 1 & 0 & 1 & 0 \\ \sqrt{\varepsilon^{(I-1)}} & 0 & -\sqrt{\varepsilon^{(I-1)}} & 0, \\ 0 & 1 & 0 & 1 \\ 0 & -\sqrt{\varepsilon^{(I-1)}} & 0 & \sqrt{\varepsilon^{(I-1)}} \end{pmatrix} \begin{pmatrix} ^{TE}g_0^{(I-1)+}(\bar{z}^{(I-1)}) \\ ^{TM}g_0^{(I-1)+}(\bar{z}^{(I-1)}) \\ ^{TE}g_0^{(I-1)-}(\bar{z}^{(I-1)}) \\ ^{TM}g_0^{(I-1)-}(\bar{z}^{(I-1)}) \end{pmatrix}$$
$$= \begin{pmatrix} 1 & 0 & 1 & 0 \\ \sqrt{\varepsilon^{(I)}} & 0 & -\sqrt{\varepsilon^{(I)}} & 0 \\ 0 & 1 & 0 & 1 \\ 0 & -\sqrt{\varepsilon^{(I)}} & 0 & \sqrt{\varepsilon^{(I)}} \end{pmatrix} \begin{pmatrix} ^{TE}g_0^{(I)+}(\bar{z}^{(I-1)}) \\ ^{TM}g_0^{(I)+}(\bar{z}^{(I-1)}) \\ ^{TE}g_0^{(I)-}(\bar{z}^{(I-1)}) \\ ^{TM}g_0^{(I)-}(\bar{z}^{(I-1)}) \end{pmatrix}, \tag{9.295}$$

where $\bar{z}^{(I-1)}$ is the coordinate specifying the interface plane between the regions $(I-1)$ and (I).

From this,

$$\mathbf{g}_0^{(I-1,u)}(\bar{z}^{(I-1)}) = \left[\mathbf{T}_0^{(I-1,u)}\right]^{-1}\mathbf{T}_0^{(I,u)}\mathbf{g}_0^{(I,u)}(\bar{z}^{(I-1)}), \tag{9.296}$$

where the interface matrix is given by

$$\mathbf{B}_0^{(I-1,I;\,u)} = \left[\mathbf{T}_0^{(I-1,u)}\right]^{-1}\mathbf{T}_0^{(I,u)}. \tag{9.297}$$

It can be obtained by considering Eq. (9.316)

$$\begin{aligned}
&\mathbf{B}_0^{(I-1,I;\,u)} \\
&= \frac{1}{2\sqrt{\varepsilon^{(I-1)}}}\begin{pmatrix} \sqrt{\varepsilon^{(I-1)}} & 1 & 0 & 0 \\ 0 & 0 & \sqrt{\varepsilon^{(I-1)}} & -1 \\ \sqrt{\varepsilon^{(I-1)}} & -1 & 0 & 0 \\ 0 & 0 & \sqrt{\varepsilon^{(I-1)}} & 1 \end{pmatrix}\begin{pmatrix} 1 & 0 & 1 & 0 \\ \sqrt{\varepsilon^{(I)}} & 0 & -\sqrt{\varepsilon^{(I)}} & 0 \\ 0 & 1 & 0 & 1 \\ 0 & -\sqrt{\varepsilon^{(I)}} & 0 & \sqrt{\varepsilon^{(I)}} \end{pmatrix} \\
&= \frac{1}{2\sqrt{\varepsilon^{(I-1)}}}\begin{pmatrix} \sqrt{\varepsilon^{(I-1)}}+\sqrt{\varepsilon^{(I)}} & 0 & \sqrt{\varepsilon^{(I-1)}}-\sqrt{\varepsilon^{(I)}} & 0 \\ 0 & \sqrt{\varepsilon^{(I-1)}}+\sqrt{\varepsilon^{(I)}} & 0 & \sqrt{\varepsilon^{(I-1)}}-\sqrt{\varepsilon^{(I)}} \\ \sqrt{\varepsilon^{(I-1)}}-\sqrt{\varepsilon^{(I)}} & 0 & \sqrt{\varepsilon^{(I-1)}}+\sqrt{\varepsilon^{(I)}} & 0 \\ 0 & \sqrt{\varepsilon^{(I-1)}}-\sqrt{\varepsilon^{(I)}} & 0 & \sqrt{\varepsilon^{(I-1)}}+\sqrt{\varepsilon^{(I)}} \end{pmatrix}.
\end{aligned} \tag{9.298}$$

The substitution into Eq. (9.296) gives

$$\begin{aligned}
&\begin{pmatrix} {}^{TE}g_0^{(I-1)+} \\ {}^{TM}g_0^{(I-1)+} \\ {}^{TE}g_0^{(I-1)-} \\ {}^{TM}g_0^{(I-1)-} \end{pmatrix} \\
&= \frac{1}{2\sqrt{\varepsilon^{(I-1)}}}\begin{pmatrix} \sqrt{\varepsilon^{(I-1)}}+\sqrt{\varepsilon^{(I)}} & 0 & \sqrt{\varepsilon^{(I-1)}}-\sqrt{\varepsilon^{(I)}} & 0 \\ 0 & \sqrt{\varepsilon^{(I-1)}}+\sqrt{\varepsilon^{(I)}} & 0 & \sqrt{\varepsilon^{(I-1)}}-\sqrt{\varepsilon^{(I)}} \\ \sqrt{\varepsilon^{(I-1)}}-\sqrt{\varepsilon^{(I)}} & 0 & \sqrt{\varepsilon^{(I-1)}}+\sqrt{\varepsilon^{(I)}} & 0 \\ 0 & \sqrt{\varepsilon^{(I-1)}}-\sqrt{\varepsilon^{(I)}} & 0 & \sqrt{\varepsilon^{(I-1)}}+\sqrt{\varepsilon^{(I)}} \end{pmatrix} \\
&\times \begin{pmatrix} {}^{TE}g_0^{(I)+} \\ {}^{TM}g_0^{(I)+} \\ {}^{TE}g_0^{(I)-} \\ {}^{TM}g_0^{(I)-} \end{pmatrix}.
\end{aligned} \tag{9.299}$$

Developing the matrix multiplication, we observe that the equations for TE and TM waves are mutually independent

$$2\sqrt{\varepsilon^{(\mathrm{I}-1)}}\;{}^{TE}g_0^{(\mathrm{I}-1)+} = \left(\sqrt{\varepsilon^{(\mathrm{I}-1)}} + \sqrt{\varepsilon^{(\mathrm{I})}}\right){}^{TE}g_0^{(\mathrm{I})+} + \left(\sqrt{\varepsilon^{(\mathrm{I}-1)}} - \sqrt{\varepsilon^{(\mathrm{I})}}\right){}^{TE}g_0^{(\mathrm{I})-} \tag{9.300a}$$

$$2\sqrt{\varepsilon^{(\mathrm{I}-1)}}\;{}^{TM}g_0^{(\mathrm{I}-1)+} = \left(\sqrt{\varepsilon^{(\mathrm{I}-1)}} + \sqrt{\varepsilon^{(\mathrm{I})}}\right){}^{TM}g_0^{(\mathrm{I})+} + \left(\sqrt{\varepsilon^{(\mathrm{I}-1)}} - \sqrt{\varepsilon^{(\mathrm{I})}}\right){}^{TM}g_0^{(\mathrm{I})-} \tag{9.300b}$$

$$2\sqrt{\varepsilon^{(\mathrm{I}-1)}}\;{}^{TE}g_0^{(\mathrm{I}-1)-} = \left(\sqrt{\varepsilon^{(\mathrm{I}-1)}} - \sqrt{\varepsilon^{(\mathrm{I})}}\right){}^{TE}g_0^{(\mathrm{I})+} + \left(\sqrt{\varepsilon^{(\mathrm{I}-1)}} + \sqrt{\varepsilon^{(\mathrm{I})}}\right){}^{TE}g_0^{(\mathrm{I})-} \tag{9.300c}$$

$$2\sqrt{\varepsilon^{(\mathrm{I}-1)}}\;{}^{TM}g_0^{(\mathrm{I}-1)-} = \left(\sqrt{\varepsilon^{(\mathrm{I}-1)}} - \sqrt{\varepsilon^{(\mathrm{I})}}\right){}^{TM}g_0^{(\mathrm{I})+} + \left(\sqrt{\varepsilon^{(\mathrm{I}-1)}} + \sqrt{\varepsilon^{(\mathrm{I})}}\right){}^{TM}g_0^{(\mathrm{I})-}. \tag{9.300d}$$

Let us now assume that there are no waves arriving at the interface between the regions (I−1) and (I) from the region (I), *i.e.*, ${}^{TE}g_0^{(\mathrm{I})+} = 0$ and ${}^{TM}g_0^{(\mathrm{I})+} = 0$. This simplifies the system (9.300a) to (9.300d) as follows

$$2\sqrt{\varepsilon^{(\mathrm{I}-1)}}\;{}^{TE}g_0^{(\mathrm{I}-1)+} = \left(\sqrt{\varepsilon^{(\mathrm{I}-1)}} - \sqrt{\varepsilon^{(\mathrm{I})}}\right){}^{TE}g_0^{(\mathrm{I})-} \tag{9.301a}$$

$$2\sqrt{\varepsilon^{(\mathrm{I}-1)}}\;{}^{TM}g_0^{(\mathrm{I}-1)+} = \left(\sqrt{\varepsilon^{(\mathrm{I}-1)}} - \sqrt{\varepsilon^{(\mathrm{I})}}\right){}^{TM}g_0^{(\mathrm{I})-} \tag{9.301b}$$

$$2\sqrt{\varepsilon^{(\mathrm{I}-1)}}\;{}^{TE}g_0^{(\mathrm{I}-1)-} = \left(\sqrt{\varepsilon^{(\mathrm{I}-1)}} + \sqrt{\varepsilon^{(\mathrm{I})}}\right){}^{TE}g_0^{(\mathrm{I})-} \tag{9.301c}$$

$$2\sqrt{\varepsilon^{(\mathrm{I}-1)}}\;{}^{TM}g_0^{(\mathrm{I}-1)-} = \left(\sqrt{\varepsilon^{(\mathrm{I}-1)}} + \sqrt{\varepsilon^{(\mathrm{I})}}\right){}^{TM}g_0^{(\mathrm{I})-}. \tag{9.301d}$$

Next, we consider the case of the wave incident from the region (I−1) of the unit amplitude and TE polarization, *i.e.*, ${}^{TE}g_0^{(\mathrm{I}-1)-} = 1$ and ${}^{TM}g_0^{(\mathrm{I}-1)-} = 0$. There is no incident wave of TM polarization. From Eqs. (9.301), we obtain

$$2\sqrt{\varepsilon^{(\mathrm{I}-1)}}\;{}^{TE}g_0^{(\mathrm{I}-1)+} = \left(\sqrt{\varepsilon^{(\mathrm{I}-1)}} - \sqrt{\varepsilon^{(\mathrm{I})}}\right){}^{TE}g_0^{(\mathrm{I})-} \tag{9.302a}$$

$$2\sqrt{\varepsilon^{(\mathrm{I}-1)}}\;{}^{TM}g_0^{(\mathrm{I}-1)+} = \left(\sqrt{\varepsilon^{(\mathrm{I}-1)}} - \sqrt{\varepsilon^{(\mathrm{I})}}\right){}^{TM}g_0^{(\mathrm{I})-} \tag{9.302b}$$

$$2\sqrt{\varepsilon^{(\mathrm{I}-1)}} = \left(\sqrt{\varepsilon^{(\mathrm{I}-1)}} + \sqrt{\varepsilon^{(\mathrm{I})}}\right){}^{TE}g_0^{-(\mathrm{I})} \tag{9.302c}$$

$$0 = \left(\sqrt{\varepsilon^{(\mathrm{I}-1)}} + \sqrt{\varepsilon^{(\mathrm{I})}}\right){}^{TM}g_0^{(\mathrm{I})-}. \tag{9.302d}$$

As a consequence, we have also ${}^{TM}g_0^{(\mathrm{I})-} = 0$ and ${}^{TM}g_0^{(\mathrm{I}-1)+} = 0$. This means that there are also no transmitted and reflected waves of the TM polarization. Using Eq. (9.302c), we obtain for the amplitude of the transmitted TE wave

$${}^{TE}t^{(\mathrm{I}-1,\mathrm{I})} = {}^{TE}g_0^{(\mathrm{I})-} = \frac{2\sqrt{\varepsilon^{(\mathrm{I}-1)}}}{\left(\sqrt{\varepsilon^{(\mathrm{I}-1)}} + \sqrt{\varepsilon^{(\mathrm{I})}}\right)}. \tag{9.303}$$

This is the well-known Fresnel amplitude transmission coefficient ${}^{TE}t^{(\mathrm{I}-1,\mathrm{I})}$ for the wave traversing the interface from the medium (I−1) to the medium

(I) at the normal incidence. We next divide Eq. (9.302a) by (9.302c) and eliminate ${}^{TE}g_0^{(\mathrm{I})-}$

$$ {}^{TE}r^{(\mathrm{I-1,I})} = {}^{TE}g_0^{(\mathrm{I})+} = \frac{(\sqrt{\varepsilon^{(\mathrm{I-1})}} - \sqrt{\varepsilon^{(\mathrm{I})}})\,{}^{TE}g_0^{(\mathrm{I})-}}{(\sqrt{\varepsilon^{(\mathrm{I-1})}} + \sqrt{\varepsilon^{(\mathrm{I})}})\,{}^{TE}g_0^{(\mathrm{I})-}}. \tag{9.304} $$

This is the classical Fresnel amplitude reflection coefficient ${}^{TE}r^{(\mathrm{I-1,I})}$ for the wave incident from the medium $(\mathrm{I}-1)$ at the interface between to the media $(\mathrm{I}-1)$ and (I) at the normal incidence. A similar procedure provides ${}^{TM}t^{(\mathrm{I-1,I})}$ and ${}^{TM}r^{(\mathrm{I-1,I})}$.

Next, we consider the case of a homogeneous isotropic region (1) sandwiched between two isotropic half spaces (0) and (2). We make use of Eq. (9.263) simplified with $\mathbf{s}^{(1)} = 0$

$$ \begin{aligned} \mathbf{g}_0^{(1)}(0) &= \left[\mathbf{T}_0^{(0u)}\right]^{-1}\mathbf{T}_0^{(1u)}\left[\mathbf{U}_0^{(1u)}(\bar{d}^{(1)})\right]^{-1}\left[\mathbf{T}_0^{(1u)}\right]^{-1}\mathbf{T}_0^{(2u)}\mathbf{g}_0^{(2)}(\bar{d}^{(1)}) \\ &= \mathbf{B}_0^{(01;u)}\mathbf{U}_0^{(1u)}(-\bar{d}^{(1)})\mathbf{B}_0^{(12;u)}\mathbf{g}_0^{(2)}(\bar{d}^{(1)}). \end{aligned} \tag{9.305} $$

Here, the matrices $\mathbf{B}_0^{(01;u)}$ and $\mathbf{B}_0^{(12;u)}$ are given in Eq. (9.297) with $\mathrm{I} = 1$ and $\mathrm{I} = 2$, respectively. The transmission matrix $\mathbf{U}_0^{(1u)}(\bar{d}^{(1)}) = [\mathbf{U}_0^{(1u)}(-\bar{d}^{(1)})]^{-1}$ is given according to Eq. (9.274) by

$$ \mathbf{U}_0^{(1u)}(\bar{d}^{(1)}) = \begin{pmatrix} \exp(\mathrm{j}\sqrt{\varepsilon^{(1)}}\bar{d}^{(1)}) & 0 & 0 & 0 \\ 0 & \exp(\mathrm{j}\sqrt{\varepsilon^{(1)}}\bar{d}^{(1)}) & 0 & 0 \\ 0 & 0 & \exp(-\mathrm{j}\sqrt{\varepsilon^{(1)}}\bar{d}^{(1)}) & 0 \\ 0 & 0 & 0 & \exp(-\mathrm{j}\sqrt{\varepsilon^{(1)}}\bar{d}^{(1)}) \end{pmatrix}. \tag{9.306} $$

The relation between amplitudes in the region (0) at the interface between the regions (0) and (1) and those in the region (2) at the interface between regions (1) and (2) is given by Eq. (9.305), *i.e.*,

$$ \begin{pmatrix} {}^{TE}g_0^{(0)+}(0) \\ {}^{TM}g_0^{(0)+}(0) \\ {}^{TE}g_0^{(0)-}(0) \\ {}^{TM}g_0^{(0)-}(0) \end{pmatrix} $$

$$ = \frac{1}{4\sqrt{\varepsilon^{(0)}}\sqrt{\varepsilon^{(1)}}}\begin{pmatrix} \sqrt{\varepsilon^{(0)}}+\sqrt{\varepsilon^{(1)}} & 0 & \sqrt{\varepsilon^{(0)}}-\sqrt{\varepsilon^{(1)}} & 0 \\ 0 & \sqrt{\varepsilon^{(0)}}+\sqrt{\varepsilon^{(1)}} & 0 & \sqrt{\varepsilon^{(0)}}-\sqrt{\varepsilon^{(1)}} \\ \sqrt{\varepsilon^{(0)}}-\sqrt{\varepsilon^{(1)}} & 0 & \sqrt{\varepsilon^{(0)}}+\sqrt{\varepsilon^{(1)}} & 0 \\ 0 & \sqrt{\varepsilon^{(0)}}-\sqrt{\varepsilon^{(1)}} & 0 & \sqrt{\varepsilon^{(0)}}+\sqrt{\varepsilon^{(1)}} \end{pmatrix} $$

$$\times \begin{pmatrix} \exp(-j\sqrt{\varepsilon^{(1)}}\bar{d}^{(1)}) & 0 & 0 & 0 \\ 0 & \exp(-j\sqrt{\varepsilon^{(1)}}\bar{d}^{(1)}) & 0 & 0 \\ 0 & 0 & \exp(j\sqrt{\varepsilon^{(1)}}\bar{d}^{(1)}) & 0 \\ 0 & 0 & 0 & \exp(j\sqrt{\varepsilon^{(1)}}\bar{d}^{(1)}) \end{pmatrix}$$

$$\times \begin{pmatrix} \sqrt{\varepsilon^{(1)}} + \sqrt{\varepsilon^{(2)}} & 0 & \sqrt{\varepsilon^{(1)}} - \sqrt{\varepsilon^{(2)}} & 0 \\ 0 & \sqrt{\varepsilon^{(1)}} + \sqrt{\varepsilon^{(2)}} & 0 & \sqrt{\varepsilon^{(1)}} - \sqrt{\varepsilon^{(2)}} \\ \sqrt{\varepsilon^{(1)}} - \sqrt{\varepsilon^{(2)}} & 0 & \sqrt{\varepsilon^{(1)}} + \sqrt{\varepsilon^{(2)}} & 0 \\ 0 & \sqrt{\varepsilon^{(1)}} - \sqrt{\varepsilon^{(2)}} & 0 & \sqrt{\varepsilon^{(1)}} + \sqrt{\varepsilon^{(2)}} \end{pmatrix}$$

$$\times \begin{pmatrix} {}^{TE}g_0^{(2)+}(\bar{d}^{(1)}) \\ {}^{TM}g_0^{(2)+}(\bar{d}^{(1)}) \\ {}^{TE}g_0^{(2)-}(\bar{d}^{(1)}) \\ {}^{TM}g_0^{(2)-}(\bar{d}^{(1)}) \end{pmatrix}. \tag{9.307}$$

The evaluation of this equation provides

$$\begin{pmatrix} {}^{TE}g_0^{(0)+}(0) \\ {}^{TM}g_0^{(0)+}(0) \\ {}^{TE}g_0^{(0)-}(0) \\ {}^{TM}g_0^{(0)-}(0) \end{pmatrix} = \frac{1}{t^{(01)}t^{(12)}}$$

$$\times \begin{pmatrix} e^{-j\sqrt{\varepsilon_1}\bar{d}^{(1)}} + r^{(01)}r^{(12)}e^{j\sqrt{\varepsilon_1}\bar{d}^{(1)}} & 0 \\ 0 & e^{-j\sqrt{\varepsilon_1}\bar{d}^{(1)}} + r^{(01)}r^{(12)}e^{j\sqrt{\varepsilon_1}\bar{d}^{(1)}} \\ r^{(01)}e^{-j\sqrt{\varepsilon_1}\bar{d}^{(1)}} + r^{(12)}e^{j\sqrt{\varepsilon_1}\bar{d}^{(1)}} & 0 \\ 0 & r^{(01)}e^{-j\sqrt{\varepsilon_1}\bar{d}^{(1)}} + r^{(12)}e^{j\sqrt{\varepsilon_1}\bar{d}^{(1)}} \end{pmatrix.$$

$$\begin{matrix} r^{(12)}e^{-j\sqrt{\varepsilon_1}\bar{d}^{(1)}} + r^{(01)}e^{j\sqrt{\varepsilon_1}\bar{d}^{(1)}} & 0 \\ 0 & r^{(12)}e^{-j\sqrt{\varepsilon_1}\bar{d}^{(1)}} + r^{(01)}e^{j\sqrt{\varepsilon_1}\bar{d}^{(1)}} \\ r^{(01)}r^{(12)}e^{-j\sqrt{\varepsilon_1}\bar{d}^{(1)}} + e^{j\sqrt{\varepsilon_1}\bar{d}^{(1)}} & 0 \\ 0 & r^{(01)}r^{(12)}e^{-j\sqrt{\varepsilon_1}\bar{d}^{(1)}} + e^{j\sqrt{\varepsilon_1}\bar{d}^{(1)}} \end{matrix}$$

$$\times \begin{pmatrix} {}^{TE}g_0^{(2)+}(\bar{d}^{(1)}) \\ {}^{TM}g_0^{(2)+}(\bar{d}^{(1)}) \\ {}^{TE}g_0^{(2)-}(\bar{d}^{(1)}) \\ {}^{TM}g_0^{(2)-}(\bar{d}^{(1)}) \end{pmatrix}, \tag{9.308}$$

where we have employed the abbreviations

$$t^{(01)} = \frac{2\sqrt{\varepsilon^{(0)}}}{\sqrt{\varepsilon^{(0)}} + \sqrt{\varepsilon^{(1)}}}, \tag{9.309a}$$

$$t^{(12)} = \frac{2\sqrt{\varepsilon^{(1)}}}{\sqrt{\varepsilon^{(1)}} + \sqrt{\varepsilon^{(2)}}}, \tag{9.309b}$$

$$r^{(01)} = \frac{\sqrt{\varepsilon^{(0)}} - \sqrt{\varepsilon^{(1)}}}{\sqrt{\varepsilon^{(0)}} + \sqrt{\varepsilon^{(1)}}}, \tag{9.309c}$$

$$r^{(12)} = \frac{\sqrt{\varepsilon^{(1)}} - \sqrt{\varepsilon^{(2)}}}{\sqrt{\varepsilon^{(1)}} + \sqrt{\varepsilon^{(2)}}}. \tag{9.309d}$$

Assuming that no waves impinge on the structure from the region (2), *i.e.*, ${}^{TE}g_0^{(2)+}(\bar{d}^{(1)}) = 0$ and ${}^{TM}g_0^{(2)+}(\bar{d}^{(1)}) = 0$, and that only incident wave is ${}^{TE}g_0^{(0)-}(\bar{d}^{(1)}) = 1$ while ${}^{TM}g_0^{(0)-}(\bar{d}^{(1)}) = 0$, we may write

$$\begin{pmatrix} {}^{TE}g_0^{(0)+}(0) \\ {}^{TM}g_0^{(0)+}(0) \\ 1 \\ 0 \end{pmatrix} = \frac{\exp(-\mathrm{j}\sqrt{\varepsilon^{(1)}}\bar{d}^{(1)})}{t^{(01)}t^{(12)}} \begin{pmatrix} 1 + r^{(01)}r^{(12)}\mathrm{e}^{2\mathrm{j}\sqrt{\varepsilon^{(1)}}\bar{d}^{(1)}} & 0 & r^{(12)} + r^{(01)}\mathrm{e}^{2\mathrm{j}\sqrt{\varepsilon^{(1)}}\bar{d}^{(1)}} & 0 \\ 0 & 1 + r^{(01)}r^{(12)}\mathrm{e}^{2\mathrm{j}\sqrt{\varepsilon^{(1)}}\bar{d}^{(1)}} & 0 & r^{(12)} + r^{(01)}\mathrm{e}^{2\mathrm{j}\sqrt{\varepsilon^{(1)}}\bar{d}^{(1)}} \\ r^{(01)} + r^{(12)}\mathrm{e}^{2\mathrm{j}\sqrt{\varepsilon^{(1)}}\bar{d}^{(1)}} & 0 & r^{(01)}r^{(12)} + \mathrm{e}^{2\mathrm{j}\sqrt{\varepsilon^{(1)}}\bar{d}^{(1)}} & 0 \\ 0 & r^{(01)} + r^{(12)}\mathrm{e}^{2\mathrm{j}\sqrt{\varepsilon^{(1)}}\bar{d}^{(1)}} & 0 & r^{(01)}r^{(12)} + \mathrm{e}^{2\mathrm{j}\sqrt{\varepsilon^{(1)}}\bar{d}^{(1)}} \end{pmatrix} \begin{pmatrix} 0 \\ 0 \\ {}^{TE}g_0^{(2)-}(\bar{d}^{(1)}) \\ {}^{TM}g_0^{(2)-}(\bar{d}^{(1)}) \end{pmatrix}. \tag{9.310}$$

As expected, ${}^{TM}g_0^{(2)-}(\bar{d}^{(1)}) = 0$. This is consistent with the absence of TE↔TM mode coupling in the isotropic structure. Then, $t^{(02)} = {}^{TE}g_0^{(2)-}(\bar{d}^{(1)})$ represents the global amplitude transmission coefficient, and $r^{(02)} = {}^{TE}g_0^{(0)+}(\bar{d}^{(1)})$ represents the global amplitude reflection coefficient of the structure. We have to consider the following two equations

$$r^{(02)} = \frac{\exp(-\mathrm{j}\sqrt{\varepsilon^{(1)}}\bar{d}^{(1)})}{t^{(01)}t^{(12)}}[r^{(12)} + r^{(01)}\exp(2\mathrm{j}\sqrt{\varepsilon^{(1)}}\bar{d}^{(1)})]t^{(02)}, \tag{9.311a}$$

$$1 = \frac{\exp(-\mathrm{j}\sqrt{\varepsilon^{(1)}}\bar{d}^{(1)})}{t^{(01)}t^{(12)}}[r^{(01)}r^{(12)} + \exp(2\mathrm{j}\sqrt{\varepsilon^{(1)}}\bar{d}^{(1)})]t^{(02)}. \tag{9.311b}$$

We arrive at Airy formulae for the case of normal light incidence

$$t^{(02)} = \frac{t^{(01)}t^{(12)}\exp(-\mathrm{j}\sqrt{\varepsilon^{(1)}}\bar{d}^{(1)})}{1 + r^{(01)}r^{(12)}\exp(-2\mathrm{j}\sqrt{\varepsilon^{(1)}}\bar{d}^{(1)})}, \tag{9.312a}$$

$$r^{(02)} = \frac{r^{(01)} + r^{(12)}\exp(-2\mathrm{j}\sqrt{\varepsilon^{(1)}}\bar{d}^{(1)})}{1 + r^{(12)}r_{23}\exp(-2\mathrm{j}\sqrt{\varepsilon^{(1)}}\bar{d}^{(1)})}. \tag{9.312b}$$

9.11 Homogeneous Isotropic Layers at Oblique Incidence

We consider a homogeneous isotropic layer at oblique light incidence $q_0 = \sin\varphi_i^{(0)}$. In this example, we shall use the TM modes normalized to unity. We are concerned with homogeneous regions where we can put $s_m = 0$. The matrix $\mathbf{C}_m^{(u)}$ and the corresponding column vector $\mathbf{f}_{\hat{t}m}^{(u)}$ are given by Eq. (9.116) and Eq. (9.114), respectively

$$\mathbf{C}_m^{(u)} = \begin{pmatrix} s_m & p_m^2/\varepsilon - 1 & 0 & -p_m q_m/\varepsilon \\ q_m^2 - \varepsilon & s_m & -p_m q_m & 0 \\ 0 & p_m q_m/\varepsilon & s_m & 1 - q_m^2/\varepsilon \\ p_m q_m & 0 & \varepsilon - p_m^2 & s_m \end{pmatrix}, \qquad \mathbf{f}_{\hat{t}m}^{(u)} = \begin{pmatrix} e_{xm} \\ h_{ym} \\ e_{ym} \\ h_{xm} \end{pmatrix}.$$

The proper values κ_m of the matrix $\mathbf{C}_m^{(u)}$ were found from the condition $\det\left(\mathbf{C}_m^{(u)} - \kappa_m \mathbf{1}\right) = 0$ for the vanishing determinant. They are given by Eq. (9.132). The matrix $\mathbf{C}_m^{(u)}$ was transformed to a diagonal one in Eq. (9.201) using Eq. (9.198) and Eq. (9.200), for $\overline{\mathbf{T}}_m^{(u)}$ and $\left[\overline{\mathbf{T}}_m^{(u)}\right]^{-1}$, respectively.

From Eq. (9.221) and Eq. (9.198) we have

$$\bar{\mathbf{f}}_{\hat{t}m}^{(u)} = \overline{\mathbf{T}}_m^{(u)}\mathbf{g}_m^{(u)},$$

$$\begin{pmatrix} \bar{e}_{xm} \\ \bar{h}_{ym} \\ \bar{e}_{ym} \\ \bar{h}_{xm} \end{pmatrix} = \begin{pmatrix} \dot{q}_m & \dot{p}_m\xi_m\varepsilon^{-1/2} & \dot{q}_m & \dot{p}_m\xi_m\varepsilon^{-1/2} \\ \dot{q}_m\xi_m & \dot{p}_m\varepsilon^{1/2} & -\dot{q}_m\xi_m & -\dot{p}_m\varepsilon^{1/2} \\ -\dot{p}_m & \dot{q}_m\xi_m\varepsilon^{-1/2} & -\dot{p}_m & \dot{q}_m\xi_m\varepsilon^{-1/2} \\ \dot{p}_m\xi_m & -\dot{q}_m\varepsilon^{1/2} & -\dot{p}_m\xi_m & \dot{q}_m\varepsilon^{1/2} \end{pmatrix} \begin{pmatrix} {}^{TE}g_m^+ \\ {}^{TM}g_m^+ \\ {}^{TE}g_m^- \\ {}^{TM}g_m^- \end{pmatrix}, \tag{9.313}$$

$${}^{TE}g_m^+(\bar{z}) = {}^{TE}g_m^+(\bar{z}^0)\exp[-\mathrm{j}\xi_m(\bar{z} - \bar{z}^{(0)})], \tag{9.314a}$$

$${}^{TM}g_m^+(\bar{z}) = {}^{TM}g_m^+(\bar{z}^0)\exp[-\mathrm{j}\xi_m(\bar{z} - \bar{z}^{(0)})], \tag{9.314b}$$

$${}^{TE}g_m^-(\bar{z}) = {}^{TE}g_m^-(\bar{z}^0)\exp[\mathrm{j}\xi_m(\bar{z} - \bar{z}^{(0)})], \tag{9.314c}$$

$${}^{TM}g_m^-(\bar{z}) = {}^{TM}g_m^-(\bar{z}^0)\exp[\mathrm{j}\xi_m(\bar{z} - \bar{z}^{(0)})], \tag{9.314d}$$

in agreement with Eq. (9.127). The equation for the transmission matrix (9.248) in homogeneous isotropic regions $\mathbf{g}_{\mathrm{I}m}(\bar{z}) = \mathbf{U}_{\mathrm{I}m}^{(u)}(\bar{z}-\bar{z}^{(0)})\mathbf{g}_{\mathrm{I}m}(\bar{z}^{0})$ can be written as

$$\begin{pmatrix} {}^{TE}g_m^{(\mathrm{I})+}(\bar{z}) \\ {}^{TM}g_m^{(\mathrm{I})+}(\bar{z}) \\ {}^{TE}g_m^{(\mathrm{I})-}(\bar{z}) \\ {}^{TM}g_m^{(\mathrm{I})-}(\bar{z}) \end{pmatrix} = \begin{pmatrix} \exp\left[-\mathrm{j}\xi_m^{(\mathrm{I})}(\bar{z}-\bar{z}^{(0)})\right] & 0 & 0 & 0 \\ 0 & \exp\left[-\mathrm{j}\xi_m^{(\mathrm{I})}(\bar{z}-\bar{z}^{(0)})\right] & 0 & 0 \\ 0 & 0 & \exp\left[\mathrm{j}\xi_m^{(\mathrm{I})}(\bar{z}-\bar{z}^{(0)})\right] & 0 \\ 0 & 0 & 0 & \exp\left[\mathrm{j}\xi_m^{(\mathrm{I})}(\bar{z}-\bar{z}^{(0)})\right] \end{pmatrix} \times \begin{pmatrix} {}^{TE}g_m^{(\mathrm{I})+}(\bar{z}^{0}) \\ {}^{TM}g_m^{(\mathrm{I})+}(\bar{z}^{0}) \\ {}^{TE}g_m^{(\mathrm{I})-}(\bar{z}^{0}) \\ {}^{TM}g_m^{(\mathrm{I})-}(\bar{z}^{0}) \end{pmatrix}. \tag{9.315}$$

We have included the index (I) distinguishing the regions (or the layers). At interfaces between homogeneous regions, the requirement of continuity for the tangential field components takes the form

$$\begin{aligned} \bar{\mathbf{f}}_{\hat{t}m}^{(\mathrm{I}-1,u)}(\bar{z}^{(\mathrm{I}-1)}) &= \bar{\mathbf{f}}_{\hat{t}m}^{(\mathrm{I},u)}(\bar{z}^{(\mathrm{I}-1)}), \\ \overline{\mathbf{T}}_m^{(\mathrm{I}-1,u)}\mathbf{g}_m^{(\mathrm{I}-1,u)}(\bar{z}^{(\mathrm{I}-1)}) &= \overline{\mathbf{T}}_m^{(\mathrm{I},u)}\mathbf{g}_m^{(\mathrm{I},u)}(\bar{z}^{(\mathrm{I}-1)}), \end{aligned} \tag{9.316}$$

$$\begin{pmatrix} \dot{q}_m & \dot{p}_m\xi_m^{(\mathrm{I}-1)}(\varepsilon^{(\mathrm{I}-1)})^{-1/2} & \dot{q}_m & \dot{p}_m\xi_m^{(\mathrm{I}-1)}(\varepsilon^{(\mathrm{I}-1)})^{-1/2} \\ \dot{q}_m\xi_m^{(\mathrm{I}-1)} & \dot{p}_m(\varepsilon^{(\mathrm{I}-1)})^{1/2} & -\dot{q}_m\xi_m^{(\mathrm{I}-1)} & -\dot{p}_m(\varepsilon^{(\mathrm{I}-1)})^{1/2} \\ -\dot{p}_m & \dot{q}_m\xi_m^{(\mathrm{I}-1)}(\varepsilon^{(\mathrm{I}-1)})^{-1/2} & -\dot{p}_m & \dot{q}_m\xi_m^{(\mathrm{I}-1)}(\varepsilon^{(\mathrm{I}-1)})^{-1/2} \\ \dot{p}_m\xi_m^{(\mathrm{I}-1)} & -\dot{q}_m(\varepsilon^{(\mathrm{I}-1)})^{1/2} & -\dot{p}_m\xi_m^{(\mathrm{I}-1)} & \dot{q}_m(\varepsilon^{(\mathrm{I}-1)})^{1/2} \end{pmatrix} \times \begin{pmatrix} {}^{TE}g_m^{(\mathrm{I}-1)+}(\bar{z}^{(\mathrm{I}-1)}) \\ {}^{TM}g_m^{(\mathrm{I}-1)+}(\bar{z}^{(\mathrm{I}-1)}) \\ {}^{TE}g_m^{(\mathrm{I}-1)-}(\bar{z}^{(\mathrm{I}-1)}) \\ {}^{TM}g_m^{(\mathrm{I}-1)-}(\bar{z}^{(\mathrm{I}-1)}) \end{pmatrix}$$

$$= \begin{pmatrix} \dot{q}_m & \dot{p}_m\xi_m^{(\mathrm{I})}\varepsilon_{\mathrm{I}}^{-1/2} & \dot{q}_m & \dot{p}_m\xi_m^{(\mathrm{I})}\varepsilon_{\mathrm{I}}^{-1/2} \\ \dot{q}_m\xi_m^{(\mathrm{I})} & \dot{p}_m\varepsilon^{(\mathrm{I})1/2} & -\dot{q}_m\xi_m^{(\mathrm{I})} & -\dot{p}_m\varepsilon^{(\mathrm{I})1/2} \\ -\dot{p}_m & \dot{q}_m\xi_m^{(\mathrm{I})}\varepsilon_{\mathrm{I}}^{-1/2} & -\dot{p}_m & \dot{q}_m\xi_m^{(\mathrm{I})}\varepsilon_{\mathrm{I}}^{-1/2} \\ \dot{p}_m\xi_m^{(\mathrm{I})} & -\dot{q}_m\varepsilon^{(\mathrm{I})1/2} & -\dot{p}_m\xi_m^{(\mathrm{I})} & \dot{q}_m\varepsilon^{(\mathrm{I})1/2} \end{pmatrix} \begin{pmatrix} {}^{TE}g_m^{(\mathrm{I})+}(\bar{z}^{(\mathrm{I}-1)}) \\ {}^{TM}g_m^{(\mathrm{I})+}(\bar{z}^{(\mathrm{I}-1)}) \\ {}^{TE}g_m^{(\mathrm{I})-}(\bar{z}^{(\mathrm{I}-1)}) \\ {}^{TM}g_m^{(\mathrm{I})-}(\bar{z}^{(\mathrm{I}-1)}) \end{pmatrix}, \tag{9.317}$$

where $\bar{z}^{(\mathrm{I}-1)}$ is the coordinate specifying the interface plane between the regions (I − 1) and (I).

From this

$$\mathbf{g}_m^{(\mathrm{I}-1,u)}(\bar{z}^{(\mathrm{I}-1)}) = \left[\mathbf{T}_m^{(\mathrm{I}-1,u)}\right]^{-1} \mathbf{T}_m^{(\mathrm{I},u)} \mathbf{g}_m^{(\mathrm{I},u)}(\bar{z}^{(\mathrm{I}-1)}), \tag{9.318}$$

where the interface matrix is given by

$$\mathbf{B}_{(m}^{(\mathrm{I}-1,\mathrm{I};\,u)} = \left[\mathbf{T}_m^{(\mathrm{I}-1,u)}\right]^{-1} \mathbf{T}_m^{(\mathrm{I},u)}. \tag{9.319}$$

It can be obtained by considering Eq. (9.316)

$$\begin{aligned}
\mathbf{B}_m^{(\mathrm{I}-1,\mathrm{I};\,u)} &= \left[\mathbf{T}_{\mathrm{I}-1,m}^{(u)}\right]^{-1} \mathbf{T}_m^{(\mathrm{I},u)} = \frac{1}{2\xi_m^{(\mathrm{I}-1)}} \\
&\times \begin{pmatrix}
\dot{q}_m \xi_m^{(\mathrm{I}-1)} & \dot{q}_m & -\dot{p}_m \xi_m^{(\mathrm{I}-1)} & \dot{p}_m \\
\dot{p}_m (\varepsilon^{(\mathrm{I}-1)})^{1/2} & \dot{p}_m \xi_m^{(\mathrm{I}-1)} (\varepsilon^{(\mathrm{I}-1)})^{-1/2} & \dot{q}_m (\varepsilon^{(\mathrm{I}-1)})^{1/2} & -\dot{q}_m \xi_m^{(\mathrm{I}-1)} (\varepsilon^{(\mathrm{I}-1)})^{-1/2} \\
\dot{q}_m \xi_m^{(\mathrm{I}-1)} & -\dot{q}_m & -\dot{p}_m \xi_m^{(\mathrm{I}-1)} & -\dot{p}_m \\
\dot{p}_m (\varepsilon^{(\mathrm{I}-1)})^{1/2} & -\dot{p}_m \xi_m^{(\mathrm{I}-1)} (\varepsilon^{(\mathrm{I}-1)})^{-1/2} & \dot{q}_m (\varepsilon^{(\mathrm{I}-1)})^{1/2} & \dot{q}_m \xi_m^{(\mathrm{I}-1)} (\varepsilon^{(\mathrm{I}-1)})^{-1/2}
\end{pmatrix} \\
&\times \begin{pmatrix}
\dot{q}_m & \dot{p}_m \xi_{\mathrm{I},m} \varepsilon_{\mathrm{I}}^{-1/2} & \dot{q}_m & \dot{p}_m \xi_{\mathrm{I},m} \varepsilon_{\mathrm{I}}^{-1/2} \\
\dot{q}_m \xi_{\mathrm{I},m} & \dot{p}_m \varepsilon^{(\mathrm{I})1/2} & -\dot{q}_m \xi_{\mathrm{I},m} & -\dot{p}_m \varepsilon^{(\mathrm{I})1/2} \\
-\dot{p}_m & \dot{q}_m \xi_{\mathrm{I},m} \varepsilon_{\mathrm{I}}^{-1/2} & -\dot{p}_m & \dot{q}_m \xi_{\mathrm{I},m} \varepsilon_{\mathrm{I}}^{-1/2} \\
\dot{p}_m \xi_{\mathrm{I},m} & -\dot{q}_m \varepsilon^{(\mathrm{I})1/2} & -\dot{p}_m \xi_{\mathrm{I},m} & \dot{q}_m \varepsilon^{(\mathrm{I})1/2}
\end{pmatrix} \\
&= \frac{1}{2\xi_m^{(\mathrm{I}-1)}} \\
&\times \begin{pmatrix}
\left(\xi_m^{(\mathrm{I}-1)} + \xi_m^{(\mathrm{I})}\right) & 0 & \left(\xi_m^{(\mathrm{I}-1)} - \xi_m^{(\mathrm{I})}\right) & 0 \\
0 & \left(\dfrac{\varepsilon^{(\mathrm{I}-1)}\xi_m^{(\mathrm{I})} + \varepsilon^{(\mathrm{I})}\xi_m^{(\mathrm{I}-1)}}{\varepsilon^{(\mathrm{I}-1)1/2}\varepsilon^{(\mathrm{I})1/2}}\right) & 0 & \left(\dfrac{\varepsilon^{(\mathrm{I}-1)}\xi_m^{(\mathrm{I})} - \varepsilon^{(\mathrm{I})}\xi_m^{(\mathrm{I}-1)}}{\varepsilon^{(\mathrm{I}-1)1/2}\varepsilon^{(\mathrm{I})1/2}}\right) \\
\left(\xi_m^{(\mathrm{I}-1)} - \xi_m^{(\mathrm{I})}\right) & 0 & \left(\xi_m^{(\mathrm{I}-1)} + \xi_m^{(\mathrm{I})}\right) & 0 \\
0 & \left(\dfrac{\varepsilon^{(\mathrm{I}-1)}\xi_m^{(\mathrm{I})} - \varepsilon^{(\mathrm{I})}\xi_m^{(\mathrm{I}-1)}}{\varepsilon^{(\mathrm{I}-1)1/2}\varepsilon^{(\mathrm{I})1/2}}\right) & 0 & \left(\dfrac{\varepsilon^{(\mathrm{I}-1)}\xi_m^{(\mathrm{I})} + \varepsilon^{(\mathrm{I})}\xi_m^{(\mathrm{I}-1)}}{\varepsilon^{(\mathrm{I}-1)1/2}\varepsilon^{(\mathrm{I})1/2}}\right)
\end{pmatrix}.
\end{aligned} \tag{9.320}$$

The substitution into Eq. (9.296) gives

$$\begin{pmatrix} {}^{TE}g_0^{(\mathrm{I}-1)+} \\ {}^{TM}g_0^{(\mathrm{I}-1)+} \\ {}^{TE}g_0^{(\mathrm{I}-1)-} \\ {}^{TE}g_0^{(\mathrm{I}-1)-} \end{pmatrix} = \frac{1}{2\xi_m^{(\mathrm{I}-1)}}$$

$$\times \begin{pmatrix} \left(\xi_m^{(\mathrm{I}-1)} + \xi_m^{(\mathrm{I})}\right) & 0 & \left(\xi_m^{(\mathrm{I}-1)} - \xi_m^{(\mathrm{I})}\right) & 0 \\ 0 & \left(\frac{\varepsilon^{(\mathrm{I}-1)}\xi_m^{(\mathrm{I})} + \varepsilon^{(\mathrm{I})}\xi_m^{(\mathrm{I}-1)}}{\varepsilon^{(\mathrm{I}-1)1/2}\varepsilon^{(\mathrm{I})1/2}}\right) & 0 & \left(\frac{\varepsilon^{(\mathrm{I}-1)}\xi_m^{(\mathrm{I})} - \varepsilon^{(\mathrm{I})}\xi_m^{(\mathrm{I}-1)}}{\varepsilon^{(\mathrm{I}-1)1/2}\varepsilon^{(\mathrm{I})1/2}}\right) \\ \left(\xi_m^{(\mathrm{I}-1)} - \xi_m^{(\mathrm{I})}\right) & 0 & \left(\xi_m^{(\mathrm{I}-1)} + \xi_m^{(\mathrm{I})}\right) & 0 \\ 0 & \left(\frac{\varepsilon^{(\mathrm{I}-1)}\xi_m^{(\mathrm{I})} - \varepsilon^{(\mathrm{I})}\xi_m^{(\mathrm{I}-1)}}{\varepsilon^{(\mathrm{I}-1)1/2}\varepsilon^{(\mathrm{I})1/2}}\right) & 0 & \left(\frac{\varepsilon^{(\mathrm{I}-1)}\xi_m^{(\mathrm{I})} + \varepsilon^{(\mathrm{I})}\xi_m^{(\mathrm{I}-1)}}{\varepsilon^{(\mathrm{I}-1)1/2}\varepsilon^{(\mathrm{I})1/2}}\right) \end{pmatrix}$$

$$\times \begin{pmatrix} {}^{TE}g_0^{(\mathrm{I})+} \\ {}^{TM}g_0^{(\mathrm{I})+} \\ {}^{TE}g_0^{(\mathrm{I})-} \\ {}^{TM}g_0^{(\mathrm{I})-} \end{pmatrix}. \tag{9.321}$$

We observe that the equations for TE and TM waves are mutually independent

$$2\xi_m^{(\mathrm{I}-1)}\,{}^{TE}g_0^{(\mathrm{I}-1)+} = \left(\xi_m^{(\mathrm{I}-1)} + \xi_m^{(\mathrm{I})}\right){}^{TE}g_0^{(\mathrm{I})+} + \left(\xi_m^{(\mathrm{I}-1)} - \xi_m^{(\mathrm{I})}\right){}^{TE}g_0^{(\mathrm{I})-} \tag{9.322a}$$

$$\begin{aligned} 2\xi_m^{(\mathrm{I}-1)}\,{}^{TM}g_0^{(\mathrm{I}-1)+} &= \left(\varepsilon^{(\mathrm{I}-1)1/2}\frac{\xi_m^{(\mathrm{I})}}{\varepsilon^{(\mathrm{I})1/2}}\frac{\xi_m^{(\mathrm{I}-1)}}{\varepsilon^{(\mathrm{I}-1)1/2}}\varepsilon^{(\mathrm{I})1/2}\right){}^{TM}g_0^{(\mathrm{I})+} \\ &\quad + \left(\varepsilon^{(\mathrm{I}-1)1/2}\frac{\xi_m^{(\mathrm{I})}}{\varepsilon^{(\mathrm{I})1/2}} - \frac{\xi_m^{(\mathrm{I}-1)}}{\varepsilon^{(\mathrm{I}-1)1/2}}\varepsilon^{(\mathrm{I})1/2}\right){}^{TM}g_0^{(\mathrm{I})-} \end{aligned} \tag{9.322b}$$

$$2\xi_m^{(\mathrm{I}-1)}\,{}^{TE}g_0^{(\mathrm{I}-1)-} = \left(\xi_m^{(\mathrm{I}-1)} - \xi_m^{(\mathrm{I})}\right){}^{TE}g_0^{(\mathrm{I})+} + \left(\xi_m^{(\mathrm{I}-1)} + \xi_m^{(\mathrm{I})}\right){}^{TE}g_0^{(\mathrm{I})-} \tag{9.322c}$$

$$\begin{aligned} 2\xi_m^{(\mathrm{I}-1)}\,{}^{TM}g_0^{(\mathrm{I}-1)-} &= \left(\varepsilon^{(\mathrm{I}-1)1/2}\frac{\xi_m^{(\mathrm{I})}}{\varepsilon^{(\mathrm{I})1/2}} - \frac{\xi_m^{(\mathrm{I}-1)}}{\varepsilon^{(\mathrm{I}-1)1/2}}\varepsilon^{(\mathrm{I})1/2}\right){}^{TM}g_0^{(\mathrm{I})+} \\ &\quad + \left(\varepsilon^{(\mathrm{I}-1)1/2}\frac{\xi_m^{(\mathrm{I})}}{\varepsilon^{(\mathrm{I})1/2}} + \frac{\xi_m^{(\mathrm{I}-1)}}{\varepsilon^{(\mathrm{I}-1)1/2}}\varepsilon^{(\mathrm{I})1/2}\right){}^{TM}g_0^{(\mathrm{I})-}. \end{aligned} \tag{9.322d}$$

Let us now assume that there are no waves arriving at the interface between the regions (I−1) and (I) from the region (I), *i.e.*, ${}^{TE}g_0^{(\mathrm{I})+} = 0$ and ${}^{TM}g_0^{(\mathrm{I})+} = 0$. This simplifies the system (9.300a) to (9.300d) as follows

$$2\xi_m^{(\mathrm{I}-1)}\,{}^{TE}g_0^{(\mathrm{I}-1)+} = \left(\xi_m^{(\mathrm{I}-1)} - \xi_m^{(\mathrm{I})}\right){}^{TE}g_0^{(\mathrm{I})-} \tag{9.323a}$$

$$2\xi_m^{(\mathrm{I}-1)}\,{}^{TM}g_0^{(\mathrm{I}-1)+} = \left(\varepsilon^{(\mathrm{I}-1)1/2}\frac{\xi_m^{(\mathrm{I})}}{\varepsilon^{(\mathrm{I})1/2}} - \frac{\xi_m^{(\mathrm{I}-1)}}{\varepsilon^{(\mathrm{I}-1)1/2}}\varepsilon^{(\mathrm{I})1/2}\right){}^{TM}g_0^{(\mathrm{I})-} \tag{9.323b}$$

$$2\xi_m^{(I-1)}\,{}^{TE}g_0^{(I-1)-} = \left(\xi_m^{(I-1)} + \xi_m^{(I)}\right) {}^{TE}g_0^{(I)-} \tag{9.323c}$$

$$2\xi_m^{(I-1)}\,{}^{TM}g_0^{(I-1)-} = \left(\varepsilon^{(I-1)1/2}\frac{\xi_m^{(I)}}{\varepsilon^{(I)1/2}} + \frac{\xi_m^{(I-1)}}{\varepsilon^{(I-1)1/2}}\varepsilon^{(I)1/2}\right) {}^{TM}g_0^{(I)-}. \tag{9.323d}$$

Next, we consider the case of the wave incident from the region $(I-1)$ of the unit amplitude and TE polarization, *i.e.*, ${}^{TE}g_0^{(I-1)-} = 1$ and ${}^{TM}g_0^{(I-1)-} = 0$. There is no incident wave of TM polarization. From Eq. (9.301a) to Eq. (9.301d), we obtain

$$2\xi_m^{(I-1)}\,{}^{TE}g_0^{(I-1)+} = \left(\xi_m^{(I-1)} - \xi_m^{(I)}\right) {}^{TE}g_0^{(I)-} \tag{9.324a}$$

$$2\xi_m^{(I-1)}\,{}^{TM}g_0^{(I-1)+} = \left(\varepsilon^{(I-1)1/2}\frac{\xi_m^{(I)}}{\varepsilon^{(I)1/2}} - \frac{\xi_m^{(I-1)}}{\varepsilon^{(I-1)1/2}}\varepsilon^{(I)1/2}\right) {}^{TM}g_0^{(I)-} \tag{9.324b}$$

$$2\xi_m^{(I-1)} = \left(\xi_m^{(I-1)} + \xi_m^{(I)}\right) {}^{TE}g_0^{(I)-} \tag{9.324c}$$

$$0 = \left(\varepsilon^{(I-1)1/2}\frac{\xi_m^{(I)}}{\varepsilon^{(I)1/2}} + \frac{\xi_m^{(I-1)}}{\varepsilon^{(I-1)1/2}}\varepsilon^{(I)1/2}\right) {}^{TM}g_0^{(I)-}. \tag{9.324d}$$

As a consequence, we have also ${}^{TM}g_0^{(I)-} = 0$ and ${}^{TM}g_0^{(I-1)+} = 0$. This means that there are also no transmitted and reflected waves of the TM polarization. Using Eq. (9.302c), we obtain for the amplitude of the transmitted TE wave

$${}^{TE}t^{(I-1,I)} = {}^{TE}g_0^{(I)-} = \frac{2\xi_m^{(I-1)}}{\xi_m^{(I-1)} + \xi_m^{(I)}}. \tag{9.325}$$

The restriction to $m = 0$ leads to the well-known TE Fresnel amplitude transmission coefficient ${}^{TE}t^{(I-1,I)}$ for the wave traversing the interface from the medium $(I-1)$ to the medium I at an arbitrary angle of incidence. We next divide Eq. (9.302a) by Eq. (9.302c) and eliminate ${}^{TE}g_0^{(I)-}$

$${}^{TE}r^{(I-1,I)} = {}^{TE}g_0^{(I-1)+} = \frac{\left(\xi_m^{(I-1)} - \xi_m^{(I)}\right) {}^{TE}g_0^{(I)-}}{\left(\xi_m^{(I-1)} + \xi_m^{(I)}\right) {}^{TE}g_0^{(I)-}}. \tag{9.326}$$

The restriction to $m = 0$ leads to the amplitude reflection coefficient ${}^{TE}r^{(I-1,I)}$ for the wave incident from the medium $(I-1)$ at the interface between the media $(I-1)$ and (I) at an arbitrary angle of incidence. It remains to consider the case of the TM wave incident from the region $(I-1)$ of the unit amplitude, *i.e.*, ${}^{TM}g_0^{(I-1)-} = 1$ and ${}^{TE}g_0^{(I-1)-} = 0$. There is no incident wave of TE polarization. From Eqs. (9.301), we obtain

$$2\xi_m^{(\mathrm{I}-1)}\,{}^{TE}g_0^{(\mathrm{I}-1)+} = \left(\xi_m^{(\mathrm{I}-1)} - \xi_m^{(\mathrm{I})}\right)\,{}^{TE}g_0^{(\mathrm{I})-} \tag{9.327a}$$

$$2\xi_m^{(\mathrm{I}-1)}\,{}^{TM}g_0^{(\mathrm{I}-1)+} = \left(\varepsilon^{(\mathrm{I}-1)1/2}\frac{\xi_m^{(\mathrm{I})}}{\varepsilon^{(\mathrm{I})1/2}} - \frac{\xi_m^{(\mathrm{I}-1)}}{\varepsilon^{(\mathrm{I}-1)1/2}}\varepsilon^{(\mathrm{I})1/2}\right)\,{}^{TM}g_0^{(\mathrm{I})-} \tag{9.327b}$$

$$0 = \left(\xi_m^{(\mathrm{I}-1)} + \xi_m^{(\mathrm{I})}\right)\,{}^{TE}g_0^{(\mathrm{I})-} \tag{9.327c}$$

$$2\xi_m^{(\mathrm{I}-1)} = \left(\varepsilon^{(\mathrm{I}-1)1/2}\frac{\xi_m^{(\mathrm{I})}}{\varepsilon^{(\mathrm{I})1/2}} + \frac{\xi_m^{(\mathrm{I}-1)}}{\varepsilon^{(\mathrm{I}-1)1/2}}\varepsilon^{(\mathrm{I})1/2}\right)\,{}^{TM}g_0^{(\mathrm{I})-}. \tag{9.327d}$$

As a consequence, we have also ${}^{TE}g_0^{(\mathrm{I})-} = 0$ and ${}^{TE}g_0^{(\mathrm{I}-1)+} = 0$. This means that there are also no transmitted and reflected waves of the TE polarization. Using Eq. (9.302d), we obtain for the amplitude of the transmitted TE wave

$${}^{TM}t^{(\mathrm{I}-1,\mathrm{I})} = {}^{TM}g_0^{(\mathrm{I})-} = \frac{2\xi_m^{(\mathrm{I}-1)}}{\left(\varepsilon^{(\mathrm{I}-1)1/2}\frac{\xi_m^{(\mathrm{I})}}{\varepsilon^{(\mathrm{I})1/2}} + \frac{\xi_m^{(\mathrm{I}-1)}}{\varepsilon^{(\mathrm{I}-1)1/2}}\varepsilon^{(\mathrm{I})1/2}\right)}. \tag{9.328}$$

The restriction to $m = 0$ leads to the well-known Fresnel amplitude transmission coefficient ${}^{TE}t^{(\mathrm{I}-1,\mathrm{I})}$ for the wave traversing the interface from the medium $(\mathrm{I}-1)$ to the medium I at an arbitrary angle of incidence. We next divide Eq. (9.302a) by Eq. (9.302c) and eliminate ${}^{TE}g_0^{(\mathrm{I})-}$

$${}^{TM}r^{(\mathrm{I}-1,\mathrm{I})} = {}^{TM}g_0^{(\mathrm{I}-1)+} = \frac{\left(\varepsilon^{(\mathrm{I}-1)1/2}\frac{\xi_m^{(\mathrm{I})}}{\varepsilon^{(\mathrm{I})1/2}} - \frac{\xi_m^{(\mathrm{I}-1)}}{\varepsilon^{(\mathrm{I}-1)1/2}}\varepsilon^{(\mathrm{I})1/2}\right)\,{}^{TM}g_0^{(\mathrm{I})-}}{\left(\varepsilon^{(\mathrm{I}-1)1/2}\frac{\xi_m^{(\mathrm{I})}}{\varepsilon^{(\mathrm{I})1/2}} + \frac{\xi_m^{(\mathrm{I}-1)}}{\varepsilon^{(\mathrm{I}-1)1/2}}\varepsilon^{(\mathrm{I})1/2}\right)\,{}^{TM}g_0^{(\mathrm{I})-}}. \tag{9.329}$$

The restriction to $m = 0$ leads to the classical TM Fresnel amplitude reflection coefficient ${}^{TM}r^{(\mathrm{I}-1,\mathrm{I})}$ for the wave incident from the medium $(\mathrm{I}-1)$ at the interface between the media $(\mathrm{I}-1)$ and I at an arbitrary angle of incidence.

Next, we consider the case of a homogeneous isotropic region (1) sandwiched between two isotropic half spaces (0) and (2). We make use of Eq. (9.263) simplified with $\mathbf{s}^{(1)} = 0$

$$\begin{aligned}\mathbf{g}_m^{(0)}(0) &= \left[\mathbf{T}_m^{(0u)}\right]^{-1}\mathbf{T}_m^{(1u)}\mathbf{U}_m^{(1)}(-\bar{d}^{(1)})\left[\mathbf{T}_m^{(1u)}\right]^{-1}\mathbf{T}_m^{(2u)}\mathbf{g}_m^{(2)}(-\bar{d}^{(1)})\\ &= \mathbf{B}_m^{(01)}\mathbf{U}_m^{(1)}(-\bar{d}^{(1)})\mathbf{B}_m^{(12)}\mathbf{g}_m^{(2)}(\bar{d}^{(1)}).\end{aligned} \tag{9.330}$$

Here, the matrices $\mathbf{B}_m^{(01)}$ and $\mathbf{B}_m^{(12)}$ are given in Eq. (9.297) with $\mathrm{I} = 1$ and $\mathrm{I} = 2$, respectively. The transmission matrix $\mathbf{U}_m^{(1)}(\bar{d}^{(1)})$ is given according to Eq. (9.274) by

$$\mathbf{U}_m^{(1)}(\bar{d}^{(1)}) = \begin{pmatrix} \exp\left(-\mathrm{j}\xi_m^{(1)}\bar{d}^{(1)}\right) & 0 & 0 & 0 \\ 0 & \exp\left(-\mathrm{j}\xi_m^{(1)}\bar{d}^{(1)}\right) & 0 & 0 \\ 0 & 0 & \exp\left(\mathrm{j}\xi_m^{(1)}\bar{d}^{(1)}\right) & 0 \\ 0 & 0 & 0 & \exp\left(\mathrm{j}\xi_m^{(1)}\bar{d}^{(1)}\right) \end{pmatrix}. \tag{9.331}$$

The relation between amplitudes in the region (0) at the interface between the regions (0) and (1) and those in the region (2) at the interface between regions (1) and (2) is given by

$$\begin{pmatrix} {}^{TE}g_m^{(0)+}(0) \\ {}^{TM}g_m^{(0)+}(0) \\ {}^{TE}g_m^{(0)-}(0) \\ {}^{TM}g_m^{(0)-}(0) \end{pmatrix} = \frac{1}{2\xi_m^{(0)}}\frac{1}{2\xi_m^{(1)}}$$

$$\times \begin{pmatrix} \left(\xi_m^{(0)}+\xi_m^{(1)}\right) & 0 & \left(\xi_m^{(0)}-\xi_m^{(1)}\right) & 0 \\ 0 & \left(\varepsilon^{(0)1/2}\frac{\xi_m^{(1)}}{\varepsilon^{(1)1/2}}+\frac{\xi_m^{(0)}}{\varepsilon^{(0)1/2}}\varepsilon^{(1)1/2}\right) & 0 & \left(\varepsilon^{(0)1/2}\frac{\xi_m^{(1)}}{\varepsilon^{(1)1/2}}-\frac{\xi_m^{(0)}}{\varepsilon^{(0)1/2}}\varepsilon^{(1)1/2}\right) \\ \left(\xi_m^{(0)}-\xi_m^{(1)}\right) & 0 & \left(\xi_m^{(0)}+\xi_m^{(1)}\right) & 0 \\ 0 & \left(\varepsilon^{(0)1/2}\frac{\xi_m^{(1)}}{\varepsilon^{(1)1/2}}-\frac{\xi_m^{(0)}}{\varepsilon^{(0)1/2}}\varepsilon^{(1)1/2}\right) & 0 & \left(\varepsilon^{(0)1/2}\frac{\xi_m^{(1)}}{\varepsilon^{(1)1/2}}+\frac{\xi_m^{(0)}}{\varepsilon^{(0)1/2}}\varepsilon^{(1)1/2}\right) \end{pmatrix}$$

$$\times \begin{pmatrix} \exp\left(-\mathrm{j}\xi_m^{(1)}\bar{d}^{(1)}\right) & 0 & 0 & 0 \\ 0 & \exp\left(-\mathrm{j}\xi_m^{(1)}\bar{d}^{(1)}\right) & 0 & 0 \\ 0 & 0 & \exp\left(\mathrm{j}\xi_m^{(1)}\bar{d}^{(1)}\right) & 0 \\ 0 & 0 & 0 & \exp\left(\mathrm{j}\xi_m^{(1)}\bar{d}^{(1)}\right) \end{pmatrix}$$

$$\times \begin{pmatrix} \left(\xi_m^{(1)}+\xi_m^{(2)}\right) & 0 & \left(\xi_m^{(1)}-\xi_m^{(2)}\right) & 0 \\ 0 & \left(\varepsilon^{(1)1/2}\frac{\xi_m^{(2)}}{\varepsilon^{(2)1/2}}+\frac{\xi_m^{(1)}}{\varepsilon^{(1)1/2}}\varepsilon^{(2)1/2}\right) & 0 & \left(\varepsilon^{(1)1/2}\frac{\xi_m^{(2)}}{\varepsilon^{(2)1/2}}-\frac{\xi_m^{(1)}}{\varepsilon^{(1)1/2}}\varepsilon^{(2)1/2}\right) \\ \left(\xi_m^{(1)}-\xi_m^{(2)}\right) & 0 & \left(\xi_m^{(1)}+\xi_m^{(2)}\right) & 0 \\ 0 & \left(\varepsilon^{(1)1/2}\frac{\xi_m^{(2)}}{\varepsilon^{(2)1/2}}-\frac{\xi_m^{(1)}}{\varepsilon^{(1)1/2}}\varepsilon^{(2)1/2}\right) & 0 & \left(\varepsilon^{(1)1/2}\frac{\xi_m^{(2)}}{\varepsilon^{(2)1/2}}+\frac{\xi_m^{(1)}}{\varepsilon^{(1)1/2}}\varepsilon^{(2)1/2}\right) \end{pmatrix}$$

$$\times \begin{pmatrix} {}^{TE}g_m^{(2)+}(\bar{d}^{(1)}) \\ {}^{TM}g_m^{(2)+}(\bar{d}^{(1)}) \\ {}^{TE}g_m^{(2)-}(\bar{d}^{(1)}) \\ {}^{TM}g_m^{(2)-}(\bar{d}^{(1)}) \end{pmatrix}. \tag{9.332}$$

The evaluation of this equation provides

$$\begin{pmatrix} {}^{TE}g_{1m}^{+}(0) \\ {}^{TM}g_{1m}^{+}(0) \\ {}^{TE}g_{1m}^{-}(0) \\ {}^{TM}g_{1m}^{-}(0) \end{pmatrix} = \frac{e^{-j\xi_m^{(1)}\bar{d}^{(1)}}}{t^{(01)}t^{(12)}}$$
$$\times \begin{pmatrix} 1+r^{(01)}r^{(12)}e^{2j\xi_m^{(1)}\bar{d}^{(1)}} & 0 & r^{(12)}+r^{(01)}e^{2j\xi_m^{(1)}\bar{d}^{(1)}} & 0 \\ 0 & 1+r^{(01)}r^{(12)}e^{2j\xi_m^{(1)}\bar{d}^{(1)}} & 0 & r^{(12)}+r^{(01)}e^{2j\xi_m^{(1)}\bar{d}^{(1)}} \\ r^{(01)}+r^{(12)}e^{2j\xi_m^{(1)}\bar{d}^{(1)}} & 0 & r^{(01)}r^{(12)}+e^{2j\xi_m^{(1)}\bar{d}^{(1)}} & 0 \\ 0 & r^{(01)}+r^{(12)}e^{2j\xi_m^{(1)}\bar{d}^{(1)}} & 0 & r^{(01)}r^{(12)}+e^{2j\xi_m^{(1)}\bar{d}^{(1)}} \end{pmatrix}$$
$$\times \begin{pmatrix} {}^{TE}g_m^{(2)+}(\bar{d}^{(1)}) \\ {}^{TM}g_m^{(2)+}(\bar{d}^{(1)}) \\ {}^{TE}g_m^{(2)-}(\bar{d}^{(1)}) \\ {}^{TM}g_m^{(2)-}(\bar{d}^{(1)}) \end{pmatrix} \tag{9.333}$$

for the TE polarization

$${}^{TE}t^{(01)} = \frac{2\xi_m^{(0)}}{\xi_m^{(0)}+\xi_m^{(1)}}, \tag{9.334a}$$

$${}^{TE}t^{(12)} = \frac{2\xi_m^{(1)}}{\xi_m^{(1)}+\xi_m^{(2)}}, \tag{9.334b}$$

$${}^{TE}r^{(01)} = \frac{\xi_m^{(0)}-\xi_m^{(1)}}{\xi_m^{(0)}+\xi_m^{(1)}}, \tag{9.334c}$$

$${}^{TE}r^{(12)} = \frac{\xi_m^{(1)}-\xi_m^{(2)}}{\xi_m^{(1)}+\xi_m^{(2)}}, \tag{9.334d}$$

and for the TM polarization

$${}^{TM}t^{(01)} = \frac{2\xi_m^{(0)}}{\varepsilon^{(0)1/2}\dfrac{\xi_m^{(1)}}{\varepsilon^{(1)1/2}} + \dfrac{\xi_m^{(0)}}{\varepsilon^{(0)1/2}}\varepsilon^{(1)1/2}}, \tag{9.335a}$$

$${}^{TM}t^{(12)} = \frac{2\xi_m^{(1)}}{\varepsilon^{(1)1/2}\dfrac{\xi_m^{(2)}}{\varepsilon^{(2)1/2}} + \dfrac{\xi_m^{(1)}}{\varepsilon^{(1)1/2}}\varepsilon^{(2)1/2}}, \tag{9.335b}$$

$${}^{TM}r^{(01)} = \frac{\varepsilon^{(0)1/2}\dfrac{\xi_m^{(1)}}{\varepsilon^{(1)1/2}} - \dfrac{\xi_m^{(0)}}{\varepsilon^{(0)1/2}}\varepsilon^{(1)1/2}}{\varepsilon^{(0)1/2}\dfrac{\xi_m^{(1)}}{\varepsilon^{(1)1/2}} + \dfrac{\xi_m^{(0)}}{\varepsilon^{(0)1/2}}\varepsilon^{(1)1/2}}, \tag{9.335c}$$

$$ {}^{TM}r^{(12)} = \frac{\varepsilon^{(1)1/2}\dfrac{\xi_m^{(2)}}{\varepsilon^{(2)1/2}} - \dfrac{\xi_m^{(1)}}{\varepsilon^{(1)1/2}}\varepsilon^{(2)1/2}}{\varepsilon^{(1)1/2}\dfrac{\xi_m^{(1)}}{\varepsilon^{(2)1/2}} + \dfrac{\xi_m^{(1)}}{\varepsilon^{(1)1/2}}\varepsilon^{(2)1/2}}. \tag{9.335d} $$

Assuming that no waves impinge on the structure from the region (2), *i.e.*, ${}^{TE}g_m^{(2)+}(\bar{d}^{(1)}) = 0$ and ${}^{TM}g_m^{(2)+}(\bar{d}^{(1)}) = 0$ and that only incident wave is ${}^{TE}g_m^{(0)-}(0) = 1$ while ${}^{TM}g_m^{(0)-}(0) = 0$, we may write

$$ \begin{pmatrix} {}^{TE}g_m^{(0)+}(0) \\ {}^{TM}g_m^{(0)+}(0) \\ 1 \\ 0 \end{pmatrix} = \frac{\exp\left(-j\xi_m^{(1)}\bar{d}^{(1)}\right)}{t^{(01)}t^{(12)}} $$

$$ \times \begin{pmatrix} 1 + r^{(01)}r^{(12)}e^{2j\xi_m^{(1)}\bar{d}^{(1)}} & 0 & r^{(12)} + r^{(01)}e^{2j\xi_m^{(1)}\bar{d}^{(1)}} & 0 \\ 0 & 1 + r^{(01)}r^{(12)}e^{2j\xi_m^{(1)}\bar{d}^{(1)}} & 0 & r^{(12)} + r^{(01)}e^{2j\xi_m^{(1)}\bar{d}^{(1)}} \\ r^{(01)} + r^{(12)}e^{2j\xi_m^{(1)}\bar{d}^{(1)}} & 0 & r^{(01)}r^{(12)} + e^{2j\xi_m^{(1)}\bar{d}^{(1)}} & 0 \\ 0 & r^{(01)} + r^{(12)}e^{2j\xi_m^{(1)}\bar{d}^{(1)}} & 0 & r^{(01)}r^{(12)} + e^{2j\xi_m^{(1)}\bar{d}^{(1)}} \end{pmatrix} $$

$$ \times \begin{pmatrix} 0 \\ 0 \\ {}^{TE}g_m^{(2)-}(\bar{d}^{(1)}) \\ {}^{TM}g_m^{(2)-}(\bar{d}^{(1)}) \end{pmatrix}. \tag{9.336} $$

From the evaluation of this matrix product, we find ${}^{TM}g_m^{(2)-}(\bar{d}^{(1)}) = 0$ and ${}^{TM}g_m^{(0)+}(0) = 0$. There are no TM reflected and transmitted waves. This is consistent with the absence of TE↔TM coupling in the isotropic structure. Then, $t^{(02)} = {}^{TE}g_m^{(2)-}(\bar{d}^{(1)})$ represents the global amplitude transmission coefficient, and $r^{(02)} = {}^{TE}g_m^{(0)+}(\bar{d}^{(1)})$ represents the global amplitude reflection coefficient of the structure. We have to consider the following two equations

$$ r^{(02)} = \frac{\exp\left(-j\xi_m^{(1)}\bar{d}^{(1)}\right)}{t^{(01)}t^{(12)}}\left(r^{(12)} + r^{(01)}e^{2j\xi_m^{(1)}\bar{d}^{(1)}}\right)t^{(02)}, $$

$$ 1 = \frac{\exp\left(-j\xi_m^{(1)}\bar{d}^{(1)}\right)}{t^{(01)}t^{(12)}}\left(r^{(01)}r^{(12)} + e^{2j\xi_m^{(1)}\bar{d}^{(1)}}\right)t^{(02)}. $$

We arrive at Airy formulae for a general angle of the plane wave incidence

$$ t^{(02)} = \frac{t^{(01)}t^{(12)}\exp\left(-j\xi_m^{(1)}\bar{d}^{(1)}\right)}{1 + r^{(01)}r^{(12)}\exp\left(-2j\xi_m^{(1)}\bar{d}^{(1)}\right)}, \tag{9.337a} $$

$$ r^{(02)} = \frac{r^{(01)} + r^{(12)}\exp\left(-2j\xi_m^{(1)}\bar{d}^{(1)}\right)}{1 + r^{(01)}r^{(12)}\exp\left(-2j\xi_m^{(1)}\bar{d}^{(1)}\right)}, \tag{9.337b} $$

employed, *e.g.*, in ellipsometry and Fabry-Perot spectroscopy.

This concludes our treatment of anisotropic multilayer diffraction gratings [9]. The procedures were illustrated on simpler situations for both homogeneous isotropic and anisotropic media including the Fresnel formulae for a single interface between homogeneous isotropic media and Airy formulae for field amplitudes in a system formed by an isotropic layer sandwiched between two isotropic half spaces. The meaning of the matrices representing the isotropic regions was compared with that pertaining to planar multilayers. The main purpose was to provide an introduction to the analytical aspects of the wave diffraction in periodic anisotropic multilayers required for an efficient use of numerical procedures [53]. More information on the subject can be found in the literature [54–63].

References

1. F. Abelès, "Recherches sur la propagation des ondes électromagnétiques sinusoïdales dans les milieux stratifiés. Application aux couches minces," Ann. Phys. Paris **5**, 596–640 (1950).
2. Š. Višňovský, R. Lopušník, M. Nývlt, V. Prosser, J. Ferré, C. Train, P. Beauvillain, D. Renard, R. Krishnan, and J. A. C. Bland, "Analytical expressions for polar magnetooptics in magnetic multilayers," Czech. J. Phys. B **50**, 857–882, 2000.
3. S. Visnovsky, K. Postava, and T. Yamaguchi, "Magneto-optic polar Kerr and Faraday effects in periodic multilayers," Optics Express **9**, 158–171, 2001, http://www.opticsexpress.org.
4. K. Postava, J. Pištora, and Š. Višňovský, "Magneto-optical effects in ultrathin structures at transversal magnetization," Czech. J. Phys. B **49**, 1185–1204, 1999.
5. C. Cohen-Tannoudji, B. Diu, F. Laloe, *Quantum Mechanics*, translation of *Mécanique quantique* (A Wiley Interscience Publication Hermann, Paris, France, 1977).
6. D. W. Berreman, "Optics in stratified and anisotropic media: 4 × 4-matrix formulation," J. Opt. Soc. Am. **62**, 502–510, 1972.
7. D. W. Berreman, "Optics in smoothly varying anisotropic planar structures: Application to liquid-crystal twist cells," J. Opt. Soc. Am. **63**, 1374–1380, 1973.
8. P. Yeh, J. Opt. Soc. Am. 69, 742–756, 1979; "Optics of anisotropic layered media: A new 4 × 4 matrix algebra," Surf. Sci. **96**, 41–53, 1980.
9. K. Rokushima and J. Yamakita, "Analysis of anisotropic dielectric gratings," J. Opt. Soc. Am. **73**, 901–908, 1983. Reprinted in *Selected Papers on Diffraction Gratings*, D. Maystre, ed., SPIE Milestone Series, Volume MS **83** (1993), pp. 519–526, SPIE Optical Engineering Press.
10. H. Kogelnik, "Coupled wave theory for thick hologram gratings," Bell. Syst. Tech. J. **48**, 2909–2947, 1969.
11. O. Francescangeli, S. Melone, and R. Deleo, "Dynamic diffraction of guided electromagnetic waves by 2-dimensional periodic dielectric gratings," Phys. Rev. A **43**, 6975–6989, 1991.

12. D. W. Berreman and A. T. Macrander, "Asymmetric x-ray diffraction by strained crystal wafers: 8 × 8-matrix dynamical theory," Phys. Rev. B **37**, 6030–6040, 1988.
13. E. N. Glytsis and T. K. Gaylord, "3-dimensional (vector) rigorous coupled-wave analysis of anisotropic grating diffraction," J. Opt. Soc. Am. A **7**, 1399–1420, 1990.
14. Š. Višňovský and K. Yasumoto, "Multilayer anisotropic bi-periodic diffraction gratings," Czech. J. Phys. B **51**, 229–247, 2001.
15. L. Li, "Fourier modal method for crossed anisotropic gratings with arbitrary permittivity and permeability tensors," J. Optics A: Pure Appl. Opt. **5**, 345–355, 2003.
16. S. P. Liu, "Exact theories for Light, X-Ray, electron, and neutron diffractions from planar media with periodic structures," Phys. Rev. B **39**, 10640–10650, 1989.
17. D. E. Aspnes, "Expanding horizons: New developments in ellipsometry and polarimetry," Thin Solid Films **455–456**, 3–13, 2004.
18. H.-T. Huang and F. L. Terry Jr., "Spectroscopic ellipsometry and reflectometry from gratings (Scatterometry) for critical dimension measurement and in-situ, real time process monitoring," Thin Solid Films (2004), in press.
19. S. Ponti, C. Oldano, and M. Becchi, "Bloch wave approach to the optics of crystals," Phys. Rev. E **64**, Art. No. 021704, 2001.
20. M. Becchi, C. Oldano, and S. Ponti, "Spatial dispersion and optics of crystals," J. Opt. A-Pure Appl. Opt. **1**, 713–718, 1999.
21. D. Ciprian and J. Pištora, "Magneto-optic periodic strip structures," Acta Phys Pol. A **99**, 33–46, 2001.
22. J. M. Jarem and P. P. Banerjee, "Application of the complex Poynting theorem to diffraction gratings," J. Opt. Soc. Am. A **16**, 1097–1107, 1999.
23. G. Montemezzani and M. Zgonik, "Light diffraction at mixed phase and absorption gratings in anisotropic media for arbitrary geometries," Phys. Rev. E **55**, 1035–1047, 1997.
24. Y. Ohkawa, Y. Tsuji, and M. Koshiba, "Analysis of anisotropic dielectric grating diffraction using the finite-element method," J. Opt. Soc. Am. A **13**, 1006–1012, 1996.
25. P. Galatola, C. Oldano, and P. B. S. Kumar, "Symmetry properties of anisotropic dielectric gratings," J. Opt. Soc. Am. A **11**, 1332–1341, 1994.
26. R. A. Depine, V. L. Brudny, and A. Lakhtakia, "T-matrix approach for calculating the electromagnetic fields diffracted by a corrugated, anisotropic grating," J. Mod. Optics **39**, 589–601, 1992.
27. A. Vial and D. Van Labeke, "Diffraction hysteresis loop modelisation in transverse magneto-optical Kerr effect," Opt. Commun. **153**, 125–133, 1998.
28. Y. Pagani, D. Van Labeke, B. Guizal, A. Vial, and F. Baida, "Diffraction hysteresis loop modeling in magneto-optical gratings," Opt. Commun. **209**, 237–244, 2002.
29. N. Bardou, B. Bartenlian, F. Rousseaux, D. Decanini, F. Carsenac, C. Chappert, P. Veillet, P. Beauvillain, R. Mégy, Y. Suzuki, and J. Ferré, "Light diffraction effects in the magneto-optical properties of 2D arrays of magnetic dots of Au/Co/Au(111) films with perpendicular magnetic anisotropy," J. Magn. Magn. Mat. **156**, 293–294, 1995.

30. Y. Suzuki, C. Chappert, P. Bruno, P. Veillet, "Simple model for the magneto-optical Kerr diffraction of a regular array of magnetic dots," J. Magn. Magn. Mat. **165**, 516–519, 1997.
31. S. Mori, K. Mukai, J. Yamakita, and K. Rokushima, "Analysis of dielectric lamellar gratings coated with anisotropic layers," J. Opt. Soc. Am. A **7**, 1661–1665, 1990.
32. S. Mitani, K. Takanashi, H. Nakajima, K. Sato, R. Schreiber, P. Grünberg, and H. Fujimori, "Structural and magnetic properties of Fe/noble metal monoatomic multilayers equivalent to $L1_0$ ordered alloys," J. Magn. Magn. Mat. **156**, 7–10, 1996.
33. K. Machida, T. Tezuka, T. Yamamoto, T. Ishibashi, Y. Morishita, A. Koukitu, and K. Sato, "Magnetic structure of cross-shaped permalloy arrays embedded in silicon wafers," J. Magn. Magn. Mat. 290, 779–782, 2004.
34. K. Yasumoto, H. Toyama, and T. Kushta, "Accurate analysis of two-dimensional electromagnetic scattering from multilayered periodic arrays of circular cylinders using lattice sums technique," IEEE Trans. Antenna Propag. **52**, 2603–2611, 2004.
35. J.-M. Lourtioz, H. Benisty, V. Berger, J.-M. Gérard, D. Maystre, and A. Tchelnokov, *Les cristaux photoniques ou la lumière en cage* (Collection Téchnique et Scientifique des Télécommunications, GET et Lavoisier, Paris 2003); *Photonic Crystals: Towards Nanoscale Photonic Devices* (Springer 2004).
36. J. Lafait, T. Yamaguchi, J. M. Frigerio, A. Bichri, and K. Driss-Khodja, "Effective medium equivalent to a symmetric multilayer at oblique incidence," Appl. Opt. **29**, 2460–2465, 1990.
37. M. Born and E. Wolf with contributions by A. B. Bhatia, P. C. Clemmow, D. Gabor, A. R. Stokes, A. M. Taylor, P. A. Wayman, and W. L. Wilcock, *Principles of Optics, Electromagnetic Theory of Propagation Interference and Diffraction of Light*, Sixth (Corrected) Edition (Cambridge University Press, Cambridge, 1997).
38. R. M. A. Azzam and N. M. Bashara, *Ellipsometry and Polarized Light* (Elsevier, Amsterdam, 1987).
39. A. K. Zvezdin and V. A. Kotov, *Modern Magnetooptics and Magnetooptical Materials*, (Institute of Physics Publishing, Bristol and Philadelphia, 1997).
40. Š. Višňovský, "Magneto-optical permittivity tensor in crystals," Czech. J. Phys. B **36**, 1424–1433, 1986.
41. Š. Višňovský, "Magneto–optical ellipsometry," Czech. J. Phys. B **36**, 625–650, 1986.
42. M. Schubert, "Polarization-dependent optical parameters of arbitrarily anisotropic homogenous layered systems," Phys. Rev. B **53** 4265-4274, 1996.
43. P. Lalanne and G. M. Morris, "Highly improved convergence of the coupled-wave method for TM polarization," J. Opt. Soc. Am. A **13** 779, 1996.
44. G. Granet and B. Guizal, "Efficient implementation of the coupled-wave method for metallic lamellar gratings in TM polarization," J. Opt. Soc. Am. A **13** 1019, 1996.
45. Lifeng Li, "Formulation and comparison of two recursive matrix algorithms for modeling layered diffraction gratings," J. Opt. Soc. Am. A **13**, 1870, 1996.
46. Lifeng Li, "New formulation of the Fourier modal method for crossed surface-relief gratings," J. Opt. Soc. Am. **14**, 2758–2767, 1997.

47. L. Li, "Reformulation of the Fourier modal method for surface-relief gratings made with anisotropic materials," J. Mod. Opt. **45** 1313–1334, 1998.
48. K. Rokushima and J. Yamakita, "Analysis of diffraction in periodic liquid-crystals – The optics of the chiral smectic-C phase," J. Opt. Soc. Am. A **4**, 27–33, 1987.
49. D. Marcuse, *Light Transmission Optics* (Bell Laboratories Series, Van Nostrand Rienhold Company, New York, 1972), Chapter 1.
50. D. Marcuse, *Theory of Dielectric Optical Waveguides* (Academic Press, New York and London, 1974), Chapter 2.
51. H. Kogelnik, in Topics in Applied Physics, vol. 9, *Integrated Optics*, ed. T. Tamir, Springer-Verlag, Berlin, 1975.
52. P. K. Tien, "Integrated optics and new wave phenomena in optical waveguides," Rev. Mod. Phys. **49** 361–420, 1977.
53. K. Rokushima, R. Antoš, J. Mistrík, Š. Višňovský, and T. Yamaguchi, "Optics of anisotropic nanostructures," to be submitted to Czech. J. Phys.
54. L. Li, "Multilayer modal method for diffraction gratings of arbitrary profile, depth, and permittivity," J. Opt. Soc. Am. A **10**, 2581–2591, 1993.
55. F. Montiel and M. Nevière, "Differential theory of gratings: extension to deep gratings of arbitrary profile and permittivity through the R-matrix propagation algorithm," J. Opt. Soc. Am. A **11**, Issue 12, 3241–3250, 1994.
56. N. P. K. Cotter, T. W. Preist, and J. R. Sambles, "Scattering-matrix approach to multilayer diffraction," J. Opt. Soc. Am. A **12**, 1097–1103, 1995.
57. L. Li, "Bremmer series, R-matrix propagation algorithm, and numerical modeling of diffraction gratings," J. Opt. Soc. Am. A **11**, 2829–2836, 1994.
58. L. Li, "Formulation and comparison of two recursive matrix algorithms for modeling layered diffraction gratings," J. Opt. Soc. Am. A **13**, 1024–1035, 1996.
59. L. Li, "Use of Fourier series in the analysis of discontinuous periodic structures," J. Opt. Soc. Am. A **13**, 1870–1876, 1996.
60. E. Popov and M. Nevière, "Grating theory: New equations in Fourier space leading to fast converging results for TM polarization," J. Opt. Soc. Am. A **17**, 1773-1784, 2000.
61. K. Watanabe, R. Petit, and M. Nevière, "Differential theory of gratings made of anisotropic materials," J. Opt. Soc. Am. A **19**, 325–334, 2002.
62. R. Antoš, J. Mistrík, T. Yamaguchi, Š. Višňovský, S. O. Demokritov, and B. Hillebrands, "Evidence of native oxides on the capping and substrate of Permalloy gratings by magneto-optical spectroscopy in the zeroeth and first-diffraction orders," Appl. Phys. Lett. **86**, 231101, 2005.
63. R. Antoš, J. Mistrík, T. Yamaguchi, Š. Višňovský, S. O. Demokritov, and B. Hillebrands, "Evaluation of the quality of Permalloy gratings by diffracted magneto-optical spectroscopy," Optics Express 2005.

Appendix A

Circular Polarizations

The distinction between right (RCP) and left (LCP) circular polarizations is a matter of convention. We adopt that employed by Azzama and Bashara [1]. The polarizations may be characterized in terms of the space-time dependent real electric wave fields. We denote the real electric field of the RCP wave (with a positive helicity) as $\widetilde{\boldsymbol{E}}_{\mathcal{R}}$ and that of the LCP wave (with a negative helicity) as $\widetilde{\boldsymbol{E}}_{\mathcal{L}}$ and define (Definition 1)

$$\begin{aligned}\widetilde{\boldsymbol{E}}_{\mathcal{R}} &= \Re\{(\hat{x} + \mathrm{j}\hat{y})\exp[\mathrm{j}(\omega t - \gamma z - \phi_{\mathcal{R}})]\} \\ &= \hat{x}\cos(\omega t - \gamma z - \phi_{\mathcal{R}}) - \hat{y}\sin(\omega t - \gamma z - \phi_{\mathcal{R}}),\end{aligned} \tag{A.1a}$$

$$\begin{aligned}\widetilde{\boldsymbol{E}}_{\mathcal{L}} &= \Re\{(\hat{x} - \mathrm{j}\hat{y})\exp[\mathrm{j}(\omega t - \gamma z - \phi_{\mathcal{L}})]\} \\ &= \hat{x}\cos(\omega t - \gamma z - \phi_{\mathcal{L}}) + \hat{y}\sin(\omega t - \gamma z - \phi_{\mathcal{L}}).\end{aligned} \tag{A.1b}$$

At a fixed time, the endpoints of the displacement from the z axis, representing the electric field vectors $\widetilde{\boldsymbol{E}}_{\mathcal{R}}$ of the RCP wave, form a right-handed helix (Definition 2). In a fixed plane perpendicular to the z axis, the endpoint of the vector representing the field $\widetilde{\boldsymbol{E}}_{\mathcal{R}}$ traces a circle rotating in the sense from the positive x axis towards the negative y axis with an angular frequency ω, provided the sense of the propagation vector is oriented parallel to the positive z axis. In other words, when looking against the propagation vector, in a fixed plane, (z = const) $\widetilde{\boldsymbol{E}}_{\mathcal{R}}$ rotates clockwise and $\widetilde{\boldsymbol{E}}_{\mathcal{L}}$ anticlockwise (Definition 3). The distinction between the RCP and LCP waves is illustrated in Figure A.1. Two opposite circularly polarized waves propagate parallel to the magnetization, $\boldsymbol{M}$, with different propagation vectors, $\boldsymbol{\gamma}_{\mathcal{R}} \neq \boldsymbol{\gamma}_{\mathcal{L}}$. Figure A.2 displays the situation where the right (RCP) and left (LCP) circularly polarized waves propagate parallel to the positive z axis with the propagation vectors $\boldsymbol{\gamma}_{\mathcal{R}} = \hat{z}\gamma_{\mathcal{R}}$ and $\boldsymbol{\gamma}_{\mathcal{L}} = \hat{z}\gamma_{\mathcal{L}}$, respectively. Here,

$$\gamma_{\mathcal{R}} = \frac{\omega}{c} N_{\mathcal{R}}$$

and

$$\gamma_{\mathcal{R}} = \frac{\omega}{c} N_{\mathcal{R}}$$

are expressed with the complex indices of refraction $N_{\mathcal{R}} = n_{\mathcal{R}} - \mathrm{j}k_{\mathcal{R}}$ and $N_{\mathcal{L}} = n_{\mathcal{L}} - \mathrm{j}k_{\mathcal{L}}$ for the RCP and LCP waves, respectively. In a special case of

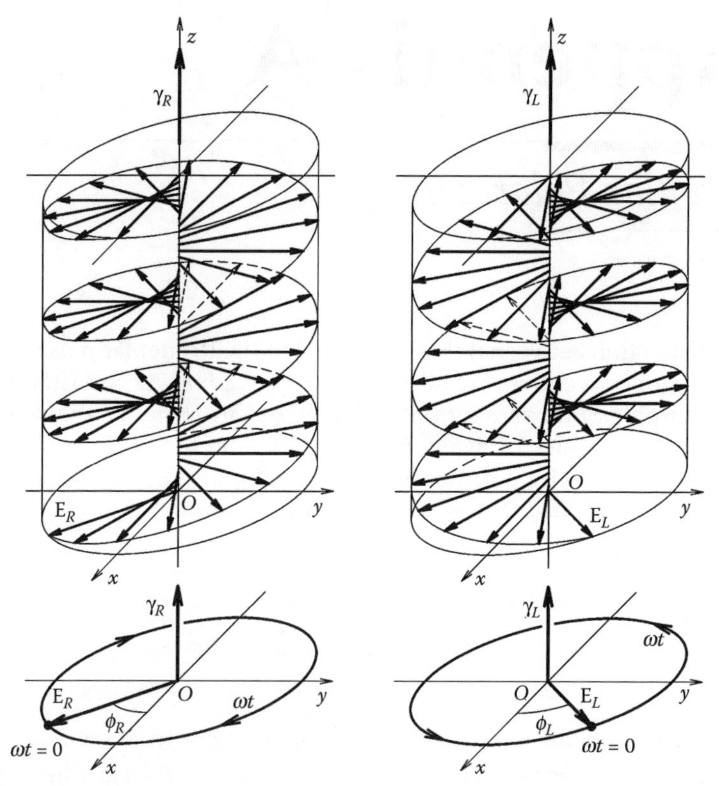

FIGURE A.1
Right (RCP) and left (LCP) circularly polarized waves propagating along the positive direction of the z axis with the electric fields $\widetilde{\boldsymbol{E}}_{\mathcal{R}}$ and $\widetilde{\boldsymbol{E}}_{\mathcal{L}}$, respectively. The propagation vectors for the RCP and LCP waves are denoted $\boldsymbol{\gamma}_{\mathcal{R}} = \hat{\boldsymbol{z}}\gamma_{\mathcal{R}}$ and $\boldsymbol{\gamma}_{\mathcal{L}} = \hat{\boldsymbol{z}}\gamma_{\mathcal{L}}$, respectively. In isotropic media, $\gamma_{\mathcal{L}} = \gamma_{\mathcal{L}}$. The upper part of the figure shows the spatial dependence at a fixed time, $t = 0$. The lower part of the figure shows the time dependence in a fixed point on the z axis, $z = 0$.

a nonabsorbing medium, considered in Figure A.2, $N_{\mathcal{R}} = n_{\mathcal{R}}$ and $N_{\mathcal{L}} = n_{\mathcal{L}}$ are the real quantities. The RCP and LCP waves propagate with different phase velocities $cn_{\mathcal{R}}^{-1} \neq cn_{\mathcal{L}}^{-1}$ without attenuation. The endpoints of the electric field vectors $\widetilde{\boldsymbol{E}}_{\mathcal{R}}$ and $\widetilde{\boldsymbol{E}}_{\mathcal{L}}$ of the RCP and LCP waves form right-handed and left-handed helices, respectively. The pitch of the RCP and LCP helices follow from

$$\lambda_{\mathcal{R}} = \frac{2\pi\, c}{n_{\mathcal{R}}\omega},$$

and

$$\lambda_{\mathcal{L}} = \frac{2\pi\, c}{n_{\mathcal{L}}\omega},$$

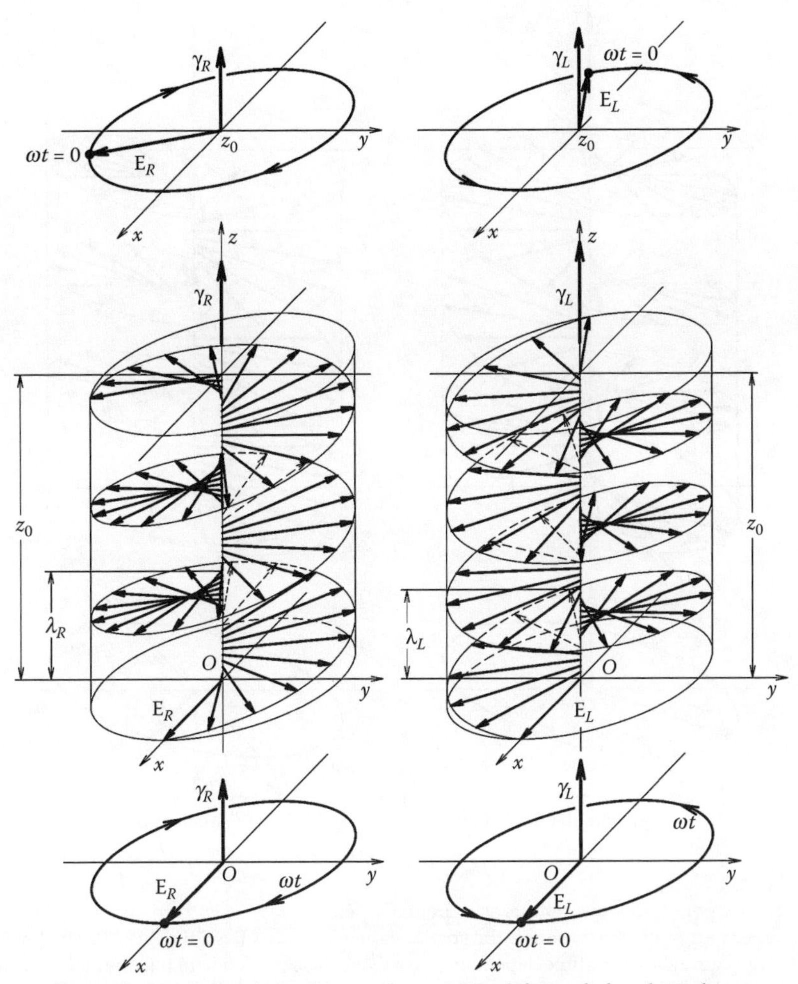

FIGURE A.2
Two opposite circularly polarized waves in a nonabsorbing medium propagate parallel to the magnetization, $\boldsymbol{M}$, with different real propagation vectors, $\gamma_{\mathcal{R}} \neq \gamma_{\mathcal{L}}$. The central part of the figure shows the spatial dependence at a fixed time, $t = 0$. The endpoint of the electric field vectors $\boldsymbol{E}_{\mathcal{R}}$ and $\boldsymbol{E}_{\mathcal{L}}$ of the RCP and LCP waves form right-handed and left-handed helices, respectively. The pitch of the RCP helix $\lambda_{\mathcal{R}}$ and that for the LCP helix $\lambda_{\mathcal{L}}$ are indicated. The lower part of the figure shows the time dependence in a plane $z = 0$. The superposition results in a linearly polarized wave of the zero azimuth. The upper part of the figure shows the time dependence in a plane $z = z_0$. The superposition results again in a linearly polarized wave with an azimuth in general different from that at $z = 0$. The figure shows the situation where $\gamma_{\mathcal{R}} < \gamma_{\mathcal{L}}$, corresponding to $n_{\mathcal{R}} < n_{\mathcal{L}}$, with the pitches $\lambda_{\mathcal{R}} > \lambda_{\mathcal{L}}$.

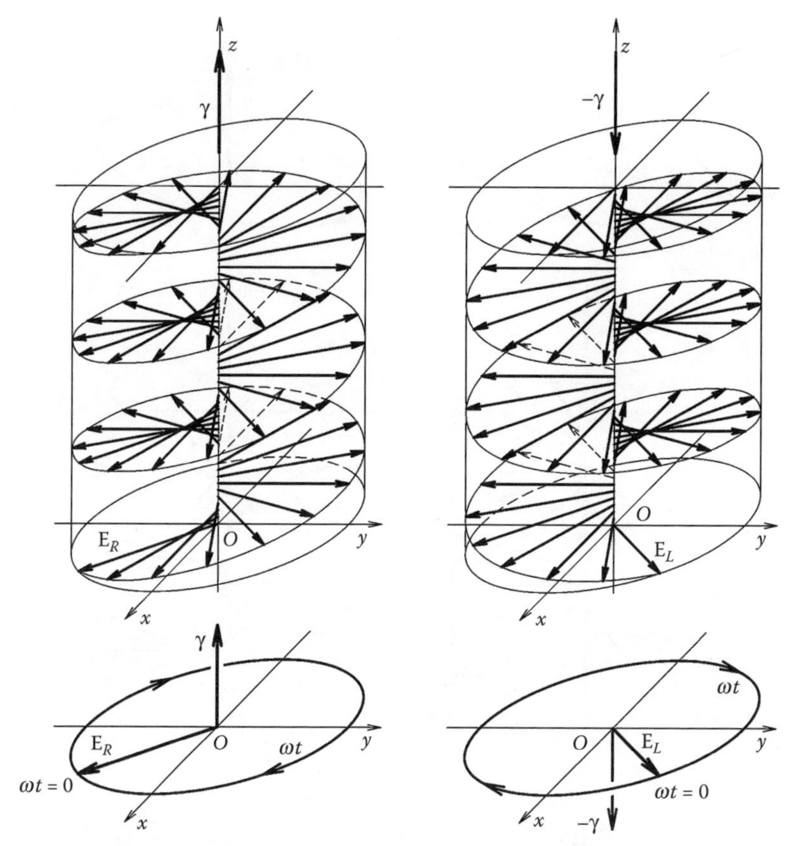

FIGURE A.3
A circularly polarized wave reflects as a circularly polarized wave with an opposite helicity. The upper part of the figure shows the spatial dependence at a fixed time, $t = 0$. The lower part of the figure shows the time dependence in a fixed plane perpendicular to the z axis, $z = 0$. The forward right circularly polarized wave is displayed on the left, whereas the retrograde left circularly polarized wave is displayed on the right.

respectively. In a plane $z = 0$, the superposition results in a linearly polarized wave of the zero azimuth with an electric field,

$$\boldsymbol{E}(t; 0) = \boldsymbol{E}_{\mathcal{R}}(t; 0) + \boldsymbol{E}_{\mathcal{L}}(t; 0) = \hat{\boldsymbol{x}} E(t; 0),$$

oscillating parallel to the x axis. In a plane $z = z_0 > 0$, the superposition of the CP wave fields again results in a linear polarization. Its azimuth θ is in general different from zero with the electric field

$$\boldsymbol{E}(t; z_0) = \boldsymbol{E}_{\mathcal{R}}(t; z_0) + \boldsymbol{E}_{\mathcal{L}}(t; z_0) = \left(\hat{\boldsymbol{x}} \cos\theta + \hat{\boldsymbol{y}} \sin\theta\right) E(t; z_0).$$

At an interface of two isotropic media, a right circularly polarized wave reflects at the normal light incidence as a left circularly polarized wave. In an analogous way, a left circularly polarized wave reflects as a right circularly polarized wave. Figure A.3 shows an RCP wave with the propagation vector $\boldsymbol{\gamma} = \hat{\mathbf{z}}\gamma$ travelling in the positive direction of the z axis. In a fixed time, the endpoints of the vectors representing $\boldsymbol{E}_{\mathcal{R}}$, the electric fields of the RCP wave, form a right handed helix in the space. Upon the reflection, the propagation vector transforms to an opposite one, $-\boldsymbol{\gamma} = -\hat{\mathbf{z}}\gamma$. The symmetry operation of a reflection in a plane normal to the z axis transforms the right-handed helix to a left-handed helix. The reflected wave is LCP; the endpoints of the vectors representing $\boldsymbol{E}_{\mathcal{L}}$, the electric fields of the LCP wave, form a left-handed helix in the space. Although the helicities of the RCP and LCP wave differ, the sense of rotation as time elapses, in a fixed plane normal to the propagation vectors, is the same for RCP and LCP waves. Both $\boldsymbol{E}_{\mathcal{R}}$ and $\boldsymbol{E}_{\mathcal{L}}$ rotate from the positive y axis to the positive x axis. In a medium magnetized parallel to the z axis, with the magnetization $\boldsymbol{M} = \hat{\mathbf{z}}\,M$, these two waves travel with the same propagation constant, either $\gamma_{+}(+\boldsymbol{M})$ or $\gamma_{-}(-\boldsymbol{M})$. When $\boldsymbol{M}$ is reversed to $-\boldsymbol{M}$, the propagation constant of both the forward and retrograde waves change respectively to either $\gamma_{-}(-\boldsymbol{M})$ or $\gamma_{+}(+\boldsymbol{M})$. However, in a medium displaying natural optical activity, the propagation constants of the two waves differ. We therefore observe $+\hat{\mathbf{z}}\gamma_{\pm}$ for the forward CP wave and $-\hat{\mathbf{z}}\gamma_{\mp}$ for the retrograde one.

Appendix B

Fresnel Formulae

The powerful 4 × 4 matrix formalism must cover simpler situations as special cases. We illustrate its use in the derivation of the classical reflection and transmission Fresnel formula for a planar interface between two linear isotropic homogenous media. In the formalism, the relation between the field in the medium (0) and the medium (1) close to the interface is represented by the matrix equation (3.77) and we can, therefore, write

$$\mathbf{E}^{(0)} = [\mathbf{D}^{(0)}]^{-1}\mathbf{D}^{(1)}\mathbf{E}^{(1)}. \tag{B.1}$$

Making use of Eq. (3.82a) and Eq. (3.82b), this relation can be expanded as

$$\begin{bmatrix} E_{0s}^{(i)} \\ E_{0s}^{(r)} \\ E_{0p}^{(i)} \\ E_{0p}^{(r)} \end{bmatrix} = (2N^{(0)}\cos\varphi^{(0)})^{-1} \begin{bmatrix} N^{(0)}\cos\varphi^{(0)} & 1 & 0 & 0 \\ N^{(0)}\cos\varphi^{(0)} & -1 & 0 & 0 \\ 0 & 0 & N^{(0)} & -\cos\varphi^{(0)} \\ 0 & 0 & N^{(0)} & \cos\varphi^{(0)} \end{bmatrix}$$
$$\times \begin{bmatrix} 1 & 1 & 0 & 0 \\ N^{(1)}\cos\varphi^{(1)} & -N^{(1)}\cos\varphi^{(1)} & 0 & 0 \\ 0 & 0 & \cos\varphi^{(1)} & \cos\varphi^{(1)} \\ 0 & 0 & -N^{(1)} & N^{(1)} \end{bmatrix} \begin{bmatrix} E_{0s}^{(t)} \\ 0 \\ E_{03}^{(t)} \\ 0 \end{bmatrix}. \tag{B.2}$$

Here, $E_{0s}^{(i)}$ and $E_{0p}^{(i)}$ are the incident s- and p-wave amplitudes, $E_{0s}^{(r)}$ and $E_{0p}^{(r)}$ denote reflected in the incident medium (0) s- and p-wave amplitudes, and $E_{0s}^{(t)}$ and $E_{0p}^{(t)}$ denote transmitted to the medium (1) s- and p-wave amplitudes. The $\mathbf{M}$ matrix takes a 2 × 2 block diagonal form, as there is no interaction between TE and TM polarizations (or no mode coupling) in isotropic media.

$$\begin{bmatrix} \mathrm{M}_{11} & \mathrm{M}_{12} \\ \mathrm{M}_{21} & \mathrm{M}_{22} \end{bmatrix} = (2N^{(0)}\cos\varphi^{(0)})^{-1}$$
$$\times \begin{bmatrix} N^{(0)}\cos\varphi^{(0)} + N^{(1)}\cos\varphi^{(1)} & N^{(0)}\cos\varphi^{(0)} - N^{(1)}\cos\varphi^{(1)} \\ N^{(0)}\cos\varphi^{(0)} - N^{(1)}\cos\varphi^{(1)} & N^{(0)}\cos\varphi^{(0)} + N^{(1)}\cos\varphi^{(1)} \end{bmatrix} \tag{B.3a}$$

$$\begin{bmatrix} M_{33} & M_{34} \\ M_{43} & M_{44} \end{bmatrix} = (2N^{(0)}\cos\varphi^{(0)})^{-1}$$
$$\times \begin{bmatrix} N^{(0)}\cos\varphi^{(1)} + N^{(1)}\cos\varphi^{(0)} & N^{(0)}\cos\varphi^{(1)} - N^{(1)}\cos\varphi^{(0)} \\ N^{(0)}\cos\varphi^{(1)} - N^{(1)}\cos\varphi^{(0)} & N^{(0)}\cos\varphi^{(1)} + N^{(1)}\cos\varphi^{(0)} \end{bmatrix}. \tag{B.3b}$$

We obtain for the reflection and transmission coefficients defined in Eqs. (3.70a) to (3.70h)

$$r_{12}^{(01)} = r_{ss}^{(01)} = \frac{M_{21}}{M_{11}} = \frac{N^{(0)}\cos\varphi^{(0)} - N^{(1)}\cos\varphi^{(1)}}{N^{(0)}\cos\varphi^{(0)} + N^{(1)}\cos\varphi^{(1)}} \tag{B.4a}$$
$$r_{14}^{(01)} = r_{32}^{(01)} = 0 \tag{B.4b}$$
$$r_{34}^{(01)} = r_{pp}^{(01)} = \frac{M_{43}}{M_{33}} = \frac{N^{(0)}\cos\varphi^{(1)} - N^{(1)}\cos\varphi^{(0)}}{N^{(0)}\cos\varphi^{(1)} + N^{(1)}\cos\varphi^{(0)}} \tag{B.4c}$$
$$t_{11}^{(01)} = t_{ss}^{(01)} = \frac{1}{M_{11}} = \frac{2N^{(0)}\cos\varphi^{(0)}}{N^{(0)}\cos\varphi^{(0)} + N^{(1)}\cos\varphi^{(1)}} \tag{B.4d}$$
$$t_{13}^{(01)} = t_{31}^{(01)} = 0 \tag{B.4e}$$
$$t_{33}^{(01)} = t_{pp}^{(01)} = \frac{1}{M_{33}} = \frac{2N^{(0)}\cos\varphi^{(0)}}{N^{(0)}\cos\varphi^{(1)} + N^{(1)}\cos\varphi^{(0)}}. \tag{B.4f}$$

Note that the subscripts 1, ..., 4 number the proper polarization modes (1-incident s, 2-reflected s, 3-incident p, and 4-reflected p). At normal wave incidence, $\varphi^{(0)} = 0$ according to Eq. (B.4a) and Eq. (B.4b), $r_{12}^{(01)} = r_{34}^{(01)}$ corresponding to $r_{ss}^{(01)} = r_{pp}^{(01)}$.

We have mentioned on page 147 that the total phase of the product $E_{0j}^{(n)} \boldsymbol{e}_j^{(n)}$ can be arbitrarily distributed between the scalar amplitudes $E_{0j}^{(n)}$ and the polarization vectors $\boldsymbol{e}_j^{(n)}$. The latter define the wave polarizations. The choice is a matter of convention [1] and was discussed in detail by Sokolov [3]. The phase relations are of importance in the multilayer magnetooptics. It is useful to consider the situation at an interface between linear isotropic homogeneous nonabsorbing media. We adopt the notation introduced by Wangsness [4] and define the interface normal $\hat{\boldsymbol{n}}$ pointing from the entrance medium (0) to the exit medium (1). The unit vector $\hat{\boldsymbol{\tau}}$ is common to the plane of incidence and to the plane of interface. Then, the plane of incidence is specified by the vector product $\hat{\boldsymbol{n}} \times \hat{\boldsymbol{\tau}}$. We choose for the real indices of refraction values $n^{(0)} = 1$ and $n^{(0)} = 1.5$ (the extinction coefficients are zero, $k^{(0)} = k^{(1)} = 0$) corresponding to a vacuum and a glass, respectively, and consider three angles of incidence, *i.e.*, $\varphi^{(i)} = 10°$, $\arg\tan 1.5 = 56, 3°$, and $80°$. Figure B.1 shows the orientation of electric field vector amplitudes, **E**, of the linearly polarized s and p proper polarization modes as computed from Eq. (B.4).

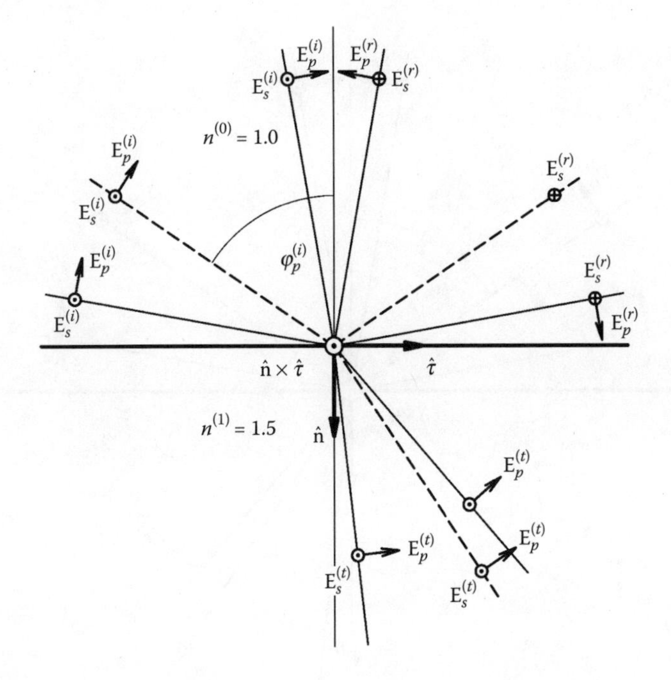

FIGURE B.1
Orientation of the electric field vector amplitudes of incident (i), reflected (r), and transmitted (t) waves polarized perpendicular (s) and parallel (p) to the plane of incidence at an interface between a vacuum and a glass with the real indices of refraction $n^{(0)} = 1$ and $n^{(0)} = 1.5$, respectively. The figure illustrates the situation close to the zero angle of incidence, at the Brewster angle, $\varphi_p^{(i)} = \arg\tan 1.5$ (where the scalar amplitude of the reflected wave polarized parallel to the plane of incidence crosses zero), and above the Brewster angle near the grazing incidence. The incident s polarizations have the orientation of the vector product $\hat{\boldsymbol{n}} \times \hat{\boldsymbol{\tau}}$. The reflected s polarizations have the opposite orientation given by $-\hat{\boldsymbol{n}} \times \hat{\boldsymbol{\tau}}$.

Figure B.2 illustrates the convention introduced by Fresnel's student Arago and applied in the classical ellipsometry of isotropic surfaces. The positive sense of the unit vectors of the electric fields polarized perpendicular to the plane of incidence (s polarized) is chosen congruent for the incident, reflected, and transmitted waves. Then, the choice of the positive sense for the unit vectors of the electric fields polarized parallel to the plane of incidence follows from a duality transformation applied to the Maxwell equations. In this special case, the Maxwell equations can be written in the form

$$\nabla \times \boldsymbol{H} = +\varepsilon \frac{\partial \boldsymbol{E}}{\partial t} \tag{B.5a}$$

$$\nabla \times \boldsymbol{E} = -\mu \frac{\partial \boldsymbol{H}}{\partial t}, \tag{B.5b}$$

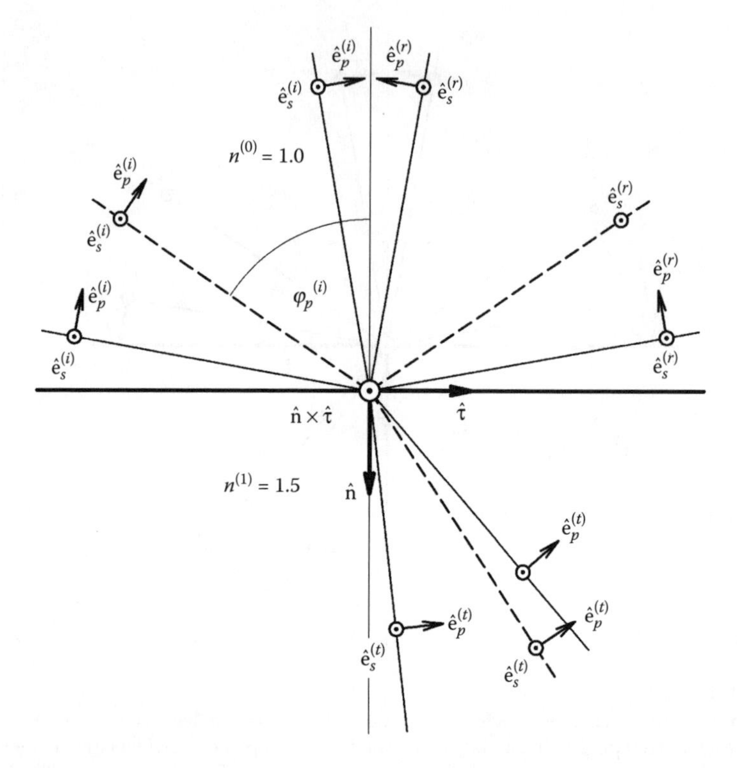

FIGURE B.2
Conventional orientation of the proper s and p polarizations of the electric fields of the incident, reflected, and transmitted waves applied in the classical ellipsometry (Arago convention). Both the incident and reflected s polarizations have the orientation of the vector product $\hat{n} \times \hat{\tau}$.

where ε and μ denote the medium permittivity and permeability, respectively. They are invariant with respect to the simultaneous exchange of $\varepsilon \leftrightarrow \mu$, $\boldsymbol{E} \rightarrow \mp \boldsymbol{H}$, and $\boldsymbol{H} \rightarrow \mp \boldsymbol{E}$. At normal wave incidence, the concept of the plane of incidence breaks down, and the waves polarized perpendicular and parallel to the plane of incidence cannot be physically distinguished. Starting from the congruent polarizations perpendicular to the plane of incidence, the duality transformation gives an antiparallel orientation for the incident and reflected polarizations of the incident and reflected waves polarized parallel to the plane of incidence. This introduces a π phase shift between the two polarizations compensated by an appropriate sign change of the field amplitudes $\mathrm{E}_{0j}^{(n)}$, which results in $r_{ss}^{(01)} = -r_{pp}^{(01)}$ for the Arago convention. This might be inconvenient and confusing in magnetooptics dealing predominantly with circularly polarized waves. We therefore adopt the original Fresnel convention illustrated in Figure B.3. The positive sense of

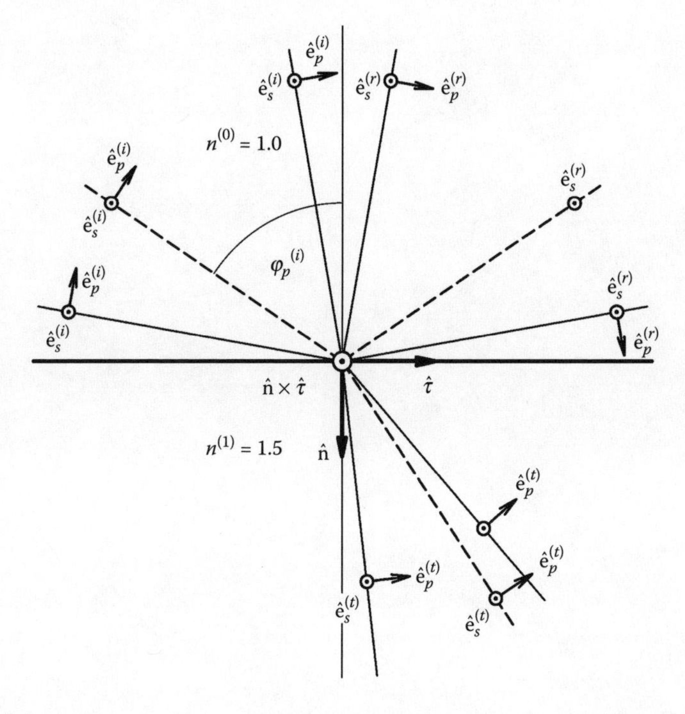

FIGURE B.3
Fresnel choice for the orientation of the proper s and p electric field polarizations of the incident (i), reflected (r), and transmitted (t) waves. Both the incident and reflected s polarizations have the orientation of the vector product $\hat{n} \times \hat{\tau}$.

the unit vectors of the electric fields polarized perpendicular to the plane of incidence (s polarized) remains congruent for the incident, reflected, and transmitted waves. At normal incidence, the p polarized electric fields of all the waves have the congruent orientation, contrary to the Arago convention of Figure B.2.

Appendix C

Isotropic Multilayers

Using the results of Section 3.4 and Section 3.5, we can construct **M** matrix for general isotropic multilayers. The relevant matrices are given in Eq. (3.60), Eq. (3.61), and Eq. (3.62). It is instructive to apply the 4×4 matrix formalism to the problems, the solutions of which are well known.

C.1 Film–Substrate System

We now obtain the classical Airy formulae for an ambient–film–substrate system, consisting of isotropic media, *i.e.*, the configuration corresponding to a Fabry–Perot resonator. The formulae give the global transmission and reflection coefficients of the system. We shall compute them separately for the TE and TM polarizations. As a simple application of the formalism, we also obtain the characteristic equation for a dielectric waveguide. The M matrix of the system has the form

$$\begin{aligned} \mathbf{M}^{(02)} &= [\mathbf{D}^{(0)}]^{-1}\mathbf{D}^{(1)}\mathbf{P}^{(1)}[\mathbf{D}^{(1)}]^{-1}\mathbf{D}^{(2)} \\ &= [\mathbf{D}^{(0)}]^{-1}\mathbf{S}^{(1)}\mathbf{D}^{(2)}. \end{aligned} \tag{C.1}$$

The characteristic matrix for an isotropic film follows from the product

$$\begin{aligned} \mathbf{S}^{(1)} &= \mathbf{D}^{(1)}\mathbf{P}^{(1)}[\mathbf{D}^{(1)}]^{-1} \\ &= \begin{pmatrix} \cos\beta^{(1)} & \mathrm{j}N_z^{(1)-1}\sin\beta^{(1)} & 0 & 0 \\ \mathrm{j}N_z^{(1)}\sin\beta^{(1)} & \cos\beta^{(1)} & 0 & 0 \\ 0 & 0 & \cos\beta^{(1)} & -\mathrm{j}\,\varepsilon_0^{(1)-1}N_z^{(1)}\sin\beta^{(1)} \\ 0 & 0 & -\mathrm{j}\varepsilon_0^{(1)}N_z^{(1)-1}\sin\beta^{(1)} & \cos\beta^{(1)} \end{pmatrix}. \end{aligned} \tag{C.2}$$

Here, $\varepsilon_0^{(1)} = N^{(1)2}$ and $N_z^{(1)} = N^{(1)}\cos\varphi^{(1)}$ $\varphi^{(1)}$ denotes the angle of refraction, and $\beta^{(1)} = \kappa^{(1)}d^{(1)}$ with $\kappa^{(1)} = \frac{\omega}{c}N_z^{(1)}$.

C.1.1 TE Polarization

We split the matrix multiplication into 2×2 blocks and write for the TE polarization

$$\begin{bmatrix} M_{11} & M_{12} \\ M_{21} & M_{22} \end{bmatrix} = (2N^{(0)} \cos\varphi^{(0)})^{-1} \begin{bmatrix} N^{(0)} \cos\varphi^{(0)} & 1 \\ N^{(0)} \cos\varphi^{(0)} & -1 \end{bmatrix}$$
$$\times \begin{bmatrix} \cos\kappa^{(1)} d^{(1)} & jN_z^{(1)-1} \sin\kappa^{(1)} d^{(1)} \\ jN_z^{(1)} \sin\kappa^{(1)} d^{(1)} & \cos\kappa^{(1)} d^{(1)} \end{bmatrix}$$
$$\times \begin{bmatrix} 1 & 1 \\ N^{(2)} \cos\varphi^{(2)} & -N^{(2)} \cos\varphi^{(2)} \end{bmatrix}. \tag{C.3}$$

The relevant matrix elements are

$$M_{11} = (2N^{(0)} \cos\varphi^{(0)})^{-1} \left[N^{(0)} \cos\varphi^{(0)} \cos\kappa^{(1)} d^{(1)} + jN_z^{(1)} \sin\kappa^{(1)} d^{(1)} \right.$$
$$\left. + \left(\cos\kappa^{(1)} d^{(1)} + jN^{(0)} \cos\varphi^{(0)} N_z^{(1)-1} \sin\kappa^{(1)} d^{(1)} \right) N^{(2)} \cos\varphi^{(2)} \right] \tag{C.4}$$

$$M_{21} = (2N^{(0)} \cos\varphi^{(0)})^{-1} \left[N^{(0)} \cos\varphi^{(0)} \cos\kappa^{(1)} d^{(1)} - jN_z^{(1)} \sin\kappa^{(1)} d^{(1)} \right.$$
$$\left. - \left(\cos\kappa^{(1)} d^{(1)} + jN^{(0)} \cos\varphi^{(0)} N_z^{(1)-1} \sin\kappa^{(1)} d^{(1)} \right) N^{(2)} \cos\varphi^{(2)} \right]. \tag{C.5}$$

We replace $\cos\kappa^{(1)} d^{(1)}$ and $\sin\kappa^{(1)} d^{(1)}$ by the exponential functions

$$M_{11} = (2N^{(0)} \cos\varphi^{(0)})^{-1} N_z^{(1)-1} \left[\left(N^{(0)} \cos\varphi^{(0)} + N^{(2)} \cos\varphi^{(2)} \right) N_z^{(1)} \cos\kappa^{(1)} d^{(1)} \right.$$
$$\left. + \left(N_z^{(1)2} + N^{(0)} \cos\varphi^{(0)} N^{(2)} \cos\varphi^{(2)} \right) j \sin\kappa^{(1)} d^{(1)} \right]$$
$$= \left(4N^{(0)} \cos\varphi^{(0)} N_z^{(1)} \right)^{-1} \left\{ \left[\left(N^{(0)} \cos\varphi^{(0)} + N^{(2)} \cos\varphi^{(2)} \right) N_z^{(1)} \right. \right.$$
$$\left. + N_z^{(1)2} + N^{(0)} N^{(2)} \cos\varphi^{(0)} \cos\varphi^{(2)} \right] \exp(j\kappa^{(1)} d^{(1)})$$
$$+ \left[\left(N^{(0)} \cos\varphi^{(0)} + N^{(2)} \cos\varphi^{(2)} \right) N_z^{(1)} - N_z^{(1)2} \right.$$
$$\left. \left. - N^{(0)} N^{(2)} \cos\varphi^{(0)} \cos\varphi^{(2)} \right] \exp(-j\kappa^{(1)} d^{(1)}) \right\}. \tag{C.6}$$

The matrix element M_{11} provides the TE transmission coefficient of the structure

$$t_{ss} = \frac{1}{M_{11}} = \frac{2N^{(0)} \cos\varphi^{(0)}}{N^{(0)} \cos\varphi^{(0)} + N_z^{(1)}} \frac{2N_z^{(1)}}{N_z^{(1)} + N^{(2)} \cos\varphi^{(2)}} \exp(-j\kappa^{(1)} d^{(1)})$$
$$\times \left[1 + \frac{N^{(0)} \cos\varphi^{(0)} - N_z^{(1)}}{N^{(0)} \cos\varphi^{(0)} + N_z^{(1)}} \frac{N_z^{(1)} - N^{(2)} \cos\varphi^{(2)}}{N_z^{(1)} + N^{(2)} \cos\varphi^{(2)}} \exp(-2j\kappa^{(1)} d^{(1)}) \right]^{-1} \tag{C.7}$$

or

$$t_{ss} = \frac{t_{ss}^{(01)} t_{ss}^{(12)} \exp(-j\kappa^{(1)} d^{(1)})}{1 + r_{ss}^{(01)} r_{ss}^{(12)} \exp(-2j\kappa^{(1)} d^{(1)})}. \tag{C.8}$$

In order to analyze the reflection case, we also need M_{21}

$$
\begin{aligned}
M_{21} = {} & \left(4N^{(0)}\cos\varphi^{(0)}N_z^{(1)}\right)^{-1}\exp(j\kappa^{(1)}d^{(1)}) \\
& \times\left\{\left[\left(N^{(0)}\cos\varphi^{(0)} - N^{(2)}\cos\varphi^{(2)}\right)N_z^{(1)} - N_z^{(1)2} + N^{(0)}\cos\varphi^{(0)}N^{(2)}\cos\varphi^{(2)}\right]\right. \\
& + \left[\left(N^{(0)}\cos\varphi^{(0)} - N^{(2)}\cos\varphi^{(2)}\right)N_z^{(1)} + N_z^{(1)2} + N^{(0)}\cos\varphi^{(0)}N^{(2)}\cos\varphi^{(2)}\right] \\
& \left.\times\exp(-j\kappa^{(1)}d^{(1)})\right\} \\
= {} & \left(4N^{(0)}\cos\varphi^{(0)}N_z^{(1)}\right)^{-1}\exp(j\kappa^{(1)}d^{(1)})\left(N^{(0)}\cos\varphi^{(0)} + N_z^{(1)}\right) \\
& \times\left(N_z^{(1)} + N^{(2)}\cos\varphi^{(2)}\right) \\
& \times\left[\frac{N^{(0)}\cos\varphi^{(0)} - N_z^{(1)}}{N^{(0)}\cos\varphi^{(0)} + N_z^{(1)}} + \frac{N_z^{(1)} - N^{(2)}\cos\varphi^{(2)}}{N_z^{(1)} + N^{(2)}\cos\varphi^{(2)}}\exp(-2j\kappa^{(1)}d^{(1)})\right].
\end{aligned}
\tag{C.9}
$$

The reflection coefficient for the TE polarization can be expressed as

$$
r_{ss} = \frac{M_{21}}{M_{11}} = \frac{r_{ss}^{(01)} + r_{ss}^{(12)}\exp(-2j\kappa^{(1)}d^{(1)})}{1 + r_{ss}^{(01)}r_{ss}^{(12)}\exp(-2j\kappa^{(1)}d^{(1)})}. \tag{C.10}
$$

The waveguiding condition follows from (3.76) and for the TE modes in this system becomes $M_{11} = 0$

$$
1 + \left(\frac{N^{(0)}\cos\varphi^{(0)} - N_z^{(1)}}{N^{(0)}\cos\varphi^{(0)} + N_z^{(1)}}\right)\left(\frac{N_z^{(1)} - N^{(2)}\cos\varphi^{(2)}}{N_z^{(1)} + N^{(2)}\cos\varphi^{(2)}}\right)\exp(-2j\kappa^{(1)}d^{(1)}) = 0 \tag{C.11}
$$

or, in a concise form,

$$
1 - r_{ss}^{(10)}r_{ss}^{(12)}\exp(-2j\kappa^{(1)}d^{(1)}) = 0. \tag{C.12}
$$

This condition can be fulfilled, provided the reflection coefficients $r_{ss}^{(01)} = -r_{ss}^{(10)}$ and $r_{ss}^{(12)}$ are of unit magnitudes, *i.e.*, provided they pertain to the total reflection case. In other words, this necessary condition for the existence of the guided modes requires that the zigzag angle of the wave in the film $\varphi^{(1)}$ is higher than the critical angles for the total reflection at both upper and lower, *i.e*,

$$
\sin\varphi^{(1)} > \frac{N^{(0)}}{N^{(1)}}, \tag{C.13a}
$$

$$
\sin\varphi^{(1)} > \frac{N^{(2)}}{N^{(1)}}. \tag{C.13b}
$$

They become

$$
r_{ss}^{(10)} = \frac{N^{(1)}\cos\varphi^{(1)} + j(N^{(1)2}\sin^2\varphi^{(1)} - N^{(0)2})^{1/2}}{N^{(1)}\cos\varphi^{(1)} - j(N^{(1)2}\sin^2\varphi^{(1)} - N^{(0)2})^{1/2}} = e^{2j\theta_{ss}^{(10)}}, \tag{C.14a}
$$

$$
r_{ss}^{(12)} = \frac{N^{(1)}\cos\varphi^{(1)} + j(N^{(1)2}\sin^2\varphi^{(1)} - N^{(2)2})^{1/2}}{N^{(1)}\cos\varphi^{(1)} - j(N^{(1)2}\sin^2\varphi^{(1)} - N^{(2)2})^{1/2}} = e^{2j\theta_{ss}^{(12)}}. \tag{C.14b}
$$

They can be expressed in an imaginary exponential form. The corresponding phase angles are given by

$$\theta_{ss}^{(10)} = \arctan\left[\frac{(N^{(1)2}\sin^2\varphi^{(1)} - N^{(0)2})^{1/2}}{N^{(1)}\cos\varphi^{(1)}}\right], \qquad (N^{(1)} > N^{(0)}), \tag{C.15a}$$

$$\theta_{ss}^{(12)} = \arctan\left[\frac{(N^{(1)2}\sin^2\varphi^{(1)} - N^{(2)2})^{1/2}}{N^{(1)}\cos\varphi^{(1)}}\right], \qquad (N^{(1)} > N^{(2)}). \tag{C.15b}$$

The equation (C.12) can be alternatively expressed as

$$\kappa^{(1)}d^{(1)} - \theta_{ss}^{(10)} - \theta_{ss}^{(12)} = \nu\pi, \tag{C.16}$$

where ν is an integer. We call

$$\kappa^{(1)} = \frac{\omega}{c}N^{(1)}\cos\varphi^{(1)} \tag{C.17a}$$

and

$$\frac{\omega}{c}N^{(1)}\sin\varphi^{(1)} \tag{C.17b}$$

the transverse and longitudinal propagation constants, respectively, and define the transverse attenuation constants in sandwiching media as follows

$$\gamma^{(0)} = \frac{\omega}{c}\left[\left(N^{(1)}\sin\varphi^{(1)}\right)^2 - N^{(0)2}\right]^{1/2}, \tag{C.18a}$$

$$\gamma^{(2)} = \frac{\omega}{c}\left[\left(N^{(1)}\sin\varphi^{(1)}\right)^2 - N^{(2)2}\right]^{1/2}. \tag{C.18b}$$

Using these definitions, the characteristic equation can be written in a more concise form [2]

$$\tan\kappa^{(1)}d^{(1)} = \frac{\kappa^{(1)}\left(\gamma^{(0)} + \gamma^{(2)}\right)}{\kappa^{(1)2} - \gamma^{(0)}\gamma^{(2)}}. \tag{C.19}$$

The transverse attenuation constants, $\gamma^{(0)}$ and $\gamma^{(2)}$, specify the penetration depth for the evanescent wave fields in the sandwiching media. For $\gamma^{(0)} \to 0$ or $\gamma^{(2)} \to 0$, *i.e.*, when the critical angles are approached from higher zigzag angles, the evanescent fields extend to infinity.

C.1.2 TM Polarization

We start again from the matrix representation (C.1). For the TM case, it is sufficient to consider lower diagonal 2×2 block

$$\begin{bmatrix} M_{33} & M_{34} \\ M_{43} & M_{44} \end{bmatrix} = (2N^{(0)}\cos\varphi^{(0)})^{-1} \begin{bmatrix} N^{(0)} & -\cos\varphi^{(0)} \\ N^{(0)} & \cos\varphi^{(0)} \end{bmatrix} \times \begin{bmatrix} \cos\kappa^{(1)}d^{(1)} & -j\,\varepsilon_0^{(1)-1} N_z^{(1)} \sin\kappa^{(1)}d^{(1)} \\ -j\varepsilon_0^{(1)} N_z^{(1)-1} \sin\kappa^{(1)}d^{(1)} & \cos\kappa^{(1)}d^{(1)} \end{bmatrix} \times \begin{bmatrix} \cos\varphi^{(2)} & \cos\varphi^{(2)} \\ -N^{(2)} & N^{(2)} \end{bmatrix}. \tag{C.20}$$

We find for M_{33} and M_{43}

$$M_{33} = (2N^{(0)}\cos\varphi^{(0)})^{-1}\big[\big(N^{(0)}\cos\kappa^{(1)}d^{(1)} + j\cos\varphi^{(0)}\varepsilon_0^{(1)} N_z^{(1)-1}\sin\kappa^{(1)}d^{(1)}\big) \times \cos\varphi^{(2)} + \big(\cos\varphi^{(0)}\cos\kappa^{(1)}d^{(1)} + jN^{(0)}\varepsilon_0^{(1)-1}N_z^{(1)}\sin\kappa^{(1)}d^{(1)}\big)N^{(2)}\big], \tag{C.21}$$

$$M_{43} = (2N^{(0)}\cos\varphi^{(0)})^{-1}\big[\big(N^{(0)}\cos\kappa^{(1)}d^{(1)} - j\cos\varphi^{(0)}\varepsilon_0^{(1)} N_z^{(1)-1}\sin\kappa^{(1)}d^{(1)}\big) \times \cos\varphi^{(2)} - \big(\cos\varphi^{(0)}\cos\kappa^{(1)}d^{(1)} - jN^{(0)}\varepsilon_0^{(1)-1}N_z^{(1)}\sin\kappa^{(1)}d^{(1)}\big)N^{(2)}\big]. \tag{C.22}$$

We find for the TM amplitude transmission and reflection coefficients, respectively

$$t_{pp} = \frac{1}{M_{33}} = \frac{t_{pp}^{(01)}t_{pp}^{(12)}\exp(-j\kappa^{(1)}d^{(1)})}{1 + r_{pp}^{(01)}r_{pp}^{(12)}\exp(-2j\kappa^{(1)}d^{(1)})} \tag{C.23}$$

$$r_{pp} = \frac{M_{43}}{M_{33}} = \frac{r_{pp}^{(01)} + r_{pp}^{(12)}\exp(-2j\kappa^{(1)}d^{(1)})}{1 + r_{pp}^{(01)}r_{pp}^{(12)}\exp(-2j\kappa^{(1)}d^{(1)})}. \tag{C.24}$$

According to Eq. (3.76), the waveguiding condition for the TM modes in this system becomes $M_{33} = 0$. From Eq. (C.21)

$$N^{(0)}\cos\varphi^{(2)}\cos\kappa^{(1)}d^{(1)} + j\cos\varphi^{(0)}\varepsilon_0^{(1)}N_z^{(1)-1}\sin\kappa^{(1)}d^{(1)}\cos\varphi^{(2)} + jN^{(0)}\varepsilon_0^{(1)-1}N_z^{(1)}\sin\kappa^{(1)}d^{(1)}N^{(2)} + \cos\varphi^{(0)}\cos\kappa^{(1)}d^{(1)}N^{(2)} = 0 \tag{C.25}$$

We denote $\varepsilon_0^{(1)}N_z^{(1)-1} = N^{(1)}\cos\varphi^{(1)-1}$ and obtain the characteristic equation

$$1 + \left(\frac{N^{(0)}\cos\varphi^{(1)} - N^{(1)}\cos\varphi^{(0)}}{N^{(0)}\cos\varphi^{(1)} + N^{(1)}\cos\varphi^{(0)}}\right)\left(\frac{N^{(1)}\cos\varphi^{(2)} - N^{(2)}\cos\varphi^{(1)}}{N^{(1)}\cos\varphi^{(2)} + N^{(2)}\cos\varphi^{(1)}}\right) \times \exp(-2j\kappa^{(1)}d^{(1)}) = 0 \tag{C.26}$$

or in a concise form analogous to Eq. (C.12)

$$1 - r_{pp}^{(10)} r_{pp}^{(12)} \exp(-2\mathrm{i}\kappa^{(1)} d^{(1)}) = 0. \tag{C.27}$$

Above the critical angles for the total reflection, the TM reflection coefficients can be written as

$$r_{pp}^{(10)} = -\frac{N^{(0)2}\cos\varphi^{(1)} + \mathrm{j}N^{(1)}\left(N^{(1)2}\sin^2\varphi^{(1)} - N^{(0)2}\right)^{1/2}}{N^{(0)2}\cos\varphi^{(1)} - \mathrm{j}N^{(1)}\left(N^{(1)2}\sin^2\varphi^{(1)} - N^{(0)2}\right)^{1/2}} = -\mathrm{e}^{2\mathrm{j}\theta_{pp}^{(10)}}, \tag{C.28a}$$

$$r_{pp}^{(12)} = -\frac{N^{(2)2}\cos\varphi^{(1)} + \mathrm{j}N^{(1)}\left(N^{(1)2}\sin^2\varphi^{(1)} - N^{(2)2}\right)^{1/2}}{N^{(2)2}\cos\varphi^{(1)} - \mathrm{j}N^{(1)}\left(N^{(1)2}\sin^2\varphi^{(1)} - N^{(2)2}\right)^{1/2}} = -\mathrm{e}^{2\mathrm{j}\theta_{pp}^{(12)}}. \tag{C.28b}$$

The reflection coefficients are now of unit magnitude and can be expressed in an imaginary exponential form. The corresponding phase angles are given by

$$\theta_{pp}^{(10)} = \arctan\left[\frac{N^{(1)2}}{N^{(0)2}} \frac{(N^{(1)2}\sin^2\varphi^{(1)} - N^{(0)2})^{1/2}}{N^{(1)}\cos\varphi^{(1)}}\right], \quad (N^{(1)} > N^{(0)}), \tag{C.29a}$$

$$\theta_{pp}^{(12)} = \arctan\left[\frac{N^{(1)2}}{N^{(2)2}} \frac{(N^{(1)2}\sin^2\varphi^{(1)} - N^{(2)2})^{1/2}}{N^{(1)}\cos\varphi^{(1)}}\right], \quad (N^{(1)} > N^{(2)}). \tag{C.29b}$$

The characteristic equation for the TM polarization is obtained in a way similar to Eq. (C.16)

$$\kappa^{(1)} d^{(1)} - \theta_{pp}^{(10)} - \theta_{pp}^{(12)} = \nu\pi. \tag{C.30}$$

This transverse resonance condition may be explicitly expressed as

$$\frac{\omega}{c} d^{(1)} N^{(1)} \cos\varphi^{(1)} - \arctan\left[\frac{N^{(1)2}}{N^{(0)2}} \frac{(N^{(1)2}\sin^2\varphi^{(1)} - N^{(0)2})^{1/2}}{N^{(1)}\cos\varphi^{(1)}}\right] - \arctan\left[\frac{N^{(1)2}}{N^{(2)2}} \frac{\left(N^{(1)2}\sin\varphi^{(1)2} - N^{(2)2}\right)^{1/2}}{N^{(1)}\cos\varphi^{(1)}}\right] = \nu\pi. \tag{C.31}$$

In terms of the transverse propagation and attenuation constants, given by Eq. (C.17a) and Eq. (C.18), this takes the form [2]

$$\tan\kappa^{(1)} d^{(1)} = \frac{\dfrac{\kappa^{(1)}}{N^{(1)2}}\left(\dfrac{\gamma^{(0)}}{N^{(0)2}} + \dfrac{\gamma^{(2)}}{N^{(2)2}}\right)}{\dfrac{\kappa^{(1)2}}{N^{(1)4}} - \dfrac{\gamma^{(0)}}{N^{(0)2}} \dfrac{\gamma^{(2)}}{N^{(2)2}}}. \tag{C.32}$$

Again, the transverse attenuation constants, $\gamma^{(0)}$ and $\gamma^{(2)}$, specify the penetration depth for the evanescent wave fields in the sandwiching media.

As in the TE case, for $\gamma^{(0)} \rightarrow 0$ or $\gamma^{(2)} \rightarrow 0$, *i.e.*, when the critical angles are approached from higher zigzag angles, the evanescent fields extend to infinity.[1]

C.2 Two-Layer System

We consider the planar structure involving four media and consisting of two films sandwiched between a semi-infinite cover and a semi-infinite substrate. The structure is represented by the matrix product

$$\begin{aligned}\mathbf{M}^{(03)} &= [\mathbf{D}^{(0)}]^{-1}\mathbf{D}^{(1)}\mathbf{P}^{(1)}[\mathbf{D}^{(1)}]^{-1}\mathbf{D}^{(2)}\mathbf{P}^{(2)}[\mathbf{D}^{(2)}]^{-1}\mathbf{D}^{(3)} \\ &= [\mathbf{D}^{(0)}]^{-1}\mathbf{S}^{(1)}\mathbf{S}^{(2)}\mathbf{D}^{(3)}. \end{aligned} \tag{C.33}$$

In view of their independence, the calculations for the TE and TM modes can be carried out separately using the 2×2 submatrices.

C.2.1 TE Polarization

For the TE case, we have

$$\begin{aligned}
\begin{bmatrix} \mathrm{M}_{11} & \mathrm{M}_{12} \\ \mathrm{M}_{21} & \mathrm{M}_{22} \end{bmatrix} &= (2N^{(0)}\cos\varphi^{(0)})^{-1} \begin{bmatrix} N^{(0)}\cos\varphi^{(0)} & 1 \\ N^{(0)}\cos\varphi^{(0)} & -1 \end{bmatrix} \\
&\times \begin{bmatrix} \cos\kappa^{(1)}d^{(1)} & \mathrm{j}N_z^{(1)-1}\sin\kappa^{(1)}d^{(1)} \\ \mathrm{j}N_z^{(1)}\sin\kappa^{(1)}d^{(1)} & \cos\kappa^{(1)}d^{(1)} \end{bmatrix} \begin{bmatrix} \cos\kappa^{(2)}d^{(2)} & \mathrm{j}N_z^{(2)-1}\sin\kappa^{(2)}d^{(2)} \\ \mathrm{j}N_z^{(2)}\sin\kappa^{(2)}d^{(2)} & \cos\kappa^{(2)}d^{(2)} \end{bmatrix} \\
&\times \begin{bmatrix} 1 & 1 \\ N^{(3)}\cos\varphi^{(3)} & -N^{(3)}\cos\varphi^{(3)} \end{bmatrix} \\
&= \left(2N_z^{(0)}\right)^{-1} \begin{bmatrix} N_z^{(0)}\cos\kappa^{(1)}d^{(1)} + \mathrm{j}N_z^{(1)}\sin\kappa^{(1)}d^{(1)} & \mathrm{j}N_z^{(0)}N_z^{(1)-1}\sin\kappa^{(1)}d^{(1)} + \cos\kappa^{(1)}d^{(1)} \\ N_z^{(0)}\cos\kappa^{(1)}d^{(1)} - \mathrm{j}N_z^{(1)}\sin\kappa^{(1)}d^{(1)} & \mathrm{j}N_z^{(0)}N_z^{(1)-1}\sin\kappa^{(1)}d^{(1)} - \cos\kappa^{(1)}d^{(1)} \end{bmatrix} \\
&\times \begin{bmatrix} \cos\kappa^{(2)}d^{(2)} + \mathrm{j}N_z^{(2)-1}N_z^{(3)}\sin\kappa^{(1)}d^{(1)} & \cos\kappa^{(2)}d^{(2)} - \mathrm{j}N_z^{(2)-1}N_z^{(3)}\sin\kappa^{(2)}d^{(2)} \\ \mathrm{j}N_z^{(2)}\sin\kappa^{(2)}d^{(2)} + N_z^{(3)}\cos\kappa^{(2)}d^{(2)} & \mathrm{j}N_z^{(2)}\sin\kappa^{(2)}d^{(2)} - N_z^{(3)}\cos\kappa^{(2)}d^{(2)} \end{bmatrix},
\end{aligned} \tag{C.34}$$

where we have set $N^{(0)}\cos\varphi^{(0)} = N_z^{(0)}$. We obtain for the relevant element M_{11}

[1] An interesting aspect of Eq. (C.32) is the existence of the solution for $d^{(1)} \rightarrow 0$, provided the permittivity $N^{(2)2}$ is approximately real and negative (plasmon condition).

$$\begin{aligned}
M_{11} = {} & \left[(2N_z^{(0)} 2N_z^{(1)} 2N_z^{(2)} \exp\left(-j\kappa^{(1)} d^{(1)} - j\kappa^{(2)} d^{(2)}\right)\right]^{-1} \\
& \times \left\{ \left(N_z^{(0)} + N_z^{(1)}\right) \left(N_z^{(1)} + N_z^{(2)}\right) \left(N_z^{(2)} + N_z^{(3)}\right) \right. \\
& + \left(N_z^{(0)} + N_z^{(1)}\right) \left(N_z^{(1)} - N_z^{(2)}\right) \left(N_z^{(2)} - N_z^{(3)}\right) \exp(-2j\kappa^{(2)} d^{(2)}) \\
& + \left(N_z^{(0)} - N_z^{(1)}\right) \left(N_z^{(1)} - N_z^{(2)}\right) \left(N_z^{(2)} + N_z^{(3)}\right) \exp(-2j\kappa^{(1)} d^{(1)}) \\
& + \left(N_z^{(0)} - N_z^{(1)}\right) \left(N_z^{(1)} + N_z^{(2)}\right) \left(N_z^{(2)} - N_z^{(3)}\right) \\
& \left. \times \exp\left[-2j\left(\kappa^{(1)} d^{(1)} + \kappa^{(2)} d^{(2)}\right)\right] \right\}.
\end{aligned} \tag{C.35}$$

The transmission coefficient of the structure is given by

$$\begin{aligned}
t_{ss}^{(03)} &= \frac{1}{M_{11}} \\
&= \frac{t_{ss}^{(01)} \left[t_{ss}^{(12)} t_{ss}^{(23)} \exp(-j\kappa^{(2)} d^{(2)})\right] \exp(-j\kappa^{(1)} d^{(1)})}{1 + r_{ss}^{(12)} r_{ss}^{(23)} \exp\left(-2j\kappa^{(2)} d^{(2)}\right) + r_{ss}^{(01)} \left[r_{ss}^{(12)} + r_{ss}^{(23)} \exp(-2j\kappa^{(2)} d^{(2)})\right] \exp(-2j\kappa^{(1)} d^{(1)})}.
\end{aligned} \tag{C.36}$$

This can be expressed in the form

$$t_{ss}^{(03)} = \frac{t_{ss}^{(01)} \left[\dfrac{t_{ss}^{(12)} t_{ss}^{(23)} \exp(-j\kappa^{(2)} d^{(2)})}{1 + r_{ss}^{(12)} r_{ss}^{(23)} \exp(-2j\kappa^{(2)} d^{(2)})}\right] \exp(-j\kappa^{(1)} d^{(1)})}{1 + r_{ss}^{(01)} \left[\dfrac{r_{ss}^{(12)} + r_{ss}^{(23)} \exp(-2j\kappa^{(2)} d^{(2)})}{1 + r_{ss}^{(12)} r_{ss}^{(23)} \exp(-2j\kappa^{(2)} d^{(2)})}\right] \exp(-2j\kappa^{(1)} d^{(1)})}. \tag{C.37}$$

We obtain for the relevant element M_{21}

$$\begin{aligned}
M_{21} = {} & \left[(2N_z^{(0)} 2N_z^{(1)} 2N_z^{(2)} \exp(-j\kappa^{(1)} d^{(1)} - j\kappa^{(2)} d^{(2)})\right]^{-1} \\
& \times \left\{ \left(N_z^{(0)} - N_z^{(1)}\right) \left(N_z^{(1)} + N_z^{(2)}\right) \left(N_z^{(2)} + N_z^{(3)}\right) \right. \\
& + \left(N_z^{(0)} - N_z^{(1)}\right) \left(N_z^{(1)} - N_z^{(2)}\right) \left(N_z^{(2)} - N_z^{(3)}\right) \exp(-2j\kappa^{(2)} d^{(2)}) \\
& + \left(N_z^{(0)} + N_z^{(1)}\right) \left(N_z^{(1)} - N_z^{(2)}\right) \left(N_z^{(2)} + N_z^{(3)}\right) \exp(-2j\kappa^{(1)} d^{(1)}) \\
& + \left(N_z^{(0)} + N_z^{(1)}\right) \left(N_z^{(1)} + N_z^{(2)}\right) \left(N_z^{(2)} - N_z^{(3)}\right) \\
& \left. \times \exp\left[-2j\left(\kappa^{(1)} d^{(1)} + \kappa^{(2)} d^{(2)}\right)\right] \right\}.
\end{aligned} \tag{C.38}$$

The reflection coefficient of the structure is given by

$$\begin{aligned}
r_{ss}^{(03)} &= \frac{M_{21}}{M_{11}} \\
&= \frac{r_{ss}^{(01)} \left[1 + r_{ss}^{(12)} r_{ss}^{(23)} \exp(-2j\kappa^{(2)} d^{(2)})\right] + \left[r_{ss}^{(12)} + r_{ss}^{(23)} \exp(-2j\kappa^{(2)} d^{(2)})\right] \exp(-2j\kappa^{(1)} d^{(1)})}{1 + r_{ss}^{(12)} r_{ss}^{(23)} \exp(-2j\kappa^{(2)} d^{(2)}) + r_{ss}^{(01)} \left[r_{ss}^{(12)} + r_{ss}^{(23)} \exp(-2j\kappa^{(2)} d^{(2)})\right] \exp(-2j\kappa^{(1)} d^{(1)})}.
\end{aligned} \tag{C.39}$$

and after rearranging

$$r_{ss}^{(03)} = \frac{r_{ss}^{(01)} + \left[\dfrac{r_{ss}^{(12)} + r_{ss}^{(23)} \exp(-2\mathrm{j}\kappa^{(2)} d^{(2)})}{1 + r_{ss}^{(12)} r_{ss}^{(23)} \exp(-2\mathrm{j}\kappa^{(2)} d^{(2)})}\right] \exp(-2\mathrm{j}\kappa^{(1)} d^{(1)})}{1 + r_{ss}^{(01)} \left[\dfrac{r_{ss}^{(12)} + r_{ss}^{(23)} \exp(-2\mathrm{j}\kappa^{(2)} d^{(2)})}{1 + r_{ss}^{(12)} r_{ss}^{(23)} \exp(-2\mathrm{j}\kappa^{(2)} d^{(2)})}\right] \exp(-2\mathrm{j}\kappa^{(1)} d^{(1)})}. \tag{C.40}$$

The comparison with the formulae for the three media structure illustrates the recursive approach for the calculation of more complicated systems.

C.2.2 TM Polarization

The structure of the expressions for the TM case is the same:

$$\begin{aligned} t_{pp}^{(03)} &= \frac{1}{\mathrm{M}_{33}} \\ &= \frac{t_{pp}^{(01)} \left[\dfrac{t_{pp}^{(12)} t_{pp}^{(23)} \exp(-\mathrm{j}\kappa^{(2)} d^{(2)})}{1 + r_{pp}^{(12)} r_{pp}^{(23)} \exp(-2\mathrm{j}\kappa^{(2)} d^{(2)})}\right] \exp(-\mathrm{j}\kappa^{(1)} d^{(1)})}{1 + r_{pp}^{(01)} \left[\dfrac{r_{pp}^{(12)} + r_{pp}^{(23)} \exp(-2\mathrm{j}\kappa^{(2)} d^{(2)})}{1 + r_{pp}^{(12)} r_{pp}^{(23)} \exp(-2\mathrm{i}\kappa^{(2)} d^{(2)})}\right] \exp(-2\mathrm{j}\kappa^{(1)} d^{(1)})} \end{aligned} \tag{C.41}$$

$$\begin{aligned} r_{pp}^{(03)} &= \frac{\mathrm{M}_{43}}{\mathrm{M}_{33}} \\ &= \frac{r_{pp}^{(01)} + \left[\dfrac{r_{pp}^{(12)} + r_{pp}^{(23)} \exp(-2\mathrm{j}\kappa^{(2)} d^{(2)})}{1 + r_{pp}^{(12)} r_{pp}^{(23)} \exp(-2\mathrm{j}\kappa^{(2)} d^{(2)})}\right] \exp(-2\mathrm{j}\kappa^{(1)} d^{(1)})}{1 + r_{pp}^{(01)} \left[\dfrac{r_{pp}^{(12)} + r_{pp}^{(23)} \exp(-2\mathrm{j}\kappa^{(2)} d^{(2)})}{1 + r_{pp}^{(12)} r_{pp}^{(23)} \exp(-2\mathrm{j}\kappa^{(2)} d^{(2)})}\right] \exp(-2\mathrm{j}\kappa^{(1)} d^{(1)})}. \end{aligned} \tag{C.42}$$

C.2.3 Waveguide

The waveguiding conditions for TE and TM modes are given by $\mathrm{M}_{11} = 0$ or $\mathrm{M}_{33} = 0$, respectively, *i.e.*,

$$\begin{aligned} &1 + r_{ss}^{(12)} r_{ss}^{(23)} \exp(-2\mathrm{j}\kappa^{(2)} d^{(2)}) + r_{ss}^{(01)} \left[r_{ss}^{(12)} + r_{ss}^{(23)} \exp(-2\mathrm{j}\kappa^{(2)} d^{(2)})\right] \\ &\quad \times \exp(-2\mathrm{j}\kappa^{(1)} d^{(1)}) = 0, \end{aligned} \tag{C.43}$$

$$\begin{aligned} &1 + r_{pp}^{(12)} r_{pp}^{(23)} \exp(-2\mathrm{j}\kappa^{(2)} d^{(2)}) + r_{pp}^{(01)} \left[r_{pp}^{(12)} + r_{pp}^{(23)} \exp(-2\mathrm{j}\kappa^{(2)} d^{(2)})\right] \\ &\quad \times \exp(-2\mathrm{j}\kappa^{(1)} d^{(1)}) = 0. \end{aligned} \tag{C.44}$$

The transverse propagation constants in the layers are defined as

$$\kappa^{(1)} = \frac{\omega}{c}\left(\varepsilon_0^{(1)} - N_y^2\right)^{1/2} = \frac{\omega}{c}\left(\varepsilon_0^{(1)} - N_y^2\right)^{1/2} \tag{C.45}$$

$$\kappa^{(2)} = \frac{\omega}{c}\left(\varepsilon_0^{(2)} - N_y^2\right)^{1/2}. \tag{C.46}$$

Let us assume that $\varepsilon_0^{(1)} > \varepsilon_0^{(2)}$. The guiding becomes restricted to a single layer, *e.g.*, to the layer (1), provided the transverse propagation constant in the layer 1 satisfies

$$\kappa^{(1)2} = \frac{\omega^2}{c^2}\left(\varepsilon_0^{(1)} - N_y^2\right) > 0, \tag{C.47}$$

while the transverse attenuation constant in the layer (2) satisfies

$$\gamma^{(2)2} = -\kappa^{(2)2} = \frac{\omega^2}{c^2}(N_y^2 - \varepsilon_0^{(2)}) > 0. \tag{C.48}$$

We, of course, assume that in the sandwiching media (0) and (3), the waves are evanescent, *i.e.*,

$$\gamma^{(0)2} = -\kappa^{(0)2} = \frac{\omega^2}{c^2}(N_y^2 - \varepsilon_0^{(0)}) > 0 \tag{C.49}$$

and

$$\gamma^{(3)2} = -\kappa^{(3)2} = \frac{\omega^2}{c^2}(N_y^2 - \varepsilon_0^{(3)}) > 0. \tag{C.50}$$

Appendix D

Single Layer at Polar Magnetization

In Section 4.3, we considered the transmission and reflection response of a multilayer at polar magnetization and normal light incidence. A considerable simplification of the problem results from the restriction to the effect of zero and of first order in magnetization. Here, we apply the procedure to simple systems, *i.e.*, a single interface between two half spaces, an isotropic one and a magnetic one, and a single magnetic layer sandwiched between two isotropic half spaces. The purpose is to make the understanding of the approach and the introduced notation easier.

D.1 Single Interface

We have obtained the **M** matrix for a single interface in the circularly polarized representation in Section 4.2.1. We start from the **M** matrix given in Eq. (4.43) and the corresponding Fresnel reflection and transmission coefficients for circularly polarized waves given in Eq. (4.44). In the absence of magnetization ($\boldsymbol{M} = 0$), where $M_{11} = M_{33}$, $M_{21} = M_{43}$, and $N_{+}^{(1)} = N_{-}^{(1)} = N^{(1)}$, these simply reduce to the Fresnel coefficients for an interface between isotropic media. The Fresnel coefficients for this special case of normal incidence are valid for any pair of the orthogonally polarized proper polarization modes. Here, we start from CP modes

$$r^{(01)} = \frac{N^{(0)} - N^{(1)}}{N^{(0)} + N^{(1)}}, \tag{D.1a}$$

and

$$t^{(01)} = \frac{2N^{(0)}}{N^{(0)} + N^{(1)}} = 1 + r^{(01)}. \tag{D.1b}$$

The elements of the Cartesian Jones transmission and reflection matrices, defined, *e.g.*, in Eq. (1.162) and Eq. (1.163),

$$\begin{pmatrix} t_{xx}^{(01)} & t_{xy}^{(01)} \\ t_{yx}^{(01)} & t_{yy}^{(01)} \end{pmatrix} \quad \text{and} \quad \begin{pmatrix} r_{xx}^{(01)} & r_{xy}^{(01)} \\ r_{yx}^{(01)} & r_{yy}^{(01)} \end{pmatrix}, \tag{D.2}$$

follow from Eqs. (4.98)

$$t_{xx}^{(01)} = t_{yy}^{(01)} = t^{(01)} = \frac{1}{\mathrm{M}_{11}^{(\mathrm{CP})}} = \frac{2N^{(0)}}{N^{(0)} + N^{(1)}}, \tag{D.3a}$$

$$\begin{aligned} t_{yx}^{(01)} &= -t_{xy}^{(01)} = \mathrm{j}\Delta t_{xx}^{(01)} = \mathrm{j}\Delta\left(\frac{1}{\mathrm{M}_{11}^{(\mathrm{CP})}}\right) \\ &= \mathrm{j}\frac{\mathrm{d}\,t^{(01)}}{\mathrm{d}\,N^{(1)}}\Delta N^{(1)} = -\mathrm{j}\frac{\Delta N^{(1)}}{2N^{(1)}}\frac{4N^{(0)}N^{(1)}}{(N^{(0)}+N^{(1)})^2} \\ &= -\mathrm{j}\frac{\Delta N^{(1)}}{2N^{(1)}}(1 - r^{(01)2}), \end{aligned} \tag{D.3b}$$

$$r_{xx}^{(01)} = r_{yy}^{(01)} = r^{(01)} = \frac{\mathrm{M}_{21}^{(\mathrm{CP})}}{\mathrm{M}_{11}^{(\mathrm{CP})}} = \frac{N^{(0)} - N^{(1)}}{N^{(0)} + N^{(1)}}, \tag{D.3c}$$

$$\begin{aligned} r_{yx}^{(01)} &= -r_{xy}^{(01)} = \mathrm{j}\Delta r_{xx}^{(01)} = \mathrm{j}\Delta\left(\frac{\mathrm{M}_{21}^{(\mathrm{CP})}}{\mathrm{M}_{11}^{(\mathrm{CP})}}\right) \\ &= \mathrm{j}\frac{\mathrm{d}\,r^{(01)}}{\mathrm{d}\,N^{(1)}}\Delta N^{(1)} = -\mathrm{j}\frac{\Delta N^{(1)}}{2N^{(1)}}\frac{4N^{(0)}N^{(1)}}{(N^{(0)}+N^{(1)})^2} \\ &= -\mathrm{j}\frac{\Delta N^{(1)}}{2N^{(1)}}(1 - r^{(01)2}), \end{aligned} \tag{D.3d}$$

where

$$\frac{\mathrm{d}\,t^{(01)}}{\mathrm{d}\,N^{(1)}} = \frac{\mathrm{d}(1 + r^{(01)})}{\mathrm{d}\,N^{(1)}} = \frac{-2N^{(0)}}{(N^{(0)}+N^{(1)})^2} = -\frac{1}{2N^{(1)}}(1 - r^{(01)2}). \tag{D.4}$$

We can characterize the transformations of the polarization states with the complex numbers in the Cartesian representation. For incident waves of zero azimuth and zero ellipticity, we obtain in transmission

$$\begin{aligned} \chi_t^{(01)} &= \frac{t_{yx}^{(01)}}{t_{xx}^{(01)}} = -\mathrm{j}\frac{\Delta N^{(1)}}{2N^{(1)}}\frac{1 - r^{(01)2}}{1 + r^{(01)}} = \frac{-\mathrm{j}\Delta N^{(1)}}{N^{(0)}+N^{(1)}}\frac{1 - r^{(01)2}}{1 + r^{(01)}} \\ &= \frac{\varepsilon_{xy}^{(1)}}{2N^{(1)}(N^{(0)}+N^{(1)})}, \end{aligned} \tag{D.5}$$

and in reflection

$$\begin{aligned} \chi_r^{(01)} &= \frac{r_{yx}^{(01)}}{r_{xx}^{(01)}} = -\mathrm{j}\frac{\Delta N^{(1)}}{2N^{(1)}}\frac{1 - r^{(01)2}}{r^{(01)}} \\ &= \frac{-\mathrm{j}\Delta N^{(1)}}{2N^{(1)}}\frac{4N^{(0)}N^{(1)}}{(N^{(0)}+N^{(1)})^2}\frac{N^{(0)} - N^{(1)}}{N^{(0)} + N^{(1)}} = \frac{-2\mathrm{j}N^{(0)}\Delta N^{(1)}}{N^{(0)2} - N^{(1)2}} \\ &= \frac{\varepsilon_{xy}^{(1)}\varepsilon_{xx}^{(0)1/2}}{\varepsilon_{xx}^{(1)1/2}\left(\varepsilon_{xx}^{(0)} - \varepsilon_{xx}^{(1)}\right)}. \end{aligned} \tag{D.6}$$

We recognize the result for the single-interface MO polar Kerr effect at normal incidence obtained in a slightly modified form in Eq. (4.91) or Eq. (1.53). In the last step of Eq. (D.5) and Eq. (D.6), we have used Eq. (4.96).

D.2 Single Layer

The response in a single magnetic layer sandwiched between two different isotropic half spaces is characterized by the matrix product given in Eq. (4.45). To obtain the linear approximation to the Jones reflection and transmission matrices of the system

$$\begin{pmatrix} t_{xx}^{(02)} & t_{xy}^{(02)} \\ t_{yx}^{(02)} & t_{yy}^{(02)} \end{pmatrix} \quad \text{and} \quad \begin{pmatrix} r_{xx}^{(02)} & r_{xy}^{(02)} \\ r_{yx}^{(02)} & r_{yy}^{(02)} \end{pmatrix}, \tag{D.7}$$

we employ Eq. (4.46). We start from the case of zero magnetization ($\boldsymbol{M} = 0$) and obtain

$$t_{xx}^{(02)} = t_{yy}^{(02)} = t^{(01)} = \frac{1}{\mathrm{M}_{11}^{(\mathrm{CP})}} = \frac{t^{(01)} t^{(12)} \mathrm{e}^{-\mathrm{j}\beta^{(1)}}}{1 + r^{(01)} r^{(12)} \mathrm{e}^{-2\mathrm{j}\beta^{(1)}}}, \tag{D.8a}$$

$$r_{xx}^{(02)} = r_{yy}^{(02)} = r^{(02)} = \frac{\mathrm{M}_{21}^{(\mathrm{CP})}}{\mathrm{M}_{11}^{(\mathrm{CP})}} = \frac{r^{(01)} + r^{(12)} \mathrm{e}^{-2\mathrm{j}\beta^{(1)}}}{1 + r^{(01)} r^{(12)} \mathrm{e}^{-2\mathrm{j}\beta^{(1)}}}, \tag{D.8b}$$

where $t^{(01)}$ and $r^{(01)}$ are given in Eq. (D.1). The Fresnel reflection and transmission coefficients for the additional interface are given by

$$r^{(12)} = \frac{N^{(1)} - N^{(2)}}{N^{(1)} + N^{(2)}}, \tag{D.9a}$$

$$t^{(12)} = \frac{2N^{(1)}}{N^{(1)} + N^{(2)}} = 1 + r^{(12)}. \tag{D.9b}$$

As before, the product of the propagation constant $\left(\omega N^{(1)}/c\right)$ with the layer thickness, $d^{(1)}$, is denoted by

$$\beta^{(1)} = N^{(1)} \frac{\omega}{c} d^{(1)}. \tag{D.10}$$

The off-diagonal coefficient of the Jones transmission and reflection matrices follow from Eq. (4.98)

$$
\begin{aligned}
t_{yx}^{(02)} &= -t_{xy}^{(02)} = \mathrm{j}\Delta t_{xx}^{(02)} = \mathrm{j}\Delta\left(\frac{1}{\mathrm{M}_{11}^{(\mathrm{CP})}}\right) = \mathrm{j}\Delta N^{(1)}\frac{\mathrm{d}t^{(02)}}{\mathrm{d}\,N^{(1)}} \\
&= \mathrm{j}\Delta N^{(1)}\frac{\mathrm{d}}{\mathrm{d}\,N^{(1)}}\left(\frac{t^{(01)}t^{(12)}\mathrm{e}^{-\mathrm{j}\beta^{(1)}}}{1+r^{(01)}r^{(12)}\mathrm{e}^{-2\mathrm{j}\beta^{(1)}}}\right) \\
&= \mathrm{j}\Delta N^{(1)}\mathrm{e}^{-\mathrm{j}\beta^{(1)}}\left(1+\mathrm{r}^{(01)}\mathrm{r}^{(12)}\mathrm{e}^{-2\mathrm{j}\beta^{(1)}}\right)^{-2} \\
&\quad\times\Big[\left(t^{(01)\prime}t^{(12)}+t^{(01)}t^{(12)\prime}-\mathrm{j}\beta^{(1)}t^{(01)}t^{(12)}\right)\left(1+r^{(01)}r^{(12)}\mathrm{e}^{-2\mathrm{j}\beta^{(1)}}\right) \\
&\quad- t^{(01)}t^{(12)}\mathrm{e}^{-2\mathrm{j}\beta^{(1)}}\left(\mathrm{r}^{(01)\prime}\mathrm{r}^{(12)}+\mathrm{r}^{(01)}\mathrm{r}^{(12)\prime}-2\mathrm{j}\beta^{(1)}\mathrm{r}^{(01)}\mathrm{r}^{(12)}\right)\Big], \qquad \text{(D.11a)}
\end{aligned}
$$

$$
\begin{aligned}
r_{yx}^{(02)} &= -r_{xy}^{(02)} = \mathrm{j}\Delta r_{xx}^{(02)} = \mathrm{j}\Delta\left(\frac{\mathrm{M}_{21}^{(\mathrm{CP})}}{\mathrm{M}_{11}^{(\mathrm{CP})}}\right) = \mathrm{j}\Delta N^{(1)}\frac{\mathrm{d}r^{(02)}}{\mathrm{d}\,N^{(1)}} \\
&= \mathrm{j}\Delta N^{(1)}\frac{\mathrm{d}}{\mathrm{d}\,N^{(1)}}\left(\frac{r^{(01)}+r^{(12)}\mathrm{e}^{-2\mathrm{j}\beta^{(1)}}}{1+r^{(01)}r^{(12)}\mathrm{e}^{-2\mathrm{j}\beta^{(1)}}}\right) \\
&= \mathrm{j}\Delta N^{(1)}\left(1+r^{(01)}r^{(12)}\mathrm{e}^{-2\mathrm{j}\beta^{(1)}}\right)^{-2}\Big[r^{(01)\prime}\left(1-r^{(12)2}\mathrm{e}^{-4\mathrm{j}\beta^{(1)}}\right) \\
&\quad+\mathrm{e}^{-2\mathrm{j}\beta^{(1)}}(1-\mathrm{r}^{(01)2})(\mathrm{r}^{(12)\prime}-2\mathrm{j}\beta^{(1)\prime}\mathrm{r}^{(12)})\Big]. \qquad \text{(D.11b)}
\end{aligned}
$$

The prime indicate the derivatives of $N^{(1)}$. We have already established $t^{(01)\prime}$ and $r^{(01)\prime}$ in Eq. (D.4), and we have in a similar way

$$
\frac{\mathrm{d}\,t^{(12)}}{\mathrm{d}\,N^{(1)}} = \frac{\mathrm{d}\left(1+r^{(12)}\right)}{\mathrm{d}\,N^{(1)}} = \frac{2N^{(2)}}{\left(N^{(1)}+N^{(2)}\right)^2} = \frac{1}{2N^{(1)}}(1-r^{(12)2}) \qquad \text{(D.12)}
$$

and $\beta^{(1)\prime} = \dfrac{\omega}{c}d^{(1)}$. We employ Eq. (D.9) and Eq. (D.12) and transform $t_{yx}^{(02)}$ to

$$
\begin{aligned}
t_{yx}^{(02)} &= \frac{\mathrm{j}\Delta N^{(1)}}{2N^{(1)}}\,\mathrm{e}^{-\mathrm{j}\beta^{(1)}}\left(1+\mathrm{r}^{(01)}\mathrm{r}^{(12)}\mathrm{e}^{-2\mathrm{j}\beta^{(1)}}\right)^{-2} \\
&\quad\times\Big\{-(1-r^{(01)2})(1+r^{(12)})+(1+r^{(01)})(1-r^{(12)2})-2\mathrm{j}\beta^{(1)}t^{(01)}t^{(12)} \\
&\quad+\Big[-(1-r^{(01)2})(1+r^{(12)})r^{(01)}r^{(12)}+(1+r^{(01)})(1-r^{(12)2})r^{(01)}r^{(12)} \\
&\quad-2\mathrm{j}\beta^{(1)}t^{(01)}t^{(12)}r^{(01)}r^{(12)}+(1+r^{(01)})(1+r^{(12)})(1-r^{(01)2})r^{(12)} \\
&\quad-(1+r^{(01)})(1+r^{(12)})r^{(01)}(1-r^{(12)2}) \\
&\quad+4\mathrm{j}\beta^{(1)}t^{(01)}t^{(12)}r^{(01)}r^{(12)}\Big]\,\mathrm{e}^{-2\mathrm{j}\beta^{(1)}}\Big\}
\end{aligned}
$$

$$= \frac{\mathrm{j}\Delta N^{(1)}}{2N^{(1)}} t^{(01)} t^{(12)} \,\mathrm{e}^{-\mathrm{j}\beta^{(1)}} \left(1 + \mathrm{r}^{(01)} \mathrm{r}^{(12)} \mathrm{e}^{-2\mathrm{j}\beta^{(1)}}\right)^{-2} \Big\{ (\mathrm{r}^{(01)} - \mathrm{r}^{(12)}) - 2\mathrm{j}\beta^{(1)}$$
$$+ \big[-(1 - r^{(01)}) r^{(01)} r^{(12)} + (1 - r^{(12)}) r^{(01)} r^{(12)} + (1 - r^{(01)2}) r^{(12)}$$
$$- r^{(01)} (1 - r^{(12)2}) + 2\mathrm{j}\beta^{(1)} r^{(01)} r^{(12)} \big] \,\mathrm{e}^{-2\mathrm{j}\beta^{(1)}} \Big\} . \qquad \text{(D.13)}$$

Finally, we have

$$t_{yx}^{(02)} = \frac{\Delta N^{(1)}}{2N^{(1)}} t^{(01)} t^{(12)} \mathrm{e}^{-\mathrm{j}\beta^{(1)}} \left(1 + \mathrm{r}^{(01)} \mathrm{r}^{(12)} \mathrm{e}^{-2\mathrm{j}\beta^{(1)}}\right)^{-2}$$
$$\times \left[2\beta^{(1)} \left(1 - r^{(01)} r^{(12)} \mathrm{e}^{-2\mathrm{j}\beta^{(1)}}\right) + \mathrm{j} \left(r^{(01)} - r^{(12)}\right) \left(1 - \mathrm{e}^{-2\mathrm{j}\beta^{(1)}}\right)\right], \qquad \text{(D.14a)}$$

$$r_{yx}^{(02)} = \frac{\Delta N^{(1)}}{2N^{(1)}} \left(1 + r^{(01)} r^{(12)} \mathrm{e}^{-2\mathrm{j}\beta^{(1)}}\right)^{-2}$$
$$\times (1 - r^{(01)2}) \left[4\beta^{(1)} r^{(12)} \mathrm{e}^{-2\mathrm{j}\beta^{(1)}} - \mathrm{j} \left(1 + \mathrm{r}^{(12)2} \mathrm{e}^{-2\mathrm{j}\beta^{(1)}}\right) \left(1 - \mathrm{e}^{-2\mathrm{j}\beta^{(1)}}\right)\right]. \qquad \text{(D.14b)}$$

When the second interface of the absorbing layer is removed to infinity, the exponential tends to zero, *i.e.*, $\mathrm{e}^{-2\mathrm{j}\beta^{(1)}} \to 0$ and Eq. (D.14) simplifies to Eq. (D.3d) corresponding to a single interface. Note that Eq. (D.14) represents a special case of Eq. (4.106) with $\mathcal{N} = 1$, $I_1^{(1)} = r^{(01)} - r^{(12)}$, $J_1^{(1)} = 1 - r^{(01)} r^{(12)} \mathrm{e}^{-2\mathrm{j}\beta^{(1)}}$, $H_1^{(1)} = r^{(12)}$, and $K_1^{(1)} = 1 + r^{(12)2} \,\mathrm{e}^{-2\mathrm{j}\beta^{(1)}}$, $\Delta N^{(2)} = 0$, and

$$\mathrm{M}_0^{(1)} = 1 + r^{(01)} r^{(12)} \mathrm{e}^{-2\mathrm{j}\beta^{(1)}}. \qquad \text{(D.15)}$$

The ultrathin film approximation is obtained by the development of $t_{yx}^{(02)}\left(\beta^{(1)}\right)$ and $r_{yx}^{(02)}\left(\beta^{(1)}\right)$ into Taylor series followed by the restriction to the term linear in $(\beta^{(1)})$. The $\beta^{(1)}$ independent term vanishes as seen from Eq. (D.14). We have

$$r_{yx}^{(02)} = \Delta N^{(1)} \frac{\omega}{c} d^{(1)} \frac{t^{(01)} t^{(10)} (1 + r^{(12)})^2}{(1 + r^{(01)} r^{(12)})^2}$$
$$= \Delta N^{(1)} \frac{\omega}{c} d^{(1)} \frac{4N^{(0)} N^{(1)}}{(N^{(0)} + N^{(2)})^2}. \qquad \text{(D.16)}$$

Appendix E

Chebyshev Polynomials

Chebyshev polynomials of the second kind:

$$p_0 = 0, \tag{E.1}$$
$$p_1 = 1, \tag{E.2}$$
$$p_2 = 2m_{11}, \tag{E.3}$$
$$p_3 = 4(m_{11})^2 - 1, \tag{E.4}$$
$$p_4 = 8(m_{11})^3 - 4m_{11}, \tag{E.5}$$
$$p_5 = 16(m_{11})^4 - 12(m_{11})^2 + 1, \tag{E.6}$$
$$p_6 = 32(m_{11})^5 - 32(m_{11})^3 + 6m_{11}, \tag{E.7}$$
$$p_7 = 64(m_{11})^6 - 80(m_{11})^4 + 24(m_{11})^2 - 1, \tag{E.8}$$
$$p_8 = 128(m_{11})^7 - 192(m_{11})^5 + 80(m_{11})^3 - 8m_{11}, \tag{E.9}$$
$$p_9 = 256(m_{11})^8 - 448(m_{11})^6 + 240(m_{11})^4 - 40(m_{11})^2 + 1, \tag{E.10}$$
$$p_{10} = 512(m_{11})^9 - 1024(m_{11})^7 + 672(m_{11})^5 - 160(m_{11})^3 + 10m_{11}, \tag{E.11}$$

Derivatives of Chebyshev polynomials of the second kind:

$$[p_0]' = 0, \tag{E.12}$$
$$[p_1]' = 0, \tag{E.13}$$
$$[p_2]' = 2p_1 = 2, \tag{E.14}$$
$$[p_3]' = 4p_2, \tag{E.15}$$
$$[p_4]' = 6p_3 + 2p_1, \tag{E.16}$$
$$[p_5]' = 8p_4 + 4p_2, \tag{E.17}$$
$$[p_6]' = 10p_5 + 6p_3 + 2p_1, \tag{E.18}$$
$$[p_7]' = 12p_6 + 8p_4 + 4p_2, \tag{E.19}$$
$$[p_8]' = 14p_7 + 10p_5 + 6p_3 + 2p_1, \tag{E.20}$$
$$[p_9]' = 16p_8 + 12p_6 + 8p_4 + 4p_2, \tag{E.21}$$
$$[p_{10}]' = 18p_9 + 14p_7 + 10p_5 + 6p_3 + 2p_1. \tag{E.22}$$

Higher orders are obtained using the recursion relations ($q \geq 0$):

$$p_{q+1}(m_{11}) = 2m_{11}\,p_q(m_{11}) - p_{q-1}(m_{11}), \tag{E.23}$$

$$[p_{q+1}(m_{11})]^2 - 2m_{11}\,p_q(m_{11}) \tag{E.24}$$

$$[p_q(m_{11})]^2 - p_{q-1}(m_{11})\,p_{q+1}(m_{11}) = 1, \tag{E.25}$$

$$[p_{q+1}(m_{11})]' = \frac{[p_q(m_{11})]'[m_{11}\,p_{q+1}(m_{11}) - p_q(m_{11})] + p_q(m_{11})\,p_{q+1}(m_{11})}{p_{q+1}(m_{11}) - m_{11}\,p_q(m_{11})}, \tag{E.26}$$

$$[p_{q+2}(m_{11})]' = 2(q+1)\,p_{q+1}(m_{11}) + p_q(m_{11})'. \tag{E.27}$$

Appendix F

Proper Value Equation

To obtain the proper value equation on page 447, we first evaluate a general 4×4 matrix and compare the result with $\det\left(\mathbf{C}_0^{(ua)} - \kappa_0 \mathbf{1}\right)$

$$\det\left(\mathbf{C}_0^{(ua)} - \kappa_0 \mathbf{1}\right) = \begin{vmatrix} a_{11} & a_{12} & a_{13} & a_{14} \\ a_{21} & a_{22} & a_{23} & a_{24} \\ a_{31} & a_{32} & a_{33} & a_{34} \\ a_{41} & a_{42} & a_{43} & a_{44} \end{vmatrix}$$

$$= a_{11} \begin{vmatrix} a_{22} & a_{23} & a_{24} \\ a_{32} & a_{33} & a_{34} \\ a_{42} & a_{43} & a_{44} \end{vmatrix} - a_{21} \begin{vmatrix} a_{12} & a_{13} & a_{14} \\ a_{32} & a_{33} & a_{34} \\ a_{42} & a_{43} & a_{44} \end{vmatrix}$$

$$+ a_{31} \begin{vmatrix} a_{12} & a_{13} & a_{14} \\ a_{22} & a_{23} & a_{24} \\ a_{42} & a_{43} & a_{44} \end{vmatrix} - a_{41} \begin{vmatrix} a_{12} & a_{13} & a_{14} \\ a_{22} & a_{23} & a_{24} \\ a_{32} & a_{33} & a_{34} \end{vmatrix}$$

$$\begin{aligned}
&= a_{11}\left(\underline{a_{22}a_{33}a_{44}} + \underline{a_{23}a_{34}a_{42}} + a_{32}a_{43}a_{24} - a_{24}a_{33}a_{42} - \underline{a_{23}a_{32}a_{44}} - \underline{a_{34}a_{43}a_{22}}\right) \\
&- a_{21}\left(\underline{a_{12}a_{33}a_{44}} + a_{13}a_{34}a_{42} + a_{14}a_{32}a_{43} - a_{14}a_{33}a_{42} - a_{13}a_{32}a_{44} - \underline{a_{12}a_{34}a_{43}}\right) \\
&+ a_{31}\left(\underline{a_{12}a_{23}a_{44}} + a_{13}a_{24}a_{42} + a_{14}a_{22}a_{43} - a_{14}a_{23}a_{42} - a_{13}a_{22}a_{44} - a_{12}a_{43}a_{24}\right) \\
&- a_{41}\left(\underline{a_{12}a_{23}a_{34}} + a_{13}a_{24}a_{32} + \underline{a_{14}a_{22}a_{33}} - a_{14}a_{23}a_{32} - \underline{a_{13}a_{22}a_{34}} - \underline{a_{12}a_{24}a_{33}}\right).
\end{aligned} \tag{F.1}$$

The substitution according to Eq. (9.245) provides

$$\begin{aligned}
a_{11}a_{22}a_{33}a_{44} &= \left(p_m \varepsilon_{zz}^{-1} \varepsilon_{zx} + s_m - \kappa_m\right)\left(\varepsilon_{xz}\varepsilon_{zz}^{-1} p_m + s_m - \kappa_m\right) \\
&\quad \times \left(q_m \varepsilon_{zz}^{-1} \varepsilon_{zy} + s_m - \kappa_m\right)\left(\varepsilon_{yz}\varepsilon_{zz}^{-1} q_m + s_m - \kappa_m\right) \\
&= \underline{(s_m - \kappa_m)^4} \; \underline{+ (s_m - \kappa_m)^3 \varepsilon_{zz}^{-1}\left(p_m \varepsilon_{zx} + p_m \varepsilon_{xz} + q_m \varepsilon_{yz} + q_m \varepsilon_{zy}\right)} \\
&\quad + (s_m - \kappa_m)^2 \varepsilon_{zz}^{-2}\left(p_m^2 \varepsilon_{zx}\varepsilon_{xz} + p_m q_m \varepsilon_{zx}\varepsilon_{yz} + p_m q_m \varepsilon_{zx}\varepsilon_{zy}\right.
\end{aligned}$$

$$
\begin{aligned}
&+p_m q_m \varepsilon_{xz}\varepsilon_{yz} + p_m q_m \varepsilon_{xz}\varepsilon_{zy} + q_m^2 \varepsilon_{yz}\varepsilon_{zy} \,) \\
&+(s_m - \kappa_m)\varepsilon_{zz}^{-3}\left(p_m q_m^2 \varepsilon_{zx}\varepsilon_{yz}\varepsilon_{zy} + p_m q_m^2 \varepsilon_{xz}\varepsilon_{yz}\varepsilon_{zy}\right. \\
&\left.+\, p_m^2 q_m \varepsilon_{zx}\varepsilon_{xz}\varepsilon_{yz} + p_m^2 q_m \varepsilon_{zx}\varepsilon_{xz}\varepsilon_{zy}\right) \\
&+p_m^2 q_m^2 \varepsilon_{zz}^{-4}\varepsilon_{zx}\varepsilon_{xz}\varepsilon_{yz}\varepsilon_{zy}
\end{aligned} \tag{F.2}
$$

$$
\begin{aligned}
a_{11}a_{23}a_{34}a_{42} &= \left(p_m\varepsilon_{zz}^{-1}\varepsilon_{zx} + s_m - \kappa_m\right)\left(\varepsilon_{xz}\varepsilon_{zz}^{-1}\varepsilon_{zy} - \varepsilon_{xy} - q_m p_m\right) \\
&\quad \times \left(1 - q_m^2\varepsilon_{zz}^{-1}\right)\left(-\varepsilon_{yz}\varepsilon_{zz}^{-1}p_m\right) \\
&= \varepsilon_{yz}\varepsilon_{zz}^{-1}p_m(s_m - \kappa_m)\left[(\varepsilon_{xy} + q_m p_m) - \varepsilon_{xz}\varepsilon_{zz}^{-1}\varepsilon_{zy}\right]\left(1 - q_m^2\varepsilon_{zz}^{-1}\right) \\
&\quad - p_m^2\varepsilon_{zz}^{-2}\varepsilon_{zx}\varepsilon_{yz}\left(1 - q_m^2\varepsilon_{zz}^{-1}\right)\left(\varepsilon_{xz}\varepsilon_{zz}^{-1}\varepsilon_{zy} - \varepsilon_{xy} - q_m p_m\right) \\
&= \underline{p_m\varepsilon_{yz}\varepsilon_{zz}^{-1}(s_m - \kappa_m)(\varepsilon_{xy} + q_m p_m)} - p_m q_m^2\varepsilon_{yz}\varepsilon_{zz}^{-2}(s_m - \kappa_m) \\
&\quad \times(\varepsilon_{xy} + q_m p_m) - p_m\varepsilon_{xz}\varepsilon_{zy}\varepsilon_{yz}\varepsilon_{zz}^{-2}(s_m - \kappa_m) + p_m q_m^2\varepsilon_{xz}\varepsilon_{yz}\varepsilon_{zy}\varepsilon_{zz}^{-3} \\
&\quad \times(s_m - \kappa_m) + p_m^2\varepsilon_{zz}^{-2}\varepsilon_{zx}\varepsilon_{yz}(\varepsilon_{xy} + q_m p_m) - p_m^2 q_m^2\varepsilon_{zz}^{-3}\varepsilon_{zx}\varepsilon_{yz} \\
&\quad \times(\varepsilon_{xy} + p_m q_m) - p_m^2\varepsilon_{zz}^{-3}\varepsilon_{zx}\varepsilon_{xz}\varepsilon_{yz}\varepsilon_{zy} + p_m^2 q_m^2\varepsilon_{zz}^{-4}\varepsilon_{zx}\varepsilon_{xz}\varepsilon_{yz}\varepsilon_{zy}
\end{aligned} \tag{F.3}
$$

$$
\begin{aligned}
a_{11}a_{32}a_{43}a_{24} &= \left(p_m\varepsilon_{zz}^{-1}\varepsilon_{zx} + s_m - \kappa_m\right)\left(q_m\varepsilon_{zz}^{-1}p_m\right) \\
&\quad \times\left(\varepsilon_{yy} - \varepsilon_{yz}\varepsilon_{zz}^{-1}\varepsilon_{zy} - p_m^2\right)\left(-\varepsilon_{xz}\varepsilon_{zz}^{-1}q_m\right) \\
&= -p_m q_m^2\varepsilon_{xz}\varepsilon_{zz}^{-2}(s_m - \kappa_m)\left(\varepsilon_{yy} - p_m^2\right) + p_m q_m^2\varepsilon_{xz}\varepsilon_{yz}\varepsilon_{zy}\varepsilon_{zz}^{-3} \\
&\quad \times(s_m - \kappa_m) - p_m^2 q_m^2\varepsilon_{zx}\varepsilon_{xz}\varepsilon_{zz}^{-3}\left(\varepsilon_{yy} - p_m^2\right) + p_m^2 q_m^2\varepsilon_{zz}^{-4}\varepsilon_{zx}\varepsilon_{xz}\varepsilon_{yz}\varepsilon_{zy}
\end{aligned} \tag{F.4}
$$

$$
\begin{aligned}
a_{11}a_{24}a_{33}a_{42} &= \left(p_m\varepsilon_{zz}^{-1}\varepsilon_{zx} + s_m - \kappa_m\right)\left(-\varepsilon_{xz}\varepsilon_{zz}^{-1}q_m\right) \\
&\quad \times\left(q_m\varepsilon_{zz}^{-1}\varepsilon_{zy} + s_m - \kappa_m\right)\left(-\varepsilon_{yz}\varepsilon_{zz}^{-1}p_m\right) \\
&= p_m q_m\varepsilon_{xz}\varepsilon_{yz}\varepsilon_{zz}^{-2}(s_m - \kappa_m)^2 + p_m q_m\varepsilon_{xz}\varepsilon_{yz}\varepsilon_{zz}^{-3}(s_m - \kappa_m) \\
&\quad \times\left(p_m\varepsilon_{zx} + q_m\varepsilon_{zy}\right) + p_m^2 q_m^2\varepsilon_{zz}^{-4}\varepsilon_{zx}\varepsilon_{xz}\varepsilon_{yz}\varepsilon_{zy}
\end{aligned} \tag{F.5}
$$

$$
\begin{aligned}
a_{11}a_{23}a_{32}a_{44} &= \left(p_m\varepsilon_{zz}^{-1}\varepsilon_{zx} + s_m - \kappa_m\right)\left(\varepsilon_{xz}\varepsilon_{zz}^{-1}\varepsilon_{zy} - \varepsilon_{xy} - q_m p_m\right) \\
&\quad \times\left(q_m\varepsilon_{zz}^{-1}p_m\right)\left(\varepsilon_{yz}\varepsilon_{zz}^{-1}q_m + s_m - \kappa_m\right) \\
&= \underline{-p_m q_m\varepsilon_{zz}^{-1}(s_m - \kappa_m)^2\left(\varepsilon_{xy} + p_m q_m\right)} + p_m q_m\varepsilon_{xz}\varepsilon_{zy}\varepsilon_{zz}^{-2}(s_m - \kappa_m)^2 \\
&\quad - p_m q_m\varepsilon_{zz}^{-2}(s_m - \kappa_m)\left(p_m\varepsilon_{zx} + q_m\varepsilon_{yz}\right)\left(\varepsilon_{xy} + q_m p_m\right) \\
&\quad + p_m q_m\varepsilon_{zz}^{-3}\left[(s_m - \kappa_m)\left(p_m\varepsilon_{zx} + q_m\varepsilon_{yz}\right)\varepsilon_{xz}\varepsilon_{zy}\right. \\
&\quad \left. -\, p_m q_m\varepsilon_{zx}\varepsilon_{yz}(\varepsilon_{xy} + p_m q_m)\right] + p_m^2 q_m^2\varepsilon_{zz}^{-4}\varepsilon_{zx}\varepsilon_{xz}\varepsilon_{yz}\varepsilon_{zy}
\end{aligned} \tag{F.6}
$$

$$
\begin{aligned}
a_{11}a_{34}a_{43}a_{22} &= \left(p_m\varepsilon_{zz}^{-1}\varepsilon_{zx} + s_m - \kappa_m\right)\left(1 - q_m^2\varepsilon_{zz}^{-1}\right) \\
&\quad \times \left(\varepsilon_{yy} - \varepsilon_{yz}\varepsilon_{zz}^{-1}\varepsilon_{zy} - p_m^2\right)\left(\varepsilon_{xz}\varepsilon_{zz}^{-1}p_m + s_m - \kappa_m\right) \\
&= (s_m - \kappa_m)^2\left[\left(1 - q_m^2\varepsilon_{zz}^{-1}\right)\left(\varepsilon_{yy} - p_m^2\right) - \varepsilon_{yz}\varepsilon_{zz}^{-1}\varepsilon_{zy} + q_m^2\varepsilon_{yz}\varepsilon_{zy}\varepsilon_{zz}^{-2}\right] \\
&\quad + (s_m - \kappa_m)p_m\varepsilon_{zz}^{-1}\left(\varepsilon_{zx} + \varepsilon_{xz}\right) \\
&\quad \times \left[\left(\varepsilon_{yy} - p_m^2\right) - \varepsilon_{yz}\varepsilon_{zz}^{-1}\varepsilon_{zy}\left(1 - q_m^2\varepsilon_{zz}^{-1}\right) - q_m^2\varepsilon_{zz}^{-1}\left(\varepsilon_{yy} - p_m^2\right)\right] \\
&\quad + p_m^2\varepsilon_{zz}^{-2}\varepsilon_{zx}\varepsilon_{xz}\left(1 - q_m^2\varepsilon_{zz}^{-1}\right)\left(\varepsilon_{yy} - \varepsilon_{yz}\varepsilon_{zz}^{-1}\varepsilon_{zy} - p_m^2\right) \\
&= \underline{(s_m - \kappa_m)^2\left(\varepsilon_{yy} - p_m^2\right) - (s_m - \kappa_m)^2\varepsilon_{zz}^{-1}\left[q_m^2\left(\varepsilon_{yy} - p_m^2\right) + \varepsilon_{yz}\varepsilon_{zy}\right]} \\
&\quad \underline{+ (s_m - \kappa_m)p_m\varepsilon_{zz}^{-1}\left(\varepsilon_{zx} + \varepsilon_{xz}\right)\left(\varepsilon_{yy} - p_m^2\right)} \\
&\quad + (s_m - \kappa_m)^2 q_m^2\varepsilon_{yz}\varepsilon_{zy}\varepsilon_{zz}^{-2} \\
&\quad - (s_m - \kappa_m)p_m\varepsilon_{zz}^{-2}\left(\varepsilon_{zx} + \varepsilon_{xz}\right)\left[\varepsilon_{yz}\varepsilon_{zy} + q_m^2\left(\varepsilon_{yy} - p_m^2\right)\right] \\
&\quad + p_m q_m^2\varepsilon_{yz}\varepsilon_{zy}(s_m - \kappa_m)\varepsilon_{zz}^{-3}\left(\varepsilon_{zx} + \varepsilon_{xz}\right) \\
&\quad + p_m^2\varepsilon_{zz}^{-2}\varepsilon_{zx}\varepsilon_{xz}\left(\varepsilon_{yy} - p_m^2\right) - p_m^2 q_m^2\varepsilon_{zz}^{-3}\varepsilon_{zx}\varepsilon_{xz}\left(\varepsilon_{yy} - p_m^2\right) \\
&\quad - p_m^2\varepsilon_{zz}^{-3}\varepsilon_{zx}\varepsilon_{xz}\varepsilon_{yz}\varepsilon_{zy} + p_m^2 q_m^2\varepsilon_{zz}^{-4}\varepsilon_{zx}\varepsilon_{xz}\varepsilon_{yz}\varepsilon_{zy}
\end{aligned} \tag{F.7}
$$

$$
\begin{aligned}
a_{21}a_{12}a_{33}a_{44} &= \left(\varepsilon_{xz}\varepsilon_{zz}^{-1}\varepsilon_{zx} - \varepsilon_{yy} + q_m^2\right)\left(p_m^2\varepsilon_{zz}^{-1} - 1\right) \\
&\quad \times \left(q_m\varepsilon_{zz}^{-1}\varepsilon_{zy} + s_m - \kappa_m\right)\left(\varepsilon_{yz}\varepsilon_{zz}^{-1}q_m + s_m - \kappa_m\right) \\
&= \left[(s_m - \kappa_m)^2 + (s_m - \kappa_m)q_m\varepsilon_{zz}^{-1}\left(\varepsilon_{yz} + \varepsilon_{zy}\right) + q_m^2\varepsilon_{zz}^{-2}\varepsilon_{yz}\varepsilon_{zy}\right] \\
&\quad \times \left[\varepsilon_{zz}^{-2}p_m^2\varepsilon_{zx}\varepsilon_{xz} + \varepsilon_{zz}^{-1}\left(p_m^2 q_m^2 - p_m^2\varepsilon_{yy} - \varepsilon_{zx}\varepsilon_{xz}\right) + \left(\varepsilon_{yy} - q_m^2\right)\right] \\
&= (s_m - \kappa_m)^2\left[\left(\varepsilon_{yy} - q_m^2\right) + \varepsilon_{zz}^{-1}\left(p_m^2 q_m^2 - p_m^2\varepsilon_{yy} - \varepsilon_{zx}\varepsilon_{xz}\right)\right. \\
&\quad \left. + \varepsilon_{zz}^{-2}p_m^2\varepsilon_{zx}\varepsilon_{xz}\right] + (s_m - \kappa_m)q_m\varepsilon_{zz}^{-1}\left(\varepsilon_{yz} + \varepsilon_{zy}\right) \\
&\quad \times \left[\left(\varepsilon_{yy} - q_m^2\right) + \varepsilon_{zz}^{-1}\left(p_m^2 q_m^2 - p_m^2\varepsilon_{yy} - \varepsilon_{zx}\varepsilon_{xz}\right) + \varepsilon_{zz}^{-2}p_m^2\varepsilon_{zx}\varepsilon_{xz}\right] \\
&\quad + q_m^2\varepsilon_{zz}^{-2}\varepsilon_{yz}\varepsilon_{zy}\left[\varepsilon_{zz}^{-2}p_m^2\varepsilon_{zx}\varepsilon_{xz} + \varepsilon_{zz}^{-1}\left(p_m^2 q_m^2 - p_m^2\varepsilon_{yy} - \varepsilon_{zx}\varepsilon_{xz}\right)\right. \\
&\quad \left. + \left(\varepsilon_{yy} - q_m^2\right)\right] \\
&= \underline{(s_m - \kappa_m)^2\left(\varepsilon_{yy} - q_m^2\right) + \varepsilon_{zz}^{-1}\left(p_m^2 q_m^2 - p_m^2\varepsilon_{yy} - \varepsilon_{zx}\varepsilon_{xz}\right)} \\
&\quad \underline{+ (s_m - \kappa_m)q_m\varepsilon_{zz}^{-1}\left(\varepsilon_{yz} + \varepsilon_{zy}\right)\left(\varepsilon_{yy} - q_m^2\right)} + (s_m - \kappa_m)^2\varepsilon_{zz}^{-2}p_m^2 \\
&\quad \times \varepsilon_{zx}\varepsilon_{xz} + (s_m - \kappa_m)q_m\varepsilon_{zz}^{-2}\left(\varepsilon_{yz} + \varepsilon_{zy}\right)\left(p_m^2 q_m^2 - p_m^2\varepsilon_{yy} - \varepsilon_{zx}\varepsilon_{xz}\right) \\
&\quad + (s_m - \kappa_m)p_m^2 q_m\varepsilon_{zx}\varepsilon_{xz}\varepsilon_{zz}^{-3}\left(\varepsilon_{yz} + \varepsilon_{zy}\right) + q_m^2\varepsilon_{zz}^{-2}\varepsilon_{yz}\varepsilon_{zy}\left(\varepsilon_{yy} - q_m^2\right) \\
&\quad + q_m^2\varepsilon_{zz}^{-3}\varepsilon_{yz}\varepsilon_{zy}\left(p_m^2 q_m^2 - p_m^2\varepsilon_{yy} - \varepsilon_{zx}\varepsilon_{xz}\right) + p_m^2 q_m^2\varepsilon_{zz}^{-4}\varepsilon_{zx}\varepsilon_{xz}\varepsilon_{yz}\varepsilon_{zy},
\end{aligned} \tag{F.8}
$$

$$\begin{aligned}
a_{21}a_{13}a_{34}a_{42} &= \left(\varepsilon_{xz}\varepsilon_{zz}^{-1}\varepsilon_{zx} - \varepsilon_{yy} + q_m^2\right)\left(p_m\varepsilon_{zz}^{-1}\varepsilon_{zy}\right)\left(1 - q_m^2\varepsilon_{zz}^{-1}\right)\left(-\varepsilon_{yz}\varepsilon_{zz}^{-1}p_m\right) \\
&= \varepsilon_{zz}^{-2}p_m^2\varepsilon_{yz}\varepsilon_{zy}\left(\varepsilon_{yy} - q_m^2 - q_m^2\varepsilon_{zz}^{-1}\varepsilon_{yy} + q_m^4\varepsilon_{zz}^{-1} - \varepsilon_{xz}\varepsilon_{zz}^{-1}\varepsilon_{zx}\right. \\
&\quad \left. + q_m^2\varepsilon_{zz}^{-2}\varepsilon_{zx}\varepsilon_{xz}\right) \\
&= \varepsilon_{zz}^{-2}p_m^2\varepsilon_{yz}\varepsilon_{zy}\left(\varepsilon_{yy} - q_m^2\right) + \varepsilon_{zz}^{-3}p_m^2\varepsilon_{yz}\varepsilon_{zy}\left(q_m^4 - q_m^2\varepsilon_{yy} - \varepsilon_{zx}\varepsilon_{xz}\right) \\
&\quad + p_m^2q_m^2\varepsilon_{zz}^{-4}\varepsilon_{zx}\varepsilon_{xz}\varepsilon_{yz}\varepsilon_{zy},
\end{aligned} \tag{F.9}$$

$$\begin{aligned}
a_{21}a_{14}a_{32}a_{43} &= \left(\varepsilon_{xz}\varepsilon_{zz}^{-1}\varepsilon_{zx} - \varepsilon_{yy} + q_m^2\right)\left(-p_m\varepsilon_{zz}^{-1}q_m\right) \\
&\quad \times \left(q_m\varepsilon_{zz}^{-1}p_m\right)\left(\varepsilon_{yy} - \varepsilon_{yz}\varepsilon_{zz}^{-1}\varepsilon_{zy} - p_m^2\right) \\
&= p_m^2q_m^2\varepsilon_{zz}^{-2}\left(\varepsilon_{yy} - q_m^2 - \varepsilon_{xz}\varepsilon_{zz}^{-1}\varepsilon_{zx}\right)\left(\varepsilon_{yy} - \varepsilon_{yz}\varepsilon_{zz}^{-1}\varepsilon_{zy} - p_m^2\right) \\
&= p_m^2q_m^2\varepsilon_{zz}^{-2}\left(\varepsilon_{yy} - q_m^2\right)\left(\varepsilon_{yy} - p_m^2\right) - p_m^2q_m^2\varepsilon_{zz}^{-3} \\
&\quad \times \left[\varepsilon_{yz}\varepsilon_{zy}\left(\varepsilon_{yy} - q_m^2\right) + \varepsilon_{xz}\varepsilon_{zx}\left(\varepsilon_{yy} - p_m^2\right)\right] + p_m^2q_m^2\varepsilon_{zz}^{-4}\varepsilon_{xz}\varepsilon_{zx}\varepsilon_{yz},
\end{aligned} \tag{F.10}$$

$$\begin{aligned}
a_{21}a_{14}a_{33}a_{42} &= \left(\varepsilon_{xz}\varepsilon_{zz}^{-1}\varepsilon_{zx} - \varepsilon_{yy} + q_m^2\right)\left(-p_m\varepsilon_{zz}^{-1}q_m\right) \\
&\quad \times \left(q_m\varepsilon_{zz}^{-1}\varepsilon_{zy} + s_m - \kappa_m\right)\left(-\varepsilon_{yz}\varepsilon_{zz}^{-1}p_m\right) \\
&= -p_m^2q_m\varepsilon_{zz}^{-2}\varepsilon_{yz}(s_m - \kappa_m)\left(\varepsilon_{yy} - q_m^2\right) + p_m^2q_m\varepsilon_{zz}^{-3}\varepsilon_{zx}\varepsilon_{xz}\varepsilon_{yz} \\
&\quad - p_m^2q_m^2\varepsilon_{zz}^{-3}\varepsilon_{yz}\varepsilon_{zy}\left(\varepsilon_{yy} - q_m^2\right) + p_m^2q_m^2\varepsilon_{zz}^{-4}\varepsilon_{zx}\varepsilon_{xz}\varepsilon_{yz}\varepsilon_{zy}
\end{aligned} \tag{F.11}$$

$$\begin{aligned}
a_{21}a_{13}a_{32}a_{44} &= \left(\varepsilon_{xz}\varepsilon_{zz}^{-1}\varepsilon_{zx} - \varepsilon_{yy} + q_m^2\right)\left(p_m\varepsilon_{zz}^{-1}\varepsilon_{zy}\right) \\
&\quad \times \left(q_m\varepsilon_{zz}^{-1}p_m\right)\left(\varepsilon_{yz}\varepsilon_{zz}^{-1}q_m + s_m - \kappa_m\right) \\
&= p_m^2q_m\varepsilon_{zy}\varepsilon_{zz}^{-2}(s_m - \kappa_m)\left(\varepsilon_{xz}\varepsilon_{zz}^{-1}\varepsilon_{zx} - \varepsilon_{yy} + q_m^2\right) \\
&\quad - p_m^2q_m^2\varepsilon_{zy}\varepsilon_{yz}\varepsilon_{zz}^{-3}\left(\varepsilon_{yy} - q_m^2\right) + p_m^2q_m^2\varepsilon_{zz}^{-4}\varepsilon_{zx}\varepsilon_{xz}\varepsilon_{yz}\varepsilon_{zy}
\end{aligned} \tag{F.12}$$

$$\begin{aligned}
a_{21}a_{12}a_{34}a_{43} &= \left(\varepsilon_{xz}\varepsilon_{zz}^{-1}\varepsilon_{zx} - \varepsilon_{yy} + q_m^2\right)\left(p_m^2\varepsilon_{zz}^{-1} - 1\right) \\
&\quad \times \left(1 - q_m^2\varepsilon_{zz}^{-1}\right)\left(\varepsilon_{yy} - \varepsilon_{yz}\varepsilon_{zz}^{-1}\varepsilon_{zy} - p_m^2\right) \\
&= \left(\varepsilon_{yy} - q_m^2 - \varepsilon_{xz}\varepsilon_{zz}^{-1}\varepsilon_{zx}\right)\left(1 - p_m^2\varepsilon_{zz}^{-1}\right) \\
&\quad \times \left(\varepsilon_{yy} - p_m^2 - \varepsilon_{yz}\varepsilon_{zz}^{-1}\varepsilon_{zy}\right)\left(1 - q_m^2\varepsilon_{zz}^{-1}\right) \\
&= \left[1 - \varepsilon_{zz}^{-1}\left(p_m^2 + q_m^2\right) + p_m^2q_m^2\varepsilon_{zz}^{-2}\right] \\
&\quad \times \left(\varepsilon_{yy} - q_m^2 - \varepsilon_{xz}\varepsilon_{zz}^{-1}\varepsilon_{zx}\right)\left(\varepsilon_{yy} - p_m^2 - \varepsilon_{yz}\varepsilon_{zz}^{-1}\varepsilon_{zy}\right)
\end{aligned}$$

$$= \underline{\left(\varepsilon_{yy}-q_m^2\right)\left(\varepsilon_{yy}-p_m^2\right)} \underline{-\left(p_m^2+q_m^2\right)\varepsilon_{zz}^{-1}\left(\varepsilon_{yy}-q_m^2\right)\left(\varepsilon_{yy}-p_m^2\right)}$$
$$\underline{-\varepsilon_{zz}^{-1}\left[\varepsilon_{yz}\varepsilon_{zy}\left(\varepsilon_{yy}-q_m^2\right)+\varepsilon_{zx}\varepsilon_{xz}\left(\varepsilon_{zz}-p_m^2\right)\right]}$$
$$+p_m^2q_m^2\varepsilon_{zz}^{-2}\left(\varepsilon_{yy}-q_m^2\right)\left(\varepsilon_{yy}-p_m^2\right)$$
$$+\left(p_m^2+q_m^2\right)\varepsilon_{zz}^{-2}\left[\left(\varepsilon_{yy}-q_m^2\right)\varepsilon_{yz}\varepsilon_{zy}+\left(\varepsilon_{yy}-p_m^2\right)\varepsilon_{xz}\varepsilon_{zx}\right]$$
$$-p_m^2q_m^2\varepsilon_{zz}^{-3}\varepsilon_{zz}^{-1}\left[\left(\varepsilon_{yy}-q_m^2\right)\varepsilon_{yz}\varepsilon_{zy}+\left(\varepsilon_{yy}-p_m^2\right)\varepsilon_{xz}\varepsilon_{zx}\right]$$
$$+\varepsilon_{zx}\varepsilon_{xz}\varepsilon_{yz}\varepsilon_{zy}\varepsilon_{zz}^{-2}-\left(p_m^2+q_m^2\right)\varepsilon_{zx}\varepsilon_{xz}\varepsilon_{yz}\varepsilon_{zy}\varepsilon_{zz}^{-3}+p_m^2q_m^2\varepsilon_{zz}^{-4}\varepsilon_{zx}\varepsilon_{xz}\varepsilon_{yz}\varepsilon_{zy} \tag{F.13}$$

$$\begin{aligned} a_{31}a_{12}a_{23}a_{44} &= \left(q_m\varepsilon_{zz}^{-1}\varepsilon_{zx}\right)\left(p_m^2\varepsilon_{zz}^{-1}-1\right) \\ &\quad\times\left(\varepsilon_{xz}\varepsilon_{zz}^{-1}\varepsilon_{zy}-\varepsilon_{xy}-p_mq_m\right)\left(\varepsilon_{yz}\varepsilon_{zz}^{-1}q_m+s_m-\kappa_m\right) \\ &= \underline{q_m\varepsilon_{zz}^{-1}\varepsilon_{zx}(s_m-\kappa_m)\left(\varepsilon_{xy}+p_mq_m\right)} \\ &\quad-q_m\varepsilon_{zz}^{-2}\varepsilon_{zx}(s_m-\kappa_m)\left(\varepsilon_{xz}\varepsilon_{zy}+p_m^2\varepsilon_{xy}+p_m^3q_m\right) \\ &\quad+p_m^2q_m\varepsilon_{zz}^{-3}\varepsilon_{zx}\varepsilon_{xz}\varepsilon_{zy}(s_m-\kappa_m) \\ &\quad+q_m^2\varepsilon_{zz}^{-2}\varepsilon_{zx}\varepsilon_{yz}\left(\varepsilon_{xy}+p_mq_m\right)-q_m^2\varepsilon_{zz}^{-3}\varepsilon_{zx}\varepsilon_{yz}\left(\varepsilon_{xz}\varepsilon_{zy}+p_m^2\varepsilon_{xy}\right. \\ &\quad\left.+\,p_m^3q_m\right)+p_m^2q_m^2\varepsilon_{zz}^{-4}\varepsilon_{zx}\varepsilon_{xz}\varepsilon_{yz}\varepsilon_{zy} \end{aligned} \tag{F.14}$$

$$\begin{aligned} a_{31}a_{13}a_{24}a_{42} &= \left(q_m\varepsilon_{zz}^{-1}\varepsilon_{zx}\right)\left(p_m\varepsilon_{zz}^{-1}\varepsilon_{zy}\right)\left(-\varepsilon_{xz}\varepsilon_{zz}^{-1}q_m\right)\left(-\varepsilon_{yz}\varepsilon_{zz}^{-1}p_m\right) \\ &= p_m^2q_m^2\varepsilon_{zz}^{-4}\varepsilon_{zx}\varepsilon_{xz}\varepsilon_{yz}\varepsilon_{zy} \end{aligned} \tag{F.15}$$

$$\begin{aligned} a_{31}a_{14}a_{22}a_{43} &= \left(q_m\varepsilon_{zz}^{-1}\varepsilon_{zx}\right)\left(-p_m\varepsilon_{zz}^{-1}q_m\right) \\ &\quad\times\left(\varepsilon_{xz}\varepsilon_{zz}^{-1}p_m+s_m-\kappa_m\right)\left(\varepsilon_{yy}-\varepsilon_{yz}\varepsilon_{zz}^{-1}\varepsilon_{zy}-p_m^2\right) \\ &= -p_mq_m^2\varepsilon_{zz}^{-2}\varepsilon_{zx}\left(\varepsilon_{xz}\varepsilon_{zz}^{-1}p_m+s_m-\kappa_m\right)\left(\varepsilon_{yy}-\varepsilon_{yz}\varepsilon_{zz}^{-1}\varepsilon_{zy}-p_m^2\right) \\ &= -p_mq_m^2\varepsilon_{zz}^{-2}\varepsilon_{zx}(s_m-\kappa_m)\left(\varepsilon_{yy}-p_m^2\right) \\ &\quad+p_mq_m^2\varepsilon_{zx}\varepsilon_{yz}\varepsilon_{zy}\varepsilon_{zz}^{-3}(s_m-\kappa_m) \\ &\quad-p_m^2q_m^2\varepsilon_{zz}^{-3}\varepsilon_{zx}\varepsilon_{xz}\left(\varepsilon_{yy}-p_m^2\right)+p_m^2q_m^2\varepsilon_{zz}^{-4}\varepsilon_{zx}\varepsilon_{xz}\varepsilon_{yz}\varepsilon_{zy} \end{aligned} \tag{F.16}$$

$$\begin{aligned} a_{31}a_{14}a_{23}a_{42} &= \left(q_m\varepsilon_{zz}^{-1}\varepsilon_{zx}\right)\left(-p_m\varepsilon_{zz}^{-1}q_m\right)\left(\varepsilon_{xz}\varepsilon_{zz}^{-1}\varepsilon_{zy}-\varepsilon_{xy}-p_mq_m\right)\left(-\varepsilon_{yz}\varepsilon_{zz}^{-1}p_m\right) \\ &= -p_m^2q_m^2\varepsilon_{zz}^{-3}\varepsilon_{zx}\varepsilon_{yz}(\varepsilon_{xy}+p_mq_m)+p_m^2q_m^2\varepsilon_{zz}^{-4}\varepsilon_{zx}\varepsilon_{xz}\varepsilon_{yz}\varepsilon_{zy} \end{aligned} \tag{F.17}$$

$$
\begin{aligned}
a_{31}a_{13}a_{22}a_{44} &= \left(q_m\varepsilon_{zz}^{-1}\varepsilon_{zx}\right)\left(p_m\varepsilon_{zz}^{-1}\varepsilon_{zy}\right) \\
&\quad\times\left(\varepsilon_{xz}\varepsilon_{zz}^{-1}p_m + s_m - \kappa_m\right)\left(\varepsilon_{yz}\varepsilon_{zz}^{-1}q_m + s_m - \kappa_m\right) \\
&= p_m q_m \varepsilon_{zz}^{-2}\varepsilon_{zx}\varepsilon_{zy}\left(s_m-\kappa_m\right)^2 + \left(s_m-\kappa_m\right)\left(p_m\varepsilon_{xz}+q_m\varepsilon_{yz}\right) \\
&\quad\times p_m q_m \varepsilon_{zz}^{-3}\varepsilon_{zx}\varepsilon_{zy} + p_m^2 q_m^2 \varepsilon_{zz}^{-4}\varepsilon_{zx}\varepsilon_{xz}\varepsilon_{yz}\varepsilon_{zy}
\end{aligned}
\tag{F.18}
$$

$$
\begin{aligned}
a_{31}a_{12}a_{43}a_{24} &= \left(q_m\varepsilon_{zz}^{-1}\varepsilon_{zx}\right)\left(p_m^2\varepsilon_{zz}^{-1}-1\right)\left(\varepsilon_{yy}-\varepsilon_{yz}\varepsilon_{zz}^{-1}\varepsilon_{zy}-p_m^2\right)\left(-\varepsilon_{xz}\varepsilon_{zz}^{-1}q_m\right) \\
&= -q_m^2\varepsilon_{zz}^{-2}\varepsilon_{zx}\varepsilon_{xz}\left(\varepsilon_{yy}-\varepsilon_{yz}\varepsilon_{zz}^{-1}\varepsilon_{zy}-p_m^2\right)\left(p_m^2\varepsilon_{zz}^{-1}-1\right) \\
&= q_m^2\varepsilon_{zz}^{-2}\varepsilon_{zx}\varepsilon_{xz}\left(\varepsilon_{yy}-p_m^2\right) - q_m^2\varepsilon_{zz}^{-3}\varepsilon_{zx}\varepsilon_{xz}\varepsilon_{yz}\varepsilon_{zy} \\
&\quad - p_m^2 q_m^2\varepsilon_{zz}^{-3}\varepsilon_{zx}\varepsilon_{xz}\left(\varepsilon_{yy}-p_m^2\right) + p_m^2 q_m^2\varepsilon_{zz}^{-4}\varepsilon_{zx}\varepsilon_{xz}\varepsilon_{yz}\varepsilon_{zy},
\end{aligned}
\tag{F.19}
$$

$$
\begin{aligned}
a_{41}a_{12}a_{23}a_{34} &= \left(\varepsilon_{zy}-\varepsilon_{yz}\varepsilon_{zz}^{-1}\varepsilon_{zx}+p_m q_m\right)\left(p_m^2\varepsilon_{zz}^{-1}-1\right) \\
&\quad\times\left(\varepsilon_{xz}\varepsilon_{zz}^{-1}\varepsilon_{zy}-\varepsilon_{xy}-p_m q_m\right)\left(1-q_m^2\varepsilon_{zz}^{-1}\right) \\
&= \left(\varepsilon_{zy}+p_m q_m-\varepsilon_{yz}\varepsilon_{zz}^{-1}\varepsilon_{zx}\right)\left(1-p_m^2\varepsilon_{zz}^{-1}\right) \\
&\quad\times\left(\varepsilon_{xy}+p_m q_m-\varepsilon_{xz}\varepsilon_{zz}^{-1}\varepsilon_{zy}\right)\left(1-q_m^2\varepsilon_{zz}^{-1}\right) \\
&= \left[\left(\varepsilon_{zy}+p_m q_m\right)-\varepsilon_{yz}\varepsilon_{zz}^{-1}\varepsilon_{zx}\right]\left[\left(\varepsilon_{xy}+p_m q_m\right)-\varepsilon_{xz}\varepsilon_{zz}^{-1}\varepsilon_{zy}\right] \\
&\quad\times\left[1-\left(p_m^2+q_m^2\right)\varepsilon_{zz}^{-1}+p_m^2 q_m^2\varepsilon_{zz}^{-2}\right] \\
&= \left[1-\left(p_m^2+q_m^2\right)\varepsilon_{zz}^{-1}+p_m^2 q_m^2\varepsilon_{zz}^{-2}\right](\varepsilon_{zy}+p_m q_m)(\varepsilon_{xy}+p_m q_m) \\
&\quad-\left[\varepsilon_{yz}\varepsilon_{zz}^{-1}\varepsilon_{zx}\left(\varepsilon_{xy}+p_m q_m\right)+\varepsilon_{xz}\varepsilon_{zz}^{-1}\varepsilon_{zy}\left(\varepsilon_{zy}+p_m q_m\right)\right] \\
&\quad+\varepsilon_{zx}\varepsilon_{xz}\varepsilon_{yz}\varepsilon_{zy}\varepsilon_{zz}^{-2}+\varepsilon_{zz}^{-2}\left[\left(p_m^2+q_m^2\right)-p_m^2 q_m^2\varepsilon_{zz}^{-1}\right] \\
&\quad\times\left[\varepsilon_{yz}\varepsilon_{zx}\left(\varepsilon_{xy}+p_m q_m\right)+\varepsilon_{xz}\varepsilon_{zy}\left(\varepsilon_{zy}+p_m q_m\right)\right. \\
&\quad\left.-\varepsilon_{zx}\varepsilon_{xz}\varepsilon_{yz}\varepsilon_{zy}\varepsilon_{zz}^{-1}\right] \\
&= \underline{(\varepsilon_{zy}+p_m q_m)(\varepsilon_{xy}+p_m q_m)} \\
&\quad\underline{-\left(p_m^2+q_m^2\right)\varepsilon_{zz}^{-1}(\varepsilon_{zy}+p_m q_m)(\varepsilon_{xy}+p_m q_m)} \\
&\quad\underline{-\left[\varepsilon_{yz}\varepsilon_{zz}^{-1}\varepsilon_{zx}\left(\varepsilon_{xy}+p_m q_m\right)+\varepsilon_{xz}\varepsilon_{zz}^{-1}\varepsilon_{zy}\left(\varepsilon_{zy}+p_m q_m\right)\right]} \\
&\quad+p_m^2 q_m^2\varepsilon_{zz}^{-2}(\varepsilon_{zy}+p_m q_m)(\varepsilon_{xy}+p_m q_m)+\varepsilon_{zx}\varepsilon_{xz}\varepsilon_{yz}\varepsilon_{zy}\varepsilon_{zz}^{-2} \\
&\quad+\varepsilon_{zz}^{-2}\left(p_m^2+q_m^2\right)\left[\varepsilon_{yz}\varepsilon_{zx}\left(\varepsilon_{xy}+p_m q_m\right)+\varepsilon_{xz}\varepsilon_{zy}\left(\varepsilon_{zy}+p_m q_m\right)\right] \\
&\quad-\varepsilon_{zz}^{-3}\left(p_m^2+q_m^2\right)\varepsilon_{zx}\varepsilon_{xz}\varepsilon_{yz}\varepsilon_{zy} \\
&\quad-p_m^2 q_m^2\varepsilon_{zz}^{-3}\left[\varepsilon_{yz}\varepsilon_{zx}\left(\varepsilon_{xy}+p_m q_m\right)+\varepsilon_{xz}\varepsilon_{zy}\left(\varepsilon_{zy}+p_m q_m\right)\right] \\
&\quad+p_m^2 q_m^2\varepsilon_{zx}\varepsilon_{xz}\varepsilon_{zz}^{-4}\varepsilon_{yz}\varepsilon_{zy}
\end{aligned}
\tag{F.20}
$$

$$a_{41}a_{13}a_{24}a_{32} = \left(\varepsilon_{zy} - \varepsilon_{yz}\varepsilon_{zz}^{-1}\varepsilon_{zx} + p_m q_m\right)\left(p_m\varepsilon_{zz}^{-1}\varepsilon_{zy}\right)$$
$$\times\left(q_m\varepsilon_{zz}^{-1}p_m\right)\left(-\varepsilon_{xz}\varepsilon_{zz}^{-1}q_m\right)$$
$$= -p_m^2q_m^2\varepsilon_{zz}^{-3}\varepsilon_{zy}\varepsilon_{xz}(\varepsilon_{zy} + p_mq_m) + p_m^2q_m^2\varepsilon_{zz}^{-4}\varepsilon_{zx}\varepsilon_{xz}\varepsilon_{yz}\varepsilon_{zy}, \tag{F.21}$$

$$a_{41}a_{14}a_{22}a_{33} = \left(\varepsilon_{zy} - \varepsilon_{yz}\varepsilon_{zz}^{-1}\varepsilon_{zx} + p_mq_m\right)\left(-p_m\varepsilon_{zz}^{-1}q_m\right)$$
$$\times\left(\varepsilon_{xz}\varepsilon_{zz}^{-1}p_m + s_m - \kappa_m\right)\left(q_m\varepsilon_{zz}^{-1}\varepsilon_{zy} + s_m - \kappa_m\right)$$
$$= -p_m\varepsilon_{zz}^{-1}q_m\left[(s_m - \kappa_m)^2\left(\varepsilon_{zy} - \varepsilon_{yz}\varepsilon_{zz}^{-1}\varepsilon_{zx} + p_mq_m\right)\right.$$
$$\left. + (s_m - \kappa_m)\,\varepsilon_{zz}^{-1}\left(q_m\varepsilon_{zy} + p_m\varepsilon_{xz}\right)\left(\varepsilon_{zy} - \varepsilon_{yz}\varepsilon_{zz}^{-1}\varepsilon_{zx} + p_mq_m\right)\right]$$
$$-p_m^2q_m^2\varepsilon_{zz}^{-3}\varepsilon_{zy}\varepsilon_{xz}(\varepsilon_{zy} + p_mq_m) + p_m^2q_m^2\varepsilon_{zz}^{-4}\varepsilon_{zx}\varepsilon_{xz}\varepsilon_{yz}\varepsilon_{zy}$$
$$= \underline{-p_mq_m\varepsilon_{zz}^{-1}\,(s_m - \kappa_m)^2\left(\varepsilon_{zy} + p_mq_m\right)} + p_mq_m\varepsilon_{yz}\varepsilon_{zx}\varepsilon_{zz}^{-2}$$
$$\times (s_m - \kappa_m)^2 - p_mq_m\varepsilon_{zz}^{-2}(s_m - \kappa_m)\left(q_m\varepsilon_{zy} + p_m\varepsilon_{xz}\right)\left(\varepsilon_{zy} + p_mq_m\right)$$
$$+p_mq_m\varepsilon_{zx}\varepsilon_{yz}\varepsilon_{zz}^{-3}\,(s_m - \kappa_m)\left(q_m\varepsilon_{zy} + p_m\varepsilon_{xz}\right)$$
$$- p_m^2q_m^2\varepsilon_{zz}^{-3}\varepsilon_{zy}\varepsilon_{xz}(\varepsilon_{zy} + p_mq_m) + p_m^2q_m^2\varepsilon_{zz}^{-4}\varepsilon_{zx}\varepsilon_{xz}\varepsilon_{yz}\varepsilon_{zy} \tag{F.22}$$

$$a_{41}a_{14}a_{23}a_{32} = \left(\varepsilon_{zy} - \varepsilon_{yz}\varepsilon_{zz}^{-1}\varepsilon_{zx} + p_mq_m\right)\left(-p_m\varepsilon_{zz}^{-1}q_m\right)$$
$$\times\left(\varepsilon_{xz}\varepsilon_{zz}^{-1}\varepsilon_{zy} - \varepsilon_{xy} - p_mq_m\right)\left(q_m\varepsilon_{zz}^{-1}p_m\right)$$
$$= -p_m^2q_m^2\varepsilon_{zz}^{-2}\left(\varepsilon_{zy} - \varepsilon_{yz}\varepsilon_{zz}^{-1}\varepsilon_{zx} + p_mq_m\right)\left(\varepsilon_{xz}\varepsilon_{zz}^{-1}\varepsilon_{zy} - \varepsilon_{xy} - p_mq_m\right)$$
$$= p_m^2q_m^2\varepsilon_{zz}^{-2}(\varepsilon_{zy} + p_mq_m)(\varepsilon_{xy} + p_mq_m)$$
$$-p_m^2q_m^2\varepsilon_{yz}\varepsilon_{zx}(\varepsilon_{xy} + p_mq_m)\varepsilon_{zz}^{-3}$$
$$-p_m^2q_m^2(\varepsilon_{zy} + p_mq_m)\varepsilon_{xz}\varepsilon_{zy}\varepsilon_{zz}^{-3} + p_m^2q_m^2\varepsilon_{zz}^{-4}\varepsilon_{zx}\varepsilon_{xz}\varepsilon_{yz}\varepsilon_{zy}, \tag{F.23}$$

$$a_{41}a_{13}a_{22}a_{34} = \left(\varepsilon_{zy} - \varepsilon_{yz}\varepsilon_{zz}^{-1}\varepsilon_{zx} + p_mq_m\right)\left(p_m\varepsilon_{zz}^{-1}\varepsilon_{zy}\right)\left(\varepsilon_{xz}\varepsilon_{zz}^{-1}p_m + s_m - \kappa_m\right)$$
$$\times\left(1 - q_m^2\varepsilon_{zz}^{-1}\right)$$
$$= p_m\varepsilon_{zz}^{-1}\varepsilon_{zy}\left[\left(\varepsilon_{zy} + p_mq_m\right) - \varepsilon_{yz}\varepsilon_{zz}^{-1}\varepsilon_{zx} - q_m^2\varepsilon_{zz}^{-1}\left(\varepsilon_{zy} + p_mq_m\right)\right.$$
$$\left.\times +q_m^2\varepsilon_{zz}^{-2}\varepsilon_{yz}\varepsilon_{zx}\right]\left[p_m\varepsilon_{zz}^{-1}\varepsilon_{xz} + (s_m - \kappa_m)\right]$$
$$= \underline{(s_m - \kappa_m)\,p_m\varepsilon_{zz}^{-1}\varepsilon_{zy}\left(\varepsilon_{zy} + p_mq_m\right)}$$
$$- (s_m - \kappa_m)\,p_m\varepsilon_{zz}^{-2}\varepsilon_{zy}\left(\varepsilon_{zx}\varepsilon_{yz} + q_m^2\varepsilon_{zy} + p_mq_m^3\right)$$
$$+ (s_m - \kappa_m)\,p_mq_m^2\varepsilon_{zz}^{-3}\varepsilon_{zx}\varepsilon_{yz}\varepsilon_{zy} + p_m^2\varepsilon_{zz}^{-2}\varepsilon_{xz}\varepsilon_{zy}(\varepsilon_{zy} + p_mq_m)$$
$$-p_m^2\varepsilon_{zz}^{-3}\varepsilon_{xz}\varepsilon_{zy}\left(\varepsilon_{zx}\varepsilon_{yz} + q_m^2\varepsilon_{zy} + p_mq_m^3\right) + p_m^2q_m^2\varepsilon_{zz}^{-4}\varepsilon_{zx}\varepsilon_{xz}\varepsilon_{yz}\varepsilon_{zy}, \tag{F.24}$$

$$
\begin{aligned}
a_{41}a_{12}a_{24}a_{33} &= \left(\varepsilon_{zy} - \varepsilon_{yz}\varepsilon_{zz}^{-1}\varepsilon_{zx} + p_m q_m\right)\left(p_m^2\varepsilon_{zz}^{-1} - 1\right) \\
&\quad \times \left(-\varepsilon_{xz}\varepsilon_{zz}^{-1}q_m\right)\left(q_m\varepsilon_{zz}^{-1}\varepsilon_{zy} + s_m - \kappa_m\right) \\
&= \left(\varepsilon_{zy} - \varepsilon_{yz}\varepsilon_{zz}^{-1}\varepsilon_{zx} + p_m q_m\right)\left(1 - p_m^2\varepsilon_{zz}^{-1}\right) \\
&\quad \times \left(\varepsilon_{xz}\varepsilon_{zz}^{-1}q_m\right)\left(q_m\varepsilon_{zz}^{-1}\varepsilon_{zy} + s_m - \kappa_m\right) \\
&= q_m\varepsilon_{zz}^{-1}\varepsilon_{xz}\left[\left(\varepsilon_{zy} + p_m q_m\right) - \varepsilon_{yz}\varepsilon_{zz}^{-1}\varepsilon_{zx} - p_m^2\varepsilon_{zz}^{-1}\left(\varepsilon_{zy} + p_m q_m\right)\right. \\
&\quad \left. + p_m^2\varepsilon_{zz}^{-2}\varepsilon_{yz}\varepsilon_{zx}\right] \\
&\quad \times \left[q_m\varepsilon_{zz}^{-1}\varepsilon_{zy} + (s_m - \kappa_m)\right] \\
&= \underline{(s_m - \kappa_m)\, q_m\varepsilon_{zz}^{-1}\varepsilon_{xz}\left(\varepsilon_{zy} + p_m q_m\right)} \\
&\quad - q_m\varepsilon_{zz}^{-2}\varepsilon_{xz}\,(s_m - \kappa_m)\left(\varepsilon_{zx}\varepsilon_{yz} + p_m^2\varepsilon_{zy} + p_m^3 q_m\right) \\
&\quad + (s_m - \kappa_m)\, p_m^2 q_m\varepsilon_{zz}^{-3}\varepsilon_{zx}\varepsilon_{xz}\varepsilon_{yz} + q_m^2\varepsilon_{zz}^{-2}\varepsilon_{xz}\varepsilon_{zy}(\varepsilon_{zy} + p_m q_m) \\
&\quad - q_m^2\varepsilon_{zz}^{-3}\varepsilon_{xz}\varepsilon_{zy}\left(\varepsilon_{zx}\varepsilon_{yz} + p_m^2\varepsilon_{zy} + p_m^3 q_m\right) + p_m^2 q_m^2\varepsilon_{zz}^{-4}\varepsilon_{zx}\varepsilon_{xz}\varepsilon_{yz}\varepsilon_{zy}.
\end{aligned}
\tag{F.25}
$$

In Eqs. (F.2) to (F.25), the terms containing ε_{xx} of the power 0 and -1 were underlined. Also underlined are the corresponding members of the 4×4 determinant in Eq. (F.1). These are the only terms that contribute into the secular equation. Each of Eqs. (F.2) through (F.25) contains the term $q_m^2 s_m^2 \varepsilon_{xx}^{-4}\varepsilon_{xy}\varepsilon_{yx}\varepsilon_{zx}\varepsilon_{xz}$. Consequently, these terms cancel. All the terms containing ε_{xx}^{-3} and ε_{xx}^{-2} also cancel. After the evaluation of the determinants, we arrive at Eq. (9.247).

References

1. R. M. A. Azzam and N. M. Bashara, *Ellipsometry and Polarized Light* (Elsevier, Amsterdam, 1987).
2. D. Marcuse, *Theory of Dielectric Optical Waveguides* (Academic Press, New York and London, 1974), Chapter 2.
3. A. V. Sokolov, *Optical properties of metals* (Blackie, Glasgow and London, 1967).
4. R. Wangsness, *Electromagnetic Fields*, (J. Wiley, 1986).

Index

D

E

F

G

H

I

J

K

L

M

Q

R

S

T

U

V

W

Y

Z